Periodic Table of the Elements with the Gmelin System Numbers

1 H 2																	2 He 1
3 Li 20	4 Be 26											5 B 13	6 C 14	7 N 4	8 O 3	9 F 5	10 Ne 1
11 Na 21	12 Mg 27											13 Al 35	14 Si 15	15 P 16	16 S 9	17 Cl 6	18 Ar 1
19 * K 22	20 Ca 28	21 Sc 39	22 Ti 41	23 V 48	24 Cr 52	25 Mn 56	26 Fe 59	27 Co 58	28 Ni 57	29 Cu 60	30 Zn 32	31 Ga 36	32 Ge 45	33 As 17	34 Se 10	35 Br 7	36 Kr 1
37 Rb 24	38 Sr 29	39 Y 39	40 Zr 42	41 Nb 49	42 Mo 53	43 Tc 69	44 Ru 63	45 Rh 64	46 Pd 65	47 Ag 61	48 Cd 33	49 In 37	50 Sn 46	51 Sb 18	52 Te 11	53 I 8	54 Xe 1
55 Cs 25	56 Ba 30	57** La 39	72 Hf 43	73 Ta 50	74 W 54	75 Re 70	76 Os 66	77 Ir 67	78 Pt 68	79 Au 62	80 Hg 34	81 Tl 38	82 Pb 47	83 Bi 19	84 Po 12	85 At 8a	86 Rn 1
87 Fr 25a	88 Ra 31	89*** Ac 40	104 71	105 71													* NH_4 23

**Lanthanides 39	58 Ce	59 Pr	60 Nd	61 Pm	62 Sm	63 Eu	64 Gd	65 Tb	66 Dy	67 Ho	68 Er	69 Tm	70 Yb	71 Lu
***Actinides	90 Th 44	91 Pa 51	92 U 55	93 Np 71	94 Pu 71	95 Am 71	96 Cm 71	97 Bk 71	98 Cf 71	99 Es 71	100 Fm 71	101 Md 71	102 No 71	103 Lr 71

A Key to the Gmelin System is given on the Inside Back Cover

Gmelin Handbook of Inorganic and Organometallic Chemistry

8th Edition

Gmelin Handbook of Inorganic and Organometallic Chemistry

8th Edition

Gmelin Handbuch der Anorganischen Chemie

Achte, völlig neu bearbeitete Auflage

PREPARED AND ISSUED BY
Gmelin-Institut für Anorganische Chemie
der Max-Planck-Gesellschaft
zur Förderung der Wissenschaften
Director: Ekkehard Fluck

FOUNDED BY
Leopold Gmelin

8TH EDITION
8th Edition begun under the auspices of the
Deutsche Chemische Gesellschaft by R. J. Meyer

CONTINUED BY
E. H. E. Pietsch and A. Kotowski, and by
Margot Becke-Goehring

Springer-Verlag Berlin Heidelberg GmbH 1944

Gmelin-Institut für Anorganische Chemie
der Max-Planck-Gesellschaft zur Förderung der Wissenschaften

GMELIN HANDBOOK

Dr. J. von Jouanne

Dr. L. Berg, Dr. H. Bergmann, Dr. J. Faust, J. Füssel, Dr. H. Katscher, Dr. R. Keim, Dipl.-Phys. D. Koschel, Dr. A. Kubny, Dr. P. Merlet, Dr. M. Mirbach, Prof. Dr. W. Petz, Dr. H. Schäfer, Dr. F. A. Schröder, Dr. A. Slawisch, Dr. W. Töpper

Dr. R. Albrecht, Dr. G. Bär, D. Barthel, Dr. N. Baumann, Dr. K. Behrends, Dr. W. Behrendt, D. Benzaid, Dr. R. Bohrer, K. D. Bonn, Dipl.-Chem. U. Boßlet, Dr. U. Busch, A.-K. Castro, Dipl.-Ing. V. A. Chavizon, E. Cloos, A. Dittmar, Dipl.-Geol. R. Ditz, R. Dowideit, Dr. H.-J. Fachmann, B. Fischer, Dr. D. Fischer, Dr. K. Greiner, Dipl.-Bibl. W. Grieser, Dr. R. Haubold, Dipl.-Min. H. Hein, H.-P. Hente, H. W. Herold, U. Hettwer, G. Hoell, Dr. G. Hönes, Dr. W. Hoffmann, G. Horndasch, Dr. W. Huisl, Dr. M. Irmler, B. Jaeger, Dr. R. Jotter, Dipl.-Chem. P. Kämpf, Dr. B. Kalbskopf, H.-G. Karrenberg, Dipl.-Phys. H. Keller-Rudek, Dipl.-Chem. C. Koeppel, R. Kolb, Dr. M. Kotowski, E. Kranz, E. Krawczyk, Dipl.-Chem. I. Kreuzbichler, Dr. W. Kurtz, M. Langer, Dr. B. Ledüc, H. Mathis, E. Meinhard, M. Meßer, C. Metz, K. Meyer, E. Mlitzke, Dipl.-Chem. B. Mohsin, Dr. U. Neu-Becker, K. Nöring, Dipl.-Min. U. Nohl, Dr. U. Ohms-Bredemann, Dr. H. Pscheidl, Dipl.-Phys. H.-J. Richter-Ditten, E. Rudolph, Dr. B. Sarbas, Dr. R. Schemm, Dr. D. Schiöberg, V. Schlicht, Dipl.-Chem. D. Schneider, E. Schneider, A. Schwärzel, Dr. B. Schwager, R. Simeone, Dr. F. Stein, Dr. C. Strametz, Dr. G. Swoboda, Dr. D. Tille, A. Tuttas, Dipl.-Phys. J. Wagner, R. Wagner, M. Walter, Dr. E. Warkentin, Dr. C. Weber, Dr. A. Wietelmann, Dr. M. Winter, Dr. B. Wöbke, K. Wolff

GMELIN ONLINE

Dr. G. Olbrich

Dr. P. Kuhn

Dr. R. Baier, Dr. K. Beesk, Dr. B. Becker, Dipl.-Chem. E. Best, Dr. H.-U. Böhmer, Dipl.-Phys. R. Bost, Dr. A. Brandl, Dr. R. Braun, Dr. T. Buck, R. Hanz, Dipl.-Phys. C. Heinrich-Sterzel, Dr. S. Kalwellis-Mohn, Dr. A. Kirchhoff, Dr. M. Körfer, Dipl.-Chem. H. Köttelwesch, Dr. V. Kruppa, Dr. M. Kunz, Dipl.-Chem. R. Maass, Dr. A. Nebel, Dipl.-Chem. R. Nohl, Dr. M. Nohlen, H. Reinhardt, Dr. B. Rempfer, Dipl.-Ing. H. Vanecek, Dipl.-Phys. M. Weiss

Organometallic Compounds in the Gmelin Handbook

The following listing indicates in which volumes these compounds are discussed or are referred to:

Ag	Silber B 5 (1975)
Au	Organogold Compounds (1980)
Be	Organoberyllium Compounds 1 (1987)
Bi	Bismut-Organische Verbindungen (1977)
Co	Kobalt-Organische Verbindungen 1, 2 (1973), Kobalt Erg.-Bd. A (1961), B 1 (1963), B 2 (1964)
Cr	Chrom-Organische Verbindungen (1971)
Cu	Organocopper Compounds 1 (1985), 2 (1983), 3 (1986), 4 (1987), Index (1987)
Fe	Eisen-Organische Verbindungen A 1 (1974), A 2 (1977), A 3 (1978), A 4 (1980), A 5 (1981), A 6 (1977), A 7 (1980), Organoiron Compounds A 8 (1986), A 9 (1989), A 10 (1991), Eisen-Organische Verbindungen B 1 (partly in English; 1976), Organoiron Compounds B 2 (1978), Eisen-Organische Verbindungen B 3 (partly in English; 1979), B 4, B 5 (1978), Organoiron Compounds B 6, B 7 (1981), B 8, B 9 (1985), B 10 (1986), B 11 (1983), B 12 (1984), B 13 (1988), B 14, B 15 (1989), B 16a, B 16b, B 17 (1990), B 18 (1991), B 19 (1992), Eisen-Organische Verbindungen C 1, C 2 (1979), Organoiron Compounds C 3 (1980), C 4, C 5 (1981), C 6a (1991), C 6b (1992), C 7 (1985), and Eisen B (1929–1932)
Ga	Organogallium Compounds 1 (1986)
Ge	Organogermanium Compounds 1 (1988), 2 (1989), 3 (1990), 4 (1994) **present volume**, 5 (1993)
Hf	Organohafnium Compounds (1973)
In	Organoindium Compounds 1 (1991)
Mo	Organomolybdenum Compounds 5 (1992), 6 (1990), 7 (1991), 8 (1992), 9 (1993), 12 (1994)
Nb	Niob B 4 (1973)
Ni	Nickel-Organische Verbindungen 1 (1975), 2 (1974), Register (1975), Nickel B 3 (1966), and C 1 (1968), C 2 (1969), Organonickel Compounds Suppl. Vol. 1 (1993)
Np, Pu	Transurane C (partly in English; 1972)
Os	Organoosmium Compounds A 1 (1992), A 2 (1993), B 6 (1993)
Pb	Organolead Compounds 1 (1987), 2 (1990), 3 (1992)
Po	Polonium Main Volume (1941)
Pt	Platin C (1939) and D (1957)
Re	Organorhenium 1, 2 (1989), 3 (1992), 5 (1994)
Ru	Ruthenium Erg.-Bd. (1970)
Sb	Organoantimony Compounds 1, 2 (1981), 3 (1982), 4 (1986), 5 (1990)
Sc, Y, La to Lu	Rare Earth Elements D 6 (1983)
Sn	Zinn-Organische Verbindungen 1 (1975) to 6 (1979); Organotin Compounds 7 (1980) to 21 (1994)
Ta	Tantal B 2 (1971)
Ti	Titan-Organische Verbindungen 1 (1977), 2 (1980), Organotitanium Compounds 3 (1984), 4 and Register (1984), 5 (1990)
U	Uranium Suppl. Vol. E 2 (1980)
V	Vanadium-Organische Verbindungen (1971), Vanadium B (1967)
Zr	Organozirconium Compounds (1973)

Gmelin Handbook of Inorganic and Organometallic Chemistry

8th Edition

Ge

Organogermanium Compounds

Part 4

Compounds with Germanium-Hydrogen Bonds

With 6 illustrations

AUTHORS — John E. Drake, University of Windsor, Ontario
Christa Siebert (Maintal)
Bernd Wöbke

FORMULA INDEX — Rainer Bohrer, Bernd Kalbskopf, Hans-Jürgen Richter-Ditten

EDITORS — Ulrich Krüerke, Christa Siebert, Bernd Wöbke

CHIEF EDITOR — Ulrich Krüerke

Springer-Verlag Berlin Heidelberg GmbH 1944

LITERATURE CLOSING DATE: END OF 1992
IN SOME CASES MORE RECENT DATA HAVE BEEN CONSIDERED

Library of Congress Catalog Card Number: Agr 25-1383

ISBN 978-3-662-06326-2 ISBN 978-3-662-06324-8 (eBook)
DOI 10.1007/978-3-662-06324-8

Originally published by Springer-Verlag Berlin Heidelberg New York Tokyo in 1994
Softcover reprint of the hardcover 8th edition 1994

Preface

The present volume in the organogermanium series describes mononuclear compounds containing only germanium-carbon and germanium-hydrogen bonds (Chapter 1.3). Germanium hydrides with other additional non-carbon ligands, such as halogen or oxygen-bonded groups, appear in later chapters according to the Gmelin principle of the last position. Compounds with Ge-H and Ge-O bonds have already been described in Volume 5, Section 1.5.1.4, pp. 50/62. The present volume covers the literature to the end of 1992 and includes many references up to 1994.

The nomenclature recommended by IUPAC has been generally adhered to. However, compound names were largely avoided, as most of the compounds are presented in tables and are only identified by their formulas. Many of the data in the tables appear in abbreviated form without units; general explanations are given on pp. X/XI.

The volume contains an empirical formula index (p. 327) and a ligand formula index (p. 341).

The editor wishes to express his gratitude to the former author, Professor J. E. Drake, and to Professor J. Satgé for his kind advice and fruitful collaboration. Thanks are due also to Dr. A. R. Pebler for editing the English text and to Mr. H.-G. Karrenberg for drawing the numerous formulas and molecular structures.

Frankfurt am Main
August 1994

Ulrich Krüerke

Explanations, Abbreviations, and Units

Most compounds in this volume are presented in tables in which numerous abbreviations are used and the units are omitted for the sake of brevity. This necessitates the following clarification:

The **formation** of a compound is briefly described at the first place of the third column of a table. Evident procedures of a preparation such as hydrolysis after an organometallic reaction, evaporation of solvent, drying, etc., are usually omitted; 20°C replaces frequently the term "room temperature" in the original. GC or TLC stand for gas-liquid or thin-layer chromatography.

Abbreviations for solvents and other substances are as follows:

AIBN	azobisisobutyronitrile	THF	tetrahydrofuran
DME	dimethoxyethane	TMEDA	tetramethylethylenediamine
DMF	dimethylformamide	acac or	
DMSO	dimethylsulfoxide	$C_5H_7O_2$	acetylacetonato
HMPT	hexamethylphosphoric triamide		

Temperatures are given in °C, otherwise K stands for Kelvin. Abbreviations used with temperatures are m.p. for melting point, b.p. for boiling point, dec. for decomposition, and subl. for sublimation. Terms like 80°C/0.1 mean the boiling or sublimation point at a pressure of 0.1 Torr. **Densities** d are given in g/cm^3; d^t is the density at t°C referred to water at 4°C.

NMR represents **nuclear magnetic resonance**. Chemical shifts δ are given in ppm and are positive to low field from the following reference substances: $Si(CH_3)_4$ for ^1H and ^{13}C, $CFCl_3$ for ^{19}F, H_3PO_4 for ^{31}P, $Ge(CH_3)_4$ for ^{73}Ge, and $Sn(CH_3)_4$ for ^{119}Sn. Reference substances for other nuclei are indicated in parentheses, e.g. ^{15}N NMR (C_6D_6; ref. neat CH_3NO_2). Multiplicities of the signals are abbreviated s, d, t, q (singlet to quartet), quint, sext, sept (quintet to septet), and m (multiplet); terms like dd (double doublet) and t's (triplets) are also used. Assignments referring to labeled structural formulas are given in the form C-4, H-3,5. Coupling constants J in Hz usually appear in parentheses behind the δ value, along with the multiplicity and the assignment, and refer to the respective nucleus. If a more precise designation is necessary, they are given as, e.g. $^nJ(C,H)$ or J(1,3) referring to labeled formulas.

Optical spectra are labeled IR (infrared), Raman, or UV (electronic spectrum including the visible region). IR bands and Raman lines are given in cm^{-1}; the band assignments are usually labeled with the symbols ν for stretching vibration and δ for deformation vibration. Intensities are added in parentheses either in the common qualitative terms (s, m, w, vs, etc.) or as numerical relative intensities along with p or dp for polarized or depolarized Raman lines. The UV absorption maxima λ_{max} are given in nm followed by the extinction coefficient ε ($L \cdot cm^{-1} \cdot mol^{-1}$) or log ε in parentheses; sh means shoulder.

Photoelectron spectra are abbreviated PE, e.g. PE/He(I) with the ionization energies in eV.

Solvents or the **physical state** of the sample and the temperature are given in parentheses immediately after the spectral symbol, e.g. Raman (solid), ^{13}C NMR (C_6D_6, 50°C), or at the end of the data if spectra for various media are reported. Common solvents are given by their formulas (C_6H_{12} = cyclohexane) or the abbreviations listed above.

The fragments of **mass spectra**, abbreviated MS (CI means chemical ionization), are given in brackets followed by the relative intensities in parentheses; $[M]^+$ is the molecular ion.

Figures of molecular structures give only selected parameters. Barred bond lengths (in Å) or angles are mean values for parameters of the same type. d_m and d_c distinguish measured and calculated densities based on X-ray diffraction analysis.

Formulas of symmetrical digermanes such as R_3GeGeR_3, HR_2GeGeR_2H, and $H_2(R)$-$GeGe(R)H_2$ are written Ge_2R_6, $Ge_2R_4H_2$, and $Ge_2R_2H_4$, respectively.

Table of Contents

Organogermanium Compounds

1.3 Organogermanium Hydrides

1.3.1 Triorganogermanium Hydrides

1.3.1.1 GeR_3H Compounds

1.3.1.1.1 Trimethylgermane, Trimethylgermanium Hydride, $Ge(CH_3)_3H$

1.3.1.1.1.1 Preparation and Formation

The preparation of $Ge(CH_3)_3H$ and $Ge(CH_3)_3D$ by reduction of $Ge(CH_3)_3Cl$ or $Ge(CH_3)_3Br$ with LiH or LiD was mentioned without experimental details [1]. The more commonly reported method involves the reduction of $Ge(CH_3)_3X$ (X = halogen) with $LiAlH_4$ or $LiAlD_4$ in $O(C_4H_9)_2$ [2], C_6H_6 [7], or $CH_3OCH_2CH_2OCH_3$ [5], resulting in excellent yields: For example, adding $Ge(CH_3)_3Br$ to $LiAlH_4$ (2:1 mole ratio) in $CH_3OCH_2CH_2OCH_3$, heating to 70°C, and distilling the mixture achieved a yield of 77% [5]; 65% were obtained from the reaction of $Ge(CH_3)_3Cl$ with $Li[Al(OC_4H_9\text{-}t)_3H]$ (ca. 1:2.5 mole ratio) in dioxane by slowly heating the solution to the boiling point and separating the volatiles by fractional condensation in a vacuum line [4, 12].

Flammable hydrides can be avoided by slowly adding aqueous $NaBH_4$ to $Ge(CH_3)_3Br$ (ca. 22:1 mole ratio) in aqueous 1 M HBr at reduced pressure and collecting the effluent gases in cold traps followed by fractional condensation; 95% yield [3]. $NaBH_4$ reduction to produce Ge-H compounds has been used to determine methylgermanium species in natural waters [13].

$Ge(CH_3)_3H$ was obtained with 44% yield by refluxing a mixture of $Ge(CH_3)_3Cl$ and $Si(CH_3)(H)Cl_2$ (ca. 1:2 mole ratio) for 5 h in the presence of 10% $AlCl_3$ [9]. It formed incompletely (16% conversion) upon cleaving of the Ge-Ge bond in $Ge_2(CH_3)_6$ with a slight excess of $LiAlH_4$ in THF for 3 d [10]; Ge-Ge or Ge-Si cleavage in $(CH_3)_3GeGe(CH_3)_2C_6H_5$, $(CH_3)_3Ge\text{-}Ge(C_6H_5)_3$, or $(CH_3)_3GeSi(CH_3)_2C_6H_5$ by photolysis in C_6H_{12} produced many products, including $Ge(CH_3)_3H$ [17, 19, 20]. It was also observed among the products of the pyrolysis of gaseous $Ge(CH_3)H_3$ at 420°C [6] and liquid $Ge_2(CH_3)_5H$ at 250°C [11]. Its formation as an intermediate from $Ge(CH_3)_3^{\bullet}$ and $t\text{-}C_4H_9^{\bullet}$ radicals was mentioned in a study of the photolysis of $Hg(Ge(CH_3)_3)_2$ [8].

Small amounts of $Ge(CH_3)_3H$ were detected in a spectroscopic study of the reaction of $Ge(CH_3)_3Sn(CH_3)_3$ with $Sn(CH_3)_3Cl$ in CH_3OH [15] and in a kinetic study of the reaction of $Ge(CH_3)_3Sn(CH_3)_3$ with HCl in CH_3OH ($Ge(CH_3)_3H$ presumably consumed by excess acid) [14]. It was formed as a decomposition product (0.5%) in the reaction of thermally generated Ca atoms with $Ge_2(CH_3)_6$ in a C_6H_5Br substrate [16].

ArF laser-induced photolysis of $Ge(CH_3)_4$ in the gas phase produced $Ge(CH_3)_3H$, CH_4, C_2H_4, and C_2H_6 along with pure Ge films on a quartz surface [21].

The heat of formation $\Delta H_f = -12.5$ kcal/mol was calculated using the AM1 procedure [18].

References:

[1] Ponomarenko, V. A.; Yzenkova, G. Ya. (Dokl. Akad. Nauk SSSR **122** [1958] 405/8; C.A. **1959** 112).
[2] Satgé, J. (Ann. Chim. [Paris] [13] **6** [1961] 519/73).
[3] Griffith, J. E. (Inorg. Chem. **2** [1963] 375/7).
[4] Van de Vondel, D. F. (J. Organomet. Chem. **3** [1965] 400/5).
[5] Fish, R. R.; Kuivila, M. G. (J. Org. Chem. **31** [1966] 2445/50).
[6] Kohanek, J. J.; Estacio, P.; Ring, M. A. (Inorg. Chem. **8** [1969] 2516/7).
[7] Durig, J. R.; Chen, M. M.; Li, Y. S.; Turner, J. B. (J. Phys. Chem. **77** [1973] 227/9).
[8] Lehnig, M.; Werner, F.; Neumann, W. P. (J. Organomet. Chem. **97** [1975] 357/87).
[9] Zueva, G. Ya.; Serezhkina, N. V.; Khaustova, V. A.; Ponomarenko, V. A. (Izv. Akad. Nauk SSSR Ser. Khim. **1976** 912/5; Bull. Acad. Sci. USSR Div. Chem. Sci. [Engl. Transl.] **1976** 890/3).
[10] Duffaut, N.; Dunoges, J.; Calas, R.; Rivière, P.; Satgé, J.; Cazes, A. (J. Organomet. Chem. **149** [1978] 57/63).
[11] Ma, E. C. L.; Paquin, D. P.; Gaspar, P. P. (J. Chem. Soc. Chem. Commun. **1980** 381/2).
[12] Imai, Y.; Aida, K. (Bull. Chem. Soc. Jpn. **54** [1981] 3323/6).
[13] Hambrick, G. A.; Froelich, P. N.; Andreae, M. O.; Lewis, B. L. (Anal. Chem. **56** [1984] 421/4).
[14] Cuthbertson, M. J.; Hawker, D. W.; Wells, P. R. (J. Organomet. Chem. **287** [1985] 7/23).
[15] Cuthbertson, M. J.; Wells, P. R. (J. Organomet. Chem. **287** [1985] 25/38).
[16] Mochida, K.; Yamanishi, T. (Bull. Chem. Soc. Jpn. **60** [1987] 3429/30).
[17] Mochida, K.; Kikkawa, H.; Nakadaira, Y. (Chem. Lett. **1988** 1089/92).
[18] Dewar, M. J. S.; Jie, C. (Organometallics **8** [1989] 1544/7).
[19] Mochida, K.; Kikkawa, H.; Nakadaira, Y. (J. Organomet. Chem. **412** [1991] 9/19).
[20] Mochida, K.; Wakasa, M.; Sakaguchi, Y.; Hayashi, H. (Bull. Chem. Soc. Jpn. **64** [1991] 1889/95).
[21] Pola, J.; Taylor, R. (J. Organomet. Chem. **437** [1992] 271/8).

1.3.1.1.1.2 The Molecule, Spectra, and Physical Properties

Structure. The Ge-H distance, r(Ge-H) = 1.532 ± 0.001 Å, has been determined from the ground-state rotational constants (obtained from the microwave spectrum between 18.0 and 40.0 GHz) of each of the five $Ge(CH_3)_3H$ and $Ge(CH_3)_3D$ isotopic species. With this value and an assumed structure for the CH_3 group, the distance r(Ge-C) = 1.947 ± 0.006 Å and the angle C-Ge-H = 109.3° ± 0.1° were calculated. The Ge-H bond is longer than those in GeH_4 and $Ge(CH_3)H_3$, whereas the Ge-C bond length is essentially the same as the ones in $Ge(CH_3)H_3$ and $Ge(CH_3)_2H_2$ [19]. Using slightly different CH_3 group parameters, derived from "isolated" ν_{is}(C-H) vibrations observed in deuterated derivatives, led to r_0(Ge-H) = 1.5361(3), r_0(Ge-C) = 1.9477(6) Å, and the angle C-Ge-H = 109.3(3)°. This revised r_0(Ge-H) gave a better linear correlation with ν_{is}(Ge-H) for the few germanes so far examined [34]. More recent calculations used the MM2 force field for obtaining the bond lengths Ge-H (1.530 Å) and Ge-C (1.949 Å) for the staggered conformation and the bond angles H-C-Ge (110.1°) and C-Ge-C (109.6°) for both the staggered and eclipsed conformations [40]. For AM1-calculated geometrical parameters, see also [41].

Force field calculations of two conformations of $M(CH_3)_nH_{4-n}$ compounds (M = Si, Ge, and Sn; n = 1 to 3) have been carried out, using parameters for the force field functions in such a way that the calculated structures and torsional barriers were consistent with the available microwave data [17]. A torsional barrier of 1.23 kcal/mol resulted from MM2 calculations [40].

Polarizability anisotropies and tensor elements of bonds and groups are presented for several MA_3B compounds of Si, Ge, and Sn, including $Ge(CH_3)_3H$ [37].

The molecule has a gas-phase **dipole moment** of μ = 0.6685 D at 25°C and 100 Torr [13]. Dipole moments and formal charge distributions were calculated for a few organogermanium compounds, μ_{calc} = 0.55 D for $Ge(CH_3)_3H$ compared with an experimental value of μ_{obs} = 0.67 D [20]. The $Ge\text{-}CH_3$ group moment, μ = 0.33 D, was determined [16] from the experimental value above and an estimated Ge-H bond moment of ca. 1.0 D [2]; see also [40, 41]. Anisotropic polarizabilities parallel and perpendicular to the Ge-H and $Ge\text{-}CH_3$ bonds were deduced from electric birefringences and dipole moments in CCl_4 solutions in conjunction with Raman band intensities [29].

The **bond dissociation energy,** $D(Ge(CH_3)_3\text{-}H)$ = 340±10 kJ/mol (81 kcal/mol), was derived from the kinetics of the reaction of $Ge(CH_3)_3H$ with I_2 [21, 22, 25]; see also p. 7. An apparently too low value (63 kcal/mol) was obtained by relating the activation energy for H-atom abstraction to the bond dissociation energy (BEBO method) [21]. Literature values of the heats of formation were combined with kinetic and appearance potential data to give self-consistent $D(M(CH_3)_3\text{-}X)$ values for a number of compounds of Group IV elements, yielding 82 kcal/mol for Ge-H [24]; see also the vibrational spectrum.

Ionization energies (in eV) from He(I) and He(II) photoelectron spectra are listed below along with the proposed assignments based on a comparison with spectra of similar compounds and semi-empirical CNDO/2 calculations:

vertical IP	10.57	10.93	13.13	13.90	14.30	16.40	18.02
assignment	4e	$4a_1$	$1a_2$	3e	2e	$3a_1$	$2a_1$

The first two bands represent predominantly Ge-C and Ge-H bonding. The $2a_1$ orbital contains a significant proportion of Ge-H bonding (spectrum illustrated along with those of $Ge(CH_3)_3F$ and $Ge(CH_3)_3Cl$) [27]. An AM1-calculated IP of 10.94 eV was given in [41]. Using the MNDO-UHF SCF method, calculations were reported of the molecular and electronic structures of a range of neutral organogermanes and the corresponding cation radicals. $Ge(CH_3)_3H$ was calculated to give upon ionization $[Ge(CH_3)_3]^+$ and hydrogen atoms [36].

The **core-electron binding energies** from the X-ray photoelectron spectrum are 36.35 eV for Ge 3d and 289.88 eV for C 1s (relative to Ne 2s at 48.45 and Ar $2p_{3/2}$ at 248.63 eV, respectively). A comparison of calculated and experimental energies for the Ge 3d level in the $Ge(CH_3)_nH_{4-n}$ series is based on both CNDO/2 calculations and the electronegativity-equalization method involving s-orbital participation [23].

Nuclear Magnetic Resonance Spectra. The ^1H NMR spectrum has been recorded by several workers [7, 9, 10, 11, 13, 15]. Data for the neat liquid: δ(ppm) = −0.14 (d, CH_3, $^1J(C, H)$ = 126.2 Hz) and 3.62 (decet, HGe, $^3J(H, H)$ = 3.4 Hz) [11, 13]; data for solutions in CCl_4: δ(ppm) = 0.21 ($^1J(C, H)$ = 127 Hz) and 3.92 ($^3J(H, H)$ = 3.40 Hz) [10]; δ(ppm) = 0.19 and 3.66 for a 1:1 volume mixture with $Si(CH_3)_4$ or C_6H_{12} [7, 9]; δ(ppm) = 0.21 (d) and 3.83 (sept) in HMPT-THF-d_8 [42]. Relationships between the Ge-H and $Ge\text{-}CH_3$ chemical shifts and the ν(Ge-H) vibration or the sum of the Taft polar constants ($\Sigma\sigma^*$) were discussed [7, 9]; for a correlation between ^1H NMR data and the Ge-X π(d-p) interaction in $Ge(CH_3)_nX_{4-n}$ compounds, see also [15].

^{13}C NMR spectrum: δ(ppm) = −4.38 for the neat liquid or in $CDCl_3$ [26] and −2.15 in HMPT-THF-d_8 [42]. The ^{13}C NMR shifts of several methylgermanium compounds were correlated with the calculated charge on the C atoms, the ^{1}H NMR shifts of the CH_3 protons, and the ^{13}C NMR shifts for analogous C, Si, and Sn compounds [26].

^{73}Ge NMR spectrum (in C_6H_{14}): $\delta = -56.9 \pm 0.2$ ppm ($W_{1/2} = 4.0 \pm 0.5$ Hz) and J(Ge,H) = 93 ± 2 Hz. The nuclear relaxation and correlation times were listed also and the ^{73}Ge nuclear quadrupole coupling constant roughly estimated [38]. A close correlation exists between the shifts for ^{73}Ge versus ^{29}Si and ^{119}Sn: $\delta(Ge) = 3.32\delta(Si) + 39.9$ and $\delta(Sn) = 1.56\delta(Ge) - 87.4$ [35]. Data on the ^{73}Ge NMR spectra of numerous organogermanium compounds (including $Ge(CH_3)_3H$) have been critically assessed in [39]. For recent ab initio calculations of ^{73}Ge chemical shifts, see [43].

The **vibrational spectra** have been reported in total or in part by several groups [3, 4, 6, 12, 28, 30, 32]. The fundamental vibrations are listed in Table 1. The first assignments [12] were essentially confirmed by considering the spectra of **$Ge(CH_3)_3D$, $Ge(CD_3)_3H$,** and **$Ge(CD_3)_3D$** and by performing normal coordinate calculations [28]. The IR and Raman spectra of all isotopic species are displayed in [28]; for $Ge(CH_3)_3H$, see also [12]. The Ge-C stretching force constant, f(GeC) = 2.67×10^{-2} N/m, is nearly equal to those in $Ge(CH_3)_4$ and $Ge(CH_3)_3Cl$ [28].

Table 1
Vibrational Spectra of $Ge(CH_3)_3H$.
Wavenumbers in cm^{-1}.

IR, gas [12]	[28]	Raman, liquid [12]	[28]	assignment	
2981(s)	2982	2981(3)	2982	$\nu_{as}(CH)$	$\nu_1, \nu_{13}, \nu_{14}$
2926(s)	2922	2916(5)	2913	$\nu_s(CH)$	ν_2, ν_{15}
2049(vs)	2040	2045(5)	2035	$\nu(GeH)$	ν_7
1410 to 1445(w)	1426	1415(0)	1419	$\delta_{as}(CH_3)$	$\nu_3, \nu_{16}, \nu_{17}$
1259(s)	1246	1248(2)	1247	$\delta_s(CH_3)$	ν_4
1251(s)	–	1240(2)	–		ν_{18}
–	850*)	–	850	$\delta(CH_3) + \rho(CH_3)$	
845(vs)	833	850(0)	830	$\rho(CH_3)$	$\nu_5, \nu_{19}, \nu_{20}$
631(nw)	624*)	627(3)	626	$\rho(CH_3) + \delta(GeH)$	ν_{23}
601(vs)	592	599(3)	597	$\nu_{as}(GeC)$	ν_{21}
–	571*)	573(10)	573	$\nu_s(GeC)$	ν_6
–	187	919(5)	189	$\rho_{as}(GeC_3) + \rho_s(GeH_3)$	ν_8, ν_{22}

*) Frequencies taken from the solid-state spectrum.

Values of $\nu(GeH)$ (2031 cm^{-1} in C_7H_{16} [3, 6] and 2032 cm^{-1} for the gas [4]) were used to establish a correlation between the inductive constants of substituents in Ge-H and Ge-D compounds and Taft's polar constants [3, 4, 6]. For a number of GeRR′R″H species, $\nu(GeH)$ correlates well with the sum of σ^* constants via $\nu(GeH) = 2008 + 16.5\ \Sigma\sigma^*$ [6]. However, a more detailed analysis led to $\nu(GeH) = 2037 + 47\ \Sigma\sigma^*$ for alkyl groups and hydrogen on germanium, including GeH_4 [32, 33]. The fundamental and first-overtone GeH stretching frequencies (2041.2 and 4014.0 cm^{-1}, respectively) were used to deduce Ge-H dissociation energies for a series of germanes with isolated Ge-H bonds, treating the Ge-H bond as a diatomic molecule. This gave D_0° = 72.9 kcal/mol and indicated that the Ge-H bond in GeH_4 is ca. 2 kcal per CH_3 group weaker than in $Ge(CH_3)_3H$ [30].

The IR spectrum of **$Ge(CD_3)_2(CHD_2)H$** yielded the frequencies of the "isolated" ν(CH) to be 2952.9 and 2961.5 cm^{-1}. Quite large differences in strength were found within a single methyl group, even when it is attached to a fairly large atom such as Ge [31].

Physical Properties. $Ge(CH_3)_3H$ is a colorless liquid with a density of d = 1.0128 g/cm^3 at 20 °C and a refractive index of n_D = 1.3890 at 20 °C [1]. Melting point: −123.1 ± 0.2 °C [8]. Boiling points of 26 °C/755.5 Torr [1, 5] or 760 Torr [14] and 28 °C [18] were reported; 27 °C/760 Torr was extrapolated from the vapor pressure below [8].

Data for **$Ge(CH_3)_3D$:** b.p. 26 °C/758 Torr, d = 1.0207 g/cm^3 at 20 °C, and n_D = 1.3893 at 20 °C [1].

The temperature dependence of the vapor pressure of $Ge(CH_3)_3H$ can be expressed by log (p/Torr) = 5.4904 − 0.00544 T + 1.75 log T − 1594.4/T, which was derived from the following experimental vapor pressures [8]:

T in K	194.7	200.1	203.2	209.4	214.3	227.9
p in Torr	1.79	2.92	3.75	6.28	9.1	24.0
T in K	237.3	242.4	249.9	263.5	273.2	283.6
p in Torr	43.3	57.8	88.1	174.3	269.7	414.0

This gives a boiling point of 27.0 °C, an enthalpy of vaporization of ΔH_v = 6097 cal/mol, and a Trouton constant of 20.3 cal · mol^{-1} · K^{-1} [8].

References:

[1] Ponomarenko, V. A.; Vzenkova, G. Ya.; Egorov, Yu. (Dokl. Akad. Nauk SSSR **122** [1958] 405/8; C.A. **1959** 112).
[2] Kartsev, G. N.; Syrkin, Ya. K.; Mironov, V. F. (Izv. Akad. Nauk SSSR Otd. Khim. Nauk **1960** 948/9; C.A. **1960** 21902).
[3] Mathis-Noel, R.; Mathis, F.; Satgé, J. (Bull. Soc. Chim. Fr. **1961** 676).
[4] Ponomarenko, V. A.; Zueva, G. Ya.; Andreev, N. S. (Izv. Akad. Nauk SSSR Otd. Khim. Nauk **1961** 1758/62; Bull. Acad. Sci. USSR Div. Chem. Sci. [Engl. Transl.] **1961** 1639/43).
[5] Satgé, J. (Ann. Chim. [Paris] [13] **6** [1961] 519/73).
[6] Mathis, R.; Satgé, J.; Mathis, F. (Spectrochim. Acta **18** [1962] 1463/72).
[7] Egorochkin, A. N.; Khidekel', M. L.; Ponomarenko, V. A.; Zueva, G. Ya.; Svirezheva, S. S.; Razuvaev, G. A. (Izv. Akad. SSSR Ser. Khim. **1963** 1865/8; Bull. Acad. Sci. USSR Div. Chem. Sci. [Engl. Transl.] **1963** 1717/9).
[8] Griffith, J. E. (Inorg. Chem. **2** [1963] 375/7).
[9] Egorochkin, A. N.; Khidekel', M. L.; Ponomarenko, V. A.; Zueva, G. Ya.; Razuvaev, G. A. (Izv. Akad. Nauk SSR Ser. Chim. **1964** 373/5; Bull. Acad. Sci. USSR Div. Chem. Sci. [Engl. Transl.] **1964** 347/8).
[10] Schmidbaur, H. (Chem. Ber. **97** [1964] 1639/48).

[11] Van der Kelen, G. P.; Verdonk, L.; Van de Vondel, D. (Bull. Soc. Chim. Belg. **73** [1964] 733/40).
[12] Van de Vondel, D. F.; Van der Kelen, G. P. (Bull. Soc. Chim. Belg. **74** [1965] 467/8).
[13] Van de Vondel, D. F. (J. Organomet. **3** [1965] 400/5).

[14] Fish, R. H.; Kuivila, M. G. (J. Org. Chem. **31** [1966] 2445/50).
[15] Egorochkin, A. N.; Burov, A. I.; Mironov, V. F.; Gar, T. K.; Vyazankin, N. S. (Dokl. Akad. Nauk SSSR **180** [1968] 861/4; Dokl. Chem. [Engl. Transl.] **178/183** [1968] 500/3).
[16] Ulbricht, K.; Vaisarova, V.; Bazant, V.; Chavlovsky, V. (J. Organomet. Chem. **13** [1968] 343/9).
[17] Quellette, R. J. (J. Am. Chem Soc. **94** [1972] 7674/9).
[18] Sakurai, H.; Mochida, K.; Hosomi, A.; Mita, F. (J. Organomet. Chem. **38** [1972] 275/80).
[19] Durig, J. R.; Chen, M. M.; Li, Y. S.; Turner, J. B. (J. Phys. Chem. **77** [1973] 227/9).
[20] Ramalingam, S. K.; Soundararajan, S. (J. Organomet. Chem. **72** [1974] 59/63).

[21] Austin, E. R.; Lampe, F. W. (J. Phys. Chem. **81** [1977] 1546/9).
[22] Doncaster, A. M.; Walsh, R. (J. Chem. Soc. Chem. Commun. **1977** 446/7).
[23] Drake, J. E.; Riddle, C.; Glavincevski, B.; Gorzelska, K.; Henderson, H. E. (Inorg. Chem. **17** [1978] 2333/6).
[24] Jackson, R. A. (J. Organomet. Chem. **166** [1979] 17/9).
[25] Doncaster, A. M.; Walsh, R. (J. Phys. Chem. **83** [1979] 578/81).
[26] Drake, J. E.; Glavinceski, B. M.; Humphries, R. E.; Majid, A. (Can. J. Chem. **57** [1979] 1426/30).
[27] Drake, J. E.; Glavincevski, B. M., Gorzelska, K. (J. Electron Spectrosc. Relat. Phenom. **16** [1979] 331/7).
[28] Imai, Y.; Aida, K.; (Bull. Chem. Soc. Jpn. **54** [1981] 3323/6).
[29] Armstrong, R. S.; Aroney, M. J.; Skamp, K. R. (J. Chem. Soc. Faraday Trans. II **78** [1982] 1641/4).
[30] McKean, D. C.; Torto, I.; Morrisson, A. R. (J. Phys. Chem. **86** [1982] 307/9).

[31] McKean, D. C.; MacKenzie, M. W.; Torto, I. (Spectrochim. Acta A **38** [1982] 113/8).
[32] McKean, D. C.; Torto, I.; MacKenzie, M. W. (Spectrochim. Acta A **39** [1983] 399/408).
[33] McKean, D. C. (J. Mol. Struct. **113** [1984] 251/66).
[34] McKean, D. C.; MacKenzie, M. W.; Morrisson, A. R. (J. Mol. Struct. **116** [1984] 331/44).
[35] Watkinson, P. J.; MacKay, K. M. (J. Organomet. Chem. **275** [1984] 39/42).
[36] Glidewell, C. (J. Organomet. Chem. **303** [1986] 337/41).
[37] Armstrong, R. S.; Aroney, M. J.; Clark, R. J. H. (J. Raman Spectrosc. **18** [1987] 141/3).
[38] Wilkins, A. L.; Watkinson, P. J.; Mackay, K. M. (J. Chem. Soc. Dalton Trans. **1987** 2365/72).
[39] Liepins, E.; Zicmane, I.; Lukevics, E. (J. Organomet. Chem. **341** [1988] 315/33).
[40] Allinger, N. L.; Quinn, M. I.; Chen, K.; Thompson, B.; Frierson, M. R. (J. Mol. Struct. **194** [1989] 1/18).

[41] Dewar, M. J. S.; Jie, C. (Organometallics **8** [1989] 1544/7).
[42] Mochida, K.; Kugita, T.; Nakadaira, Y. (Polyhedron **9** [1990] 2263/6).
[43] Nakatsuji, H.; Nakao, T. (Intern. J. Quantum Chem. **49** [1994] 279/90).

1.3.1.1.1.3 Mass Spectrum, Thermal Decomposition, and Chemical Reactions

The **mass spectra** of many organogermanes, including $Ge(CH_3)_3H$, were generated by electron impact (70 eV) at high resolution. Their analysis gave the following fragmentation scheme (relative abundances in parentheses) [8]:

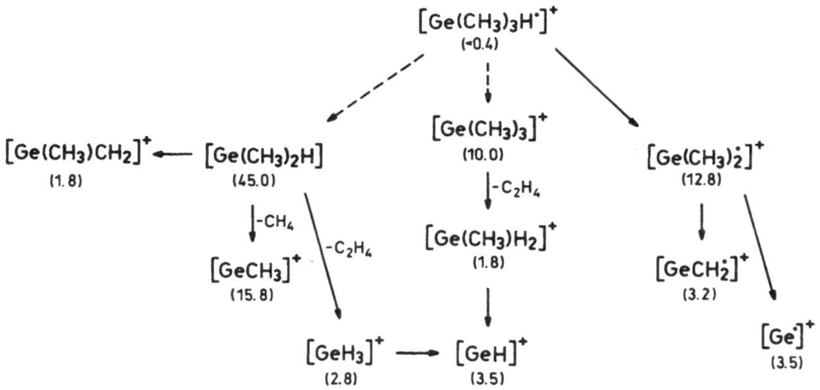

Thermal decomposition in a flow system at 470 °C (pressure not given) produced CH_4 and H_2 in a 30:1 mole ratio and a mixture of $Ge_2(CH_3)_5H$ and $Ge_2(CH_3)_6$ in about a 4:1 ratio. This reaction is much less complex than that of $Si(CH_3)_3H$ and is consistent with the following reaction steps:

$Ge(CH_3)_3H$	$\rightarrow Ge(CH_3)_3^{\cdot} + H^{\cdot}$
	$\rightarrow Ge(CH_3)_2H^{\cdot} + CH_3^{\cdot}$
$CH_3^{\cdot}/H^{\cdot} + Ge(CH_3)_3H$	$\rightarrow Ge(CH_3)_3^{\cdot} + CH_4/H_2$
$Ge(CH_3)_3^{\cdot}$	$\rightarrow Ge(CH_3)_2 + CH_3^{\cdot}$
$Ge(CH_3)_2 + Ge(CH_3)_3H$	$\rightarrow Ge_2(CH_3)_5H$
$3\ Ge(CH_3)_3^{\cdot}$	$\rightarrow Ge_2(CH_3)_6$

Decomposition in a static system (IR cell) begins at 400 °C at an appreciable rate to give CH_4 and $Ge_2(CH_3)_6$. Fourier transform IR monitoring of the integrated intensity of ν(GeH) for time-resolved spectra in the range 400 to 600 °C showed that the rate of reaction is half-order with respect to $Ge(CH_3)_3H$. Time-resolved IR spectra (displayed) of the decomposition of $Ge(CH_3)_3D$ at 460 °C showed the formation of CH_3D and transient $Ge(CH_3)_3H$ and thus indicated that abstraction of hydrogen by $Ge(CH_3)_3^{\cdot}$ radicals from CH_3 groups occurs as well as abstraction of hydridic hydrogen from Ge. The observed activation energy (260 kJ/mol) is lower than estimated from bond dissociation energies suggesting an interaction with the metallic film coating of the cell walls [49]. Pyrolysis in the presence of an equimolar amount of $CH_2{=}CHCH{=}CH_2$ at 500 °C gave CH_4, H_2, and minor amounts of $Ge(CH_3)_3CH_2CH{=}CHCH_3$ and $(CH_3)_2\overline{GeCH_2CH{=}CHCH_2}$ (with the latter in higher yield) based on a mass spectral analysis [21].

Reactions with Elements. Absolute rate constants for the reaction of H atoms (formed by Hg photosensitation of H_2-substrate mixtures at 32 °C) with $Ge(CH_3)_3H$ (at 30 to 60 Torr total pressure) were determined in experiments involving the competitive reaction of H atoms with Si_2H_6. The Arrhenius parameters $A = 0.7 \times 10^{-11}$ cm^3/s and $E_a = 0.26$ kcal/mol and the dissociation energy D(Ge-H) = 63 kcal/mol were calculated [48]; see also $Ge(CH_3)_2H_2$, p. 220.

The gas phase reaction with iodine at 420 to 474 K yielded equal amounts of $Ge(CH_3)_3I$ and HI as the only products. The results of a kinetic study were consistent with the reaction steps: $Ge(CH_3)_3H + I^{\cdot} \rightarrow Ge(CH_3)_3^{\cdot} + HI$ and $Ge(CH_3)_3^{\cdot} + I_2 \rightarrow Ge(CH_3)_3I + I^{\cdot}$. The rate of the first reaction, $\log (k_1/dm^3 \cdot mol^{-1} \cdot s^{-1}) = (9.94 \pm 0.10) - (45.3 \pm 0.9)/RT \ln 10$ (E_a in kJ/mol), was deduced from the overall, surface-independent rate law, and this gave the bond dissociation energy D(Ge-H) = 340 ± 10 kJ/mol (81 kcal/mol). This value is comparable with other estimates ranging from 338 to 386 kJ/mol [30, 34].

References on p. 12

$Ge(CH_3)_3H$ decomposes in the presence of palladium black to give $Ge_2(CH_3)_6$ and H_2 [9].

Insertion reactions into the Ge-H bond were studied with GeH_2, generated by flow pyrolysis of Ge_2H_6 at 280 °C in the presence of $Ge(CH_3)_3H$ (1:1.64 mole ratio). This reaction yielded $(CH_3)_3GeGeH_3$ and Ge_3H_8 in a 1:2 mole ratio. The relative rates were found to decrease (on a per-bond basis) in the order $Ge(CH_3)_3H > Ge(CH_3)_2H_2 > Ge_2H_6 > Ge(CH_3)H_3$. SiH_2 (from Si_2H_6 at 350 °C) could also be inserted to give $(CH_3)_3GeSiH_3$ and a 2.5-fold excess of Si_3H_8 [20]. Thermally generated Ge atoms inserted into the Ge-H bond to produce $Ge(Ge(CH_3)_3)_2H_2$ (0.2%) and $Ge_2(CH_3)_6$ [46].

Hydrogermylation Reactions. $Ge(CH_3)_3H$ can be easily added with or without catalyst across the C=C or C≡C bond of various compounds. The catalysts generally used are H_2PtCl_6 or $(Pt(Ge(CH_3)_3)(\mu\text{-}H)(P(C_6H_{11}\text{-}c)_3))_2$. These reactions are listed in Table 2 where the abbreviations "Pt" and "PtGe" designate the two catalysts. More details are given in the preparation section for each of the products in "Organogermanium Compounds" 1 (1989), 2 (1989), and 3 (1990).

$Ge(CH_3)_3H$ also adds across the carbonyl group of activated ketones, e.g. CF_3COCF_3 at 20 °C for 15 h without catalyst to yield $Ge(CH_3)_3OCH(CF_3)_2$ (ca. 25% conversion) [3, 6].

Table 2
Hydrogermylation Reactions with $Ge(CH_3)_3H$.
For the catalysts "Pt" and "PtGe", see the text above.

reactant	reaction conditions, catalyst → products (yield)	Ref.
$CH_2{=}CHC_4H_9$	20 °C, 24 h, PtGe → $Ge(CH_3)_3C_6H_{13}$ (80%)	[29, 31]
$CH_2{=}CHCH_2Si(CH_3)_3$	20 °C, 24 h, Pt → $Ge(CH_3)_3CH_2CH_2CH_2Si(CH_3)_3$ (70%)	[38]
$CH_2{=}CHCOOC_2H_5$	boiling for 2 h, Pt → $Ge(CH_3)_3CH_2CH_2COOC_2H_5$	[2]
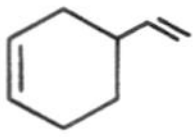	150 °C, 8 h in a sealed tube, Pt → $Ge(CH_3)_3CH_2CH_2\overline{CHCH_2CH{=}CHCH_2CH_2}$ (47%)	[5]
$CH_2{=}CHC_6H_5$	20 °C, 48 h, PtGe → $Ge(CH_3)_3CH_2CH_2C_6H_5$ (76%)	[31, 36]
$CH_2{=}CHCH{=}CH_2$	140 °C, 10 h in a sealed tube, Pt → cis- and trans-$Ge(CH_3)_3CH_2CH{=}CHCH_3$ in a 3:2 ratio (65%)	[5]
$CH_2{=}C{=}CH_2$	115 °C, 8.5 h in a sealed tube, Pt → $Ge(CH_3)_3CH_2CH{=}CH_2$ and $Ge(CH_3)_3C(CH_3){=}CH_2$ in a ca. 3:2 mole ratio (48%)	[5]
	230 °C, 72 h, none → $Ge(CH_3)_3\overline{CFCHFCF_2CF_2}$ (86%)	[4]

Table 2 (continued)

reactant	reaction conditions, catalyst → products (yield)	Ref.
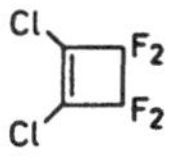	190 °C, 36 h, none (20% conversion) → $Ge(CH_3)_3C{=}C(Cl)CF_2CF_2$, $Ge(CH_3)_3C(Cl)CH(Cl)CF_2CF_2$, cis- and trans-$Ge(CH_3)_3CHCH(Cl)CF_2CF_2$, $Ge(CH_3)_3Cl$ (ca. 40:27:12:20 mole ratio) and traces of $Ge(CH_3)_3C{=}CHCF_2CF_2$	[4]
	175 °C, 5 h, Pt → $Ge(CH_3)_3C{=}CHCH_2CH_2CH_2$ and $Ge(CH_3)_3CHCH{=}CHCH_2CH_2$ in a 63:37 mole ratio (25%)	[5]
	140 °C, 10 h, Pt → $Ge(CH_3)_3CHCH{=}CHCH_2CH_2CH_2$ (60%)	[5]
	140 °C, 10 h, Pt → $Ge(CH_3)_3CHCH{=}CHCH_2CH_2CH_2$ (49%)	[5]
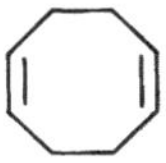	160 °C, 10 h, Pt → $Ge(CH_3)_3CHCH{=}CH(CH_2)_4CH_2$ (30%)	[5]
	150 °C, 10 h, Pt → $Ge(CH_3)_3CHCH{=}CH(CH_2)_4CH_2$ (42%)	[5]
	140 °C, 8 h, Pt → endo- and exo-$Ge(CH_3)_3C_7H_8$ (80%) (see Ge 2, Nos. 60 and 61, p. 40) and small amounts of 	[5]
$CH{\equiv}CCH_2OCH_2OCH_2C{\equiv}CCH_2OH$	65 °C, 5 h in THF → $Ge(CH_3)_3CH{=}CHCH_2OCH_2OCH_2C{\equiv}CCH_2OH$ (58%) for similar reactions with other propargyl ethers, see Ge 1, Nos. 46 and 47, pp. 8/9	[28]
$(CH_3)_3SiOCH_2C{\equiv}CCH_2OSi(CH_3)_3$	100 °C, 2 h in a sealed tube, Pt in THF → $Ge(CH_3)_3C(CH_2OSi(CH_3)_3){=}CHCH_2$-$OSi(CH_3)_3$ (89%)	[40]
$CF_2{=}PCF_3$	low temperature in $C_6H_5CH_3$ (NMR tube) → $Ge(CH_3)_3P(CF_3)CHF_2$	[44]

Reactions with Other Organic Compounds. Unlike SiR_3H, GeR_3H compounds are capable of reducing organic halides by heating in the absence of free-radical initiators; cf. $Ge(C_2H_5)_3H$, p. 53. Reductions with $Ge(CH_3)_3H$ were carried out in the presence of

References on p. 12

t-$C_4H_9OOC_4H_9$-t (DTBP) or C_6H_5CO-OO-OCC_6H_5 (DBP) at 80 or 135°C in a sealed tube for 20 h [18] or by gas-phase irradiation in the range 50 to 200°C [32]. Using either DTBP or DBP, the following reductions were performed ($Ge(CH_3)_3H$ was converted to $Ge(CH_3)_3Cl$): $CCl_4 \rightarrow CHCl_3$, $CH_2ClCHClCH_3 \rightarrow CH_2ClCH_2CH_3$ and $CH_3CHClCH_3$, cyclo-$C_6H_{11}Cl \rightarrow$ cyclo-C_6H_{12}, $Cl(CH_2)_4Cl \rightarrow C_4H_9Cl$, $CHCl_2CH_2Cl \rightarrow CH_2ClCH_2Cl$, and $C_6H_5CH_2Cl \rightarrow C_6H_5CH_3$; other reductions performed in the presence of DTBP at 135°C were $Br(CH_2)_3Cl \rightarrow C_3H_7Cl$ (and $Ge(CH_3)_3Br$), $Br(CH_2)_4Cl \rightarrow C_4H_9Cl$, cyclo-$C_6H_{11}I \rightarrow$ cyclo-C_6H_{12} (and $Ge(CH_3)_3I$). The reactivity of the halogen atom decreases in the order I > Br > Cl. The same trend was found for $C_6H_5X \rightarrow C_6H_6 + Ge(CH_3)_3X$ at 135°C with DTBP. C_6H_5F was not reduced under these conditions, but yielded some $Ge_2(CH_3)_6$; in the case of C_6H_5Cl the condensation product $Ge(CH_3)_3C_6H_5$ was formed in comparable quantities [18]. The gas-phase reactions were carried out to less than 10% conversion in order to compare the reactivities and selectivities of $Ge(CH_3)_3H$ and $Sn(CH_3)_3H$ with those of a large variety of halogenoalkanes, such as chlorobutanes, dichlorobutanes, chlorocyclohexane, and several $RCCl_3$ compounds, including CCl_4, $CHCl_3$, and $CFCl_3$ [32].

For the photoreduction of carbonyl triplets with various X-H substrates, including $Ge(CH_3)_3H$, a modified bond-energy/bond-order method was used to calculate the kinetic parameters [26].

The reaction of $Ge(CH_3)_3H$ with t-$C_4H_9OOC_4H_9$-t (2:1 mole ratio) at 135°C after 20 h yielded $Ge_2(CH_3)_6$ (40%), $Ge(CH_3)_3OC_4H_9$-t (16 to 19%), and small amounts of $(CH_3)_3GeO$-$Ge(CH_3)_3$ and $Ge(CH_3)_3OC_3H_7$-i [22]. The action of either the t-$C_4H_9^{\cdot}$ [25, 27] or t-$C_4H_9O^{\cdot}$ [27] radical on $Ge(CH_3)_3H$ leads to the $Ge(CH_3)_3^{\cdot}$ radical. These processes were used to study the reaction of $Ge(CH_3)_3^{\cdot}$ with butadiene [17] or cyclopentadiene [19]; see also Ge 3, pp. 348/54 and p. 358. $Ge(CH_3)_3H$ was used to produce the $CH(CH_2Ge(CH_3)_3)_2^{\cdot}$ radical by reaction with $Ge(CH_3)_3CH_2CH{=}CH_2$ [42].

Hydride abstraction from $Ge(CH_3)_3H$ to give $[Ge(CH_3)_3]^+$ was achieved using $[C(C_6H_5)_3]ClO_4$ in CH_2Cl_2 or sulfolane [45].

$Ge(CH_3)_3H$ was reacted with $(PC_6H_5)_5$ (5:1 mole ratio) at 210°C to give $Ge(CH_3)_3P(C_6H_5)H$ (65% yield), $P(C_6H_5)H_2$ (12%), and $P(C_6H_5)(Ge(CH_3)_3)_2$ (13%) [35]. The cleavage of element-element bonds by $Ge(CH_3)_3H$ was studied by NMR spectroscopy. Cleavage occurred at 150°C with $P_2(CF_3)_4$ and at 20°C with $As_2(CF_3)_4$ to give $E(CF_3)_2H$ and $Ge(CH_3)_3E(CF_3)_2$ (E = P or As). Reactions with $E(CF_3)_2SCF_3$ and $E(CF_3)_2SeCF_3$ were observed at 100°C for E = P and at 80°C for E = As, yielding $E(CF_3)_2H$ and either $Ge(CH_3)_3SCF_3$ or $Ge(CH_3)_3SeCF_3$. $As(CH_3)_2SCF_3$ gave at 50°C $As(CH_3)_2H$ and $Ge(CH_3)_3SCF_3$ [37].

α-Monoolefins have been polymerized with a $TiCl_3$-$Ge(CH_3)_3H$ catalyst [1].

Reactions with transition metal compounds are summarized in Table 3.

Table 3
Reactions with Transition Metal Compounds.

reactant	reaction conditions; → products (yield)	Ref.
$Ru_3(CO)_{12}$	80°C in C_6H_{14}, 19 h in a sealed tube; → $Ru(CO)_4(Ge(CH_3)_3)_2$ (80%); small amounts of $Ru_2(CO)_6(Ge(CH_3)_3)_4$ also found after prolonged heating above 80°C	[16]

Table 3 (continued)

reactant	reaction conditions; → products (yield)	Ref.
	UV irradiation in C_6H_{14} for 40 h; → $Ru(CO)_4(Ge(CH_3)_3)_2$ (19%) and $Ru_2(CO)_8(Ge(CH_3)_3)_2$ (66%); similar irradiation for 75 h gave $Ru(CO)_4(Ge(CH_3)_3)_2$ as the main product (85%)	
$Ru_2(CO)_8(Si(CH_3)_3)_2$	80°C in C_6H_{12}, 24 h in a sealed tube; → $Ru(CO)_4(Ge(CH_3)_3)_2$ (8%)	[16]
$Ru_2(CO)_8(Ge(CH_3)_3)_2$	UV irradiation in C_6H_{12} for 24 h → $Ru(CO)_4(Ge(CH_3)_3)_2$ (100%)	[16]
$Os_3(CO)_{12}$	140°C in C_6H_{14}, ca. 8 d in a sealed tube → $Os(CO)_4(Ge(CH_3)_3)_2$ (85%) and small amounts of $Os_2(CO)_6(Ge(CH_3)_3)_4$; at 140°C for 1 d $Os(CO)_4(Ge(CH_3)_3)H$ also formed (7%) UV irradiation in C_6H_{14}, ca. 8 d → $Os(CO)_4(Ge(CH_3)_3)H$ (16%) and $Os(CO)_4(Ge(CH_3)_3)_2$ (15%)	[16]
$Os_2(CO)_8(Si(CH_3)_3)_2$	80°C in C_6H_{12}, 2 d in a sealed tube → $Os(CO)_4(Ge(CH_3)_3)H$ (18%) and $Os(CO)_4(Ge(CH_3)_3)_2$ (64%)	[16]
$Rh(P(C_6H_5)_3)_3Cl$	20°C in C_6H_6 → $Rh(P(C_6H_5)_3)_2(Ge(CH_3)_3)(H)Cl$	[13]
$Rh(As(C_6H_5)_3)_3Cl$	20°C in C_6H_6 → $Rh(As(C_6H_5)_3)_2(Ge(CH_3)_3)(H)Cl$	[13]
$Rh(C_5H_5)(CO)_2$	80°C in C_6H_{14}, 5 d; → $Rh(C_5H_5)(CO)(Ge(CH_3)_3)_2$ (8%) UV irradiation in C_6H_{14} for 5 d → $Rh(C_5H_5)(CO)(Ge(CH_3)_3)_2$ (54%)	[24]
$Rh_2(C_5H_5)_2(CO)_3$	80°C in C_6H_{14}, 1 h → $Rh(C_5H_5)(CO)_2$, $Rh(C_5H_5)(CO)(Ge(CH_3)_3)_2$, and $Rh(C_5H_5)(CO)(Ge(CH_3)_3)H$	[24]
$Ir(CO)(P(C_2H_5)_3)_2Cl$ cis and trans	25°C, 6 d → $Ir(CO)(P(C_2H_5)_3)_2(H)_2Cl$ and $Ge_2(CH_3)_6$	[10, 14]
$Ir(CO)(P(C_6H_5)_3)_2Cl$ trans	80°C in C_6H_6 for 12 h and 20°C for several weeks → $Ir(CO)(P(C_6H_5)_3)_2(H)_2Ge(CH_3)_3$	[10, 14, 39]
$Ir(CO)(P(CH_3)_2C_6H_5)_2Cl$	60°C, 5 d → $Ir(CO)(P(CH_3)_2C_6H_5)_2(H)_2Ge(CH_3)_3$	[14]
$Pd(P(C_2H_5)_3)_2X_2$ X = Cl, Br; trans	40°C in a sealed system → trans-$Pd(P(C_2H_5)_3)_2(H)X$	[9]

References on p. 12

Table 3 (continued)

reactant	reaction conditions; → products (yield)	Ref.
$(Pd(\eta^3\text{-}C_3H_5)Cl)_2$	−196 to +20 °C → $Ge(CH_3)_3Cl$, CH_4, $CH_3CH{=}CH_2$, and probably $Ge(CH_3)_2(C_3H_7)H$ and $Ge(CH_3)(CH_2CH{=}CH_2)H_2$ (analyzed by GLC and MS)	[43]
$Pt(P(C_6H_5)_2CH_2CH_2\text{-}P(C_6H_5)_2)(Si(CH_3)_3)Cl$	60 °C in C_6H_6, 3 d → $Pt(P(C_6H_5)_2CH_2CH_2P(C_6H_5)_2)(Ge(CH_3)_3)Cl$ (95%) + $Si(CH_3)_3H$	[12, 15]
$Pt(P(C_6H_5)_2CH_2CH_2\text{-}P(C_6H_5)_2)(Si(CH_3)_3)_2$	50 °C, 3 d; → $Pt(P(C_6H_5)_2CH_2CH_2P(C_6H_5)_2)(Ge(CH_3)_3)_2$ (84%) + $Si(CH_3)_3H$	[12, 15]
$Pt(CH_3)_2P(C_6H_5)_2CH_2CH_2\text{-}P(C_6H_5)_2$	50 °C in C_6H_6, ca. 12 h → $Pt(CH_3)(P(C_6H_5)_2CH_2CH_2P(C_6H_5)_2)Ge(CH_3)_3$ (58%)	[23]
$Pt(C_2H_4)_2P(C_6H_{11}\text{-}c)_3$	20 °C in petroleum ether → $Pt(P(C_6H_{11}\text{-}c)_3)(H)Ge(CH_3)_3$	[29]
$Pt_3(CNC_4H_9\text{-}t)_6$	20 °C in ether, 1 h → $Pt_2(CH{=}NC_4H_9\text{-}t)_2(CNC_4H_9\text{-}t)_2(Ge(CH_3)_3)_2$ (75%)	[33]
$Hg(C_2H_5)_2$	60 °C, 11 h → $Hg(Ge(CH_3)_3)_2$	[7]
	UV irradiation, in C_6H_{14} for 1.5 h → $Hg(Ge(CH_3)_3)_2$ (83%) and C_2H_6 (95%)	[47]
$Hg(C_4H_9\text{-}t)_2$	20 °C → $Hg(Ge(CH_3)_3)_2$ (90 to 95%)	[11]

Methylgermanium species, including $Ge(CH_3)_3X$ compounds, were detected at the parts-per-trillion level in aqueous matrices via hydride generation with $NaBH_4$, graphite-furnace atomization of the germane formed, and atomic absorption spectrometry [41].

References:

[1] Shearer, N. H.; Coover, H. W.; Eastman Kodak Co. (U. S. 2925409 [1960] from C.A. **1960** 13732).

[2] Rijkens, F.; Janssen, M. J.; Drenth, W.; van der Kerk, G. J. M. (J. Organomet. Chem. **2** [1964] 347/56).

[3] Cullen, W. R.; Styan, G. E. (Inorg. Chem. **4** [1965] 1437/40).

[4] Cullen, W. R.; Styan, G. E. (J. Organomet. Chem. **6** [1966] 633/44).

[5] Fish, R. H.; Kuivila, M. G. (J. Org. Chem. **31** [1966] 2445/50).

[6] Styan, G. E. (Diss. Univ. Brit. Columbia 1966; Abstr. B **27** [1966] 1078/9).

[7] Eaborn, C.; Dutton, W. A.; Glockling, F.; Hooton, K. A. (J. Organomet. Chem. **9** [1975] 175/6).

[8] Glockling, F.; Light, J. R. C. (J. Chem. Soc. A **1968** 717/24).

[9] Glockling, F.; Brooks, E. H. (Reprints Am. Chem. Soc. Div. Petrol. Chem. **14** [1969] B135/B137).
[10] Glockling, F.; Wilbey, M. D. (J. Chem. Soc. Chem. Commun. **1969** 286/7).

[11] Neumann, W. P.; Blaukat, U. (Angew. Chem. **81** [1969] 625/6; Angew. Chem. Int. Ed. Engl. **8** [1969] 611/2).
[12] Clemmit, A. F.; Glockling, F. (J. Chem. Soc. Chem. Commun. **1970** 705/6).
[13] Glockling, F.; Hill, G. C. (J. Organomet. Chem. **22** C48/C50).
[14] Glockling, F.; Wilbey, M. D. (J. Chem. Soc. A **1970** 1675/81).
[15] Clemmit, A. F.; Glockling, F. (J. Chem. Soc. A **1971** 1164/9).
[16] Knox, S. A. R.; Stone, F. G. A. (J. Chem. Soc. A **1971** 2874/80).
[17] Kawamura, T.; Maekin, P.; Kochi, S. K. (J. Am. Chem. Soc. **94** [1972] 8065/72).
[18] Sakurai, H.; Mochida, A.; Hosomi, A.; Mita, F. (J. Organomet. Chem. **38** [1972] 275/80).
[19] Kawamura, T.; Kochi, J. K. (J. Organomet. Chem. **47** [1973] 79/88).
[20] Sefcik, M. D.; Ring, M. A. (J. Organomet. Chem. **59** [1973] 167/173).

[21] Paquin, D. P.; O'Connor, R. J.; Ring, M. A. (J. Organomet. Chem. **80** [1974] 341/8).
[22] Sakurai, H.; Nozue, I.; Hosomi, A. (J. Organomet. Chem. **80** [1974] 71/8).
[23] Glockling, F.; Pollock, R. J. I. (J. Chem. Soc. Dalton Trans. **1975** 497/8).
[24] Hill, R.; Knox, S. A. R. (J. Chem. Soc. Dalton Trans. **1975** 2622/7).
[25] Lehnig, M.; Werner, F.; Neumann, W. P. (J. Organomet. Chem. **97** [1975] 375/87).
[26] Previtali, C. M.; Scaiano, J. C. (J. Chem. Soc. Perkin Trans. II **1975** 934/8).
[27] Sakurai, H.; Mochida, K.; Kira, M. (J. Am. Chem. Soc. **97** [1975] 929/31).
[28] Dadasheva, Ya. A. (Azerb. Khim. Zh. **1976** No. 3, pp. 60/3 from C.A. **86** [1977] No. 190112).
[29] Green, M.; Howard, J. A. K.; Proud, J.; Spencer, J. L.; Stone, C. A.; Tsipis, C. A. (J. Chem. Soc. Commun. **1976** 671/2).
[30] Doncaster, A. M.; Walsh, R. (J. Chem. Soc. Chem. Commun. **1977** 446/7).

[31] Green, M.; Spencer, J. L.; Stone, F. G. A.; Tsipis, C. A. (J. Chem. Soc. Dalton Trans. **1977** 1519/25).
[32] Coates, D. A.; Tedder, J. M. (J. Chem. Soc. Perkin Trans. II **1978** 725/8).
[33] Ciriano, M.; Green, M.; Gregson, D.; Howard, J. A. K.; Spencer, J. L.; Stone, F. G. A.; Woodward, P. (J. Chem. Soc. Dalton Trans. **1979** 1294/300).
[34] Doncaster, A. M.; Walsh, R. (J. Phys. Chem. **83** [1979] 578/81).
[35] Escudié, J.; Couret, C.; Satgé, J. (Recl. J. Roy. Neth. Chem. Soc. **98** [1979] 461/6).
[36] Green, M.; Spencer, J. L.; Stone, F. G. A.; Tsipis, C. A. (Brit. Appl. 2013207 from C.A. **93** [1980] No. 150364).
[37] Dehnert, P.; Grobe, J.; Le Van, D. (Z. Naturforsch. **36b** [1981] 48/54).
[38] Taylor, T. G.; Koermer, G. S. (J. Org. Chem. **46** [1981] 3651/7).
[39] Bell, N. A.; Glockling, F.; Schneider, M. L.; Shearer, H. M. M.; Wilbey, M. D. (Acta Cryst. C **40** [1984] 625/8).
[40] Lukevics, E.; Gevorgyan, V. N.; Goldbert, Y. S.; Skymanska, M. V. (J. Organomet. Chem. **263** [1984] 283/96).

[41] Hambrick, G. A.; Froelich, P. N.; Andreae, M. O.; Lewis, B. L. (Anal. Chem. **56** [1984] 421/4).
[42] Kira, M.; Akiyama, M.; Sakurai, H. (J. Organomet. Chem. **271** [1984] 23/31).
[43] Ssebuwufu, P. J.; Glockling, F.; Harriot P. (Inorg. Chim. Acta **98** [1986] L35/L38).
[44] Grobe, J.; Le Van, D.; Nientied, J. (Z. Naturforsch. **42b** [1987] 984/92).
[45] Lambert, J. B.; Schilf, W. (Organometallics **7** [1988] 1659/60).
[46] Mochida, K.; Yoshida, Y. (Bull. Chem. Soc. Jpn. **61** [1988] 1789/90).

[47] Rybin, L. I.; Gendin, D. V.; Vyazankina, O. A.; Vyazankin, N. S. (Metalloorg. Khim. **2** [1989] 334/6; Organometal. Chem. USSR [Engl. Transl.] **2** [1989] 159/61).
[48] Austin, E. R.; Lampe, F. W. (J. Phys. Chem. **81** [1977] 1546/9).
[49] Harrison, P. G.; McManus, J.; Podesta, D. M. (J. Chem. Soc. Chem. Commun. **1992** 291/3).

1.3.1.1.2 Triethylgermane, Triethylgermanium Hydride, $Ge(C_2H_5)_3H$

1.3.1.1.2.1 Preparation and Formation

$Ge(C_2H_5)_3H$ was first prepared in a three-step process consisting of the reaction of $Ge(C_2H_5)_3Br$ with Na to give $Ge_2(C_2H_5)_6$, treatment of the latter with Li to give $Ge(C_2H_5)_3Li$ followed by ammonolysis to yield $Ge(C_2H_5)_3H$ [1].

$Ge(C_2H_5)_3H$ was obtained with 79% yield by refluxing $Ge(C_2H_5)_3Cl$ and $LiAlH_4$ in ether for 2 h followed by acid hydrolysis of the reaction mixture and usual workup of the ether layer [2]; see also [5, 55]. Reduction at room temperature within 3 d and neutral hydrolysis of the decanted ether solution is described in [59]. $LiAlH_4$ reduction in other ether solvents such as $O(CH_3)_2$ or tetraglyme are mentioned without detail in [13]. $Ge(C_2H_5)_3Cl$ and $LiAlH_4$ in C_5H_{12} at 40°C required ultrasonic agitation for 4.5 h to yield more than 95% $Ge(C_2H_5)_3H$ [45]; see also [60]. Yields above 82% resulted from the reaction of $Ge(C_2H_5)_3Cl$ with solid $LiAlH_4$ in C_6H_6, if 15-crown-5 ether or $[N(C_2H_5)_3CH_2C_6H_5]Cl$ were present as the phase transfer catalysts [48].

Reductions of $Ge(C_2H_5)_3Cl$ with LiH in $O(C_5H_{11}\text{-i})_2$ (24% yield) [4] and with LiH, NaH, $LiBH_4$, and $NaBH_4$ in C_6H_6 at 80°C in the presence of 18-crown-6 ether [60] were also reported. A quantitative yield was obtained with an LiH-containing product prepared from $Ge(C_6H_5)_2H_2$ and $LiC_4H_9\text{-t}$ (1:1.2 mole ratio) in $N(C_2H_5)_3\text{-}C_5H_{12}$ at −40 to +20°C and its subsequent reaction with $Ge(C_2H_5)_3Cl$ in THF at 60°C (15 h) in a sealed tube [57].

Reducing $Ge(C_2H_5)_3Br$ with $LiAlH_4$ in ether at 0 to 20°C (procedure described for $Sn(C_6H_5)_3H$) yielded 82% $Ge(C_2H_5)_3H$ [41]; see also [3, 5, 13]. Reduction with $NaBH_4$ in aqueous acidic medium gave $Ge(C_2H_5)_3H$ with a low yield [12].

Reduction of $Ge(C_2H_5)_3I$ with $LiH\text{-}Al(C_4H_9\text{-i})_2H$ in $C_{12}H_{26}$ at 100°C gave a 96% yield of $Ge(C_2H_5)_3H$. Its formation from $Ge(C_2H_5)_3I$ and $NaBH_4$ (in ether or refluxing THF), $Al(C_4H_9\text{-i})_2H$ (at 90°C), or $LiAlH_4$ (in refluxing ether) with 10 to 60% yield was always accompanied by isomerization to $Ge(C_2H_5)_4$ and $Ge(C_2H_5)_2H_2$. LiH in $C_6H_5CH_3$ at 100°C gave no $Ge(C_2H_5)_3H$ [44].

$Ge(C_2H_5)_3H$ resulted in 95% yield from the cleavage of $(C_2H_5)_3GeGe(C_6H_5)_3$ with $LiAlH_4$ in THF at room temperature after 24 h, whereas only traces were formed from $Ge_2(C_2H_5)_6$ under the same conditions (5% conversion after 1 d) [36]. Other $Ge(C_2H_5)_3X$ compounds (X = $OCH{=}C(C_6H_5)_2$ [16], SR [10], or $P(C_6H_5)_2$ [6]) also produced $Ge(C_2H_5)_3H$ when treated with $LiAlH_4$.

$GeLi_4$ (supposed to be formed from $GeCl_4$ and gaseous Li) yielded ca. 20% $Ge(C_2H_5)_3H$ (along with $Ge(C_2H_5)_4$) when treated with C_2H_5Br in THF [35]. It was formed in competitive reactions of $Ge(C_2H_5)_3Li$ and $Ge(C_2H_5)_2(R)Li$ compounds with CH_3OH in THF-C_6H_6 at −78 to +20°C [54, 56].

$Ge(C_2H_5)_3H$ was prepared with 45% yield from $K_2[Ge(1,2\text{-}O_2C_6H_4)_3]$ and C_2H_5MgBr (1:5 mole ratio) in refluxing ether (4 h) in the presence of 5 mol% $Ti(C_5H_5)_2Cl_2$ [51]; see also [58].

Other reports on the formation of $Ge(C_2H_5)_3H$ are summarized in Table 4. They involve decomposition and solvolysis reactions of main-group and transition metal compounds containing the $Ge(C_2H_5)_3$ group.

Table 4
Other Formations of $Ge(C_2H_5)_3H$.

reactants	conditions → products (yield)	Ref.
GeF_2 + C_2H_5Br, C_2H_5I or $GeCl_4$, $Ge(C_6H_5)_3Cl$, $Ge(C_2H_5)_2(OCH_3)_2$	insertion into C-X bonds at 100 °C or Ge-X bonds at 20 °C and characterization of the unstable products by alkylation with C_2H_5MgBr and hydrolysis → $Ge(C_2H_5)_3H$ (5 to 15%) among other fully alkylated germanes	[32]
GeF_2 + $Ge(C_2H_5)_3SCH_3$, $Ge(C_2H_5)_3N(CH_3)_2$	insertion as above followed by C_6H_5MgBr → $Ge(C_2H_5)_3H$ (10 to 14%) along with $Ge(C_2H_5)_3C_6H_5$ and $Ge_2(C_2H_5)_6$	[32]
$GeCl_2$ + $GeCl_4$	at 120 °C probably equilibrium with Ge_2Cl_6; alkylation with C_2H_5MgBr → $Ge(C_2H_5)_3H$, $Ge(C_2H_5)_4$, $(Ge(C_2H_5)_2)_n$, and traces of $Ge_2(C_2H_5)_6$	[23]
$Ge(C_2H_5)Cl$ + GeX_4 (X = Cl, Br, I)	insertion at 20 °C followed by C_2H_5MgBr → $Ge(C_2H_5)_3H$, $Ge(C_2H_5)_4$, $(Ge(C_2H_5)_2)_n$, and $Ge_2(C_2H_5)_6$ (20%)	[23]
Ge_2H_6 + C_2H_4	160 °C for about 3 d in a sealed tube → $Ge(C_2H_5)_3H$ identified among other products	[15]
$(Ge(C_2H_5)_3)_2O$ + K	1:2 mole ratio in HMPT; hydrolysis → $Ge(C_2H_5)_3H$ (via $Ge(C_2H_5)_3K$) and $(Ge(C_2H_5)_3)_2O$ (via $Ge(C_2H_5)_3OK$)	[21]
$Ge(C_2H_5)_3Cl$ + Ca atoms	in the vapor phase and condensation at 77 K; then hydrolysis → $Ge(C_2H_5)_3H$ (26%) via $Ge(C_2H_5)_3CaCl$	[46]
$Ge(C_2H_5)_2H_2$ + Na or K	in HMPT-THF, then reaction with C_2H_5Br and hydrolysis → $Ge(C_2H_5)_3H$	[56]
$Ge(C_2H_5)_3CH_2CN$ + excess $LiAlH_4$	in ether; followed by hydrolysis → $Ge(C_2H_5)_3H$ and $C_2H_5NH_2$ (probably via $Ge(C_2H_5)_3CH_2CH{=}NLi$)	[34]
$Ge(C_2H_5)_3CH_2CH_2CH_2$-CO-OOC_4H_9-t	dec. at 125 °C in a dilute solution of $CHCl_3$ → $Ge(C_2H_5)_3H$ (small amount) among many other products	[40]

References on p. 19

Table 4 (continued)

reactants	conditions → products (yield)	Ref.
$Ge(C_2H_5)_3Li + CH_3CON(CH_3)_2$	in THF at −78°C → $Ge(C_2H_5)_3H$ and $LiCH_2CON(CH_3)_2$; in C_6H_{14} at 20°C and hydrolysis → $Ge(C_2H_5)_3H$, $Ge(C_2H_5)_3COCH_3$, and $Ge(C_2H_5)_3C(CH_3)(OH)CH_2CON(CH_3)_2$	[43]
$Ge(C_2H_5)_3M + CH_3COC_6H_5$ (M = Li, K, Rb)	in C_6H_6 at 20°C → $Ge(C_2H_5)_3H$ (27, 74, or 99%) and $Ge(C_2H_5)_3C(CH_3)(C_6H_5)OH$ with $Ge(C_2H_5)_3Li$ in C_6H_6-HMPT → $Ge(C_2H_5)_3H$ (48%)	[28]
$Ge(C_2H_5)_3M + CH_2{=}CH_2$ (M = Li, Na, K)	in C_6H_6 at 20°C, hydrolysis → $Ge(C_2H_5)_3H$ (5, 44, or 83%) and $Ge(C_2H_5)_4$ (81, 37, 13%)	[27]
$Ge(C_2H_5)_3M + C_6H_5C{\equiv}CH$ (M = Li, K)	in C_6H_6 at 20°C → $Ge(C_2H_5)_3H$ (57, 82%) and $C_6H_5C{\equiv}CM$	[28]
$Ge(C_2H_5)_3Li + C_6H_5OC{\equiv}CH$	in THF at 60°C, hydrolysis → $Ge(C_2H_5)_3H$ (23%), $Ge_2(C_2H_5)_6$ (19%), and ca. 3% germylphenoxyethylenes	[33]
$Ge(C_2H_5)_3Cs + C_6H_5CH_3$	in $C_6H_5CH_3$ at 25°C (5 d) → equilibrium mixture containing $Ge(C_2H_5)_3H$ and $C_6H_5CH_2Cs$	[50, 56]
$Ge(C_2H_5)_3Li + Ge(C_2H_5)_2(R)H$ (R = t-C_4H_9, C_6H_5)	in HMPT at 25°C (3 d) → equilibrium mixtures containing $Ge(C_2H_5)_3H$	[50, 56]
$Ge(C_2H_5)_3M + Ge(C_2H_5)_2(R)H$ (M = Li, Cs; R = H, Cl, $Si(CH_3)_3$)	→ $Ge(C_2H_5)_3H$ and various other products resulting from a series of metallation experiments, dealing with the reactivity of organogermanium anions	[50, 56]
$Te(Ge(C_2H_5)_3)_2 + Sn(C_2H_5)_3H$	170°C (1 h) → $Ge(C_2H_5)_3H$ (57%) and $Te(Sn(C_2H_5)_3)_2$ (72%)	[17]
$Sb(Ge(C_2H_5)_3)_3 + Ge(C_6H_5)_3H$	230°C (15 h) → $Ge(C_2H_5)_3H$ (52%), Sb (11%), and $Sb(Ge(C_6H_5)_3)_3$ (identified by reaction with $C_2H_4Br_2$ at 100°C)	[19]
$Sb(Ge(C_2H_5)_3)_3 + Sn(C_2H_5)_3H$	180°C (10 h) → $Ge(C_2H_5)_3H$ (88%) and $Sb(Sn(C_2H_5)_3)_3$ (81%)	[14]

Table 4 (continued)

reactants	conditions → products (yield)	Ref.
$Bi(Ge(C_2H_5)_3)_3$ + CH_3COOH (excess)	170 °C (2 h) → $Ge(C_2H_5)_3H$ (53%), Bi (86%), and $Ge(C_2H_5)_3OOCCH_3$ (69%)	[9, 14]
$Bi(Ge(C_2H_5)_3)_3$ + $Sn(C_2H_5)_3H$	170 °C (12 h) → $Ge(C_2H_5)_3H$ (62%) and $Bi(Sn(C_2H_5)_3)_3$, decomposing to Bi (89%) and $Sn_2(C_2H_5)_6$ (84%)	[9, 14]
$Yb(Ge(C_2H_5)_3)Br \cdot 2\,THF$ + C_6H_5OH	in THF at 20 °C → $Ge(C_2H_5)_3H$ (50%)	[49]
$(YbH_2 \cdot (THF)_m)_n$ (m = 1 or 2) + $Ge(C_2H_5)_3Br$	in THF at 20 °C, sealed tube → $Ge(C_2H_5)_3H$ (77 to 86%) and $YbBr_2$	[53]; see also [52]
$Ti(C_5H_5)_2Cl_2$ + $Cd(Ge(C_2H_5)_3)_2$	in $C_6H_5CH_3$, 20 °C (3 d) → $Ge(C_2H_5)_3H$, $Ti(C_5H_5)_2(Cl)Ge(C_2H_5)_3$ (the main product), $Ge(C_2H_5)_3Cl$, $Ge_2(C_2H_5)_6$, Cd, and $Ti(C_5H_5)_2Cl_2 \cdot Cd(Ge(C_2H_5)_3)_2$	[31]
$Ti(C_5H_5)_2Cl_2 \cdot Cd(Ge(C_2H_5)_3)_2$	dec. in $C_6H_5CH_3$ at 100 °C (6 h) via $Ti(C_5H_5)_2(Cl)Ge(C_2H_5)_3$ → $Ge(C_2H_5)_3H$ (19%), $Ge(C_2H_5)_3Cl$, $Ge_2(C_2H_5)_6$, and $Ti(C_5H_5)_2Cl$	[31]
$Ti(C_5H_5)_2Cl_2 \cdot Cd(Ge(C_2H_5)_3)_2$ + CH_3COOH	in $C_6H_5CH_3$ at 20 °C (2d) → $Ge(C_2H_5)_3H$ (79%), $Ge(C_2H_5)_3Cl$, $Ti(C_5H_5)_2OOCCH_3$, and Cd salts	[31]
$Ti(C_5H_5)_2(Cl)Ge(C_2H_5)_3$ (solid)	in vacuum at 380 °C → $Ge(C_2H_5)_3H$ (30%), $Ge(C_2H_5)_3C_5H_5$ (43%), and $Ti(C_5H_5)_2Cl$ (26%)	[38]
$Zr(C_5H_5)_2(Cl)Ge(C_2H_5)_3$ + HCl	in $C_6H_5CH_3$, 20 °C (1 d) → $Ge(C_2H_5)_3H$ (71%), $Ge(C_2H_5)_3Cl$, and $Zr(C_5H_5)_2Cl_2$	[29]
$V(C_5H_5)_2Ge(C_2H_5)_3$	150 °C in vacuum or heating in $C_6H_5CH_3$ or THF → $Ge(C_2H_5)_3H$ (ca. 30%), $Ge(C_2H_5)_3C_5H_5$, $V(C_5H_5)_2$, and $V(C_5H_5)C_5H_4Ge(C_2H_5)_3$	[42]
$V(C_5H_5)_2Ge(C_2H_5)_3$ + HCl	in $C_6H_5CH_3$ or dioxane → $Ge(C_2H_5)_3H$ (5%), $Ge(C_2H_5)_3Cl$, H_2, and $V(C_5H_5)_2Cl$	[30]
$V(C_5H_5)_2Ge(C_2H_5)_3$ + CH_3COOH	in $C_6H_5CH_3$ (24, 48, and 78 h) → $Ge(C_2H_5)_3H$ (45, 60, and 85%)	[30]
$V(C_5H_5)_2$ + $Cd(Ge(C_2H_5)_3)_2$	in $C_6H_5CH_3$, 20 °C (2 d) → $Ge(C_2H_5)_3H$ (60%; hydrogen abstracted from C_5H_5), $Ge_2(C_2H_5)_6$, Cd, and $V(C_5H_5)_2Ge(C_2H_5)_3$	[30]

References on p. 19

Table 4 (continued)

reactants	conditions → products (yield)	Ref.
$V(C_5H_5)_2Cl_2 \cdot Cd(Ge(C_2H_5)_3)_2$	in $C_6H_5CH_3$ at 100°C (2 h) → $Ge(C_2H_5)_3H$ (24%), $Ge(C_2H_5)_3Cl$, $V(C_5H_5)_2$, $V(C_5H_5)_2Cl$, and Cd	[37]
$V(C_5H_5)_2Cl_2 \cdot Cd(Ge(C_2H_5)_3)_2$ + HCl	in $C_6H_5CH_3$ or dioxane, 20°C → $Ge(C_2H_5)_3H$ (20%), $Ge(C_2H_5)_3Cl$, $V(C_5H_5)_2Cl_2$, and $CdCl_2$	[37]
$Nb(C_5H_5)_2(Cl)Ge(C_2H_5)_3$ (solid)	in vacuum at 400°C → $Ge(C_2H_5)_3H$ (47%), $Ge(C_2H_5)_3C_5H_5$ (34%), and C_5H_6 (38%)	[38]
$Ir(CO)(H)_2Ge(C_2H_5)_3$-$(P(C_6H_5)_3)_2$ + HCl or H_2	in C_6H_6 → $Ge(C_2H_5)_3H$ along with other products	[20]
$(C_2H_5)_3GeZn(Ge(C_2H_5)_2)_n$-$Ge(C_2H_5)_3$, hydrolysis	starting material obtained from $Ge(C_2H_5)_3H$ and $Zn(C_2H_5)_2$ at 125°C → $Ge(C_2H_5)_3H$, $Ge_2(C_2H_5)_5H$, $Ge_2(C_2H_5)_6$, and $Zn(OH)_2$	[8]
$Cd(Ge(C_2H_5)_3)_2$ + HX	X = OH, C_3H_7O [11], C_6H_5O [22], CH_3COO [11, 22], C_6H_5COO [22] → $Ge(C_2H_5)_3H$ (75 to 90% [22])	[11, 22]
$Cd(Ge(C_2H_5)_3)_2$ + $Sn(C_2H_5)_3H$	20 to 25°C (3 d) → $Ge(C_2H_5)_3H$ (95%), $Sn_2(C_2H_5)_6$ (59%), and Cd (100%)	[11]
$Hg(Ge(C_2H_5)_3)_2$ + CH_3COOH	heating → $Ge(C_2H_5)_3H$ (ca. 50%), $Ge(C_2H_5)_3OOCCH_3$ and metallic Hg	[9, 14]
$Hg(Ge(C_2H_5)_3)_2$ + $Si(C_2H_5)_3EH$ (E = S, Se, or Te)	→ $Ge(C_2H_5)_3H$ along with $(C_2H_5)_3Si$-$EHgGe(C_2H_5)_3$ (E = S, Se, or Te) or $Hg(ESi(C_2H_5)_3)_2$ (E = Se or Te)	[24]
$Hg(Ge(C_2H_5)_3)_2$ + $Si(OCH_3)_3(CH_2)_nSH$ (n = 1 or 2)	→ $Ge(C_2H_5)_3H$ along with $Ge(C_2H_5)_3$-$S(CH_2)_nSi(OCH_3)_3$	[47]
$Hg(Ge(C_2H_5)_3)_2$ + $Ge(C_6F_5)_3H$	1:2 mole ratio, in C_6H_6 → $Ge(C_2H_5)_3H$ and $Hg(Ge(C_6F_5)_3)_2$	[25]

Chromatographic characteristics of $Ge(C_2H_5)_3H$ in GLC [7, 26] and GSC [26] were described in connection with equivalent data for alkyl derivatives of Si, Ge [7, 26], and Sn [26].

Standard enthalpies of formation of $Ge(C_2H_5)_3H$ based on the heat of combustion are ΔH°_f(liquid) = −40.1 ± 2 kcal/mol and ΔH°_f(gas) = −30.2 ± 2.5 kcal/mol [18].

$Ge(C_2H_5)_3D$ was obtained with 31% yield by reducing $Ge(C_2H_5)_3Cl$ with LiD in $O(C_5H_{11}\text{-i})_2$ [4] and by reducing $Ge(C_2H_5)_3Br$ with $(YbD_2 \cdot THF)_n$ [53].

$Ge(C_2H_5)_3T$ was obtained by base-catalysed tritiation of $Ge(C_2H_5)_3H$ with tritiated water in refluxing CH_3OH containing CH_3ONa [39].

References:

[1] Kraus, C.A.; Flood, E. A. (J. Am. Chem. Soc. **54** [1932] 1635/44).
[2] Anderson, H. H. (J. Am. Chem. Soc. **79** [1957] 326/8).
[3] Lesbre, M.; Satgé, J. (C. R. Hebd. Seances Acad. Sci. **247** [1958] 471/4).
[4] Ponomarenko, V. A.; Zueva, G. Ya.; Andreev, N. S. (Izv. Akad. Nauk SSSR Ser. Khim. **1961** 1758/62; Bull. Acad. Sci. USSR Div. Chem. Sci. [Engl. Transl.] **1961** 1639/43).
[5] Satgé, J. (Ann. Chim. [Paris] [13] **6** [1961] 519/73).
[6] Satgé, J.; Lesbre, M.; Baudet, M. (C. R. Hebd. Seances Acad. Sci. **259** [1964] 4733/6).
[7] Semlyen, J. A.; Walker, G. R.; Blofeld, R. E.; Phillips, C. S. G. (J. Chem. Soc. **1964** 4948/53).
[8] Vyazankin, N. S.; Razuvaev, G. A.; Korneva, S. P.; Kruglaya, O. A.; Galiulina, R. F. (Dokl. Akad. Nauk SSSR **158** [1964] 884/7; Dokl. Chem. [Engl. Transl.] **154/159** [1964] 1002/4).
[9] Kruglaya, O. A.; Vyazankin, N. S.; Razuvaev, G. A. (Zh. Obshch. Khim. **35** [1965] 394; J. Gen. Chem. USSR [Engl. Transl.] **35** [1965] 392).
[10] Satgé, J.; Lesbre, M. (Bull. Soc. Chim. Fr. **1965** 2578/81).

[11] Vyazankin, N. S.; Razuvaev, G. A.; Bychkov, V. T. (Izv. Akad. Nauk SSSR Ser. Khim. **1965** 1665/7; Bull. Acad. Sci. USSR Div. Chem. Sci. [Engl. Transl.] **1965** 1624/5).
[12] Cullen, W. R.; Styan, G. E. (J. Organomet. Chem. **6** [1966] 117/25).
[13] Mackay, K. M.; Watt, R. (J. Organomet. Chem. **6** [1966] 336/51).
[14] Vyazankin, N. S.; Razuvaev, G. A.; Kruglaya, O. A.; Semchikova, G. S. (J. Organomet. Chem. **6** [1966] 474/83).
[15] Mackay, K. M.; Watt, R. (J. Organomet. Chem. **14** [1968] 123/9).
[16] Rivière, P.; Satgé, J. (C. R. Seances Acad. Sci. C **267** [1968] 267/9).
[17] Bochkarev, M. N.; Sanina, L. P.; Vyazankin, N. S. (Zh. Obshch. Khim. **39** [1969] 135/41; J. Gen. Chem. USSR [Engl. Transl.] **39** [1969] 122/7).
[18] Tel'noi, V. I.; Kol'yakova, G. M.; Rabinovich, I. B.; Vyazankin, N. S. (Dokl. Akad. Nauk SSSR **185** [1969] 374/6; Dokl. Chem. [Engl. Transl.] **184/189** [1969] 214/6).
[19] Vyazankin, N. S.; Kalinina, G. S.; Kruglaya, O. A.; Razuvaev, G. A. (Zh. Obshch. Khim. **39** [1969] 2005/11; J. Gen. Chem. USSR [Engl. Transl.] **39** [1969] 1964/8).
[20] Glockling, F.; Wilbey, M. D. (J. Chem. Soc. A **1970** 1675/81).

[21] Bulten, E. J.; Noltes, J. G. (J. Organomet. Chem. **29** [1971] 397/407).
[22] Bychkov, V. T.; Linzina, O. V.; Razuvaev, G. A.; Vyazankin, N. S. (Izv. Akad. Nauk SSSR Ser. Khim. **1971** 2403/6; Bull. Acad. Sci. USSR Div. Chem. Sci. [Engl. Transl.] **1971** 2283/6).
[23] Massol, M.; Barrau, J.; Satgé, J. (Inorg. Nucl. Chem. Lett. **7** [1971] 895/9).
[24] Bochkarev, M. N.; Maiorova, L. P.; Charov, A. I.; Vyazankin, N. S. (Izv. Akad. Nauk SSSR Ser. Khim. **1972** 1375/9; Bull. Acad. Sci. USSR Div. Chem. Sci. [Engl. Transl.] **1972** 1324/7).
[25] Bochkarev, M. N.; Maiorova, L. P.; Vyazankin, N. S. (J. Organomet. Chem. **55** [1973] 89/96).
[26] Bortnikov, G. N.; Vyazankin, N. S.; Nikulina, N. P.; Yashin, Ya. I. (Izv. Akad. Nauk SSSR Ser. Khim. **1973** 21/4; Bull. Acad. Sci. USSR Div. Chem. Sci. [Engl. Transl.] **1973** 19/21).

[27] Gladyshev, E. N.; Fedorova, E. A.; Vyazankin, N. S.; Razuvaev, G. A. (Zh. Obshch. Khim. **43** [1973] 1315/9; J. Gen. Chem. USSR [Engl. Transl.] **43** [1973] 1306/10).
[28] Gladyshev, E. N.; Vyazankin, N. S.; Fedorova, E. A.; Yuntila, L. O.; Razuvaev, G. A. (J. Organomet. Chem. **64** [1974] 307/14).
[29] Bychkov, V. T.; Lomakova, I. V.; Domrachev, G. A. (Izv. Akad. Nauk SSSR Ser. Khim. **1975** 2115/6; Bull. Acad. Sci. USSR Div. Chem. Sci. [Engl. Transl.] **1975** 2001/2).
[30] Razuvaev, G. A.; Bychkov, V. T.; Vyshinskaya, L. I.; Latyaeva, V. N.; Spiridonova, N. N. (Dokl. Akad. Nauk SSSR **220** [1975] 854/5; Dokl. Chem. [Engl. Transl.] **220/225** [1975] 126/7).

[31] Razuvaev, G. A.; Latyaeva, V. N.; Vishinskaya, L. I.; Bytchkov, V. T.; Vasilyeva, G. A. (J. Organomet. Chem. **87** [1975] 93/9).
[32] Rivière, P.; Satgé, J.; Boy, A. (J. Organomet. Chem. **96** [1975] 25/40).
[33] Kruglaya, O. A.; Lyashenko, G. S.; Filippova, A. Kh.; Keiko, V. V.; Kalikhman, I. D.; Vyazankin, N. S. (Izv. Akad. Nauk SSSR Ser. Khim. **1976** 682/3; Bull. Acad. Sci. USSR Div. Chem. Sci. [Engl. Transl.] **1976** 669/71).
[34] Rivière-Baudet, M.; Rivière, P. (J. Organomet. Chem. **116** [1976] C49/C52).
[35] Morrison, J. A.; Lagow, R. J. (Inorg. Chem. **16** [1977] 2972/4).
[36] Duffaut, N.; Dunogues, J.; Calas, R.; Rivière, P.; Satgé, J.; Cazes, A. (J. Organomet. Chem. **149** [1978] 57/63).
[37] Vyshinskaya, L. I.; Korneva, S. P.; Cherkasov, V. K. (Izv. Akad. Nauk SSSR Ser. Khim. **1978** 1405/6; Bull. Acad. Sci. USSR Div. Chem. Sci. [Engl. Transl.] **1978** 1224/5).
[38] Domrachev, G. A.; Khamylov, V. K.; Lomakova, I. V.; Bychkov, V. T. (Zh. Obshch. Khim. **49** [1979] 2533/6; J. Gen. Chem. USSR [Engl. Transl.] **49** [1979] 2239/41).
[39] Eaborn, C.; Singh, B. (J. Organomet. Chem. **177** [1979] 333/48).
[40] Razuvaev, G. A.; Brevnova, T. N.; Chesnokova, T. A. (Zh. Obshch. Khim. **49** [1979] 2537/41; J. Gen. Chem. USSR [Engl. Transl.] **49** [1979] 2242/6).

[41] Stang, P. J.; White, M. R. (J. Am. Chem. Soc. **103** [1981] 5429/33).
[42] Razuvaev, G. A.; Latyaeva, V. N.; Mar'in, V. P.; Vyshinskaya, L. I.; Korneva, S. P.; Andrianov, Yu. A.; Krasil'nikova, E. V. (J. Organomet. Chem. **225** [1982] 233/44).
[43] Bravo-Zhivotovskii, D. A.; Pigarev, S. D.; Kalikhman, I. D.; Vyazankina, O. A.; Vyazankin, N. S. (J. Organomet. Chem. **248** [1983] 51/60).
[44] Salimgareeva, I. M.; Bogatova, N. G.; Panasenko, A. A.; Khalilov, L. M.; Purlei, I. I.; Mavrodiev, V. K.; Yur'ev, V. P. (Izv. Akad. Nauk SSSR Ser. Khim. **1983** 1605/12; Bull. Acad. Sci. USSR Div. Chem. Sci. [Engl. Transl.] **1983** 1456/61).
[45] Lukevics, E.; Gevorgyan, V. N.; Goldberg, Y. S. (Tetrahedron Lett. **25** [1984] 1415/6).
[46] Mochida, K.; Manishi, M. (Chem. Lett. **1984** 1077/80).
[47] Voronkov, M. G.; Chernov, N. F.; Gendin, D. V.; Rybin, L. I.; Vyazankina, O. A.; Vyazankin, N. S. (Izv. Akad. Nauk SSSR Ser. Khim. **1984** 2401/2; Bull. Acad. Sci. USSR Div. Chem. Sci. [Engl. Transl.] **1984** 2194/5).
[48] Gevorgyan, V. N.; Ignatovich, L. M.; Lukevics, E. (J. Organomet. Chem. **284** [1985] C31/C32).
[49] Bochkarev, L. N.; Fedyushkin, I. L.; Kholodilova, M. N.; Zhil'tsov, S. F.; Bochkarev, M. N.; Razuvaev, G. A. (Izv. Akad. Nauk SSSR Ser. Khim. **1987** 658/9; Bull. Acad. Sci. USSR Div. Chem. Sci. [Engl. Transl.] **1987** 600/2).
[50] Bravo-Zhivotovskii, D. A.; Pigarev, S. D.; Vyazankina, O. A.; Vyazankin, N. S. (Zh. Obshch. Khim. **57** [1987] 2735/8; J. Gen. Chem. USSR [Engl. Transl.] **57** [1987] 2440/2).

[51] Cerveau, G.; Chult, C.; Corriu, R. J. P.; Reyé, C. (Organometallics **7** [1988] 786/7).

[52] Bochkarev, M. N.; Fedorov, E. A.; Penyagina, I. M.; Vasina, O. A.; Khorshev, S. Ya.; Protchenko, A. V. (Metalloorg. Khim. **2** [1989] 703; Organomet. Chem. USSR [Engl. Transl.] **2** [1989] 363).
[53] Bochkarev, M. N.; Fedorova, E. A.; Penyagina, I. M.; Vasina, O. A.; Protchenko, A. V.; Khorshev, S. Ya. (Metalloorg. Khim. **2** [1989] 1317/9; Organomet. Chem. USSR [Engl. Transl.] **2** [1989] 696/7).
[54] Bravo-Zhivotovskii, D. A.; Pigarev, S. D.; Voronkov, M. G.; Vyazankin, N. S. (Zh. Obshch. Khim. **59** [1989] 863/5; J. Gen. Chem. USSR [Engl. Transl.] **59** [1989] 761/3).
[55] Kaushal, P.; Roberts, B. P. (J. Chem. Soc. Perkin Trans. II **1989** 1559/68).
[56] Pigarev, S. D.; Bravo-Zhivotovskii, D. A.; Kalikhman, I. D.; Vyazankin, N. S.; Voronkov, M. G. (J. Organomet. Chem. **369** [1989] 29/41).
[57] Castel, A.; Rivière, P.; Satgé, J.; Ko, H. Y. (Organometallics **9** [1990] 205/10).
[58] Cerveau, G.; Chuit, C.; Corriu, R. J. P.; Reyé, C. (Organometallics **10** [1991] 1510/5).
[59] Clark, K. B.; Griller, D. (Organometallics **10** [1991] 746/50).
[60] Lukevics, E.; Gevorgyan, V. (Chem. Technol. Silicon Tin. Proc. Asian Network Anal. Inorg. Chem. 1st Int. Chem. Conf. Silicon Tin, Kuala Lumpur 1989 [1992], pp. 165/77).

1.3.1.1.2.2 The Molecule, Spectra, and Physical Properties

A Ge-C distance of 1.96 Å was calculated based on the EXAFS spectrum (extended X-ray absorption fine-structure) and its Fourier transform; both illustrated and compared with the spectra of GeR_3M compounds (M = alkali metal), suggesting an increased Ge-C bond lengths (by ca. 5%) in the $[GeR_3]^-$ anions [29].

The bond dissociation energy, D(Ge-H) = 74 ± 5 kcal/mol, and the enthalpy of atomization, ΔH°_{298} = 1979.2 ± 3.5 kcal/mol, were derived from the heat of combustion [12]. D(Ge-H) = 82.3 ± 0.6 kcal/mol was determined in solution (C_6H_6 or iso-C_8H_{18}) at 23 °C by a photoacoustic technique [32].

^{1}H NMR spectrum (in C_6H_6): δ(ppm) = 0.79 (CH_2, J(C, H) = 123.5 Hz), 1.05 (CH_3, J(C, H) = 127 Hz), and 3.93 (GeH, ^{3}J(H, H) = 2.6 Hz) [10]; A_3B_2-type signals of the C_2H_5 group [11] with ^{3}J(H, H) = 7.8 Hz [10]. The ethyl protons appear as multiplets at 1.17 ppm in HMPT-C_4D_8O [30] and 1.08 ppm in $CDCl_3$ [32]. Other values for the GeH resonance: δ(ppm) = 3.64 (in HMPT-C_4D_8O) [30], 3.71 (in $CDCl_3$ [32]), 3.83 (50% in C_6H_6) [10], and 4.70 (?) (in CCl_4) [18]; for δ(GeH) of several germanium hydrides and a comparison with corresponding Si and C compounds, see also [7].

^{13}C NMR spectrum: δ(ppm) = 4.2 and 10.5 (neat?) [26]; 5.09 (CH_2) and 11.45 (CH_3) in HMPT-C_4D_8O [30]. ^{73}Ge NMR spectrum (neat): δ(ppm) = 15.7 ± 0.8, $W_{1/2}$ = 20 ± 2 Hz [26] (see also [24]), and J(Ge, H) = 88 ± 2 Hz [26].

The IR spectrum of $Ge(C_2H_5)_3H$ was first reported in some detail as part of a study on the ^{1}H NMR spectra of some ethylgermanium compounds [10]; see Table 5. The value given for ν(GeH) in [10] (2016 cm^{-1}) differs slightly from other values reported to be 2003 to 2005 [3], 2006 [6], and 2010 cm^{-1} [4] for the pure liquid, 2008 cm^{-1} in C_7H_{16}, 2002 cm^{-1} in CCl_4 [6], and 2004 cm^{-1} (no details given) [18]. Two ν(GeH) bands at 2012 and 2018 cm^{-1} were observed in the gas phase and explained by the presence of conformers with different substituent effects of CH_3 (bands and possible conformers are illustrated) [21]. The complete IR spectrum of $Ge(C_2H_5)_3H$ is illustrated in [25]. The compound was part of studies on various organogermanium compounds dealing with the dependence of the GeH bond frequency [6, 8, 13, 14, 15, 22] and their integrated intensity [16, 17] on the nature of the substituents. Measurements of the integrated intensity of ν(GeH) in C_7H_{16} in the presence of electron-donating solvents indicated donor-acceptor complexes with the Ge atom as the acceptor [23].

Table 5
IR Spectrum of $Ge(C_2H_5)_3H$; Characteristic Bands [10].

wavenumber in cm^{-1}	assignment	wavenumber in cm^{-1}	assignment
2016(vs)	ν(GeH)	1019(s)	ν(CC)
1470(m)	$\delta_{as}(CH_3)$	969(ms)	ν(CC)
1435(sh)		751(vs)	δ(GeH) and $\rho(GeC_2H_5)$
1432(w)	$\delta_{as}(CH_2)$	700(sh)	δ(GeH) and $\rho(GeC_2H_5)$
1384(w)	$\delta_s(CH_3)$	665(sh)	δ(GeH) and $\rho(GeC_2H_5)$
1235(wb)	$\delta_s(CH_2)$	568(vs)	ν(GeC)
1135(vvw)			

The He(I) photoelectron spectrum showed that the lowest ionization potentials are 9.6 (e, GeC_3) and 10.5 (a_1, GeH) eV (illustrated and compared to the spectra of $Ge(C_2H_5)_4$, $Ge(C_2H_5)_2H_2$, and $Ge(C_2H_5)H_3$) [19].

The compound is a colorless, mobile liquid with a camphor-like odor [1]. Reported boiling points in °C/Torr: 62 to 70/96 [9], 120/750.5 [4], 124.4/751 [1], 124/760 [2]; other data are in the range 120 to 123°C/760 Torr [18, 28, 31]; see also [20]. Density d^{20} = 1.009 g/cm^3 [2] (1.0043 g/cm^3 in [4]); refractive index n_D^{20} = 1.4382 [2] (1.4361 [9], 1.4338 [4], and 1.4340 (99.5% purity) [12]).

$Ge(C_2H_5)_3D$. ^1H NMR spectrum (in C_6H_6): δ(ppm) = 0.82 (CH_2) and 1.08 (CH_3), J(H, H) = 7.8 Hz [10]. IR spectrum (in liquid paraffin): ν(GeD) 1440 cm^{-1} [27] (1452 cm^{-1} [4, 10]) and δ(GeD) 490 cm^{-1} [27].

Boiling point 120°C/740 Torr, d^{20} = 1.0097 g/cm^3, and n_D^{20} = 1.4333 [4].

References:

[1] Kraus, C.A.; Flood, E. A. (J. Am. Chem. Soc. **54** [1932] 1635/44).
[2] Anderson, H. H. (J. Am. Chem. Soc. **79** [1957] 326/8).
[3] Mathis-Noël, R.; Mathis, F.; Satgé, J. (Bull. Soc. Chim. Fr. **1961** 676).
[4] Ponomarenko, V. A.; Zueva, G. Ya.; Andreev, N. S. (Izv. Akad. Nauk SSSR Ser. Khim. **1961** 1758/62; Bull. Acad. Sci. USSR Div. Chem. Sci. [Engl. Transl.] **1961** 1639/43).
[5] Satgé, J. (Ann. Chim. [Paris] [13] **6** [1961] 519/73).
[6] Mathis, R.; Satgé, J.; Mathis, F. (Spectrochim. Acta **18** [1962] 1463/72).
[7] Egorochkin, A. N.; Khidekel', M. L.; Ponomarenko, V. A.; Zueva, G. Ya.; Svirezheva, S. S.; Razuvaev, G. A. (Izv. Akad. Nauk SSSR Ser. Khim. **1963** 1865/8; Bull. Acad. Sci. USSR Div. Chem. Sci. [Engl. Transl.] **1963** 1717/9).
[8] Mathis, R.; Constant, M.; Satgé, J.; Mathis, F. (Spectrochim. Acta **20** [1964] 515/21).
[9] Vyazankin, N. S.; Razuvaev, G. A.; Korneva, S. P.; Kruglaya, O. A.; Galiulina, R. F. (Dokl. Akad. Nauk SSSR **158** [1964] 884/7; Dokl. Chem. [Engl. Transl.] **154/159** [1964] 1002/4).
[10] Mackay, K. M.; Watt, R. (J. Organomet. Chem. **6** [1966] 336/51).
[11] Egorochkin, A. N.; Burov, A. I.; Mironov, V. F.; Gar, T. K.; Vyazankin, N. S. (Dokl. Akad. Nauk SSSR **180** [1968] 861/4; Dokl. Chem. [Engl. Transl.] **178/183** [1968] 500/3).

[12] Tel'noi, V. I.; Kol'yakova, G. M.; Rabinovich, I. B.; Vyazankin, N. S. (Dokl. Akad. Nauk SSSR **185** [1969] 374/6; Dokl. Chem. [Engl. Transl.] **184/189** [1969] 214/6).
[13] Mathis, R.; Barthelat, M.; Mathis, F. (Spectrochim. Acta A **26** [1970] 1993/2000).
[14] Egorochkin, A. N.; Khorshev, S. Ya.; Ostasheva, N. S.; Satgé, J.; Rivière, P.; Barrau, J.; Massol, M. (J. Organomet. Chem. **76** [1974] 29/36).
[15] Egorochkin, A. N.; Khorshev, S. Ya.; Ostasheva, N. S.; Satgé , J.; Rivière, P.; Barrau, J.; Massol, M. (Dokl. Akad. Nauk SSSR **215** [1974] 858/60; Dokl. Chem. [Engl. Transl.] **214/219** [1974] 204/6).
[16] Egorochkin, A. N.; Khorshev, S. Ya.; Ostasheva, N. S.; Sevastyanova, E. I.; Satgé, J.; Rivière, P.; Barrau, J. (J. Organomet. Chem. **105** [1976] 311/20).
[17] Egorochkin, A. N.; Sevast'yanova, E. I.; Khorshev, S. Ya.; Ratushnaya, S. Kh.; Satgé, J.; Rivière, P.; Barrau, J.; Richelme, S. (J. Organomet. Chem. **162** [1978] 25/35).
[18] Eaborn, C.; Singh, B. (J. Organomet. Chem. **177** [1979] 333/48).
[19] Beltram, G.; Fehlner, T. P.; Mochida, K.; Kochi, J. K. (J. Electron Spectrosc. Relat. Phenom. **18** [1980] 153/9).
[20] Stang, P. J.; White, M. R. (J. Am. Chem. Soc. **103** [1981] 5429/33).

[21] McKean, D. C.; Torto, I.; Mackenzie, M. W.; Morrisson, A. R. (Spectrochim. Acta A **39** [1983] 387/98).
[22] McKean, D. C.; Torto, I.; Mackenzie, M. W. (Spectrochim. Acta A **39** [1983] 399/408).
[23] Khorshev, S. Ya.; Tsvetkova, V. L.; Egorochkin, A. N. (J. Organomet. Chem. **264** [1984] 169/78).
[24] Watkinson, P. J.; Mackay, K. M. (J. Organomet. Chem. **275** [1984] 39/42).
[25] Stanley, A. E.; Johnson, R. A.; Turner, J. B.; Roberts, A. H. (Appl. Spectrosc. **40** [1986] 374/8).
[26] Wilkins, A. L.; Watkinson, P. J.; Mackay, K. M. (J. Chem. Soc. Dalton Trans. **1987** 2365/72).
[27] Bochkarev, M. N.; Fedorova, E. A.; Penyagina, I. M.; Vasina, O. A.; Protchenko, A. V.; Khorshev, S. Ya. (Metalloorg. Khim. **2** [1989] 1317/9; Organometal. Chem. USSR [Engl. Transl.] **2** [1989] 696/7).
[28] Kaushal, P.; Roberts, B. P. (J. Chem. Soc. Perkin Trans. II **1989** 1559/68).
[29] Kugita, T.; Mochida, K.; Tohji, K.; Udagawa, Y. (Chem. Lett. **1989** 501/4).
[30] Mochida, K.; Kugita, T.; Nakadaira, Y. (Polyhedron **9** [1990] 2263/6).

[31] Cerveau, G.; Chuit, C.; Corriu, R. J. P.; Reyé, C. (Organometallics **10** [1991] 1510/5).
[32] Clark, K. B.; Griller, D. (Organometallics **10** [1991] 746/50).

1.3.1.1.2.3 Solutions, Mass Spectrum, Decomposition, and Chemical Reactions

Solutions. The compound is soluble in organic solvents and insoluble in H_2O and liquid NH_3 [1]. The solvating abilities of various solvents for $Ge(C_2H_5)_3H$ (and $Si(C_2H_5)_3H$) were determined by measuring the integrated ν(E-H) band intensity of the IR spectrum in C_7H_{16} solution and in C_7H_{16} with the donor added (ca. 0.5 mol/L). The donor-acceptor interaction of some selected solvents follows the order HMPT > $N(C_2H_5)_3$ = DMF > THF > C_5H_5N > ether > CH_3COCH_3 > CH_3CN. $Ge(C_2H_5)_3H$ and $Si(C_2H_5)_3H$ are equally solvated by donors with sterically more accessible oxygen atoms. The solvating action of nitrogen donors is strongly influenced by steric hindrance at the donor and acceptor centers [181].

From the equilibrium concentrations of the metalation reaction $Ge(C_2H_5)_3Cs + C_6H_5CH_3 \rightleftharpoons Ge(C_2H_5)_3H + C_6H_5CH_2Cs$ (at 25°C for several days), $pK_a = 39.7 \pm 0.1$ was derived for $Ge(C_2H_5)_3H$ [187] (based on pK_a = 40.9 for $C_6H_5CH_3$ according to [114]). Corresponding values in the MSAD scale [25] are pK_a = 33.3 for the title compound vs. 35 for $C_6H_5CH_3$ [199]; cf. $Ge(C_2H_5)_2(C_4H_9\text{-t})H$, p. 165, and $Ge(C_2H_5)_2(C_6H_5)H$, p. 166.

Mass Spectrum. Fragmentation by electron impact (70 eV) showed the main ions in the following order of abundance: $[C_2H_5GeH_2]^+$ > $[(C_2H_5)_2GeH]^+$ > $[(C_2H_5)_2Ge]^+$ > $[GeH]^+$ > $[C_2H_5Ge]^+$ > $[Ge]^+$ > $[(C_2H_5)_3Ge]^+$ > $[(C_2H_5)_3GeH]^+$. Metastable-supported alkene elimination ($[(C_2H_5)_2GeH]^+ \rightarrow [C_2H_5GeH_2]^+ + C_2H_4$) and cleavage of two Ge-H bonds ($[C_2H_5GeH_2]^+ \rightarrow [C_2H_5Ge]^+ + H_2$) was observed [60]. The relative amounts of $[(C_2H_5)_3GeH]^+$, $[(C_2H_5)_2GeH]^+$, and $[(C_2H_5)_3Ge]^+$ are more or less insensitive to ionizing at voltages between 20 and 70 eV. The abundance of $[C_2H_5GeH_2]^+$ strongly depends on the ionizing voltage; this fragment becomes the principal ion above 40 eV. Ge-C fragmentation ($[M]^+ \rightarrow [(C_2H_5)_2GeH]^+$) was estimated to be preferred over Ge-H fragmentation ($[M]^+ \rightarrow [(C_2H_5)_3Ge]^+$) by at least a factor of 10 [155]. For m/e values and intensities, see also [201]. The monoisotopic mass spectra of several ER_3H compounds (E = Si, Ge, Sn) including $Ge(C_2H_5)_3H$ were reported, and the main reactions discussed in terms of the physical-organic theory of mass spectra and of quasi-equilibrium theory [161].

Irradiation with a CO_2 laser (ν = 1075 cm^{-1}) in the gas phase (3.5 Torr pressure of $Ge(C_2H_5)_3H$) produced only trace quantities of a Ge deposite along with GeH_4, C_2H_4, and H_2 [185]. ArF laser photolysis of a $C_{12}H_{26}$ solution of $Ge(C_2H_5)_3H$ yielded carbon-free Ge films; C_2H_4 was identified as the gaseous product [220].

Gas-phase **pyrolysis** in a He carrier at low flow rates showed that C_2H_4 is evolved at a minimum temperature of 350 °C. The decomposition was proposed to proceed by stepwise elimination of C_2H_4 via the $Ge(C_2H_5)_2H_2$ and $Ge(C_2H_5)H_3$ intermediates to yield eventually Ge, C_2H_4, and H_2. At all temperatures of decomposition (435 to 500 °C), the Ge film was found to be crystalline (based on Raman spectroscopic and micrographic evidence) containing embedded carbon, particularly at low temperatures [189].

Reactions with Elements. Combustion of $Ge(C_2H_5)_3H$ with O_2 at 35 atm gave ΔH°_{comb} = -1203 ± 2 kcal/mol [86]. Ozonation in $C_6D_5CD_3$ or CD_3COCD_3 at −78 °C yielded $Ge(C_2H_5)_3OOOH$ [192] (cf. "Organogermanium Compounds" 5, 1993, p. 381). Reactions of $Ge(C_2H_5)_3H$ with S, Se (ca. 1:1 mole ratio), or Te (ca. 2:1 mole ratio) were conducted in an evacuated and sealed tube under the following conditions: with S at 140 to 170 °C for ca. 4 h $\rightarrow$ $Ge(C_2H_5)_3SH$ (50% yield), $(Ge(C_2H_5)_3)_2S$ (20%), and H_2S [38, 39, 40]; with Se at 200 °C for 14 to 18 h $\rightarrow$ $Ge(C_2H_5)_3SeH$ (63%) and $(Ge(C_2H_5)_3)_2Se$ (ca. 20%) [38, 39, 40]; and with Te at 210 °C for 18 h $\rightarrow$ $(Ge(C_2H_5)_3)_2Te$ (60 to 75%) and H_2 [38, 56]. $Ge(C_2H_5)_3H$ readily decolorizes a solution of Br_2 in C_2H_5Br [1].

$Ge(C_2H_5)_3H$ does not appear to react with Na in liquid NH_3 [1]. Thermally generated Ge atoms (at ca. 5×10^{-3} Torr) insert into the Ge-H bond when cocondensed with $Ge(C_2H_5)_3H$ at 77 K followed by warming to room temperature; $(Ge(C_2H_5)_3)_2GeH_2$ (0.2% yield), $Ge_2(C_2H_5)_6$ (0.4%), and small quantities of several unidentified products were formed [193].

Reactions with Inorganic Compounds. Interaction of $Ge(C_2H_5)_3H$ with H_2O in the gas phase causes a pressure lineshift and line broadening of certain transitions in the H_2O microwave spectrum [211, 213].

$Ge(C_2H_5)_3H$ reacts quantitatively with H_2O at 100 °C in the presence of Cu powder to yield $(Ge(C_2H_5)_3)_2O$ and H_2 [13]. Refluxing with 100% H_2SO_4 for 5 min gave $(Ge(C_2H_5)_3)_2SO_4$ as the main product (70% yield) along with $(Ge(C_2H_5)_3)_2O$ and H_2 [2]. Solvolysis in an acidic medium occurs approximately to the same extent as with $Si(C_2H_5)_3H$. But in an alkaline medium, $Ge(C_2H_5)_3H$ is surprisingly stable, resisting a refluxing 0.5 N KOH solution in $C_2H_5OH\text{-}H_2O$ (ca. 94:6 v/v) for several hours, whereas $Si(C_2H_5)_3H$ and $Sn(C_2H_5)_3H$ are decomposed with evolution of H_2 [10].

Refluxing $Ge(C_2H_5)_3H$ with $B(OH)_3$ in the presence of Cu powder yielded 50% $B(OGe(C_2H_5)_3)_3$ [13]. $Ge(C_2H_5)_3Cl$, SO_2, and H_2 were formed quantitatively in a vigorous reaction with SO_2Cl_2 at room temperature [8]. Electrochemical halogenation of $Ge(C_2H_5)_3H$ at a Pt cathode was attempted in moist CH_3CN in the presence of $[N(C_4H_9)_4][ClO_4]$ or $[N(C_4H_9)_4][BF_4]$, but the expected products, $Ge(C_2H_5)_3Cl$ or $Ge(C_2H_5)_3F$, were readily hydrolyzed under the experimental conditions [202]; cf. $Ge(C_6H_5)_3H$, p. 112.

The reaction of $Ge(C_2H_5)_3H$ with $GeCl_4$ (1.2:1) in ether at 35°C yielded $GeHCl_3 \cdot 2\,O(C_2H_5)_2$ and $Ge(C_2H_5)_3Cl$ (80%) [34]. A neat mixture of $GeCl_4$ and $Ge(C_2H_5)_3H$ (1:4 mole ratio) precipitated $GeCl_2$ at room temperature; subsequent heating to 130°C gave $Ge(C_2H_5)_3Cl$ (87% yield), Ge (87%), and H_2 [8, 9]; the room-temperature reaction was used to prepare $GeCl_2$ which actually had the composition $GeCl_{1.49}$ owing to decomposition during workup [66]. Equimolar amounts of $Ge(C_2H_5)_3H$, $GeCl_4$, and $CH(=N_2)COOC_2H_5$ in C_5H_{12} (48 h) gave $Ge(C_2H_5)_3Cl$, $GeCl_2$ (57%), $CH_2ClCOOC_2H_5$, and N_2 [166]. $Ge(C_2H_5)_3F$ (20% yield) was formed from $Ge(C_2H_5)_3H$ and GeF_2 in dioxane, presumable via the unstable, nondetectable $(C_2H_5)_3GeGe(H)F_2$ [120, 130]. $Ge(C_2H_5)_3H$ did not react with $Ge(OH)(OCH_2CH_2)_3N$ in the presence of H_2PtCl_6 [216]; see also $Ge(C_6H_5)_3H$, p. 112.

Reactions of $Ge(C_2H_5)_3H$ with various transition metal compounds are listed in Table 6. The Rh compound I (R = H or OCH_3) did not react with $Ge(C_2H_5)_3H$ in contrast to $Ge(C_2H_5)_2H_2$ [212].

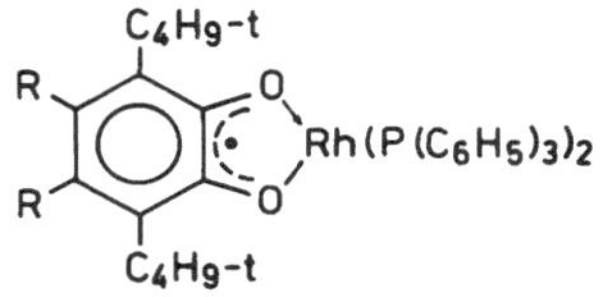

I

Table 6
Reactions of $Ge(C_2H_5)_3H$ with Transition Metal Compounds.

reactant	conditions → products (yield)	Ref.
$TiCl_4$	refluxing, 40 min → $Ge(C_2H_5)_3Cl$ (70%), $TiCl_3$, $TiCl_2$, H_2	[2]
	130°C, 2 h → Ti^{3+} (95%), Ti^{2+} (traces)	[55]
$VOCl_3$	refluxing, 2 min → $Ge(C_2H_5)_3Cl$ (95%), $VOCl_2$, VOCl, H_2	[2]
CrO_2Cl_2	refluxing, 5 min → $Ge(C_2H_5)_3Cl$ (45%), Cr_2O_3, H_2O, H_2	[2]
WCl_6	2:1 mole ratio, in C_6H_6, C_6H_5Cl, or CCl_4, 20°C, 3 min → $Ge(C_2H_5)_3Cl$, WCl_{6-n} ($n \leq 4$), H_2; reaction rate similar to that of $Si(C_2H_5)_3H$ and $Sn(C_2H_5)_3H$	[129]
$KMnO_4$	in CH_3COCH_3 → $(Ge(C_2H_5)_3)_2O$ (28%), MnO_2, KOH	[2]
$Fe(L)_3(ClO_4)_3$*)	in CH_3CN, 25°C → $[Ge(C_2H_5)_3]^+$, $[Fe(L)_3]^{2+}$, H^+; kinetics and mechanism studied	[160]
$PdCl_2$	refluxing, 10 min → $Ge(C_2H_5)_3Cl$ (88%), Pd, H_2	[2]
K_2PtCl_6	refluxing, 2 h → $Ge(C_2H_5)_3Cl$ (95%), Pt, KCl, H_2	[2]

References on p. 63

Table 6 (continued)

reactant	conditions → products (yield)	Ref.
$CuBr_2$	refluxing, 1 h → $Ge(C_2H_5)_3Br$ (90%), Cu_2Br_2, H_2	[2]
$KAuCl_4$	refluxing, 15 min → $Ge(C_2H_5)_3Cl$ (95%), Au, KCl, H_2	[2]
$CdCl_2$	not reduced to Cd metal	[2]
$HgCl_2$	refluxing, 10 min → $Ge(C_2H_5)_3Cl$ (55%), Hg, HCl, H_2	[2]
Hg_2Cl_2	refluxing, 35 min → $Ge(C_2H_5)_3Cl$ (90%), Hg, HCl, H_2	[2]
$HgBr_2$	refluxing, 45 min → $Ge(C_2H_5)_3Br$ (98%), Hg, HBr, H_2	[2]
HgI_2	refluxing, 15 min → $Ge(C_2H_5)_3I$ (99%), Hg_2I_2, H_2	[2]
$HgSO_4$	refluxing, 15 min → $(Ge(C_2H_5)_3)_2SO_4$ (70%), $(Ge(C_2H_5)_3)_2O$, Hg, H_2	[2]
$Hg(OOCCH_3)_2$	refluxing, 15 min → $Ge(C_2H_5)_3OOCCH_3$ (60%), CH_3COOH, Hg, H_2	[2]

*) L = 1,10-phenanthroline and its derivatives $C_{12}H_7N_2Cl$-5, $C_{12}H_7N_2NO_2$-5, and $C_{12}H_6N_2(C_6H_5)_2$-4,7, or 2,2′-bipyridine.

Hydrogermylation Reactions. Reports on the addition of $Ge(C_2H_5)_3H$ to unsaturated organic compounds are numerous. These reactions are summarized in Table 7. They include a large variety of alkenes (pp. 27/34) and alkynes (pp. 34/52) and a few other unsaturated systems such as C=C=O, C=N, and N=N (p. 52). A review on hydrogermylations has recently been published [222].

H_2PtCl_6 is the mostly used catalyst and is abbreviated "cat." in Table 7. Other Ni-based catalyst systems were investigated using butadiene as the substrate, e.g. $Ni(CO)_4$-$P(C_6H_5)_3$, $Ni(C_5H_7O_2)$-$P(C_6H_5)_3$ ($C_5H_7O_2$ = acetylacetonato), $Ni(C_5H_7O_2)$-$P(C_6H_5)_3$-$GeHCl_3$, Ni-$(C_5H_7O_2)$-$P(C_6H_5)_3$-AlR_3 (R = C_2H_5 and i-C_4H_9), and the products were analyzed for the cis:trans ratio of $Ge(C_2H_5)_3CH_2CH{=}CHCH_3$ and the proportions of simultaneously formed oligomers such as $Ge(C_2H_5)_3CH_2CH{=}CH(CH_2)_3CH{=}CH_2$ and $Ge(C_2H_5)_3CH_2CH{=}CH$-$(CH_2)_2CH{=}CHCH_3$ [177]; see also p. 31. Recent investigations of the hydrogermylation of $C_6H_5C{\equiv}CH$ in THF at 20 to 50°C to give cis- and trans-$Ge(C_2H_5)_3CH{=}CHC_6H_5$ and minor amounts of $Ge(C_2H_5)_3C(C_6H_5){=}CH_2$ involved the following catalysts: the chlorometallates $[N(C_2H_5)_3CH_2C_6H_5]^+_m[MCl_n]^{m-}$ (M = Pt, Pd, Rh, Ir, Fe, Co, Cu; m = 1 to 3, n = 3 to 6), polymer-anchored chlorometallates, $[polymer\text{-}CH_2P(C_4H_9)_3]^+_m[MCl_n]^{m-}$, H_2PtCl_6, and some other neutral Rh and Ru complexes. The complexes containing Pt, Pd, and Rh were found to be the most effective catalysts. The cis:trans ratio of the β adduct depends on the transition metal and decreases in the order Ir > Rh ≫ Pd > Pt, whilst the ratio of a:β adducts is less influenced by the metal involved. With H_2PtCl_6 as the catalyst, the reaction proceeds much faster in $C_2H_4Cl_2$ and THF than in nonpolar solvents such as C_6H_{14} or $C_6H_5CH_3$ [214].

NMR- and IR-spectroscopic studies were carried out with several SiR_3H and GeR_3H compounds (including $Ge(C_2H_5)_3H$) for examining the effects of the reagent type [154, 156], the nature and amount of catalyst [156], the mixing order, temperature [154, 156], and solvent

[154] on the hydrosilylation and hydrogermylation of $C_6H_5CH{=}CH_2$ [154] and $C_6H_5CH_2$-$CH{=}CH_2$ [156].

A study of the mechanism of hydrogen transfer from $Ge(C_2H_5)_3H$ and other R_3EH compounds (E = Si, Ge, and Sn) to $(NC)_2C{=}C(CN)_2$ revealed the formation of characteristic but labile 1:1 adducts of unknown structure. After reacting $Ge(C_2H_5)_3H$ with $(NC)_2C{=}C(CN)_2$ in H_2O-CH_2Cl_2 at room temperature for 10 min, $(NC)_2CHCH(CN)_2$ and $Ge(C_2H_5)_3OH$ (or the oxide) were detected with 90% yield. Under rigorously anhydrous conditions a red material was observed but no $(NC)_2CHCH(CN)_2$ [155].

(Text continues on p. 53)

Table 7
Hydrogermylation Reactions with $Ge(C_2H_5)_3H$.

reactant	conditions (cat. = H_2PtCl_6) → products (yield) and remarks	Ref.
reactants with one C=C bond		
$CH_2{=}CHOC_4H_9$	γ irradiated (^{60}Co source) → $Ge(C_2H_5)_3CH_2CH_2OC_4H_9$	[162]
$CH_2{=}CHOCH_2\overline{CHCH_2O}$	heating in C_6H_6, 20 h, cat. → $Ge(C_2H_5)_3CH_2CH_2OCH_2\overline{CHCH_2O}$ (60%)	[133]
$CH_2{=}CHO(CH_2)_2OCH_2\overline{CHCH_2O}$	heating in C_6H_6, 20 h, cat. $Ge(C_2H_5)_3CH_2CH_2O(CH_2)_2OCH_2$-$\overline{CHCH_2O}$ (70%)	[133]
$CH_2{=}CHO(CH_2)_4OCH_2\overline{CHCH_2O}$	heating in C_6H_6, 20 h, cat. → $Ge(C_2H_5)_3CH_2CH_2O(CH_2)_4OCH_2$-$\overline{CHCH_2O}$ (68%)	[133]
$CH_2{=}CHOOCCH_3$	refluxing, 5 h, cat. → $Ge(C_2H_5)_3CH_2CH_2OOCCH_3$ (39%)	[169]
$CH_2{=}CHSC_6H_5$	refluxing, 3 h, cat., → $Ge(C_2H_5)_3CH_2CH_2SC_6H_5$	[122]
$CH_2{=}CHSC_6H_4CH_3$-4	refluxing, 3 h, cat. → $Ge(C_2H_5)_3CH_2CH_2SC_6H_4CH_3$-4	[122]
$CH_2{=}CHSC_6H_4(C_4H_9$-t)-4	refluxing, 3 h, cat. → $Ge(C_2H_5)_3CH_2CH_2SC_6H_4(C_4H_9$-t)-4	[122]
$CH_2{=}CHP(O)Cl_2$	150 to 160°C, 10 h, cat. → $Ge(C_2H_5)_3CH_2CH_2P(O)Cl_2$ (17%)	[59]
$CH_2{=}CHP(O)(OC_2H_5)_2$	150 to 160°C, 10 h, cat. → $Ge(C_2H_5)_3CH_2CH_2P(O)(OC_2H_5)_2$ (51%)	[59]
$CH_2{=}CHC_6H_5$	cat. → $Ge(C_2H_5)_3CH(CH_3)C_6H_5$ and $Ge(C_2H_5)_3CH_2CH_2C_6H_5$	[154]

References on p. 63

Table 7 (continued)

reactant	conditions (cat. = H_2PtCl_6) → products (yield) and remarks	Ref.
CH_2=$CHCH_2Cl$	refluxing, 2 h, cat. → $Ge(C_2H_5)_3CH_2CH_2CH_2Cl$ (5%) $Ge(C_2H_5)_3CH_2CH$=CH_2 (26%), CH_2=$CHCH_3$, and $Ge(C_2H_5)_3Cl$ (55%)	[14]
CH_2=$C(CH_3)CH_2Cl$	refluxing, 2 h, cat. → $Ge(C_2H_5)_3CH_2CH(CH_3)CH_2Cl$ (36%), CH_2=$C(CH_3)_2$, and $Ge(C_2H_5)_3Cl$ (41%)	[14]
CH_2=$CHCH_2OH$	ca. 90 °C, 4 h, cat. → $Ge(C_2H_5)_3CH_2CH_2CH_2OH$ (73%) 125 to 140 °C, 1 to 3 h, cat. → $Ge(C_2H_5)_3CH_2CH_2CH_2OH$ (65%) γ irradiated (^{60}Co source) → $Ge(C_2H_5)_3CH_2CH_2CH_2OH$	[7] [128] [162]
CH_2=$CHCH_2OCH_2CHCH_2O$	refluxing in C_6H_6, 16 h, cat. → $Ge(C_2H_5)_3CH_2CH_2CH_2OCH_2CHCH_2O$ (68%); the same product formed at 80 °C in the presence of heptanoyl-peroxide	[109, 119] [217]
CH_2=$CHCH_2OCH_2CHCH_2OOCCH_3$ OCH_2CHCH_2O	refluxing in C_6H_6, 1 d, cat. → $Ge(C_2H_5)_3(CH_2)_3OCH_2$- $CHCH_2OOCCH_3$ (72%) OCH_2CHCH_2O	[132]
CH_2=$CHCH_2OOCH$	125 to 140 °C, 1 to 3 h, cat. → $Ge(C_2H_5)_3CH_2CH_2CH_2OOCH$ (67%)	[128]
CH_2=$CHCH_2OOCCl$	90 °C, 5 h, cat. → $Ge(C_2H_5)_3CH_2CH_2CH_2OOCCl$ (traces), $Ge(C_2H_5)_3Cl$ (main product), CH_2=$CHCH_2OOCH$, CH_2=$CHCH_3$, and CO_2	[128]
CH_2=$CHCH_2OOCCH_3$	→ $Ge(C_2H_5)_3CH_2CH_2CH_2OOCCH_3$ (50%)	[14]
CH_2=$CHCH_2OOCCH_2NHSi(CH_3)_3$	135 °C in the presence of DTBP → $Ge(C_2H_5)_3CH_2CH_2CH_2OOCCH_2NHSi$-$(CH_3)_3$ (65%)	[217]
CH_2=$CHCH_2OCOOCH_3$	125 to 140 °C, 1 to 3 h, cat. → $Ge(C_2H_5)_3CH_2CH_2CH_2OCOOCH_3$ (82%)	[128]
CH_2=$CHCH_2OSi(CH_3)_3$	125 to 140 °C, 1 to 3 h, cat. → $Ge(C_2H_5)_3CH_2CH_2CH_2OSi(CH_3)_3$ (59%)	[128]
CH_2=$CHCH_2OC_6H_4OCH_3$-2	cat. → $Ge(C_2H_5)_3CH_2CH_2CH_2OC_6H_4OCH_3$-2	[157]

Table 7 (continued)

reactant	conditions (cat. = H_2PtCl_6) → products (yield) and remarks	Ref.
$CH_2{=}CHCH_2NCO$	109 to 134 °C, 23 h, cat. → $Ge(C_2H_5)_3CH_2CH_2CH_2NCO$ (40%)	[105, 128]
$CH_2{=}CHCH_2NH_2$	→ $Ge(C_2H_5)_3CH_2CH_2CH_2NH_2$ (80%)	[30]
$CH_2{=}CHCH_2NHCH_2COOC_2H_5$	5:1 mole ratio, 135 °C, 5 h, in the presence of DTBP → $Ge(C_2H_5)_3CH_2CH_2CH_2NHCH_2$-$COOC_2H_5$ (80%)	[217]
$CH_2{=}CHCH_2NHSi(CH_3)_3$	125 to 140 °C, 1 to 3 h, cat. → $Ge(C_2H_5)_3CH_2CH_2CH_2NHSi(CH_3)_3$ (40%)	[128]
$CH_2{=}CHCH_2N(Si(CH_3)_3)_2$	125 to 140 °C, 1 to 3 h, cat. → $Ge(C_2H_5)_3CH_2CH_2CH_2N(Si(CH_3)_3)_2$ (46%)	[105, 128]
$CH_2{=}CHCH_2P(O)Cl_2$	160 to 165 °C, 10 h, cat. → $Ge(C_2H_5)_3(CH_2)_3P(O)Cl_2$ (52%)	[59]
$CH_2{=}CHCH_2P(O)(OC_2H_5)_2$	160 to 165 °C, 10 h, cat. → $Ge(C_2H_5)_3(CH_2)_3P(O)(OC_2H_5)_2$ (60%)	[59]
$CH_2{=}CHCH_2P(O)(OC_4H_9)_2$	200 to 220 °C, 10 h, no catalyst → $Ge(C_2H_5)_3(CH_2)_3P(O)(OC_4H_9)_2$ (37%)	[59]
$CH_2{=}CHCH_2B(OCH_3)_2$	reflux temperature rising from 110 to 190 °C, 17.5 h → $Ge(C_2H_5)_3CH_2CH_2CH_2B(OCH_3)_2$ (78%)	[29]
$CH_2{=}CHCHO$	Pt asbestos and $C_6H_4(OH)_2$-1,4 inhibitor → $Ge(C_2H_5)_3CH_2CH_2CHO$	[4, 9]
$CH_2{=}CHCOOH$	refluxing, 8 h → $Ge(C_2H_5)_3CH_2CH_2COOH$ (49%)	[4, 9]
$CH_2{=}C(CH_3)COOCH_3$	150 °C, 80 h, cat., sealed evacuated tube → $Ge(C_2H_5)_3CH_2CH(CH_3)COOCH_3$ (60%)	[172]
$CH_2{=}CHCOOC_2H_5$	90 °C, 0.5 h, then 125 °C, 1.5 h, cat. → $Ge(C_2H_5)_3CH_2CH_2COOC_2H_5$	[20]
$CH_2{=}CHCOOSi(CH_3)_3$	→ $Ge(C_2H_5)_3CH_2CH_2COOSi(CH_3)_3$ (58%)	[30]
$CH_2{=}CHCN$	refluxing → $Ge(C_2H_5)_3CH_2CH_2CN$ (main product) and $Ge(C_2H_5)_3CH_2CH(CN)CH_2CH_2CN$	[4, 9]

References on p. 63

Table 7 (continued)

reactant	conditions (cat. = H_2PtCl_6) → products (yield) and remarks	Ref.
$CH_2{=}CHCH_2C_6H_5$	cat. → $Ge(C_2H_5)_3CH_2CH_2CH_2C_6H_5$	[156]
$CH_2{=}CHCOCH_3$	reflux, 10 h, no catalyst → $Ge(C_2H_5)_3CH_2CH_2COCH_3$ (31%)	[36]
	with cat. → $Ge(C_2H_5)_3OC(CH_3){=}CHCH_3$	[50]
$CH_2{=}CHCH_2COOH$	125 to 140°C, 1 to 3 h, cat. → $Ge(C_2H_5)_3CH_2CH_2CH_2COOH$ (70%)	[128]
$CH_2{=}CHCH_2COOC_2H_5$	70 to 80°C, 2 h, cat. and $C_6H_4(OH)_2$-1,4 inhibitor → $Ge(C_2H_5)_3CH_2CH_2CH_2COOC_2H_5$ (93%)	[20]
$CH_2{=}CHCH_2COOSi(CH_3)_3$	125 to 140°C, 1 to 3 h, cat. → $Ge(C_2H_5)_3CH_2CH_2CH_2COOSi(CH_3)_3$ (65%)	[128]
$CH_2{=}CHCH_2CH(COOR)_2$ R = C_2H_5?	140 to 145°C, 100 h, cat., sealed evacuated tube → $Ge(C_2H_5)_3CH_2(CH_2)_2CH(COOR)_2$ (not isolated); hydrolyzed and decarboxylated to give $Ge(C_2H_5)_3(CH_2)_4COOH$	[168]
$CH_2{=}CHCH_2CH_2COCH_3$	125 to 140°C, 1 to 3 h, cat. → $Ge(C_2H_5)_3CH_2(CH_2)_2CH_2COCH_3$ (76%)	[128]
$CH_2{=}CH(CH_2)_5CH_3$	→ $Ge(C_2H_5)_3CH_2(CH_2)_6CH_3$	[9]
$Ge(C_2H_5)_3CH{=}CH_2$	refluxing, several hours, cat. [6] or Pt asbestos [9] → $Ge(C_2H_5)_3CH_2CH_2Ge(C_2H_5)_3$	[6, 9]; see also [5]
$Ge(C_2H_5)_3CH_2CH{=}CH_2$	refluxing, several hours, Pt asbestos → $Ge(C_2H_5)_3CH_2CH_2CH_2Ge(C_2H_5)_3$	[9]; see also [5]
$Ge(C_2H_5)_3OCH_2CH{=}CH_2$	150°C, 15 h, cat., sealed tube → $Ge(C_2H_5)_3OCH_2CH_2CH_2Ge(C_2H_5)_3$ (43%)	[21]
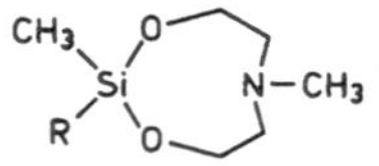 R = $CH{=}CH_2$	100°C, 8 h, $Rh(C_5H_7O_2)(CO)_2$ catalyst, sealed tube → 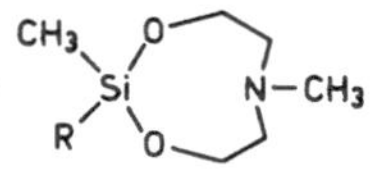 R = $Ge(C_2H_5)_3CH_2CH_2$ (98%)	[191]

Table 7 (continued)

reactant	conditions (cat. = H_2PtCl_6) → products (yield) and remarks	Ref.
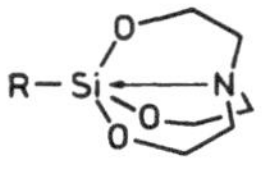 R = $CH{=}CH_2$	80 °C in C_6H_6, 8 h, $Rh(C_5H_7O_2)(CO)_2$ catalyst, sealed tube → 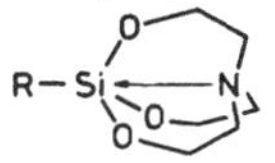 R = $Ge(C_2H_5)_3CH_2CH_2$ (100%)	[186]
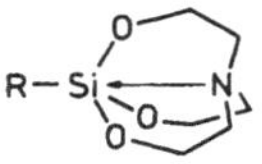 R = $CH_2CH{=}CH_2$	80 °C in C_6H_6, 8 h, $Rh(C_5H_7O_2)(CO)_2$ catalyst, sealed tube → 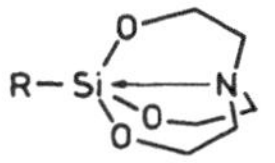 R = $Ge(C_2H_5)_3(CH_2)_2CH_2$ (100%)	[186]
reactants with conjugated C=C bonds		
$CH_2{=}CHCH{=}CH_2$	120 °C, 15 h, cat., autoclave → $Ge(C_2H_5)_3CH_2CH{=}CHCH_3$ (36%, cis:trans ≈ 1.1:1), $Ge(C_2H_5)_3CH_2CH_2C_6H_9$-c, $CH_2{=}CHC_6H_9$-c (trace amount)	[36]
	1:2 mole ratio, 60 °C, 4 h, $Ni(CO)_4$-$P(C_6H_5)_3$ catalyst, autoclave → $Ge(C_2H_5)_3CH_2CH{=}CHCH_3$ (97%, cis:trans = 9:1)	[177]
	1:1 mole ratio, 60 °C, 4 h, $Ni(C_5H_7O_2)_2$-$P(C_6H_5)_3$-$Al(C_4H_9\text{-}i)_3$ catalyst, autoclave → $Ge(C_2H_5)_3CH_2CH{=}CHCH_3$ (46%, cis:trans ≈ 22:1), $Ge(C_2H_5)_3CH_2CH{=}CH(CH_2)_3CH{=}CH_2$ (39%, cis:trans ≈ 2:1), $Ge(C_2H_5)_3CH_2$-$CH{=}CH(CH_2)_2CH{=}CHCH_3$ (3%) for the study of other catalyst systems, see the text, p. 26	
$CD_2{=}CDCD{=}CD_2$	80 °C, 4 h, Ni-P-Al catalyst as for butadiene, autoclave → $Ge(C_2H_5)_3CD_2CD{=}CDCD_2H$ (24%), $Ge(C_2H_5)_3CD_2CD{=}CDCD_2CD_2CD{=}CD$-$CD_2H$ (59%), and $Ge(C_2H_5)_3CD_2$-$CD{=}CDCDHCD_2CD{=}CDCD_3$	[177]
$CH_2{=}C(CH_3)CH{=}CH_2$	100 °C, 4 h, Ni-P-Al catalyst as for butadiene, autoclave	[177]

References on p. 63

Table 7 (continued)

reactant	conditions (cat. = H_2PtCl_6) → products (yield) and remarks	Ref.
	→ $Ge(C_2H_5)_3CH_2C(CH_3)=CHCH_3$ (62%) and $Ge(C_2H_5)_3CH_2CH=C(CH_3)_2$ (ca. 15%)	
$CH_2=CHCH=CHCH_3$	60 °C, 4 h, Ni-P-Al catalyst as for butadiene, autoclave → $Ge(C_2H_5)_3CH_2CH=CHCH_2CH_3$ (50%) and $Ge(C_2H_5)_3CH_2CH_2CH=CHCH_3$ (ca. 13%)	[177]
$CH_2=CHCH=CHCH(CH_3)CH=CH_2$	100 °C, 4 h, Ni-P-Al catalyst as for butadiene, autoclave → $Ge(C_2H_5)_3CH_2CH=CHCH_2$-$CH(CH_3)CH=CH_2$ (76%)	[177]
$CH_2=C(CH_3)CH=CHCH_2CH_2$-$C(CH_3)=CH_2$	100 °C, 4 h, Ni-P-Al catalyst as for butadiene, autoclave → $Ge(C_2H_5)_3CH_2C(CH_3)=CH(CH_2)_3$-$C(CH_3)=CH_2$ (67%)	[177]
	80 °C, 4 h, Ni-P-Al catalyst as for butadiene, autoclave → 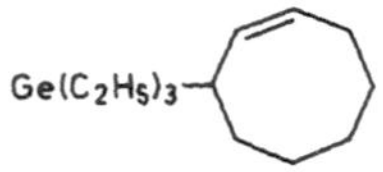 (45%)	[177]
	80 °C, 4 h, Ni-P-Al catalyst as for butadiene, autoclave → 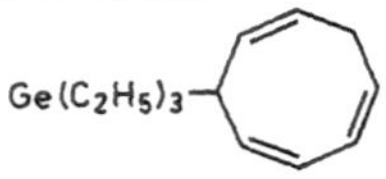 (25%)	[177]
reactants with cumulated C=C bonds		
$CH_2=C=CHCH_3$	110 °C, 10 h, cat. → $Ge(C_2H_5)_3C(C_2H_5)=CH_2$, $Ge(C_2H_5)_3CH_2CH=CHCH_3$ (only trans isomer), and $Ge(C_2H_5)_3C(CH_3)=CHCH_3$ in a 6:7:7 ratio (40% total yield)	[99]
$CH_2=C=CHC_6H_5$	70 °C, 1 h, cat. → $Ge(C_2H_5)_3CH_2CH=CHC_6H_5$ (trans isomer and traces of cis isomer), $Ge(C_2H_5)_3C(CH_2C_6H_5)=CH_2$, and $Ge(C_2H_5)_3C(CH_3)=CHC_6H_5$ in a 2:1:2 ratio (79% total yield)	[99]
$CH(Cl)=C=C(CH_3)_2$	after adding the cat. the temperature rose to 100 °C	[99]

Table 7 (continued)

reactant	conditions (cat. = H_2PtCl_6) → products (yield) and remarks	Ref.
	→ $Ge(C_2H_5)_3CH_2CH{=}C(CH_3)_2$ (25%) (formed via $CH_2{=}C{=}C(CH_3)_2$ before the cat. was added), $Ge(C_2H_5)_3C$-$(C_3H_7$-i$){=}CH_2$ (traces), $Ge(C_2H_5)_3C(CH_3){=}C(CH_3)_2$ (traces), and $Ge(C_2H_5)_3Cl$ (38%)	
$CH_2{=}C{=}CHOC_6H_5$	cat. → $Ge(C_2H_5)_3C(CH_2OC_6H_5){=}CH_2$, $Ge(C_2H_5)_3C(CH_3){=}CHOC_6H_5$, and cis-$Ge(C_2H_5)_3CH_2CH{=}CHOC_6H_5$	[163]
reactants with two isolated C=C bonds		
$CH_2{=}CHSi(CH_3)_2CH_2CH{=}CH_2$	40 to 50 °C, 0.5 h, cat. → $Ge(C_2H_5)_3CH_2CH_2Si(CH_3)_2$-$CH_2CH{=}CH_2$ (79%)	[96]
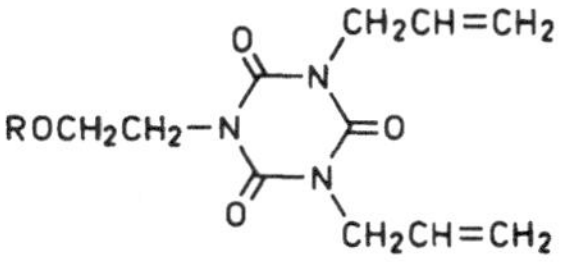 R = H	3:1 mole ratio, 120 °C, 12 h, cat. 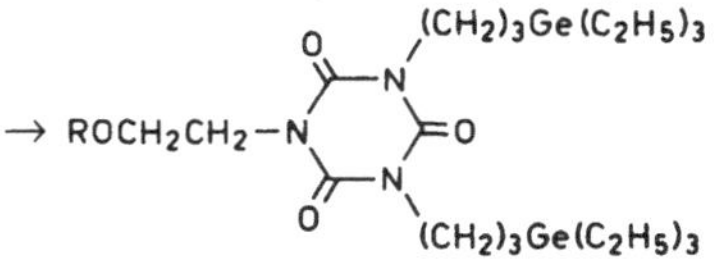 R = H (88%)	[197]
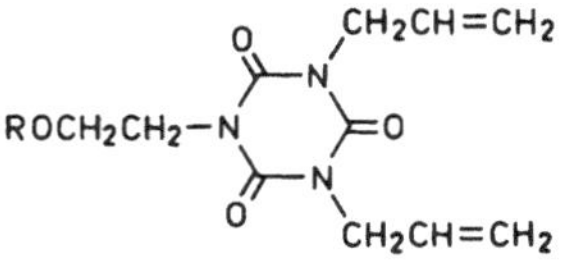 R = $Si(CH_3)_3$	2:1 mole ratio, 100 to 120 °C, 12 h, cat. 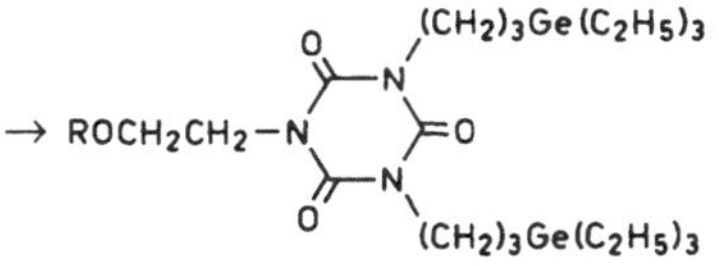 R = $Si(CH_3)_3$ (87%)	[197]
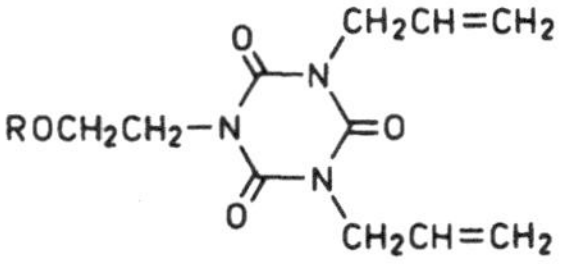 R = $Ge(C_2H_5)_3$	2:1 mole ratio, 120 °C, 12 h, cat. 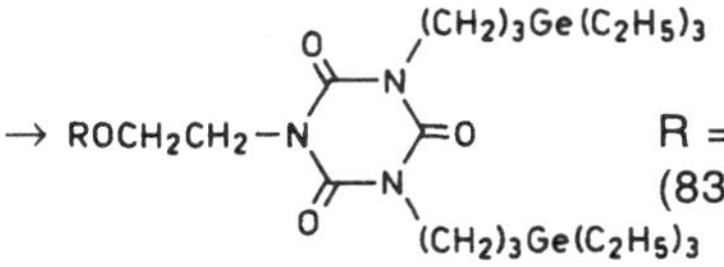 R = $Ge(C_2H_5)_3$ (83%)	[197]
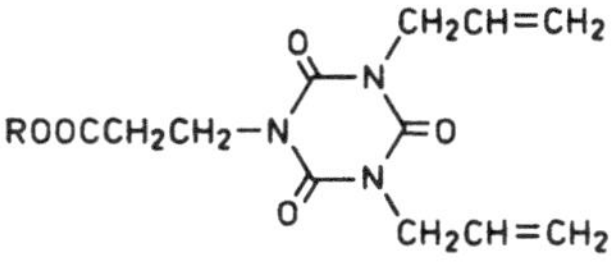 R = H	3:1 mole ratio, 160 to 170 °C, 8 to 10 h, cat. 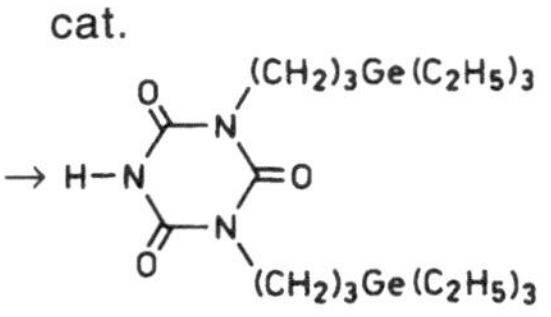 (85%) and $Ge(C_2H_5)_3$-$OOCCH{=}CH_2$ (89%)	[197]

References on p. 63

Table 7 (continued)

reactant	conditions (cat. = H_2PtCl_6) → products (yield) and remarks	Ref.
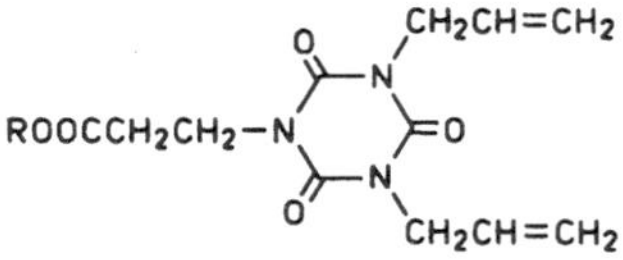 R = $Si(CH_3)_3$	3:1 mole ratio, 100 °C, 8 h, cat. → 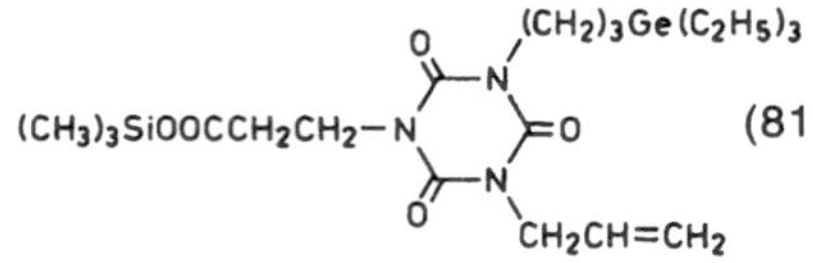 (81%)	[197]
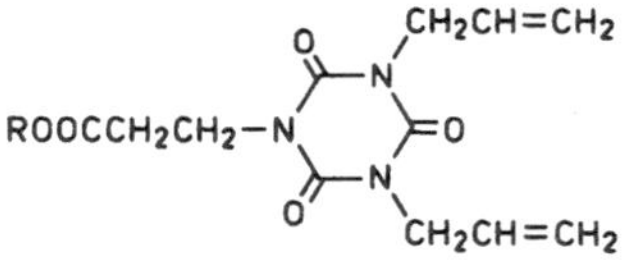 R = $Ge(C_2H_5)_3$	cat. → 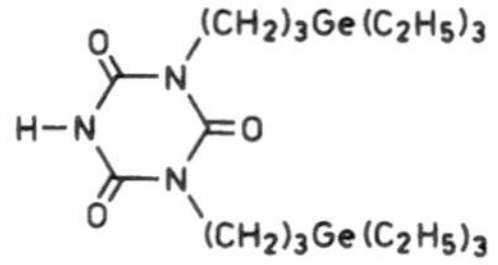 and $Ge(C_2H_5)_3$-$OOCCH{=}CH_2$	[197]
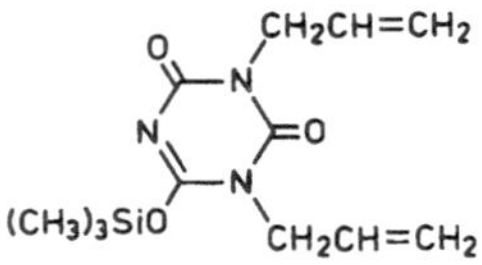	2:1 mole ratio, 120 °C, 20 h, cat., evacuated tube → 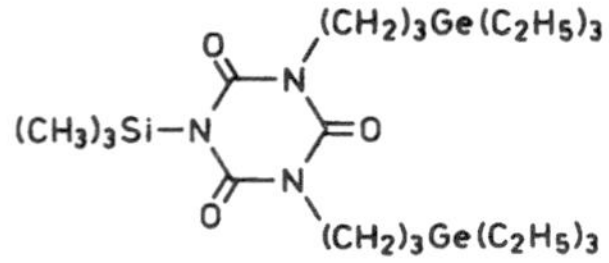 (91%)	[190]
reactants with one C≡C bond		
CH≡CH	100 °C, cat. → $Ge(C_2H_5)_3CH{=}CH_2$ and $Ge(C_2H_5)_3CH_2CH_2Ge(C_2H_5)_3$	[6, 9]
$CH{\equiv}COC_6H_5$	95 °C, 6 h, $Rh(P(C_6H_5)_3)Cl$ catalyst → cis- and trans- $Ge(C_2H_5)_3CH{=}CH$-OC_6H_5 and $Ge(C_2H_5)_3C(OC_6H_5){=}CH_2$ in a 3:6:1 ratio	[171]
	80 °C, 2 h, cat. → cis- and trans-$Ge(C_2H_5)_3CH{=}CH$-OC_6H_5 and $Ge(C_2H_5)_3C(OC_6H_5){=}CH_2$ in a 9:9:2 ratio	[137]
$CH{\equiv}CSC_6H_5$	70 °C, in THF, cat. → cis- and trans-$Ge(C_2H_5)_3CH{=}CHSC_6H_5$ and $Ge(C_2H_5)_3C(SC_6H_5){=}CH_2$ in a 29:23:10 ratio	[140]
	90 °C, no solvent, 6 h, cat. → cis- and trans-$Ge(C_2H_5)_3CH{=}CHSC_6H_5$ (46% and 31%), $Ge(C_2H_5)_3C(SC_6H_5){=}CH_2$ (4%), and $Ge(C_2H_5)_3SC_6H_5$ (20%)	
	95 °C, 6 h, $Rh(P(C_6H_5)_3)Cl$ catalyst → cis- and trans-$Ge(C_2H_5)_3CH{=}CHSC_6H_5$	[171]

Table 7 (continued)

reactant	conditions (cat. = H_2PtCl_6) → products (yield) and remarks	Ref.
	and $Ge(C_2H_5)_3C(SC_6H_5)=CH_2$ in a 10:9:1 ratio	
$CH\equiv CC_6H_5$	100°C, 6 h, cat. → cis- and trans-$Ge(C_2H_5)_3CH=CHC_6H_5$ (traces of the cis isomer) and $Ge(C_2H_5)_3C(C_6H_5)=CH_2$ for studies of other catalyst systems, see p. 26	[140]
	→ 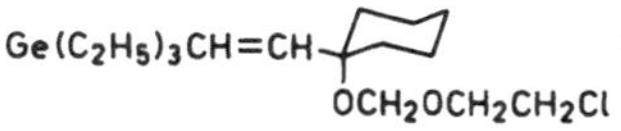 (34%)	[100]
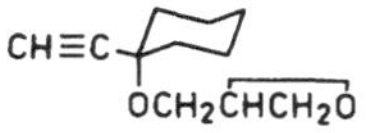	80°C, 5 h, cat. → 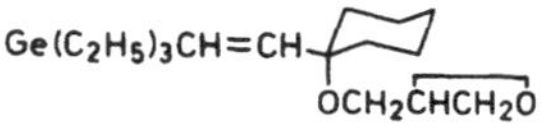 (72%)	[117]
	ca. 80°C, cat. → 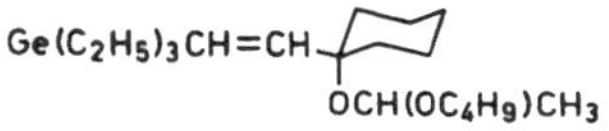	[72]
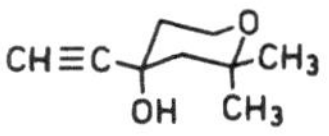	heating in C_6H_6, cat. → 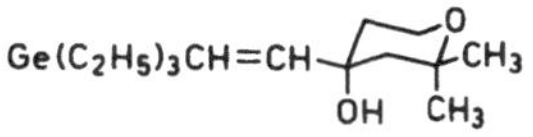 (trans) and 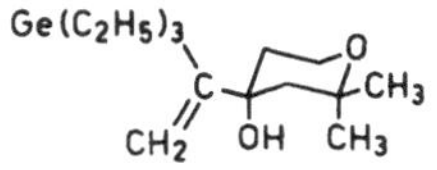	[102]
$CH\equiv CCH_2Cl$	25°C, 10 min, cat., no solvent → trans-$Ge(C_2H_5)_3CH=CHCH_2Cl$ (ca. 22%), $Ge(C_2H_5)_3C(CH_2Cl)=CH_2$ (ca. 40%), $CH\equiv CCH_3$ (33%), and $Ge(C_2H_5)_3Cl$ (33%) (exothermic reaction) 80°C in CH_3CN, 1 h, cat. → trans-$Ge(C_2H_5)_3CH=CHCH_2Cl$ (ca. 6%), $Ge(C_2H_5)_3C(CH_2Cl)=CH_2$ (ca. 8%), $CH\equiv CCH_3$ (56%), and $Ge(C_2H_5)_3Cl$ (56%)	[81]; see also [35]
$CH\equiv CCH_2OH$	ca. 90°C, 4 h, cat. → $Ge(C_2H_5)_3CH=CHCH_2OH$ (37%)	[7]
$CH\equiv CCH_2OCH_2OCH_2CH_2Cl$	70°C, 4 h, cat. → cis- and trans-$Ge(C_2H_5)_3$-$CH=CHCH_2OCH_2OCH_2CH_2Cl$ (55%)	[44]

References on p. 63

Table 7 (continued)

reactant	conditions (cat. = H_2PtCl_6) → products (yield) and remarks	Ref.
$CH{\equiv}CCH_2OCH_2OC_6H_{13}$	cat. → $Ge(C_2H_5)_3CH{=}CHCH_2OCH_2OC_6H_{13}$ (48%)	[65]
$CH{\equiv}CCH_2OCH(CH_3)OC_6H_{13}$	cat. → $Ge(C_2H_5)_3CH{=}CHCH_2OCH(CH_3)OC_6H_{13}$ (50%)	[65]
$CH{\equiv}CCH_2OC_6H_5$	90 °C, 6 h, cat. → cis-$Ge(C_2H_5)_3CH{=}CHCH_2OC_6H_5$ and $Ge(C_2H_5)_3C(CH_2OC_6H_5){=}CH_2$	[140]
	95 °C, 6 h, $Rh(P(C_6H_5)_3)Cl$ catalyst → trans-$Ge(C_2H_5)_3CH{=}CHCH_2OC_6H_5$ and $Ge(C_2H_5)_3C(CH_2OC_6H_5){=}CH_2$ in a 4:1 ratio	[171]
$CH{\equiv}CCH_2OC_6H_3(CH_3)_2$-2,5	95 °C, 6 h, $Rh(P(C_6H_5)_3)Cl$ catalyst → trans-$Ge(C_2H_5)_3CH{=}CHCH_2OC_6H_3(CH_3)_2$-2,5 and $Ge(C_2H_5)_3C(CH_2OC_6H_3(CH_3)_2){=}CH_2$ in a 9:1 ratio	[171]
$CH{\equiv}CCH_2OC_6H_4Cl$-4	95 °C, 6 h, $Rh(P(C_6H_5)_3)Cl$ catalyst → trans-$Ge(C_2H_5)_3CH{=}CHCH_2OC_6H_4$-Cl-4 and $Ge(C_2H_5)_3C(CH_2OC_6H_4$-Cl-4)$=CH_2$ in a 7:3 ratio	[171]
$CH{\equiv}CCH_2OC_6H_4Br$-2	95 °C, 6 h, $Rh(P(C_6H_5)_3)Cl$ catalyst → trans-$Ge(C_2H_5)_3CH{=}CHCH_2OC_6H_4$-Br-2 and $Ge(C_2H_5)_3C(CH_2OC_6H_4$-Br-2)$=CH_2$ in a 7:3 ratio	[171]
$CH{\equiv}CCH_2OSi(CH_3)_2C_6H_4CH_3$-4	100 to 120 °C, 3 to 4 h, cat. → $Ge(C_2H_5)_3CH{=}CHCH_2OSi(CH_3)_2C_6H_4CH_3$-4 (70%)	[93]
$CH{\equiv}CCOOCH_3$	20 °C in THF, 18 h, cat. → $Ge(C_2H_5)_3CH{=}CHCOOCH_3$ and $Ge(C_2H_5)_3C({=}CH_2)COOCH_3$ in an 11:89 ratio (99% total) 50 °C in C_6H_{14}, 3.5 h, cat. → products as above in a 43:57 ratio (99% total) reactions also carried out in $CHCl_3$, $C_2H_4Cl_2$, or C_6H_6	[219]
$CH{\equiv}CCOOC_2H_5$	in refluxing THF, 2 h, cat. → $Ge(C_2H_5)_3CH{=}CHCOOC_2H_5$ and $Ge(C_2H_5)_3C({=}CH_2)COOC_2H_5$ in an 18:82 ratio (67% total)	[219]

Table 7 (continued)

reactant	conditions (cat. = H_2PtCl_6) → products (yield) and remarks	Ref.
	no solvent, cat. → products as above in a 49:51 ratio (68% total)	
$CH{\equiv}CCH_2SC_6H_5$	100 °C, 6 h, cat. → trans-$Ge(C_2H_5)_3CH{=}CHCH_2SC_6H_5$ and $Ge(C_2H_5)_3C(CH_2SC_6H_5){=}CH_2$ in a 1:1 ratio	[163]
	95 °C, 6 h, $Rh(P(C_6H_5)_3)Cl$ catalyst → trans-$Ge(C_2H_5)_3CH{=}CHCH_2SC_6H_5$ and $Ge(C_2H_5)_3C(CH_2SC_6H_5){=}CH_2$ in a 2:3 ratio	[171]
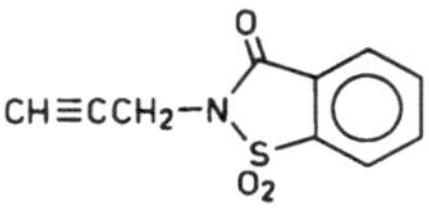	70 °C, 6 h, in THF, cat. → 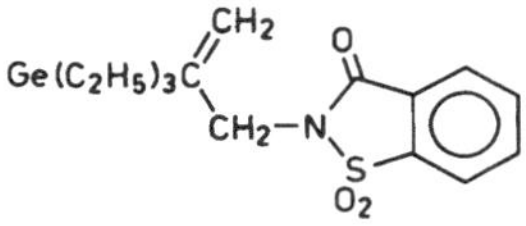 and 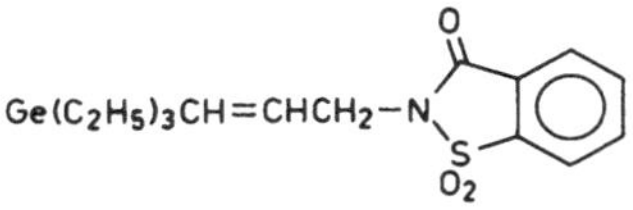 in a 5:1 ratio; similar results even at ca. 20 °C or in C_6H_{14}-C_6H_6	[215]
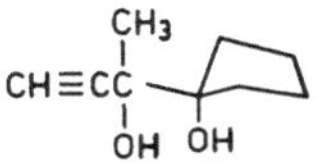	80 °C, 4 h, cat. → 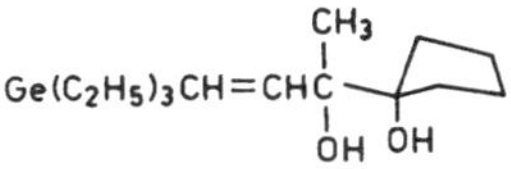 (cis,trans isomers) and 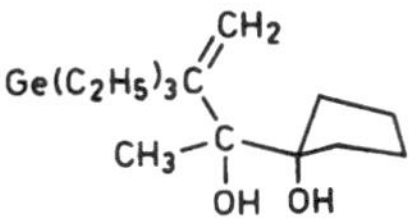 (86% total)	[175]
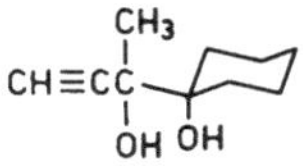	analogous to the preceding reaction → 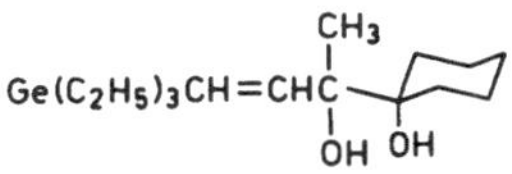 (97%; only trans isomer) and 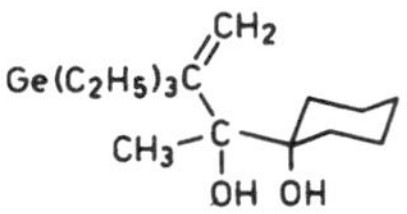 (small amounts)	[175]
$CH{\equiv}CC(CH_3)_2OCH(OH)CCl_3$	70 °C, 16 h, then 120 °C, 0.5 h, cat. → $Ge(C_2H_5)_3CH{=}CHC(CH_3)_2$-$OCH(OH)CH_3$ (58%)	[101]

References on p. 63

Table 7 (continued)

reactant	conditions (cat. = H_2PtCl_6) → products (yield) and remarks	Ref.
$CH{\equiv}CC(CH_3)_2OCH_2\overline{CHCH_2O}$	80 °C, 5 h, cat. → $Ge(C_2H_5)_3CH{=}CHC(CH_3)_2OCH_2\text{-}\overline{CHCH_2O}$ (76%)	[117]
$CH{\equiv}CC(CH_3)_2OSi(CH_3)_2\text{-}C_6H_4CH_3\text{-}4$	100 to 120 °C, 3 to 4 h, cat. → $Ge(C_2H_5)_3CH{=}CHC(CH_3)_2OSi(CH_3)_2\text{-}C_6H_4CH_3\text{-}4$ (64%)	[93]
$CH{\equiv}CC(OH)(CH_3)C_{10}H_7\text{-}1$	90 °C, 4 h, cat.; addition to the C≡C bond of the starting compound (60%)	[141]
$CH{\equiv}CC(CH_3)(C_{10}H_7\text{-}1)OSi(CH_3)_3$	90 °C, 6 h, cat.; addition to the C≡C bond of the starting compound (62%)	[141]
$CH{\equiv}CCH(OH)C_3H_7\text{-}i$	80 to 100 °C, 6 h, cat. → $Ge(C_2H_5)_3CH{=}CHCH(OH)C_3H_7\text{-}i$	[32]
$CH{\equiv}CC(OH)(CH_3)C_4H_9\text{-}t$	refluxing, 2 h, cat. → $Ge(C_2H_5)_3CH{=}CHC(OH)(CH_3)C_4H_9\text{-}t$ (65%)	[37]
$CH{\equiv}CC(OH)(CH_3)CH_2CH_2OCH_3$	cat., 20 min, temperature rising to 125 °C → $Ge(C_2H_5)_3CH{=}CHC(OH)(CH_3)CH_2\text{-}CH_2OCH_3$ (cis,trans isomers, 7 and 9%) and $Ge(C_2H_5)_3C({=}CH_2)C(OH)(CH_3)\text{-}CH_2CH_2OCH_3$ (74%)	[98]
$CH{\equiv}CC(CH_3)(C_2H_5)\text{-}OCH_2\overline{CHCH_2O}$	80 °C, 5 h, cat. → $Ge(C_2H_5)_3CH{=}CHC(CH_3)\text{-}(C_2H_5)OCH_2\overline{CHCH_2O}$ (70%)	[117]
$CH{\equiv}CCH(C_3H_7\text{-}i)\text{-}OCH_2\overline{CHCH_2O}$	80 °C, 5 h, cat. → $Ge(C_2H_5)_3CH{=}CHCH(C_3H_7\text{-}i)\text{-}OCH_2\overline{CHCH_2O}$ (77%)	[117]
$CH{\equiv}CC(CH_3)(C_2H_5)\text{-}OCH_2OCH_2CH_2Cl$	cat. → $Ge(C_2H_5)_3CH{=}CHC(CH_3)(C_2H_5)OCH_2\text{-}OCH_2CH_2Cl$	[54]
$CH{\equiv}CC(CH_3)(C_4H_9\text{-}t)\text{-}OCH(OH)CCl_3$	110 to 120 °C, 30 h, cat. → $Ge(C_2H_5)_3CH{=}CHC(CH_3)(C_4H_9\text{-}t)\text{-}OCH(OH)CCl_3$ (47%) and $Ge(C_2H_5)_3CH{=}CH\text{-}C(CH_3)(C_4H_9\text{-}t)\text{-}OCH(OH)CH_2Cl$ (43%)	[101]
$CH{\equiv}CC(CH_3)(C_2H_5)\text{-}OSi(CH_3)_2C_6H_4CH_3\text{-}4$	100 to 120 °C, 3 to 4 h, cat. → $Ge(C_2H_5)_3CH{=}CHC(CH_3)(C_2H_5)O\text{-}Si(CH_3)_2C_6H_4CH_3\text{-}4$ (60%)	[93]

Table 7 (continued)

reactant	conditions (cat. = H_2PtCl_6) → products (yield) and remarks	Ref.
$CH{\equiv}CCH_2C_3H_7$	100 °C, 2 h, cat. (very exothermic reaction) → $Ge(C_2H_5)_3CH{=}CHCH_2C_3H_7$ (100%)	[6, 9]
$CH{\equiv}CCH(OH)C_3H_7$	80 to 100 °C, 6 h, cat. → $Ge(C_2H_5)_3CH{=}CHCH(OH)C_3H_7$	[32]
$CH{\equiv}CC(OH)(CH_3)C_3H_7$	reflux, 2 h, cat. → $Ge(C_2H_5)_3CH{=}CHC(OH)(CH_3)C_3H_7$	[37]
$CH{\equiv}CCH(C_3H_7)$-$OCH(OH)CCl_3$	50 to 55 °C, 6 h, cat. → $Ge(C_2H_5)_3CH{=}CHCH(C_3H_7)OCH$-$(OH)CH_2Cl$	[88]
$CH{\equiv}CCH(C_3H_7)$-$OCH_2OCH_2CH_2Cl$	cat. → $Ge(C_2H_5)_3CH{=}CHCH(C_3H_7)OCH_2$-$OCH_2CH_2Cl$	[54]
$CH{\equiv}CC(CH_3)(C_3H_7)$-$OCH(OC_4H_9)CH_3$	ca. 80 °C, cat. → $Ge(C_2H_5)_3CH{=}CHC(CH_3)(C_3H_7)OCH$-$(OC_4H_9)CH_3$	[72]
$CH{\equiv}CC(OH)(CH_3)C_4H_9$	refluxing, 2 h, cat. → $Ge(C_2H_5)_3CH{=}CHC(OH)(CH_3)C_4H_9$	[37]
$CH_3C{\equiv}COCH_3$	80 to 100 °C, 20 to 24 h → $Ge(C_2H_5)_3C(OCH_3){=}CHCH_3$ and $Ge(C_2H_5)_3C(CH_3){=}CHOCH_3$ in a 2.3 : 1 ratio (50% total yield); Ge trans to H in both products	[127]
$CH_3C{\equiv}COC_2H_5$	80 to 100 °C, 20 to 24 h → $Ge(C_2H_5)_3C(OC_2H_5){=}CHCH_3$ and $Ge(C_2H_5)_3C(CH_3){=}CHOC_2H_5$ in a 2.3 : 1 ratio (60% total yield); Ge trans to H in both products	[127]
$CH_3C{\equiv}COC_6H_5$	cat. → $Ge(C_2H_5)_3C(OC_6H_5){=}CHCH_3$ and $Ge(C_2H_5)_3C(CH_3){=}CHOC_6H_5$ in a 1 : 1 ratio	[163]
$CH_3C{\equiv}CSC_6H_5$	100 °C, 6 h, cat. → $Ge(C_2H_5)_3C(SC_6H_5){=}CHCH_3$ (main product)	[163]
$CH_3C{\equiv}CN(C_2H_5)_2$	100 °C, 10 to 15 h, UV irradiated → $Ge(C_2H_5)_3C(CH_3){=}CHN(C_2H_5)_2$ (cis,trans isomers; 90%)	[126]

References on p. 63

Table 7 (continued)

reactant	conditions (cat. = H_2PtCl_6) → products (yield) and remarks	Ref.
$ClCH_2C{\equiv}CCH_2Cl$	50 °C, in THF, 4 h, cat. → $Ge(C_2H_5)_3C(CH_2Cl){=}CHCH_2Cl$ (65%) along with $Ge(C_2H_5)_3Cl$	[221]
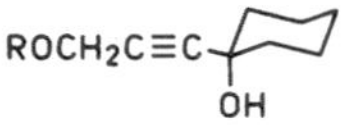 R = CH_3	100 to 120 °C, 4 h, cat. → 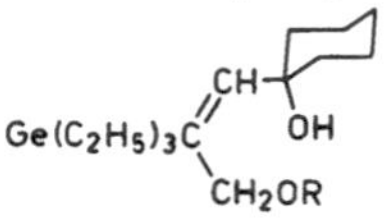 (48%)	[71]
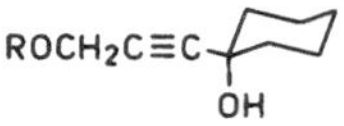 R = $ClCH_2CH_2$	→ 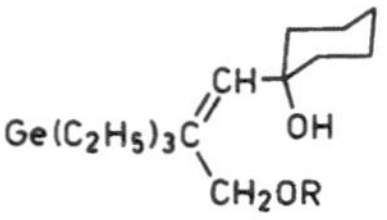 (40%)	[100]
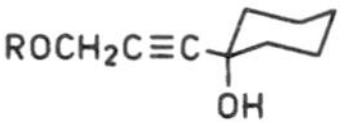 R = C_3H_7	100 to 120 °C, 4 h, cat. → 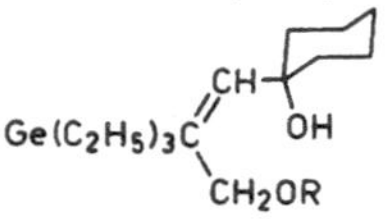 (49%)	[71]
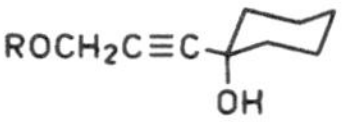 R = C_4H_9	100 to 120 °C, 4 h, cat. → 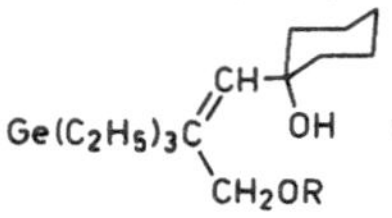 (46%)	[71]
$C_6H_5C{\equiv}CCH(OH)C_6H_5$	50 to 55 °C, 1 h, cat. → $Ge(C_2H_5)_3C(CH(OH)C_6H_5){=}CHC_6H_5$ (60%)	[97]
$CF_3C{\equiv}CCF_3$	UV-irradiated, 7 d → trans-$Ge(C_2H_5)_3C(CF_3){=}CHCF_3$	[33]
$CH_3OOCC{\equiv}CCOOCH_3$	100 to 110 °C, 15 h, cat. → $Ge(C_2H_5)_3C(COOCH_3){=}CHCOOCH_3$ (65%)	[145]
$HOCH_2C{\equiv}CC(OH)(CH_3)C_{10}H_7$-2	95 °C, 4 h, cat. → 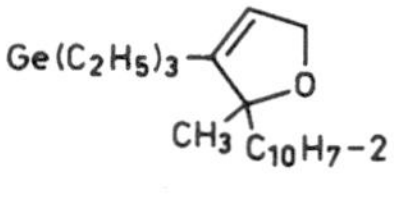 and 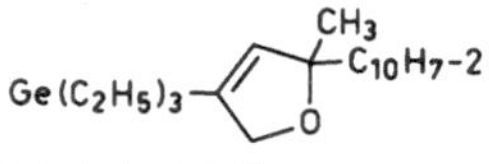 (68% total yield)	[104]
$CH_3OCH_2C{\equiv}CC(OH)(CH_3)_2$	150 °C, 2 h, cat. → $Ge(C_2H_5)_3C(CH_2OCH_3){=}CHC(OH)$-$(CH_3)_2$ (57%)	[94]

Table 7 (continued)

reactant	conditions (cat. = H_2PtCl_6) → products (yield) and remarks	Ref.
$ClCH_2C{\equiv}CCH(OH)CCl_3$	80 to 85°C, 2 h, cat. → $Ge(C_2H_5)_3C(CH_2Cl){=}CHCH(OH)CCl_3$ (50%)	[84]
$C_2H_5OCH_2C{\equiv}CC(OH)(CH_3)C_2H_5$	cat. → $Ge(C_2H_5)_3C(CH_2OC_2H_5){=}CHC(OH)(CH_3)C_2H_5$	[94]
$CH_3OCH_2C{\equiv}CCH(OH)C_3H_7$	cat. → $Ge(C_2H_5)_3C(CH_2OCH_3){=}CHCH(OH)C_3H_7$	[94]
$C_2H_5OCH_2C{\equiv}CCH(OH)C_3H_7$	cat. → $Ge(C_2H_5)_3C(CH_2OC_2H_5){=}CHCH(OH)C_3H_7$	[94]
$C_3H_7OCH_2C{\equiv}CCH(OH)C_3H_7$	100°C, cat. → $Ge(C_2H_5)_3C(CH_2OC_3H_7){=}CHCH(OH)C_3H_7$	[70]
i-$C_3H_7OCH_2C{\equiv}CCH(OH)C_3H_7$	100°C, 2 h, cat. → $Ge(C_2H_5)_3C(CH_2OC_3H_7$-i)$=CHCH(OH)C_3H_7$ (51%)	[70]
$C_4H_9OCH_2C{\equiv}CCH(OH)C_3H_7$	100°C, cat. → $Ge(C_2H_5)_3C(CH_2OC_4H_9){=}CHCH(OH)C_3H_7$	[70]
i-$C_4H_9OCH_2C{\equiv}CCH(OH)C_3H_7$	100°C, cat. → $Ge(C_2H_5)_3C(CH_2OC_4H_9$-i)$=CHCH(OH)C_3H_7$	[70]
$C_2H_5C{\equiv}COCH_3$	80 to 100°C, 20 to 24 h → $Ge(C_2H_5)_3C(OCH_3){=}CHC_2H_5$ and $Ge(C_2H_5)_3C(C_2H_5){=}CHOCH_3$ in a 1.5:1 ratio (71% total yield); Ge trans to H in both products	[127]
$C_2H_5C{\equiv}COC_2H_5$	80 to 100°C, 20 to 24 h → $Ge(C_2H_5)_3C(OC_2H_5){=}CHC_2H_5$ and $Ge(C_2H_5)_3C(C_2H_5){=}CHOC_2H_5$ in a 3:2 ratio (54% total yield); Ge trans to H in both products 80°C, 12 h, in the presence of AIBN → the same products as above in a 3:2 ratio (50% total yield)	[127]
i-$C_3H_7C{\equiv}COC_2H_5$	80 to 100°C, 20 to 24 h → $Ge(C_2H_5)_3C(OC_2H_5){=}CHC_3H_7$-i and $Ge(C_2H_5)_3C(C_3H_7$-i)$=CHOC_2H_5$ in a	[127]

References on p. 63

Table 7 (continued)

reactant	conditions (cat. = H_2PtCl_6) → products (yield) and remarks	Ref.
	1.1:1 ratio (61% total yield); Ge trans to H in both products	
$C_2H_5C{\equiv}CN(C_2H_5)_2$	100 °C, 10 to 15 h, UV-irradiated → $Ge(C_2H_5)_3C(C_2H_5){=}CHN(C_2H_5)_2$ (72%); cis,trans isomers	[126]
i-$C_3H_7C{\equiv}CN(C_2H_5)_2$	100 °C, 10 to 15 h, UV-irradiated → $Ge(C_2H_5)_3C(C_3H_7\text{-i}){=}CHN(C_2H_5)_2$ (60%); cis,trans isomers	[126]
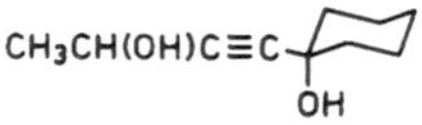	115 to 120 °C, 30 h, Pt-C catalyst → 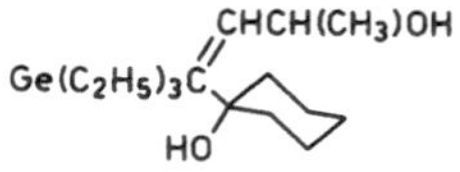 (15%) and 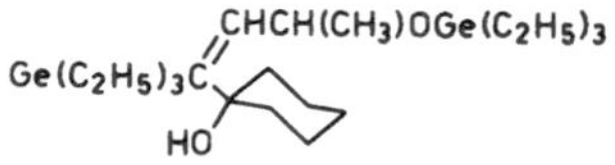 (8%) H_2PtCl_6 cat. → 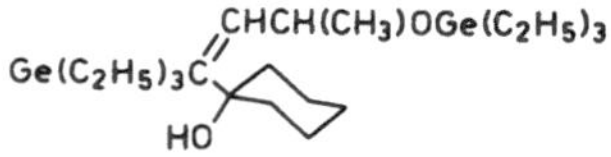 (29%)	[26]
$C_6H_5C{\equiv}CCH(OH)CH_3$	50 to 55 °C, 1 h, cat. → $Ge(C_2H_5)_3C(CH(OH)CH_3){=}CHC_6H_5$ (70%)	[97]
$C_6H_5C{\equiv}CC(OH)(CH_3)_2$	30 min, cat. → $Ge(C_2H_5)_3C(C(OH)(CH_3)_2){=}CHC_6H_5$ (85%)	[46]
$C_6H_5C{\equiv}CC(CH_3)_2C_6H_4OH\text{-}4$	100 °C, 1.5 to 2 h, cat. → $Ge(C_2H_5)_3C(C(CH_3)_2C_6H_4OH\text{-}4){=}CH\text{-}C_6H_5$ (73%)	[142]
$C_6H_5C{\equiv}CC(CH_3)_2C_6H_4OCH_3\text{-}4$	100 °C, 1.5 to 2 h, cat. → $Ge(C_2H_5)_3C(C(CH_3)_2C_6H_4OCH_3\text{-}4){=}CH\text{-}C_6H_5$ (90%)	[142]
$C_6H_5C{\equiv}CC(OH)(CH_3)C_{10}H_7\text{-}2$	90 °C, cat. → $Ge(C_2H_5)_3C(C_6H_5){=}CHC(OH)(CH_3)\text{-}C_{10}H_7\text{-}2$ and $Ge(C_2H_5)_3C(C(OH)\text{-}(CH_3)C_{10}H_7\text{-}2){=}CHC_6H_5$ in a 1:1 ratio (86% total yield)	[167]
$(CH_3)_2(OH)CC{\equiv}CC(OH)(CH_3)_2$	140 to 150 °C, 5 h, cat. → $Ge(C_2H_5)_3C(C(OH)\text{-}(CH_3)_2){=}CH(OH)(CH_3)_2$ (40%)	[6, 9]

Table 7 (continued)

reactant	conditions (cat. = H_2PtCl_6) → products (yield) and remarks	Ref.
$(CH_3)_2(C_2H_5O)CC{\equiv}CC(OC_2H_5)$-$(CH_3)_2$	100°C, 1.5 h, cat. → $Ge(C_2H_5)_3C(C(OC_2H_5)$-$(CH_3)_2)$=CH-$C(OC_2H_5)(CH_3)_2$ (45%)	[47]
$CH_3(OH)CHC{\equiv}CC(OH)(CH_3)_2$	100°C, 20 h, Pt-C catalyst, 35 min → $Ge(C_2H_5)_3C(C(OH)$-$(CH_3)_2)$=$CHCH(OH)CH_3$ (40%) 35 min, cat. → $Ge(C_2H_5)_3C(C(OH)$-$(CH_3)_2)$=$CHCH(CH_3)OGe(C_2H_5)_3$ (28%)	[26]
$(CH_3)_2(OH)CC{\equiv}C$-$C(OH)(CH_3)C_{10}H_7$-1	90°C, 4 h, cat. → 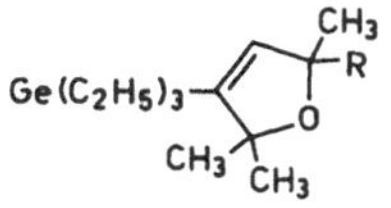 and 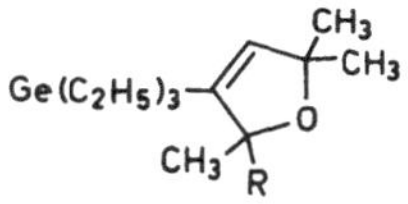 (R = 1-$C_{10}H_7$) in a 2:3 ratio (72% total yield)	[103]
$(CH_3)_2(OH)CC{\equiv}C$-$C(OH)(CH_3)C_{10}H_7$-2	90°C, 4 h, cat. → 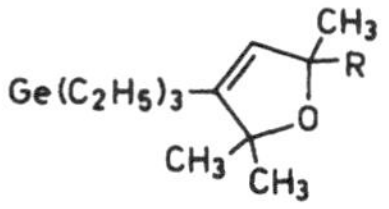 and 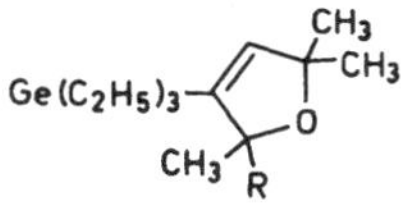 (R = 2-$C_{10}H_7$) in a 3:2 ratio (76% total yield)	[103]
2-$C_{10}H_7(CH_3)(OH)CC{\equiv}C$-$C(OH)(CH_3)C_{10}H_7$-2	90°C, 5 h, cat. → 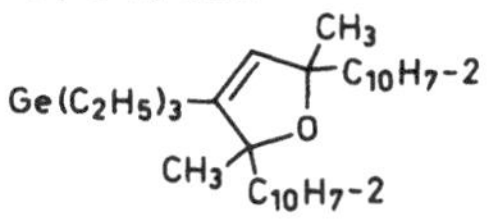 (47%) and 2-$C_{10}H_7(CH_3)C$=C=$CHC(OH)$-$(CH_3)C_{10}H_7$-2 (21%)	[103]
$CH_3(OH)CHC{\equiv}C$-$C(OH)(CH_3)C_{10}H_7$-2	85 to 90°C, 4 h, cat. → 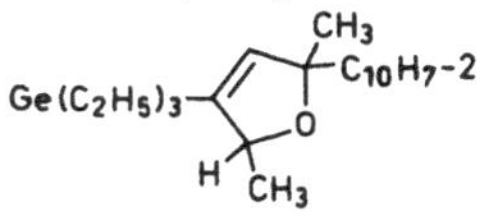	[135]

References on p. 63

Table 7 (continued)

reactant	conditions (cat. = H_2PtCl_6) → products (yield) and remarks	Ref.
	and 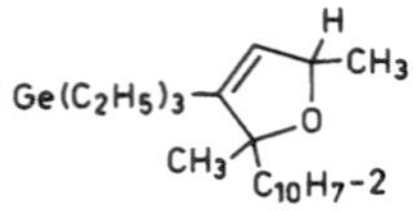 (58% total yield)	
$C_4H_7S(CH_3)(OH)CC{\equiv}C$- $C(OH)(CH_3)C_4H_7S$ C_4H_7S = 	90 °C, cat. → 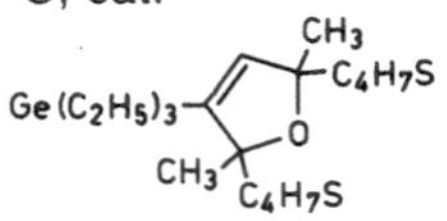 (72%)	[123]
$(CH_3)_2(OH)CC{\equiv}C$- $C(OH)(CH_3)C_2H_4OCH_3$	55 to 60 °C, 3 h, cat. → $Ge(C_2H_5)_3C(C(OH)$- $(CH_3)_2)$=$CHC(OH)(CH_3)C_2H_4OCH_3$ (62%) and 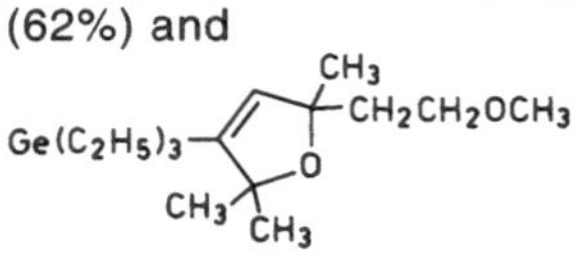 (20%)	[62]
$(CH_3)_2(OH)CC{\equiv}CC(OH)(C_3H_7)_2$	60 to 70 °C, 4 h, cat. → $Ge(C_2H_5)_3C(C(OH)$- $(CH_3)_2)$=$CHC(OH)(C_3H_7)_2$ (13%)	[63]
2-$C_{10}H_7(CH_3)(OH)CC{\equiv}C$- $C(OH)(CH_3)C_3H_7$	85 to 90 °C, 4 h, cat. → 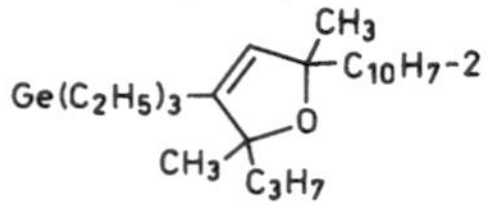 and 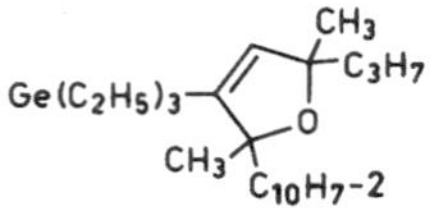 in a 10:1 ratio (71% total yield)	[103]
2-$C_{10}H_7(CH_3)(OH)CC{\equiv}C$- $CH(OH)C_3H_7$	85 to 90 °C, 4 h, cat. → $Ge(C_2H_5)_3C(CH(OH)C_3H_7)$=CH- $C(C_{10}H_7$-2)=CH_2 and $Ge(C_2H_5)_3C(C(C_{10}H_7$-2)=$CH_2)$=CH- $CH(OH)C_3H_7$ (42% total yield) along with 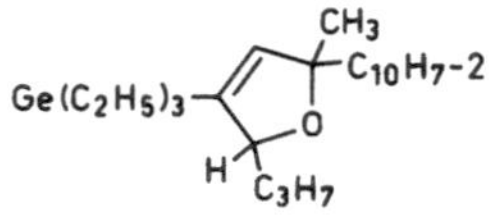 and 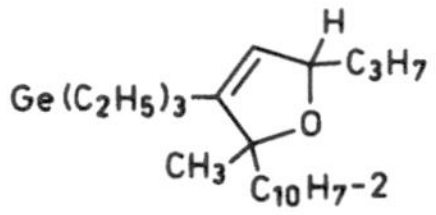 (33% total yield)	[135]

Table 7 (continued)

reactant	conditions (cat. = H_2PtCl_6) → products (yield) and remarks	Ref.
$C_6H_5C{\equiv}CC(OH)(CH_3)C_2H_5$	30 min, cat. → $Ge(C_2H_5)_3C(C(OH)(CH_3)C_2H_5){=}CHC_6H_5$ (80%)	[46]
$CH_3OCH_2CH_2(CH_3)(OH)CC{\equiv}C$-$C(OH)(CH_2)_4CH_2$	55 to 60°C, 3 h, cat. → $Ge(C_2H_5)_3C(C(OH)(CH_2)_4CH_2){=}CH$-$C(OH)(CH_3)C_2H_4OCH_3$ (28%) and 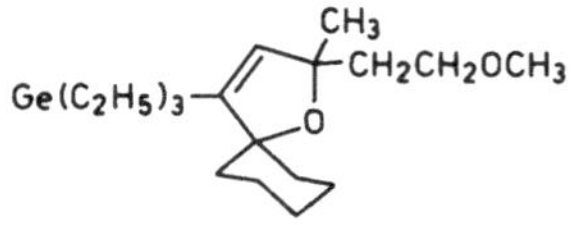	[62]
$CH_3OCH_2CH_2(CH_3)(OH)CC{\equiv}C$-$C(OH)(CH_3)C_2H_5$	55 to 60°C, 3 h, cat. → $Ge(C_2H_5)_3C(C(OH)(CH_3)C_2H_5){=}CH$-$C(OH)(CH_3)C_2H_4OCH_3$ (90%) and 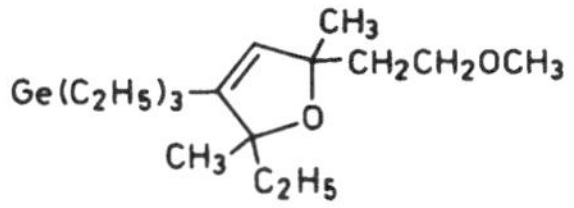	[62]
$CH_3OC_2H_4(CH_3)(OH)CC{\equiv}C$-$C(OH)(CH_3)C_2H_4OCH_3$	55 to 60°C, 3 h, cat. → $Ge(C_2H_5)_3C(C(OH)(CH_3)$-$C_2H_4OCH_3){=}CHC(OH)(CH_3)$-$C_2H_4OCH_3$ (63%) and 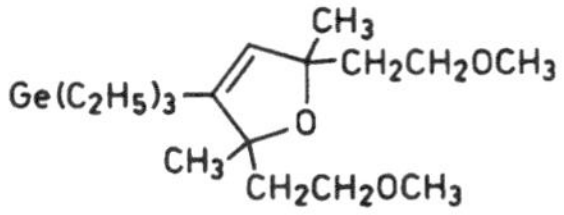	[62]
$C_2H_5(CH_3)(OH)CC{\equiv}C$-$C(OH)(C_3H_7)_2$	elevated temperature (molten alkyne), cat. → $Ge(C_2H_5)_3C(C(OH)(CH_3)C_2H_5){=}CH$-$C(OH)(C_3H_7)_2$ (8%)	[63]
$C_6H_5C{\equiv}CCH(OH)C_3H_7$	50 to 55°C, 1 h, cat. → $Ge(C_2H_5)_3C(CH(OH)C_3H_7){=}CHC_6H_5$ (70%)	[97]
$C_6H_5C{\equiv}CSi(OCH_3)_3$	100°C, 8 h, sealed tube → $Ge(C_2H_5)_3C(Si(OCH_3)_3){=}CHC_6H_5$ (73%) and $Ge(C_2H_5)_3C(C_6H_5){=}CHSi(OCH_3)_3$ (14%)	[191]
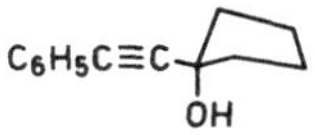	30 min, cat. (very exothermic reaction) → 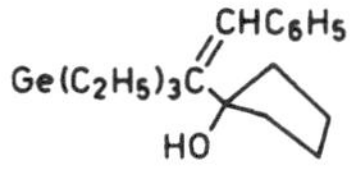 (80%)	[61]

References on p. 63

Table 7 (continued)

reactant	conditions (cat. = H_2PtCl_6) → products (yield) and remarks	Ref.
	and 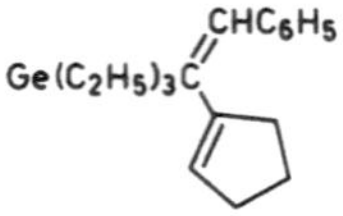	
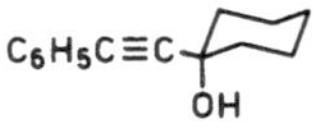	30 min, cat. → 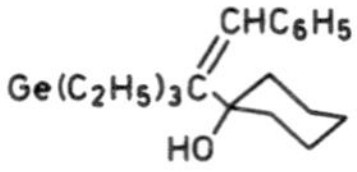 (80%)	[61]
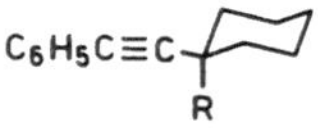 R = C_6H_4OH-4	100 °C, 1.5 to 2 h, cat. → 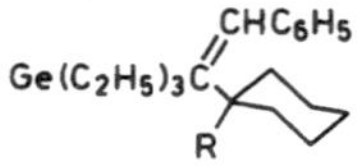 (92%)	[142]
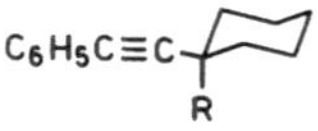 R = $C_6H_4OCH_3$-4	100 °C, 1.5 to 2 h, cat. → 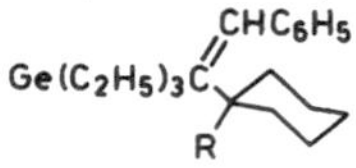 (90%)	[142]
$Ge(C_2H_5)_3OCH_2C{\equiv}CH$	140 °C, 12 h, cat. → $Ge(C_2H_5)_3OCH_2CH{=}CHGe(C_2H_5)_3$ (70%)	[21]
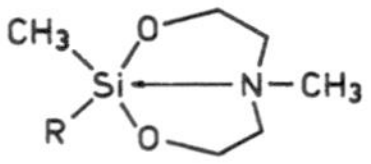 R = $C{\equiv}CC_6H_5$	100 °C, 8 h, $Rh(C_5H_7O_2)(CO)_2$ catalyst, in CH_3CN in a sealed tube → 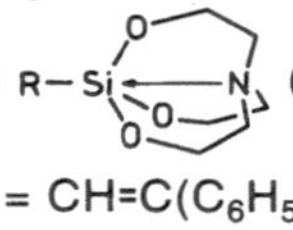 (96%) R = $CH{=}C(C_6H_5)Ge(C_2H_5)_3$	[191]
reactants with one C≡C and one or two C=C bonds		
$CH{\equiv}CCH{=}CH_2$	ca. −80 °C, 3 to 4 h, then at room temperature (1 h) and 80 to 90 °C for 2 h, cat., in autoclave → $Ge(C_2H_5)_3CH{=}CHCH{=}CH_2$ (64%)	[85]
$CH{\equiv}CC(CH_3){=}CH_2$	cat. → trans-$Ge(C_2H_5)_3CH{=}CHC(CH_3){=}CH_2$ and $Ge(C_2H_5)_3C({=}CH_2)C(CH_3){=}CH_2$ in a 86 : 14 ratio (50% total yield)	[80]
	conditions like those of the previous reaction → $Ge(C_2H_5)_3CH{=}CHC(CH_3){=}CH_2$ (59%)	[85]
$CH{\equiv}CCH{=}CHCH_3$	conditions like those of the previous reaction → $Ge(C_2H_5)_3CH{=}CHCH{=}CHCH_3$ (78%)	[85]

Table 7 (continued)

reactant	conditions (cat. = H_2PtCl_6) → products (yield) and remarks	Ref.
$CH{\equiv}CCH_2OCH_2OCH_2CH{=}CH_2$	cat. → $Ge(C_2H_5)_3CH{=}CHCH_2OCH_2$- $OCH_2CH{=}CH_2$ (28%) and $Ge(C_2H_5)_3CH{=}CHCH_2OCH_2O(CH_2)_3$- $Ge(C_2H_5)_3$ (10%)	[65]
$CH{\equiv}CCH_2OCH(CH_3)OCH_2CH{=}CH_2$	cat. → $Ge(C_2H_5)_3CH{=}CHCH_2OCH(CH_3)$- $OCH_2CH{=}CH_2$ (33%) and $Ge(C_2H_5)_3CH{=}CHCH_2OCH(CH_3)O$- $(CH_2)_3Ge(C_2H_5)_3$ (10%)	[65]
cis-$CH{\equiv}CCH_2OOCCH{=}CHCOOH$	80 to 85°C, 7 h, cat. → $Ge(C_2H_5)_3CH{=}CHCH_2OOCCH{=}CH$- COOH (48%)	[143]
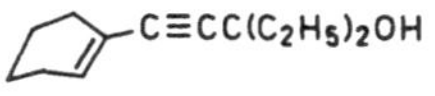	→ 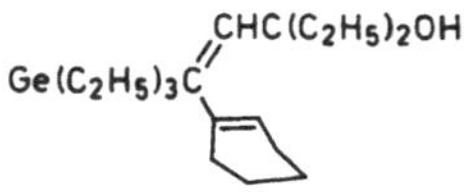 (74%)	[125]
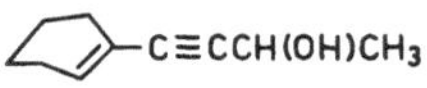	cat. → 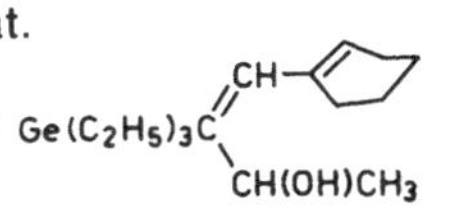 and 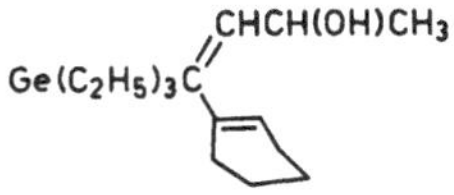 in a ca. 1:1 ratio (76% total yield)	[124]
	cat. → 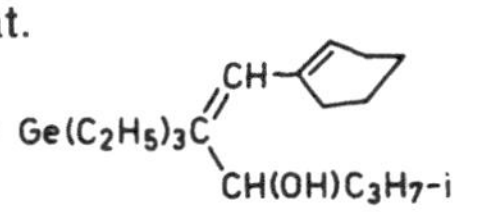 and 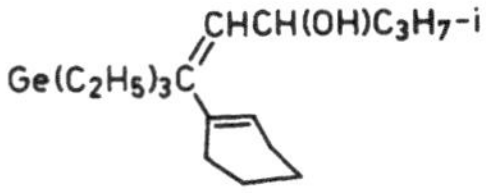 in a ca. 1:2 ratio (74% total yield)	[124]
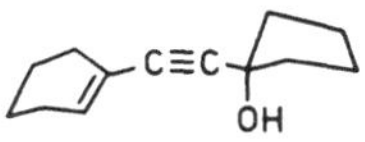	→ 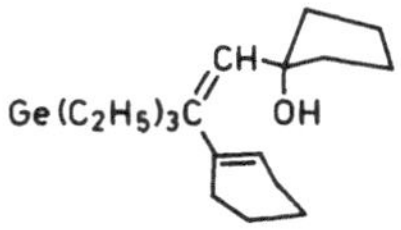 (73%)	[125]

References on p. 63

Table 7 (continued)

reactant	conditions (cat. = H_2PtCl_6) → products (yield) and remarks	Ref.
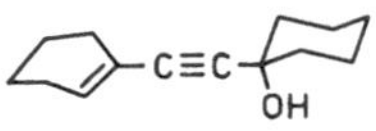	→ 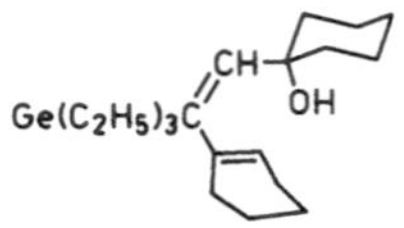 (66%)	[125]
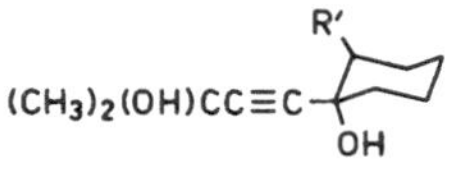 $R' = CH_2CH{=}CH_2$	80 to 90°C, 4 h, cat. → 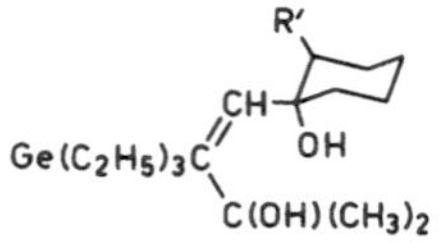 (26%), 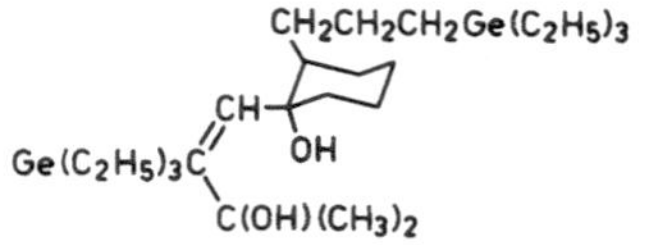 and 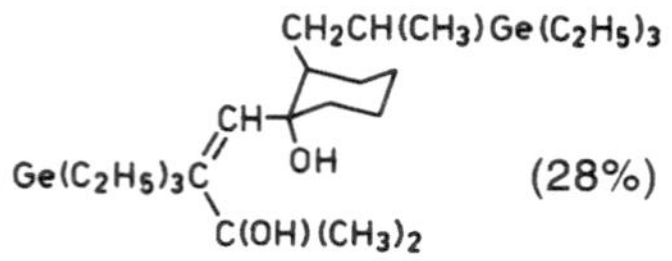 (28%)	[144]
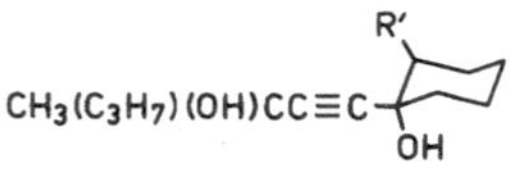 $R' = CH_2CH{=}CH_2$	cat. → products analogous to those of the previous reaction	[144]
$CH_2{=}CHC{\equiv}CCH_2CH_2OH$	1 h, cat. (exothermic reaction) → $Ge(C_2H_5)_3C(CH_2CH_2OH){=}CHCH{=}CH_2$ (40%), $Ge(C_2H_5)_3CH_2CH{=}CHCH(CH_2$-$CH_2OH)Ge(C_2H_5)_3$ (7%), and polymeric material	[19]
$CH_2{=}CHC{\equiv}CCH(OH)CH_3$	1 h, cat. (exothermic reaction) → $Ge(C_2H_5)_3C(CH(OH)CH_3){=}CHCH{=}CH_2$ (54%), $Ge(C_2H_5)_3CH_2CH{=}CHCH$-$(CH(OH)CH_3)Ge(C_2H_5)_3$ (5%), and polymeric material	[19]
$CH_2{=}CHC{\equiv}CCH(OH)C_3H_7$	1 h, cat. (exothermic reaction) → $Ge(C_2H_5)_3C(CH(OH)$-$C_3H_7){=}CHCH{=}CH_2$ (56%), $Ge(C_2H_5)_3CH_2CH{=}CHCH$-$(CH(OH)C_3H_7)Ge(C_2H_5)_3$ (11%), and polymeric material	[19]
$CH_2{=}CHC{\equiv}CCH(OH)C_6H_5$	100°C, 1 h, cat. (exothermic reaction after initial warming)	[19]

Table 7 (continued)

reactant	conditions (cat. = H_2PtCl_6) → products (yield) and remarks	Ref.
	→ $Ge(C_2H_5)_3C$-($CH(OH)C_6H_5$)=CHCH=CH_2 (41%), $Ge(C_2H_5)_3CH_2CH$=CHCH-($CH(OH)C_6H_5$)$Ge(C_2H_5)_3$ (9%), and polymeric material	
CH_2=CHC≡CC(OH)$(CH_3)_2$	ca. 1 h, cat. (exothermic reaction) → $Ge(C_2H_5)_3C$(C(OH)-$(CH_3)_2$)=CHCH=CH_2 (70%)	[15]
CH_2=CHC≡CC(OH)$(CH_3)C_2H_5$	cat. → $Ge(C_2H_5)_3C$(C(OH)-$(CH_3)C_2H_5$)=CHCH=CH_2 (79%)	[15]
CH_2=CHC≡CC(OH)$(CH_3)C_3H_7$	heating on a water bath, 3 h, cat. (temperature rising to 170°C) → $Ge(C_2H_5)_3C$(CH=CH_2)=CHC-(OH)$(CH_3)C_3H_7$ (40%)	[63]
CH_2=CHC≡CC(OH)$(CH_3)C_6H_5$	cat. → $Ge(C_2H_5)_3C$(C(OH)-$(CH_3)C_6H_5$)=CHCH=CH_2 (68%)	[64]
CH_2=CHC≡CC$(CH_3)_2OCH_3$	100°C, 6 h, cat. → $Ge(C_2H_5)_3C$(CH=CH_2)=CHC$(CH_3)_2$-OCH_3, $Ge(C_2H_5)_3C$(=CHCH=CH_2)C-$(CH_3)_2OCH_3$, and $Ge(C_2H_5)_3C$(=C=$CHCH_3$)C$(CH_3)_2OCH_3$ in a 64:26:10 ratio (77% total yield)	[91]; see also [77]
CH_2=CHC≡CC$(CH_3)_2C_6H_4OCH_3$-4	100°C, 6 h, cat. → $Ge(C_2H_5)_3C$(CH=CH_2)=CHC$(CH_3)_2$-$C_6H_4OCH_3$-4 and $Ge(C_2H_5)_3C$(=C=$CHCH_3$)C$(CH_3)_2$-$C_6H_4OCH_3$-4 in a 4:1 ratio (79% total yield)	[92]; see also [78]
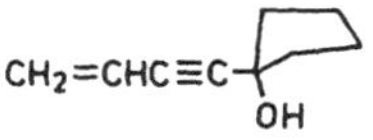	cat. → 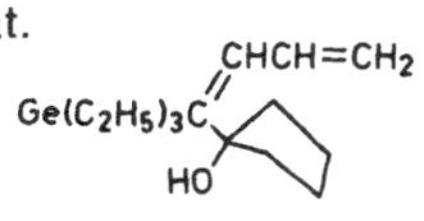 (73%)	[15]
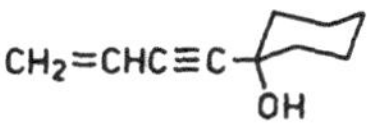	cat. → 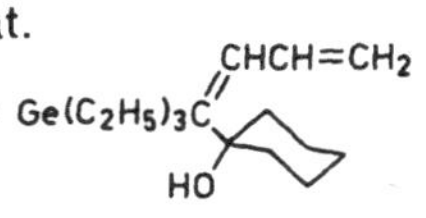 (69%)	[15]
$Ge(C_2H_5)_3C$(C(OH)$(CH_3)_2$)=CH-C≡CC(OH)$(CH_3)_2$	120°C, 1 h, cat. → $Ge(C_2H_5)_3C$(C(OH)$(CH_3)_2$)=CHCH=C-(C(OH)$(CH_3)_2Ge(C_2H_5)_3$ (13%)	[18]

References on p. 63

Table 7 (continued)

reactant	conditions (cat. = H_2PtCl_6) → products (yield) and remarks	Ref.
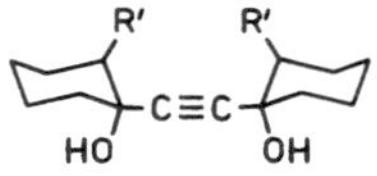 $R' = CH_2CH{=}CH_2$	cat. → 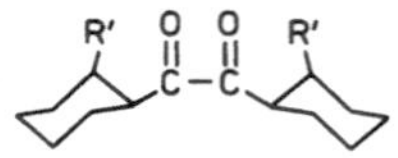 two products with $R' = CH_2CH_2CH_2Ge(C_2H_5)_3$ or $CH_2CH(CH_3)Ge(C_2H_5)_3$ (23% total yield)	[144]
reactants with two C≡C bonds		
$(CH{\equiv}CCH_2OSi(CH_3)(C_2H_5)\text{-})_2O$	refluxing, 2 h, cat. → $(Ge(C_2H_5)_3CH{=}CHCH_2OSi\text{-}(CH_3)(C_2H_5)\text{-})_2O$ (44%)	[69]
$(CH{\equiv}CCH_2OSi(CH_3)(C_3H_7)\text{-})_2O$	refluxing, 2 h, cat. → $(Ge(C_2H_5)_3CH{=}CHCH_2OSi\text{-}(CH_3)(C_3H_7)\text{-})_2O$ (44%)	[69]
$(CH{\equiv}CC(CH_3)(C_4H_9)O\text{-}Si(CH_3)(C_3H_7)\text{-})_2O$	refluxing, 2 h, cat. → $(Ge(C_2H_5)_3CH{=}CHC(CH_3)(C_4H_9)O\text{-}Si(CH_3)(C_3H_7)\text{-})_2O$ (37%)	[69]
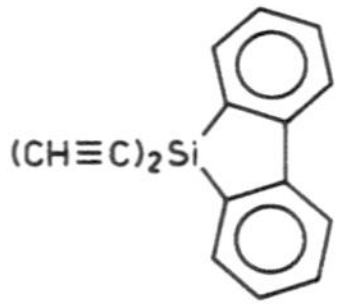	60 °C, 6 h, cat. → 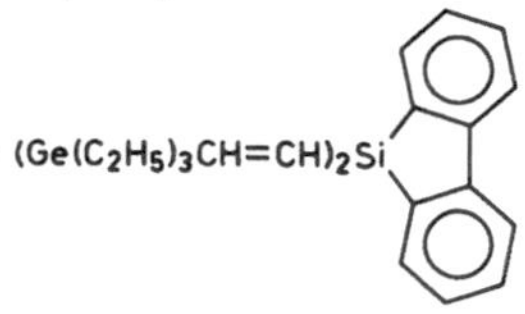 (56%)	[116]
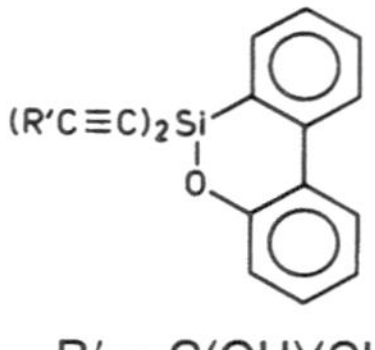 $R' = C(OH)(CH_3)_2$	95 °C, 8 h, cat. → 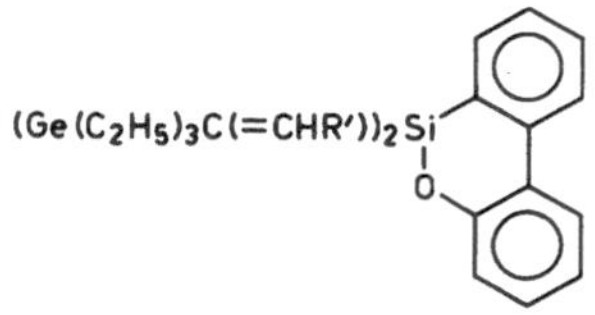 (13%) and 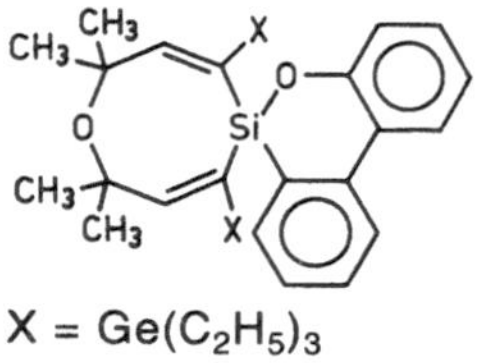 (10%) $X = Ge(C_2H_5)_3$	[115]

Table 7 (continued)

reactant	conditions (cat. = H_2PtCl_6) → products (yield) and remarks	Ref.
$(CH_2(OH)C{\equiv}C)_2Si(C_6H_5)_2$	cat. → 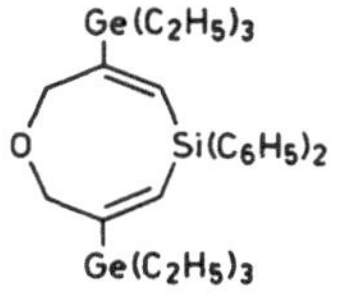 (43%)	[170]
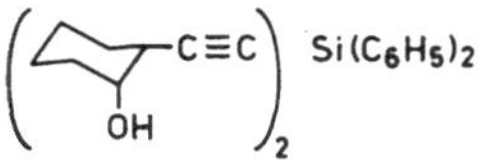	cat. → 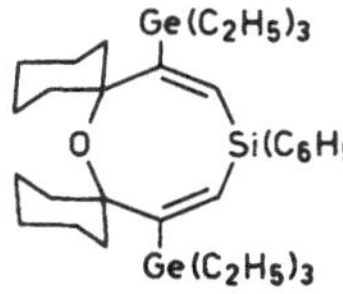 (61%)	[170]
$(CH{\equiv}CCH_2O)_2CHCH_3$	75 °C, 10 to 12 h, cat. in C_6H_6 → $(Ge(C_2H_5)_3CH{=}CHCH_2O)_2CHCH_3$ (55% ?), $Ge(C_2H_5)_3CH{=}CHCH_2OCH(CH_3)OCH_2C{\equiv}CH$ (51% ?)	[139]
$CH{\equiv}CCH_2OCH(CH_3)OCH_2C{\equiv}CC(OH)(CH_3)_2$	cat. → $Ge(C_2H_5)_3CH{=}CHCH_2OCH(CH_3)OCH_2C{\equiv}CC(OH)(CH_3)_2$ (52%)	[139]
$(CH_3)_2(OH)CC{\equiv}CC{\equiv}C{-}C(OH)(CH_3)_2$	118 °C, 30 min, cat. → $Ge(C_2H_5)_3C(C(OH)(CH_3)_2){=}CHC{\equiv}CC(OH)(CH_3)_2$ (80%)	[18]
$C_2H_5(CH_3)(OH)CC{\equiv}C{-}C{\equiv}CC(OH)(CH_3)C_2H_5$	not heated, 1 h, cat. (exothermic reaction) → $Ge(C_2H_5)_3C(C(CH_3){=}CHCH_3){=}CHC{\equiv}C{-}C(OH)(CH_3)C_2H_5$ (56%) and $(Ge(C_2H_5)_3)_2O$ 120 °C, 1 h, small amount of cat. → $Ge(C_2H_5)_3C(C(OH)(CH_3)C_2H_5){=}CHC{\equiv}CC(OH)(CH_3)C_2H_5$ (70%)	[18]
$i{-}C_3H_7(CH_3)(OH)CC{\equiv}C{-}C{\equiv}CC(OH)(CH_3)C_3H_7{-}i$	120 °C, 4 h, small amount of cat. → $Ge(C_2H_5)_3C(C(OH)(CH_3){-}C_3H_7{-}i){=}CHC{\equiv}CC(OH)(CH_3)C_3H_7{-}i$ (50%) 110 °C, 1.5 h, twofold amount of cat. → same product (only 10%)	[27]
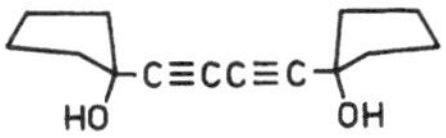	60 °C, 90 h, small amount of cat. → 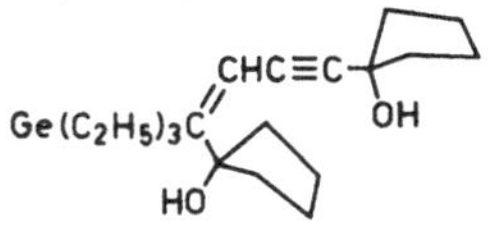 (23%)	[27]

References on p. 63

Table 7 (continued)

reactant	conditions (cat. = H_2PtCl_6) → products (yield) and remarks	Ref.
	120 °C, 12 h, twofold amount of cat. → 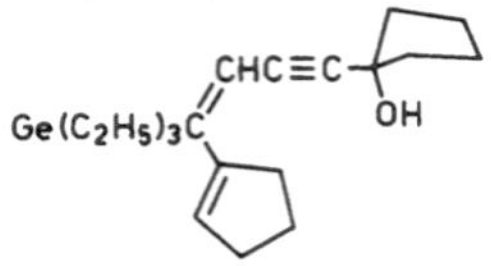 (9%)	[27]
	130 °C, 1 h, small amount of cat. → 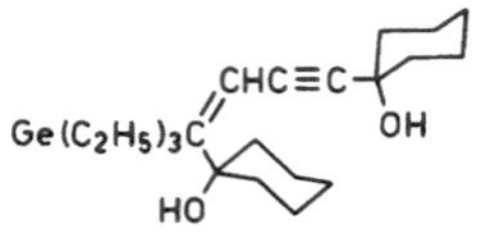 (70%) and $(Ge(C_2H_5)_3)_2O$	[27]
	20 °C, 1 h, twofold amount of cat. → 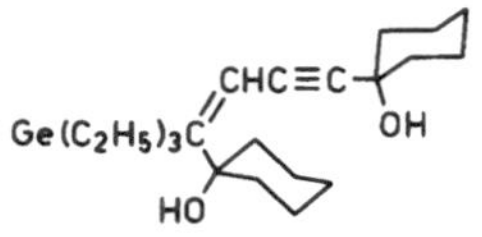 (10%) and 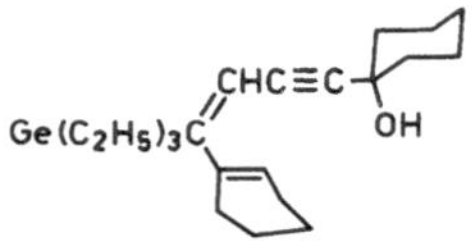 (7%)	[27]
ketene reactants		
CH_2=C=O	UV-irradiated, 2 to 4 h → $Ge(C_2H_5)_3COCH_3$ (60%)	[79]
$C(C_6H_5)_2$=C=O	30 min → $Ge(C_2H_5)_3OCH=C(C_6H_5)_2$ (80%)	[68, 83]
reactants with N=N and C=NO bonds		
$CH_3OOCN=NCOOCH_3$	refluxing in C_6H_6, 2 d → $Ge(C_2H_5)_3N(COOCH_3)N(COOCH_3)H$ (97%)	[113]
$C_2H_5OOCN=NCOOC_2H_5$	refluxing in C_6H_6, 1 d → $Ge(C_2H_5)_3N(COOC_2H_5)N(COOC_2H_5)H$ (95%); slowly reacting at room temperature, accelerated in the presence of Pt/C	[113]
$C_6H_5CH=N(C_6H_5)O$	100 °C, 3 d, in THF → $Ge(C_2H_5)_3ON(C_6H_5)CH_2C_6H_5$ (25%)	[108, 148]

Table 7 (continued)

reactant	conditions (cat. = H_2PtCl_6) → products (yield) and remarks	Ref.
$C_6H_5CH{=}N(C_4H_9\text{-t})O$	30 °C, UV-irradiated for 1 h, in C_6D_6 → no reaction 80 °C in C_6D_6, with or without traces of AIBN → no reaction	[148]
$(CH_3)_2\overline{CCH_2CH_2CH{=}N}O$	100 °C, 1 d, in C_6D_6 → $Ge(C_2H_5)_3O\overline{NCH_2CH_2CH_2C}(CH_3)_2$ (17%)	[148]
$(CH_3)_2\overline{CCH_2CH_2C(CH_3){=}N}O$	200 °C, 2 d, in C_6D_6 → $Ge(C_2H_5)_3O\overline{NCH(CH_3)CH_2CH_2C}$-$(CH_3)_2$ (11%)	[148]
$(CH_3)_2$-$\overline{CCH(C_6H_5)CH_2C(C_6H_5){=}N}O$	100 °C, 1 d, in C_6D_6 → no reaction	[148]

Reactions with Organic, Halogen-Containing Compounds. $Ge(C_2H_5)_3H$ reduces organic halides (RX with X = Cl, Br, or I) to the corresponding hydrocarbons RH while it is converted into $Ge(C_2H_5)_3X$. Thus reacting $Ge(C_2H_5)_3H$ with excess CCl_4 at 80 °C for 1 h yielded $Ge(C_2H_5)_3Cl$ and $CHCl_3$ in an uncatalyzed free-radical chain process. This reaction did not occur below room temperature or at 80 °C in the presence of galvinoxyl (Formula I) as a free-radical scavenger [106]. Refluxing $Ge(C_2H_5)_3H$ with CH_3OCH_2Cl in $O(CH_3)_2$ as the solvent gave $Ge(C_2H_5)_3Cl$ and $O(CH_3)_2$ quantitatively [8]. Reduction of $C(C_6H_5)_3Cl$ to $C(C_6H_5)_3H$ by $Ge(C_2H_5)_3H$ in C_6H_6 at 80 °C proceeds by a free-radical mechanism, but it apparently involves an ionic mechanism in polar solvents such as CH_2Cl_2 or CH_3NO_2 at room temperature [138].

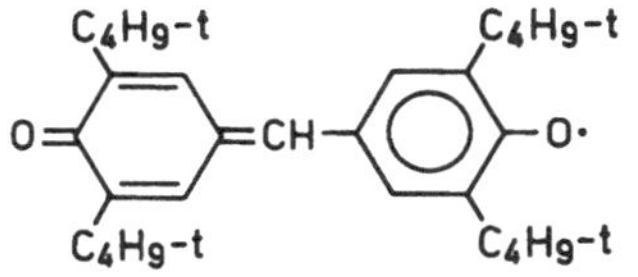

I

The reactivity of some aryl-substituted benzyl chlorides, $XC_6H_4CH_2Cl$ (X = 3-CN, 4-CN, 3-CF_3, 3-Cl, 4-Cl, 3-CH_3, or 4-CH_3), towards $Ge(C_2H_5)_3H$ containing $C_6H_5CO\text{-}OO\text{-}COC_6H_5$ (in C_6H_6 at 80 °C) compared to the reactivity of $C_6H_5CH_2Cl$ increases when X is an electron-withdrawing substituent. The reaction rates can be correlated with the Hammett equation with a ρ value of 0.312. No reduction of the aryl-bonded chlorine of 3- or 4-$ClC_6H_4CH_2Cl$ was observed [107].

The stereoselectivity of the abstraction of the halogen atom from cis- and trans-$XC_6H_{10}C_4H_9$-t-4 (X = Cl or Br; Formulas II) and from exo- and endo-$C_7H_{11}Cl$-2 (Formulas III) by $Ge(C_2H_5)_3H$ was studied [182].

References on p. 63

cis trans

II

exo endo

III

Reactions between $Ge(C_2H_5)_3H$ and unsaturated organic halides are more complex. Treating $CH{\equiv}CCH_2Cl$ with $Ge(C_2H_5)_3H$ at 150 °C for 15 h in the absence of any catalyst or solvent yielded $CH{\equiv}CCH_3$ as the sole product with 45% yield [81]; see also [35]. For this reaction in the presence of the H_2PtCl_6 catalyst, see Table 7. In contrast to $CH_2{=}CHCH_2Cl$ and $CH_2{=}C(CH_3)CH_2Cl$ (see Table 7, p. 28), refluxing the $CH_2{=}CHCH_2X$ analogues with X = Br or I and $Ge(C_2H_5)_3H$ for 2 h in the presence of H_2PtCl_6 gave no hydrogermylation product but $Ge(C_2H_5)_3CH_2CH{=}CH_2$ (37 and 27% yield, respectively), $Ge(C_2H_5)_3X$ (40 and 63% yield, respectively), and $CH_2{=}CHCH_3$ [14].

Slightly warming a 1:1 mixture of $Ge(C_2H_5)_3H$ and $(CH_3)_2C{=}C{=}CHCl$ in the absence of a catalyst resulted partly in the conversion to $Ge(C_2H_5)_3Cl$ and $(CH_3)_2C{=}C{=}CH_2$ [99]. For reactions with $CH{\equiv}CCH(C_3H_7)OCH(OH)CCl_3$ [88], $CH{\equiv}CC(CH_3)_2OCH(OH)CCl_3$, and $CH{\equiv}CC(CH_3)(C_4H_9\text{-t})OCH(OH)CCl_3$ [101] in the presence of H_2PtCl_6, see Table 7, pp. 37/9.

The reaction with $CH_2{=}CHCH_2OOCCl$ in the presence of H_2PtCl_6 yielded $Ge(C_2H_5)_3Cl$ (main product) but only traces of the hydrogermylation product; see Table 7, p. 28.

Reactions with Alcohols and Phenols. Refluxing $Ge(C_2H_5)_3H$ with C_6H_5OH in C_6H_6 or $C_6H_4(CH_3)_2$ in the presence of colloidal Ni or C_6H_5ONa, respectively, gave small amounts of $Ge(C_2H_5)_3OC_6H_5$ [176].

Reactions of $Ge(C_2H_5)_3H$ with various alcohols and phenols in the presence of Cu powder as the catalyst are listed in Table 8. H_2 is evolved in these reactions [13, 21]. $CH{\equiv}CCH_2OH$ is hydrogermylated by $Ge(C_2H_5)_3H$ in the presence of H_2PtCl_6 (Table 7, p. 35). Reactions with $(CH_3)_2(OH)CC{\equiv}CCH(OH)CH_3$ and compound IV (hydrogermylation and evolution of H_2) are also listed in Table 7. For reactions of the Ge-containing alcohols $Ge(C_2H_5)_3$-$CH_2CH_2CH_2OH$, $Ge(C_2H_5)_3CH{=}CHCH_2OH$, and $Ge(C_2H_5)_3C{\equiv}CCH_2OH$ in the presence of Cu, see Table 9, p. 59/60.

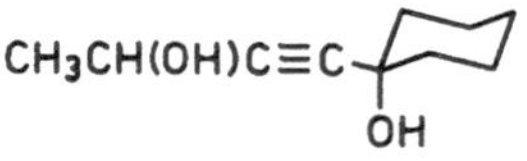

IV

Table 8
Reactions of $Ge(C_2H_5)_3H$ with Alcohols or Phenols in the Presence of Cu Powder.

reactant	conditions	products (yield)	Ref.
CH_3OH	reflux, 10 h	$Ge(C_2H_5)_3OCH_3$ (90%), H_2 (97%)	[21]
C_2H_5OH	reflux, 15 h	$Ge(C_2H_5)_3OC_2H_5$ (89%), H_2 (95%)	[21]
$n\text{-}C_3H_7OH$	95 °C	$Ge(C_2H_5)_3O(CH_2)_2CH_3$ (100%), H_2	[13]
$i\text{-}C_3H_7OH$	reflux, 3 d	$Ge(C_2H_5)_3OCH(CH_3)_2$ (81%), H_2 (95%)	[21]

Table 8 (continued)

reactant	conditions	products (yield)	Ref.
n-C_4H_9OH	95 °C	$Ge(C_2H_5)_3O(CH_2)_3CH_3$ (100%), H_2	[13]
t-C_4H_9OH	reflux, 3 d	$Ge(C_2H_5)_3OC(CH_3)_3$ (58%), H_2 (85%)	[21]
n-$C_7H_{15}OH$	95 °C [13], 120 °C [12]	$Ge(C_2H_5)_3O(CH_2)_6CH_3$, H_2	[12, 13]
$C_6H_5CH_2OH$	95 °C	$Ge(C_2H_5)_3OCH_2C_6H_5$, H_2	[13]
$CH_2{=}CHCH_2OH$	reflux, 36 h	$Ge(C_2H_5)_3OCH_2CH{=}CH_2$ (51%), H_2 (54%)	[21]
$CH_3CH{=}CHCH_2OH$	reflux, 12 h	$Ge(C_2H_5)_3OCH_2CH{=}CHCH_3$ (62%), H_2 (79%)	[21]
$HOCH_2\overline{C{=}CHCH{=}CHO}$	reflux, 5 h	$Ge(C_2H_5)_3OCH_2$-$\overline{C{=}CHCH{=}CHO}$ (89%), H_2 (92%)	[21]
$HOCH_2CH_2OH$	reflux	$Ge(C_2H_5)_3O(CH_2)_2OGe(C_2H_5)_3$ (82%), H_2	[13]
$HO(CH_2)_4OH$	reflux	$Ge(C_2H_5)_3O(CH_2)_4OGe(C_2H_5)_3$ (90%), H_2	[13]
C_6H_5OH	80 °C	$Ge(C_2H_5)_3OC_6H_5$ (90%), H_2	[13]
$C_6H_4(OH)_2$-1,4		$C_6H_4(OGe(C_2H_5)_3)_2$-1,4 (88%), H_2	[13]

The first-order rate constant for the loss of tritium from **$Ge(C_2H_5)_3T$** by T/H exchange in CH_3OH (1.5 M CH_3ONa) at 30 °C relative to $k(Ge(C_6H_5)_3T) = 1$ (in CH_3OH, 0.421 M CH_3ONa) was found to be $k_{rel} = 0.006$. Since considerable solvolysis with generation of hydrogen and $Ge(C_2H_5)_3OCH_3$ took place, the k_{rel} value is estimated to be only half as large [152].

Reactions with Ketones. Heating a 1:1 mixture of $Ge(C_2H_5)_3H$ and cyclohexanone at 150 °C for 2 d in the presence of Cu powder gave $Ge(C_2H_5)_3OC_6H_{11}$ quantitatively [12]. Heating equimolar amounts of $Ge(C_2H_5)_3H$ and quinone V in C_6H_6 at 130 °C for 2 h yielded compounds VI and VII (61 and 34% yield, respectively); only traces of the products were obtained at 60 °C in C_6H_6, $N(C_2H_5)_3$, or dioxane [194].

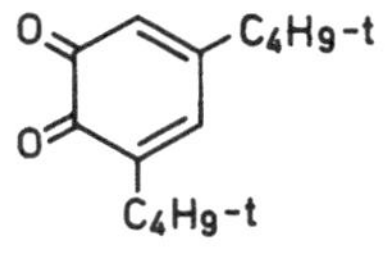

V VI VII

Irradiating a mixture of $Ge(C_2H_5)_3H$, quinone VIII, and t-$C_4H_9OOC_4H_9$-t in C_6H_6 in an ESR cavity immediately produced the ESR signal of the radical IX [151]. Photolysis of $Ge(C_2H_5)_3H$ and t-$C_4H_9OOC_4H_9$-t with CH_3COCH_3 (at 0 °C), $CH_3COCOCH_3$ (at −20 °C), and C_6H_5CO-COC_6H_5 (at 0 to 40 °C) gave the radicals $Ge(C_2H_5)_3OC^{\cdot}(CH_3)_2$, $Ge(C_2H_5)_3OC^{\cdot}(CH_3)COCH_3$, and $Ge(C_2H_5)_3OC^{\cdot}(C_6H_5)COC_6H_5$, respectively [206].

References on p. 63

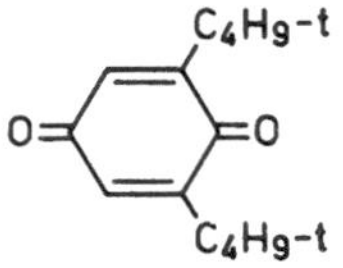

VIII

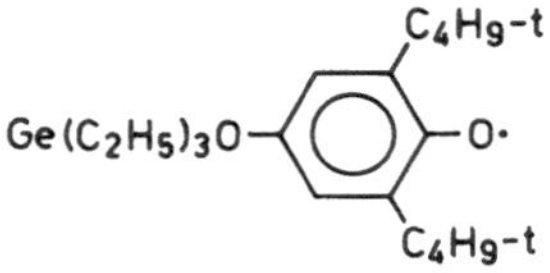

IX

The yields of the ketyl radicals escaping from the radical pairs, formed by the reactions of $Ge(C_2H_5)_3H$ with the triplet states of $C_6H_5COCH_3$ and xanthone (Formula X), were found to be higher in a magnetic field [180]. Abstraction of H or D from $Ge(C_2H_5)_3H$ or $Ge(C_2H_5)_3D$, respectively, by a benzene ring with formation of a cyclohexadienyl-type radical (instead of the ketyl radical) was confirmed by chemically induced, dynamic electron polarization (CIDEP) measurements to occur when they react with triplet xanthone [184]. Studies of this reaction with a laser photolysis-ESR technique and using partially deuteriated xanthone revealed that hydrogen addition occurs at the 4-position of the xanthone [195].

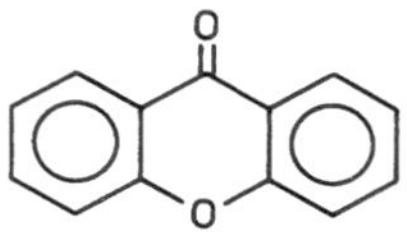

X

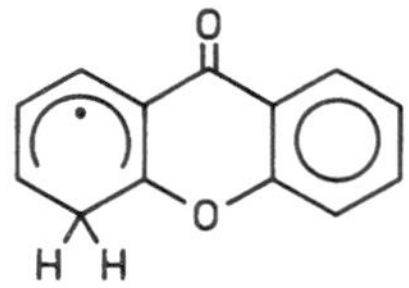

XI

Laser flash photolysis of a mixture of $Ge(C_2H_5)_3H$ and xanthone in i-C_3H_7OH yielded a cyclohexadienyl-type radical XI [218].

Reactions with Organic Acids and Esters. The following reactions of $Ge(C_2H_5)_3H$ with carboxylic acids were reported, all at reflux temperature for 1 to 3 h: $CF_3COOH \rightarrow Ge(C_2H_5)_3OOCCF_3$ (95%); $C_2F_5COOH \rightarrow Ge(C_2H_5)_3OOCC_2F_5$ (99%); $C_3F_7COOH \rightarrow Ge(C_2H_5)_3OOCC_3F_7$ (97%); $CCl_3COOH \rightarrow Ge(C_2H_5)_3Cl$ (80%), CH_3COOH, and $Ge(C_2H_5)_3OOCCH_2Cl$; $CBr_3COOH \rightarrow Ge(C_2H_5)_3Br$ (94%) and CH_3COOH; $CH_2ICOOH \rightarrow Ge(C_2H_5)_3I$ (94%) and CH_3COOH [2]. CH_2=CHCOOH was hydrogermylated but yielded 60% $Ge(C_2H_5)_3OOCCH{=}CH_2$ in the presence of Cu powder [13].

Photolysis of $Ge(C_2H_5)_3H$ and $C_2H_5OOCCOOC_2H_5$ in the presence of t-C_4H_9O-OC_4H_9-t at 0 and 20 °C gave the $Ge(C_2H_5)_3OC^{\cdot}(OC_2H_5)COOC_2H_5$ radical [206].

Reactions with Aldehydes. Reacting $Ge(C_2H_5)_3H$ with $C_6H_{13}CHO$ or C_6H_5CHO at 150 °C in the presence of Cu powder (24 and 40 h) gave $Ge(C_2H_5)_3OC_7H_{15}$ (100% yield) and $Ge(C_2H_5)_3OCH_2C_6H_5$ (80% yield), respectively [12]. $CH_2{=}CH(CH_2)_8CHO$ yielded $Ge(C_2H_5)_3O(CH_2)_9CH{=}CH_2$ under similar conditions [205]; see also [203].

Reactions with Epoxides. Treating $\overline{OCH_2C}HCH{=}CH_2$, $\overline{OCH_2C}(CH_3)CH{=}CH_2$, or $\overline{OCH_2C}HC(Cl){=}CH_2$ with $Ge(C_2H_5)_3H$ (equimolar amounts) in the presence of H_2PtCl_6 (in sealed tubes in a water bath for 18 to 20 h or refluxing the neat reactants for 3 h) gave $Ge(C_2H_5)_3OCH_2CH{=}CHCH_3$, $Ge(C_2H_5)_3OCH_2C(CH_3){=}CHCH_3$, and $Ge(C_2H_5)_3OCH_2CH{=}C(Cl)CH_3$, respectively [45]. $Ge(C_2H_5)_3OCH(CH_3)CH{=}CHCH_3$ was formed from $\overline{OCH(CH_3)C}HCH{=}CH_2$ under similar conditions [67].

Reactions with Sulfur Compounds. $Ge(C_2H_5)_3H + C_4H_9SH$ (refluxing for 3 d in the presence of Pt as the catalyst) $\rightarrow Ge(C_2H_5)_3SC_4H_9$ (75% yield) with evolution of H_2 [13] (similar yield with a Ni/SiO_2 catalyst after 10 h refluxing). $Ge(C_2H_5)_3H + HSCH_2CH_2SH$ (2:1 mole ratio; refluxing in the presence of Ni/SiO_2) → $Ge(C_2H_5)_3SCH_2CH_2SGe(C_2H_5)_3$ (46% yield) and H_2 (86%). $Ge(C_2H_5)_3H$ + CH_3SSCH_3 (equimolar amounts; refluxing) → $Ge(C_2H_5)_3SCH_3$ (92%) and CH_3SH [31].

$Ge(C_2H_5)_3H$ + $RC_6H_4SO_2SC_6H_4R$ (R = H or CH_3-4) in the presence of AIBN (2:1:0.4 mole ratio) in C_6H_6 for 17 h in air → $Ge(C_2H_5)_3OSO_2C_6H_4R$ (82 and 96% yield, respectively) and $Ge(C_2H_5)_3SC_6H_4R$ (82 and 92% yield, respectively). Under deoxygenated conditions in C_6H_6 at 80°C for 38 h in the presence of AIBN, interaction between equimolar amounts of $Ge(C_2H_5)_3H$ and 4-$CH_3C_6H_4SO_2SC_6H_4CH_3$-4 yielded $Ge(C_2H_5)_3OSOC_6H_4CH_3$-4, Ge-$(C_2H_5)_3SC_6H_4CH_3$-4, and 4-$CH_3C_6H_4SO_2C(CH_3)_2CN$. In the absence of AIBN, $RC_6H_4SO_2$-SC_6H_4R compounds did not react with $Ge(C_2H_5)_3H$. $Ge(C_2H_5)_3H$ + $RC_6H_4SO_3H$ (equimolar amounts) refluxing in $C_6H_5CH_3$ for 1 d → $Ge(C_2H_5)_3OSO_2C_6H_4R$ [183].

$Ge(C_2H_5)_3H$ + 4-$CH_3C_6H_4SO_2SeC_6H_5$ in the presence of AIBN → $Ge(C_2H_5)_3OSO_2$-$C_6H_4CH_3$-4 (93% yield) and $C_6H_5SeSeC_6H_5$ (98% yield) when carried out in air, but Ge-$(C_2H_5)_3OSOC_6H_4CH_3$-4 (35%), $C_6H_5SeSeC_6H_5$ (58%), and 4-$CH_3C_6H_4SO_2C(CH_3)_2CN$ (31% yield) in an evacuated tube [183].

Reactions with Se and Te Compounds. $Ge(C_2H_5)_3H$ + $(C_2H_5)_2Se$ (2:1 mole ratio) at 200°C for 20 h → $(Ge(C_2H_5)_3)_2Se$ (45%) [38, 40] with liberation of C_2H_6 (74% yield) [40]. $Ge(C_2H_5)_3H$ + $(C_2H_5)_2Te$ (ca. 1.2:1 mole ratio) at 140°C for 7 h → $Ge(C_2H_5)_3TeC_2H_5$ (39%), $(Ge(C_2H_5)_3)_2Te$ (58%), and C_2H_6 (99% yield) [56]; see also [39]. Interaction between $Ge(C_2H_5)_3H$ and $Te(Si(C_2H_5)_3)_2$ (equimolar amounts) at 230°C for 36 h → $(Ge(C_2H_5)_3)_2Te$ (79%) and $Si(C_2H_5)_3H$ (49% yield) [75].

Reactions with Nitrogen Compounds. For the reactions with RN=NR compounds and several nitrones (containing C=NO groups), see the hydrogermylation in Table 7, p. 52.

$Ge(C_2H_5)_3H$ + CH_2N_2 (UV-irradiated) → $Ge(CH_3)(C_2H_5)_3$ (9% yield) [11, 16]; none was formed in the presence of Cu powder [11]. $Ge(C_2H_5)_3H$ + $CH(=N_2)COOC_2H_5$ (2:3 mole ratio) in the presence of Cu powder (exothermic) → $Ge(C_2H_5)_3CH_2COOC_2H_5$ (53% yield) [20]. $RCOCH(=N_2)$ (R = CH_3 or C_6H_5) in ether or C_6H_6 in the presence of Cu powder → $Ge(C_2H_5)_3CH_2COR$ with liberation of N_2 [4]. For reactions with $Ge(C_2H_5)_3C(=N_2)COCH_3$ and $Ge(C_2H_5)_3$-$C(=N_2)COOC_2H_5$ [136], see Table 9.

$Ge(C_2H_5)_3H$ + $BrCH_2CH_2CONCO$ in C_6H_{12} at 55°C in the presence of t-C_4H_9O-$N=NOC_4H_9$-t → $Ge(C_2H_5)_3N(COR)CHO$ and/or $Ge(C_2H_5)_3CONHCOR$ (both with R = C_2H_4Br or C_2H_5), C_2H_5CONCO (forming $C_2H_5CONHCOOCH_3$ in a secondary reaction with CH_3OH), and succinimide XII. $Ge(C_2H_5)_3H$ + $BrCH_2C(CH_3)_2CONCO$ (7:2 mole ratio) under the same conditions followed by treatment with CH_3OH → 2,2-dimethylsuccinimide XIII (87%), t-$C_4H_9CONHCOOCH_3$ (0.6%), and s-$C_4H_9CONHCOOCH_3$ (9% yield); $BrCH_2C(CH_3)_2$-$CONHCOOCH_3$ was not detected among the products [198].

XII XIII

For reactions of $Ge(C_2H_5)_3H$ with $CH_2=CHCH_2NCO$ [105, 128] and some diallylisocyanurates [197], see the hydrogermylation reactions in Table 7, pp. 29 and 33.

$Ge(C_2H_5)_3H$ + $C_6H_5SO_2NCO$ in dry $C_6H_5CH_3$ at 120 to 130°C for 3 h → $Ge(C_2H_5)_3N$-$(CHO)SO_2C_6H_5$ (85% yield) in a tautomeric equilibrium with $Ge(C_2H_5)_3OCH=NSO_2C_6H_5$ (de-

References on p. 63

tected by IR). $Ge(C_2H_5)_3H$ + $C_6H_5SO_2N_3$ in $C_6H_5CH_3$ at 120 °C in the presence of H_2PtCl_6 → $(Ge(C_2H_5)_3)_2O$, $C_6H_5SO_2NH_2$ (89%), and N_2 (98% yield) via hydrolytically unstable $Ge(C_2H_5)_3NHSO_2C_6H_5$ [121].

$Ge(C_2H_5)_3H$ + $\overline{CH(C_6H_5)N(C_4H_9\text{-t})O}$ (oxaziridine; equimolar amounts) in C_6D_6 at 90 °C for 5 h → $(Ge(C_2H_5)_3)_2O$, $C_6H_5CH{=}NC_4H_9$-t (both 90% yield), and C_6H_5CHO (5 to 6% yield); a reaction mechanism involving the unstable $Ge(C_2H_5)_3OCH(C_6H_5)NHC_4H_9$-t was discussed [149].

Reactions with Carbenes, Radicals, and Carbenium Ions. Dropwise addition of t-C_4H_9OK in DME to a solution of $CH{\equiv}CC(OSO_2CF_3){=}CR_2$ (R = CH_3 or C_6H_5) and excess $Ge(C_2H_5)_3H$ in DME at −50 °C in the presence of t-C_4H_9NO as a radical trap yielded $Ge(C_2H_5)_3CH{=}C{=}C{=}CR_2$ (84 or 81%, respectively) via the carbene intermediates $[R_2C{=}C{=}C{=}C{:}]$ [164]. $CH{\equiv}CC{\equiv}CC(OSO_2CF_3){=}C(CH_3)_2$, t-$C_4H_9OK$, and $Ge(C_2H_5)_3H$ (1 : 1.2 : 4.2 mole ratio) in glyme gave $Ge(C_2H_5)_3CH{=}C{=}C{=}C{=}C{=}C(CH_3)_2$, $Ge(C_2H_5)_3C{\equiv}CCH{=}C{=}C{=}C(CH_3)_2$, and $Ge(C_2H_5)_3C{\equiv}CC{\equiv}CCH{=}C(CH_3)_2$ (1, 12, and 19% yield, respectively) via the unstable carbene $[(CH_3)_2C{=}C{=}C{=}C{=}C{=}C{:}]$ [165]. For the reaction with the carbene complex $Cr(CO)_5C(OCH_3)C_6H_5$ and pyridine, see Table 10, p. 62.

Photochemically generated t-$C_4H_9O^{\bullet}$ reacts with $Ge(C_2H_5)_3H$ to form the $Ge(C_2H_5)_3^{\bullet}$ radical [131, 150] which was used for generating new radicals by abstraction of halogen (X = Cl or Br) from $Ge(CH_3)_2(CH_2X)H$, $Si(CH_3)_2(CH_2X)(H)$, $Si(CH_3)(C_6H_5)(CH_2X)H$, $Si(C_6H_5)_2(CH_2X)H$, or $Si(CH_3)_2(C(CH_3)_2X)H$ [150].

A study of the second-order rate constant of the reactions of $Ge(C_2H_5)_3H$ with $[(C_6H_5)_3C]X$ (X = $[SbF_6]$, $[AsF_6]$, $[FeCl_4]$) or $[cyclo\text{-}C_7H_7][BF_4]$ in CH_2Cl_2 (yielding $Ge(C_2H_5)_3Y$ (Y = F or Cl) and $CH(C_6H_5)_3$ or cyclo-C_7H_8 as the final products) showed that the rate does not depend on the anion. The reaction of $Ge(C_2H_5)_3H$ is 25-times faster than that of $Si(C_2H_5)_3H$ for $[(C_6H_5)_3C]^+$ and 60-times faster for $[cyclo\text{-}C_7H_7]^+$ [188].

Reactions with Organometallic Compounds; Main Group Elements. Reactions of $Ge(C_2H_5)_3H$ with organogermanium compounds and other organometallic compounds of Main Group elements are listed in Table 9.

Table 9
Reactions of $Ge(C_2H_5)_3H$ with Organometallic Compounds of the Main Group Elements.

reactant	conditions products (yield)	Ref.
LiC_4H_9	in ether → $Ge(C_2H_5)_3Li$ (<10%) (slow reaction)	[3]
LiC_6H_5	in ether → $Ge(C_2H_5)_3Li$ (<10%) (slow reaction)	[3]
KC_2H_5	in C_6H_{14}, 0 °C → $Ge(C_2H_5)_3K$ (11%), $Ge(C_2H_5)_4$ (47%), $Ge_2(C_2H_5)_6$ (traces), C_2H_6 (ca. 50%), and 1-hexene (35%)	[207]
$Tl(C_2H_5)_3$	100 °C, 2 h → $(Ge(C_2H_5)_3)_3Tl$ (91%), C_2H_6 (100%)	[42, 48]

Table 9 (continued)

reactant	conditions products (yield)	Ref.
$Si(CH_3)_3Cl$	in THF-C_6H_6, room temperature, 1 h → $Ge(C_2H_5)_3Si(CH_3)_3$	[196]
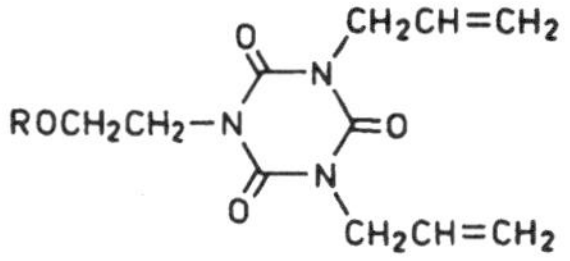 R = $Si(CH_3)_3$	hydrogermylation; see Table 7, p. 33	
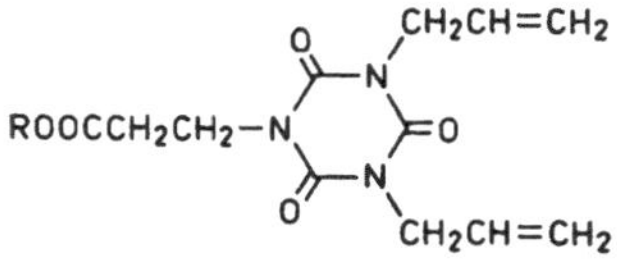 R = $Si(CH_3)_3$	hydrogermylation; see Table 7, p. 34	
$Ge(C_2H_5)_3CH_2CH_2CH_2OH$	refluxing, 10 h, Cu catalyst → $Ge(C_2H_5)_3O(CH_2)_3Ge(C_2H_5)_3$ (72%), H_2 (85%)	[21]
$Ge(C_2H_5)_3CH{=}CHCH_2OH$	refluxing, 12 h, Cu catalyst → $Ge(C_2H_5)_3OCH_2CH{=}CHGe(C_2H_5)_3$ (56%), H_2 (60%)	[21]
$\overline{Ge(C_4H_9)_2CH_2CH_2CH_2}$	H_2PtCl_6 catalyst → $Ge(C_2H_5)_3CH_2CH_2CH_2Ge(C_4H_9)_2H$ and polymers	[51, 52]
$Ge(C_2H_5)_3C({=}N_2)COCH_3$	in C_6H_{14}, 100 to 110°C, 15 h, Cu catalyst, evacuated tube → $(Ge(C_2H_5)_3)_2CHCOCH_3$ (13%), $Ge(C_2H_5)_3CH{=}C(CH_3)OGe(C_2H_5)_3$ (15% cis and 35% trans), and N_2 (100%)	[136]
$Ge(C_2H_5)_3C({=}N_2)COOC_2H_5$	in C_6H_6, 100°C, 10 h, Cu catalyst, evacuated tube → $(Ge(C_2H_5)_3)_2CHCOOC_2H_5$ (66%) and N_2 (100%)	[136]
$Ge_2(C_6F_5)_6$	in THF, 100°C, 3 h → no reaction in contrast to $Sn(C_2H_5)_3H$	[209]
$Ge(C_6H_5)Cl$	in C_6H_6, 100°C, 2 d → $Ge(C_2H_5)_3Cl$ (27%), $(Ge(C_6H_5)H)_n$, Ge	[118]
$Ge(C_2H_5)_2Cl_2$	refluxing, 2 h, $AlCl_3$ catalyst, exothermic reaction → $Ge(C_2H_5)_3Cl$ and $Ge(C_2H_5)_2(H)Cl$ (both ca. 100%)	[8]

References on p. 63

Table 9 (continued)

reactant	conditions products (yield)	Ref.
$Ge(C_6F_5)_3Br$	refluxing, 1 h → $Ge(C_6F_5)_3H$ (89%) and $Ge(C_2H_5)_3Br$	[110]
$Ge(C_2H_5)_2Br_2$	refluxing, 2 h, $AlBr_3$ catalyst, exothermic reaction → $Ge(C_2H_5)_3Br$ and $Ge(C_2H_5)_2(H)Br$ (both ca. 100%)	[8]
$Ge(C_6F_5)_2Br_2$	ca. 4:1 mole ratio, 120°C, 6 h → $Ge(C_2H_5)_3Br$ (77%), $Ge(C_6F_5)_2(H)Br$ (68%), and $Ge(C_6F_5)_2H_2$ (24%)	[208]
$Ge_2(C_6F_5)_4Br_2$	ca. 14:1 mole ratio, 150°C, 8 h → $Ge(C_2H_5)_3Br$ (66%) and $Ge_2(C_6F_5)_4H_2$ (35%)	[208]
$Ge(CF_3)_3I$	60°C, 6 h, sealed tube → $Ge(CF_3)_3H$ (98%) and $Ge(C_2H_5)_3I$	[178]
$Ge(C_2H_5)_3C{\equiv}CCH_2OH$	Cu catalyst → $Ge(C_2H_5)_3C{\equiv}CCH_2OGe(C_2H_5)_3$ (20%) and H_2	[21]
$Ge(C_2H_5)_3OCH_2C{\equiv}CH$	hydrogermylation	[21]
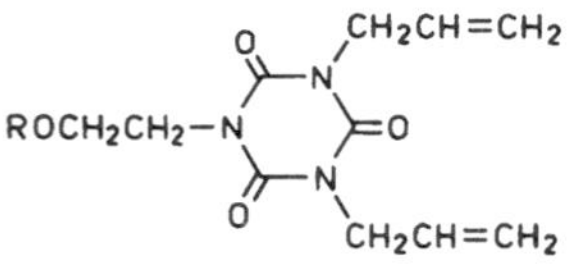 R = $Ge(C_2H_5)_3$	hydrogermylation; see Table 7, p. 33	
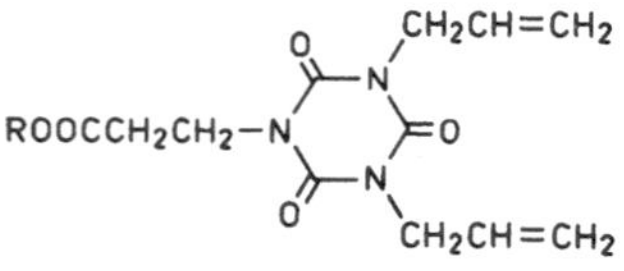 R = $Ge(C_2H_5)_3$	hydrogermylation; see Table 7, p. 34	
$Ge(C_2H_5)_3Li$	in THF, 60 to 70°C, 7 to 8 h, → $Ge_2(C_2H_5)_6$ (36% [73], 42% [58]) and LiH	[58, 73]
$Ge(C_2H_5)_3K$	in HMPT, 1 h, room temperature → $Ge_2(C_2H_5)_6$ (40%) and KH	[95]
$Sn(C_2H_5)_3OCH_3$	ca. 180°C, evacuated sealed tube → $Ge(C_2H_5)_3OCH_3$ (48%), $Sn_2(C_2H_5)_6$ (39%), and CH_3OH (34%)	[57]
$Pb(C_2H_5)_4$	165 to 170°C, 44 h → $Ge_2(C_2H_5)_6$ (18%), $Ge(C_2H_5)_4$ (13%), $(Ge(C_2H_5)_3)_nPb(C_2H_5)_{4-n}$ (n = 1 to 4) (18%), C_2H_6 (63%), and Pb (46%)	[74]

Table 9 (continued)

reactant	conditions products (yield)	Ref.
$(PC_6H_5)_5$	210°C, 36 h → $Ge(C_2H_5)_3P(H)C_6H_5$ (62%), $(Ge(C_2H_5)_3)_2PC_6H_5$ (12%), and $PH_2C_6H_5$ (15%)	[153]
$Sb(C_2H_5)_3$	200°C, 15 h → $(Ge(C_2H_5)_3)_3Sb$ (75%) and C_2H_6 (100%)	[41, 43]
$Sb(Si(C_2H_5)_3)_3$	230°C, 16 h → $(Ge(C_2H_5)_3)_3Sb$ (59%), $Si(C_2H_5)_3H$ (68%)	[41, 43]
$Sb(Si(C_6H_5)_3)_3$	230°C, 15 h → $(Ge(C_2H_5)_3)_3Sb$ (>83%), $Si(C_6H_5)_3H$ (72%), and Sb (7%)	[87]
$Bi(C_2H_5)_3$	3.25:1 mole ratio, 140 to 145°C, 8.5 h → $(Ge(C_2H_5)_3)_3Bi$ (55%) and C_2H_6 (80%)	[41, 43]; see also [28, 87]
	2:1 mole ratio, 130 to 135°C, 7 h → $(Ge(C_2H_5)_3)_2BiC_2H_5$ (86%) and C_2H_6 (100%)	[28]
	1:1 mole ratio, 145 to 150°C, 6 h → $(Ge(C_2H_5)_3)_3Bi$ (main product) and $Ge(C_2H_5)_3Bi(C_2H_5)_2$ (8%)	[28]
$Bi(Si(C_2H_5)_3)_3$	180°C, 16 h → $(Ge(C_2H_5)_3)_3Bi$ (73%), $Si(C_2H_5)_3H$ (98%)	[41, 43]
$Bi(C_2H_5)_2Ge(C_6F_5)_3$ $Bi(C_2H_5)(Ge(C_6F_5)_3)_2$	no reaction at 150°C	[210]

Reactions with Organometallic Compounds; Transition Metals. Reactions of $Ge(C_2H_5)_3H$ with various organometallic species containing transition metals are summarized in Table 10.

Table 10
Reactions of $Ge(C_2H_5)_3H$ with Organometallic Transition Metal Compounds.

reactant	conditions products (yield) and remarks	Ref.
$Ti(C_5H_5)_2(CH_3)_2$	in THF, 50°C, 6 h → $Ge_2(C_2H_5)_6$ (83%), $Ti(C_5H_5)_2$ (95%), and CH_4 (96%)	[147]
$Zr(C_5H_5)_2(CH_3)_2$	in DME, UV-irradiated at room temperature, 20 h → $Ge_2(C_2H_5)_6$ (70%), $Zr(C_5H_5)_2 \cdot$ DME (65%), and CH_4 (93%)	[147]

References on p. 63

Table 10 (continued)

reactant	conditions products (yield) and remarks	Ref.
	in DME, 110 °C, 15 h → $Ge_2(C_2H_5)_6$ (100%), $(Zr(CH_3)(C_5H_5)_2)_2$ (70%), and CH_4 (100%)	
$V(C_5H_5)_2$	in $C_6H_5CH_3$, 20 °C, 20 h → no reaction	[173]
$V(C_5H_5)_2(CH_3)_2$	in $C_6H_5CH_3$, 20 °C, 700 h → $Ge(C_2H_5)_3CH_3$ (20%), $Ge_2(C_2H_5)_6$ (18%), $V(C_5H_5)_2$ (29%), $V(C_5H_5)_2Ge(C_2H_5)_3$ (60%), and CH_4 (67%)	[158]
$V(C_5H_5)_2CH_2Si(CH_3)_3$	100 °C, 10 to 15 min → $V(C_5H_5)_2Ge(C_2H_5)_3$ (91%), $Si(CH_3)_4$	[146]
$Cr(CO)_5C(OCH_3)C_6H_5$	1:1 mole ratio, in refluxing C_6H_{14} in the presence of pyridine, 10 min → $Ge(C_2H_5)_3CH(OCH_3)C_6H_5$ (66%); kinetics and mechanism studied	[112] [111, 134]
$Na[HRu_3(CO)_{11}]$	in THF, 45 °C, 4 h → $[HRu_3(CO)_{10}(Ge(C_2H_5)_3)_2]^-$ (62%), CO, and H_2	[174]
$Co_3(CO)_9(\mu_3\text{-}CH)$	in C_6H_6, refluxing, 15 h → $Co_3(CO)_9(\mu_3\text{-}CGe(C_2H_5)_3)$ (30%) and H_2	[159]
$Ir(CO)(P(C_6H_5)_3)_2Cl$	in C_6H_6, refluxing, 24 h → $Ir(CO)(P(C_6H_5)_3)_2(H)_2Ge(C_2H_5)_3$ (88%), $Ge(C_2H_5)_3Cl$, $Ge_2(C_2H_5)_6$ (traces), and H_2	[76, 90]
$Zn(C_2H_5)_2$	125 °C, 15 h → undistillable product (55%; proposed composition: $Ge(C_2H_5)_3Zn(Ge(C_2H_5)_2)_nGe(C_2H_5)_3$), $Ge(C_2H_5)_4$ (9%), Zn (30%), and C_2H_6 (94%)	[24]
$Cd(C_2H_5)_2$	80 to 85 °C, 3 h → $(Ge(C_2H_5)_3)_2Cd$ (79%), C_2H_6 (90%), and Cd (traces)	[23]
$Hg(C_2H_5)_2$	2:1 mole ratio, 100 to 120 °C → $(Ge(C_2H_5)_3)_2Hg$ (66%), C_2H_6 (97%); the reaction proceeds via a radical-chain mechanism and is inhibited by $(C_6H_5)_2CO$ [179]	[17]
	2:1 mole ratio, 80 °C, 7 h → $(Ge(C_2H_5)_3)_2Hg$ (68%), C_2H_6 (91%), Hg (1%), no C_2H_4 and C_4H_{10}	[73]
	ca. 1.3:1 mole ratio, 80 °C, 3 h → $(Ge(C_2H_5)_3)_2Hg$ (84%), $Ge(C_2H_5)_3HgC_2H_5$ (8%), C_2H_6 (71%), and Hg (4%)	

Table 10 (continued)

reactant	conditions products (yield) and remarks	Ref.
	at 25 °C after 1 h practically no reaction, only 1% C_2H_6 formed	[199]
	20 °C, then 120 °C → $(Ge(C_2H_5)_3)_2Hg$ and C_2H_6	[49]
	2:1 mole ratio, UV-irradiated (λ = 253.7 nm), 2 h → $(Ge(C_2H_5)_3)_2Hg$ (88%), C_2H_6 (95%)	[200]
$Hg(C_4H_9\text{-}t)_2$	25 °C, 18 h → $(Ge(C_2H_5)_3)_2Hg$ (98%) and iso-C_4H_{10}; reaction inhibited by $(C_6H_5)_2CO$ [179]	[82]
$Hg(C_6H_5)CHBr_2$	in C_6H_5Cl, 130 °C, 66 h → $Ge(C_2H_5)_3CH_2Br$ (28%), $Hg(C_6H_5)Br$ (70%), and $Ge(C_2H_5)_3Br$ (30%, formed in a secondary reaction between $Ge(C_2H_5)_3H$ and $Hg(C_6H_5)Br$)	[204]
$Hg(C_6H_5)CCl_2Br$	in C_6H_6, 85 to 90 °C, 2 h → $Ge(C_2H_5)_3CCl_2H$ (83%) and $Hg(C_6H_5)Br$	[53]
$Hg(CH_2COCH_3)_2$	60 °C, 3 h, in C_6H_6 → $Ge(C_2H_5)_3OC(CH_3){=}CH_2$ (30%) and $Ge(C_2H_5)_3CH_2COCH_3$ (traces)	[89]
$Hg(CH_2CHO)_2$	40 °C, 1.5 h, in $CHCl_3$ → $Ge(C_2H_5)_3OCH{=}CH_2$ (76%) and $Ge(C_2H_5)_3CH_2$-CHO (trace)	[89]
$Hg(C_2H_5)Si(C_2H_5)_3$	100 °C, 1 to 2 h → $Ge(C_2H_5)_3HgSi(C_2H_5)_3$ (26%) and C_2H_6 (73%)	[22, 73]
$Hg(C_2H_5)Si_2(C_2H_5)_5$	100 °C, 1 to 2 h → $Ge(C_2H_5)_3HgSi(C_2H_5)_2Si(C_2H_5)_3$ and C_2H_6	[22]
$Hg(C_2H_5)Ge(C_2H_5)_3$	100 °C, 20 h, sealed tube → $(Ge(C_2H_5)_3)_2Hg$ (80%) and C_2H_6 (84%)	[73]

References:

[1] Kraus, C. A.; Flood, E. A. (J. Am. Chem. Soc. **54** [1932] 1635/44).
[2] Anderson, H. H. (J. Am. Chem. Soc. **79** [1957] 326/8).
[3] Hughes, M. B. (Diss. Iowa State College 1958; Diss. Abstr. **19** [1958/59] 1921).
[4] Lesbre, M.; Satgé, J. (C. R. Hebd. Seances Acad. Sci. **247** [1958] 471/4).
[5] Mazerolles, P.; Lesbre, M. (C. R. Hebd. Seances Acad. Sci. **248** [1959] 2018/20).
[6] Lesbre, M.; Satgé, J. (C. R. Hebd. Seances Acad. Sci. **250** [1960] 2220/2).
[7] Dzhurinskaya, N. G.; Mironov, V. F.; Petrov, A. D. (Dokl. Akad. Nauk SSSR **138** [1961] 1107/10; Dokl. Chem. [Engl. Transl.] **136/141** [1961] 574/7).

[8] Lesbre, M.; Satgé, J. (C. R. Hebd. Seances Acad. Sci. **252** [1961] 1976/8).
[9] Satgé, J. (Ann. Chim. [Paris] [13] **6** [1961] 519/73).
[10] Schott, G.; Harzdorf, C. (Z. Anorg. Allg. Chem. **307** [1961] 105/8).

[11] Kramer, K.; Wright, A. (Angew. Chem. **74** [1962] 468/9; Angew. Chem. Int. Ed. Engl. **1** [1962] 402/3).
[12] Lesbre, M.; Satgé, J. (C. R. Hebd. Seances Acad. Sci. **254** [1962] 1453/5).
[13] Lesbre, M.; Satgé, J. (C. R. Hebd. Seances Acad. Sci. **254** [1962] 4051/3).
[14] Mironov, V. F.; Dzhurinskaya, N. G.; Gar, T. K.; Petrov, A. D. (Izv. Akad. Nauk SSSR Ser. Khim. **1962** 460/5; Bull. Acad. Sci. USSR Div. Chem. Sci. [Engl. Transl.] **1962** 425/30).
[15] Gverdtsiteli, I. M.; Guntsadze, T. P.; Petrov, A. D. (Dokl. Akad. Nauk SSSR **153** [1963] 107/10; Dokl. Chem. [Engl. Transl.] **148/153** [1963] 881/4).
[16] Kramer, K. A. W.; Wright, A. N. (J. Chem. Soc. **1963** 3604/8).
[17] Vyazankin, N. S.; Razuvaev, G. A.; Gladyshev, E. N. (Dokl. Akad. Nauk SSSR **151** [1963] 1326/8; Dokl. Chem. [Engl. Transl.] **148/153** [1963] 653/5).
[18] Gverdtsiteli, I. M.; Buachidze, M. A. (Dokl. Akad. Nauk SSSR **158** [1964] 147/50; Dokl. Chem. [Engl. Transl.] **154/159** [1964] 840/3).
[19] Gverdtsiteli, I. M.; Guntsadze, T. P.; Petrov, A. D. (Dokl. Akad. Nauk SSSR **157** [1964] 607/10; Dokl. Chem. [Engl. Transl.] **154/159** [1964] 711/4).
[20] Rijkens, F.; Janssen, M. J.; Drenth, W.; van der Kerk, G. J. M. (J. Organomet. Chem. **2** [1964] 347/56).

[21] Satgé, J. (Bull. Soc. Chim. Fr. **1964** 630/4).
[22] Vyazankin, N. S.; Razuvaev, G. A.; Gladyshev, E. N.; Gurikova, T. G. (Dokl. Akad. Nauk SSSR **155** [1964] 1108/10; Dokl. Chem. [Engl. Transl.] **154/159** [1964] 360/2).
[23] Vyazankin, N. S.; Razuvaev, G. A.; Bychkov, V. T. (Dokl. Akad. Nauk SSSR **158** [1964] 382/4; Dokl. Chem. [Engl. Transl.] **154/159** [1964] 877/9).
[24] Vyazankin, N. S.; Razuvaev, G. A.; Korneva, S. P.; Kruglaya, O. A.; Galiulina, R. F. (Dokl. Akad. Nauk SSSR **158** [1964] 884/7; Dokl. Chem. [Engl. Transl.] **154/159** [1964] 1002/4).
[25] Cram, D. J. (Fundamentals of Carbanion Chemistry, Academic, New York-London 1965, p. 19).
[26] Gverdtsiteli, I. M.; Buachidze, M. A. (Soobshch. Akad. Nauk Gruz. SSR **37** [1965] 59/64; C.A. **62** [1965] 14716).
[27] Gverdtsiteli, I. M.; Buachidze, M. A. (Soobshch. Akad. Nauk Gruz. SSR **37** [1965] 323/30; C.A. **62** [1965] 14719).
[28] Kruglaya, O. A.; Vyazankin, N. S.; Razuvaev, G. A. (Zh. Obshch. Khim. **35** [1965] 394; J. Gen. Chem. [Engl. Transl.] **35** [1965] 392).
[29] Mikhailov, B. M.; Bubnov, Yu. N.; Kiselev, V. G. (Izv. Akad. Nauk SSSR Ser. Khim. **1965** 68/72; Bull. Acad. Sci. USSR Div. Chem. Sci. [Engl. Transl.] **1965** 58/61).
[30] Mironov, V. F.; Gar, T. K. (Izv. Akad. Nauk SSSR Ser. Khim. **1965** 291/300; Bull. Acad. Sci. USSR Div. Chem. Sci. [Engl. Transl.] **1965** 273/80).

[31] Satgé, J.; Lesbre, M. (Bull. Soc. Chim. Fr. **1965** 2578/81).
[32] Shikhiev, I. A.; Aslanov, I. A.; Mekhmandarova, N. T.; Verdieva, S. Sh. (Azerb. Khim. Zh. **1965** No. 4, pp. 42/4; C.A. **64** [1966] 9760).
[33] Cullen, W. R.; Styan, G. E. (J. Organomet. Chem. **6** [1966] 117/25).
[34] Massol, M.; Satgé, J. (Bull. Soc. Chim. Fr. **1966** 2737/43).
[35] Massol, M.; Satgé, J.; Lesbre, M. (C. R. Hebd. Seances Acad. Sci. C **262** [1966] 1806/9).
[36] Satgé, J.; Massol, M.; Lesbre, M. (J. Organomet. Chem. **5** [1966] 241/53).

[37] Shikhiev, I. A.; Aslanov, I. A.; Mekhmandarova, N. T. (Zh. Obshch. Khim. **36** [1966] 1295/7; J. Gen. Chem. USSR [Engl. Transl.] **36** [1966] 1310/1).
[38] Vyazankin, N. S.; Bochkarev, M. N.; Sanina, L. P. (Zh. Obshch. Khim. **36** [1966] 166; J. Gen. Chem. USSR [Engl. Transl.] **36** [1966] 175).
[39] Vyazankin, N. S.; Bochkarev, M. N.; Sanina, L. P. (Zh. Obshch. Khim. **36** [1966] 1154/5; J. Gen. Chem. USSR [Engl. Transl.] **36** [1966] 1169).
[40] Vyazankin, N. S.; Bochkarev, M. N.; Sanina, L. P. (Zh. Obshch. Khim. **36** [1966] 1961/4; J. Gen. Chem. USSR [Engl. Transl.] **36** [1966] 1954/7).

[41] Vyazankin, N. S.; Kruglaya, O. A.; Razuvaev, G. A.; Semchikova, G. S. (Dokl. Akad. Nauk SSSR **166** [1966] 99/102; Dokl. Chem. [Engl. Transl.] **166/171** [1966] 8/11).
[42] Vyazankin, N. S.; Mitrofanova, E. V.; Kruglaya, O. A.; Razuvaev, G. A. (Zh. Obshch. Khim. **36** [1966] 160; J. Gen. Chem. USSR [Engl. Transl.] **36** [1966] 166).
[43] Vyazankin, N. S.; Razuvaev, G. A.; Kruglaya, O. A.; Semchikova, G. S. (J. Organomet. Chem. **6** [1966] 474/83).
[44] Ali-zade, I. G.; Shikhieva, M. I.; Salimov, M. A.; Abdullaev, N. D.; Shikhiev, I. A. (Dokl. Akad. Nauk SSSR **173** [1967] 89/92; Dokl. Chem. [Engl. Transl.] **172/177** [1967] 201/4).
[45] Bryskovskaya, A. V.; Al'bitskaya, V. M. (Zh. Obshch. Khim. **37** [1967] 1553/8; J. Gen. Chem. USSR [Engl. Transl.] **37** [1967] 1474/8).
[46] Gverdtsiteli, I. M.; Baramidze, L. V.; Chelidze, M. V. (Zh. Obshch. Khim. **37** [1967] 2654/6; J. Gen. Chem. USSR [Engl. Transl.] **37** [1967] 2526/7).
[47] Gverdtsiteli, I. M.; Buachidze, M. A. (Soobshch. Akad. Nauk Gruz. SSR **48** [1967] 571/4; C.A. **68** [1968] No. 105315).
[48] Kruglaya, O. A.; Vyazankin, N. S.; Razuvaev, G. A.; Mitrofanova, E. V. (Dokl. Akad. Nauk SSSR **173** [1967] 834/6; Dokl. Chem. [Engl. Transl.] **172/177** [1967] 310/2).
[49] Kühlein, K. (Dipl.-Arbeit, Univ. Gießen 1963 from Kühlein, K.; Neumann, W. P.; Becker, H. P.; Angew. Chem. **79** [1967] 870/1; Angew. Chem. Int. Ed. Engl. **6** [1967] 876).
[50] Massol, M. (Diss. Toulouse 1967 from [83]).

[51] Mazerolles, P.; Dubac, J. (C. R. Hebd. Seances Acad. Sci. C **265** [1967] 403/6).
[52] Mazerolles, P.; Dubac, J.; Lesbre, M. (Tetrahedron Lett. **1967** 255/8).
[53] Seyferth, D.; Burlitch, J. M.; Dertouzos, H.; Simmons, H. D., Jr. (J. Organomet. Chem. **7** [1967] 405/13).
[54] Shikhiev, I. A.; Mustafaev, R. M.; Abdullaev, N. D. (Uch. Zap. Azerb. Gos. Univ. Ser. Khim. Nauk **1967** No. 2, pp. 56/9; C.A. **70** [1969] No. 106624).
[55] Sorokin, G. V.; Pozdnyakova, M. V.; Ter-Asaturova, N. I.; Perchenko, V. N.; Nametkin, N. S. (Dokl. Akad. Nauk SSSR **174** [1967] 376/7; Dokl. Chem. [Engl. Transl.] **172/177** [1967] 465/6).
[56] Vyazankin, N. S.; Bochkarev, M. N.; Sanina, L. P. (Zh. Obshch. Khim. **37** [1967] 1037/40; J. Gen. Chem. USSR [Engl. Transl.] **37** [1967] 980/3).
[57] Vyazankin, N. S.; Gladyshev, E. N.; Korneva, S. P. (Zh. Obshch. Khim. **37** [1967] 1736/8; J. Gen. Chem. USSR [Engl. Transl.] **37** [1967] 1655/6).
[58] Vyazankin, N. S.; Razuvaev, G. A.; Gladyshev, E. N.; Korneva, S. P. (J. Organomet. Chem. **7** [1967] 353/7).
[59] Dzhurinskaya, N. G.; Mikhailyants, S. A.; Evdakov, V. P. (Zh. Obshch. Khim. **38** [1968] 1267/9; J. Gen. Chem. USSR [Engl. Transl.] **38** [1968] 1220/2).
[60] Glockling, F.; Light, J. R. C. (J. Chem. Soc. A **1968** 717/34).

[61] Gverdtsiteli, I. M.; Baramidze, L. V. (Zh. Obshch. Khim. **38** [1968] 1598/601; J. Gen. Chem. USSR [Engl. Transl.] **38** [1968] 1547/50).
[62] Gverdtsiteli, I. M.; Gelashvili, E. S. (Soobshch. Akad. Nauk Gruz. SSR **52** [1968] 69/74; C.A. **70** [1969] No. 78108).

[63] Gverdtsiteli, I. M.; Guntsadze, T. P.; Gudavadze, M. I. (Soobshch. Akad. Nauk Gruz. SSR **50** [1968] 609/12; C.A. **69** [1968] No. 87124).

[64] Gverdtsiteli, I. M.; Guntsadze, T. P.; Kalandarishvili, A. A. (Tr. Tbilis Univ. **126** [1968] 209/13; Ref. Zh. Khim. **1969** No. 12Zh496; C.A. **73** [1970] No. 4002).

[65] Mamedov, Sh. M.; Shikhieva, M. I.; Shikhiev, I. A. (Azerb. Khim. Zh. **1968** No. 2, pp. 85/91; C.A. **69** [1968] No. 106834).

[66] Mendelsohn, J.-C.; Metras, F.; Lahournère, J.-C.; Valade, J. (J. Organomet. Chem. **12** [1968] 327/40).

[67] Nekhorosheva, E. V.; Al'bitskaya, V. M. (Zh. Obshch. Khim. **38** [1968] 1511/7; J. Gen. Chem. USSR [Engl. Transl.] **38** [1968] 1461/6).

[68] Rivière, P.; Satgé, J. (C. R. Seances Acad. Sci. C **267** [1968] 267/9).

[69] Shikhiev, I. A.; Askerov, G. F.; Garaeva, Sh. V. (Zh. Obshch. Khim. **38** [1968] 639/43; J. Gen. Chem. USSR [Engl. Transl.] **38** [1968] 616/9).

[70] Shikhiev, I. A.; Isaev, E. M.; Verdieva, S. Sh.; Aslanov, I. A. (Uch. Zap. Azerb. Gos. Univ. Ser. Khim. Nauk **1968** No. 2, pp. 83/6; C.A. **72** [1970] No. 90595).

[71] Shikhiev, I. A.; Nasirova, M. M. (Azerb. Khim. Zh. **1968** No. 1, pp. 23/5; C.A. **70** [1969] No. 47561).

[72] Shikhiev, I. A.; Nasirova, M. M. (Uch. Zap. Azerb. Gos. Univ. Ser. Khim. Nauk **1968** No. 3, pp. 61/5; C.A. **73** [1970] No. 3980).

[73] Vyazankin, N. S.; Gladyshev, E. N.; Korneva, S. P.; Razuvaev, G. A.; Arkhangel'skaya, E. A. (Zh. Obshch. Khim. **38** [1968] 1803/9; J. Gen. Chem. USSR [Engl. Transl.] **38** [1968] 1757/61).

[74] Vyazankin, N. S.; Kalinina, G. S.; Kruglaya, O. A.; Razuvaev, G. A. (Zh. Obshch. Khim. **38** [1968] 906/11; J. Gen. Chem. USSR [Engl. Transl.] **38** [1968] 870/4).

[75] Bochkarev, M. N.; Sanina, L. P.; Vyazankin, N. S. (Zh. Obshch. Khim. **39** [1969] 135/41; J. Gen. Chem. USSR [Engl. Transl.] **39** [1969] 122/7).

[76] Glockling, F.; Wilbey, M. D. (J. Chem. Soc. Chem. Commun. **1969** 286/7).

[77] Kakhniashvili, A. I.; Ioramashvili, D. Sh.; Fedin, E. I.; Petrovskii, P. V.; Rubin, I. D. (Soobshch. Akad. Nauk Gruz. SSR **53** [1969] 573/6; C.A. **71** [1969] No. 60353).

[78] Kakhniashvili, A. I.; Ioramashvili, D. Sh.; Fedin, E. I.; Petrovskii, P. V.; Rubin, I. D. (Soobshch. Akad. Nauk Gruz. SSR **54** [1969] 337/40; C.A. **71** [1969] No. 60453).

[79] Kazankova, M. A.; Lutsenko, I. F. (Zh. Obshch. Khim. **39** [1969] 926; J. Gen. Chem. USSR [Engl. Transl.] **39** [1969] 891).

[80] Massol, M.; Satgé, J.; Cabadi, Y. (C. R. Seances Acad. Sci. C **268** [1969] 1814/6).

[81] Massol, M.; Satgé, J.; Lesbre, M. (J. Organomet. Chem. **17** [1969] 25/39).

[82] Neumann, W. P.; Blaukat, U. (Angew. Chem. **81** [1969] 625/6; Angew. Chem. Int. Ed. Engl. **8** [1969] 611).

[83] Satgé, J.; Rivière, P. (J. Organomet. Chem. **16** [1969] 71/82).

[84] Shikhiev, I. A.; Karaev, S. F. (Uch. Zap. Azerb. Gos. Univ. Ser. Khim. Nauk **1969** No. 2, pp. 81/4; Ref. Zh. Khim. **1970** No. 9Zh498; C.A. **75** [1971] No. 6038).

[85] Stadnichuk, M. D.; Petrov, A. A. (Zh. Obshch. Khim. **39** [1969] 2597/8; J. Gen. Chem. USSR [Engl. Transl.] **39** [1969] 2537).

[86] Tel'noi, V. I.; Kol'yakova, G. M.; Rabinovich, I. B.; Vyazankin, N. S. (Dokl. Akad. Nauk SSSR **185** [1969] 374/6; Dokl. Chem. [Engl. Transl.] **184/189** [1969] 214/6).

[87] Vyazankin, N. S.; Kalinina, G. S.; Kruglaya, O. A.; Razuvaev, G. A. (Zh. Obshch. Khim. **39** [1969] 2005/11; J. Gen. Chem. USSR [Engl. Transl.] **39** [1969] 1964/8).

[88] Aslanov, I. A.; Verdieva, S. Sh. (Zh. Obshch. Khim. **40** [1970] 1266/8 ; J. Gen. Chem. USSR [Engl. Transl.] **40** [1970] 1257/9).

[89] Belavin, I. Yu.; Nguyen, D. H.; Tvorogov, A. N.; Baukov, Yu. I.; Lutsenko, I. F. (Zh. Obshch. Khim. **40** [1970] 1065/75; J. Gen. Chem. USSR [Engl. Transl.] **40** [1970] 1052/61).
[90] Glockling, F.; Wilbey, M. D. (J. Chem. Soc. A **1970** 1675/81).
[91] Kakhniashvili, A. I.; Ioramashvili, D. Sh. (Zh. Obshch. Khim. **40** [1970] 1552/5; J. Gen. Chem. USSR [Engl. Transl.] **40** [1970] 1539/41).
[92] Kakhniashvili, A. I.; Ioramashvili, D. Sh. (Zh. Obshch. Khim. **40** [1970] 1556/9; J. Gen. Chem. USSR [Engl. Transl.] **40** [1970] 1542/5).
[93] Shikhiev, I. A.; Gasanova, R. Yu.; Askerov, G. F.; Rzaeva, S. A. (Zh. Obshch. Khim. **40** [1970] 817/9; J. Gen. Chem. USSR [Engl. Transl.] **40** [1970] 795/7).
[94] Shikhiev, I. A.; Askerov, G. F.; Isaev, E. M.; Ramazanzade, Z. M.; Shakhverdieva, F. M.; Mustafaev, R. M. (Uch. Zap. Azerb. Univ. Ser. Khim. Nauk **1970** No. 4, pp. 75/9; Ref. Zh. Khim. **1971** No. 16Zh354; C.A. **77** [1972] No. 101811).
[95] Bulten, E. J.; Noltes, J. G. (J. Organomet. Chem. **29** [1971] 397/407).
[96] Gar, T. K.; Buyakov, A. A.; Kisin, A. V.; Mironov, V. F. (Zh. Obshch. Khim. **41** [1971] 1589/94; J. Gen. Chem. USSR [Engl. Transl.] **41** [1971] 1596/600).
[97] Gverdtsiteli, I. M.; Baramidze, L. V.; Tsikaridze, N. V. (Zh. Obshch. Khim. **41** [1971] 139/41; J. Gen. Chem. USSR [Engl. Transl.] **41** [1971] 134/6).
[98] Gverdtsiteli, I. M.; Gelashvili, E. S. (Zh. Obshch. Khim. **41** [1971] 2061/6; J. Gen. Chem. USSR [Engl. Transl.] **41** [1971] 2080/5).
[99] Massol, M.; Cabadi, Y.; Satgé, J. (Bull. Soc. Chim. Fr. **1971** 3235/44).
[100] Nasirova, M. M.; Shikhiev, I. A.; Rzaeva, Sh. M. (Uch. Zap. Azerb. Univ. Ser. Khim. Nauk **1971** No. 2, pp. 64/7; C.A. **78** [1973] No. 29296).
[101] Shikhiev, I. A.; Rzaeva, S. A. (Uch. Zap. Azerb. Inst. Nefti Khim. Ser. 9 **1971** No. 2, pp. 116/9; C.A. **77** [1972] No. 88612).
[102] Azerbaev, I. N.; Erzhanov, K. B.; Polatbekov, K. B.; Bazalitskaya, V. S. (Khim. Atsetilena Tekhnol. Karbida Kal'tsiya Dokl. Vses Nauchno Tekh. Konf., Temirtau, Kaz., 1969 [1972] pp. 177/80; C.A. **79** [1973] No. 115687).
[103] Gverdtsiteli, I. M.; Chanturiya, M. D. (Zh. Obshch. Khim. **42** [1972] 1773/7; J. Gen. Chem. USSR [Engl. Transl.] **42** [1972] 1760/2).
[104] Gverdtsiteli, I. M.; Chanturiya, M. D. (Soobshch. Akad. Nauk Gruz. SSR **65** [1972] 73/6; C.A. **76** [1972] No. 140281).
[105] Mironov, V. F.; Tsotadze, M. V.; Gar, T. K. (Soobshch. Akad. Nauk Gruz. SSR **68** [1972] 77/9; C.A. **78** [1973] No. 29948).
[106] Sakurai, H.; Mochida, K.; Hosomi, A.; Mita, F. (J. Organomet. Chem. **38** [1972] 275/80).
[107] Sakurai, H.; Mochida, K. (J. Organomet. Chem. **42** [1972] 339/43).
[108] Satgé, J.; Lesbre, M.; Rivière, P.; Richelme, S. (J. Organomet. Chem. **34** [1972] C18/C20).
[109] Sultanov, R. A.; Aronova, L. L.; Sadykh-zade, S. I. (Zh. Obshch. Khim. **42** [1972] 1872; J. Gen. Chem. USSR [Engl. Transl.] **42** [1972] 1862).
[110] Bochkarev, M. N.; Maiorova, L. P.; Vyazankin, N. S. (J. Organomet. Chem. **55** [1973] 89/96).
[111] Connor, J. A.; Day, J. P.; Turner, R. M. (J. Chem. Soc. Chem. Commun. **1973** 578/9).
[112] Connor, J. A.; Rose, P. D.; Turner, R. M. (J. Organomet. Chem. **55** [1973] 111/9).
[113] Linke, K.-H.; Göhausen, H. J. (Chem. Ber. **106** [1973] 3438/49).
[114] Streitwieser, A., Jr.; Granger, M. R.; Mares, F.; Wolf, R. A. (J. Am. Chem. Soc. **95** [1973] 4257/61).
[115] Gverdtsiteli, I. M.; Chernyshev, E. A.; Dzotsenidze, L. A. (Soobshch. Akad. Nauk Gruz. SSR **74** [1974] 93/6; C.A. **81** [1974] No. 63717).

[116] Gverdtsiteli, I. M.; Édiberidze, D. A.; Chernyshev, E. A. (Zh. Obshch. Khim. **44** [1974] 2449/52; J. Gen. Chem. USSR [Engl. Transl.] **44** [1974] 2409/11).
[117] Gverdtsiteli, I. M.; Gelashvili, E. S.; Topchiashvili, É. E. (Zh. Obshch. Khim. **44** [1974] 2452/6; J. Gen. Chem. USSR [Engl. Transl.] **44** [1974] 2412/5).
[118] Rivière, P.; Satgé, J.; Dousse, G.; Rivière-Baudet, M.; Couret, C. (J. Organomet. Chem. **72** [1974] 339/50).
[119] Sadykh-zade, S. I.; Sultanov, R. A.; Aronova, L. L.; Pestunovich, V. A. (Zh. Obshch. Khim. **44** [1974] 1787/9; J. Gen. Chem. USSR [Engl. Transl.] **44** [1974] 1753/5).
[120] Satgé, J.; Rivière, P.; Boy, A. (C. R. Hebd. Seances Acad. Sci. C **278** [1974] 1309/12).

[121] Dergunov, Yu. I.; Mysin, N. I.; Mushkin, Yu. I. (Zh. Obshch. Khim. **45** [1975] 1280/5; J. Gen. Chem. USSR [Engl. Transl.] **45** [1975] 1257/60).
[122] Dzhafarov, A. A.; Aslanov, I. A.; Kochkin, D. A. (Zh. Obshch. Khim. **45** [1975] 2023/5; J. Gen. Chem. USSR [Engl. Transl.] **45** [1975] 1986/8).
[123] Gverdtsiteli, I. M.; Chanturiya, M. D. (Zh. Obshch. Khim. **45** [1975] 2349; J. Gen. Chem. USSR [Engl. Transl.] **45** [1975] 2309).
[124] Gverdtsiteli, I. M.; Talakvadze, T. G. (Soobshch. Akad. Nauk Gruz. SSR **79** [1975] 89/92; C.A. **83** [1975] No. 193462).
[125] Gverdtsiteli, I. M.; Talakvadze, T. G. (Soobshch. Akad. Nauk Gruz. SSR **79** [1975] 601/4; C.A. **84** [1976] No. 59690).
[126] Kazankova, M. A.; Zverkova, T. I.; Levin, M. Z.; Lutsenko, I. F. (Zh. Obshch. Khim. **45** [1975] 73/81; J. Gen. Chem. USSR [Engl. Transl.] **45** [1975] 66/72).
[127] Kazankova, M. A.; Zverkova, T. I.; Lutsenko, I. F. (Zh. Obshch. Khim. **45** [1975] 2044/51; J. Gen. Chem. USSR [Engl. Transl.] **45** [1975] 2006/12).
[128] Mironov, V. F.; Tsotadze, M. V.; Gar, T. K.; Gverdtsiteli, I. M. (Zh. Obshch. Khim. **45** [1975] 2185/9; J. Gen. Chem. USSR [Engl. Transl.] **45** [1975] 2148/51).
[129] Nametkin, N. S.; Vdovin, V. M.; Karel'skii, V. N.; Silkina, I. V.; Kacharmin, B. V.; Babich, E. D. (Dokl. Akad. Nauk SSSR **220** [1975] 601/4; Dokl. Chem. [Engl. Transl.] **220/225** [1975] 97/9).
[130] Rivière, P.; Satgé, J.; Boy, A. (J. Organomet. Chem. **96** [1975] 25/40).

[131] Sakurai, H.; Mochida, K.; Kira, M. (J. Am. Chem. Soc. **97** [1975] 929/31).
[132] Sultanov, R. A.; Askerov, O. V.; Aronova, L. A.; Khudayarov, I. A.; Sadykh-zade, S. I. (Zh. Obshch. Khim. **45** [1975] 2102/3; J. Gen. Chem. USSR [Engl. Transl.] **45** [1975] 2068).
[133] Aronova, L. L.; Sadykh-zade, S. I.; Sultanov, R. A. (Azerb. Khim. Zh. **1976** No. 5, pp. 43/5; C.A. **87** [1977] No. 39613).
[134] Connor, J. A.; Day, J. P.; Turner, R. M. (J. Chem. Soc. Dalton Trans. **1976** 283/5).
[135] Gverdtsiteli, M. I.; Chanturiya, M. D. (Zh. Obshch. Khim. **46** [1976] 865/8; J. Gen. Chem. USSR [Engl. Transl.] **46** [1976] 863/5).
[136] Kruglaya, O. A.; Fedot'eva, I. B.; Fedot'ev, B. V.; Kalikhman, I. D.; Brodskaya, É. I.; Vyazankin, N. S. (Izv. Akad. Nauk SSSR Ser. Khim. **1976** 1887/9; Bull. Acad. Sci. USSR Div. Chem. Sci. [Engl. Transl.] **1976** 1777/9).
[137] Kruglaya, O. A.; Lyashenko, G. S.; Filippova, A. Kh.; Keiko, V. V.; Kalikhman, I. D.; Vyazankin, N. S. (Izv. Akad. Nauk SSSR Ser. Khim. **1976** 682/3; Bull. Acad. Sci. USSR Div. Chem. Sci. [Engl. Transl.] **1976** 669/71).
[138] Sakurai, H.; Mochida, K. (Bull. Chem. Soc. Jpn. **49** [1976] 3703/4).
[139] Shikhiev, I. A.; Rzaeva, S. A.; Dadasheva, Ya. A.; Guseinova, M. A. (Khim. Elementoorg. Soedin. **1976** 71/4; C.A. **85** [1976] No. 177573).
[140] Filippova, A. Kh.; Lyashenko, G. S.; Kruglaya, O. A.; Keiko, V. V.; Kalikhman, I. D.; Vyazankin, N. S. (Izv. Akad. Nauk SSSR Ser. Khim. **1977** 660/3; Bull. Acad. Sci. USSR Div. Chem. Sci. [Engl. Transl.] **1977** 596/9).

[141] Gverdtsiteli, I. M.; Chanturiya, M. D. (Soobshch. Akad. Nauk Gruz. SSR **88** [1977] 593/6; C.A. **88** [1978] No. 190443).
[142] Kakhniashvili, A. I.; Ioramashvili, D. Sh.; Nadirashvili, M. D. (Soobshch. Akad. Nauk Gruz. SSR **87** [1977] 81/4; C.A. **88** [1978] No. 7000).
[143] Aslanov, I. A.; Dzhafarov, E. F.; Gadzhieva, M. A.; Kulibekova, A. M. (Azerb. Khim. Zh. **1978** No. 5, pp. 42/3; C.A. **91** [1979] No. 38881).
[144] Gverdtsiteli, I. M.; Baramidze, L. V.; Dzhananashvili, E. V. (Soobshch. Akad. Nauk Gruz. SSR **89** [1978] 593/6; C.A. **89** [1978] No. 43655).
[145] Kruglaya, O. A.; Fedot'eva, I. B.; Fedot'ev, B. V.; Kalikhman, I. D.; Vyazankin, N. S. (Zh. Obshch. Khim. **48** [1978] 1431; J. Gen. Chem. USSR [Engl. Transl.] **48** [1978] 1315/6).
[146] Razuvaev, G. A.; Latyaeva, V. N.; Gladyshev, E. N.; Krasilnikova, E. V.; Lineva, A. N.; Kozina, A. P. (Inorg. Chim. Acta **31** [1978] L357/L360).
[147] Razuvaev, G. A.; Vyshinskaya, L. I.; Vasil'eva, G. A.; Latyaeva, V. N.; Timoshenko, S. Ya.; Ermolaev, N. L. (Izv. Akad. Nauk SSSR Ser. Khim. **1978** 2584/8; Bull. Acad. Sci. USSR Div. Chem. Sci. [Engl. Transl.] **1978** 2310/4).
[148] Rivière, P.; Richelme, S.; Rivière-Baudet, M.; Satgé, J.; Gynane, M. J. S.; Lappert, M. F. (J. Chem. Res. Synop. **1978** 218/9; J. Chem. Res. Miniprint **1978** 2801/16).
[149] Rivière, P.; Rivière-Baudet, M.; Richelme, S.; Satgé, J. (Bull. Soc. Chim. Fr. **1978** II 193/6).
[150] Sakurai, H.; Mochida, K. (J. Organomet. Chem. **154** [1978] 353/68).

[151] Chen, K. S.; Foster, T.; Wan, J. K. S. (J. Chem. Soc. Perkin Trans. II **1979** 1288/92).
[152] Eaborn, C.; Singh, B. (J. Organomet. Chem. **177** [1979] 333/48).
[153] Escudié, J.; Couret, C.; Satgé, J. (Recl. Trav. Chim. Pays-Bas **98** [1979] 461/6).
[154] Ioramashvili, D. Sh.; Shudra, O. S. (Soobshch. Akad. Nauk Gruz. SSR **94** [1979] 81/4; C.A. **91** [1979] No. 91727).
[155] Klingler, R. J.; Mochida, K.; Kochi, J. K. (J. Am. Chem. Soc. **101** [1979] 6626/37).
[156] Ioramashvili, D. Sh. (Soobshch. Akad. Nauk Gruz. SSR **97** [1980] 357/60; C.A. **93** [1980] No. 131601).
[157] Ioramashvili, D. Sh. (Izv. Akad. Nauk Gruz. SSR Ser. Khim. **6** [1980] 129/35 from C.A. **93** [1980] No. 239555).
[158] Razuvaev, G. A.; Korneva, S. P.; Vyshinskaya, L. I.; Sorokina, L. A. (Zh. Obshch. Khim. **50** [1980] 891/4; J. Gen. Chem. USSR [Engl. Transl.] **50** [1980] 719/21).
[159] Seyferth, D.; Withers, H. P., Jr. (J. Organomet. Chem. **188** [1980] 329/33).
[160] Wong, C. L.; Klingler, R. J.; Kochi, J. K. (Inorg. Chem. **19** [1980] 423/30).

[161] Hottmann, K. (J. Prakt. Chem. **323** [1981] 399/419).
[162] Lopatina, V. S.; Kocheshkov, K. A.; Fomina, N. V.; Rodionov, A. I.; Shapet'ko, N. N.; Yankelevich, A. Z. (Zh. Obshch. Khim. **51** [1981] 2580/2; J. Gen. Chem. USSR [Engl. Transl.] **51** [1981] 2225/7).
[163] Lyashenko, G. S.; Filippova, A. Kh.; Kalikhman, I. D.; Keiko, V. V.; Kruglaya, O. A.; Vyazankin, N. S. (Izv. Akad. Nauk SSSR Ser. Khim. **1981** 874/8; Bull. Acad. Sci. USSR Div. Chem. Sci. [Engl. Transl.] **1981** 654/8).
[164] Stang, P. J.; White, M. R. (J. Am. Chem. Soc. **103** [1981] 5429/33).
[165] Stang, P. J.; Ladika, M. (J. Am. Chem. Soc. **103** [1981] 6437/43).
[166] Castel, A.; Rivière. P.; Satgé, J. (J. Organomet. Chem. **232** [1982] 137/46).
[167] Chanturiya, M. D. (Soobshch. Akad. Nauk Gruz. SSR **108** [1982] 565/8; C.A. **99** [1983] No. 53305).
[168] Chesnokova, T. A.; Razuvaev, G. A.; Brevnova, T. N. (Zh. Obshch. Khim. **52** [1982] 2754/62; J. Gen. Chem. USSR [Engl. Transl.] **52** [1982] 2428/35).

[169] Eaborn, C.; Mahmoud, F. M. S.; Taylor, R. (J. Chem. Soc. Perkin Trans. II **1982** 1313/9).
[170] Ediberidze, D. A.; Chernyshev, E. A. (Izv. Akad. Nauk Gruz. SSR Ser. Khim. **8** [1982] 116/20; C.A. **97** [1982] No. 216299).

[171] Lyashenko, G. S.; Filippova, A. Kh.; Kalikhman, I. D.; Naumova, G. G.; Vyazankin, N. S. (Izv. Akad. Nauk SSSR Ser. Khim. **1982** 2618/20; Bull. Acad. Sci. USSR Div. Chem. Sci. [Engl. Transl.] **1982** 2312/4).
[172] Razuvaev, G. A.; Brevnova, T. N.; Chesnokova, T. A.; Troitskaya, L. S.; Shevtsova, M. P. (Zh. Obshch. Khim. **52** [1982] 1832/7; J. Gen. Chem. USSR [Engl. Transl.] **52** [1982] 1624/8).
[173] Razuvaev, G. A.; Latyaeva, V. N.; Mar'in, V. P.; Vyshinskaya, L. I.; Korneva, S. P.; Andrianov, Yu. A.; Krasil'nikova, E. V. (J. Organomet. Chem. **225** [1982] 233/44).
[174] Süss-Fink, G.; Ott, J.; Schmidkonz, B.; Guldner, K. (Chem. Ber. **115** [1982] 2487/93).
[175] Cherkezishvili, K. I.; Gelashvili, K. Sh.; Kublashvili, R. I. (Soobshch. Akad. Nauk Gruz. SSR **109** [1983] 297/300; C.A. **99** [1983] No. 70885).
[176] Gar, T. K.; Khromova, N. Yu.; Tandura, S. N.; Mironov, V. F. (Zh. Obshch. Khim. **53** [1983] 1800/7; J. Gen. Chem. USSR [Engl. Transl.] **53** [1983] 1619/26).
[177] Salimgareeva, I. M.; Bogatova, N. G.; Panasenko, A. A.; Khalilov, L. M.; Purlei, I. I.; Mavrodiev, V. K.; Yur'ev, V. P. (Izv. Akad. Nauk SSSR Ser. Khim. **1983** 1605/12; Bull. Acad. Sci. USSR Div. Chem. Sci. [Engl. Transl.] **1983** 1456/61).
[178] Ermolaev, N. L.; Bochkarev, M. N.; Razuvaev, G. A.; Grishin, Yu. K.; Ustynyuk, Yu. A. (Zh. Obshch. Khim. **54** [1984] 96/100; J. Gen. Chem. USSR [Engl. Transl.] **54** [1984] 83/6).
[179] Gendin, D. V.; Rybin, L. I.; Vakul'skaya, T. I.; Vyazankina, O. A.; Vyazankin, N. S. (Izv. Akad. Nauk SSSR Ser. Khim. **1984** 2381/3; Bull. Acad. Sci. USSR Div. Chem. Sci. [Engl. Transl.] **1984** 2176/8).
[180] Hayashi, H.; Sakaguchi, Y.; Mochida, K. (Chem. Lett. **1984** 79/82).

[181] Khorshev, S. Ya.; Tsvetkova, V. L.; Egorochkin, A. N. (J. Organomet. Chem. **264** [1984] 169/78).
[182] Dneprovskii, A. S.; Pertsikov, B. Z. (Zh. Org. Khim. **21** [1985] 250/6; J. Org. Chem. USSR [Engl. Transl.] **21** [1985] 222/7).
[183] Kobayashi, Mi.; Kobayashi, Ma.; Yoshida, M. (Bull. Chem. Soc. Jpn. **58** [1985] 473/6).
[184] Sakaguchi, Y.; Hayashi, H.; Murai, H.; I'Haya, Y. J.; Mochida, K. (Chem. Phys. Lett. **120** [1985] 401/5).
[185] Stanley, A. E.; Johnson, R. A.; Turner, J. B.; Roberts, A. H. (Appl. Spectrosc. **40** [1986] 374/8).
[186] Voronkov, M. G.; Adamovich, S. N.; Kudyakov, N. M.; Khramtsova, S. Yu.; Rakhlin, V. I.; Mirskov, R. G. (Izv. Akad. Nauk SSSR Ser. Khim. **1986** 488/9; Bull. Acad. Sci. USSR Div. Chem. Sci. [Engl. Transl.] **1986** 451/2).
[187] Bravo-Zhivotovskii, D. A.; Pigarev, S. D.; Vyazankina, O. A.; Vyazankin, N. S. (Zh. Obshch. Khim. **57** [1987] 2735/8; J. Gen. Chem. USSR [Engl. Transl.] **57** [1987] 2440/2).
[188] Chojnowski, J.; Fortuniak, W.; Stańczyk, W. (J. Am. Chem. Soc. **109** [1987] 7776/81).
[189] Morancho, R.; Reynes, A.; El Boucham, J.; Sepiani, N.; Mazerolles, P.; Grégoire, P. (Proc. 6th Eur. Conf. Chem. Vap. Deposition, Jerusalem 1987, pp. 381/8).
[190] Razuvaev, G. A.; Gordetsov, A. S.; Latyaeva, V. N.; Cherepennikova, N. F.; Brevnova, T. N.; Skobeleva, S. E.; Konkina, T. N.; Kozina, A. P. (Zh. Obshch. Khim. **57** [1987] 464/8; J. Gen. Chem. USSR [Engl. Transl.] **57** [1987] 405/8).

[191] Voronkov, M. G.; Adamovich, S. N.; Khramtsova, S. Yu.; Shternberg, B. Z.; Rakhlin, V. I.; Mirskov, R. G. (Izv. Akad. Nauk SSSR Ser. Khim. **1987** 1424/7; Bull. Acad. Sci. USSR Div. Chem. Sci. [Engl. Transl.] **1987** 1317/9).
[192] Koenig, M.; Barrau, J.; Ben Hamida, N. (J. Organomet. Chem. **356** [1988] 133/9).
[193] Mochida, K.; Yoshida, Y. (Bull. Chem. Soc. Jpn. **61** [1988] 1789/90).
[194] Rivière, P.; Castel, A.; Satgé, J.; Guyot, D.; Ko, Y. H. (J. Organomet. Chem. **339** [1988] 51/60).
[195] Sakaguchi, Y.; Hayashi, H.; Murai, H.; I'Haya, Y. J. (J. Am. Chem. Soc. **110** [1988] 7479/84).
[196] Bravo-Zhivotovskii, D. A.; Pigarev, S. D.; Voronkov, M. G.; Vyazankin, N. S. (Zh. Obshch. Khim. **59** [1989] 863/5; J. Gen. Chem. USSR [Engl. Transl.] **59** [1989] 761/3).
[197] Gordetsov, A. S.; Cherepennikova, T. N.; Kozina, A. P.; Karlik, V. M.; Tikhonova, T. N.; Dergunov, Yu. I. (Zh. Obshch. Khim. **59** [1989] 1595/600; J. Gen. Chem. USSR [Engl. Transl.] **59** [1989] 1418/22).
[198] Kaushal, P.; Roberts, B. P. (J. Chem. Soc. Perkin Trans. II **1989** 1559/68).
[199] Pigarev, S. D.; Bravo-Zhivotovskii, D. A.; Kalikhman, I. D.; Vyazankin, N. S.; Voronkov, M. G. (J. Organomet. Chem. **369** [1989] 29/41).
[200] Rybin, L. I.; Gendin, D. V.; Vyazankina, O. A.; Vyazankin, N. S. (Metalloorg. Khim. **2** [1989] 334/6; Organomet. Chem. USSR [Engl. Transl.] **2** [1989] 159/61).

[201] Clark, K. B.; Griller, D. (Organometallics **10** [1991] 746/50).
[202] Okano, M.; Mochida, K. (Bull. Chem. Soc. Jpn. **64** [1991] 1381/2).
[203] Satgé, J.; Lesbre, M. (Bull. Soc. Chim. Fr. **1962** 703).
[204] Seyferth, D.; Andrews, S. B.; Simmons, H. D., Jr. (J. Organomet. Chem. **17** [1969] 9/15).
[205] Lesbre, M.; Mazerolles, P.; Satgé, J. (The Organic Compounds of Germanium, Wiley, London-New York-Sydney-Toronto 1971, p. 301).
[206] Cooper, J.; Hudson, A.; Jackson, R. A. (J. Chem. Soc. Perkin Trans. II **1973** 1933/7).
[207] Gladyshev, E. N.; Fedorova, E. A.; Vyazankin, N. S.; Razuvaev, G. A. (Zh. Obshch. Khim. **43** [1973] 1315/9; J. Gen. Chem. USSR [Engl. Transl.] **43** [1973] 1306/10).
[208] Bochkarev, M. N.; Maiorova, L. P.; Korneva, S. P.; Bochkarev, L. N.; Vyazankin, N. S. (J. Organomet. Chem. **73** [1974] 229/36).
[209] Bochkarev, M. N.; Razuvaev, G. A.; Vyazankin, N. S. (Izv. Akad. Nauk SSSR Ser. Khim. **1975** 1820/5; Bull. Acad. Sci. USSR Div. Chem. Sci. [Engl. Transl.] **1975** 1701/5).
[210] Bochkarev, M. N.; Gur'ev, N. I.; Razuvaev, G. A. (J. Organomet. Chem. **162** [1978] 289/95).

[211] Belov, S. P.; Krupnov, A. F.; Markov, V. N.; Mel'nikov, A. A.; Skvortsov, V. A.; Tret'yakov, M. Yu. (J. Molec. Spectrosc. **101** [1983] 258/70).
[212] Abakumov, G. A.; Cherkasov, V. K.; Nevodchikov, V. I.; Razuvaev, G. A. (Dokl. Akad. Nauk SSSR **282** [1985] 1402/5; Dokl. Phys. Chem. [Engl. Transl.] **280/285** [1985] 579/81).
[213] Skvortsov, V. A. (Izv. Vyssh. Uchebn. Zaved. Radiofiz. **33** [1990] 1349/56; Radiophys. Quantum Electron. [Engl. Transl.] **33** [1990] 991/7).
[214] Lukevics, E.; Barabanov, D. I.; Ignatovich, L. M. (Appl. Organomet. Chem. **5** [1991] 379/83).
[215] Lyashchenko, G. S.; Medvedeva, A. S.; Safronova, L. P.; Bannikova, O. B.; Voronkov, M. G. (Izv. Akad. Nauk SSSR Ser. Khim. **1991** 2889/91; Bull. Acad. Sci. USSR Div. Chem. Sci. [Engl. Transl.] **1991** 2520/3).

[216] Lukevics, E.; Ignatovich, L.; Shilina, N.; Germane, S. (Appl. Organomet. Chem. **6** [1992] 261/6).
[217] Rakhimov, A. I.; Gnatyuk, V. P.; Putilin, V. A. (Metalloorg. Khim. **5** [1992] 467/9; Organomet. Chem. USSR [Engl. Transl.] **5** [1992] 224/6).
[218] Sakaguchi, Y.; Hayashi, H.; I'Haya, Y. J. (J. Photochem. Photobiol. A **65** [1992] 183/90).
[219] Dirnens, V.; Barabanov, D. I.; Liepins, E.; Ignatovich, L. M.; Lukevics, E. (J. Organomet. Chem. **435** [1992] 257/63).
[220] Pola, J.; Parsons, J. P.; Taylor, R. (J. Mater. Chem. **2** [1992] 1289/92).
[221] Boukherroub, R.; Manuel, G. (J. Organomet. Chem. **460** [1993] 155/61).
[222] Wolfsberger, W. (J. Prakt. Chem. **334** [1992] 453/64).

1.3.1.1.2.4 Applications

The activity of an aluminomolybdenum catalyst used for the disproportionation of 1-heptene at 60 °C was increased by impregnating the catalyst with 2% $Ge(C_2H_5)_3H$ [3].

C_2H_4 was polymerized with a $TiCl_4$-$Ge(C_2H_5)_3H$ catalyst in C_7H_{16} at room temperature [1]. Effective copolymerization constants for the copolymerization of 2-methyl-5-vinylpyridin and vinyl acetate in the presence of AIBN as the initiator and $Ge(C_2H_5)_3H$ as the regulator have been reported [6]. A mixture of $Ge(C_2H_5)_3H$ and $B(C_2H_5)_3$ effects the quantitative isomerization of Z-$Ge(C_2H_5)_3CH{=}CHSi(CH_3)_3$ to the E-isomer in C_6H_6 at 60 °C [5].

The thermal decomposition of poly(vinyl chloride) was delayed, and the formation of a crosslinked polymer as well as a color change of the polymer were prevented by adding $Ge(C_2H_5)_3H$ [2]. The kinetics of the initial stage of the thermal dehydrochlorination was examined on chlorinated poly(vinyl chloride) samples and on poly(vinyl chloride) samples treated at 150 °C with $Ge(C_2H_5)_3H$ or $Sn(C_4H_9)_2Cl_2$-$Ge(C_2H_5)_3H$ for 4 h. In the latter case the initial rate of dehydrochlorination under vacuum decreased considerably [4].

References:

[1] Shearer, N. H.; Coover, H. W., Jr.; Eastman Kodak Co. (U. S. 2925409 [1960]; C.A. **1960** 13732).
[2] Myakov, V. N.; Troitskii, B. B.; Razuvaev, G. A. (Sint. Issled. Eff. Khim. Polim. Mater. No. 3 [1969] 184/91; C.A. **75** [1971] No. 77460).
[3] Bashkirov, A. N.; Fridman, R. A.; Nosakova, S. M.; Liberov, L. G.; Babich, E. D.; Vdovin, V. M. (Kinet. Katal. **16** [1975] 1353; Kinet. Catal. [Engl. Transl.] **16** [1975] 1180).
[4] Troitskii, B. B.; Troitskaya, L. S.; Denisova, V. N.; Luzinova, Z. B. (Polym. J. [Tokyo] **10** [1978] 377/85; C.A. **89** [1978] No. 147370).
[5] Ichinose, Y.; Nozaki, K.; Wakamatsu, K.; Oshima, K.; Utimoto, K. (Tetrahedron Lett. **28** [1987] 3709/12).
[6] Semchikov, Yu. D.; Gromov, V. F.; Teleshov, E. N. (Vysokomol. Soedin. A **33** [1991] 1428/41; Polym. Sci. USSR [Engl. Transl.] **33** [1991] 1322/35).

1.3.1.1.3 Tripropylgermane, Tripropylgermanium Hydride, $Ge(C_3H_7)_3H$

$Ge(C_3H_7)_3H$ was obtained with low yield along with $Ge(C_3H_7)H_3$ by reducing with $LiAlH_4$ a mixture of propylchlorogermanes which had been formed from $GeCl_4$ and LiC_3H_7 (1.1 : 1 mole ratio) in isopropyl ether [1]; see also [4, 9].

Chromatographic characteristics of $Ge(C_3H_7)_3H$ in GLC were described in connection with data of several alkyl derivatives of Si and Ge [9].

The compound is a colorless liquid, b.p. 65 °C/20 Torr, 183 °C/742 Torr, and distilling without decomposition at atmospheric pressure [1]; d^{20} = 0.9694 g/cm^3 and n_D^{20} = 1.4441 [4]. It could be stored in a closed vial for three months without change [1].

IR spectrum (neat): ν(GeH) 2003 cm^{-1} [6, 16]; see also [3] (2008 cm^{-1} was incorrectly cited in [18, 19, 32]). For a relationship between ν(GeH) and the sum of the Taft σ^* constants of the three C_3H_7 groups, see [6, 16, 18, 19].

The monoisotopic mass spectrum (70 eV) shows fragmentation with elimination of olefins (C_3H_6, C_2H_4). It was reported together with the mass spectra of several other ER_3H compounds (E = Si, Ge, or Sn), whose main reactions and the correlations between the molecular properties and mass spectrometric ion abundances were discussed [28].

Several hydrogermylation reactions of $Ge(C_3H_7)_3H$ with ethylenic or acetylenic compounds have been reported. The results are summarized in Table 11; "cat." stands for the H_2PtCl_6 catalyst, preferentially used unless otherwise stated. A review on hydrogermylations has recently been published [36].

Table 11
Hydrogermylation Reactions of $Ge(C_3H_7)_3H$.

reactant	conditions (cat. = H_2PtCl_6) → products (yield)	Ref.
reactants with a C=C bond		
$CH_2{=}CHOCH_2\overline{CHCH_2O}$	heating in C_6H_6, 20 h, cat. → $Ge(C_3H_7)_3CH_2CH_2OCH_2\overline{CHCH_2O}$ (62%)	[24]
$CH_2{=}CHO(CH_2)_2OCH_2\overline{CHCH_2O}$	heating in C_6H_6, 20 h, cat. → $Ge(C_3H_7)_3CH_2CH_2O(CH_2)_2OCH_2\overline{CHCH_2O}$ (65%)	[24]
$CH_2{=}CHO(CH_2)_4OCH_2\overline{CHCH_2O}$	heating in C_6H_6, 20 h, cat. → $Ge(C_3H_7)_3CH_2CH_2O(CH_2)_4OCH_2\overline{CHCH_2O}$ (60%)	[24]
$CH_2{=}CHC_6H_5$	cat. → $Ge(C_3H_7)_3CH(CH_3)C_6H_5$ and $Ge(C_3H_7)_3CH_2CH_2C_6H_5$	[25]
$CH_2{=}CHCH_2Cl$	100 °C, 6 h, cat. → $Ge(C_3H_7)_3(CH_2)_3Cl$ (5%)	[31]
$CH_2{=}CHCH_2OH$	refluxing, 10 h, in the presence of $C_6H_5CO\text{-}OO\text{-}OCC_6H_5$ → $Ge(C_3H_7)_3(CH_2)_3OH$ (14%)	[4]; see also [2]
$CH_2{=}CHCH_2OC_4H_9$-t	100 °C, 7 h, cat. → $Ge(C_3H_7)_3(CH_2)_3OC_4H_9$-t (50%)	[31]
$CH_2{=}CHCH_2OCH_2\overline{CHCH_2O}$	refluxing in C_6H_6, 16 h, cat. → $Ge(C_3H_7)_3(CH_2)_3OCH_2\overline{CHCH_2O}$ (64%)	[17, 20]

References on p. 76

Table 11 (continued)

reactant	conditions (cat. = H_2PtCl_6) → products (yield)	Ref.
CH_2=$CHCH_2OC_6H_4OCH_3$-2	cat. → $Ge(C_3H_7)_3(CH_2)_3OC_6H_4OCH_3$-2	[27]
CH_2=$CHCOOCH_3$	100 °C, 8 h → $Ge(C_3H_7)_3CH_2CH_2COOCH_3$ (34%) and $Ge(C_3H_7)_3CH_2CH(COOCH_3)CH_2CH_2$-$COOCH_3$	[4]; see also [2]
CH_2=$CHCOOC_2H_5$	cat. → $Ge(C_3H_7)_3CH_2CH_2COOC_2H_5$	[29]
CH_2=$C(CH_3)COOCH_3$	145 to 150 °C, 90 h, cat., sealed tube → $Ge(C_3H_7)_3CH_2CH(CH_3)COOCH_3$ (60%)	[30]
CH_2=$CHCN$	150 °C, 24 h, sealed tube → $Ge(C_3H_7)_3CH_2CH_2CN$ (24%), $Ge(C_3H_7)_3CH_2CH(CN)CH_2CH_2CN$, and $Ge(C_3H_7)_3(CH_2CH(CN))_2CH_2CH_2CN$	[4]
CH_2=$CHCH_2C_6H_5$	cat. → $Ge(C_3H_7)_3CH_2CH_2CH_2C_6H_5$	[26]
CH_2=$CHCH_2COOC_2H_5$	100 to 110 °C, 8 h, sealed tube → $Ge(C_3H_7)_3(CH_2)_3COOC_2H_5$ (40%)	[31]
reactants with a C≡C bond		
CH≡COR R = C_2H_5, C_4H_9	heating for several hours → $Ge(C_3H_7)_3CH$=$CHOR$ (65%); cis isomers are the main products	[10]
$CH≡CC_{10}H_{21}$	25 °C, 1 h, cat. → $Ge(C_3H_7)_3C(=CH_2)C_{10}H_{21}$ and trans-$Ge(C_3H_7)_3CH$=$CHC_{10}H_{21}$ in a 13:87 ratio (93% total yield) Na_2PtCl_6 catalyst → $Ge(C_3H_7)_3C(=CH_2)C_{10}H_{21}$ and trans-$Ge(C_3H_7)_3CH$=$CHC_{10}H_{21}$ in a 20:80 ratio H_2PtCl_4 catalyst → $Ge(C_3H_7)_3C(=CH_2)C_{10}H_{21}$ and trans-$Ge(C_3H_7)_3CH$=$CHC_{10}H_{21}$ in a 15:85 ratio K_2PtCl_4 catalyst → trans-$Ge(C_3H_7)_3CH$=$CHC_{10}H_{21}$ and cis-$Ge(C_3H_7)_3CH$=$CHC_{10}H_{21}$ in a 68:32 ratio (78% total yield) $Pt(P(C_6H_5)_3)_2Cl_2$ catalyst → trans-$Ge(C_3H_7)_3CH$=$CHC_{10}H_{21}$ and cis-$Ge(C_3H_7)_3CH$=$CHC_{10}H_{21}$ in a 10:90 ratio (47% total yield)	[33]

Table 11 (continued)

reactant	conditions (cat. = H_2PtCl_6) → products (yield)	Ref.
	80 °C, 3 h, no catalyst → trans-$Ge(C_3H_7)_3CH{=}CHC_{10}H_{21}$ and cis-$Ge(C_3H_7)_3CH{=}CHC_{10}H_{21}$ in a 13:87 ratio (74% total yield)	
	25 °C, 5 h, in THF, $Pd(P(C_6H_5)_3)_4$ catalyst → $Ge(C_3H_7)_3C({=}CH_2)C_{10}H_{21}$ and trans-$Ge(C_3H_7)_3CH{=}CHC_{10}H_{21}$ in a 20:80 ratio (83% total yield)	[34]
	60 °C, in the presence of $B(C_2H_5)_3$ → trans-$Ge(C_3H_7)_3CH{=}CHC_{10}H_{21}$ and cis-$Ge(C_3H_7)_3CH{=}CHC_{10}H_{21}$ in a 2:1 ratio (79% total yield)	[35]
$C_2H_5C{\equiv}COR$ $R = CH_3, C_2H_5$	80 to 100 °C, 20 to 24 h → $Ge(C_3H_7)_3C(OR){=}CHC_2H_5$ and $Ge(C_3H_7)_3C(C_2H_5){=}CHOR$ in a 1.5:1 ratio (50 and 54% total yields); Ge trans to H in both products	[22]
i-$C_3H_7C{\equiv}CN(C_2H_5)_2$	100 °C, 36 h, in the presence of AIBN → $Ge(C_3H_7)_3C(C_3H_7\text{-i}){=}CHN(C_2H_5)_2$ (78%); cis and trans isomers	[21]
reactants with C≡C and C=C bonds		
$CH_2{=}CHC{\equiv}CC(CH_3)_2OCH_3$	100 °C, 6 h, cat. → $Ge(C_3H_7)_3C(CH{=}CH_2){=}CHC(CH_3)_2OCH_3$, $Ge(C_3H_7)_3C({=}CHCH{=}CH_2)C(CH_3)_2OCH_3$ and $Ge(C_3H_7)_3C({=}C{=}CHCH_3)C(CH_3)_2OCH_3$ in a 74:13:13 ratio (40% total yield)	[14]; see also [11]
$CH_2{=}CHC{\equiv}CC(CH_3)_2C_6H_4OCH_3$-4	100 °C, 6 h, cat. → $Ge(C_3H_7)_3C(CH{=}CH_2){=}CHC(CH_3)_2C_6H_4$-$OCH_3$-4 and $Ge(C_3H_7)_3C({=}C{=}CHCH_3)$-$C(CH_3)_2C_6H_4OCH_3$-4 in a 91:9 ratio (41% total yield)	[15]; see also [12]
ketenes		
$CH_2{=}C{=}O$	UV irradiation, 2.5 h → $Ge(C_3H_7)_3COCH_3$ (70%)	[13]

Reacting the compound under UV irradiation with CH_2N_2 yielded 5% $Ge(CH_3)(C_3H_7)_3$ [7]. Photochemically generated t-$C_4H_9O^{\bullet}$ abstracts hydrogen to produce the $Ge(C_3H_7)_3^{\bullet}$ radical [23].

Treating cis-$Ge(C_6H_5)_3CH{=}CHC_{10}H_{21}$ with equimolar amounts of $Ge(C_3H_7)_3H$ and $B(C_2H_5)_3$ at 60 °C yielded a 2:5 mixture of trans-$Ge(C_3H_7)_3CH{=}CHC_{10}H_{21}$ and trans-

References on p. 76

$Ge(C_6H_5)_3CH{=}CHC_{10}H_{21}$ [35]. Reacting $Ge(C_3H_7)_3H$ with $Hg(CH_2CHO)_2$ (in boiling $CHCl_3$) or $Hg(CH_2COOC_3H_7)_2$ gave $Ge(C_3H_7)_3OCH{=}CH_2$ with 60% [5] and $Ge(C_3H_7)_3CH_2COOC_3H_7$ with 41% yield, respectively [8].

References:

[1] Johnson, O. H.; Jones, L. V. (J. Org. Chem. **17** [1952] 1172/6).
[2] Lesbre, M.; Satgé, J. (C. R. Hebd. Séance Acad. Sci. **247** [1958] 471/4).
[3] Mathis-Noël, R.; Mathis, F.; Satgé, J. (Bull. Soc. Chim. Fr. **1961** 676).
[4] Satgé, J. (Ann. Chim. [Paris] [13] **6** [1961] 519/73).
[5] Baukov, Yu. I.; Lutsenko, I. F. (Zh. Obshch. Khim. **32** [1962] 3838/9; J. Gen. Chem. USSR [Engl. Transl.] **32** [1962] 3767).
[6] Mathis, R.; Satgé, J.; Mathis, F. (Spectrochim. Acta **18** [1962] 1463/72).
[7] Kramer, K. A. W.; Wright, A. N. (J. Chem. Soc. **1963** 3604/8).
[8] Lutsenko, N. F.; Baukov, Yu. I.; Khasapov, B. N. (Zh. Obshch. Khim. **33** [1963] 2724/7; J. Gen. Chem. USSR [Engl. Transl.] **33** [1963] 2653/5).
[9] Semlyen, J. A.; Walker, G. R.; Blofeld, R. E.; Phillips, C. S. G. (J. Chem. Soc. **1964** 4948/53).
[10] Kazankova, M. A.; Protsenko, N. P.; Lutsenko, I. F. (Zh. Obshch. Khim. **38** [1968] 106/8; J. Gen. Chem. USSR [Engl. Transl.] **38** [1968] 104/6).

[11] Kakhniashvili, A. I.; Ioramashvili, D. Sh.; Fedin, E. I.; Petrovskii, P. V.; Rubin, I. D. (Soobshch. Akad. Nauk Gruz. SSR **53** [1969] 573/6; C. A. **71** [1969] No. 60353).
[12] Kakhniashvili, A. I.; Ioramashvili, D. Sh.; Fedin, E. I.; Petrovskii, P. V.; Rubin, I. D. (Soobshch. Akad. Nauk Gruz. SSR **54** [1969] 337/40; C. A. **71** [1969] No. 60453).
[13] Kazankova, M. A.; Lutsenko, I. F. (Zh. Obshch. Khim. **39** [1969] 926; J. Gen. Chem. USSR [Engl. Transl.] **39** [1969] 891).
[14] Kakhniashvili, A. I.; Ioramashvili, D. Sh. (Zh. Obshch. Khim. **40** [1970] 1552/5; J. Gen. Chem. USSR [Engl. Transl.] **40** [1970] 1539/41).
[15] Kakhniashvili, A. I.; Ioramashvili, D. Sh. (Zh. Obshch. Khim. **40** [1970] 1556/9; J. Gen. Chem. USSR [Engl. Transl.] **40** [1970] 1542/5).
[16] Mathis, R.; Barthelat, M.; Mathis, F. (Spectrochim. Acta A **26** [1970] 1993/2000).
[17] Sultanov, R. A.; Aronova, L. L.; Sadykh-zade, S. I. (Zh. Obshch. Khim. **42** [1972] 1872; J. Gen. Chem. USSR [Engl. Transl.] **42** [1972] 1862).
[18] Egorochkin, A. N.; Khorshev, S. Ya.; Ostasheva, N. S.; Satgé, J.; Rivière, P.; Barrau, J.; Massol, M. (J. Organomet. Chem. **76** [1974] 29/36).
[19] Egorochkin, A. N.; Khorshev, S. Ya.; Ostasheva, N. S.; Satgé, J.; Rivière, P.; Barrau, J.; Massol, M. (Dokl. Akad. Nauk SSSR **215** [1974] 858/60; Dokl. Chem. [Engl. Transl.] **214/219** [1974] 204/6).
[20] Sadykh-zade, S. I.; Sultanov, R. A.; Aronova, L. L.; Pestunovich, V. A. (Zh. Obshch. Khim. **44** [1974] 1787/9; J. Gen. Chem. USSR [Engl. Transl.] **44** [1974] 1753/5).

[21] Kazankova, M. A.; Zverkova, T. I.; Levin, M. Z.; Lutsenko, I. F. (Zh. Obshch. Khim. **45** [1975] 73/81; J. Gen. Chem. USSR [Engl. Transl.] **45** [1975] 66/72).
[22] Kazankova, M. A.; Zverkova, T. I.; Lutsenko, I. F. (Zh. Obshch. Khim. **45** [1975] 2044/51; J. Gen. Chem. USSR [Engl. Transl.] **45** [1975] 2006/12).
[23] Sakurai, H.; Mochida, K.; Kira, M. (J. Am. Chem. Soc. **97** [1975] 929/31).
[24] Aronova, L. L.; Sadykh-zade, S. I.; Sultanov, R. A. (Azerb. Khim. Zh. **1976** No. 5, pp. 43/5; C. A. **87** [1977] No. 39613).
[25] Ioramashvili, D. Sh.; Shudra, O. S. (Soobshch. Akad. Nauk Gruz. SSR **94** [1979] 81/4; C. A. **91** [1979] No. 91727).

[26] Ioramashvili, D. Sh. (Soobshch. Akad. Nauk Gruz. SSR **97** [1980] 357/60; C. A. **93** [1980] No. 131601).
[27] Ioramashvili, D. Sh. (Izv. Akad. Nauk Gruz. SSR Ser. Khim. **6** [1980] 129/35 from C. A. **93** [1980] No. 239555).
[28] Hottmann, K. (J. Prakt. Chem. **323** [1981] 399/419).
[29] Razuvaev, G. A.; Brevnova, T. N.; Chesnokova, T. A. (Zh. Obshch. Khim. **51** [1981] 813/7; J. Gen. Chem. USSR [Engl. Transl.] **51** [1981] 672/5).
[30] Chesnokova, T. A.; Razuvaev, G. A.; Brevnova, T. N. (Zh. Obshch. Khim. **52** [1982] 2754/62; J. Gen. Chem. USSR [Engl. Transl.] **52** [1982] 2428/35).

[31] Razuvaev, G. A.; Brevnova, T. N.; Chesnokova, T. A. (Zh. Obshch. Khim. **52** [1982] 617/21; J. Gen. Chem. USSR [Engl. Transl.] **52** [1982] 541/5).
[32] McKean, D. C.; Torto, I.; Mackenzie, M. W. (Spectrochim. Acta A **39** [1983] 399/408).
[33] Oda, H.; Morizawa, Y.; Oshima, K.; Nozaki, H. (Tetrahedron Lett. **25** [1984] 3221/4).
[34] Ichinose, Y.; Oda, H.; Oshima, K.; Utimoto, K. (Bull. Chem. Soc. Jpn. **60** [1987] 3468/70).
[35] Ichinose, Y.; Nozaki, K.; Wakamatsu, K.; Oshima, K.; Utimoto, K. (Tetrahedron Lett. **28** [1987] 3709/12).
[36] Wolfsberger, W. (J. Prakt. Chem. **334** [1992] 453/64).

1.3.1.1.4 Triisopropylgermane, Triisopropylgermanium Hydride, $Ge(C_3H_7\text{-}i)_3H$

$Ge(C_3H_7\text{-}i)_3H$ is one of the main products of the reaction of $GeCl_4$ with $i\text{-}C_3H_7MgX$ (X = Cl or Br) and subsequent hydrolysis. The reaction probably proceeds via a $Ge(C_3H_7\text{-}i)_3MgX$ intermediate [1, 9, 11]. Using excess $i\text{-}C_3H_7MgCl$ in C_6H_6-ether resulted in a faster reaction and yielded more title compound (35%) when free Mg was present. Treating the reaction mixture with $LiAlH_4$ before hydrolysis increased the yield to 40% [9]. The title compound was also present in a reaction mixture obtained from $GeCl_4$, CH_3MgI, and $i\text{-}C_3H_7MgBr$ in ether, but it could not be separated from $Ge_2(CH_3)_5C_3H_7\text{-}i$ (one of the numerous products) by gas chromatography [8].

Refluxing $Ge(C_3H_7\text{-}i)_3Cl$ with Na-K in $C_6H_5CH_3$ and subsequently treating the mixture with H_2O gave the compound along with $Ge_2(C_3H_7\text{-}i)_6$ and $(Ge(C_3H_7\text{-}i)_2H)_2$ [9]. Hydrolysis of the reaction mixture obtained from $Ge(C_3H_7\text{-}i)_3M$ (M = Li or K) and C_2H_4 yielded 16 and 45% $Ge(C_3H_7\text{-}i)_3H$, respectively [15].

The title compound was one of several products generated by photolyzing compound I in $C_6H_5CH_3$ at 5°C for 18 h [19].

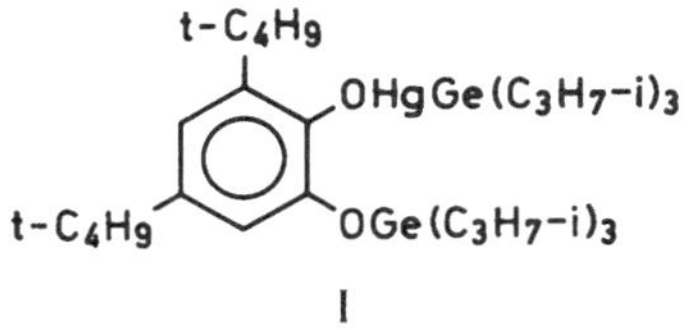

I

Boiling point: 88 to 89°C/42 Torr [11], 174°C/760 Torr [3], 176°C [9]; d^{20} = 0.975 [11] or 0.9770 g/cm^3 [3], n_D^{20} = 1.4500 [11] or 1.4505 [3].

^{1}H NMR spectrum: δ(ppm) = 1.2 (d's, CH_3), 3.64 [9] or 3.55 (GeH) [5]; the CH multiplets overlap the doublets on the low-field side [9]. IR spectrum: ν(GeH) 1989(s), δ($CH(CH_3)_2$) 1211(w) [7], δ(GeH) 710(s), ν_{as}(GeC) 565(m), ν_s(GeC) 556(sh,m) cm^{-1}; other bands at

522(w), 420(s), 313(s), and 239(w) cm^{-1} [7, 9]; other ν(GeH) assignments are 1994 (no details) [5], 1989 (neat) [4, 14] (see also [2, 3, 11]), 1988 (in C_7H_{16}) [4, 6], and 1985 (in CCl_4) cm^{-1} [4] (erroneous value in [9, 13]). Two ν(GeH) bands of conformers were predicted [20] on the basis of substituent effects [21]. The relationship between the sum of the Taft σ^* constants of the three i-C_3H_7 groups and δ(GeH) [5], ν(GeH) [4, 5, 16, 17, 21], and the integrated absorption intensity A(GeH) [6] was discussed in several papers.

Reacting the $Ge(C_3H_7\text{-}i)_3H$ with Se at 200 °C for 15 h gave $Ge(C_3H_7\text{-}i)_3SeH$ with 67% yield [10]. Interaction with $C_6H_5SCH{=}C(C_6H_5)_2$ in the presence of AIBN for 8 h yielded 28% $Ge(C_3H_7\text{-}i)_3CH{=}C(C_6H_5)_2$ [22]. $Hg(Ge(C_3H_7\text{-}i)_3)_2$ was obtained along with C_2H_6 by treating the compound with $Hg(C_2H_5)_2$ at 115 to 125 °C for 3 h in the absence of O_2 [12]. Photochemically generated $t\text{-}C_4H_9O^{\bullet}$ converted $Ge(C_3H_7\text{-}i)_3H$ into the $(C_3H_7\text{-}i)_3Ge^{\bullet}$ radical [18].

$Ge(C_3H_7\text{-}i)_3H$ did not react with $Ir(CO)(P(C_6H_5)_3)_2Cl$, even after prolonged refluxing in C_6H_6 [13].

$Ge(C_3H_7\text{-}i)_3D$. IR spectrum: ν(GeD) 1433(s) [7, 9], $\delta(CH(CH_3)_2)$ 1224(w) [7], $\nu_{as}(GeC)$ 564(s), $\nu_s'(GeC)$ 542(s), δ(GeD) 508(s) cm^{-1}; other bands at 420(s) and 311(s) cm^{-1} [7, 9].

The deposition of Ge films uncontaminated by C has been achieved, using an ArF laser, by photolysis of a $C_{12}H_{26}$ solution of $Ge(C_3H_7\text{-}i)_3H$. C_3H_6 was identified as the gaseous product [23].

References:

[1] Mazerolles, P. (Diss. Toulouse 1959 from Lesbre, M.; Mazerolles, P.; Satgé, J.; The Organic Compounds of Germanium, London-New York-Sydney-Toronto 1971, p. 262; also cited in [9]).
[2] Mathis-Noël, R.; Mathis, F.; Satgé, J. (Bull. Soc. Chim. Fr. **1961** 676).
[3] Satgé, J. (Ann. Chim. [Paris] [13] **6** [1961] 519/73).
[4] Mathis, R.; Satgé, J.; Mathis, F. (Spectrochim. Acta **18** [1962] 1463/72).
[5] Egorochkin, A. N.; Khidekel', M. L.; Ponomarenko, V. A.; Zueva, G. Ya.; Svirezheva, S. S.; Razuvaev, G. A. (Izv. Akad. Nauk SSSR Ser. Khim. **1963** 1865/8; Bull. Acad. Sci. USSR Div. Chem. Sci. [Engl. Transl.] **1963** 1717/9).
[6] Mathis, R.; Constant, M.; Satgé, J.; Mathis, F. (Spectrochim. Acta **20** [1964] 515/21).
[7] Cross, R. J.; Glockling, F. (J. Organomet. Chem. **3** [1965] 146/55).
[8] Semlyen, J. A.; Walker, G. R.; Phillips, C. S. G. (J. Chem. Soc. **1965** 1197/203).
[9] Carrick, A.; Glockling, F. (J. Chem. Soc. A **1966** 623/9).
[10] Vyazankin, N. S.; Bochkarev, M. N.; Sanina, L. P. (Zh. Obshch. Khim. **37** [1967] 1037/40; J. Gen. Chem. USSR [Engl. Transl.] **37** [1967] 980/3).
[11] Mendelsohn, J.-C.; Métras, F.; Lahournère, J.-C.; Valade, J. (J. Organomet. Chem. **12** [1968] 327/40).
[12] Glushakova, V. N.; Aleksandrov, Yu. A.; Anfilov, N. V. (Zh. Obshch. Khim. **39** [1969] 1896; J. Gen. Chem. USSR [Engl. Transl.] **39** [1969] 1859).
[13] Glockling, F.; Wilbey, M. D. (J. Chem. Soc. A **1970** 1675/81).
[14] Mathis, R.; Barthelat, M.; Mathis, F. (Spectrochim. Acta A **26** [1970] 1993/2000).
[15] Gladyshev, E. N.; Fedorova, E. A.; Vyazankin, N. S.; Razuvaev, G. A. (Zh. Obshch. Khim. **43** [1973] 1315/9; J. Gen. Chem. USSR [Engl. Transl.] **43** [1973] 1306/10).
[16] Egorochkin, A. N.; Khorshev, S. Ya.; Ostasheva, N S.; Satgé, J.; Rivière, P.; Barrau, J.; Massol, M. (J. Organomet. Chem. **76** [1974] 29/36).
[17] Egorochkin, A. N.; Khorshev, S. Ya.; Ostasheva, N. S.; Satgé, J.; Rivière, P.; Barrau, J.; Massol, M. (Dokl. Akad. Nauk SSSR **215** [1974] 858/60; Dokl. Chem. [Engl. Transl.] **214/219** [1974] 204/6).

[18] Sakurai, H.; Mochida, K.; Kira, M. (J. Am. Chem. Soc. **97** [1975] 929/31).
[19] Bayushkin, P. Ya.; Gladyshev, E. N.; Razuvaev, G. A.; Emelina, E. A. (Izv. Akad. Nauk SSSR Ser. Khim. **1976** 447/9 ; Bull. Acad. Sci. USSR Div. Chem. Sci. [Engl. Transl.] **1976** 431/3).
[20] McKean, D. C.; Torto, I.; Mackenzie, M. W.; Morrisson, A. R. (Spectrochim. Acta A **39** [1983] 387/98).
[21] McKean, D. C.; Torto, I.; Mackenzie, M. W. (Spectrochim. Acta A **39** [1983] 399/408).
[22] Ichinose, Y.; Oshima, K.; Utimoto, K. (Chem. Lett. **1988** 669/72).
[23] Pola, J.; Parsons, J. P.; Taylor, R. (J. Mater. Chem. **2** [1992] 1289/92).

1.3.1.1.5 Tributylgermane, Tributylgermanium Hydride, $Ge(C_4H_9)_3H$

1.3.1.1.5.1 Preparation and Formation

The compound was obtained with high yields by reducing $Ge(C_4H_9)_3Cl$ (91% yield) [8], $Ge(C_4H_9)_3Br$ [1, 10], $Ge(C_4H_9)_3I$ (86%) [5], or $Ge(C_4H_9)_3OC_6H_{11}$-c (ca. 100%) [3] with $LiAlH_4$ [1, 3, 8] in ether [1, 3, 5] or in C_6H_6 containing $[N(C_2H_5)_3CH_2C_6H_5]Cl$ [10]. It was quantitatively formed along with smaller amounts of $Ge(C_6H_5)_3H$ and $Li[(Ge(C_6H_5)_3)AlH_3]$ by cleaving $(C_4H_9)_3GeGe(C_6H_5)_3$ with excess $LiAlH_4$ in THF at room temperature for 1 d [7]. Refluxing $Ge(C_4H_9)_3I$ and $NaBH_4$ in THF for 4 h gave a 91% yield [2].

$Ge(C_4H_9)_3H$ was prepared with 75 to 83% yield from $K_2[Ge(1,2\text{-}O_2C_6H_4)_3]$ and excess C_4H_9MgBr in refluxing ether for 4 h in the presence of $Ti(C_5H_5)_2Cl_2$ [9]. Treatment of $Li_2[Ge(1,2\text{-}O_2C_6H_4)_3]$ with C_4H_9MgBr in ether or THF followed by reduction with $LiAlH_4$ gave mixtures of $Ge(C_4H_9)_4$ (as the main product) and $Ge(C_4H_9)_3H$ [11]. A 58% yield was obtained from $K_2[Ge(\text{-}OCH(CH_3)CH(CH_3)O\text{-})_3]$ and C_4H_9MgBr in ether followed by treatment with $LiAlH_4$ [9]; similar procedure used in [11].

$Ge(C_4H_9)_3H$ was a by-product (10 to 17% yield) when $Ge(C_4H_9)_3CH_2CH_2CN$ or $Ge(C_4H_9)_3CH_2CH_2COOCH_3$ were converted with $LiAlH_4$ in refluxing ether into $Ge(C_4H_9)_3CH_2CH_2CH_2NH_2$ or $Ge(C_4H_9)_3CH_2CH_2CH_2OH$, possibly due to the presence of branched components in the starting material [2].

For chromatographic characteristics of $Ge(C_4H_9)_3H$ in GLC in the context of various alkyl derivatives of Si and Ge, see [4].

$Ge(C_4H_9)_3D$ was synthesized by reducing $Ge(C_4H_9)_3Br$ with $LiAlD_4$ [6].

References:

[1] Lesbre, M.; Satgé, J. (C. R. Hebd. Séances Acad. Sci. **247** [1958] 471/4).
[2] Satgé, J. (Ann. Chim. [Paris] [13] **6** [1961] 519/73).
[3] Lesbre, M.; Satgé, J. (C. R. Hebd. Séances Acad. Sci. **254** [1962] 1453/5).
[4] Semlyen, J. A.; Walker, G. R.; Blofeld, R. E.; Phillips, C. S. G. (J. Chem. Soc. **1964** 4948/53).
[5] Nesmeyanov, A. N.; Borisov, A. E. (Dokl. Akad. Nauk SSSR **174** [1967] 96/9; Dokl. Chem. [Engl. Transl.] **172/177** [1967] 424/7).
[6] Corriu, R. J. P.; Moreau, J. J. E. (J. Organomet. Chem. **40** [1972] 73/96).
[7] Duffaut, N.; Dunogues, J.; Calas, R.; Rivière, P.; Satgé, J.; Cazes, A. (J. Organomet. Chem. **149** [1978] 57/63).
[8] Stang, P. J.; White, M. R. (J. Am. Chem. Soc. **103** [1981] 5429/33).
[9] Cerveau, G.; Chult, C.; Corriu, R. J. P.; Reyé, C. (Organometallics **7** [1988] 786/7).

[10] Wilt, J. W.; Lusztyk, J.; Peeran, M.; Ingold, K. U. (J. Am. Chem. Soc. **110** [1988] 281/7).

[11] Cerveau, G.; Chuit, C.; Corriu, R. J. P.; Reyé, C. (Organometallics **10** [1991] 1510/5).

1.3.1.1.5.2 The Molecule, Spectra, and Physical Properties

The strength of the germanium-hydrogen bond was estimated to be D(Ge-H) ≈ 84 kcal/mol [6]; D(Ge-H) = 82.6 ± 0.6 kcal/mol was obtained in solution (C_6H_6 or iso-C_8H_{18}) at 23 °C by a photoacoustic technique [18]. The photo-electron spectrum has the lowest ionization potentials at 9.37 (e, GeC_3) and 10.2 (a_1, GeH) eV [13].

^{1}H NMR spectrum: δ(ppm) = 3.67 (sept, GeH, J ≈ 3 Hz) [17]; 1.17 (m, C_4H_9), and 3.75 (m, GeH) [18] in $CDCl_3$; see also [12].

IR spectrum (wavenumber in cm^{-1}): The ν(GeH) band was found at 2000 to 2005 for the neat liquid [3, 7, 17] ; 1998 to 1999 in polar solvents such as dioxane, CH_3COCH_3, CH_3CN, HMPT, or pyridine [8]; 2000 to 2002 in C_6H_6, CS_2, $CHCl_3$, or CCl_4 [8]; 2004 cm^{-1} in ether, 2005 in Freon 113, 2008 in C_6H_{14} [8] or C_7H_{16} [3, 4, 8] (complete IR spectrum displayed in [2]). Nonconsistent values (1980 and 2013 cm^{-1}, solvent not given) were reported in [11, 12]. The compound was part of studies correlating the GeH bond frequency of various organogermanium compounds [3, 7, 8, 9, 10] and their integrated absorption intensity [4, 11] with the nature of the substituents [3, 4, 7, 9, 10, 11]; see also [15].

The following boiling points were reported: 61 to 63 °C/0.3 Torr [14], 80 °C/0.1 Torr [17], 117 to 118 °C/20 Torr [5], 121 °C/19 Torr [2], 123 to 125 °C/20 Torr [1, 2, 16], and 232 to 233 °C/760 Torr [2].

The compound is a liquid with d^{20} = 0.9455 [2] or 0.9490 g/cm^3 [1]; n_D^{20} is in the range 1.4502 to 1.4520 [1, 2, 5].

References:

[1] Lesbre, M.; Satgé, J. (C. R. Hebd. Séances Acad. Sci. **247** [1958] 471/4).
[2] Satgé, J. (Ann. Chim. [Paris] [13] **6** [1961] 519/73).
[3] Mathis, R.; Satgé, J.; Mathis, F. (Spectrochim. Acta **18** [1962] 1463/72).
[4] Mathis, R.; Constant, M.; Satgé, J.; Mathis, F. (Spectrochim. Acta **20** [1964] 515/21).
[5] Nesmeyanov, A. N.; Borisov, A. E. (Dokl. Akad. Nauk SSSR **174** [1967] 96/9; Dokl. Chem. [Engl. Transl.] **172/177** [1967] 424/7).
[6] Carlsson, D. J.; Ingold, K. U.; Bray, L. C. (Int. J. Chem. Kinet. **1** [1969] 315/23).
[7] Mathis, R.; Barthelat, M.; Mathis, F. (Spectrochim. Acta A **26** [1970] 1993/2000).
[8] Mathis, R.; Barthelat, M.; Mathis, F. (Spectrochim. Acta A **26** [1970] 2001/5).
[9] Egorochkin, A. N.; Khorshev, S. Ya.; Ostasheva, N. S.; Satgé, J.; Rivière, P.; Barrau, J.; Massol, M. (J. Organomet. Chem. **76** [1974] 29/36).
[10] Egorochkin, A. N.; Khorshev, S. Ya.; Ostasheva, N. S.; Satgé, J.; Rivière, P.; Barrau, J.; Massol, M. (Dokl. Akad. Nauk SSSR **215** [1974] 858/60; Dokl. Chem. [Engl. Transl.] **214/219** [1974] 204/6).
[11] Egorochkin, A. N.; Sevast'yanova, E. I.; Khorshev, S. Ya.; Ratushnaya, S. Kh.; Satgé, J.; Rivière, P.; Barrau, J.; Richelme, S. (J. Organomet. Chem. **162** [1978] 25/35).
[12] Wong, C. L.; Klingler, R. J.; Kochi, J. K. (Inorg. Chem. **19** [1980] 423/30).
[13] Gin, A.; Weiner, M. A.; Wong, C. L.; Kochi, J. K. (unpublished results from [12, p. 425]).
[14] Stang, P. J.; White, M. R. (J. Am. Chem. Soc. **103** [1981] 5429/33).

[15] McKean, D. C.; Torto, I.; Mackenzie, M. W. (Spectrochim. Acta A **39** [1983] 399/408).
[16] Cerveau, G.; Chult, C.; Corriu, R. J. P.; Reyé, C. (Organometallics **7** [1988] 786/7).
[17] Wilt, J. W.; Lusztyk, J.; Peeran, M.; Ingold, K. U. (J. Am. Chem. Soc. **110** [1988] 281/7).
[18] Clark, K. B.; Griller, D. (Organometallics **10** [1991] 746/50).

1.3.1.1.5.3 Chemical Behavior

Mass Spectrum. The m/e values and their intensities are listed in [66]. The monoisotopic mass spectrum at 70 eV showed fragmentation with elimination of olefins (C_4H_8, C_3H_6, C_2H_4). It was reported together with the mass spectra of several other ER_3H compounds (E = Si, Ge, or Sn), whose main reactions and correlations between the molecular properties and mass-spectrometric ion abundances were discussed [48].

Reactions with Elements and Inorganic Compounds. Low-temperature photolysis of $Ge(C_4H_9)_3H$ in CH_3CN in the presence of I_2 or $[IrCl_6]^{2-}$ produced hydrogen atoms [41]. $Ge(C_4H_9)_3H$ in DME is quantitatively cleaved by NaH at 40 °C or KH at room temperature to yield $Ge(C_4H_9)_3Na$ and $Ge(C_4H_9)_3K$, respectively [40]. The reaction with $AlCl_3$ at 130 °C for 6 h gave $Ge(C_4H_9)_3Cl$ (75% yield), Al, and H_2 [4, 5]. $Ge(C_4H_9)_3F$ was formed with GeF_2 [30, 32].

Oxidative cleavage of $Ge(C_4H_9)_3H$ by $Fe(L)_3(ClO_4)_3$ compounds in CH_3CN at 25 °C yields $[Ge(C_2H_5)_3]^+$, $[Fe(L)_3]^{2+}$, and H^+ (L = 1,10-phenanthroline, its derivatives $C_{12}H_7N_2Cl$-5, $C_{12}H_7N_2NO_2$-5, and $C_{12}H_6N_2(C_6H_5)_2$-4,7, or 2,2′-bipyridine); the mechanism of this reaction was studied in [47].

Hydrogermylation Reactions. $Ge(C_4H_9)_3H$ was involved in NMR- and IR-spectroscopic studies of several GeR_3H and SiR_3H compounds to examine the effects of reagent [37, 42], nature and amount of catalyst [42], mixing order, temperature [37, 42], and solvent [37] on the hydrogermylation and hydrosilylation of $C_6H_5CH{=}CH_2$ [37] and $C_6H_5CH_2CH{=}CH_2$ [42]. It was included in a study of the mechanism of hydrogen transfer from ER_3H compounds (E = Si, Ge, and Sn) to $(NC)_2C{=}C(CN)_2$ which leads to characteristic, but labile 1 : 1 adducts of uncertain structure [38].

The γ-radiation-initiated hydrogermylation of unsymmetrical olefins proceeds regioselectively with formation of the β-adducts $Ge(C_4H_9)_3CH_2CH_2R$ regardless of the polarization of the double bond [49]. Hydrogermylation of 2-cyclohexen-1-one was not observed, even at high olefin concentration in the presence of $C_{11}H_{23}I$ [62].

The main hydrogermylation reactions are listed in Table 12. A review on hydrogermylations has recently been published [69].

Table 12
Hydrogermylation Reactions with $Ge(C_4H_9)_3H$.

reactant	conditions (cat. = H_2PtCl_6) → products (yield)	Ref.
reactants with a C=C bond		
$CH_2{=}CHOC_2H_5$	γ radiation → $Ge(C_4H_9)_3CH_2CH_2OC_2H_5$	[49]

Table 12 (continued)

reactant	conditions (cat. = H_2PtCl_6) → products (yield)	Ref.
$CH_2{=}CHOC_4H_9$	130°C, 8 h → $Ge(C_4H_9)_3CH_2CH_2OC_4H_9$ (72%) γ radiation, 25°C → $Ge(C_4H_9)_3CH_2CH_2OC_4H_9$ (94%)	[5] [39, 44]
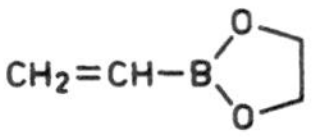	100°C, 2 h, sealed tube → 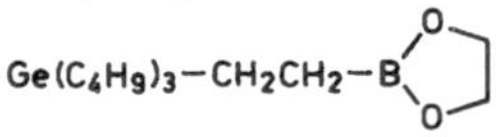 (95%) 100°C, 5 h, sealed tube, $C_6H_4(OH)_2$-1,4 → 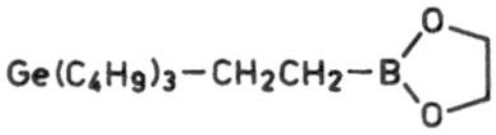 (71%)	[13]
$CH_2{=}CHC_6H_5$	cat. → $Ge(C_4H_9)_3CH(CH_3)C_6H_5$ and $Ge(C_4H_9)_3CH_2CH_2C_6H_5$ γ radiation, 25°C → $Ge(C_4H_9)_3CH_2CH_2C_6H_5$ (46 to 50%)	[37] [39, 44]
$CH_2{=}CHCH_2OH$	reflux for 15 h, Pt-asbestos catalyst → $Ge(C_4H_9)_3CH_2CH_2CH_2OH$ (57%) γ radiation, 25°C → $Ge(C_4H_9)_3CH_2CH_2CH_2OH$ (76%)	[5]; see also [1] [39, 44, 49]
$CH_2{=}C(CH_3)CH_2OH$	120 to 145°C, 10 h, cat. → $Ge(C_4H_9)_3CH_2CH(CH_3)CH_2OH$ (47%) and small amount of H_2	[12]
$CH_2{=}CHCH_2OC_6H_4OCH_3$-2	cat. → $Ge(C_4H_9)_3(CH_2)_3OC_6H_4OCH_3$-2	[43]
$CH_2{=}CHCHO$	90°C, 10 h, Pt-asbestos catalyst and $C_6H_4(OH)_2$-1,4 inhibitor → $Ge(C_4H_9)_3CH_2CH_2CHO$ (60%)	[5]; see also [1]
$CH_2{=}CHCOOCH_3$	100°C, 8 h → $Ge(C_4H_9)_3CH_2CH_2$-$COOCH_3$ (40%) and $Ge(C_4H_9)_3CH(CH_3)COOCH_3$ (?) γ radiation, 25°C → $Ge(C_4H_9)_3CH_2CH_2COOCH_3$ (92%)	[5]; see also [25, p. 294] [44, 49]
$CH_2{=}C(CH_3)COOCH_3$	γ radiation, 25°C → $Ge(C_4H_9)_3CH_2CH(CH_3)COOCH_3$ (92%) 25°C, 60 h, no γ radiation → $Ge(C_4H_9)_3CH_2CH(CH_3)COOCH_3$ (19%)	[39, 44, 49] [39]
$CH_2{=}CHCOOC_2H_5$	100°C, 10 h, Pt-asbestos catalyst and $C_6H_4(OH)_2$-1,4 inhibitor	[5]

Table 12 (continued)

reactant	conditions (cat. = H_2PtCl_6) → products (yield)	Ref.
	→ $Ge(C_4H_9)_3CH_2CH_2COOC_2H_5$ (34%) and $Ge(C_4H_9)_3CH_2CH(COOC_2H_5)$-$CH_2CH_2COOC_2H_5$	
	refluxing, no catalyst → $Ge(C_4H_9)_3CH_2CH_2COOC_2H_5$	[1]
CH_2=$CHCH_2SH$	reflux for 15 h, cat. → $Ge(C_4H_9)_3CH_2CH_2CH_2SH$ (30%)	[5]
CH_2=$CHCH_2NH_2$	80 to 130°C (bath), 14 h, cat. → $Ge(C_4H_9)_3CH_2CH_2CH_2NH_2$ (74%)	[5]
	γ radiation, 25°C → $Ge(C_4H_9)_3CH_2CH_2CH_2NH_2$ (80%)	[39, 44]
CH_2=$CHCN$	refluxing for 5 h → $Ge(C_4H_9)_3CH_2CH_2CN$ (42%), $Ge(C_4H_9)_3CH_2CH(CN)CH_2CH_2CN$, and $Ge(C_4H_9)_3CH(CH_3)CN$ (?)	[1, 5]
	γ radiation, 25°C → $Ge(C_4H_9)_3CH_2CH_2CN$ (65%); see also reactions with halohydrocarbons and acid halides, p. 86	[39, 44]
CH_2=$CHCH_2C_6H_5$	cat. → $Ge(C_4H_9)_3CH_2CH_2CH_2C_6H_5$	[42]
CH_2=CHC_4H_9	γ radiation, 25°C → $Ge(C_4H_9)_3C_6H_{13}$ (98%)	[44, 49]
CH_2=CHC_5H_{11}	γ radiation, 25°C → $Ge(C_4H_9)_3C_7H_{15}$ (86%)	[39]
CH_2=CHC_6H_{13}	γ radiation, 25°C → $Ge(C_4H_9)_3C_8H_{17}$ (64%)	[39, 44]
$CH_2CHC_7H_{15}$	80°C; reaction used for determining the the rate constant for S_{H^2} attack of the $Ge(C_4H_9)_3^{\bullet}$ radical; products not given	[59]
c-C_6H_{10}	γ radiation, 25°C → $Ge(C_4H_9)_3C_6H_{11}$-c (75%)	[39, 44]
	80°C; reaction used for determining the rate constant for S_{H^2} attack of the $Ge(C_4H_9)_3^{\bullet}$ radical	[59]
reactants with a C≡C bond		
$CH{\equiv}CSi(C_2H_5)_3$	70 to 80°C, 22 h, cat., sealed tube → $Ge(C_4H_9)_3CH$=$CHSi(C_2H_5)_3$ and $Ge(C_4H_9)_3CH$=$CHGe(C_4H_9)_3$	[18]

References on p. 89

Table 12 (continued)

reactant	conditions (cat. = H_2PtCl_6) → products (yield)	Ref.
$CH{\equiv}CGe(C_4H_9)_3$	200 °C, 10 h, cat. → $Ge(C_4H_9)_3CH{=}CHGe(C_4H_9)_3$ (90%)	[3]; see also [2, 5]
$CH{\equiv}CC_6H_5$	20 °C, 1.5 h, $Rh(P(C_6H_5)_3)_3Cl$ catalyst → cis-$Ge(C_4H_9)_3CH{=}CHC_6H_5$, trans-$Ge(C_4H_9)_3CH{=}CHC_6H_5$, and $Ge(C_4H_9)_3C(C_6H_5){=}CH_2$ in a 14:85:1 ratio (total 80%)	[27], see also [2, 5]
	20 °C, 0.5 h, cat. → cis-$Ge(C_4H_9)_3CH{=}CHC_6H_5$, trans-$Ge(C_4H_9)_3CH{=}CHC_6H_5$, and $Ge(C_4H_9)_3C(C_6H_5){=}CH_2$ in a 2:75:23 ratio (total 68%)	
	50 °C, 2 h, cis-$Pt(P(C_6H_5)_3)_2Cl_2$ catalyst → cis-$Ge(C_4H_9)_3CH{=}CHC_6H_5$, trans-$Ge(C_4H_9)_3CH{=}CHC_6H_5$, and $Ge(C_4H_9)_3C(C_6H_5){=}CH_2$ in a 7:73:20 ratio (total 59%)	
	40 °C, 1 h, in CH_2Cl_2, $Rh(hfa)(C_2H_4)_2$ cat. (hfa = $(CH(COCF_3)_2)$), minor amount of $Ge(C_4H_9)_3H$ → cis-$Ge(C_4H_9)_3CH{=}CHC_6H_5$, trans-$Ge(C_4H_9)_3CH{=}CHC_6H_5$, and $Ge(C_4H_9)_3C(C_6H_5){=}CH_2$ in a 1:4:95 ratio (total 100%); other Rh complexes also catalyze this reaction, but with less regioselectivity	[67]
	γ radiation, 25 °C → $Ge(C_4H_9)_3CH{=}CHC_6H_5$ (51%)	[39, 44]
	25 °C, 60 h, no γ radiation → $Ge(C_4H_9)_3CH{=}CHC_6H_5$ (9%)	[39]
$CH{\equiv}CCH_2OH$	100 °C, 6 h, cat. → $Ge(C_4H_9)_3CH{=}CHCH_2OH$ (80%)	[2, 5]; see also [25, p. 297]
	γ radiation, 25 °C → $Ge(C_4H_9)_3CH{=}CHCH_2OH$ (60%)	[39, 44]
	25 °C, 60 h, no γ radiation → $Ge(C_4H_9)_3CH{=}CHCH_2OH$ (11%)	[39]
$CH{\equiv}CC(OH)(CH_3)_2$	cat., exothermic → $Ge(C_4H_9)_3CH{=}CHC(OH)(CH_3)_2$ (90%)	[2, 5]; see also [25, p. 297]
$CH{\equiv}CC(OH)(CH_3)C_2H_5$	60 to 65 °C, 2 h, cat. → $Ge(C_4H_9)_3CH{=}CHC(OH)(CH_3)C_2H_5$	[16]

Table 12 (continued)

reactant	conditions (cat. = H_2PtCl_6) → products (yield)	Ref.
$CH{\equiv}CCH(OH)C_3H_7$	60 to 65 °C, 2 h, cat. → $Ge(C_4H_9)_3CH{=}CHCH(OH)C_3H_7$ (58%)	[16]
$C_2H_5C{\equiv}COCH_3$	80 to 100 °C, 20 to 24 h → $Ge(C_4H_9)_3C(OCH_3){=}CHC_2H_5$ and $Ge(C_4H_9)_3C(C_2H_5){=}CHOCH_3$ in a 1.5 : 1 ratio (75% total yield) (Ge trans to H in both products)	[31]
$C_6H_5C{\equiv}CC_6H_5$	γ radiation, 25 °C → $Ge(C_4H_9)_3C(C_6H_5){=}CHC_6H_5$ (80%)	[39]
reactants with a C≡C and a C=C bond		
$CH_2{=}CHC{\equiv}CC(CH_3)_2OCH_3$	100 °C, 6 h, cat. → $Ge(C_4H_9)_3C(CH{=}CH_2){=}CHC(CH_3)_2$-$OCH_3$, $Ge(C_4H_9)_3C({=}CHCH{=}CH_2)C$-$(CH_3)_2OCH_3$, and $Ge(C_4H_9)_3C({=}C{=}CHCH_3)C(CH_3)_2OCH_3$ in a 80 : 16 : 4 ratio (total 52%)	[23]; see also [20]
$CH_2{=}CHC{\equiv}CC(CH_3)_2C_6H_4OCH_3$-4	100 °C, 6 h, cat. → $Ge(C_4H_9)_3C(CH{=}CH_2){=}CHC(CH_3)_2$-$C_6H_4OCH_3$-4 and $Ge(C_4H_9)_3C({=}C{=}CHCH_3)C(CH_3)_2$-$C_6H_4OCH_3$-4 in a 19 : 1 ratio (total 55%)	[24]; see also [21]
reactants with two C≡C bonds		
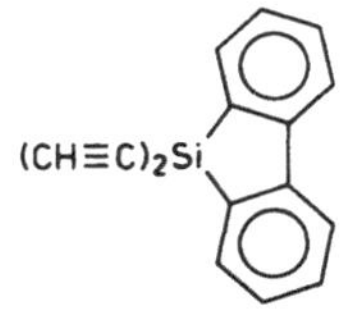	90 to 100 °C, 8 h, cat. → 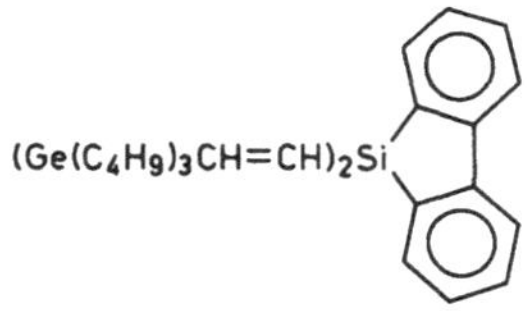 (20%)	[28]
ketenes and reactants with conjugated C=C and C=O bonds		
$CH_2{=}C{=}O$	UV radiation, 2 to 4 h → $Ge(C_4H_9)_3COCH_3$ (84%)	[22]
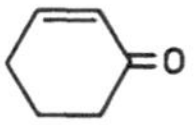	80 °C; reaction used for determining the rate constant for S_{H2} attack of the $Ge(C_4H_9)_3^{\cdot}$ radical; products not given	[59]

Reactions with Halogen-Containing Compounds. $Ge(C_4H_9)_3H$ reduces RX compounds (R = alkyl or alkenyl; X = Cl, Br, or I) to the corresponding hydrocarbons RH [4, 5, 19, 62, 63].

References on p. 89

With CH_2=$CHCH_2X$ (X = Cl or Br), no hydrogermylation was observed, even in the presence of a catalyst like Pt-asbestos or H_2PtCl_6. Acid chlorides are reduced to the corresponding aldehydes [4, 5, 64]. The reactions are listed in Table 13.

The kinetics of photo-initiated reductions of CCl_4 and CH_3I by $Ge(C_4H_9)_3H$ was studied in C_6H_{12} at 25°C with α,α'-azobiscyclohexylnitrile as the photoinitiator. The rate constants for hydrogen abstraction by $CCl_3^{\cdot}$ and $CH_3^{\cdot}$ radicals are both (1 to 2)$\times10^5$ $M^{-1} \cdot sec^{-1}$. $Sn(C_4H_9)_3H$ is a better hydrogen donor to the alkyl radical (by the factor 10 to 20) as is $Ge(C_4H_9)_3H$ [19]. Rate constants were determined for the S_{H2} attack of the $Ge(C_4H_9)_3^{\cdot}$ radical (generated from $Ge(C_4H_9)_3H$) on $C_7H_{15}Cl$, $C_8H_{17}Br$, $C_6H_5CH_2Cl$, $C_3H_7COOCH_2Cl$, and $C_2H_5OOCCH_2Cl$ at 80°C [59].

$Ge(C_4H_9)_3H$ was also used for the reductive alkylation of an activated double bond (CH_2=CHCN, 2-cyclohexen-1-one) by an alkyl halide RX in the presence of AIBN (RX = t-C_4H_9I, $C_{11}H_{23}I$, c-$C_6H_{11}I$, $C_6H_5CH_2I$; only low yields of olefin-alkyl adducts with $C_{11}H_{23}Cl$ or $C_{11}H_{23}Br$). After heating the mixture in CH_3CN or C_6H_6 at ca. 80°C for about 8 h, the title compound was largely transformed to $Ge(C_4H_9)_3X$ (X = halogen), but with a high concentration of CH_2=CHCN, hydrogermylation was also observed to yield $Ge(C_4H_9)_3CH_2CH_2CN$. The performance of the $Ge(C_4H_9)_3H$ reagent was compared to that of the more commonly employed $Sn(C_4H_9)_3H$ and of other triorganogermanium and triorganotin hydrides [62]; see also [58].

After reacting **$Ge(C_4H_9)_3D$** with CH_2=$CH(CH_2)_3C(CH_3)_2Br$ at 120°C, only ca. 70% of the CH_2=$CH(CH_2)_3CH(CH_3)_2$ produced contained a deuterium atom. Therefore, the intermediate tertiary alkyl radical must have abstracted some hydrogen from the C_9H_{20} solvent or from C_4H_9 groups of the germane. Similar experiments with CH_2=$CH(CH_2)_3CH(Br)CH_3$ revealed essentially 100% labeling of the hydrocarbon products with a one deuterium atom [61].

Table 13
Reactions of $Ge(C_4H_9)_3H$ with Halogenated Hydrocarbons and Acid Halides.
AIBN = 2,2'-azobisisobutyronitrile, DMVN = 2,2'-azobis(2,4-dimethylvaleronitrile).

reactant	conditions → products (yield)	Ref.
CCl_4	220°C, 8 h, sealed tube → $Ge(C_4H_9)_3Cl$ (90%) and $CHCl_3$ (90%)	[5]; see also [4]
	photo-initiated in C_6H_{12} → $Ge(C_4H_9)_3Cl$ and $CHCl_3$; $\Delta H = -40$ kcal/mol	[19]
CH_3I	photo-initiated in C_6H_{12} → $Ge(C_4H_9)_3I$ and CH_4; $\Delta H = -27$ kcal/mol	[19]
CF_3I	photo-initiated in C_6H_{12}; no further information given	[19]
C_3H_7I	refluxing, 3 h → $Ge(C_4H_9)_3I$ (95%), C_3H_8	[5]
t-C_4H_9Br	photo-initiated in C_6H_{12}; no further information given	[19]
$C_7H_{15}Cl$	170°C, 4 h → $Ge(C_4H_9)_3Cl$ (80%)	[4, 5]
$C_7H_{15}Br$	125°C; then 150°C, 4 h → $Ge(C_4H_9)_3Br$ (95%)	[5]

Table 13 (continued)

reactant	conditions → products (yield)	Ref.
$C_8H_{17}Br$	80 °C, 4 h, in C_6H_6, AIBN → C_8H_{18} (80%)	[59]
$C_{11}H_{23}I$	80 °C, 8 h, in C_6H_6, AIBN → $C_{11}H_{24}$ (76%)	[62]
$Si(CH_3)_3CH_2CH_2Br$	33 °C, in C_6H_6, DMVN → $Ge(C_4H_9)_3Br$ and $Si(CH_3)_3C_2H_5$	[63]
$Si(CH_3)_3CH_2CH_2CH_2Br$	33 °C, in C_6H_6, DMVN → $Ge(C_4H_9)_3Br$ and $Si(CH_3)_3C_3H_7$	[63]
$CH_2{=}CHCH_2Cl$	without catalyst → $Ge(C_4H_9)_3Cl$ and C_3H_6	[4, 5]
$CH_2{=}CHCH_2Br$	without catalyst → $Ge(C_4H_9)_3Br$ and C_3H_6	[5]
$C_3H_7COOCH_2Cl$	80 °C, 4 h, in C_6H_6, AIBN → $C_3H_7COOCH_3$ (?) (85%)	[59]
C_6H_5COCl	200 °C → $Ge(C_4H_9)_3Cl$ (70%) and C_6H_5CHO 80 °C, 1 h, in HMPT, $Pd(P(C_6H_5)_3)_4$ cat. → $Ge(C_4H_9)_3Cl$ and C_6H_5CHO (ca. 93%, contaminated with $(Ge(C_4H_9)_3)_2O$)	[4, 5] [64]
$2\text{-}FC_6H_4COCl$	80 °C, 0.5 h, in HMPT, $Pd(P(C_6H_5)_3)_4$ cat. → $Ge(C_4H_9)_3Cl$ and $2\text{-}FC_6H_4CHO$ (ca. 80%, contaminated with $(Ge(C_4H_9)_3)_2O$)	[64]
$4\text{-}NO_2C_6H_4COCl$	80 °C, 1 h, in HMPT, $Pd(P(C_6H_5)_3)_4$ cat. → $Ge(C_4H_9)_3Cl$ and $4\text{-}NO_2C_6H_4CHO$ (80%)	[64]
$2,4\text{-}Cl_2C_6H_3COCl$	80 °C, 1 h, in HMPT, $Pd(P(C_6H_5)_3)_4$ cat. → $Ge(C_4H_9)_3Cl$ and $2,4\text{-}Cl_2C_6H_3CHO$ (85%)	[64]
$C_6H_5CH{=}CHCOCl$	90 °C, 1.5 h, in HMPT, $Pd(P(C_6H_5)_3)_4$ cat., small excess of $Ge(C_4H_9)_3H$ → $Ge(C_4H_9)_3Cl$ and $C_6H_5CH{=}CHCHO$ (88%) 100 °C, 2 h, in HMPT, $Pd(P(C_6H_5)_3)_4$ cat., large excess of $Ge(C_4H_9)_3H$ → $C_6H_5CH_2CH_2CHO$ (37%)	[64]

Reactions with Alcohols. $Ge(C_4H_9)_3H$ undergoes dehydrocondensation with alcohols in the presence of Cu powder to yield alkoxygermanes: with CH_3OH (refluxing for 10 d) → $Ge(C_4H_9)_3OCH_3$ (94%) and H_2 (98%) [12]; with C_2H_5OH (refluxing for 15 h) → $Ge(C_4H_9)_3$-OC_2H_5 (90%) and H_2 (96%) [12]; with $CH_2{=}C(CH_3)CH_2OH$ (at 113 to 114 °C for 1 d) → $Ge(C_4H_9)_3OCH_2C(CH_3){=}CH_2$ (35%) and H_2 (37%) [12]; and with $CH_3(CH_2)_3C{\equiv}CCH_2CH_2OH$ (140 °C) → $Ge(C_4H_9)_3OCH_2CH_2C{\equiv}C(CH_2)_3CH_3$ (25%) [12].

References on p. 89

Reactions with Ketones. Reactions of $Ge(C_4H_9)_3H$ with ketones yield alkoxygermanes (all reactions at 150 to 160°C for ca. 2 days in the presence of Cu powder): $CH_3COC_6H_{13} \rightarrow Ge(C_4H_9)_3OCH(CH_3)C_6H_{13}$ (50%); $C_2H_5COC_4H_9 \rightarrow Ge(C_4H_9)_3OCH(C_2H_5)C_4H_9$ (50%) [7]; $CH_3COCH_2CH_2CH{=}C(CH_3)_2 \rightarrow Ge(C_4H_9)_3OCH(CH_3)CH_2CH_2CH{=}C(CH_3)_2$ (70%) [25, p. 301]; and cyclohexanone $\rightarrow Ge(C_4H_9)_3OC_6H_{11}$-c (100% or 60% without catalyst [25, p. 301] (see also [7]). UV irradiation of $Ge(C_4H_9)_3H$ and $C_6H_5COCOC_6H_5$ in t-$C_4H_9OOC_4H_9$-t gave the radical $Ge(C_4H_9)_3OC^{\bullet}(C_6H_5)COC_6H_5$ [35]. Similarly, melting a mixture of $Ge(C_4H_9)_3H$ and acenaphthoquinone yielded the radical I [36]:

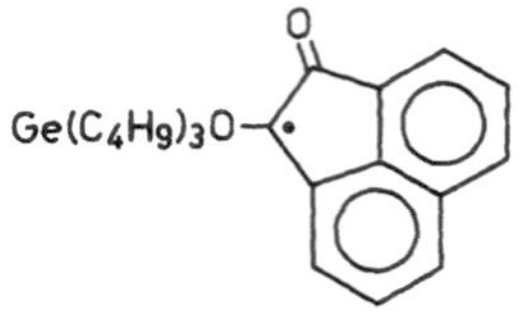

I

Reactions with Organic Acids. In the presence of Cu powder, $Ge(C_4H_9)_3H$ and CH_3COOH formed $Ge(C_4H_9)_3OOCCH_3$ and H_2 quantitatively. No reaction occurred in the absence of the catalyst [8]. Reactions with CH_3COOH, $CH_2ClCOOH$, C_6H_5COOH, and $C_6H_4(NO_2\text{-}3)COOH$ in several solvents (2-picoline, pyridine, quinoline, quinaldine) at 110°C in the presence of Cu were studied kinetically. Electron-donating solvents have a catalytic influence on these reactions [34].

Reactions with Sulfur and Selenium Compounds. Reacting $Ge(C_4H_9)_3H$ in the presence of AIBN with $C_4H_9OCH_2SC_6H_5$ in $C_6H_5CH_3$ at 110°C or with $C_4H_9COOCH_2SeC_6H_5$ in C_6H_6 at 80°C (both for 4 h) gave $C_4H_9OCH_3$ (50% yield) or $C_4H_9COOCH_3$ (55 to 95% yield) depending on the proportions of the starting materials. Rate constants were determined for the S_H2 attack of the $Ge(C_4H_9)_3^{\bullet}$ radical (generated from $Ge(C_4H_9)_3H$) on the following compounds: $C_3H_7COOCH_2SC_6H_5$, $C_4H_9COOCH_2SC_6H_4CN$-4, $C_4H_9OCH_2SC_6H_5$, C_2H_5OOC-$CH_2SC_6H_5$, $C_4H_9COOCH_2SeC_6H_5$, $C_4H_9OCH_2SeC_6H_5$, $C_2H_5OOCCH_2SeC_6H_5$, and compound II at 80°C [59]. $CH_3(CH_2)_{16}C(CH_3)_2OCH_2SCH_3$ was not reduced by various tin hydrides, but $Ge(C_4H_9)_3H$ in $C_6H_5CH_3$ at 110°C reduced it cleanly to yield $CH_3(CH_2)_{16}C(CH_3)_2OCH_3$ without fragmentation [51].

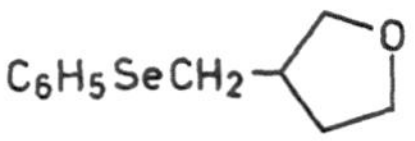

II

Attempts to use $Ge(C_4H_9)_3H$ as a reagent for the reductive alkylation of $CH_2{=}CHCN$ by $C_{11}H_{23}SC_6H_5$ or $C_{11}H_{23}SeC_6H_5$ in the presence of AIBN yielded only low amounts of $C_{13}H_{27}CN$ and $C_{11}H_{24}$ (each less than 5% yield) [62].

Reactions with Nitrogen Compounds. In the presence of Cu powder, $Ge(C_4H_9)_3H$ reacts with $CH(N_2)COOC_2H_5$ in ether, C_6H_6, or $C_6H_5CH_3$ to give $Ge(C_4H_9)_3CH_2COOC_2H_5$ with 40 to 50% yield [1, 5]. However, with CH_2N_2 under UV irradiation the yield of $Ge(CH_3)(C_4H_9)_3$ was only 2% [9].

Reactions with Carbenes, Radicals, and Ions. Slowly adding t-C_4H_9OK in DME to a solution of $CH{\equiv}CC(OSO_2CF_3){=}C(CH_3)_2$ and excess $Ge(C_4H_9)_3H$ in DME at −50°C in the presence of t-C_4H_9NO as a radical trap led to $Ge(C_4H_9)_3CH{=}C{=}C{=}C(CH_3)_2$ (40% yield) via the carbene intermediate $[(CH_3)_2C{=}C{=}C{=}C:]$ [50].

At ambient temperatures, $Ge(C_4H_9)_3H$ is less reactive than $Sn(C_4H_9)_3H$ towards radicals, e.g. ca. 20-times less reactive towards primary alkyl radicals [54] and ca. 80-times less reactive towards the secondary 1-methyl-5-hexenyl radical. The hydrogen-donating ability towards a tertiary alkyl radical is rather poor (see also the reactions with halogen-containing compounds and acid halides above) [61]. The utility of $Ge(C_4H_9)_3H$ as a hydrogen donor in special radical chain reactions involving a very slow β-scission or rearrangement was compared with that of $Si_2(CH_3)_5H$ and $Sn(C_4H_9)_3H$ [60].

The action of photochemically generated t-$C_4H_9O^{\bullet}$ on $Ge(C_4H_9)_3H$ leads to the $(C_4H_9)_3Ge^{\bullet}$ radical [33, 56], the rate constants for hydrogen abstraction being $(9.2 \pm 1.4) \times 10^7$ and $(8.0 \pm 0.6) \times 10^7$ $L \cdot mol^{-1} \cdot s^{-1}$ at 300 K (measured by different methods) [53, 66]. A rate constant was also measured for the reaction with the 5-hexenyl radical in C_8H_{18} at temperatures varying from 233 to 393 K [54]; see also [65]. Rate constants for H atom abstraction by $Si(CH_3)_3(CH_2)_n^{\bullet}$ (n = 1 to 3) were determined at ambient temperatures. The order of decreasing radical reactivity was found to be $Si(CH_3)_3CH_2^{\bullet}$ > n-alkyl$^{\bullet}$ > $Si(CH_3)_3CH_2CH_2CH_2^{\bullet}$ > $Si(CH_3)_3CH_2CH_2^{\bullet}$ [63]. Rate constants were also measured for the reactions with $CH_3^{\bullet}$ or $CCl_3^{\bullet}$ in C_6H_{12} [19], with $C_6H_5^{\bullet}$ in C_6H_6, with $(CH_3)_2C{=}CH^{\bullet}$ or c-$C_3H_5^{\bullet}$ in C_5H_{12} [57], with $(CF_3)_2NO^{\bullet}$ in Freon solvents [55], and with the 1-methyl-5-hexenyl radical [61].

The kinetics of H^- transfer from $Ge(C_4H_9)_3H$ to $[4\text{-}XC_6H_4CHC_6H_4Y\text{-}4]^+$ ions (X = Y = $N(CH_3)_2$; X = OCH_3, Y = OCH_3, OC_6H_5, CH_3, or H) was studied in CH_2Cl_2 at −70 °C. In the case of $[4\text{-}CH_3OC_6H_4CHC_6H_4OCH_3\text{-}4]^-$, the rate constant was independent of the nature of the counterion ($[CF_3SO_2O]^-$ or $[TiCl_5]^-$) [68].

Reactions with Organometallic Compounds. Isotope exchange between $Ge(C_4H_9)_3H$ and $Ge(C_6H_5)_3D$ in the presence of hydrogermylation catalysts was studied using $Rh(P(C_6H_5)_3)_3Cl$ in C_6D_6 and cis-$Pt(P(C_6H_5)_3)_2Cl_2$ or H_2PtCl_6 in $CDCl_3$ [26]. Small amounts of $Ge(C_4H_9)_3H$ turned out to be useful in transalkylations at 200 °C between germanium compounds such as $Ge(C_4H_9)_4$-$GeCl_4$, $Ge(C_4H_9)_4$-$Ge(C_4H_9)Cl_3$, $Ge(C_4H_9)_4$-$Ge(C_4H_9)Cl_3$-$GeCl_4$, $Ge(C_4H_9)_4$-$Ge(C_4H_9)_2Cl_2$, $Ge(C_4H_9)_3Cl$-$GeCl_4$, $Ge(C_4H_9)_2Cl_2$-$GeCl_4$). It was assumed that in these reactions a germanium dihalide is formed from $Ge(C_4H_9)_3H$ and $GeCl_4$ and acts as the common, catalytically active species [15].

In the reaction of $Ge(CH_2C_6H_5)_3H$ with excess LiC_4H_9, $Ge(C_4H_9)_3H$ may be an intermediate (not detected) and the precursor of $Ge(C_4H_9)_4$ which was found among the final products [11]. It did not react with $Sn(CH_3)_3N(C_2H_5)_2$ at 100 °C within 65 h [14]. UV irradiation of a mixture of $Ge(C_4H_9)_3H$ and $Si(CH_3)_2C(CH_3)_2C(CH_3)_2$ in C_5H_{12} (2 h) [46] or THF-C_6H_6 [52] yielded $Ge(C_4H_9)_3Si(CH_3)_2C(CH_3)_2CH(CH_3)_2$ (70%); heating in THF-C_6H_6 at 68 °C (18 h) only gave 28% yield [46, 52]. For reactions with $Si(CH_3)_3CH_2CH_2Br$ and $Si(CH_3)_3CH_2CH_2CH_2Br$, see Table 13. For the reaction with $Ge(C_4H_9)_3C{\equiv}CH$, see hydrogermylation in Table 12, p. 84. Treating $Sb(C_4H_9)_2C{\equiv}CH$ with $Ge(C_4H_9)_3H$ at 80 °C (23 h) and then at 97 °C (11 h) gave $Ge(C_4H_9)_3C{\equiv}CH$ and $Sb(C_4H_9)_2H$ in the first step and finally $(C_4H_9)_2SbCH{=}CHSb(C_4H_9)_2$ [17].

The following reactions of $Ge(C_4H_9)_3H$ with organometallic compounds of transition metals have been reported: $Co_3(CO)_9(\mu_3\text{-}CH)$ (in refluxing C_6H_6 for 8 h) → $Co_3(CO)_9(\mu_3\text{-}CGe(C_4H_9)_3)$ (44%) and H_2 [45]; $Hg(CH_2CHO)_2$ (in refluxing $CHCl_3$ for 30 min) → $Ge(C_4H_9)_3\text{-}OCH{=}CH_2$ (54%), Hg, and CH_3CHO [6]; $Hg(CH_2COOCH_3)_2$ (in refluxing THF for 1 h → $Ge(C_4H_9)_3CH_2COOCH_3$ and Hg [10]; $Hg(CH{=}CHCl)_2$ (gentle heating in ether) → $Ge(C_4H_9)_3CH{=}CHCl$, Hg, and H_2 [29].

References:

[1] Lesbre, M.; Satgé, J. (C. R. Hebd. Séances Acad. Sci. **247** [1958] 471/4).

[2] Lesbre, M.; Satgé, J. (C. R. Hebd. Séances Acad. Sci. **250** [1960] 2220/2).

[3] Mazerolles, P. (Bull. Soc. Chim. Fr. **1960** 856/60).

[4] Lesbre, M.; Satgé, J. (C. R. Hebd. Séances Acad. Sci. **252** [1961] 1976/8).
[5] Satgé, J. (Ann. Chim. [Paris] [13] **6** [1961] 519/73).
[6] Baukov, Yu. I.; Lutsenko, I. F. (Zh. Obshch. Khim. **32** [1962] 3838/9; J. Gen. Chem. [USSR] **32** [1962] 3767).
[7] Lesbre, M.; Satgé, J. (C. R. Hebd. Séances Acad. Sci. **254** [1962] 1453/5).
[8] Lesbre, M.; Satgé, J. (C. R. Hebd. Séances Acad. Sci. **254** [1962] 4051/3).
[9] Kramer, K. A. W.; Wright, A. N. (J. Chem. Soc. **1963** 3604/8).
[10] Lutsenko, N. F.; Baukov, Yu. I.; Khasapov, B. N. (Zh. Obshch. Khim. **33** [1963] 2724/7; J. Gen. Chem. USSR [Engl. Transl.] **33** [1963] 2653/5).

[11] Cross, R. J.; Glockling, F. (J. Chem. Soc. **1964** 4125/33).
[12] Satgé, J. (Bull. Soc. Chim. Fr. **1964** 630/4).
[13] Braun, J. (C. R. Hebd. Séances Acad. Sci. **260** [1965] 218/20).
[14] Neumann, W. P.; Schneider, B.; Sommer, R. (Liebigs Ann. Chem. **692** [1966] 1/11).
[15] Rijkens, F.; Bulten, E. J.; Drenth, W.; van der Kerk, G. J. M. (Recl. Trav. Chim. Pays-Bas **85** [1966] 1223/9).
[16] Shikhiev, I. A.; Abdullaev, N. D.; Aliev, M. I. (Zh. Obshch. Khim. **36** [1966] 942/3; J. Gen. Chem. USSR [Engl. Transl.] **36** [1966] 956/7).
[17] Nesmeyanov, A. N.; Borisov, A. E.; Novikova, N. V. (Dokl. Akad. Nauk SSSR **172** [1967] 1329/32; Dokl. Chem. [Engl. Transl.] **172/177** [1967] 172/5).
[18] Nesmeyanov, A. N.; Borisov, A. E. (Dokl. Akad. Nauk SSSR **174** [1967] 96/9; Dokl. Chem. [Engl. Transl.] **172/177** [1967] 424/7).
[19] Carlsson, D. J.; Ingold, K. U.; Bray, L. C. (Int. J. Chem. Kinet. **1** [1969] 315/23).
[20] Kakhniashvili, A. I.; Ioramashvili, D. Sh.; Fedin, E. I.; Petrovskii, P. V.; Rubin, I. D. (Soobshch. Akad. Nauk Gruz. SSR **53** [1969] 573/6; C. A. **71** [1969] No. 60353).

[21] Kakhniashvili, A. I.; Ioramashvili, D. Sh.; Fedin, E. I.; Petrovskii, P. V.; Rubin, I. D. (Soobshch. Akad. Nauk Gruz. SSR **54** [1969] 337/40; C. A. **71** [1969] No. 60453).
[22] Kazankova, M. A.; Lutsenko, I. F. (Zh. Obshch. Khim. **39** [1969] 926; J. Gen. Chem. USSR [Engl. Transl.] **39** [1969] 891).
[23] Kakhniashvili, A. I.; Ioramashvili, D. Sh. (Zh. Obshch. Khim. **40** [1970] 1552/5; J. Gen. Chem. USSR [Engl. Transl.] **40** [1970] 1539/41).
[24] Kakhniashvili, A. I.; Ioramashvili, D. Sh. (Zh. Obshch. Khim. **40** [1970] 1556/9; J. Gen. Chem. USSR [Engl. Transl.] **40** [1970] 1542/5).
[25] Lesbre, M.; Mazerolles, P.; Satgé, J. (The Organic Compounds of Germanium; Wiley, London-New York-Sydney-Toronto 1971).
[26] Corriu, R. J. P.; Moreau, J. J. E. (J. Organomet. Chem. **40** [1972] 55/72).
[27] Corriu, R. J. P.; Moreau, J. J. E. (J. Organomet. Chem. **40** [1972] 73/96).
[28] Gverdtsiteli, I. M.; Édiberidze, D. A.; Chernyshev, E. A. (Zh. Obshch. Khim. **44** [1974] 2449/52; J. Gen. Chem. USSR [Engl. Transl.] **44** [1974] 2409/11).
[29] Nesmeyanov, A. N.; Borisov, A. E. (Izv. Akad. Nauk SSSR Ser. Khim. **1974** 1667/8; Bull. Acad. Sci. USSR Div. Chem. Sci. [Engl. Transl.] **1974** 1596).
[30] Satgé, J.; Rivière, P.; Boy, A. (C. R. Hebd. Séances Acad. Sci. C **278** [1974] 1309/12).

[31] Kazankova, M. A.; Zverkova, T. I.; Lutsenko, I. F. (Zh. Obshch. Khim. **45** [1975] 2044/51; J. Gen. Chem. USSR [Engl. Transl.] **45** [1975] 2006/12).
[32] Rivière, P.; Satgé, J.; Boy, A. (J. Organomet. Chem. **96** [1975] 25/40).
[33] Sakurai, H.; Mochida, K.; Kira, M. (J. Am. Chem. Soc. **97** [1975] 929/31).
[34] Saratov, I. E.; Ivanov, V. A.; Reikhsfel'd, V. O. (Zh. Obshch. Khim. **46** [1976] 1052/7; J. Gen. Chem. USSR [Engl. Transl.] **46** [1976] 1048/52).
[35] Alberti, A.; Hudson, A. (Chem. Phys. Lett. **48** [1977] 331/3).
[36] Alberti, A.; Hudson, A. (J. Chem. Soc. Perkin Trans. II **1978** 1098/102).

[37] Ioramashvili, D. Sh.; Shudra, O. S. (Soobshch. Akad. Nauk Gruz. SSR **94** [1979] 81/4; C. A. **91** [1979] No. 91727).
[38] Klingler, R. J.; Mochida, K.; Kochi, J. K. (J. Am. Chem. Soc. **101** [1979] 6626/37).
[39] Lopatina, V. S.; Sheverdina, N. I.; Kocheshkov, K. A.; Fomina, N. V. (Dokl. Akad. Nauk SSSR **246** [1979] 620/2; Dokl. Chem. [Engl. Transl.] **244/249** [1979] 263/4).
[40] Corriu, R. J. P.; Guérin, C. (J. Organomet. Chem. **197** [1980] C19/C21).

[41] Fukuzumi, S.; Kochi, J. K. (Inorg. Chem. **19** [1980] 3022/6).
[42] Ioramashvili, D. Sh. (Soobshch. Akad. Nauk Gruz. SSR **97** [1980] 357/60; C. A. **93** [1980] No. 131601).
[43] Ioramashvili, D. Sh. (Izv. Akad. Nauk Gruz. SSR Ser. Khim. **6** [1980] 129/35 from C. A. **93** [1980] No. 239555).
[44] Lopatina, V. S.; Sheverdina, N. I.; Fomina, N. V.; Kocheshkov, K. A.; Panov, E. M. (Izv. Akad. Nauk SSSR Ser. Khim. **1980** 378/82; Bull. Acad. Sci. USSR Div. Chem. Sci. [Engl. Transl.] **1980** 291/4).
[45] Seyferth, D.; Withers, H. P., Jr. (J. Organomet. Chem. **188** [1980] 329/33).
[46] Seyferth, D.; Escudié, J.; Shannon, M. L.; Satgé, J. (J. Organomet. Chem. **198** [1980] C51/C54).
[47] Wong, C. L.; Klingler, R. J.; Kochi, J. K. (Inorg. Chem. **19** [1980] 423/30).
[48] Hottmann, K. (J. Prakt. Chem. **323** [1981] 399/419).
[49] Lopatina, V. S.; Kocheshkov, K. A.; Fomina, N. V.; Rodionov, A. I.; Shapet'ko, N. N.; Yankelevich, A. Z. (Zh. Obshch. Khim. **51** [1981] 2580/2; J. Gen. Chem. USSR [Engl. Transl.] **51** [1981] 2225/7).
[50] Stang, P. J.; White, M. R. (J. Am. Chem. Soc. **103** [1981] 5429/33).

[51] Barton, D. H. R.; Hartwig, W.; Motherwell, R. S. H.; Motherwell, W. B.; Stange, A. (Tetrahedron Lett. **23** [1982] 2019/22).
[52] Seyferth, D.; Annarelli, D. C.; Shannon, M. L.; Escudié, J.; Duncan, D. P. (J. Organomet. Chem. **225** [1982] 177/91).
[53] Chatgilialoglu, C.; Ingold, K. U.; Lusztyk, J.; Nazran, A. S.; Scaiano, J. C. (Organometallics **2** [1983] 1332/5).
[54] Lusztyk, J.; Maillard, B.; Lindsay, D. A.; Ingold, K. U. (J. Am. Chem. Soc. **105** [1983] 3578/80).
[55] Doba, T.; Ingold, K. U. (J. Am. Chem. Soc. **106** [1984] 3958/63).
[56] Ingold, K. U.; Lusztyk, J.; Scaiano, J. C. (J. Am. Chem. Soc. **106** [1984] 343/8).
[57] Johnston, L. J.; Lusztyk, J.; Wayner, D. D. M.; Abeywickreyma, A. N.; Beckwith, A. L. J.; Scaiano, J. C.; Ingold, K. U. (J. Am. Chem. Soc. **107** [1985] 4594/6).
[58] Pike, P.; Hershberger, S.; Hershberger, J. (Tetrahedron Lett. **26** [1985] 6289/90).
[59] Beckwith, A. L. J.; Pigou, P. E. (Austral. J. Chem. **39** [1986] 1151/5).
[60] Lusztyk, J.; Maillard, B.; Ingold, K. U. (J. Org. Chem. **51** [1986] 2457/60).

[61] Lusztyk, J.; Maillard, B.; Deycard, S.; Lindsay, D. A.; Ingold, K. U. (J. Org. Chem. **52** [1987] 3509/14).
[62] Pike, P.; Hershberger, S.; Hershberger, J. (Tetrahedron **44** [1988] 6295/304).
[63] Wilt, J. W.; Lusztyk, J.; Peeran, M.; Ingold, K. U. (J. Am. Chem. Soc. **110** [1988] 281/7).
[64] Geng, L.; Lu, X. (J. Organomet. Chem. **376** [1989] 41/3).
[65] Kaushal, P.; Roberts, B. P. (J. Chem. Soc. Perkin Trans. II **1989** 1559/68, 1565).
[66] Clark, K. B.; Griller, D. (Organometallics **10** [1991] 746/50).
[67] Wada, F.; Abe, S.; Yonemaru, N.; Kikukawa, K.; Matsuda, T. (Bull. Chem. Soc. Jpn. **64** [1991] 1701/3).
[68] Mayr, H.; Basso, N. (Angew. Chem. **104** [1992] 1103/5; Angew. Chem. Int. Ed. Engl. **31** [1992] 1046/8).
[69] Wolfsberger, W. (J. Prakt. Chem. **334** [1992] 453/64).

1.3.1.1.5.1 Uses

Stoichiometric quantities of $Ge(C_4H_9)_3H$ were used as a source of free radicals for converting the esters I (R′ = H or CH_3; R = $CH_2SeC_6H_5$), II (R = CH_2I), and III (R = CH_2I) into lactones via regio- and stereoselective ring closure of alkenoyloxymethyl radicals. The direct reduction products (I to III with R = CH_3) were simultaneously formed in variable amounts. Reactions mediated by $Ge(C_4H_9)_3H$ give higher relative yields of cyclized products than do reactions of $Sn(C_4H_9)_3H$, but in the case of I (with R′ = H) it was difficult to sustain the chain resulting in poor overall yield [2].

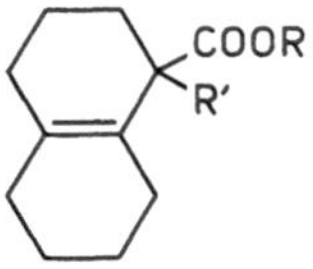

I

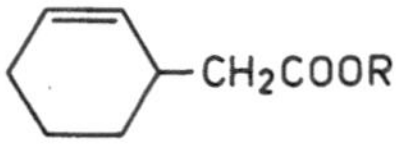

II

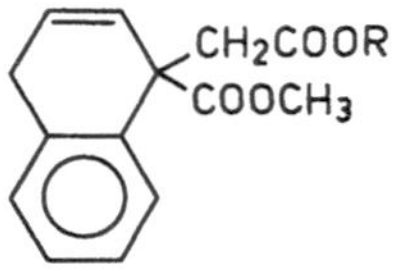

III

Similarly, bicyclic esters with less than 5% yield were obtained by treating compounds IV to VII with $Ge(C_4H_9)_3H$ [3].

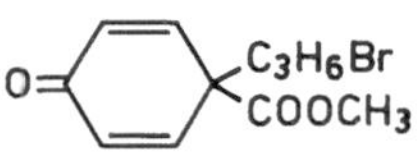

IV

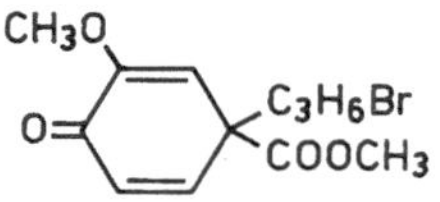

V

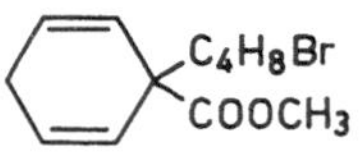

VI

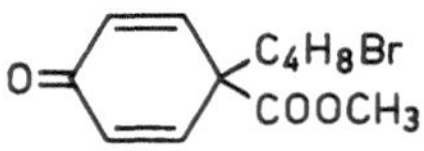

VII

Reacting the trienyl bromide VIII with $Ge(C_4H_9)_3H$ and a trace of AIBN in C_6H_6 at 80 °C for 20 h gave a mixture containing the two stereoisomers IX (27% yield) and six related products [1].

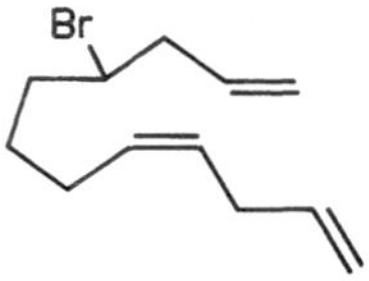

VIII

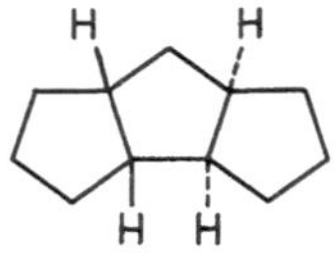

IX

References:

[1] Beckwith, A. L. J.; Roberts, D. H.; Schiesser, C. H.; Wallner, A. (Tetrahedron Lett. **26** [1985] 3349/52).

[2] Beckwith, A. L. J.; Pigou, P. E. (J. Chem. Soc. Chem. Commun. **1986** 85/6).

[3] Beckwith, A. L. J.; Roberts, D. H. (J. Am. Chem. Soc. **108** [1986] 5893/901).

1.3.1.1.6 GeR_3H Compounds with R Larger than Butyl, Cycloalkyl, and Substituted Alkyl

$Ge(C_5H_{11})_3H$

$Ge(C_5H_{11})_3H$ was prepared by reducing $Ge(C_5H_{11})_3Br$ with $LiAlH_4$ in ether [3]; see also [6]. Physical data: b.p. 150 °C/11 Torr, d^{20} = 0.9310 g/cm^3, and n_D^{20} = 1.4542 [3, 6].

Refluxing with CH_2=CHCN gave $Ge(C_5H_{11})_3CH_2CH_2CN$ [3, 6] along with $Ge(C_5H_{11})_3$-$CH_2CH(CN)CH_2CH_2CN$. In the presence of Cu powder, $Ge(C_5H_{11})_3H$ reacted with CH(=N_2)-$COOC_2H_5$ in C_6H_6 to give $Ge(C_5H_{11})_3CH_2COOC_2H_5$ with 35% yield [6].

$Ge(C_5H_{11}$-i$)_3H$

$Ge(C_5H_{11}$-i$)_3H$ was prepared by reducing $Ge(C_5H_{11}$-i$)_3Br$ with $LiAlH_4$ in ether [3]; see also [6]. Physical data: b.p. 140 °C/17 Torr, d^{20} = 0.9238 g/cm^3, and n_D^{20} = 1.4517 [3, 6].

Refluxing with CH_2=$CHCH_2CN$ for 20 h in the presence of Pt-asbestos as catalyst yielded $Ge(C_5H_{11}$-i$)_3CH_2CH_2CH_2CN$ with 50% yield [6]; see also [5]. Refluxing with C_3H_7I gave $Ge(C_5H_{11}$-i$)_3I$ and C_3H_8 [6].

$Ge(C_5H_9$-c$)_3H$

The reaction of $GeCl_4$ with c-C_5H_9MgCl in ether at 4 °C followed by hydrolysis gave the compound with 35% yield along with $Ge(C_5H_9$-c$)_3Cl$ as the major product. IR spectrum: ν(GeH) 1996 and δ(GeH) ca. 700 cm^{-1}. The compound boils at 121 to 122 °C/0.4 Torr; d^{20} = 1.0954 g/cm^3 and n_D^{20} = 1.5162 [16].

$Ge(C_6H_{13})_3H$

The compound was prepared with 80% yield by treating $K_2[Ge(1,2\text{-}O_2C_6H_4)_3]$ with C_6H_{13}-MgBr-$MgBr_2$ in ether at 0 °C followed by reduction with $LiAlH_4$ and hydrolysis by 25% H_2SO_4 [42]. It was also obtained with 26% yield by reducing $Ge(C_6H_{13})_3Br$ with $LiAlH_4$ in refluxing ether for 4 h [2].

IR spectrum: ν(GeH) 1980 (solvent not given) [2], 1997.5 (HMPT), 1998.5 (pyridine), 2000 (CH_3CN), 2001 ($CHCl_3$ and CH_3COCH_3), 2002.5 (CCl_4), 2005 (neat), 2005.5 (ether), 2008 (C_6H_{14} or C_7H_{16}) cm^{-1} [17]. The compound was one of several organogermanes included in investigations on the isolated GeH stretching frequency [30].

Its boiling point was quoted to be 105 to 107 °C/13.3 Pa [42], 122 to 125 °C/0.5 Torr [2], 169 to 170 °C/9 Torr; d^{20} = 0.9228 [6] and d^{25} = 0.917 g/cm^3 [2]; n_D^{20} = 1.4582 [6] and n_D^{21} = 1.4565 [2].

The pure compound does not evolve hydrogen at a noticeable rate when it is treated with dilute alcoholic KOH, although $Ge(C_6H_5)_3H$ reacts readily [2].

$Ge(C_6H_{11}$-c$)_3H$

The compound was prepared by gently refluxing $Ge(C_6H_{11}$-c$)_3Cl$ and $LiAlH_4$ in ether for 1 h; 87% yield. It was obtained practically quantitatively from $Ge(C_6H_{11}$-c$)_3OH$ and a large excess of $LiAlH_4$ in refluxing ether after 2 h [1]. Hydrolysis of the products obtained in the reaction of excess c-$C_6H_{11}MgCl$ with $GeCl_4$ in ether at 0 °C yielded at most ca. 30% [10]; similar results in [16].

^{1}H NMR spectrum: δ = 3.60 ppm (GeH). IR spectrum: ν(GeH) 1986 [7], 1995(s), δ(GeH) 723(s) cm^{-1} [16]. The compound melts at 24 to 25 °C [1]; b.p. 141 to 143 °C/0.3 Torr [16], 145 °C/9 Torr [11]; d^{25} = 1.0838 g/cm^3 and n_D^{25} = 1.5223 [16].

Thermal decomposition of $Ge(C_6H_{11}$-c$)_3H$ begins at ca. 360 °C and leads at 400 to 450 °C after 6 h to Ge, H_2, C_6H_{12}, C_6H_{10}, C_6H_6, and highly condensed compounds containing cyclohexyl rings as the main decomposition products. GeH_4 and C_6H_{10} are probably formed

References on p. 98

initially during the decomposition process. It was shown that pyrolysis of C_6H_{10} between 350 and 450 °C, yielding C_6H_{12}, C_6H_6, and olefins, is greatly accelerated in the presence of catalytic amounts of the title compound [11].

The title compound is oxidized fairly rapidly when stored in contact with the air. Bubbling a stream of air through a refluxing solution of $Ge(C_6H_{11}\text{-c})_3H$ in CCl_4 for 2 h yielded $Ge(C_6H_{11}\text{-c})_3OH$ [1]. Heating with Se at 200 °C for 7 h yielded 31% $Ge(C_6H_{11}\text{-c})_3SeH$ [13]. Reactions with Br_2 in C_2H_5Br and with I_2 in CCl_4 gave $Ge(C_6H_{11}\text{-c})_3Br$ and $Ge(C_6H_{11}\text{-c})_3I$, respectively, along with evolution of HX (X = Br or I) [1].

The reaction with $C_{11}H_{23}I$ in C_6H_6 at 80 °C for 8 h in the presence of AIBN gave $C_{11}H_{24}$ with 64% yield. Under the same conditions, the reaction with $C_{11}H_{23}I$ and 2-cyclohexen-1-one in CH_3CN led to $C_{11}H_{24}$ (37% yield) and the adduct I (16%) [40].

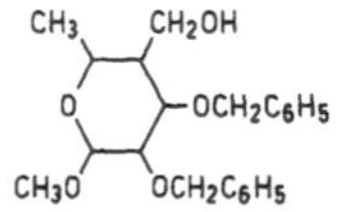

I

Treatment with $Te(C_2H_5)_2$ at 200 °C for 3.5 h led to $(Ge(C_6H_{11}\text{-c})_3)_2Te$ (78% yield) with evolution of C_2H_6 [13]. No H/D exchange was found after successive treatment with excess $c\text{-}C_6H_{11}MgCl$ and D_2O [16].

$CH_2{=}CHCH_2C_3H_7\text{-i}$ was polymerized with a catalyst mixture containing $MoCl_5$ and $Ge(C_6H_{11}\text{-c})_3H$ [4].

$Ge(C_6H_{11}\text{-c})_3D$ was obtained by reacting $GeCl_4$ with $c\text{-}C_6H_{11}MgCl$ followed by treatment of the products with D_2O [10, 16]. IR spectrum: ν(GeD) 1428(vs) and δ(GeD) at 515(vs) cm^{-1}. The compound boils at 143 to 144 °C/0.1 Torr; d^{20} = 1.0913 g/cm^3 and n_D^{20} = 1.5250 [16].

$Ge(C_7H_{15})_3H$

The compound was prepared from $Ge(C_7H_{15})_3Cl$ and $LiAlH_4$; it boils at 182 °C/1.7 Torr; d^{20} = 0.9108 g/cm^3 and n_D^{20} = 1.4600 [6].

$Ge(C_8H_{17})_3H$

$Ge(C_8H_{17})_3H$ was prepared by refluxing $K_2[Ge(1,2\text{-}O_2C_6H_4)_3]$ and $C_8H_{17}MgBr$ in ether in the presence of some $Ti(C_5H_5)_2Cl_2$ for 1 h with 77% yield. It was also obtained similarly using $Li_2[Ge(1,2\text{-}O_2C_6H_4)_3]$ or $K_2[Ge(\text{-}OCH(CH_3)CH(CH_3)O\text{-})_3]$ instead of $K_2[Ge(1,2\text{-}O_2C_6H_4)_3]$ as the starting material and ether or THF as solvent [42].

Boiling point: 155 to 160 °C/53.3 Pa [42] and 179 to 180 °C/0.4 Torr (identical pressure); d^{20} = 0.9061 g/cm^3 and n_D^{20} = 1.4610 [6].

$CH_2{=}CHCH_2C_4H_9\text{-t}$ was polymerized with a catalyst mixture containing $Ge(C_8H_{17})_3H$ and CrO_2Cl_2 [4].

$Ge(CH_2C_6H_5)_3H$

The compound was prepared by refluxing crude $Ge(CH_2C_6H_5)_3Br$ (obtained from $Ge(CH_2C_6H_5)_4$ and Br_2 in $C_2H_4Br_2$-1,2) and $LiAlH_4$ in C_6H_6-ether (1 : 1) for 2 h with 85% yield.

It was also formed (76% yield) by treating $Ge(CH_2C_6H_5)_4$ with Li shot in DME for ca. 1 h followed by hydrolysis of the $Ge(CH_2C_6H_5)_3Li$ formed. Longer reaction times favor the formation of $Ge(CH_2C_6H_5)_2Li_2$. Cleavage of $Ge_2(CH_2C_6H_5)_6$ with Li followed by hydrolysis gave only a 25% yield of $Ge(CH_2C_6H_5)_3H$ [8].

^1H NMR spectrum (in CCl_4): δ(ppm) = 2.19 (CH_2), 4.18 (GeH), and 6.93 (C_6H_5) [8]. IR spectrum: ν(GeH) 2034(m) [8, 9], 2034(m) or 2038 cm^{-1} (in C_7H_{16} or CCl_4) [21], ν(GeC) 541(w) cm^{-1}, and further bands at 334(w, complex) and 239(m) cm^{-1} [9]. The integrated intensity of the ν(GeH) band in C_7H_{16} and CCl_4 was measured and used for establishing a correlation with the sum of the inductive Taft σ* constants of the three $CH_2C_6H_5$ substituents [21]. The resonance constant σ_R^0 of the $CH_2Ge(CH_2\text{-}C_6H_5)_2H$ unit was determined from the integrated intensity value of the IR bands in the 1600 cm^{-1} region [23].

The compound forms colorless needles melting at 80 to 82 °C (crystallized from CH_3OH). It boils at 164 °C/10^{-3} Torr [8].

Slight decomposition occurred on heating to 370 °C for 1 d in an evacuated, sealed tube; at 390 °C complete decomposition yielded $C_6H_5CH_3$, $C_6H_5CH_2CH_2C_6H_5$, a trace of trans-$C_6H_5CH{=}CHC_6H_5$, and a dark brown polymer which did not contain Ge [8].

In a study of electron-impact-induced fragmentation at 70 eV, the metastable-supported cleavages of one and two Ge-(C,H) bonds via $[(C_6H_5CH_2)_3GeH]^{+\cdot} \rightarrow [(C_6H_5CH_2)_2GeH]^+ + C_6H_5CH_2^{\cdot}$ and $[(C_6H_5CH_2)_2GeH]^+ \rightarrow [C_6H_5CH_2Ge]^+ + C_6H_5CH_3$ and the metastable transition $[C_6H_5CH_2Ge]^+ \rightarrow [C_5H_5Ge]^+ + C_2H_2$ were confirmed. A complete fragmentation pattern is presented [15].

In the reaction between $Ge(CH_2C_6H_5)_3H$ and Li, both Ge-H and Ge-$CH_2C_6H_5$ bonds are cleaved. Treating the title compound with Li in DME and then with CH_3I in ether gave $Ge(CH_3)_2(CH_2C_6H_5)_2$, $Ge(CH_3)(CH_2C_6H_5)_3$, $C_6H_5CH_3$, and probably $Ge(CH_3)(CH_2C_6H_5)_2H$. $Ge(CH_2C_6H_5)_3H$ reacts with LiC_4H_9 by hydrogen-metal exchange to yield $Ge(CH_2C_6H_5)_3Li$ and also with displacement of either $[H]^-$ or $[C_6H_5CH_2]^-$ via nucleophilic attack by $[C_4H_9]^-$. This was confirmed by successively treating the title compound with LiC_4H_9 and CH_3I at −10 to −12 °C, resulting in a mixture of products containing $Ge_2(CH_2C_6H_5)_6$, $Ge(CH_2C_6H_5)_4$, $Ge(C_4H_9)(CH_2C_6H_5)_3$, $Ge(CH_3)(CH_2C_6H_5)_3$, and probably $Ge(C_4H_9)_2(CH_2C_6H_5)H$ and $Ge(C_4H_9)(CH_2C_6H_5)_2H$. A similar experiment at room temperature yielded $Ge_2(CH_2C_6H_5)_6$ and $Ge(C_4H_9)_n(CH_2C_6H_5)_{4-n}$ (n = 0 to 4) but no $Ge(CH_3)(CH_2C_6H_5)_3$ and no volatile products containing Ge-H bonds. Reaction with $C_6H_5CH_2Li$ in THF-ether at room temperature followed by addition of CH_3I yielded $Ge(CH_2C_6H_5)_4$ as the main product along with $Ge(CH_3)(CH_2C_6H_5)_3$ and a trace of $Ge(CH_3)_2(CH_2C_6H_5)_2$ [8].

The compound forms a charge-transfer complex with $(NC)_2C{=}C(CN)_2$ which was used for calculating the σ_P value of the $CH_2Ge(CH_2C_6H_5)_2H$ substituent from the value of the electronic absorption [23].

$Ge(CH_2C_6H_5)_3D$ was obtained with 77% yield by treating $Ge(CH_2C_6H_5)_3Li$ (crude product prepared from $Ge(CH_2C_6H_5)_4$ and Li shot) with D_2O [8]. IR spectrum: ν(GeD) 1464(s) [8, 9], ν(GeC) 563(m), 541(w), and δ(GeD) 472(m) cm^{-1}; further bands at 334(w, complex) and 238(m) cm^{-1} [9].

The compound melts at 81 °C (crystallized from CH_3OH) and boils at 170 to 176 °C/10^{-3} Torr [8]. In a study of electron-impact-induced fragmentation at 70 eV, the metastable-supported cleavages of one and two Ge-(C,D) bonds via $[(C_6H_5CH_2)_3GeD]^{+\cdot} \rightarrow [(C_6H_5CH_2)_2GeD]^+ + C_6H_5CH_2^{\cdot}$ and $(C_6H_5CH_2)_2GeD]^+ \rightarrow [C_6H_5CH_2Ge]^+ + C_7H_7D$ were confirmed [15].

References on p. 98

$Ge(CH_2C_4H_9\text{-}t)_3H$

The compound was prepared by reducing crude $Ge(CH_2C_4H_9\text{-}t)_3Cl$ (obtained from $GeCl_4$ and $t\text{-}C_4H_9CH_2MgBr$) with $LiAlH_4$ in refluxing ether for 5 h followed by hydrolysis [33].

^1H NMR spectrum (in $CDCl_3$): δ(ppm) = 0.94 (d, CH_2, J = 4.0 Hz), 0.97 (s, CH_3), and 4.03 (m, GeH). IR spectrum (neat): ν(GeH) 2040 cm^{-1}. The compound boils at 140 °C/16 Torr.

Photochemically generated $t\text{-}C_4H_9O^{\cdot}$ abstracts hydrogen to yield the corresponding germyl radical [33].

$Ge(CH_2C(CH_3)_2C_6H_5)_3H$

$Ge(CH_2C(CH_3)_2C_6H_5)_3H$ was prepared by reducing $Ge(CH_2C(CH_3)_2C_6H_5)_3Cl$ with $LiAlH_4$ in refluxing ether for 1 h followed by hydrolysis and recrystallization from C_6H_{14}; 97% yield [34].

^1H NMR spectrum (CS_2): δ(ppm) = 0.71 (d, CH_2), 1.21 (s, CH_3), 3.40 to 3.70 (sept, GeH), and 7.0 to 7.20 (m, C_6H_5). IR spectrum (KBr): ν(GeH) 1980 cm^{-1}. The compound melts at 58 °C [34].

The mass spectrum (70 eV, 270 °C) showed the molecular ion $[M]^+$ (2) and the fragments $[Ge(CH_2C(CH_3)_2C_6H_5)_2]^+$ (92), $[GeCH_2C(CH_3)_2C_6H_5]^+$ (16), $[CH_2C(CH_3)_2C_6H_5]^+$ (5), and $[CH_2C_6H_5]^+$ (100) [34]. Irradiating a mixture of $Ge(CH_2C(CH_3)_2C_6H_5)_3H$ and $t\text{-}C_4H_9OOC_4H_9\text{-}t$ at 80 °C produced the ESR signal of the $Ge(CH_2C(CH_3)_2C_6H_5)_3^{\cdot}$ radical which immediately disappeared when the lamp was switched off [34]; see also [39].

$Ge(CH_2Si(CH_3)_3)_3H$

The compound was obtained with 85% yield by refluxing $Ge(CH_2Si(CH_3)_3)_3Br$ and excess $LiAlH_4$ in ether for 20 h followed by hydrolysis with H_2O-HCl [19]; see also [33]. It was formed with 76% yield from $Ge(CH_2Si(CH_3)_3)_3Li$ and $Si(CH_3)_3OOC_4H_9\text{-}t$ in THF at 20 °C during a few minutes along with $CH_2{=}C(CH_3)_2$ and $Si(CH_3)_3OOLi$. It was also obtained by hydrolysis of $Ge(CH_2Si(CH_3)_3)_3Li$ or $Ge(CH_2Si(CH_3)_3)_3OSi(CH_3)_3$ [19].

^1H NMR spectrum (in CCl_4): δ(ppm) = −0.19 (d, CH_2, J = 3.8 Hz), −0.05 (s, CH_3), and 4.00 (sept, GeH, J = 3.8 Hz). IR spectrum (neat): 1985 (GeH) and 1235 ($SiCH_3$) cm^{-1} [33]. Reported boiling points range from 78 to 80 °C/1 Torr, 79 to 81 °C/1 Torr, 80 to 81 °C/1 Torr [19], and 117 to 118 °C/18 Torr [33]; n_D^{20} = 1.4610, 1.4615, and 1.4619 [19], n^{16} = 1.4620 [33].

$Ge(CH_2Si(CH_3)_3)_3H$ was reacted with freshly distilled $Hg(C_2H_5)_2$ (2:1 mole ratio) at 80 °C for 5 h to give $Hg(Ge(CH_2Si(CH_3)_3)_3)_2$ (89% yield) and C_2H_6 (100%) [19]; similar in [18]. Treating it with $Cd(C_2H_5)_2$ at room temperature yielded after 2 h $Cd(Ge(CH_2Si(CH_3)_3)_3)_2$ (45%) [18]. With $Tl(C_2H_5)_3$ at 50 °C after 5 h, $Tl(Ge(CH_2Si(CH_3)_3)_3)_3$ (67%) and C_2H_6 (100%) were formed [20]. Abstraction of hydrogen by photochemically generated $t\text{-}C_4H_9O^{\cdot}$ gave the corresponding germyl radical [33].

$Ge(CH_2COOCH_3)_3H$

The compound was prepared with 70% yield from $Ge(CH_2COOCH_3)_3Br$ and $Sn(C_2H_5)_3H$ [14]. ^1H NMR spectrum at 34 °C: δ(ppm) = 2.22 (d, CH_2, ^3J = 2.2 Hz), 3.61 (CH_3), and 4.55 (t, GeH, ^3J = 2.2 Hz). IR spectrum: ν(GeH) 2086 cm^{-1} [12, 14] and ν(C=O) 1730 cm^{-1}. The compound boils at 110 to 111 °C/0.7 Torr; d^{20} = 1.3580 g/cm^3 and n_D^{20} = 1.4752 [14].

$Ge(CF_3)_3H$

The compound was first prepared by reducing $Ge(CF_3)_3I$ with $NaBH_4$ in 1 M aqueous HBr (93% yield) [22] or in aqueous H_3PO_4 [26, 37]. This method also produces some $Ge(CF_3)_2H_2$ (less than 5%) [37]. The compound was also isolated from the exchange reaction between $Ge(CF_3)_3I$ and $Ge(C_2H_5)_3H$ in a sealed tube at 60°C for 6 h (98% yield) [32]; see also [24]. Treating $Hg(Ge(CF_3)_3)_2$ with H_2O readily gave $Ge(CF_3)_3H$, $(Ge(CF_3)_3)_2O$, and Hg [24]; the same compound gave $Ge(CF_3)_3H$, $Ge(CF_3)_3Cl$, and Hg when treated with HCl in C_6H_6 at 100°C for 5 h [24, 32]. $(CF_3)_3GeGe(C_2H_5)_3$ readily reacted with HCl in THF to yield $Ge(CF_3)_3H$ and $Ge(C_2H_5)_3Cl$ [32]. The title compound was also found among the products formed from $Zn(Ge(CF_3)_3)_2 \cdot 2THF$ or $Ti(C_5H_5)_2Cl \cdot Zn(Ge(CF_3)_3)_2 \cdot DME$ and HCl in DME at 20°C [35] or from $V(C_5H_5)_2(Ge(CF_3)_3)_2$ and HCl along with $V(C_5H_5)_2(Cl)Ge(CF_3)_3$ [41].

$Ge(CF_3)_3H$, $(CF_3)_3GeM(P(C_6H_5)_3)_2H$, and Hg were obtained from the reaction of $(CF_3)_3GeHgM(P(C_6H_5)_3)_2Ge(CF_3)_3$ (M = Pt, Pd) with H_2 [36].

^1H NMR spectrum: δ = 5.23 [22] or 5.65 [37] ppm (GeH, $^3J(H, F)$ = 6.7 Hz) [22, 37]. ^{13}C NMR spectrum (in C_6D_6; ref. C_6D_6): δ = 127.5 ppm ($^2J(C, H)$ = 15.1 Hz) [37]. ^{19}F NMR spectrum (neat): δ(ppm) = −26.4 (ref. external CF_3COOH) [22], −50.1 (ref. internal $CFCl_3$) [28, 37]; $^1J(C, F)$ = 329.5 [37] or 330.2 Hz [28], $^3J(C, F)$ = 4.8, $^4J(F, F)$ = 4.10 [37] or 4.1(2) Hz [28].

IR spectrum (gas): ν(GeH) 2155 cm^{-1}(s) [22, 35]; other bands at 1167 (vvs) and 734(s) cm^{-1} [22]. The gas-phase IR and liquid-phase Raman spectra are completely listed and assigned (Raman polarization spectrum illustrated) [26]; for the fundamental vibrations, see Table 14. The force constants were calculated to be f(GeC) = 2.27, f(GeH) = 2.762, and f(CF) = 5.82 N/cm. These values are all slightly higher than those of $Ge(CF_3)_2H_2$ [26]. The compound was one of several organogermanes included in studies of isolated GeH stretching frequencies [29, 30].

The He(I) and He(II) photoelectron spectra were plotted to compare them with the spectra of $Ge(CF_3)_2H_2$ and $Ge(CF_3)H_3$. The vertical ionization energies for the range 27 to 10 eV and their assignments are listed [25]; the lowest ionization energies are 12.80 (13e) and 13.55 ($9a_1$) eV. The average positions of all spectral features are slightly shifted to higher (more stable) energies with an increasing number of CF_3 groups. The same trend was observed for the core levels in the series $Ge(CF_3)_nH_{4-n}$ (n = 1 to 3). The core-level binding energies of the title compound from the X-ray photoelectron spectrum are (in eV) 38.7 for Ge3d, 131.05 for $Ge3p_{3/2}$, 135.05 for $Ge3p_{1/2}$, 298.15 for C1s, and 694.65 for F1s [25].

Table 14
The Fundamental Modes of the Vibrational Spectra of $Ge(CF_3)_3H$ [26].
Wavenumbers in cm^{-1}.

IR gas	Raman liquid	assignment
2160.5(s)	2168 (m, polarized)	ν(GeH)
1206(s)	1205 (vw, polarized)	$\nu_s(CF_3)$
	ca. 1180 (sh)	$\nu_s(CF_3)$
1167(vs)	1160 (w, broad)	$\nu_{as}(CF_3)$
1155(s)		$\nu_{as}(CF_3)$
1131(s)		$\nu_{as}(CF_3)$
735(m)		$\delta_s(CF_3)$
731.5(m)	731 (s, polarized)	$\delta_s(CF_3)$

References on p. 98

Table 14 (continued)

IR gas	Raman liquid	assignment
679(vs)	676(m)	δ(CGeH)
525(sh)	527(w)	$\delta_{as}(CF_3)$
512(m)	515(vw)	$\delta_{as}(CF_3)$
343(vs)	341(m)	$\nu(GeC_3)$
323(s)	320(m, polarized)	$\rho(CF_3)$
251(vw)	253(m)	$\rho(CF_3)$
238(w)	241(vs, polarized)	$\nu(GeC_3)$
202(m)	205(vw)	$\rho(CF_3)$
82(m, broad)		$\delta(GeC_3)$
73(sh)	75(m)	$\delta(GeC_3)$

The compound melts at −102°C [22]; b. p. 30°C [32], 31.6°C (extrapolated) [22], or 31.7°C. Vapor pressure data: log (p/mbar) = 7.861 − 1480/T; ΔH_v = 28.33 kJ/mol and ΔS_v = 92.9 kJ · mol^{-1} · K^{-1} [37].

Mass spectrum: $[Ge(CF_3)_2(CF_2)H]^+$ (10), $[Ge(CF_3)_2H]^+$ (60) [37], see also [22], $[Ge(CF_3)F_2]^+$ (5), $[Ge(CF_3)HF]^+$ (90), $[GeCF_3]^+$ (20), $[C_2F_5]^+$ (7), $[GeF_2H]^+$ (68), $[C_2F_4H]^+$ (20), $[GeF]^+$ (100), $[GeH]^+$ (6), $[Ge]^+$ (15), $[CF_3]^+$ (60), and $[CF_2H]^+$ (90) [37].

The reaction with $Hg(C_2H_5)_2$ in C_6H_6 at room temperature for 24 h and then at 50°C for 0.5 h in a sealed, evacuated tube gave $Hg(Ge(CF_3)_3)_2$ (78% yield) and C_2H_6 [32]; similar results obtained in [24]. In other solvents, such as C_6H_{14} and THF, or in the absence of a solvent, this reaction is complicated by side processes precipitating Hg and decreasing the yield between 3 and 10% [24].

Photolysis of $Mn_2(CO)_{10}$ and excess $Ge(CF_3)_3H$ at ca. 40°C gave $Ge(CF_3)_3Mn(CO)_5$ (86% yield) along with H_2. Similarly, $Ge(CF_3)_3Co(CO)_4$ was obtained by irradiating $Co_2(CO)_8$ in an excess of $Ge(CF_3)_3H$ [27]. The reaction with $Zn(C_6H_5)_2$ in C_6H_6-THF (7:2) at 20°C for 2 h yielded $Zn(Ge(CF_3)_3)_2$ · 2 THF with 82% yield [35].

Reacting an excess of $Ge(CF_3)_3H$ with $V(C_5H_5)_2$ in $C_6H_5CH_3$ at 20°C gave $V(C_5H_5)_2(Ge(CF_3)_3)_2$ (56% yield) and hydrogen [38, 41]. $V(C_5H_5)_2Ge(CF_3)_3$ was formed with 81% yield when equimolar amounts of $Ge(CF_3)_3H$ and $V(C_5H_5)_2$ were used. $V(C_5(CH_3)_5)_2Ge(CF_3)_3$ was similarly obtained, but attempts to prepare $V(C_5(CH_3)_5)_2$-$(Ge(CF_3)_3)_2$ from excess $Ge(CF_3)_3H$ and $V(C_5(CH_3)_5)_2$ failed [41].

Treating the compound with $Pt(P(C_6H_5)_3)_4$ in THF for 1 h led to $Ge(CF_3)_3Pt(P(C_6H_5)_3)_2H$ with 87% yield [31]. $Ge(CF_3)_3Pd(P(C_6H_5)_3)_2H$ was formed with $Pd(P(C_6H_5)_3)_4$ [36].

Slow H/D exchange was observed in D_3PO_4 (5% at 25°C in 24 h) [37].

$Ge(CF_3)_3D$. NMR data: $^3J(D, F)$ = 1.00 Hz [37]. IR spectrum (gas): ν(GeD) 1556(s) cm^{-1}; Raman spectrum (liquid): ν(GeD) 1562 (w, polarized) cm^{-1}. All gas-phase IR and liquid-phase Raman bands were reported and assigned in [26].

References:

[1] Johnson, O. H.; Nebergall, W. H. (J. Am. Chem. Soc. **71** [1949] 1720/2).
[2] Fuchs, R.; Gilman, H. (J. Org. Chem. **23** [1958] 911/3).
[3] Lesbre, M.; Satgé, J. (C. R. Hebd. Séances Acad. Sci. **247** [1958] 471/4).

[4] Shearer, N. H.; Coover, H. W., Jr.; Eastman Kodak Co. (U. S. 2925409 [1960]; C. A. **1960** 13732).
[5] Lesbre, M.; Satgé, J. (C. R. Hebd. Séances Acad. Sci. **252** [1961] 1976/8).
[6] Satgé, J. (Ann. Chim. [Paris] [13] **6** [1961] 519/73).
[7] Egorochkin, A. N.; Khidekel', M. L.; Ponomarenko, V. A.; Zueva, G. Ya.; Svirezheva, S. S.; Razuvaev, G. A. (Izv. Akad. Nauk SSSR Ser. Khim. **1963** 1865/8; Bull. Acad. Sci. USSR Div. Chem. Sci. [Engl. Transl.] **1963** 1717/9).
[8] Cross, R. J.; Glockling, F. (J. Chem. Soc. **1964** 4125/33).
[9] Cross, R. J.; Glockling, F. (J. Organomet. Chem. **3** [1965] 146/55).
[10] Mendelsohn, J.-C.; Métras, F.; Valade, J. (C. R. Hebd. Séances Acad. Sci. **261** [1965] 756/8).

[11] Petukhov, G. G.; Svirezheva, S. S.; Druzhkov, O. N. (Zh. Obshch. Khim. **36** [1966] 914/6; J. Gen. Chem. USSR [Engl. Transl.] **36** [1966] 929/30).
[12] Petrovskaya, L. I.; Belavin, I. Yu.; Burlachenko, G. S.; Fedin, E. I.; Baukov, Yu. I.; Lutsenko, I. F. (Zh. Strukt. Khim. **8** [1967] 168/9; J. Struct. Chem. USSR [Engl. Transl.] **8** [1967] 141/2).
[13] Vyazankin, N. S.; Bochkarev, M. N.; Sanina, L. P. (Zh. Obshch. Khim. **37** [1967] 1037/40; J. Gen. Chem. USSR [Engl. Transl.] **37** [1967] 980/3).
[14] Baukov, Yu. I.; Burlachenko, G. S.; Belavin, I. Yu.; Lutsenko, I. F. (Zh. Obshch. Khim. **38** [1968] 1899/900; J. Gen. Chem. USSR [Engl. Transl.] **38** [1968] 1846).
[15] Glockling, F.; Light, J. R. C. (J. Chem. Soc. A **1968** 717/34).
[16] Mendelsohn, J.-C.; Métras, F.; Lahournère, J.-C.; Valade, J. (J. Organomet. Chem. **12** [1968] 327/40).
[17] Mathis, R.; Barthelat, M.; Mathis, F. (Spectrochim. Acta A **26** [1970] 2001/5).
[18] Kalinina, G. S.; Kruglaya, O. A.; Petrov, B. I.; Shchupak, E. A.; Vyazankin, N. S. (Zh. Obshch. Khim. **43** [1973] 2224/8; J. Gen. Chem. USSR [Engl. Transl.] **43** [1973] 2215/8).
[19] Kalinina, G. S.; Basalgina, T. A.; Vyazankin, N. S.; Razuvaev, G. A.; Yablokov, V. A.; Yablokova, N. V. (J. Organomet. Chem. **96** [1975] 213/23).
[20] Kalinina, G. S.; Shchupak, E. A.; Vyazankin, N. S.; Razuvaev, G. A. (Izv. Akad. Nauk SSSR Ser. Khim. **1976** 1342/6; Bull. Acad. Sci. USSR Div. Chem. Sci. [Engl. Transl.] **1976** 1289/92).

[21] Egorochkin, A. N.; Sevast'yanova, E. I.; Khorshev, S. Ya.; Ratushnaya, S. Kh.; Satgé, J.; Rivière, P.; Barrau, J.; Richelme, S. (J. Organomet. Chem. **162** [1978] 25/35).
[22] Lagow, R. J.; Eujen, R.; Gerchman, L. L.; Morrison, J. A. (J. Am. Chem. Soc. **100** [1978] 1722/6).
[23] Sennikov, P. G.; Skobeleva, S. E.; Kuznetsov, V. A.; Egorochkin, A. N.; Rivière, P.; Satgé, J.; Richelme, S. (J. Organomet. Chem. **201** [1980] 213/9).
[24] Bochkarev, M. N.; Ermolaev, N. L.; Razuvaev, G. A.; Grishin, Yu. K.; Ustynyuk, Yu. A. (J. Organomet. Chem. **229** [1982] C1/C4).
[25] Drake, J. E.; Gorzelska, K.; Helbing, R.; Eujen, R. (J. Electron Spectrosc. Relat. Phenom. **26** [1982] 19/30).
[26] Eujen, R.; Mellies, R. (Spectrochim. Acta A **38** [1982] 533/40).
[27] Brauer, D. J.; Eujen, R. (Organometallics **2** [1983] 263/7).
[28] Eujen, R.; Mellies, R. (J. Fluor. Chem. **22** [1983] 263/80).
[29] McKean, D. C.; Torto, I.; Mackenzie, M. W.; Morrisson, A. R. (Spectrochim. Acta A **39** [1983] 387/98).
[30] McKean, D. C.; Torto, I.; Mackenzie, M. W. (Spectrochim. Acta A **39** [1983] 399/408).

[31] Bochkarev, M. N.; Ermolaev, N. L.; Zakharov, L. N.; Safyanov, Yu. N.; Razuvaev, G. A.; Struchkov, Yu. T. (J. Organomet. Chem. **270** [1984] 289/300).

[32] Ermolaev, N. L.; Bochkarev, M. N.; Razuvaev, G. A.; Grishin, Yu. K.; Ustynyuk, Yu. A. (Zh. Obshch. Khim. **54** [1984] 96/100; J. Gen. Chem. USSR [Engl. Transl.] **54** [1984] 83/6).
[33] Mochida, K. (Bull. Chem. Soc. Jpn. **57** [1984] 796/801).
[34] Neumann, W. P.; Schultz, K.-D.; Vieler, R. (J. Organomet. Chem. **264** [1984] 179/91).
[35] Ermolaev, N. L.; Vasil'eva, G. A.; Vyshinskaya, L. I.; Bochkarev, M. N.; Razuvaev, G. A. (Zh. Obshch. Khim. **55** [1985] 617/21; J. Gen. Chem. USSR [Engl. Transl.] **55** [1985] 547/9).
[36] Ermolaev, L. N.; Bochkarev, M. N.; Razuvaev, G. A. (Dokl. Akad. Nauk SSSR **290** [1986] 1123/5; Dokl. Chem. [Engl. Transl.] **286/291** [1986] 405/7).
[37] Eujen, R.; Mellies, R.; Petrauskas, E. (J. Organomet. Chem. **299** [1986] 29/40).
[38] Ermolaev, N. L.; Mar'in, V. P.; Vishinskaya, L. I.; Razuvaev, G. A. (Izv. Akad. Nauk SSSR Ser. Khim. **1987** 1689; Bull. Acad. Sci. USSR Div. Chem. Sci. [Engl. Transl.] **1987** 1571).
[39] Lehnig, M.; Neumann, W. P.; Wallis, E. (J. Organomet. Chem. **333** [1987] 17/24).
[40] Pike, P.; Hershberger, S.; Hershberger, J. (Tetrahedron **44** [1988] 6295/304).

[41] Ermolaev, N. L.; Mar'in, V. P.; Cherkasov, V. K.; Razuvaev, G. A. (Metalloorg. Khim. **2** [1989] 631/6; Organomet. Chem. USSR [Engl. Transl.] **2** [1989] 323/6).
[42] Cerveau, G.; Chuit, C.; Corriu, R. J. P.; Reyé, C. (Organometallics **10** [1991] 1510/5).

1.3.1.1.7 Triphenylgermane, Triphenylgermanium Hydride, $Ge(C_6H_5)_3H$

1.3.1.1.7.1 Preparation and Formation

The compound was prepared with 79% yield by refluxing $Ge(C_6H_5)_3Br$ with $LiAlH_4$ in anhydrous ether for 1 h [5]; see also [3, 8, 43]. The procedure was slightly modified in [15]. $LiAlH_4$ reduction of a mixture of bromides obtained from $Ge_2(C_6H_5)_6$ and excess Br_2 gave $Ge(C_6H_5)_3H$ with 10% yield together with $Ge(C_6H_5)_2H_2$ as the main product [4]. Similar $LiAlH_4$ reduction of a reaction mixture obtained from $GeCl_4$ and C_6H_5MgBr yielded $Ge(C_6H_5)_3H$ as a by-product [13]. $LiAlH_4$ reduction of $Ge(C_6H_5)_3Cl$ is mentioned in [43, 56]. A 66% yield of $Ge(C_6H_5)_3H$ was obtained by adding C_6H_5MgBr-$MgBr_2$ to compound I (3:4:1 mole ratio) in ether at 0°C followed by refluxing with excess $LiAlH_4$ and hydrolysis with aqueous 25% H_2SO_4 [55].

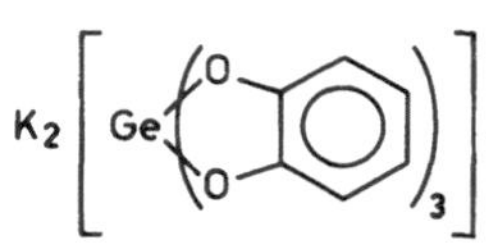

I

Only traces of $Ge(C_6H_5)_3H$ were formed upon refluxing $Ge_2(C_6H_5)_6$ and $LiAlH_4$ in THF for 3 d. Treatment of the unsymmetrical digermanes GeR_3-$Ge(C_6H_5)_3$ (R_3 = $(C_2H_5)_3$, $(C_4H_9)_3$, or $(CH_3)_2C_6H_5$) with $LiAlH_4$ in THF at room temperature for 1 d and subsequent hydrolysis yielded 96, 78, or 89% $Ge(C_6H_5)_3H$, respectively. Similar treatment of $(C_4H_9)_3GeGe(C_6H_5)_3$ for 3 d followed by deuterolysis gave $Ge(C_6H_5)_3H$ (30%), $Ge(C_6H_5)_3D$ (70%), and $Ge(C_4H_9)_3H$ (100%) [41]. Electrolysis of $(CH_3)_3GeGe(C_6H_5)_3$ in CH_3CN (Pt electrodes; $[N(C_4H_9)_4][BF_4]$ or $[N(C_4H_9)_4][ClO_4]$ electrolytes) yielded $Ge(C_6H_5)_3H$ quantitatively along with $(Ge(CH_3)_3)_2O$ in the catholyte by an electrochemically initiated chain reaction [60].

$Ge(C_6H_5)_3H$ was also obtained by $LiAlH_4$ reduction of $(Ge(C_6H_5)_3)_2O$ (in refluxing ether) [3], $Ge(C_6H_5)_3OCH=C(C_6H_5)_2$ [20], $Ge(C_6H_5)_3OC_6H_4CH_3$-4 [62], or $Ge(C_6H_5)_3ON(CH_2$-$C_6H_5)C_6H_5$ [28].

The reaction of $Ge(C_6H_2(CH_3)_3$-2,4,6$)_2(OCH_3)_2$ with $Ge(C_6H_5)_3Li$ in ether-THF followed by treatment with aqueous 6 N HCl, reduction with $LiAlH_4$, and hydrolysis with aqueous 10% HCl yielded $Ge(C_6H_5)_3H$ as a minor product (9%) along with $Ge(C_6H_2(CH_3)_3$-2,4,6$)_2H_2$ (50%) and $(C_6H_5)_3GeGe(C_6H_2(CH_3)_3$-2,4,6$)_2H$ (19%) [58].

Reduction of $Ge(C_6H_5)_3Br$ in C_2H_5OH-ether (1 : 1) with amalgamated Zn dust and aqueous 12 N HCl gave $Ge(C_6H_5)_3H$ with 16% yield, smaller amounts of $Ge(C_6H_5)_4$, and probably $Ge(C_6H_5)_3OH$. Workup included chromatography of the crude $Ge(C_6H_5)_3H$ in C_6H_{12} through an Al_2O_3 column [6].

$Ge(C_6H_5)_3H$ was quantitatively prepared by reacting $Ge(C_6H_5)_3Na$ with NH_4Br in liquid NH_3 or with H_2O in C_6H_6 [1]. It was also formed along with $Ge_2(C_6H_5)_6$ and N_2 by electrolyzing $Ge(C_6H_5)_3Na$ in liquid NH_3 [2]. It was a minor product (11% yield) of the hydrolysis of $Ge(C_6H_5)_3Li$ after refluxing (5 d) in THF [10]. Refluxing GeI_2 and excess LiC_6H_5 in ether for 5 d also yielded $Ge(C_6H_5)_3H$ as a by-product, probably via intermediate $Ge(C_6H_5)_3Li$ and its reaction with H_2O [12]; $Ge(C_6H_5)_3H$ and $Ge(C_6H_5)_4$ were the main products of a similar reaction using $GeI_2P(C_6H_5)_3$ [52].

Treatment of $Sn(CH_3)(C_3H_7$-i$)(C_6H_5)OC_{10}H_{19}$ ($C_{10}H_{19}$ = menthyl) with $Ge(C_6H_5)_3Li$ followed by hydrolysis gave a ca. 2:1 mixture of $Ge(C_6H_5)_3Sn(CH_3)(C_3H_7$-i$)C_6H_5$ and $Ge(C_6H_5)_3H$ [40]. $Ge(C_6H_5)_3H$ was also found as a by-product in the reaction of $Ge(C_6H_5)H_2Cl$ with $Ge(C_6H_5)_3Li$ followed by hydrolysis, which yielded mainly $(C_6H_5)_3Ge$-$Ge(C_6H_5)H_2$ [49].

$Ge(C_6H_5)_3H$ was found among the products of the reaction of $Ge(C_6H_5)_3Li$ with the quinone II in THF at 20 °C, along with $Ge_2(C_6H_5)_6$ and compounds III and IV. The product obtained from II and Na gave with $Ge(C_6H_5)_3Li$ under the same conditions $Ge(C_6H_5)_3H$, $Ge_2(C_6H_5)_6$, and compound III. The latter reacted with $Ge(C_6H_5)_3Li$ (2:1 mole ratio) to yield $Ge(C_6H_5)_3H$, $Ge_2(C_6H_5)_6$, and compound V [54]. Reactions between $Ge(C_6H_5)_3Li$ and several compounds favoring single-electron transfer reactions (methyl violet, compounds VI to VIII) followed by treatment with 10% HCl led to variable amounts of $Ge(C_6H_5)_3H$, $Ge_2(C_6H_5)_6$, and $Ge(C_6H_5)_3Cl$ via the $Ge(C_6H_5)_3^{\cdot}$ radical and germyl adducts [61].

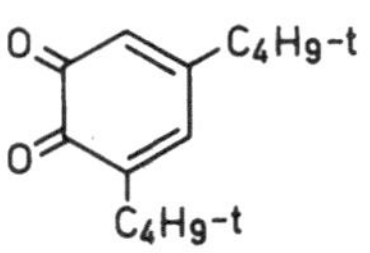

II

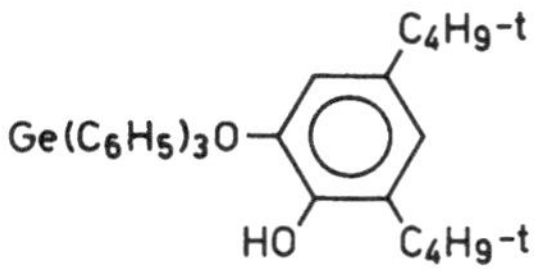

III

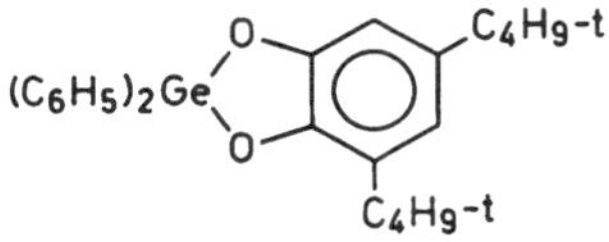

IV

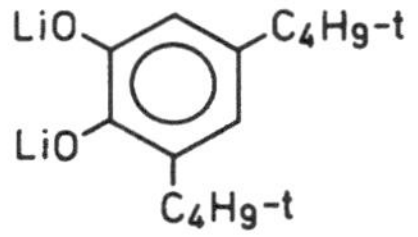

V

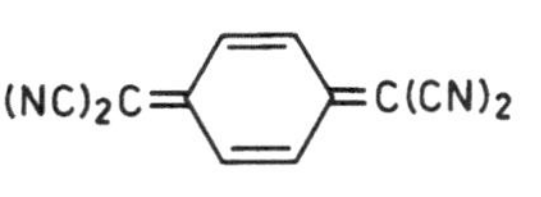

VI

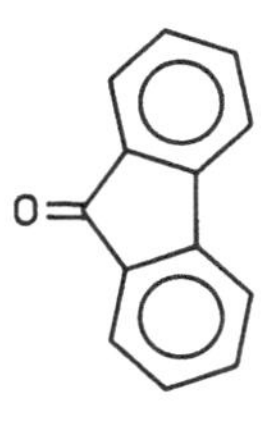

VII

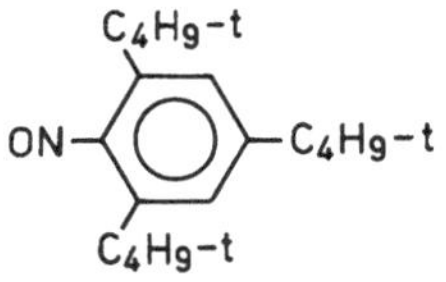

VIII

References on p. 107

$Ge(C_6H_5)_3H$ is formed in many reactions involving cleavage of the Ge-Si bond [44, 46]. Storage of $Ge(C_6H_5)_3SiR_3$ (R = CH_3 or C_2H_5) in 0.1 M $NaOCH_3$ in CH_3OH/CH_3OD at 30 or 50°C yielded $Ge(C_6H_5)_3H/Ge(C_6H_5)_3D$ [44]. Treating $Ge(C_6H_5)_3Si(CH_3)(C_6H_5)OCH_3$ with $LiAlH_4$ in ether followed by deuterolysis gave $Ge(C_6H_5)_3H$, $Si(CH_3)(C_6H_5)H_2$, and $Ge(C_6H_5)_3$-$Si(CH_3)(C_6H_5)H$ but no compounds containing Ge-D or Si-D bonds. $Ge(C_6H_5)_3$-$Si(CH_3)(C_6H_5)OCH_3$ reacted with $LiAlD_4/D_2O$ under the same conditions to yield $Ge(C_6H_5)_3H$, $Ge(C_6H_5)_3D$, $Si(CH_3)(C_6H_5)D_2$, and $Ge(C_6H_5)_3Si(CH_3)(C_6H_5)D$. Similar results were obtained with $Ge(C_6H_5)_3Si(CH_3)(C_6H_5)H$ and $LiAlH_4/D_2O$ or $LiAlD_4/D_2O$ [46].

$Ge(C_6H_5)_3H$ was also one of the products of the following reactions: $Ge(C_6H_5)_3$-$Si(CH_3)(C_6H_5)R$ with LiC_2H_5 (R = H, F, OCH_3, or $OC_{10}H_{19}$), with LiC_2H_5/C_2H_5Br (R = H or F), or with LiC_2H_5/CH_3Br (R = OCH_3); $Ge(C_6H_5)_3Si(CH_3)(C_6H_5)OCH_3$ with Li, with C_2H_5Br in the presence of traces of Li, or with C_2H_5Br in the presence of traces of LiC_2H_5; and $Ge(C_6H_5)_3$-$Si(CH_3)(C_6H_5)H$ with Li in the presence of t-$C_4H_9CH_2Cl$ [46]. The mechanism of the Ge-Si cleavage is discussed in [44, 46]. $Ge(C_6H_5)_3SnR_3$ compounds are expected to be cleaved very readily by bases to give $Ge(C_6H_5)_3H$ [44].

Thermal (150°C, 5 h) or photochemical (4 h) α-elimination from $(C_6H_5)_3GeGe(C_6H_5)Cl_2$ in excess CH_2=$C(CH_3)C(CH_3)$=CH_2 yielded small amounts of $Ge(C_6H_5)_3H$ [49]. Photolysis of $(C_6H_5)_3GeGe(CH_3)_3$ in C_6H_{12} at room temperature gave 32% $Ge(C_6H_5)_3H$ along with $Ge(CH_3)_3C_6H_5$ (19%) and $Ge(CH_3)_3H$ (12%) [57].

Treatment of $Ge(C_6H_5)_3Sb(C_6H_5)_4$ with CH_3COOH in $C_6H_5CH_3$ at 20°C for 1 d gave $Ge(C_6H_5)_3H$ with 73% yield [51]. $Ge(C_6H_5)_3H$ (10%), $Ge(C_6H_5)_3OH$ (90%), and Yb_2O_3 were obtained from $(Ge(C_6H_5)_3H)_2Yb(THF)_4$ and oxygen in THF at −78°C [53].

Formation of $Ge(C_6H_5)_3H$ by chemical cleavage or thermal decomposition of several compounds with Ge-transition metal bonds is listed in Table 15.

Table 15
Formation of $Ge(C_6H_5)_3H$ from Ge-Transition Metal Compounds.
R = C_6H_5.

starting materials	conditions → products (yield)	Ref.
$(GeR_3)_2Ti(C_5H_5)_2$	100°C, in DME, $C_6H_5CH_3$, or THF, 40 h → GeR_3H, Ge_2R_6, and $GeR_3Ti(C_5H_5)_2$	[27]
$(GeR_3)_2Ti(C_5H_5)_2$ + HCl gas	→ GeR_3H and $Ti(C_5H_5)_2Cl_2$	[27]
$GeR_3Ti(C_5H_5)_2$ + excess HCl gas	in dioxane → GeR_3H (via $GeR_3Ti(C_5H_5)_2Cl$) and $Ti(C_5H_5)_2Cl_2$	[27]
$GeR_3Ti(C_5H_5)_2$ + $NaOCH_3$	20°C, in alcohol → GeR_3H and C_5H_6	[27]
$GeR_3Ti(C_5H_5)_2Cl$ + $NaOCH_3$	→ GeR_3H and C_5H_6	[27]
$GeR_3Zr(C_5H_5)(C_5(CH_3)_5)Cl$ + H_2	8 h, ca. 7 atm H_2 → GeR_3H and $Zr(C_5H_5)(C_5(CH_3)_5)(H)Cl$	[59]

Table 15 (continued)

starting materials	conditions → products (yield)	Ref.
$GeR_3Zr(C_5(CH_3)_5)_2Cl$ + H_2	5 h, ca. 7 atm H_2 → GeR_3H and $Zr(C_5(CH_3)_5)_2(H)Cl$	[59]
$GeR_3Zr(C_5(CH_3)_5)_2Cl$ + HCl gas	1 h, in C_6D_6 → GeR_3H and $Zr(C_5(CH_3)_5)_2Cl_2$	[59]
$GeR_3Zr(C_5(CH_3)_5)_2Cl$ + $SiRH_3$	in C_6D_6, ambient room light → GeR_3H, $Zr(C_5(CH_3)_5)_2(H)Cl$, $H_2(R)SiSi(R)H_2$, $H_2(R)SiSi(R)(H)Si(R)H_2$, $H_2(R)Si(Si(R)H)_2Si(R)H_2$	[59]
$GeR_3Hf(C_5H_5)(C_5(CH_3)_5)Cl$ + $SiRH_3$	in C_6D_6, ambient room light → GeR_3H and $Hf(C_5H_5)(C_5(CH_3)_5)(Cl)Si(R)H_2$	[59]
$GeR_3V(C_5H_5)_2$ + HCl (excess)	in $C_6H_5CH_3$ or dioxane → GeR_3H (83%), Ge_2R_6, and $V(C_5H_5)_2Cl$	[36]
$GeR_3V(C_5H_5)_2$ + CH_3COOH	20 °C, in $C_6H_5CH_3$, 78 h → GeR_3H (85%), $V(C_5H_5)(OOCCH_3)_2$, and C_5H_6	[36]
$GeR_3Mn(C_5H_4CH_3)(CO)_2H$ + PR_3	20 °C, in $C_6H_5CH_3$, 6 d → GeR_3H (90%) and $Mn(C_5H_4CH_3)(CO)_2PR_3$	[47]
$GeR_3Mn(C_5H_4CH_3)(CO)_2CH_3$ + PR_3	refluxing in C_6H_6 → GeR_3H (6 to 15%), $Ge(CH_3)R_3$ (85 to 94%), $Mn(C_5H_4CH_3)(CO)_2PR_3$, and $Mn(C_5H_3(CH_3)_2)(CO)_2PR_3$	[48]
$GeR_3Os_3(CO)_{11}H$ + PR_3	20 °C, in $C_6H_5CH_3$, 5 d → GeR_3H and $Os_3(CO)_{11}PR_3$	[50]
$GeR_3Co(CO)_4$ + $LiAlH_4$	refluxing in THF, 6 h → GeR_3H (48%) 20 °C, in ether, 2 d → GeR_3H (74%)	[39]; see also [38] [43]
$GeR_3Co(CO)_4$ + CH_2=$CHCH_2MgBr$, then + H_2O	no details given → GeR_3H (52%; via GeR_3MgBr)	[39]; see also [38]
$GeR_3Co(CO)_3PR_3$ + $LiAlH_4$	20 °C, in ether, 2 d → GeR_3H (93%)	[43]
$GeR_3Co(CO)_3P(OR)_3$ + $LiAlH_4$	20 °C, in ether, 2 d → GeR_3H (92%)	[43]
$GeR_3Co(CO)_3C(OC_2H_5)C_4H_9$ + $LiAlH_4$	20 °C, in ether, 3 d → GeR_3H (64%)	[43]
$GeR_3Pd(P(C_2H_5)_3)_2$ + HCl (excess)	→ GeR_3H (100%), trans-$Pd(P(C_2H_5)_3)_2Cl_2$ and probably $Pd(P(C_2H_5)_3)_2(H)Cl$	[23]

References on p. 107

Table 15 (continued)

starting materials	conditions → products (yield)	Ref.
$GeR_3Pt(P(C_2H_5)_3)_2H$ + HCl	20 °C, in ether, 4 d → GeR_3H, $Pt(P(C_2H_5)_3)_2(H)Cl$, and H_2	[19]
$GeR_3Pt(P(C_2H_5)_3)_2H$ + $PR_2C_2H_4PR_2$	20 °C, in C_6H_6, 3.5 h → GeR_3H (76%), $Pt(PR_2C_2H_4PR_2)_2$, and $P(C_2H_5)_3$	[19]
$(GeR_3)_2Pt(P(C_2H_5)_3)_2$ + H_2	in $CH_3COOC_2H_5$ or $C_6H_5CH_3$ → GeR_3H and $GeR_3Pt(P(C_2H_5)_3)_2H$	[14]
$(GeR_3)_2Pt(P(C_2H_5)_3)_2$ + dry HCl (excess)	in C_6H_6 → GeR_3H, GeR_3Cl, trans-$Pt(P(C_2H_5)_3)_2(H)Cl$, trans-$Pt(P(C_2H_5)_3)_2Cl_2$, and traces of $(GeR_3)_2O$	[14]
$(GeR_3)_2Pt(P(C_3H_7)_3)_2$ + H_2	in $CH_3COOC_2H_5$ or $C_6H_5CH_3$ → GeR_3H and $GeR_3Pt(P(C_3H_7)_3)_2H$	[14]
$(GeR_3)_2Pt(P(C_3H_7)_3)_2$ + LiC_6H_5	20 °C, in C_6H_6, 3 d, hydrolysis → GeR_3H (via $Ge(C_6H_5)_3Li$) and cis-$Pt(P(C_3H_7)_3)_2R_2$	[14]
$(GeR_3)_2Pt(P(C_3H_7)_3)_2$ + $LiAlH_4$	20 °C, in ether-C_6H_6, 2 d → GeR_3H (98%) and an insoluble yellow Pt compound	[14]
$(GeR_3)_2Pt(P(C_3H_7)_3)_2$ + MgI_2	20 °C, in ether-C_6H_6, 46 h; then hydrolysis with air-free H_2O → GeR_3H (via GeR_3MgI), GeR_3I, trans-$Pt(P(C_3H_7)_3)_2I_2$, a hydroplatinum complex, and traces of Ge_2R_6 and $(GeR_3)_2O$	[14]
cis-$(GeR_3)_2Pt(PR_2C_2H_4PR_2)$ + HCl	in C_6H_6, 1 h → GeR_3H and cis-$Pt(PR_2C_2H_4PR_2)Cl_2$ (no reaction in ether at 20 to 25 °C over 20 d)	[19]
cis-$(GeR_3)_2Pt(PR_2C_2H_4PR_2)$ + H_2	50 °C, in C_6H_6, 100 atm, 2 d → GeR_3H and $GeR_3Pt(PR_2C_2H_4PR_2)H$ (no reaction at 20 °C and 1 atm)	[19]; see also [18]
$GeR_3CdNi(PR_3)_2GeR_3$ + HCl	20 °C, in $C_6H_5CH_3$, 1 d → GeR_3H (45%), $Ni(PR_3)_2Cl_2 \cdot CdCl_2 \cdot 4\,HCl$, $GeR_2(H)Cl$, $GeCl_4$, and H_2	[45]
$GeR_3CdNi(PR_3)_2GeR_3$ + glacial CH_3COOH	80 °C, in $C_6H_5CH_3$, 4 h → GeR_3H (71%), $Cd(OOCCH_3)_2$, $Ni(OOCCH_3)_2$, and PR_3	[45]
$GeR_3HgNi(PR_3)_2GeR_3$ + HCl	20 °C, in $C_6H_5CH_3$, 1 d → GeR_3H (51%), $Ni(PR_3)_2Cl_2 \cdot HgCl_2$, and H_2	[45]

Table 15 (continued)

starting materials	conditions → products (yield)	Ref.
$GeR_3CdPt(PR_3)_2GeR_3$ + glacial CH_3COOH	80°C, in $C_6H_5CH_3$, 4 h → GeR_3H (78%), $Pt(PR_3)_2(H)GeR_3$ (84%), and $Cd(OOCCH_3)_2 \cdot 3\,H_2O$	[45]
$GeR_3HgPt(PR_3)_2GeR_3$ + HCl	20°C, in $C_6H_5CH_3$, 1 d → GeR_3H (40%), $Pt(PR_3)_2(H)Cl_2GeR_3$ (77%), $HgCl_2$, and H_2	[45]
$Zn(C_2H_5)GeR_3 \cdot (CH_3OC_2H_4)_2O$ + glacial CH_3COOH (equimolar amounts)	20°C, in $C_6H_5CH_3$, 10 h, then 100°C, 30 min → GeR_3H (98%), $Zn(C_2H_5)OOCCH_3 \cdot (CH_3OC_2H_4)_2O$	[30]
$Zn(C_2H_5)GeR_3 \cdot (CH_3OC_2H_4)_2O$ + glacial CH_3COOH (excess)	100°C, in $C_6H_5CH_3$, 30 min, → GeR_3H (93%), $Zn(OOCCH_3)_2$, and C_2H_6	[30]
$Zn(GeR_3)_2 \cdot (CH_3OC_2H_4)_2O$ + glacial CH_3COOH (equimolar amounts)	20°C, in $C_6H_5CH_3$, 10 h → GeR_3H (91%) and $Zn(OOCCH_3)GeR_3$	[30]
$Zn(GeR_3)_2 \cdot (CH_3OC_2H_4)_2O$ + glacial CH_3COOH (excess)	100°C, in $C_6H_5CH_3$, 2 h → GeR_3H (91%) and $Zn(OOCCH_3)_2$	[30]
$Cd(GeR_3)_2$ + H_2O	100°C, in dioxane, 2 h → GeR_3H (71%), $(GeR_3)_2O$ (via GeR_3OH), and Cd	[25]
$Cd(GeR_3)_2$ + C_2H_5OH	100°C, 3 h → GeR_3H (82%), $GeR_3OC_2H_5$, and Cd	[25]
$Cd(GeR_3)_2$ + C_6H_5OH	100°C, in $C_6H_5CH_3$, 2 h → GeR_3H (60%) and $GeR_3CdOC_6H_5$	[25]
$Cd(GeR_3)_2$ + glacial CH_3COOH	100°C, in $C_6H_5CH_3$, 2 h GeR_3H (98%) and $GeR_3CdOOCCH_3$	[25]; see also [32, 45]
$Cd(GeR_3)_2$ + C_6H_5COOH	20°C, in $C_6H_5CH_3$, ca. 12 h → GeR_3H (83%), $GeR_3CdOOCC_6H_5$	[25]
$Cd(GeR_3)_2$ + CF_3COOH (excess)	90°C, in $C_6H_5CH_3$, 3 h → GeR_3H (66%) and $Cd(OOCCF_3)_2$	[25]
$Cd(GeR_3)_2$ + $V(C_5H_5)_2$	20°C, in $C_6H_5CH_3$, 2 d, → GeR_3H (30%), Ge_2R_6, $GeR_3V(C_5H_5)_2$, and Cd	[36]
$Cd(GeR_3)_2 \cdot N(C_2H_5)_2H$	180°C, in $C_6H_5CH_3$, sealed tube, 14 h → GeR_3H (78%), $GeR_3N(C_2H_5)_2$, and Cd	[25]
$GeR_3CdC_2H_5 \cdot TMED$ + H_2O (excess)	110°C, in $C_6H_5CH_3$, 5 h → GeR_3H (75%), $Cd(OH)_2$, and C_2H_6	[32]

References on p. 107

Table 15 (continued)

starting materials	conditions → products (yield)	Ref.
$GeR_3CdOOCCH_3$ + CH_3COOH	130 °C, in $C_6H_5CH_3$, 2 h → GeR_3H (68%) and $Cd(OOCCH_3)_2$	[25]; see also [32]
$Hg(GeR_3)_2$ + Li	20 °C, in THF; then hydrolysis → GeR_3H (66%; via GeR_3Li)	[24]
$Hg(GeR_3)_2$ + glacial CH_3COOH	140 to 145 °C, 6 h → GeR_3H (84%), GeR_3OOCCH_3, and Hg	[24]
$Hg(GeR_3)_2$ + HCl	−30 to −20 °C, in $C_6H_5CH_3$, 10 to 15 min → GeR_3H (86%), GeR_3Cl, and Hg	[24]

Hydrolysis of $M(Ge(C_6H_5)_3)_2$ (M = Sr or Ba) with moist CH_3OH led to $M(OH)_2$, $Ge(C_6H_5)_3H$, some $Ge(C_6H_5)_3OH$ (by hydrolysis of the hydride), and H_2 [16]. Hydrolysis of the products obtained from $Ge(CH(OH)CH(CH_3)_2)(C_6H_5)Cl_2$ and C_6H_5MgBr in ether yielded 10% $Ge(C_6H_5)_3H$ [33]; see also [26]. Treatment of $(C_2H_5)_3GeGe(C_6H_5)_2Cl$ with C_6H_5MgBr in ether gave $Ge(C_6H_5)_3H$, $Ge(C_2H_5)_3C_6H_5$, and $(C_6H_5)_3GeGe(C_2H_5)_3$ after hydrolysis [34]. $Ge(C_6H_5)_3H$ was a minor product when GeF_2 was reacted with $Si(CH_3)_3Cl$, $Si(C_6H_5)_2Cl_2$, $Ge(CH_3)_3Cl$, or $Ge(CH_3)_3OCH_3$ followed by treatment of the reaction mixtures with C_6H_5MgBr [37]; see also [35].

$Ge(C_6H_5)_3H$ resulted from the exhaustive electrolysis of $Ge(C_6H_5)_3Cl$ [17, 29] or $Ge(C_6H_5)_3Br$ [29] in DME [17, 29]. It was also obtained by electrolysis of $Ge(C_6H_5)_3$-$SeSn(C_6H_5)_3$, $Ge_2(C_6H_5)_6$ in the presence of $Si(C_6H_5)_3Cl$, $Ge(C_6H_5)_3Cl$, or $Sn(C_6H_5)_3Cl$, and by electrolysis of $Sn_2(C_6H_5)_6$ in the presence of $Ge(C_6H_5)_3Cl$. The source of the abstracted H atom could conceivably be the DME solvent known to be a good H atom donor [17].

Li-H exchange between $Ge(C_6H_5)_3Li$ and the germole IX (being at least 10^6 times more acidic than $Ge(C_6H_5)_3H$) in THF gave $Ge(C_6H_5)_3H$ and X [21]; exchange between $Ge(C_6H_5)_3Li$ and fluorene ($C_{13}H_{10}$) to yield $Ge(C_6H_5)_3H$ and $C_{13}H_9Li$ was also reported [9, 11].

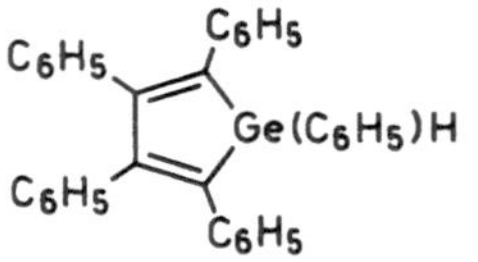

IX

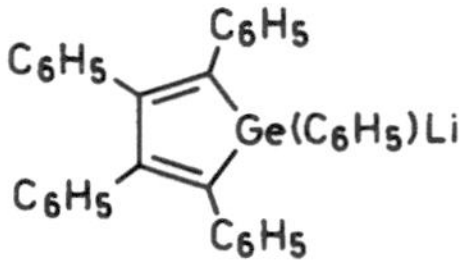

X

Crude $Ge(C_6H_5)_3H$ was purified by distillation at 138 °C/0.6 Torr [7] or 150 °C/0.01 Torr [1]. It was separated from traces of $Ge(C_6H_5)_4$ by recrystallization from CH_3OH [5].

$Ge(C_6H_5)_3D$ was obtained by reducing $Ge(C_6H_5)_3Br$ with $LiAlD_4$ in ether [22]. It was also formed by H-D exchange between $Ge(C_6H_5)_3H$ and excess CH_3OD containing CH_3ONa at ca. 21 °C [31]. Treatment of $Ge(C_6H_5)_3Co(CO)_4$ with $CH_2{=}CHCH_2MgBr$ and hydrolysis with D_2O gave $Ge(C_6H_5)_3D$ (42% yield) via $Ge(C_6H_5)_3MgBr$ [38, 39].

$Ge(C_6H_5)_3T$ was obtained by converting $Ge(C_6H_5)_3H$ with LiC_4H_9 in ether into $Ge(C_6H_5)_3Li$ and hydrolyzing the latter with tritiated water [42].

References:

[1] Kraus, C. A.; Foster, L. S. (J. Am. Chem. Soc. **49** [1927] 457/67).
[2] Foster, L. S.; Hooper, G. S. (J. Am. Chem. Soc. **57** [1935] 76/8).
[3] Johnson, O. H.; Nebergall, W. H. (J. Am. Chem. Soc. **71** [1949] 1720/2).
[4] Johnson, O. H.; Harris, D. M. (J. Am. Chem. Soc. **72** [1950] 5564/6).
[5] Johnson, O. H.; Harris, D. M. (J. Am. Chem. Soc. **72** [1950] 5566/8).
[6] West, R. (J. Am. Chem. Soc. **75** [1953] 6080/1).
[7] Hearn, D.; Fish, A. (from [8, footnote on p. 78]).
[8] Johnson, O. H.; Nebergall, W. H.; Harris, D. M. (Inorg. Synth. **5** [1957] 76/8).
[9] Gilman, H.; Gerow, C. W. (J. Org. Chem. **23** [1958] 1582/4).
[10] Gilman, H.; Zuech, E. A. (J. Org. Chem. **26** [1961] 3035/7).

[11] Gilman, H.; Marrs, O. L.; Trepka, W. J.; Diehl, J. W. (J. Org. Chem. **27** [1962] 1260/5).
[12] Glockling, F.; Hooton, K. A. (J. Chem. Soc. **1963** 1849/54).
[13] Brooks, E. H.; Glockling, F.; Hooton, K. A. (J. Chem. Soc. **1965** 4283/8).
[14] Cross, R. J.; Glockling, F. (J. Chem. Soc. **1965** 5422/32).
[15] Nicholson, D. A.; Allred, A. L. (Inorg. Chem. **4** [1965] 1747/50).
[16] Amberger, E.; Stoeger, W.; Hönigschmid-Grossich, R. (Angew. Chem. **78** [1966] 549; Angew. Chem. Int. Ed. Engl. **5** [1966] 522).
[17] Dessy, R. E.; Kitching, W.; Chivers, T. (J. Am. Chem. Soc. **88** [1966] 453/9).
[18] Brooks, E. H.; Cross, R. J.; Glockling, F. (unpublished results from: Glockling, F.; Hooton, K. A.; J. Chem. Soc. A **1967** 1066/75, 1068).
[19] Brooks, E. H.; Cross, R. J.; Glockling, F. (Inorg. Chim. Acta **2** [1968] 17/21).
[20] Rivière, P.; Satgé, J. (C. R. Seances Acad. Sci. C **267** [1968] 267/9).

[21] Curtis, M. D. (J. Am. Chem. Soc. **91** [1969] 6011/8).
[22] Durig, J. R.; Sink, C. W.; Turner, J. B. (Spectrochim. Acta A **25** [1969] 629/45).
[23] Glockling, F.; Brooks, E. H. (Prepr. Am. Chem. Soc. Div. Pet. Chem. **14** [1969] B135/B137).
[24] Vyazankin, N. S.; Bychkov, V. T.; Linzina, O. V.; Razuvaev, G. A. (Zh. Obshch. Khim. **39** [1969] 979/82; J. Gen. Chem. USSR [Engl. Transl.] **39** [1969] 950/2).
[25] Vyazankin, N. S.; Bychkov, V. T.; Linzina, O. V.; Razuvaev, G. A. (J. Organomet. Chem. **21** [1970] 107/13).
[26] Rivière, P.; Satgé, J. (C. R. Seances Acad. Sci. C **272** [1971] 413/6).
[27] Razuvaev, G. A.; Latyaeva, V. N.; Vasil'eva, G. A.; Vyshinskaya, L. I.; Mal'kova, G. Ya. (Dokl. Akad. Nauk SSSR **206** [1972] 1127/9; Dokl. Chem. [Engl. Transl.] **202/207** [1972] 805/7).
[28] Satgé, J.; Lesbre, M.; Rivière, P.; Richelme, S. (J. Organomet. Chem. **34** [1972] C18/C20).
[29] Boczkowski, R. J.; Bottei, R. S. (J. Organomet. Chem. **49** [1973] 389/407).
[30] Bychkov, V. T.; Vyazankin, N. S.; Razuvaev, G. A. (Zh. Obshch. Khim. **43** [1973] 793/8; J. Gen. Chem. USSR [Engl. Transl.] **43** [1973] 792/6).

[31] Eaborn, C.; Jenkins, I. D. (J. Chem. Soc. Chem. Commun. **1973** 780).
[32] Razuvaev, G. A.; Bychkov, V. T.; Vyazankin, N. S. (Dokl. Akad. Nauk SSSR **211** [1973] 116/9; Dokl. Chem. [Engl. Transl.] **208/213** [1973] 545/8).
[33] Rivière, P.; Satgé, J. (J. Organomet. Chem. **49** [1973] 157/72).
[34] Rivière, P.; Rivière-Baudet, M.; Couret, C.; Satgé , J. (Synth. React. Inorg. Met.-Org. Chem. **4** [1974] 295/307).
[35] Satgé, J.; Rivière, P.; Boy, A. (C. R. Seances Acad. Sci. C **278** [1974] 1309/12).

[36] Razuvaev, G. A.; Bychkov, V. T.; Vyshinskaya, L. I.; Latyaeva, V. N.; Spiridonova, N. N. (Dokl. Akad. Nauk SSSR **220** [1975] 854/5; Dokl. Chem. [Engl. Transl.] **220/225** [1975] 126/7).
[37] Rivière, P.; Satgé, J.; Boy, A. (J. Organomet. Chem. **96** [1975] 25/40).
[38] Colomer, E.; Corriu, R. (J. Chem. Soc. Chem. Commun. **1976** 176/7).
[39] Colomer, E.; Corriu, R. J. P. (J. Organomet. Chem. **133** [1977] 159/68).
[40] Gielen, M.; Simon, S. (Bull. Soc. Chim. Belg. **86** [1977] 589/94).

[41] Duffaut, N.; Dunogues, J.; Calas, R.; Rivière, P.; Satgé, J.; Cazes, A. (J. Organomet. Chem. **149** [1978] 57/63).
[42] Eaborn, C.; Singh, B. (J. Organomet. Chem. **177** [1979] 333/48).
[43] Cerveau, G.; Colomer, E.; Corriu, R. J. P.; Young, J. C. (J. Organomet. Chem. **205** [1981] 31/45).
[44] Eaborn, C.; Mahmoud, F. M. S. (J. Organomet. Chem. **205** [1981] 47/51).
[45] Titova, S. N.; Bychkov, V. T.; Domrachev, G. A.; Sorokin, Yu. A.; Konkina, T. N.; Razuvaev, G. A. (Zh. Obshch. Khim. **52** [1982] 1580/5; J. Gen. Chem. USSR [Engl. Transl.] **52** [1982] 1396/400).
[46] Corriu, R. J. P.; Ould-Kada, S.; Lanneau, G. F. (J. Organomet. Chem. **248** [1983] 39/50).
[47] Carré, F.; Colomer, E.; Corriu, R. J. P.; Vioux, A. (Organometallics **3** [1984] 1272/8).
[48] Colomer, E.; Corriu, R. J. P.; Vioux, A. (J. Organomet. Chem. **267** [1984] 107/19).
[49] Rivière, P.; Castel, A.; Satgé, J.; Guyot, D. (J. Organomet. Chem. **264** [1984] 193/206).
[50] Burgess, K.; Guerin, C.; Johnson, B. F. G.; Lewis, J. (J. Organomet. Chem. **295** [1985] C3/C6).

[51] Bychkov, V. T.; Sharutin, V. V.; Bolotova, O. P.; Ermoshkin, A. E.; Konkina, T. N. (Zh. Obshch. Khim. **55** [1985] 2398; J. Gen. Chem. USSR [Engl. Transl.] **55** [1985] 2131/2).
[52] Inoguchi, Y.; Okui, S.; Mochida, K.; Itai, A. (Bull. Chem. Soc. Jpn. **58** [1985] 974/7).
[53] Bochkarev, M. N.; Penyagina, I. M.; Rad'kov, Yu. F.; Zakharov, L. N.; Struchkov, Yu. T. (Metalloorg. Khim. **1** [1988] 718/9; Organometal. Chem. USSR [Engl. Transl.] **1** [1988] 402).
[54] Rivière, P.; Castel, A.; Ko, Y. H.; Desor, D. (J. Organomet. Chem. **386** [1990] 147/56).
[55] Cerveau, G.; Chuit, C.; Corriu, R. J. P.; Reyé, C. (Organometallics **10** [1991] 1510/5).
[56] Clark, K. B.; Griller, D. (Organometallics **10** [1991] 746/50).
[57] Mochida, K.; Wakasa, M.; Sakaguchi, Y.; Hayashi, H. (Bull. Chem. Soc. Jpn. **64** [1991] 1889/95).
[58] Rivière, P.; Rivière-Baudet, M.; Castel, A.; Desor, D.; Abdennadher, C. (Phosphorus Sulfur Silicon Relat. Elem. **61** [1991] 189/99).
[59] Woo, H.-G.; Freeman, W. P.; Tilley, T. D. (Organometallics **11** [1992] 2198/205).
[60] Okano, M.; Mochida, K. (Denki Kagaku oyobi Kogyo Butsuri Kagaku **61** [1993] 772/3; C.A. **119** [1993] No. 203519).

[61] Rivière, P.; Castel, A.; Desor, D.; Abdennadher, C. (J. Organomet. Chem. **443** [1993] 51/60).
[62] Cerveau, G.; Colomer, E.; Corriu, R. J. P. (J. Organomet. Chem. **236** [1982] 33/52).

1.3.1.1.7.2 The Molecule, Spectra, and Physical Properties

The bond dissociation energy D(Ge-H) = 80.2 ± 0.8 kcal/mol was determined in solution (C_6H_6 or iso-C_8H_{18}) at 23 °C using the laser-initiated formation of t-$C_4H_9O^{\cdot}$ and its reaction with $Ge(C_6H_5)_3H$ to give $Ge(C_6H_5)_3^{\cdot}$ and measuring the combined heats of reactions by photoacustic calorimetry [41].

^{1}H NMR spectrum: δ(ppm) = 5.85 (s, GeH), 7.14 (m, H-3,4,5), 7.51 (m, H-2,6) in C_6D_6 [42]; see also [14]; 5.61(s) to 5.75 (GeH) and 7.33 (m, C_6H_5) in $CDCl_3$ [15, 19, 41], similar data in CCl_4 [11, 25, 31]; 5.48 (GeH) and 7.30 (m, C_6H_5) in liquid NH_3 [16].

^{13}C NMR spectrum: δ(ppm) = 129.05 (C-3,5), 129.82 (C-4), 135.78 (C-2,6), and 136.36 (C-1) in THF [33]; see also [43]; 129.4 (C-3,5), 130.2 (C-4), 135.6 (C-2,6), 135.8 (C-1) in DMF-CD_3CN (spectrum displayed) [43].

The vibrational spectra (IR of the liquid and Raman of the liquid and solid) were recorded in detail, completely assigned from 3200 to 50 cm^{-1} (based on local C_{2v} symmetry of the C_6H_5 groups), and discussed together with the spectra of the halides $Ge(C_6H_5)_3X$ where X = F, Cl, and Br (IR spectrum illustrated from 4000 to 200 cm^{-1}) [20].

IR spectrum (wavenumbers in cm^{-1}): The ν(GeH) band is in the range 2036 to 2040 for commonly used solvents such as C_6H_{14}, C_6H_6, CCl_4, or CS_2 and even for polar solvents such as ether, dioxane, acetone, or pyridine; 2029 for a HMPT solution [23]; see also [22]. The rather high wavenumbers for solutions in $CHCl_3$ (2043) or CH_3CN (2044) could not be interpreted [23]; 2031 to 2033 was reported for the neat liquid or solid compound [22, 23], but 2046 for the liquid in [20]. For other ν(GeH) data, see also [10, 13, 14, 17, 18, 31, 32]. The δ(GeH) band appears at 708(vs) [20] or 709(s) [13, 31]. Bands at 1093, 323(s, br), and 303(m) cm^{-1} are characteristic for the C_6H_5Ge group [13]; 1093(s), 320(mw), and 297(mw) is given in [20]. (Spectrum in $CHCl_3$ listed down to 1000 cm^{-1} without assignment in [4].)

Raman spectrum: ν(GeH) 2041(14) for the liquid and 2033(59) cm^{-1} for the solid; δ(GeH) 712(2) for the liquid and 711(2) cm^{-1} for the solid. Other vibrations, their intensities, and partly the degree of depolarization are listed down to 170 cm^{-1}, including lattice modes of the solid at 107(<1), 83(10), 69(7), and 48(4) cm^{-1} [20].

$Ge(C_6H_5)_3H$ was involved in studies dealing with the dependence of the GeH vibration of various organogermanium compounds [9, 26, 32] and their intensity [28, 30, 32] on the nature of the substituents.

The ionization potentials in the He(I) photoelectron spectrum at 9.15 (ring), 10.11 (GeC_3), 10.9 (GeH), 11.8, 12.6, 13.1, 13.9, 14.6, 15.7, and 16.4 eV are very similar to those of $Si(C_6H_5)_3H$ and $Sn(C_6H_5)_3H$ (all spectra illustrated) and indicate no sizable interaction among the ring π orbitals and the central atom as found for $C(C_6H_5)_3H$ [27]; for illustrations of the PE spectrum, see also [39]. The first IP was used to calculate the Brown-Okamoto σ_p^+ constant of the $Ge(C_6H_5)_2H$ substituent on a phenyl ring [36]. He(I) PE spectra of $Ge(C_6H_5)_3H$, $Mn(C_5H_5)(CO)_3$, and $Mn(^5L)(CO)_2(HGe(C_6H_5)_3)$ complexes were monitored for elucidating the nature and extent of Ge-H bond interaction with the Mn atom [39].

The EXAFS (Extended X-ray Absorption Fine Structure) spectrum and the associated Fourier transform are illustrated [35, 37]. A calculation including the Fourier transform and phase factors gave a Ge-C distance of 1.96 Å [37].

The compound exists in two monotropic forms: the stable α form melting above 40 °C (see below) and the less stable β form melting at 27 °C [1] (24 °C in [10]) and forming transparent hexagonal plates on crystallization from CH_3OH [4] or cooling to 0 °C the oil obtained by evaporation of an ether solution under vacuum [2]. The melt of the β form, maintained at 30 °C, freezes immediately and completely when seeded with the α form [1].

For the α form, many variable melting points were reported ranging from 39.5 °C [29] to 47 to 49 °C [40]. Most of the melting points are between 43 and 47 °C [5, 6, 8, 21, 24, 34]; for lower values, see also [20, 31, 41]. Translucent, flaky crystals of the α form were isolated by crystallization from CH_3OH, melting at 41.0 to 41.5 °C [3, 7].

Published boiling points (in °C/Torr) are 128 to 136/0.01 [12], 137 to 145/0.01 [10], 143 to 151/0.25 [5], 151 to 152/1 [24], 155 to 157/1.5 [24], 156 to 157/1 [21], 161 to 164/2 [24]. The hot vapors are extremely penetrating and irritating [1].

The electric birefringence of $Ge(C_6H_5)_3H$ in CCl_4 was measured at 298 K and 589 nm: molar Kerr constant $_{inf}(_mK_2) = (11.6 \pm 0.8) \cdot 10^{-27}\ m^5 \cdot V^{-2} \cdot mol^{-1}$). Agreement with calculated molar Kerr constants was obtained when the angle ϕ between the C_6H_5 planes and the C_3 axis was $33 \pm 3°$, very similar to the same angle in $Si(C_6H_5)_3H$. The molar polarization and refraction ($_{inf}P_2 = 106$ and $R_D = 88.9\ cm^3$) were determined for the same conditions, and the dipole moment was calculated to be $\mu = (2.60 \pm 0.10) \cdot 10^{-30}$ Cm (ca. 8.7 D) [34]. Molecular mechanics calculations of the ground-state conformation showed the molecule to have propeller-like (C_3) geometry with the bond lengths Ge-C = 1.939 and Ge-H = 1.553 Å, the bond angles C-Ge-H = 109.0° and C-Ge-C = 110.0°, and $\phi = 33°$. This geometry resembles the "effective" conformation derived from the Kerr effect and constitutes the conformational average over a broad low-energy region where C_6H_5 rotations are relatively unhindered [38].

$Ge(C_6H_5)_3D$. IR spectrum: ν(GeD) 1473(m) [13] or 1470(w, sh) cm^{-1} (liquid phase) [20]; δ(GeD) 526(s) [13] or 530(s) cm^{-1} (liquid phase) [20]; characteristic bands at 1093, 322(s, broad), and 303(m) cm^{-1} [13]. Raman spectrum: ν(GeD) 1469 (liquid) and 1464 (solid) cm^{-1}; δ(GeD) 534 cm^{-1} (liquid) [20].

Like in the case of $Ge(C_6H_5)_3H$ two forms of $Ge(C_6H_5)_3D$ were isolated: one that melts just above room temperature (ca. 30°C) and was obtained by distillation under vacuum and crystallization of the liquid in an ice bath; the second form melts at 42°C and was isolated without distillation by dissolving the product in a small amount of ether and adding CH_3OH [20].

References:

[1] Kraus, C. A.; Foster, L. S. (J. Am. Chem. Soc. **49** [1927] 457/67).
[2] Johnson, O. H.; Nebergall, W. H. (J. Am. Chem. Soc. **71** [1949] 1720/2).
[3] Johnson, O. H.; Harris, D. M. (J. Am. Chem. Soc. **72** [1950] 5566/8).
[4] West, R. (J. Am. Chem. Soc. **75** [1953] 6080/1).
[5] Fuchs, R.; Gilman, H. (J. Org. Chem. **22** [1957] 1009/11).
[6] Gilman, H.; Gerow, C. W. (J. Am. Chem. Soc. **79** [1957] 342/5).
[7] Johnson, O. H.; Nebergall, W. H.; Harris, D. M. (Inorg. Synth. **5** [1957] 76/8).
[8] Gilman, H.; Zuech, E. A. (J. Org. Chem. **26** [1961] 3035/7).
[9] Mathis, R.; Satgé, J.; Mathis, F. (Spectrochim. Acta **18** [1962] 1463/72).
[10] Glockling, F.; Hooton, K. A. (J. Chem. Soc. **1963** 1849/54).

[11] Cross, R. J.; Glockling, F. (J. Chem. Soc. **1964** 4125/33).
[12] Brooks, E. H.; Glockling, F.; Hooton, K. A. (J. Chem. Soc. **1965** 4283/8).
[13] Cross, R. J.; Glockling, F. (J. Organomet. Chem. **3** [1965] 146/55).
[14] Cross, R. J.; Glockling, F. (J. Chem. Soc. **1965** 5422/32).
[15] Ryan, M. T.; Lehn, W. L. (J. Organomet. Chem. **4** [1965] 455/60).
[16] Birchall, T.; Jolly, W. L. (Inorg. Chem. **5** [1966] 2177/80).
[17] Creemers, H. M. J. C.; Noltes, J. G. (J. Organomet. Chem. **7** [1967] 237/47).
[18] Brooks, E. H.; Cross, R. J.; Glockling, F. (Inorg. Chim. Acta **2** [1968] 17/21).
[19] Carey, F. A.; Tremper, H. S. (Tetrahedron Lett. **1969** 1645/8).
[20] Durig, J. R.; Sink, C. W.; Turner, J. B. (Spectrochim. Acta A **25** [1969] 629/45).

[21] Vyazankin, N. S.; Bychkov, V. T.; Linzina, O. V.; Razuvaev, G. A. (Zh. Obshch. Khim. **39** [1969] 979/82; J. Gen. Chem. USSR [Engl. Transl.] **39** [1969] 950/2).

[22] Mathis, R.; Barthelat, M.; Mathis, F. (Spectrochim. Acta A **26** [1970] 1993/2000).
[23] Mathis, R.; Barthelat, M.; Mathis, F. (Spectrochim. Acta A **26** [1970] 2001/5).
[24] Vyazankin, N. S.; Bychkov, V. T.; Linzina, O. V.; Razuvaev, G. A. (J. Organomet. Chem. **21** [1970] 107/13).
[25] Rivière, P.; Satgé, J. (J. Organomet. Chem. **49** [1973] 157/72).
[26] Egorochkin, A. N.; Khorshev, S. Ya.; Ostasheva, N. S.; Satgé, J.; Rivière, P.; Barrau, J.; Massol, M. (J. Organomet. Chem. **76** [1974] 29/36).
[27] Distefano, G.; Pignataro, S.; Szepes, L.; Borossay, J. (J. Organomet. Chem. **104** [1976] 173/8).
[28] Egorochkin, A. N.; Khorshev, S. Ya.; Ostasheva, N. S.; Sevastyanova, E. I.; Satgé, J.; Rivière, P.; Barrau, J. (J. Organomet. Chem. **105** [1976] 311/20).
[29] Sakurai, H.; Mochida, K.; Kira, M. (J. Organomet. Chem. **124** [1977] 235/52).
[30] Egorochkin, A. N.; Sevast'yanova, E. I.; Khorshev, S. Ya.; Ratushnaya, S. Kh.; Satgé, J.; Rivière, P.; Barrau, J.; Richelme, S. (J. Organomet. Chem. **162** [1978] 25/35).

[31] Eaborn, C.; Singh, B. (J. Organomet. Chem. **177** [1979] 333/48).
[32] Skobeleva, S. E.; Egorochkin, A. N.; Khorshev, S. Ya.; Ratushnaya, S. Kh.; Rivière, P.; Satgé, J.; Richelme, S.; Cazes, A. (J. Organomet. Chem. **182** [1979] 1/7).
[33] Batchelor, R. J.; Birchall, T. (J. Am. Chem. Soc. **105** [1983] 3848/52).
[34] Allen, G. W.; Armstrong, R. S.; Aroney, M. J.; Skamp, K. R. (J. Mol. Struct. **129** [1985] 145/9).
[35] Mochida, K.; Fujii, A.; Tsuchiya, N.; Tohji, K.; Udagawa, Y. (Organometallics **6** [1987] 1811/2).
[36] Egorochkin, A. N.; Razuvaev, G. A.; Lopatin, M. A. (J. Organomet. Chem. **344** [1988] 49/60).
[37] Kugita, T.; Mochida, K.; Tohji, K.; Udagawa, Y. (Chem. Lett. **1989** 501/4).
[38] Allen, G. W.; Aroney, M. J.; Hambley, T. W. (J. Mol. Struct. **216** [1990] 227/40).
[39] Lichtenberger, D. L.; Rai-Chaudhuri, A. (J. Chem. Soc. Dalton Trans. **1990** 2161/6).
[40] Cerveau, G.; Chuit, C.; Corriu, R. J. P.; Reyé, C. (Organometallics **10** [1991] 1510/5).

[41] Clark, K. B.; Griller, D. (Organometallics **10** [1991] 746/50).
[42] Woo, H.-G.; Freeman, W. P.; Tilley, T. D. (Organometallics **11** [1992] 2198/205).
[43] Okano, M.; Kugita, T.; Mochida, K. (J. Electroanal. Chem. **356** [1993] 303/7).

1.3.1.1.7.3 Solutions, Mass Spectrum, Decomposition, and Chemical Reactions

Solutions. $Ge(C_6H_5)_3H$ is extremely soluble in C_6H_6, $C_6H_5CH_3$, $CHCl_3$, and petroleum ether, very soluble in CCl_4 and ether, and slightly soluble in liquid NH_3. It is fairly soluble in CH_3OH from which it is precipitated on addition of H_2O [1].

The ^1H NMR spectra of $Ge(C_6H_5)_3H$ and various dissolved hydrides of Ge, P, and As were used to determine the relative acidity of these compounds in a suspension of $NaNH_2$ in liquid NH_3. Unexpectedly, $Ge(C_6H_5)_3H$ was found to be a much weaker acid than GeH_4 [24]. Different explanations were given for this low acidity [24, 47, 51]. A spectrophotometric determination of the acidity of $Ge(C_6H_5)_3H$ in DMSO gave $pk_a = 23.0$ [36]; $pK_a = 23.2$ (referred to fluorene, $pK_a = 20.6$) was derived from an NMR study in liquid NH_3 at 30°C [51].

Mass Spectrum. The m/e values and their intensities are listed in [121]. Fragmentation by electron impact (70 eV) showed that C_6H_6 elimination is a major process leading to $[Ge(C_6H_5)_2^{\cdot}]^+$, $[GeC_6H_5]^+$, and $[Ge(C_6H_5)C_6H_4]^+$. The metastable transitions $[Ge(C_6H_5)C_6H_4]^+ \rightarrow [Ge(C_6H_5)C_4H_2]^+ + C_2H_2$ and $[Ge(C_6H_5)C_4H_2]^+ \rightarrow [Ge(C_6H_5)C_2]^+ + C_2H_2$ were confirmed. A complete fragmentation scheme including the ion abundances is presented in [33].

No sign of **thermal decomposition** was found at 250 °C [1]. Heating in an He atmosphere at 300°C for 0.5 h resulted in partial disproportionation into $Ge(C_6H_5)_4$ and $Ge(C_6H_5)_2H_2$ [4, 9].

Irradiation of $Ge(C_6H_5)_3H$ in a matrix (?) with a low-pressure Hg arc lamp at 77 K gave the $Ge(C_6H_5)_3^{\bullet}$ radical [118].

Constant-current **electrolysis** of $Ge(C_6H_5)_3H$ in a two-compartment cell in CH_3CN ($[N(C_4H_9)_4][ClO_4]$ electrolyte) yielded 60% $Ge(C_6H_5)_3Cl$ and 40% $(Ge(C_6H_5)_3)_2O$ in the cathode compartment. With $[N(C_4H_9)_4][BF_4]$ as the electrolyte, $(Ge(C_6H_5)_3)_2O$ (70% yield) and $Ge(C_6H_5)_3F$ were formed. Only traces of $Ge(C_6H_5)_3Cl$ were found after heating $Ge(C_6H_5)_3H$ in CH_3CN in the presence of $[N(C_4H_9)_4][ClO_4]$ to 80 °C [122]. Constant-current electrolysis of $Ge(C_6H_5)_3H$ in DMF ($[N(C_4H_9)_4][BF_4]$ electrolyte) at −40 °C produced at the cathode the yellow color of the $[Ge(C_6H_5)_3]^-$ anion which was identified by its ^{13}C NMR spectrum [130].

Reactions with Elements. $Ge(C_6H_5)_3H$ is stable in air under ordinary conditions [1], but on very long standing there were indications that oxidation occurred [1 to 3]. Pt complexes appear to catalyze the aerial oxidation of $Ge(C_6H_5)_3H$ in C_2H_5OH to $(Ge(C_6H_5)_3)_2O$ [20].

Treatment with halogens X_2 immediately gives $Ge(C_6H_5)_3X$ [1]. $Ge(C_6H_5)_3I$ was obtained by refluxing $Ge(C_6H_5)_3H$ and I_2 in CCl_4 [4]. $Ge(C_6H_5)_3Na$ and H_2 were the main products of the reaction of $Ge(C_6H_5)_3H$ with Na in liquid NH_3 [1].

Reactions with Inorganic Compounds. The compound is relatively stable towards hydrolysis in moist CH_3OH, but some $Ge(C_6H_5)_3OH$ and H_2 are formed slowly [23]. In the reaction of $Ge(C_6H_5)_3Li$ with fluorene $(Ge(C_6H_5)_3)_2O$ was obtained as a by-product and was supposed to result from the hydrolysis of $Ge(C_6H_5)_3H$ during workup [12]. Treatment with strong halogen acids HX gave $Ge(C_6H_5)_3X$ [1]; see also [6]. However, in solutions of $C_6H_5CH_3$ or dioxane $Ge(C_6H_5)_3H$ did not react with HCl [67]. Hydrogen was reported to be readily evolved when treating $Ge(C_6H_5)_3H$ with dilute alcoholic KOH [11], but no significant solvolysis was observed in CH_3OH containing 2.3 M CH_3ONa at ca. 21 °C after 2 d. In a solution of $\mathbf{Ge(C_6H_5)_3D}$ in C_2H_5OH containing 1.7 M KOH at ca. 21 °C, D-H exchange occurred between the germane and the alcohol (ca. one-third complete after 15 min) [61].

$Ge(C_6H_5)_3H$ reacted with NaH at 40 °C (4 h) or with KH at room temperature (1 h) in DME to give the corresponding $Ge(C_6H_5)_3M$ with good yield [82]. The equilibrium with $[Ge(C_6H_5)_3]^-$ formed in $NaNH_2$-NH_3 mixtures was used to determine the relative acidity of $Ge(C_6H_5)_3H$ [24]. $Ge(C_6H_5)_3H$ reacted with $Ge(OH)(OCH_2CH_2)_3N$ in refluxing $C_6H_4(CH_3)_2$ in the presence of Amberlyst 15 or H_2PtCl_6 to give $Ge(C_6H_5)_3OGe(OCH_2CH_2)_3N$ [124].

trans-$Pd(P(C_2H_5)_3)_2Cl_2$ reacts with $Ge(C_6H_5)_3H$ to give $Ge(C_6H_5)_3Cl$ and probably $Pd(P(C_2H_5)_3)_2(H)Cl$ [37]. The reaction with $Pt(P(CH_3)_3)_2H_2$ in THF yielded 34% cis,trans,-cis-$Pt(P(CH_3)_3)_2(Ge(C_6H_5)_3)_2H_2$ along with H_2 [112]. Photolysis of $Ge(C_6H_5)_3H$ and $[Pt_2(P_2O_5H_2)_4]^{4-}$ in CH_3CN resulted in the formation of $[Pt_2(P_2O_5H_2)_4(Ge(C_6H_5)_3)_2]^{4-}$ and $[Pt_2(P_2O_5H_2)_4H_2]^{4-}$; a rate constant of 2.9×10^7 $M^{-1} \cdot s^{-1}$ for the H atom transfer was measured [104, 105]. $Ge(C_6H_5)_3H$ did not react with cis-$Pt(P(C_2H_5)_3)_2Br_2$ in the presence of $N(C_2H_5)_3$ in refluxing C_6H_6 [20].

Hydrogermylation Reactions. In the reactions of $Ge(C_6H_5)_3H$ with CH_2=CH_2Y compounds (Y = $OOCCH_3$, CN, $CONH_2$, $COCH_3$, $COOCH_3$, and $CH_2CH_2CH_2OH$) or $CH \equiv CC(OH)(CH_3)_2$, no reducing properties of $Ge(C_6H_5)_3H$ upon the functional groups were observed [14].

The γ radiation-initiated hydrogermylation of asymmetrical olefins proceeded regioselectively to yield the β adducts regardless of the double bond polarization [90]; see also [79].

A charge-transfer complex was formed with $(NC)_2C{=}C(CN)_2$ in CH_2Cl_2 and used for determining the σ_p-Hammet and σ_p^+-Brown-Okamoto constant of the $Ge(C_6H_5)_2H$ substituent [85].

The hydrogermylation reactions are summarized in Table 16. A review on hydrogermylations has recently been published [131].

Table 16
Hydrogermylation Reactions with $Ge(C_6H_5)_3H$.
$GeR_3 = Ge(C_6H_5)_3$.

reactant	conditions → products (yield)	Ref.
reactants with one C=C bond		
$CH_2{=}C(C_6H_5)_2$	75 °C (bath), in C_7H_{16}, 2 d $C_6H_5CO\text{-}OO\text{-}COC_6H_5$ initiator → unidentified yellow oil (olefin polymer?); no hydrogermylation product	[7]
$CH_2{=}CHOOCCH_3$	50 to 60 °C, 6 h → $GeR_3CH_2CH_2OOCCH_3$ (54%)	[14]
$CH_2{=}CHC_6H_5$	120 °C, ca. 12 h → $GeR_3CH_2CH_2C_6H_5$ (40%)	[14]
	50 °C, in $C_2H_4Cl_2$, 10 h, $Rh(P(C_6H_5)_3)_3Cl$ catalyst → $GeR_3CH_2CH_2C_6H_5$ and $GeR_3CH(C_6H_5)CH_3$ in a 94 : 6 ratio (75% total yield)	[52]
	50 °C, in $C_2H_4Cl_2$, 1 d, cis-$Pt(P(C_6H_5)_3)_2Cl_2$ catalyst → $GeR_3CH_2CH_2C_6H_5$ and $GeR_3CH(C_6H_5)CH_3$ in a 89 : 11 ratio (52% total yield)	[52]
	20 °C, in $C_2H_4Cl_2$, 6 h, H_2PtCl_6 catalyst → $GeR_3CH_2CH_2C_6H_5$ and $GeR_3CH(C_6H_5)CH_3$ in a 7 : 3 ratio (60% total yield)	[52]
$CH_2{=}CHCH_2B(OCH_3)_2$	in ether, 2 h → $GeR_3(CH_2)_3B(OCH_3)_2$ (100%)	[21]
$CH_2{=}CHCH_2NH_2$	25 °C, γ radiation → $GeR_3(CH_2)_3NH_2$ (95%)	[83]; see also [90]
$CH_2{=}CHCH_2OH$	25 °C, γ radiation → $GeR_3(CH_2)_3OH$ (94%)	[83]
$CH_2{=}CHCN$	50 to 60 °C, 6 h → $GeR_3CH_2CH_2CN$ (83%)	[14]
	25 °C, γ radiation → $GeR_3CH_2CH_2CN$ (98%)	[83]
$CH_2{=}CHCONH_2$	50 to 60 °C, 21 h → $GeR_3CH_2CH_2CONH_2$	[14]

References on p. 128

Table 16 (continued)

reactant	conditions → products (yield)	Ref.
	25 °C, γ radiation → $GeR_3CH_2CH_2CONH_2$ (82%)	[83]; see also [90]
CH_2=$CHCOOCH_3$	60 to 70 °C, 67 h → $GeR_3CH_2CH_2COOCH_3$ (50%)	[14]
CH_2=$CHCOCH_3$	60 to 70 °C, 18 h [14] or AIBN initiator [39] → $GeR_3CH_2CH_2COCH_3$ (53%) [14], (80%) [39]	[14, 39]
CH_2=$CH\overline{CHCH_2O}$	heating in a sealed tube, H_2PtCl_6 catalyst → $GeR_3CH(CH_3)CH_2CH_2OGeR_3$	[27]
CH_2=$CH\overline{C(CH_3)CH_2O}$	heating in a sealed tube, H_2PtCl_6 catalyst → $GeR_3CH(CH_3)CH(CH_3)CH_2OGeR_3$	[27]
CH_2=$CH(CH_2)_3OH$	50 °C, 4 h → $GeR_3(CH_2)_4CH_2OH$ (41%)	[14]
CH_2=CHC_4H_9	80 °C, 3 h → $GeR_3C_6H_{13}$ (90%)	[29]
	60 °C, in $C_2H_4Cl_2$, 2 d → $GeR_3C_6H_{13}$ (56%) (no reaction at 20 °C for 30 d)	[52]
	80 °C, 1 h, AIBN initiator, sealed tube → $GeR_3C_6H_{13}$ (95%)	[29]
	50 °C, in $C_6H_4Cl_2$-1,2, 12 h, $Pd(P(C_6H_5)_3)_2Cl_2$ catalyst → $GeR_3C_6H_{13}$ (78%)	[52]
	50 °C, in $C_2H_4Cl_2$, 7 h, cis-$Pt(P(C_6H_5)_3)_2Cl_2$ catalyst → $GeR_3C_6H_{13}$ (68%; 51% yield at 20 °C for 7 d)	[52]
	50 °C, in $C_2H_4Cl_2$, 8 h, $Rh(P(C_6H_5)_3)_3Cl$ catalyst → $GeR_3C_6H_{13}$ (60%; 48% yield at 20 °C for 10 d)	[52]
	30 °C, 1 h 20 min, UV irradiated → $GeR_3C_6H_{13}$ (95%)	[29]
	25 °C, γ irradiated → $GeR_3C_6H_{13}$ (82%)	[83]
CH_2=$CHCH_2CH_2COCH_3$	90 °C, 1 d, AIBN initiator → $GeR_3(CH_2)_4COCH_3$ (92%)	[30, 39]
CH_2=CHC_5H_{11}	similar to the conditions used for hydrogermylation of CH_2=CHC_4H_9 → $GeR_3C_7H_{15}$	[52]
CH_2=CHC_6H_{13}	75 °C (bath), in petroleum ether, 1 d, C_6H_5CO-OO-COC_6H_5 initiator	[7]

Table 16 (continued)

reactant	conditions → products (yield)	Ref.
	→ $GeR_3C_8H_{17}$ (91%)	
	in C_7H_{16}, 2 d, UV-irradiated → $GeR_3C_8H_{17}$ (80%) and $Ge(C_6H_5)_4$	
	110 to 115°C, 5 d → $GeR_3C_8H_{17}$ (29%)	[14]
	25°C, γ-irradiated → $GeR_3C_8H_{17}$ (81%)	[83]
$CH_2{=}CH(CH_2)_8CHO$	120°C, 2 d → $GeR_3(CH_2)_{10}CHO$ (78%)	[64]
	80°C, 2 d, AIBN initiator → $GeR_3(CH_2)_{10}CHO$ (80%) and $GeR_3(CH_2)_{11}OGeR_3$ (20%)	
$CH_2{=}CHC_{10}H_{21}$	4:1 mole ratio, 25°C, in C_6H_6, 15 h, $B(C_2H_5)_3$ → $GeR_3C_{12}H_{25}$ (96%)	[109]
$CH_2{=}CHC_{16}H_{33}$	85°C, in C_7H_{16}, 17.5 h, $C_6H_5CO\text{-}OO\text{-}COC_6H_5$ initiator → $GeR_3C_{18}H_{37}$ (67%) and $(GeR_3)_2O$	[8]
$C_{10}H_{21}CH{=}CHSC_6H_5$	see reaction with sulfur compounds, p. 121	[108]
$c\text{-}C_6H_{10}$	75°C (bath), in C_7H_{16}, 2 d, $C_6H_5CO\text{-}OO\text{-}COC_6H_5$ initiator → $GeR_3C_6H_{11}\text{-}c$ and $Ge(C_6H_5)_4$	[7]
	in C_7H_{16}, UV-irradiated, 2 d → $GeR_3C_6H_{11}\text{-}c$, $Ge(C_6H_5)_4$	[7]
	60 to 70°C, 5 d → $GeR_3C_6H_{11}\text{-}c$ (39%)	[14]
	50°C, in $C_2H_4Cl_2$, 1 d, cis-$Pt(P(C_6H_5)_3)_2Cl_2$ catalyst → $GeR_3C_6H_{11}\text{-}c$ (12%)	[52]
	50°C, in $C_2H_4Cl_2$, 1 d, $Rh(P(C_6H_5)_3)_3Cl$ catalyst → $GeR_3C_6H_{11}\text{-}c$ (10%)	[52]
	25°C, γ-irradiated → $GeR_3C_6H_{11}\text{-}c$ (76%)	[83]; see also [90]
$Si(C_6H_5)_3CH_2CH{=}CH_2$	75°C, in C_6H_{14}, 1 d, $C_6H_5CO\text{-}OO\text{-}COC_6H_5$ initiator → $GeR_3CH_2CH_2CH_2Si(C_6H_5)_3$ (76%)	[10]
$GeR_3CH_2CH{=}CH_2$	refluxing in C_6H_{14}, 18.5 h, $C_6H_5CO\text{-}OO\text{-}COC_6H_5$ initiator → $GeR_3CH_2CH_2CH_2GeR_3$ (86%)	[8]
$GeR_3OCH_2CH{=}CHCH_3$	heating for 3 h, H_2PtCl_6 catalyst → $GeR_3CH(CH_3)CH_2CH_2OGeR_3$	[27]

References on p. 128

Table 16 (continued)

reactant	conditions → products (yield)	Ref.
reactants with cumulated C=C bonds		
CH_2=C=CH_2	25 °C, in THF, 5 h, $Pd(P(C_6H_5)_3)_4$ catalyst → GeR_3CH_2CH=CH_2 (88%)	[110]
	25 °C, in C_6H_6-C_6H_{14}, 2 h, $B(C_2H_5)_3$ → GeR_3CH_2CH=CH_2 (57%)	
CH_2=C=$CHC_{10}H_{21}$	25 °C, in THF, 5 h, $Pd(P(C_6H_5)_3)_4$ catalyst → $GeR_3CH(C_{10}H_{21})CH$=CH_2 (53%)	[110]
	25 °C, in C_6H_6-C_6H_{14}, 2 h, $B(C_2H_5)_3$ → GeR_3CH_2CH=$CHC_{10}H_{21}$ (82%)	
CH_2=C=$CHSi(CH_3)_2C_6H_5$	25 °C, in THF, 5 h, $Pd(P(C_6H_5)_3)_4$ catalyst → GeR_3CH_2CH=$CHSi(CH_3)_2C_6H_5$ (57%) and $GeR_3CH(Si(CH_3)_2C_6H_5)CH$=$CH_2$ (38%)	[110]
	25 °C, in C_6H_6-C_6H_{14}, 2 h, $B(C_2H_5)_3$ → GeR_3CH_2CH=$CHSi(CH_3)_2C_6H_5$ (69%) and $GeR_3C(CH_3)$=$CHSi(CH_3)_2C_6H_5$ (28%)	
$C_5H_{11}CH$=C=CHC_5H_{11}	25 °C, in THF, 5 h, $Pd(P(C_6H_5)_3)_4$ catalyst → $GeR_3CH(C_5H_{11})CH$=CHC_5H_{11} (79%)	[110]
reactants with one C≡C bond		
CH≡$CSi(CH_3)_3$	25 °C, in THF, 5 h, $Pd(P(C_6H_5)_3)_4$ catalyst → (E)-GeR_3CH=$CHSi(CH_3)_3$ (98%)	[103]
	−78 °C, in $C_6H_5CH_3$, 6 h, $B(C_2H_5)_3$ → (Z)- and (E)-GeR_3CH=$CHSi(CH_3)_3$ in a 56:44 ratio (72% total yield)	[120]
	60 °C, in C_6H_6, 3 h, $B(C_2H_5)_3$ → (E)-GeR_3CH=$CHSi(CH_3)_3$ (87%)	
CH≡$CC_5H_4FeC_5H_5$	120 to 130 °C, 31 h → cis- and trans-GeR_3CH=$CHC_5H_4FeC_5H_5$ (25%)	[55]
CH≡CC_6H_5	135 °C, 4 d → GeR_3CH=CHC_6H_5 (12%)	[14]
	50 °C, 4 to 6 d → trans-GeR_3CH=CHC_6H_5 and cis-GeR_3CH=CHC_6H_5, ratio varied from 9:91 to 100:0 (38% total yield)	[53], see also [46]
	20 °C, 3 h, $Rh(P(C_6H_5)_3)_3Cl$ catalyst → trans-GeR_3CH=CHC_6H_5, cis-GeR_3CH=CHC_6H_5, and $GeR_3C(C_6H_5)$=CH_2 in a 82:15.5:2.5 ratio (85% total yield)	[46, 53]
	50 °C, 10 h, cis-$Pt(P(C_6H_5)_3)_2Cl_2$ catalyst	

Table 16 (continued)

reactant	conditions → products (yield)	Ref.
	→ trans-$GeR_3CH{=}CHC_6H_5$, cis-$GeR_3CH{=}CHC_6H_5$, and $GeR_3C(C_6H_5){=}CH_2$ in a 85:10:5 ratio (70% total yield)	
	20°C, 0.5 h, H_2PtCl_6 catalyst → trans-$GeR_3CH{=}CHC_6H_5$, cis-$GeR_3CH{=}CHC_6H_5$, and $GeR_3C(C_6H_5){=}CH_2$ in a 70:0.5:29.5 ratio (78% total yield)	
	25°C, in THF, 5 h, $Pd(P(C_6H_5)_3)_4$ catalyst → $GeR_3C(C_6H_5){=}CH_2$ and (E)-$GeR_3CH{=}CHC_6H_5$ in a 9:91 ratio (89% total yield)	[103]
$CH{\equiv}CCH_3$	−78°C, 2 h, $B(C_2H_5)_3$ → (Z)- and (E)-$GeR_3CH{=}CHCH_3$ in a >20:1 ratio (65% total yield)	[102, 120]
$CH{\equiv}CC(OH)(CH_3)_2$	50 to 60°C, 77 h → $GeR_3CH{=}CHC(OH)(CH_3)_2$ (49%)	[14]
$CH{\equiv}CCH_2CH_2OH$	−78°C, in $C_6H_5CH_3$, 5 h, $B(C_2H_5)_3$, → (Z)- and (E)-$GeR_3CH{=}CHCH_2CH_2OH$ in a >20:1 ratio (80% total yield)	[102, 120]
	60°C, in C_6H_6, 15 h, $B(C_2H_5)_3$ → (Z)- and (E)-$GeR_3CH{=}CHCH_2CH_2OH$ in a <1:20 ratio (75% total yield)	
	25°C, in THF, 5 h, $Pd(P(C_6H_5)_3)_4$ catalyst → (E)-$GeR_3CH{=}CHCH_2CH_2OH$ (57%)	[103]
$CH{\equiv}CCH_2CH_2OCH_2C_6H_5$	25°C, in THF, 5 h, $Pd(P(C_6H_5)_3)_4$ catalyst → $GeR_3C({=}CH_2)CH_2CH_2OCH_2C_6H_5$, (E)- and (Z)-$GeR_3CH{=}CHCH_2CH_2OCH_2C_6H_5$ in a 7:91:<2 ratio (52% total yield)	[103]
$CH{\equiv}C(CH_2)_4OH$	−78°C, in $C_6H_5CH_3$, 5 h, $B(C_2H_5)_3$ → (Z)- and (E)-$GeR_3CH{=}CH(CH_2)_4OH$ in a >20:1 ratio (80% total yield)	[102]
$CH{\equiv}CC_{10}H_{21}$	100°C, 0.5 h, H_2PtCl_6 catalyst → $GeR_3C({=}CH_2)C_{10}H_{21}$ and trans-$GeR_3CH{=}CHC_{10}H_{21}$ in a 25:75 ratio (99% total yield)	[98]
	25°C, in THF, 5 h, $Pd(P(C_6H_5)_3)_4$ catalyst → $GeR_3C(C_{10}H_{21}){=}CH_2$ and (E)-$GeR_3CH{=}CHC_{10}H_{21}$ in a 14:86 ratio (97% total yield)	[103]
	−78°C, in $C_6H_5CH_3$, 3 h, $B(C_2H_5)_3$ → (Z)- and (E)-$GeR_3CH{=}CHC_{10}H_{21}$ in a >20:1 ratio (76% total yield)	[102, 120]
	−20°C, in C_6H_6, 2 h, $B(C_2H_5)_3$	

References on p. 128

Table 16 (continued)

reactant	conditions → products (yield)	Ref.
	→ (Z)- and (E)-$GeR_3CH{=}CHC_{10}H_{21}$ in a 2:1 ratio (78% total yield) 25 °C, in C_6H_6, 2 h, $B(C_2H_5)_3$ → (Z)- and (E)-$GeR_3CH{=}CHC_{10}H_{21}$ in a 1:9 ratio (77% total yield) 60 °C, in C_6H_6, 2 h, $B(C_2H_5)_3$, → (Z)- and (E)-$GeR_3CH{=}CHC_{10}H_{21}$ in a <1:20 ratio (99% total yield) 0 °C, in THF, 2 h, $B(C_2H_5)_3$ → (Z)- and (E)-$GeR_3CH{=}CHC_{10}H_{21}$ in a 8:1 ratio (84% total yield) 0 °C, in $C_6H_5CH_3$-CH_3OH, 2 h, $B(C_2H_5)_3$ → (Z)- and (E)-$GeR_3CH{=}CHC_{10}H_{21}$ in a 10:1 ratio (80% total yield)	
$CH{\equiv}C(CH_2)_9COOC_2H_5$	−78 °C, in $C_6H_5CH_3$, 12 h, $B(C_2H_5)_3$ → (Z)- and (E)-$GeR_3CH{=}CH(CH_2)_9COOC_2H_5$ in a >10:1 ratio (64% total yield) 60 °C, in C_6H_6, 15 h, $B(C_2H_5)_3$ → (Z)- and (E)-$GeR_3CH{=}CH(CH_2)_9COOC_2H_5$ in a <1:20 ratio (93% total yield)	[102, 120]
$C_5H_{11}C{\equiv}CC_5H_{11}$	−78 °C, in $C_6H_5CH_3$, 8 h, $B(C_2H_5)_3$ → (Z)- and (E)-$GeR_3C(C_5H_{11}){=}CHC_5H_{11}$ in a >20:1 ratio (65% total yield)	[102, 120]
	25 °C, in THF, 5 h, $Pd(P(C_6H_5)_3)_4$ catalyst → (E)-$GeR_3C(C_5H_{11}){=}CHC_5H_{11}$ and (E)-$GeR_3CH(C_5H_{11})CH{=}CHC_4H_9$ in a 1:2 ratio (80% total yield)	[103]
$C_3H_5C{\equiv}CC_6H_5$	130 °C, 1 d, Pt on charcoal catalyst → cis-$Ge(C_6H_5)_3C(C_6H_5){=}CHC_6H_5$ (48%)	[32]
ketenes		
$(C_6H_5)_2C{=}C{=}O$	160 °C, 12 h → $GeR_3OCH{=}C(C_6H_5)_2$ (68%)	[34, 39]
reactants with conjugated C=C and C=O bonds		
$CH_2{=}CHCHO$	UV-irradiated or 80 °C, 4 h, AIBN → $GeR_3CH_2CH_2CHO$ (68%); no reaction in CH_3NO_2 or in the presence of the galvinoxyl radical	[49, 64]
	120 °C, 4 h → $GeR_3CH_2CH_2CHO$ (18%)	[64]
$CH_3CH{=}CHCHO$	150 °C, 6 h, AIBN initiator → $GeR_3CH(C_2H_5)CHO$ (12%)	[64]

Table 16 (continued)

reactant	conditions → products (yield)	Ref.
$(CH_3)_2C{=}CHCOCH_3$	180 °C, 2 d, AIBN initiator → $GeR_3OC(CH_3){=}CHCH(CH_3)_2$ (60%) (cis : trans = 2 : 3)	[30, 39]
reactants with N=N bonds		
$C_6H_5N{=}NC_6H_5$	130 °C, 20 h $GeR_3N(C_6H_5)NHC_6H_5$ (68%)	[62]
$CH_3OOCN{=}NCOOCH_3$	refluxing in C_6H_6, 11 d → $GeR_3N(COOCH_3)NHCOOCH_3$ (96%)	[54]; see also [62]
$C_2H_5OOCN{=}NCOOC_2H_5$	refluxing in C_6H_6, 1 d → $GeR_3N(COOC_2H_5)NHCOOC_2H_5$ (90%) in C_6H_6, UV-irradiated, 3.5 h → $GeR_3N(COOC_2H_5)NHCOOC_2H_5$ (93%)	[54]; see also [62]
reactants with C=N bonds		
$C_6H_5CH{=}NC_4H_9$-t	100 °C, in C_6H_6, 1 d, t-$C_4H_9OOC_4H_9$-t initiator, sealed tube → $GeR_3N(C_4H_9\text{-t})CH_2C_6H_5$ (8%)	[80]
$C_6H_5CH{=}NC_6H_5$	100 °C, in C_6H_6, 1 d, t-$C_4H_9OOC_4H_9$-t initiator, sealed tube → $GeR_3N(C_6H_5)CH_2C_6H_5$ (57%)	[80]
reactants with C=NO bonds		
$C_6H_5CH{=}N(C_6H_5)O$	80 °C, in THF, 1 d → $GeR_3ON(C_6H_5)CH_2C_6H_5$ (40%)	[56, 76]
	20 °C, UV-irradiated, 4 h → $GeR_3ON(C_6H_5)CH_2C_6H_5$ (7%) and $GeR_3OCH_2C_6H_5$ (via decomposition of the nitrone to C_6H_5CHO)	[76]
$C_6H_5CH{=}N(C_4H_9\text{-t})O$	60 °C, in C_6D_6, UV-irradiated, 7 h → $GeR_3ON(C_4H_9\text{-t})CH_2C_6H_5$ (7 to 8%), $GeR_3OCH_2C_6H_5$, $(GeR_3)_2O$, GeR_3OH, and $C_6H_5CH{=}NC_4H_9$-t	[76]
	90 to 100 °C, 12 h, no reaction in C_6D_6 with or without AIBN	[76]
	160 °C, 1 d, no reaction in C_6D_6 in the presence of traces of t-$C_4H_9OOC_4H_9$-t	[76]
$(CH_3)_2\overline{CCH_2CH_2CH{=}N}O$	95 °C, in C_6H_6, 1.5 d → $GeR_3O\overline{NCH_2CH_2CH_2C}(CH_3)_2$ (88%)	[76]
$(CH_3)_2\overline{CCH_2CH_2C(CH_3){=}N}O$	190 to 200 °C, in C_6D_6, 12 h → $GeR_3O\overline{NCH(CH_3)CH_2CH_2C}(CH_3)_2$ (23%)	[76]

References on p. 128

Table 16 (continued)

reactant	conditions → products (yield)	Ref.
$(CH_3)_2$-$\overline{CCH(C_6H_5)CH_2C(C_6H_5)=NO}$	100°C, in C_6D_6, 1 d no reaction	[76]

Reactions with Organic Halides. Refluxing with excess C_6H_5COCl at a bath-temperature of 210°C for 45 min yielded $Ge(C_6H_5)_3Cl$ [4]. Stirring with $CHBr_3$ in C_6H_6 and slowly adding solid t-$C_4H_9OK \cdot$ t-C_4H_9OH gave $Ge(C_6H_5)_3CHBr_2$ with 78% yield (significantly reduced yields in t-C_4H_9OH as the solvent) [42].

The reaction of $Ge(C_6H_5)_3H$ with $ICH_2CH_2CH_2I$ (2:1 mole ratio) in C_6H_6 in the presence of AIBN (at 71 and 99°C) or (t-$C_4H_9O)_2$ (at 134°C) or without an initiator (at 189 and 225°C) gave both cyclization and reduction products with the latter being preferred (cycl./red. = 0.040 at 71°C, 0.074 at 99°C, 0.14 at 134°C, 0.36 at 189°C, and 0.45 at 225°C). With $ICH_2(CH_2)_3CH_2I$ between 71 and 225°C only the reduction product was observed [38].

The reaction of $Ge(C_6H_5)_3H$ with $C_{11}H_{23}I$ in C_6H_6 at 80°C (8 h) in the presence of AIBN gave $C_{11}H_{24}$ with 98% yield. $Ge(C_6H_5)_3H$ was used as a reagent for the reductive alkylation of active olefins, e.g., $C_{11}H_{23}I$ and 2-cyclohexen-1-one in CH_3CN reacted under the conditions above to produce $C_{11}H_{24}$ (83% yield) and the adduct I (13%) [113]. A 3-iodotetrahydropyran derivative was converted into the corresponding 3-hydroxymethyl compound II using $NaBH_3CN$ and catalytic amounts of $Ge(C_6H_5)_3H$ in C_6H_6-THF under CO pressure at 105°C [128].

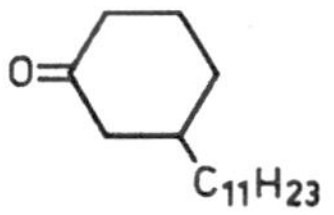

I

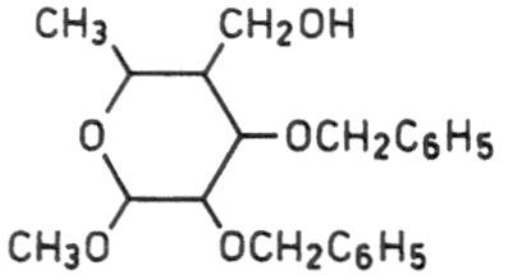

II

Reactions with Alcohols and Phenols. $Ge(C_6H_5)_3H$ does not undergo significant solvolysis at ca. 21°C in CH_3OH containing 2.3 M CH_3ONa. After 2 d, 98% of the starting material could be recovered [61]. In a similar experiment, after keeping $Ge(C_6H_5)_3H$ at 30°C in CH_3OH containing 1.5 M CH_3ONa for 1 d, no hydrogen was detectable in the gases above the liquid by mass spectroscopy [78]; H-D exchange between $Ge(C_6H_5)_3H$ and CH_3OD occurred under these conditions [61]. First-order rate constants for the loss of tritium from $Ge(C_6H_5)_3T$ during the T-H or T-D exchange with CH_3OH-CH_3ONa or CH_3OD-CH_3ONa were determined at 20 to 40°C [78]; see also the reactions with inorganic compounds on p. 112.

No dehydrocondensation was observed when $Ge(C_6H_5)_3H$ was heated with alcohols in the presence of Cu powder [48]. But boiling with $CH_3CH{=}CHCH_2OH$ in the presence of hydrogen-reduced Cu yielded $Ge(C_6H_5)_3OCH_2CH{=}CHCH_3$ [27].

Reactions with Ketones and Quinones. The following reactions of $Ge(C_6H_5)_3H$ with ketones in the presence of AIBN were reported (yield): $CH_3COCH_2C_3H_7$-i (120°C, 10 h) → $Ge(C_6H_5)_3OCH(CH_3)CH_2C_3H_7$-i (54%) [29]; $CH_3COCH_2CH_2CH{=}C(CH_3)_2$ (100°C, 10 h) → $Ge(C_6H_5)_3OCH(CH_3)CH_2CH_2CH{=}C(CH_3)_2$ (82%) [30, 39]; $C_6H_5COC_6H_5$ (120°C, 10 h) → $Ge(C_6H_5)_3OCH(C_6H_5)_2$ (58%) [48]; and cyclohexanone (70°C, 3 h) → $Ge(C_6H_5)_3OC_6H_{11}$ (92%) [29]. Heating equimolar amounts of $Ge(C_6H_5)_3H$ and compound III in C_6H_6 at 130°C (12 h) gave compounds IV (64%) and V (12%) [114].

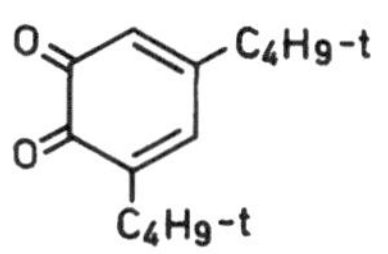

III

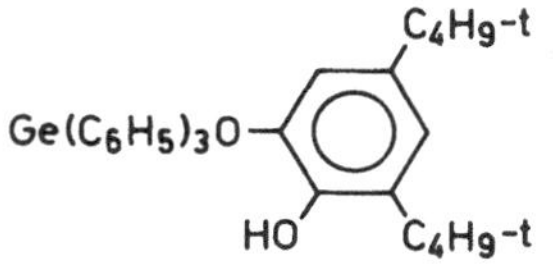

IV

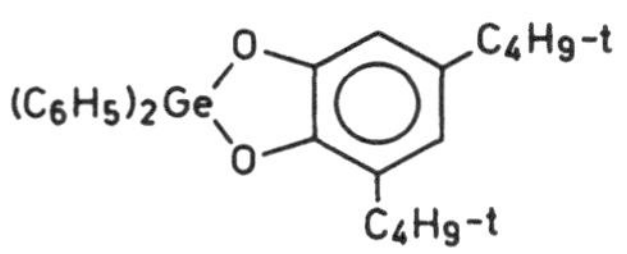

V

Reactions with Carboxylic Acids and Aldehydes. It was shown by ^{1}H NMR that 50% of $Ge(C_6H_5)_3H$ had reacted with 6.7 M CF_3COOH in $CDCl_3$ after 30 min [35].

$Ge(C_6H_5)_3H$ and C_6H_5CHO reacted in a sealed tube at 180 °C (1 d) in the presence of AIBN to yield $Ge(C_6H_5)_3OCH_2C_6H_5$ (65%); in the presence of $ZnCl_2$, this reaction led to $(Ge(C_6H_5)_3)_2O$ (70% yield) and $(C_6H_5CH_2)_2O$ (60%). With i-C_3H_7CHO in a sealed tube at 80 °C (4 h) in the presence of AIBN, $Ge(C_6H_5)_3OCH_2C_3H_7$-i was obtained with 93% yield [29]. For the reaction with $CH_2{=}CH(CH_2)_8CHO$, see Table 16, p. 115.

Reactions with Epoxides. The reaction of $Ge(C_6H_5)_3H$ with $CH_2{=}CH\overline{CHCH_2O}$, $CH_2{=}CH\overline{CHCH(C_{10}H_{21})O}$, or $CH_2{=}C(C_7H_{15})\overline{CHCH_2O}$ (4:1 mole ratio) in CH_3CN at 25 °C in the presence of $B(C_2H_5)_3$ gave trans-$Ge(C_6H_5)_3CH_2CH{=}CHCH_2OH$, trans-$Ge(C_6H_5)_3CH_2CH{=}CHCH(OH)C_{10}H_{21}$, or cis- and trans-$Ge(C_6H_5)_3CH_2C(C_7H_{15}){=}CHCH_2OH$, respectively. $Ge(C_6H_5)_3H$ and $CH_2{=}C(CH_3)CH{=}CH\overline{CHCH(C_{12}H_{25})O}$ in C_6H_6 similarly yielded $Ge(C_6H_5)_3CH_2C(CH_3){=}CHCH{=}CHCH(OH)C_{12}H_{25}$ as the major product and $C_2H_5CH_2C$-$(CH_3){=}CHCH{=}CHCH(OH)C_{12}H_{25}$ [109].

Reactions with Sulfur Compounds. The reactions of 1-alkenylsulfides $(C_6H_5)_2C{=}CHSCH_3$, $(C_6H_5)_2C{=}CHSC_6H_5$, $(C_2H_5)_2C{=}CHSC_6H_5$, VI, and VII with $Ge(C_6H_5)_3H$-$B(C_2H_5)_3$ in C_6H_6-C_6H_{14} at 60 °C (some hours) provided the corresponding 1-alkenyltriphenylgermanes in good yields. Similar treatment of a 1:1 mixture of (E)- and (Z)-$C_{10}H_{21}CH{=}CHSC_6H_5$ gave (E)-$Ge(C_6H_5)_3CH{=}CHC_{10}H_{21}$ (44% yield) and $Ge(C_6H_5)_3CH(C_{10}H_{21})CH_2SC_6H_5$ (50%). Ge-$(C_6H_5)_3CH_2CH{=}CHCH_3$ was obtained from $Ge(C_6H_5)_3H$-$B(C_2H_5)_3$ and $CH_3CH{=}CHCH_2$-SC_6H_5 [108].

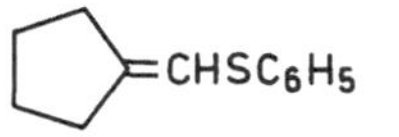

VI

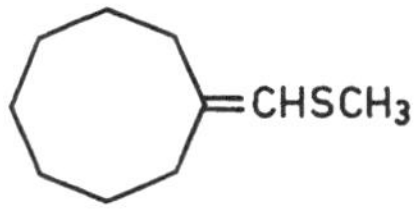

VII

Reactions with Nitrogen Compounds. $Ge(C_6H_5)_3H$ did not react with CH_2N_2 under UV irradiation, whereas similar reactions of trialkylgermanes, GeR_3H (R = C_2H_5, C_3H_7, or C_4H_9), afforded small yields of $Ge(CH_3)R_3$ [16]; see also [20].

Equimolar amounts of $Ge(C_6H_5)_3H$ and the oxaziridine $C_6H_5\overline{CHN(C_4H_9\text{-t})O}$ in C_6D_6 at 80 °C (1.5 h) gave $(Ge(C_6H_5)_3)_2O$, $Ge(C_6H_5)_3OH$, $C_6H_5CH{=}NC_4H_9$-t, C_6H_5CHO, and traces of t-$C_4H_9NH_2$. A reaction mechanism including the instable $Ge(C_6H_5)_3OCH(C_6H_5)NHC_4H_9$-t was discussed. Reacting $Ge(C_6H_5)_3H$ with an equimolar amount of compound VIII in C_6D_6 at 60 °C (12 h) gave $(Ge(C_6H_5)_3)_2O$, $Ge(C_6H_5)_3OH$, and compound IX [77].

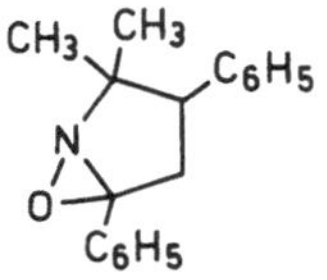

VIII

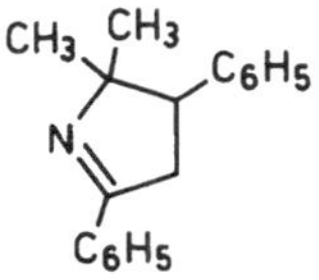

IX

References on p. 128

Reactions with Radicals and Carbenium Ions. The photochemically generated t-$C_4H_9O^•$ radical abstracts hydrogen from $Ge(C_6H_5)_3H$ at −30°C yielding the $Ge(C_6H_5)_3^•$ radical [69, 73]; see also [87, 93, 94, 123]; the rate constants for hydrogen abstraction $k = (9.2 \pm 0.4) \cdot 10^7$ and $(8.9 \pm 1.7) \times 10^7$ L · mol^{-1} · s^{-1} at 300 K were measured by different methods [95]; the value $(1.2 \pm 0.1) \times 10^8$ L · mol^{-1} · s^{-1} was given for 28.5°C [96]. The $Ge(C_6H_5)_3^•$ radical was used for reactions with aromatic nitrile oxides (formation of oxygen-centered GeR_4-derived radicals) [87], 1,4-diaza-1,3-butadienes [93], pyrazines [94], and compounds X (R = H or OCH_3) and XI [123] (formation of $Ge(C_6H_5)_3N$ radicals) [93, 94, 123]. Reactions of $Ge(C_6H_5)_3H$ with t-$C_4H_9O^•$ were also carried out in the presence of t-$C_4H_9NO_2$ (→ $Ge(C_6H_5)_3ON(C_4H_9\text{-t})O^•$) [126], C_{60} (→ $Ge(C_6H_5)_3C_{60}^•$) [127], and $P(X)(OC_2H_5)_2C(S)SCH_3$ (→ $P(X)(OC_2H_5)_2C^•(SGe(C_6H_5)_3)SCH_3$ with X = O or S) [129]. For the formation of several carbon-centered GeR_4-derived radicals and $Ge(C_6H_5)_3O$ radicals from $Ge(C_6H_5)_3H$, see "Organogermanium Compounds" 3, 1990, pp. 349/52, and "Organogermanium Compounds", 5, 1993, pp. 403/10, respectively. The galvinoxyl radical was reported to react with $Ge(C_6H_5)_3H$ in C_6H_6 at 80°C or in the presence of AIBN at 120°C to yield $Ge(C_6H_5)_3CHC_6H_2(OH\text{-}4)(R_2\text{-}3,5)$ as the major product (R = t-C_4H_9) and minor amounts of 4-$Ge(C_6H_5)_3OC_6H_2(R_2\text{-}3,5)CH_2C_6H_2(OH\text{-}4)(R_2\text{-}3,5)$ [132].

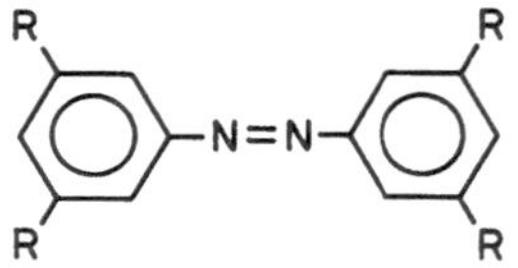

X

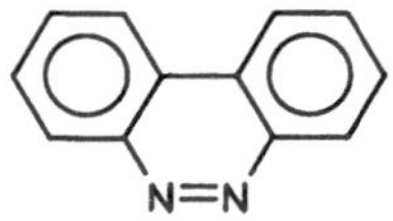

XI

Hydride transfer from $Ge(C_6H_5)_3H$ to the cations generated from the alcohols XIIa or XIIb in CH_2Cl_2-CF_3COOH gave compound XIIIa with 29%, compound XIIIb with 59 to 61%, and compound XIV with 10 to 12% yield. c-$C_3H_5C(C_6H_5)_2OH$ similarly yielded 45% c-$C_3H_5C(C_6H_5)_2H$ and 55% $(C_6H_5)_2C{=}CHCH_2CH_2OOCCF_3$. $Ge(C_6H_5)_3H$ was shown to be a better hydride donor than $Si(C_6H_5)_3H$ [35].

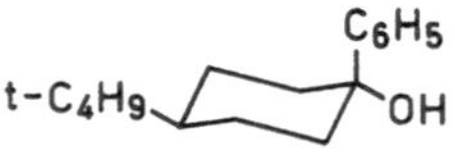

XIIa

XIIb

XIIIa

XIIIb

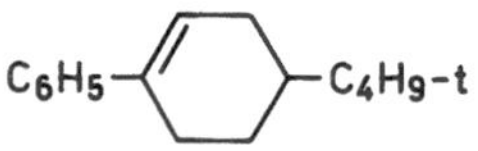

XIV

Hydride transfer from $Ge(C_6H_5)_3H$ to $[C(C_6H_5)_3][ClO_4]$ in CH_2Cl_2 or sulfolane gave $[Ge(C_6H_5)_3][ClO_4]$ and $(C_6H_5)_3CH$ [111]. The kinetics of hydride transfer from $Ge(C_6H_5)_3H$ to $[4\text{-}XC_6H_4CHC_6H_4Y\text{-}4]^+$ ions (X = Y = CH_3 or OCH_3; X = H, Y = OCH_3 or OC_6H_5) was studied in CH_2Cl_2 at −70°C [125].

Reactions with Organometallic Compounds of the Main Group Elements. In the reactions of $Ge(C_6H_5)_3H$ with LiR, variable amounts of $Ge(C_6H_5)_3Li$ (not isolated) and $Ge(C_6H_5)_3R$ are formed depending on the conditions of mixing and the nature of the group R [4, 5]; see Table 17.

First attempts to prepare $Ge(C_6H_5)_3MgBr$ from $Ge(C_6H_5)_3H$ and Grignard reagents such as C_4H_9MgBr or CH_2=$CHCH_2MgBr$ by refluxing in ether were unsuccessful [5]. However, in refluxing THF $Ge(C_6H_5)_3H$ was metalated by CH_2=$CHCH_2MgCl$, CH_2=$CHCH_2MgBr$, or C_6H_5MgBr within 2 d. The formation of the Ge-containing Grignard reagent was confirmed by carbonation (followed by hydrolysis to give $Ge(C_6H_5)_3COOH$) and by identifying Ge-$(C_6H_5)_3(CH_2)_4OH$ as the THF cleavage product. But even under these conditions, Ge-$(C_6H_5)_3H$ and C_4H_9MgBr did not appear to react [13]. The reaction between $Ge(C_6H_5)_3H$ and CH_3MgBr in the presence of $NiCl_2$ (1:8:1.2 mole ratio) in ether at 23°C (3 h) gave a quantitative yield of $Ge(CH_3)(C_6H_5)_3$. But treatment of $Ge(C_6H_5)_3D$ with C_2H_5MgBr in the presence of catalytic amounts of $Ni(P(C_6H_5)_2C_2H_4P(C_6H_5)_2)Cl_2$ (1:9:0.1 mole ratio) in ether resulted in formation of $Ge(C_6H_5)_3H$. A mechanism for this D-H exchange, including evolution of C_2H_4, was proposed [65].

Treatment of Z-$Ge(C_2H_5)_3CH$=$CHC_{10}H_{21}$ with equimolar amounts of $Ge(C_6H_5)_3H$ and $B(C_2H_5)_3$ at 60°C gave a 2:5 mixture of E-$Ge(C_2H_5)_3CH$=$CHC_{10}H_{21}$ and E-$Ge(C_6H_5)_3$-CH=$CHC_{10}H_{21}$ [102]. The reaction between $Ge(C_6H_5)_3H$ and $Sn(C_2H_5)_3N(C_2H_5)_2$ in C_6H_{12} at 20°C was approximately 1000 times slower than that between $Sn(C_6H_5)_3H$ and $Sn(C_2H_5)_3$-$N(C_2H_5)_2$. Electrophilic attack of the $Ge(C_6H_5)_3H$ hydrogen at the nitrogen atom was suggested to be involved in the reaction mechanism [28].

$Ge(C_6H_5)_3H$ did not react with a measurable rate with $Sn(C_2H_5)_3N(C_6H_{13})CHO$ or $Sn(C_4H_9)_2(N(C_6H_5)CHO)_2$ in refluxing C_3H_7CN [28], nor with $Ge(C_6H_5)_3OC_6H_{11}$ at 160°C (1 d) [29], GeF_2 [66, 68], or $Sn(C_6H_5)_3Li$ [15].

H-D exchange between $\mathbf{Ge(C_6H_5)_3D}$ and $Ge(C_4H_9)_3H$ or $Ge(C_6H_4CH_3\text{-}4)_3H$ was studied in the presence of the hydrogermylation catalysts $Rh(P(C_6H_5)_3)_3Cl$ in C_6D_6 and cis-$Pt(P(C_6H_5)_3)_2Cl_2$ or H_2PtCl_6 in $CDCl_3$ [52].

Other reactions of $Ge(C_6H_5)_3H$ with organometallic compounds of the main group elements are listed in Table 17.

Table 17
Reactions of $Ge(C_6H_5)_3H$ with Organometallic Compounds of the Main Group Elements.

reactant	conditions → products (yield)	Ref.
$LiCH_3$ + CO_2, + H_2O-HCl	20°C, in ether, 3 d → $Ge_2(C_6H_5)_6$ (10%) and $(Ge(C_6H_5)_3)_2O$ (12%) refluxing in ether, 1 d → $Ge(C_6H_5)_3COOH$ (70 to 77%) via the $Ge(C_6H_5)_3Li$ intermediate and $Ge(C_6H_5)_3CH_3$ (9 to 16%)	[5]
LiC_4H_9	−23°C, in THF, 0.5 h → $Ge(C_6H_5)_3Li$ (not isolated)	[22]; see also [36, 116]
LiC_4H_9 + CO_2, + H_2O-HCl	in ether, refluxing during the addition → $Ge(C_6H_5)_3COOH$ (97 to 100%) via the $Ge(C_6H_5)_3Li$ intermediate [5]; $GeR_3C_4H_9$ also formed with low yield [19]	[5]; see also [20, 75, 89]

References on p. 128

Table 17 (continued)

reactant	conditions → products (yield)	Ref.
LiC_6H_5	slowly adding $Ge(C_6H_5)_3H$ to excess LiC_6H_5; refluxing in ether, 12 h → $Ge(C_6H_5)_4$ (main product) and $Ge_2(C_6H_5)_6$ (1.4%)	[4]; see also [9]
	slowly adding excess LiC_6H_5 to $Ge(C_6H_5)_3H$; refluxing in ether, 12 h → $Ge_2(C_6H_5)_6$ (54 to 60%) possibly via $Ge(C_6H_5)_3Li$, and some $Ge(C_6H_5)_4$	[4]; see also [5, 9]
LiC_6H_5 + CO_2, + H_2O-HCl	slowly adding LiC_6H_5 to $Ge(C_6H_5)_3H$ (1:1 mole ratio) in ether; refluxing during the addition → $Ge(C_6H_5)_3COOH$ (83%) (via the $Ge(C_6H_5)_3Li$); no $Ge_2(C_6H_5)_6$ found	[5]
$NaCH_2SOCH_3$	20°C, in CH_3SOCH_3 → $Ge(C_6H_5)_3Na$; at 85 to 90°C (4 h) → $Ge(C_6H_5)_3CH_3$, $Ge(C_6H_5)_3SCH_3$, and some $(Ge(C_6H_5)_3)_2O$ formed in a secondary reaction	[50]
$\overline{Si(CH_3)_2C(CH_3)_2C}(CH_3)_2$	20°C, in THF, 15 h → $Ge(C_6H_5)_3Si(CH_3)_2C(CH_3)_2C_3H_7$-i (80%)	[86, 91]
$Si(C_6H_5)_3Li$	conditions not reported → $Ge(C_6H_5)_3Li$, $Si(C_6H_5)_3H$	[15]
$Si(C_6H_5)_3N_3$	200°C, 2 h → $Ge(C_6H_5)_3NHSi(C_6H_5)_3$ (79%), N_2	[45]
$Ge(C_6H_5)_3Li$	stirring, in DME, ca. 12 h → $Ge_2(C_6H_5)_6$ (1%) refluxing in DME-ether (1 d), then at 20°C (3 d) → $Ge_2(C_6H_5)_6$ (12%) refluxing in ether, 40 h → $Ge_2(C_6H_5)_6$ (12%)	[5]; see also [12]
$Ge(C_6H_5)_3N_3$	240°C, 4 h → $(Ge(C_6H_5)_3)_2NH$ (80%), N_2, some NH_3	[45]
$Sn(CH_3)_3N(C_2H_5)_2$	70°C, 2 h → $Ge(C_6H_5)_3Sn(CH_3)_3$ (92%)	[25]
$Sn(CH_3)(CH_2C(CH_3)_2C_6H_5)$-$(C_6H_5)N(C_2H_5)_2$	20°C, in ether, 1 d → $Ge(C_6H_5)_3Sn(CH_3)(CH_2C(CH_3)_2C_6H_5)$-$C_6H_5$ (70%)	[72]
$Sn(C_2H_5)_3N(C_2H_5)_2$	60°C, 2.5 h → $Ge(C_6H_5)_3Sn(C_2H_5)_3$ (52%)	[28]

Table 17 (continued)

reactant	conditions → products (yield)	Ref.
$Sn(C_4H_9)_3N(C_2H_5)_2$	60°C, 6 h → $Ge(C_6H_5)_3Sn(C_4H_9)_3$ (83%)	[25]; see also [18]
$Sn(C_6H_5)_3N(C_2H_5)_2$	→ $Ge(C_6H_5)_3Sn(C_6H_5)_3$ (17%)	[28]
$Sn(C_6H_5)_3N_3$	230°C, 2 h → $(Ge(C_6H_5)_3)_2NH$ (small amount), $Sn(C_6H_5)_4$, and N_2 (no Ge-N-Sn compound formed)	[45]
$Sn(C_2H_5)_2(N(C_2H_5)_2)_2$	1:1 mole ratio, 80 to 90°C, 1 to 1.5 h → $Ge(C_6H_5)_3Sn(C_2H_5)_2N(C_2H_5)_2$ (72%) 2:1 mole ratio, 80 to 90°C, 2 h, → $(Ge(C_6H_5)_3)_2Sn(C_2H_5)_2$ (70%)	[28]
$Sn(C_6H_5)_2(N(C_2H_5)_2)_2$	1:1 mole ratio → $Ge(C_6H_5)_3Sn(C_6H_5)_2N(C_2H_5)_2$	[28]
$Sn(C_2H_5)(N(C_2H_5)_2)_3$	1:1 mole ratio → $Ge(C_6H_5)_3Sn(C_2H_5)(N(C_2H_5)_2)_2$ 2:1 mole ratio → $(Ge(C_6H_5)_3)_2Sn(C_2H_5)N(C_2H_5)_2$ 3:1 mole ratio → $(Ge(C_6H_5)_3)_3SnC_2H_5$ (3%)	[28]
$Sn(C_6H_5)_3CH_2CH{=}CHCH_3$	60°C, in C_6H_6-C_6H_{14}, $B(C_2H_5)_3$, 9 h → $Ge(C_6H_5)_3CH_2CH{=}CHCH_3$ (98%)	[108]
(E)-$Sn(C_6H_5)_3CH{=}CHC_6H_5$	3:1 mole ratio, in C_6H_6-C_6H_{14}, $B(C_2H_5)_3$, 60°C, 12 h → (E)-$Ge(C_6H_5)_3CH{=}CHC_6H_5$ (93%)	[108]
(E)-$Sn(C_6H_5)_3$-$CH{=}C(C_2H_5)C_6H_5$	conditions like those in the previous reaction → (E)-$Ge(C_6H_5)_3CH{=}C(C_2H_5)C_6H_5$ (81%)	[108]
$PbR_3N(C_2H_5)_2$ R = n-C_4H_9, i-C_4H_9, c-C_6H_{11}, C_6H_5	exothermic reactions → $Ge(C_6H_5)_3PbR_3$	[26]
$Sb(C_2H_5)_3$	200°C, 12 h → $Sb(Ge(C_6H_5)_3)_3$ (62%) and C_2H_6 (99%)	[41]
$Sb(Ge(C_2H_5)_3)_3$	230°C, 15 h → $Sb(Ge(C_6H_5)_3)_3$ (ca. 52%), $Ge(C_2H_5)_3H$ (52%), and Sb (11%)	[41]
$Bi(C_2H_5)_3$	140 to 145°C, 4.5 h → $Bi(Ge(C_6H_5)_3)_3$ (71%) and C_2H_6 (97%)	[41]

Reactions with organometallic compounds of the transition metals are summarized in Table 18.

References on p. 128

Table 18
Reactions of $Ge(C_6H_5)_3H$ with Organometallic Compounds of the Transition Metals.

reactant	conditions → products (yield)	Ref.
$Yb(C_{10}H_8)(THF)_3$ $C_{10}H_8$ = naphthalene	4.5:1 mole ratio, 20 °C, in THF, 20 h → $(Ge(C_6H_5)_3H)_2Yb(THF)_4$ (61%); μ_2-H bridges between Ge and Yb	[115]; see also [106]
$M(N(Si(CH_3)_3)_2)_3$ M = Pr or Nd	in DME → reaction complicated by cleavage of an ether bond to give $(Ge(C_6H_5)_3)_2MOC_2H_4OCH_3 \cdot DME$	[84]
$Ti(C_5H_5)_2(CH_3)_2$	65 °C, in C_6H_{14}, 2 h → $Ti(CH_3)(C_5H_5)_2Ge(C_6H_5)_3$ (84%) and CH_4	[101, 107]
$Cr(CO)_5C(OCH_3)C_6H_5$ + C_5H_5N	1:1:6 mole ratio, refluxing in C_6H_{14}, 15 min → $Ge(C_6H_5)_3CH(OCH_3)C_6H_5$ (92%) mechanism and kinetics studied	[60] [59, 70]
$Cr(CO)_5C(NC_4H_8)C_6H_5$ + C_5H_5N	refluxing in C_6H_{14}, 7 h → $Ge(C_6H_5)_3CH(NC_4H_8)C_6H_5$ (60%)	[60]
$Mn(CO)_5CH_3$	25 to 30 °C, in CD_3CN, 50 min → CH_3CHO (98%); organometallic products not identified	[100]
$Mn(CO)_2(C_5H_5)(THF)$	in THF → $Mn(CO)_2(C_5H_5)(H)Ge(C_6H_5)_3$	[119]
$Mn(CO)_3(C_5H_4CH_3)$	irradiated in THF, ca. 12 h → $Mn(CO)_2(C_5H_4CH_3)(H)Ge(C_6H_5)_3$ (31%)	[97]; see also [119]
$Mn(CO)_2(C_5(CH_3)_5)(THF)$	in THF → $Mn(CO)_2(C_5(CH_3)_5)(H)Ge(C_6H_5)_3$	[119]
$Ru_3(CO)_{11}(CH_3CN)$	−40 °C, in CH_2Cl_2, 2 h → $Ru_3(CO)_{11}(H)(Ge(C_6H_5)_3)$ (60%)	[99]
$Na[Ru_3(CO)_{11}H]$	45 °C, in THF, 4 h → $[Ru_3(CO)_{10}(H)(Ge(C_6H_5)_3)_2]^-$ (73%), CO, and H_2	[92]
$Os_3(CO)_{11}(CH_3CN)$	20 °C, in $C_6H_5CH_3$, 12 h → $Os_3(CO)_{11}(H)(Ge(C_6H_5)_3)$ (70 to 80%)	[99]
$Os_3(CO)_{10}(CH_3CN)_2$	20 °C, in $C_6H_5CH_3$, 5 h → $Os_3(CO)_{10}(H)(CH_3CN)(Ge(C_6H_5)_3)$ (70 to 75%) and $Os_3(CO)_{11}(H)(Ge(C_6H_5)_3)$ (5%)	[99]
$Os_4(CO)_{14}$	25 °C, in C_6H_{14} → $Os_4(CO)_{14}(H)(Ge(C_6H_5)_3)$	[117]
$Co_2(CO)_8$	in C_6H_{14}, 2 h → $Co(CO)_4Ge(C_6H_5)_3$ and H_2	[71]

Table 18 (continued)

reactant	conditions → products (yield)	Ref.
$Co_3(CO)_9CH$	ca. 105 °C, in $C_6H_5CH_3$, 30 min → $Ge(C_6H_5)_3CCo_3(CO)_9$ (30 to 33%)	[74, 81]
$(Co(CO)_3P(C_6H_5)_3)_2$	in refluxing C_6H_6, 3 d → $Ge(C_6H_5)_3Co(CO)_3P(C_6H_5)_3$ (44%) and H_2	[88]
$(Co(CO)_3P(OC_6H_5)_3)_2$	in $C_6H_5CH_3$ similar to the preceding procedure → $Ge(C_6H_5)_3Co(CO)_3P(OC_6H_5)_3$ (63%) and H_2	[88]
$Ir(CO)(P(C_6H_5)_3)_2Cl$	refluxing, in C_6H_6, 7 d → $Ge(C_6H_5)_3Ir(CO)(H)(Cl)P(C_6H_5)_3$ (83%), $Ge(C_6H_5)_3Cl$, and $P(C_6H_5)_3$	[43]
$Zn(C_2H_5)_2$	mixed at −196 °C, then 20 to 110 °C; in $(CH_3OC_2H_4)_2O$ → $Zn(C_2H_5)Ge(C_6H_5)_3 \cdot (CH_3OC_2H_4)_2O$ (71%) and C_2H_6 (90%) up to 155 °C, in $(CH_3OC_2H_4)_2O$ → $Zn(Ge(C_6H_5)_3)_2 \cdot (CH_3OC_2H_4)_2O$ (76%) and C_2H_6 (90%) heating in HMPT → $Zn(Ge(C_6H_5)_3)_2 \cdot$ 2 HMPT (72%) and C_2H_6 (85%) up to 150 °C, in DMF → $Zn(Ge(C_6H_5)_3)_2 \cdot$ 2 DMF (81%) and C_2H_6 (95%)	[58]
$Zn(C_2H_5)_2 \cdot$ TMEDA	110 to 115 °C, in $CH_3CON(CH_3)_2$ → $Zn(Ge(C_6H_5)_3)_2 \cdot$ TMEDA (35%) and C_2H_6	[57]
$Zn(C_2H_5)_2 \cdot C_{10}H_8N_2$ $C_{10}H_8N_2$ = 2,2′-bipyridine	100 to 130 °C → $Zn(Ge(C_6H_5)_3)_2 \cdot C_{10}H_8N_2$ (81%) and C_2H_6	[57]
$Cd(CH_3)_2 \cdot$ TMEDA	100 to 112 °C → $Cd(Ge(C_6H_5)_3)_2 \cdot$ TMEDA (48%) and CH_4	[57]
$Cd(CH_3)_2 \cdot C_{10}H_8N_2$	80 to 127 °C, 45 min → $Cd(Ge(C_6H_5)_3)_2 \cdot C_{10}H_8N_2$ (53%) and CH_4	[57]
$Cd(C_2H_5)_2$	80 to 90 °C, in$(CH_3OC_2H_4)_2O$, 40 min → $Cd(Ge(C_6H_5)_3)_2 \cdot (CH_3OC_2H_4)_2O$ (70%) and C_2H_6 (90%)	[58, 63]
	40 °C, in TMEDA, 2 h → $Cd(C_2H_5)Ge(C_6H_5)_3 \cdot$ TMEDA (62%) and C_2H_6 (100%); at 90 °C → $Cd(Ge(C_6H_5)_3)_2 \cdot$ TMEDA (77%) 25 °C, in HMPT, 2 h → $Cd(Ge(C_6H_5)_3)_2 \cdot$ HMTP (64%) and C_2H_6 (82%)	[63]

References on p. 128

Table 18 (continued)

reactant	conditions → products (yield)	Ref.
$Hg(C_2H_5)_2$	110 to 120°C, 3 h → $Hg(Ge(C_6H_5)_3)_2$ (73%) and C_2H_6 (96%) (error in [40]?)	[40]
$Hg(C_6H_5)CCl_2Br$	ca. 80°C, in C_6H_6, 1.3 h → $Ge(C_6H_5)_3CCl_2H$ (88%) and $Hg(C_6H_5)Br$	[17, 31]
$Hg(C_6H_5)CClBr_2$	heated in C_6H_6 → $Ge(C_6H_5)_3CClBrH$ (73%)	[44]
$Hg(C_6H_5)CBr_3$	refluxing in C_6H_6, 2 h → $Ge(C_6H_5)_3CBr_2H$ (51%)	[42]

References:

[1] Kraus, C. A.; Foster, L. S. (J. Am. Chem. Soc. **49** [1927] 457/67).
[2] Foster, L. S.; Hooper, G. S. (J. Am. Chem. Soc. **57** [1935] 76/8).
[3] Johnson, O. H.; Nebergall, W. H. (J. Am. Chem. Soc. **71** [1949] 1720/2).
[4] Johnson, O. H.; Harris, D. M. (J. Am. Chem. Soc. **72** [1950] 5566/8).
[5] Gilman, H.; Gerow, C. W. (J. Am. Chem. Soc. **78** [1956] 5435/8).
[6] Anderson, H. H. (J. Am. Chem. Soc. **79** [1957] 326/8).
[7] Fuchs, R.; Gilman, H. (J. Org. Chem. **22** [1957] 1009/11).
[8] Gilman, H.; Gerow, C. W. (J. Am. Chem. Soc. **79** [1957] 342/5).
[9] Johnson, O. H.; Nebergall, W. H.; Harris, D. M. (Inorg. Synth. **5** [1957] 76/8).
[10] Meen, R. H.; Gilman, H. (J. Org. Chem. **22** [1957] 684/5).

[11] Fuchs, R.; Gilman, H. (J. Org. Chem. **23** [1958] 911/3).
[12] Gilman, H.; Gerow, C. W. (J. Org. Chem. **23** [1958] 1582/4).
[13] Gilman, H.; Zuech, E. A. (J. Org. Chem. **26** [1961] 3035/7).
[14] Henry, M. C.; Downey, M. F. (J. Org. Chem. **26** [1961] 2299/300).
[15] Gilman, H.; Zuech, E. A. (unpublished studies from Gilman, H.; Marrs, O. L.; Trepka, W. J.; Diehl, J. W.; J. Org. Chem. **27** [1962] 1260/5, 1261).
[16] Kramer, K. A. W.; Wright, A. N. (J. Chem. Soc. **1963** 3604/8).
[17] Seyferth, D.; Burlitch, J. M. (J. Am. Chem. Soc. **85** [1963] 2667/8).
[18] Sommer, R.; Neumann, W. P.; Schneider, B. (Tetrahedron Lett. **1964** 3875/8).
[19] Cross, R. J.; Glockling, F. (J. Organomet. Chem. **3** [1965] 146/55).
[20] Cross, R. J.; Glockling, F. (J. Chem. Soc. **1965** 5422/32).

[21] Mikhailov, B. M.; Bubnov, Yu. N.; Kiselev, V. G. (Izv. Akad. Nauk SSSR Ser. Khim. **1965** 68/72; Bull. Acad. Sci. USSR Div. Chem. Sci. [Engl. Transl.] **1965** 58/61).
[22] Nicholson, D. A.; Allred, A. L. (Inorg. Chem. **4** [1965] 1747/50).
[23] Amberger, E.; Stoeger, W.; Hönigschmid-Grossich, R. (Angew. Chem. **78** [1966] 549; Angew. Chem. Int. Ed. Engl. **5** [1966] 522).
[24] Birchall, T.; Jolly, W. L. (Inorg. Chem. **5** [1966] 2177/80).
[25] Neumann, W. P.; Schneider, B.; Sommer, R. (Liebigs Ann. Chem. **692** [1966] 1/11).
[26] Neumann, W. P.; Kühlein, K. (Tetrahedron Lett. **1966** 3419/21).
[27] Bryskovskaya, A. V.; Al'bitskaya, V. M. (Zh. Obshch. Khim. **37** [1967] 1553/8; J. Gen. Chem. USSR [Engl. Transl.] **37** [1967] 1474/8).

[28] Creemers, H. M. J. C.; Noltes, J. G. (J. Organomet. Chem. **7** [1967] 237/47).
[29] Rivière, P.; Satgé, J. (Bull. Soc. Chim. Fr. **1967** 4039/46).
[30] Satgé, J.; Rivière, P.; Lesbre, M. (C. R. Seances Acad. Sci. C **265** [1967] 494/6).

[31] Seyferth, D.; Burlitch, J. M.; Dertouzos, H.; Simmons, H. D., Jr. (J. Organomet. Chem. **7** [1967] 405/13).
[32] Brook, A. G.; Pannell, K. H.; Anderson, D. G. (J. Am. Chem. Soc. **90** [1968] 4374/7).
[33] Glockling, F.; Light, J. R. C. (J. Chem. Soc. A **1968** 717/34).
[34] Rivière, P.; Satgé, J. (C. R. Seances Acad. Sci. C **267** [1968] 267/9).
[35] Carey, F. A.; Tremper, H. S. (Tetrahedron Lett. **1969** 1645/8).
[36] Curtis, M. D. (J. Am. Chem. Soc. **91** [1969] 6011/8).
[37] Glockling, F.; Brooks, E. H. (Prepr. Am. Chem. Soc. Div. Pet. Chem. **14** [1969] B135/B137).
[38] Kaplan, L. (J. Chem. Soc. Chem. Commun. **1969** 106/7).
[39] Satgé, J.; Rivière, P. (J. Organomet. Chem. **16** [1969] 71/82).
[40] Vyazankin, N. S.; Bychkov, V. T.; Linzina, O. V.; Razuvaev, G. A. (Zh. Obshch. Khim. **39** [1969] 979/82; J. Gen. Chem. USSR [Engl. Transl.] **39** [1969] 950/2).

[41] Vyazankin, N. S.; Kalinina, G. S.; Kruglaya, O. A.; Razuvaev, G. A. (Zh. Obshch. Khim. **39** [1969] 2005/11; J. Gen. Chem. USSR [Engl. Transl.] **39** [1969] 1964/8).
[42] Brook, A. G.; Duff, J. M.; Anderson, D. G. (Can. J. Chem. **48** [1970] 561/9).
[43] Glockling, F.; Wilbey, M. D. (J. Chem. Soc. A **1970** 1675/81).
[44] Seyferth, D.; Hopper, S. P. (J. Organomet. Chem. **23** [1970] 99/104).
[45] Tsai, T.-T.; Lehn, W. L.; Marshall, C. J., Jr. (J. Organomet. Chem. **22** [1970] 387/93).
[46] Corriu, R. J. P.; Moreau, J. J. E. (J. Chem. Soc. Chem. Commun. **1971** 812/3).
[47] Jolly, W. L. (Inorg. Chem. **10** [1971] 2364/5).
[48] Satgé, J.; Rivière, P. (unpublished work from: Lesbre, M.; Mazerolles, P.; Satgé, J.; The Organic Compounds of Germanium, London-New York-Sydney-Toronto 1971; pp. 305, 310, 311).
[49] Rivière, P.; Satgé, J. (Angew. Chem. **83** [1971] 286/7; Angew. Chem. Int. Ed. Engl. **10** [1971] 267/8).
[50] Sandman, D. J.; West, R. (J. Organomet. Chem. **30** [1971] C61/C63).

[51] Birchall, T.; Drummond, I. (Inorg. Chem. **11** [1972] 250/2).
[52] Corriu, R. J. P.; Moreau, J. J. E. (J. Organomet. Chem. **40** [1972] 55/72).
[53] Corriu, R. J. P.; Moreau, J. J. E. (J. Organomet. Chem. **40** [1972] 73/96).
[54] Linke, K.-H.; Göhausen, H. J.; Wrobel, G. (Chem. Ber. **105** [1972] 1780/2).
[55] Nesmeyanov, A. N.; Borisov, A. E.; Novikova, N. V. (Izv. Akad. Nauk SSSR Ser. Khim. **1972** 1372/5 ; Bull. Acad. Sci. USSR Div. Chem. Sci. [Engl. Transl.] **1972** 1321/3).
[56] Satgé, J.; Lesbre, M.; Rivière, P.; Richelme, S. (J. Organomet. Chem. **34** [1972] C18/C20).
[57] des Tombe, F. J. A.; van der Kerk, G. J. M.; Creemers, H. M. J. C.; Carey, N. A. D.; Noltes, J. G. (J. Organomet. Chem. **44** [1972] 247/52).
[58] Bychkov, V. T.; Vyazankin, N. S.; Razuvaev, G. A. (Zh. Obshch. Khim. **43** [1973] 793/8; J. Gen. Chem. USSR [Engl. Transl.] **43** [1973] 792/6).
[59] Connor, J. A.; Day, J. P.; Turner, R. M. (J. Chem. Soc. Chem. Commun. **1973** 578/9).
[60] Connor, J. A.; Rose, P. D.; Turner, R. M. (J. Organomet. Chem. **55** [1973] 111/9).

[61] Eaborn, C.; Jenkins, I. D. (J. Chem. Soc. Chem. Commun. **1973** 780).
[62] Linke, K.-H.; Göhausen, H. J. (Chem. Ber. **106** [1973] 3438/49).
[63] Razuvaev, G. A.; Bychkov, V. T.; Vyazankin, N. S. (Dokl. Akad. Nauk SSSR **211** [1973] 116/9; Dokl. Chem. [Engl. Transl.] **208/213** [1973] 545/8).
[64] Rivière, P.; Satgé, J. (J. Organomet. Chem. **49** [1973] 173/89).

[65] Carré, F. H.; Corriu, R. J. P. (J. Organomet. Chem. **74** [1974] 49/58).
[66] Satgé, J.; Rivière, P.; Boy, A. (C. R. Seances Acad. Sci. C **278** [1974] 1309/12).
[67] Razuvaev, G. A.; Bychkov, V. T.; Vyshinskaya, L. I.; Latyaeva, V. N.; Spiridonova, N. N. (Dokl. Akad. Nauk SSSR **220** [1975] 854/5; Dokl. Chem. [Engl. Transl.] **220/225** [1975] 126/7).
[68] Rivière, P.; Satgé, J.; Boy, A. (J. Organomet. Chem. **96** [1975] 25/40).
[69] Sakurai, H.; Mochida, K.; Kira, M. (J. Am. Chem. Soc. **97** [1975] 929/31).
[70] Connor, J. A.; Day, J. P.; Turner, R. M. (J. Chem. Soc. Dalton Trans. **1976** 283/5).

[71] Colomer, E.; Corriu, R. J. P. (J. Organomet. Chem. **133** [1977] 159/68).
[72] Gielen, M.; Simon, S. (Bull. Soc. Chim. Belg. **86** [1977] 589/94).
[73] Sakurai, H.; Mochida, K.; Kira, M. (J. Organomet. Chem. **124** [1977] 235/52).
[74] Seyferth, D.; Nivert, C. L. (J. Am. Chem. Soc. **99** [1977] 5209/10).
[75] Duffaut, N.; Dunogues, J.; Calas, R.; Rivière, P.; Satgé, J.; Cazes, A. (J. Organomet. Chem. **149** [1978] 57/63).
[76] Rivière, P.; Richelme, S.; Rivière-Baudet, M.; Satgé, J.; Gynane, M. J. S.; Lappert, M. F. (J. Chem. Res. Synop. **1978** 218/9; J. Chem. Res. Miniprint **1978** 2801/16).
[77] Rivière, P.; Rivière-Baudet, M.; Richelme, S.; Satgé, J. (Bull. Soc. Chim. Fr. **1978** II 193/6).
[78] Eaborn, C.; Singh, B. (J. Organomet. Chem. **177** [1979] 333/48).
[79] Lopatina, V. S.; Sheverdina, N. I.; Kocheshkov, K. A.; Fomina, N. V. (Dokl. Akad. Nauk SSSR **246** [1979] 620/2; Dokl. Chem. [Engl. Transl.] **244/249** [1979] 263/4).
[80] Rivière, P.; Rivière-Baudet, M.; Richelme, S.; Castel, A.; Satgé, J. (J. Organomet. Chem. **168** [1979] 43/52).

[81] Seyferth, D.; Rudie, C. N.; Nestle, M. O. (J. Organomet. Chem. **178** [1979] 227/47).
[82] Corriu, R. J. P.; Guerin, C. (J. Organomet. Chem. **197** [1980] C19/C21).
[83] Lopatina, V. S.; Sheverdina, N. I.; Fomina, N. V.; Kocheshkov, K. A.; Panov, E. M. (Izv. Akad. Nauk SSSR Ser. Khim. **1980** 378/82; Bull. Acad. Sci. USSR Div. Chem. Sci. [Engl. Transl.] **1980** 291/4).
[84] Razuvaev, G. A.; Kalinina, G. S.; Fedorova, E. A. (J. Organomet. Chem. **190** [1980] 157/65).
[85] Sennikov, P. G.; Skobeleva, S. E.; Kuznetsov, V. A.; Egorochkin, A. N.; Rivière, P.; Satgé, J.; Richelme, S. (J. Organomet. Chem. **201** [1980] 213/9).
[86] Seyferth, D.; Escudié, J.; Shannon, M. L.; Satgé, J. (J. Organomet. Chem. **198** [1980] C51/C54).
[87] Alberti, A.; Barbaro, G.; Battaglia, A.; Guerra, M.; Bernardi, F.; Dondoni, A.; Pedulli, G. F. (J. Org. Chem. **46** [1981] 742/50).
[88] Cerveau, G.; Colomer, E.; Corriu, R. J. P.; Young, J. C. (J. Organomet. Chem. **205** [1981] 31/45).
[89] Eaborn, C.; Mahmoud, F. M. S. (J. Organomet. Chem. **205** [1981] 47/51).
[90] Lopatina, V. S.; Kocheshkov, K. A.; Fomina, N. V.; Rodionov, A. I.; Shapet'ko, N. N.; Yankelevich, A. Z. (Zh. Obshch. Khim. **51** [1981] 2580/2; J. Gen. Chem. USSR [Engl. Transl.] **51** [1981] 2225/7).

[91] Seyferth, D.; Annarelli, D. C.; Shannon, M. L.; Escudié, J.; Duncan, D. P. (J. Organomet. Chem. **225** [1982] 177/91).
[92] Süß-Fink, G.; Ott, J.; Schmidkonz, B.; Guldner, K. (Chem. Ber. **115** [1982] 2487/93).
[93] Alberti, A.; Hudson, A. (J. Organomet. Chem. **241** [1983] 313/9).
[94] Alberti, A.; Pedulli, G. F. (J. Organomet. Chem. **248** [1983] 261/7).
[95] Chatgilialoglu, C.; Ingold, K. U.; Lusztyk, J.; Nazran, A. S.; Scaiano, J. C. (Organometallics **2** [1983] 1332/5).

[96] Hayashi, H.; Mochida, K. (Chem. Phys. Lett. **101** [1983] 307/11).
[97] Carré, F.; Colomer, E.; Corriu, R. J. P.; Vioux, A. (Organometallics **3** [1984] 1272/8).
[98] Oda, H.; Morizawa, Y.; Oshima, K.; Nozaki, H. (Tetrahedron Lett. **25** [1984] 3221/4).
[99] Burgess, K.; Guerin, C.; Johnson, B. F. G.; Lewis, J. (J. Organomet. Chem. **295** [1985] C3/C6).
[100] Warner, K. E.; Norton, J. R. (Organometallics **4** [1985] 2150/60).

[101] Harrod, J. F.; Malek, A.; Rochon, F. D.; Melanson, R. (Organometallics **6** [1987] 2117/20).
[102] Ichinose, Y.; Nozaki, K.; Wakamatsu, K.; Oshima, K.; Utimoto, K. (Tetrahedron Lett. **28** [1987] 3709/12).
[103] Ichinose, Y.; Oda, H.; Oshima, K.; Utimoto, K. (Bull. Chem. Soc. Jpn. **60** [1987] 3468/70).
[104] Vlček, A., Jr.; Gray, H. B. (Inorg. Chem. **26** [1987] 1997/2001).
[105] Vlček, A., Jr.; Gray, H. B. (J. Am. Chem. Soc. **109** [1987] 286/7).
[106] Bochkarev, M. N.; Penyagina, I. M.; Rad'kov, Yu. F.; Zakharov, L. N.; Struchkov, Yu. T. (Metalloorg. Khim. **1** [1988] 718/9; Organomet. Chem. USSR [Engl. Transl.] **1** [1988] 402).
[107] Harrod, J. F. (ACS Symposium Series **1988** No. 360, pp. 99/100).
[108] Ichinose, Y.; Oshima, K.; Utimoto, K. (Chem. Lett. **1988** 669/72).
[109] Ichinose, Y.; Oshima, K.; Utimoto, K. (Chem. Lett. **1988** 1437/40).
[110] Ichinose, Y.; Oshima, K.; Utimoto, K. (Bull. Chem. Soc. Jpn. **61** [1988] 2693/5).

[111] Lambert, J. B.; Schilf, W. (Organometallics **7** [1988] 1659/60).
[112] Packett, D. L.; Syed, A.; Trogler, W. C. (Organometallics **7** [1988] 159/66).
[113] Pike, P.; Hershberger, S.; Hershberger, J. (Tetrahedron **44** [1988] 6295/304).
[114] Rivière, P.; Castel, A.; Satgé, J.; Guyot, D.; Ko, Y. H. (J. Organomet. Chem. **339** [1988] 51/60).
[115] Bochkarev, M. N.; Penyagina, I. M.; Zakharov, L. N.; Rad'kov, Yu. F.; Fedorova, E. A.; Khorshev, S. Ya.; Struchkov, Yu. T. (J. Organomet. Chem. **378** [1989] 363/73).
[116] Kiyooka, S.; Shibuya, T.; Shiota, F.; Fujiyama, R. (Bull. Chem. Soc. Jpn. **62** [1989] 1361/3).
[117] Lu, C.-Y.; Einstein, F. W. B.; Johnston, V. J.; Pomeroy, R. K. (Inorg. Chem. **28** [1989] 4212/6).
[118] Wakasa, M.; Horiuchi, K.; Mochida, K. (Nippon Kagaku Kaishi **1989** 1469/71; C. A. **112** [1990] No. 77402).
[119] Lichtenberger, D. L.; Rai-Chaudhuri, A. (J. Chem. Soc. Dalton Trans. **1990** 2161/6).
[120] Nozaki, K.; Ichinose, Y.; Wakamatsu, K.; Oshima, K.; Utimoto, K. (Bull. Chem. Soc. Jpn. **63** [1990] 2268/72).

[121] Clark, K. B.; Griller, D. (Organometallics **10** [1991] 746/50).
[122] Okano, M.; Mochida, K. (Bull. Chem. Soc. Jpn. **64** [1991] 1381/2).
[123] Alberti, A.; Bedogni, N.; Benaglia, M.; Leardini, R.; Nanni, D.; Pedulli, G. F.; Tundo, A.; Zanardi, G. (J. Org. Chem. **57** [1992] 607/13).
[124] Lukevics, E.; Ignatovich, L.; Shilina, N.; Germane, S. (Appl. Organomet. Chem. **6** [1992] 261/6).
[125] Mayr, H.; Basso, N. (Angew. Chem. **104** [1992] 1103/5; Angew. Chem. Int. Ed. Engl. **31** [1992] 1046/8).
[126] Lucarini, M.; Pedulli, G. F.; Alberti, A.; Benaglia, M. (J. Am. Chem. Soc. **114** [1992] 9603/7).
[127] Morton, J. R.; Preston, K. F.; Krusic, P. J.; Wasserman, E. (J. Chem. Soc. Perkin Trans. II **1992** 1425/9).

[128] Gupta, V.; Kahne, D. (Tetrahedron Lett. **34** [1993] 591/4).
[129] Levillain, J.; Masson, S.; Hudson, A.; Alberti, A. (J. Am. Chem. Soc. **115** [1993] 8444/6).
[130] Okano, M.; Kugita, T.; Mochida, K. (J. Electroanal. Chem. **356** [1993] 303/7).
[131] Wolfsberger, W. (J. Prakt. Chem. **334** [1992] 453/64).
[132] Rivière, P.; Castel, A.; Abdennadher, C. (Phosphorus Sulfur Silicon Relat. Elem. **82** [1993] 181/93).

1.3.1.1.7.4 Applications

CH_2=CHC_3H_7-i was polymerized with a catalyst mixture containing VCl_4 and $Ge(C_6H_5)_3H$ [1]. The (Z)-isomers of $C_5H_{11}CH$=CHC_5H_{11}, t-C_4H_9CH=CHC_8H_{17}, $C_6H_{13}CH$=CHC_6H_5, C_6H_5CH=CHC_6H_5, $Si(CH_3)_2(C_6H_5)CH$=CHC_6H_{13}, $Ge(C_6H_5)_3CH$=CH-$C_{10}H_{21}$, $Ge(C_6H_5)_3$-CH=$CHCH_2CH_2OH$, $Ge(C_6H_5)_3CH$=$CH(CH_2)_9COOC_2H_5$ [2, 4], $Si(CH_3)_2(C_6H_5)C(CH_3)$=$CH$-$CH_3$, and $Ge(C_6H_5)_3CH$=$CHCH_3$ [2] were converted to their (E)-isomers on heating in C_6H_6 at 60 °C for several hours in the presence of 10 mol% $Ge(C_6H_5)_3H$ and 10 mol% $B(C_2H_5)_3$ [2, 4].

Co carbonyl compounds were proposed to be used as catalysts and $Ge(C_6H_5)_3H$ as a promoter in the hydroformylation of CH_2=$CHCH_2OOCR$ (R = H or alkyl groups from CH_3 to C_6H_{13}) [3]. Values of effective copolymerization constants were reported for the copolymerization of C_6H_5CH=CH_2 and CH_2=$C(CH_3)COOH$ in the presence of AIBN or CH_2=$CHCN$ in the presence of c-$C_6H_{11}OOCOOCOOC_6H_{11}$-c as the initiators and $Ge(C_6H_5)_3H$ as a regulator [5]. Organic glass substitutes with improved heat resistance were obtained by bulk copolymerization of CH_2=$C(CH_3)COOCH_3$ with CH_2=$C(CH_3)COOH$ in the presence of c-$C_6H_{11}OOCOO$-$COOC_6H_{11}$-c and $Ge(C_2H_5)_3H$, the latter as a modifier [6].

References:

[1] Shearer, N. H.; Coover, H. W., Jr.; Eastman Kodak Co. (U.S. 2925409 [1960]; C.A. **1960** 13732).
[2] Ichinose, Y.; Nozaki, K.; Wakamatsu, K.; Oshima, K.; Utimoto, K. (Tetrahedron Lett. **28** [1987] 3709/12).
[3] Lin, J. J.; Texaco Inc. (U.S. 4806678 [1983/89] from C.A. **110** [1989] No. 192266).
[4] Nozaki, K.; Ichinose, Y.; Wakamatsu, K.; Oshima, K.; Utimoto, K. (Bull. Chem. Soc. Jpn. **63** [1990] 2268/72).
[5] Semchikov, Yu. D.; Gromov, V. F.; Teleshov, E. N. (Vysokomol. Soedin. A **33** [1991] 1428/41; Polym. Sci. USSR [Engl. Transl.] **33** [1991] 1322/35).
[6] Smirnova, L. A.; Semchikov, Yu. D.; Knyazeva, T. E.; Modeva, Sh. I.; Rudin, A. A.; Ryabov, S. A. (U.S.S.R. 1668369 A1 [1991], Appl. 4604815 [1988] from C.A. **116** [1992] No. 256721).

1.3.1.1.8 Other GeR_3H Compounds with R = Aryl or Heterocycle

1.3.1.1.8.1 $Ge(C_6F_5)_3H$

Preparation and Formation. $Ge(C_6F_5)_3H$ was obtained with 90% yield by refluxing $Ge(C_6F_5)_3Br$ with excess $Ge(C_2H_5)_3H$ for 1 h and in 78% yield by reducing $Ge(C_6F_5)_3Br$ with

$LiAlH_4$ in ether-$C_6H_5CH_3$ at room temperature for 30 min followed by treatment with H_2O-HCl. It was purified by recrystallization from C_6H_{14} or sublimation at 100 °C/1 Torr [2].

The reaction of $Ge_2(C_6F_5)_6$ with H_2O in THF at 70 °C for 15 min yielded 94% $Ge(C_6F_5)_3H$ along with $Ge(C_6F_5)_3OH$. A similar cleavage of $Ge_2(C_6F_5)_6$ with CH_3OH (60 °C, 15 min), HCl (70 °C, 30 min), CH_3COOH (100 °C, 1 h), CF_3COOH (100 °C, 1 h) [4, 7], or $Sn(C_2H_5)_3H$ (70 °C, 8 h) [7] in THF gave $Ge(C_6F_5)_3H$ in 97, 98, 92, 70% [4, 7], and 84% yield [7], respectively. More vigorous conditions (100 °C, 6 h) were required for reacting $Ge_2(C_6F_5)_6$ with H_2O in ether, whereas in C_6H_{14}, C_6H_6, or $C_6H_5CH_3$ the digermane did not react with the above mentioned compounds under identical conditions (all reactions in an evacuated, sealed tube) [4, 7]. $Ge_2(C_6F_5)_6$ also reacted with $Ge(C_6F_5)_2H_2$ to form a mixture containing $Ge(C_6F_5)_3H$, $Ge_2(C_6F_5)_5H$, and $Ge_2(C_6F_5)_4H_2$ [29].

$(C_6F_5)_3GeGe(C_2H_5)_3$ was cleaved by H_2O in THF (room temperature, 15 min) to give $Ge(C_6F_5)_3H$ (82% yield together with $(Ge(C_2H_5)_3)_2O$) and by HCl (20 °C, 20 min) to yield 100% $Ge(C_6F_5)_3H$ and $Ge(C_2H_5)_3Cl$. Reactions of $(C_6F_5)_3GeGe(C_2H_5)_3$ with $CuCl_2$ (room temperature to 50 °C), AgCl (100 °C, 3 h, in the dark), or $AuCl_3$ (100 °C, 45 min) in THF gave $Ge(C_6F_5)_3H$ with 50, 56, and 46% yield, respectively. The title compound originates from initially generated (but not detectable) $Ge(C_6F_5)_3^{\bullet}$ radicals, which react with the solvent. When the reaction of $(C_6F_5)_3GeGe(C_2H_5)_3$ with $CuCl_2$ was carried out in the presence of the phenol I, $Ge(C_6F_5)_3H$ was obtained together with the radical II [8].

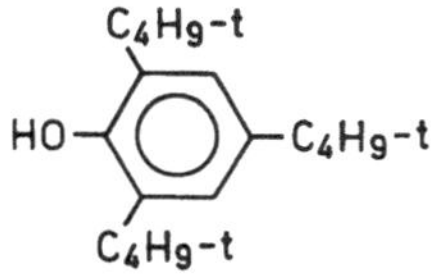

I

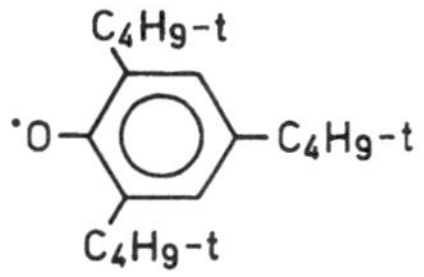

II

$Ge(C_6F_5)_3H$ appeared in the following reactions: preparation of $Ge_2(C_6F_5)_5H$ from $(C_6F_5)_3GeGe(C_2H_5)_3$ and $Ge(C_6F_5)_2(H)Br$ in THF at 20 °C (as a by-product); decomposition of $Ge_2(C_6F_5)_5H$ in THF at 95 °C for 1 h; reaction of $Ge_2(C_6F_5)_5H$ with $N(C_2H_5)_3$ (1 : 1 mole ratio) in THF at 20 °C for 5 min in the absence of oxygen and moisture (39% yield along with $Ge_2(C_6F_5)_6$ and a mixture of polymeric products); $Ge_2(C_6F_5)_5H$ with HCl (1 : 2 mole ratio) in ether (100% yield along with $Ge(C_6F_5)_2(H)Cl$); $Ge_2(C_6F_5)_5H$ with H_2O and subsequent heating at 140 °C for 5 h (along with $(\text{-}Ge(C_6F_5)_2O\text{-})_n$) [29].

Reacting $Ge(C_6F_5)_3Bi(C_2H_5)_2$ or $(Ge(C_6F_5)_3)_2BiC_2H_5$ with dry HCl in THF at room temperature yielded 95% $Ge(C_6F_5)_3H$ along with $Bi(C_2H_5)_2Cl$ or $Bi(C_2H_5)Cl_2$ [10].

Transition metal complexes containing metal-bonded $Ge(C_6F_5)_3$ groups produce $Ge(C_6F_5)_3H$ in a variety of reactions. Bubbling dry HCl through a solution of $Ge(C_6F_5)_3Pt(P(C_6H_5)_3)_2H$ in C_6H_6 at 80 °C (17 h) yielded $Ge(C_6F_5)_3H$ (64% along with $Pt(P(C_6H_5)_3)_2Cl_2$ and H_2) [12]. $V(C_5H_5)_2Ge(C_6F_5)_3$ reacted with HCl in $C_6H_5CH_3$ at 65 °C (2 h) to yield 80% $Ge(C_6F_5)_3H$ and $Ge(C_6F_5)_3Cl$ (12%) [20].

Other reactions of compounds containing Hg-bonded $Ge(C_6F_5)_3$ units are summarized in Table 19, including a $CuGe(C_6F_5)_3$ compound at the end.

References on p. 138

Table 19
Formation of $Ge(C_6F_5)_3H$ from $Ge(C_6F_5)_3$-Containing Hg and Cu Compounds.

starting material	reactant and conditions → products
$Hg(Ge(C_6F_5)_3)_2$	UV irradiation in $C_6H_5CH_3$ at 50 to 60 °C (2.5 d) → $Ge(C_6F_5)_3H$ (16%), $Ge_2(C_6F_5)_6$ (52%), $Ge(CH_2C_6H_5)(C_6F_5)_3$ (12%), and Hg (94%) [2] HCl in C_6H_6 at 100 °C (20 h) in a sealed tube → $Ge(C_6F_5)_3H$ (5%), $Ge(C_6F_5)_3Cl$ as the major product [2]
$Li[Hg(Ge(C_6F_5)_3)_3]$ [a)]	HCl → $Ge(C_6F_5)_3H$ (93%) [19] thermolysis (in evacuated tube) at 230 °C (1 h) → traces of $Ge(C_6F_5)_3H$ [19]
$Tl[Hg(Ge(C_6F_5)_3)_3]$ [c)]	HCl in THF → $Ge(C_6F_5)_3H$ (100%), $Hg(Ge(C_6F_5)_3)_2$, and TlCl [13] $Ge(C_6F_5)_3Br$ in THF → $Ge(C_6F_5)_3H$ in traces (obviously due to traces of moisture) [13] $HgCl_2$ in THF → $Ge(C_6F_5)_3H$ in traces [13] $CuCl_2$ (2:1 mole ratio) in DME at 20 °C (15 min) → $Ge(C_6F_5)_3H$ (58%) (probably via $Ge(C_6F_5)_3^{\cdot}$ and its reaction with the solvent), $Hg(Ge(C_6F_5)_3)_2$ (95%), $Ge_2(C_6F_5)_6$ (14%), TlCl, and Cu_2Cl_2 [19] $Sb(C_6H_5)_3Br_2$ → $Ge(C_6F_5)_3H$ (96%) [19] $Nb(C_5H_5)_2Cl_2$ → $Ge(C_6F_5)_3H$ (68%) [19] ZnI_2, $CoCl_2$, or $NiCl_2$ → $Ge(C_6F_5)_3H$ [19]
$M[Hg_2(Ge(C_6F_5)_3)_6]$ [b)] (M = Sm, Eu, or Yb)	HCl (1:2 mole ratio) in THF at 20 °C → $Ge(C_6F_5)_3H$ with 87 to 99% yield [22]
$M[Hg_2(Ge(C_6F_5)_3)_7]$ [a)] (M = Nd or Ho)	HCl → $Ge(C_6F_5)_3H$ [15]
$[PrX][Hg(Ge(C_6F_5)_3)_4]$ [a)] (X = Cl or Br)	HCl (1:2 mole ratio) in DME at 20 °C → $Ge(C_6F_5)_3H$ [15]
$Pr[Hg_2(Ge(C_6F_5)_3)_7]$ [a)] (wrong formula given in [9], corrected in [14])	HCl (large excess) in THF → $Ge(C_6F_5)_3H$, $Ge(C_6F_5)_3Cl$, $PrCl_3$, and Hg [9]; see also [15] HCl (small excess) in THF → $Ge(C_6F_5)_3H$, $Hg(Ge(C_6F_5)_3)_2$, $Pr(Ge(C_6F_5)_3)_2Cl$ [9] $Ge(C_6F_5)_3Br$ in THF → $Ge(C_6F_5)_3H$, $Hg(Ge(C_6F_5)_3)_2$, $Ge_2(C_6F_5)_6$ and $Pr(Ge(C_6F_5)_3)_2Br$ (probably via $Ge(C_6F_5)_3^{\cdot}$ and its reaction with the solvent) [9]
$Pr[Hg_2(Ge(C_6F_5)_3)_7]$ [a)] $[PrN(Si(CH_3)_3)_2]$-$[Hg_2(Ge(C_6F_5)_3)_6]$ [a)] or $Pr[Hg_2(Si(C_6F_5)_3)_4$-$(Ge(C_6F_5)_3)_3]$ [a)]	dry HCl (3 equivalents) in DME at 20 °C (30 min) → $Ge(C_6F_5)_3H$ with 65, 99, and 62% yield, respectively [17]
$[Cr(C_6H_6)_2][Hg(Ge(C_6F_5)_3)_3]$	HCl → $Ge(C_6F_5)_3H$ (68%) [19]

Table 19 (continued)

starting material	reactant and conditions → products
$[(C_5H_5)_2Co][(Ge(C_6F_5)_3)_2Hg$-$SGe(C_6F_5)_3]$	HCl in THF (or DME?) for 1 h → $Ge(C_6F_5)_3H$ (23%), $Ge(C_6F_5)_3Cl$, $Ge(C_6F_5)_3SH$, $Hg(Ge(C_6F_5)_3)_2$, H_2S, and an unidentified Co-containing compound [23]
$M[Cu(Ge(C_6F_5)_3)_2]_3$ [a] (M = Sm or Yb)	HCl (1:6 mole ratio) in DME at 20°C → $Ge(C_6F_5)_3H$ (71 to 84%), Cu_2Cl_2, and MCl_3 [25]

[a] 3-DME solvate. – [b] 2-DME solvate. – [c] 1.5-DME solvate.

Spectra and Physical Properties. IR spectrum (in Vaseline oil): ν(GeH) at 2223 cm^{-1} (low intensity); other bands characteristic for the C_6F_5 group are at 1650, 1525, 1480, 1390, 1290, 1090, 1025, 820, and 625 cm^{-1} [2]; an additional band was found at 780 cm^{-1} [3]. In C_7H_{16} the ν(GeH) band appears at 2153 cm^{-1} [6]; see also [11]. Studies of the dependence of the ν(GeH) and their integrated intensity on the nature of the substituents of various organogermanium compounds also involved $Ge(C_6F_5)_3H$ [6, 11].

$Ge(C_6F_5)_3H$ melts at 128 to 130 [7], 129 to 131 [3], 129 to 132 [8], or 130 to 132°C [2].

Chemical Behavior. Treating $Ge(C_6F_5)_3H$ with excess Br_2 in C_6H_6 in a sealed tube at 100°C (6 h) gave $Ge(C_6F_5)_3Br$ with 71% yield along with HBr [2]. Heating with sulfur in a sealed tube at 135 to 150°C (4 h) yielded 58% $Ge(C_6F_5)_3SH$ [5].

$Ge(C_6F_5)_3H$ reacted in THF or DME with NH_3, $N(C_2H_5)_3$, or $NH(Si(CH_3)_3)_2$ to give a complex mixture containing F^- ions [17]. Reactions with NR_3 (e.g. $N(C_2H_5)_3$), NHR_2 (e.g. $NH(C_2H_5)_2$), NH_3, $LiAlH_4$, LiH, Li, or other Lewis bases in donor solvents such as THF, DME, CH_3COCH_3, or ether at or below room temperature yielded a perfluorinated polyphenylenegermane ($(Ge(C_6F_5)_2C_6F_4)_n$ = PPG) having a star-branched macromolecular structure. The best results in regards the yield and quality of PPG were obtained when $Ge(C_6F_5)_3H$ was treated with $N(C_2H_5)_3$ in deoxygenated anhydrous THF [24]; see also [26]. In polar aromatic solvents, $Ge(C_6F_5)_3H$ did not react with $NH(C_2H_5)_2$ [29]. In the presence of $Hg(Ge(C_6F_5)_3)_2$, equimolar amounts of $Ge(C_6F_5)_3H$ and $NH(Si(CH_3)_3)_2$ in DME at 50°C gave $[N(Si(CH_3)_3)_2H_2][Hg(Ge(C_6F_5)_3)_3]$ with 44% yield within 30 min [17].

Other reactions of $Ge(C_6F_5)_3H$ are summarized in Table 20.

Table 20
Chemical Reactions of $Ge(C_6F_5)_3H$.
Mole ratios refer to the $Ge(C_6F_5)_3H$:reactants ratio.

reactant	conditions → products (yield); remarks
$Bi(C_2H_5)_3$	1:1 mole ratio, heating up to 110°C → $Ge(C_6F_5)_3Bi(C_2H_5)_2$ (75%) [10] 2:1 mole ratio, heating up to 170°C → $(Ge(C_6F_5)_3)_2BiC_2H_5$ (72%); attempts failed to obtain the $Ge(C_6F_5)_3Bi(C_2H_5)Ge(C_6F_5)_2Bi(C_2H_5)$-

References on p. 138

Table 20 (continued)

reactant	conditions → products (yield); remarks
	$Ge(C_6F_5)_3$ oligomer from $Ge(C_6F_5)_3H$ and $Ge(C_6F_5)_2(Bi(C_2H_5)_2)_2$ at 70 to 140 °C [10]
$Ge(C_2H_5)_3OCH_3$	at 100 °C (2 h) → no formation of $(C_6F_5)_3GeGe(C_2H_5)_3$ [8]
$Ge(C_2H_5)_3N(C_2H_5)_2$	in C_6H_{14} at 100 °C (1 h) → $(C_6F_5)_3GeGe(C_2H_5)_3$ (73%) and $N(C_2H_5)_2H$ (91%) [8]
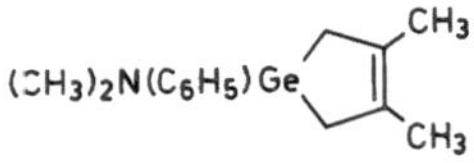	in THF, $Ge(C_6F_5)_3H$ slowly added → 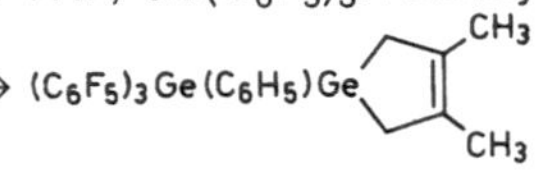 and $N(CH_3)_2H$ [16]
$Sn(C_2H_5)_3N(C_2H_5)_2$	in C_6H_6 at 20 °C (1 h) and 100 °C (4 h) → $Ge(C_6F_5)_3Sn(C_2H_5)_3$ (45%) with elimination of $N(C_2H_5)_3H$ [3]
$Tl(C_2H_5)_3$	1:1 mole ratio, in C_6H_{14} at 20 °C to 40 °C (40 min) → $Tl(Ge(C_6F_5)_3)(C_2H_5)_2$ (70%) 2:1 mole ratio in C_6H_{14} at 40 °C (2 h) → $Tl(Ge(C_6F_5)_3)_2C_2H_5$ (62%); $Tl(Ge(C_6F_5)_3)_3$ could not be prepared [18]
PrX_3, SmX_3, EuX_3 X = OC_4H_9-t or $N(Si(CH_3)_3)_2$	mixture of products containing the GeC_6F_4Ge group; F^- ions were observed [17]
$V(C_5H_5)_2$	in THF at 95 °C (10 h) → no reaction [20]
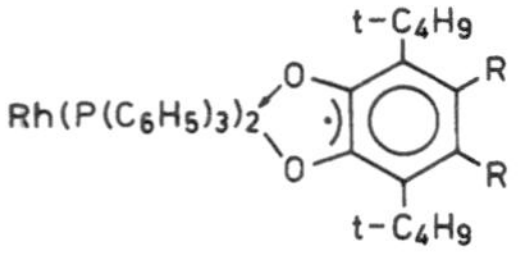 R = H or OCH_3	no reaction in contrast to $Ge(C_6F_5)_2H_2$ and $Ge(C_6F_5)H_3$ [27]
$Ni(P(C_6H_5)_3)_4$	in C_6H_6 at ca. 20 °C (3 h) → $Ge(C_6F_5)_3Ni(P(C_6H_5)_3)_2H$, indicated by the IR spectrum, but not isolated [12]
$Ni(C_5H_5)_2$	in THF at 70 °C → $[Ni_2(C_5H_5)_3][Ni(C_5H_5)(Ge(C_6F_5)_3)_2]$ after 4 h (30%) [28]
$Pd(P(C_6H_5)_3)_3$	→ $Ge(C_6F_5)_3Pd(P(C_6H_5)_3)_2H \cdot 2\ C_6H_6$ (17%) [12]
$Pt(P(C_6H_5)_3)_3$	→ $Ge(C_6F_5)_3Pt(P(C_6H_5)_3)_2H$ (62%); attempts to prepare a digermylplatinum complex from $Ge(C_6F_5)_3Pt(P(C_6H_5)_3)_2H$ and $Ge(C_6F_5)_3H$ failed [12]

Table 20 (continued)

reactant	conditions → products (yield); remarks
$Zn(C_2H_5)_2$	2.2:1 mole ratio, in C_6H_{14} at 20 to 70°C (1 h) → $Zn(Ge(C_6F_5)_3)_2$ (69%) and C_2H_6 (97%) [3]
$Cd(C_2H_5)_2$	2:1 mole ratio, in $C_6H_5CH_3$ at 20 to 100°C (3.5 h) → $Cd(Ge(C_6F_5)_3)_2$ (69%) and C_2H_6 (97%) [2]; see also [1]
$Cd(Ge(C_6F_5)_3)_2$ + $Pr(OC_4H_9\text{-}t)_3$	3:2:1 mole ratio, in DME at 50°C (1 h) → $Pr[Cd_2(Ge(C_6F_5)_3)_7] \cdot 3$ DME (72%) along with $t\text{-}C_4H_9OH$ [17]
$Hg(N(Si(CH_3)_3)_2)_2$	2:1 mole ratio, in $C_6H_5CH_3$ at 50°C (15 min) → $Hg(Ge(C_6F_5)_3)_2$ (ca. 90%) and $NH(Si(CH_3)_3)_2$ [1, 2]
$Hg(C_2H_5)_2$	2:1 mole ratio, in $C_6H_5CH_3$ at 100°C (2.5 h) → $Hg(Ge(C_6F_5)_3)_2$ and C_2H_6 (75% each) 1:1 mole ratio, under the same conditions → $Hg(C_2H_5)Ge(C_6F_5)_3$ (40%) along with C_2H_6 [1]; see also [2]
$Hg(CH_3)(m\text{-}C_2B_{10}H_{11})$	in THF at 90 to 100°C (3 h) → $Hg(Ge(C_6F_5)_3)(m\text{-}C_2B_{10}H_{11})$ (65%) and CH_4 (89%) [21]
$Hg(Si(C_6F_5)_3)_2$ + $Pr(OC_4H_9\text{-}t)_3$	3:2:1 mole ratio, in DME at 50°C (1 h) → $Pr[Hg_2(Si(C_6F_5)_3)_4(Ge(C_6F_5)_3)_3] \cdot 3$ DME (33%) along with $t\text{-}C_4H_9OH$ [17]
$Hg(Ge(C_2H_5)_3)_2$	2:1 mole ratio, in C_6H_6 at 100°C (20 min) and then without solvent at 140°C (1 h) → $Hg(Ge(C_6F_5)_3)_2$ (40%) along with $Ge(C_2H_5)_3H$ [2]
$Hg(Ge(C_6F_5)_3)_2$ + $Pr(N(Si(CH_3)_3)_2)_3$	1:1:1 mole ratio, in DME at 50°C (30 min) → $[Pr(N(Si(CH_3)_3)_2)_2][Hg(Ge(C_6F_5)_3)_3] \cdot 3$ DME (57%) along with $NH(Si(CH_3)_3)_2$ [17] 3:2:1 mole ratio → $[PrN(Si(CH_3)_3)_2][Hg_2(Ge(C_6F_5)_3)_6] \cdot 3$ DME (82%) [14]
$Hg(Ge(C_6F_5)_3)_2$ + $Pr(OC_4H_9\text{-}t)Cl_2 \cdot THF$	1.3:1.3:1 mole ratio, in DME at 50°C (30 min) → $[PrCl_2][Hg(Ge(C_6F_5)_3)_3] \cdot 3$ DME (78%) [14]
$Hg(Ge(C_6F_5)_3)_2$ + $Pr(OC_4H_9\text{-}t)_3$	3:2:1 mole ratio, as above → $Pr[Hg_2(Ge(C_6F_5)_3)_7] \cdot 3$ DME (68%) [14]
$Hg(Sn(C_6F_5)_3)_2$ + $Pr(OC_4H_9\text{-}t)_3$	3:2:1 mole ratio, in DME at 50°C (1 h) → $Pr[Hg_2(Sn(C_6F_5)_3)_4(Ge(C_6F_5)_3)_3] \cdot 3$ DME (55%) [17]
$CuOC_4H_9\text{-}t$ + $M(OC_4H_9\text{-}t)_3$ (M = Sm or Yb)	6:3:1 mole ratio, in DME at 50°C → $M[Cu(Ge(C_6F_5)_3)_2]_3 \cdot 3$ DME (61 to 81%) [25]

References on p. 138

References:

[1] Bochkarev, M. N.; Maiorova, L. P.; Bochkarev, L. N.; Vyazankin, N. S. (Izv. Akad. Nauk SSSR Ser. Khim. **1971** 2353; Bull. Acad. Sci. USSR Div. Chem. Sci. [Engl. Transl.] **1971** 2241).

[2] Bochkarev, M. N.; Maiorova, L. P.; Vyazankin, N. S. (J. Organomet. Chem. **55** [1973] 89/96).

[3] Bochkarev, M. N.; Maiorova, L. P.; Korneva, S. P.; Bochkarev, L. N.; Vyazankin, N. S. (J. Organomet. Chem. **73** [1974] 229/36).

[4] Bochkarev, M. N.; Razuvaev, G. A.; Vyazankin, N. S.; Semenov, O. Yu. (J. Organomet. Chem. **74** [1974] C4/C6).

[5] Bochkarev, M. N.; Maiorova, L. P.; Vyazankin, N. S.; Razuvaev, G. A. (J. Organomet. Chem. **82** [1974] 65/71).

[6] Egorochkin, A. N.; Khorshev, S. Ya.; Ostasheva, N. S.; Satgé, J.; Rivière, P.; Barrau, J.; Massol, M. (J. Organomet. Chem. **76** [1974] 29/36).

[7] Bochkarev, M. N.; Razuvaev, G. A.; Vyazankin, N. S. (Izv. Akad. Nauk SSSR Ser. Khim. **1975** 1820/5; Bull. Acad. Sci. USSR Div. Chem. Sci. [Engl. Transl.] **1975** 1701/5).

[8] Bochkarev, M. N.; Vyazankin, N. S.; Bochkarev, L. N.; Razuvaev, G. A. (J. Organomet. Chem. **110** [1976] 149/57).

[9] Razuvaev, G. A.; Bochkarev, L. N.; Kalinina, G. S.; Bochkarev, M. N. (Inorg. Chim. Acta **24** [1977] L40/L42).

[10] Bochkarev, M. N.; Gur'ev, N. I.; Razuvaev, G. A. (J. Organomet. Chem. **162** [1978] 289/95).

[11] Egorochkin, A. N.; Sevast'yanova, E. I.; Khorshev, S. Ya.; Ratushnaya, S. Kh.; Satgé, J.; Rivière, P.; Barrau, J.; Richelme, S. (J. Organomet. Chem. **162** [1978] 25/35).

[12] Bochkarev, M. N.; Maiorova, L. P.; Skobeleva, S. E.; Razuvaev, G. A. (Izv. Akad. Nauk SSSR Ser. Khim. **1979** 1854/8; Bull. Acad. Sci. USSR Div. Chem. Sci. [Engl. Transl.] **1979** 1717/22).

[13] Bochkarev, M. N.; Gur'ev, N. I.; Pankratov, L. V.; Razuvaev, G. A. (Inorg. Chim. Acta **44** [1980] L59/L60).

[14] Bochkarev, L. N.; Bochkarev, M. N.; Radkov, Yu. F.; Kalinina, G. S.; Razuvaev, G. A. (Inorg. Chim. Acta **45** [1980] L261/L262).

[15] Bochkarev, L. N.; Bochkarev, M. N.; Kalinina, G. S.; Razuvaev, G. A. (Izv. Akad. Nauk SSSR Ser. Khim. **1981** 2589/94; Bull. Acad. Sci. USSR Div. Chem. Sci. [Engl. Transl.] **1981** 2149/53).

[16] Castel, A.; Escudié, J.; Rivière, P.; Satgé, J.; Bochkarev, M. N.; Maiorova, L. P.; Razuvaev, G. A. (J. Organomet. Chem. **210** [1981] 37/42).

[17] Bochkarev, L. N.; Rad'kov, Yu. F.; Kalinina, G. S.; Bochkarev, M. N.; Razuvaev, G. A. (Zh. Obshch. Khim. **52** [1982] 1381/5; J. Gen. Chem. USSR [Engl. Transl.] **52** [1982] 1217/21).

[18] Bochkarev, M. N.; Basalgina, T. A.; Kalinina, G. S.; Razuvaev, G. A. (J. Organomet. Chem. **243** [1983] 405/10).

[19] Razuvaev, G. A.; Bochkarev, M. N.; Pankratov, L. V. (J. Organomet. Chem. **250** [1983] 135/43).

[20] Bochkarev, M. N.; Pankratov, L. V.; Cherkasov, V. K.; Razuvaev, G. A.; Latyaeva, V. N.; Lineva, A. N. (J. Organomet. Chem. **263** [1984] 21/7).

[21] Bochkarev, M. N.; Fedorova, E. A.; Razuvaev, G. A.; Bregadze, V. I.; Kampel, V. Ts. (J. Organomet. Chem. **265** [1984] 117/22).

[22] Bochkarev, L. N.; Orlov, N. A.; Zhil'tsov, S. F.; Bochkarev, M. N. (Zh. Obshch. Khim. **57** [1987] 2802/3; J. Gen. Chem. USSR [Engl. Transl.] **57** [1987] 2500/1).

[23] Pankratov, L. V.; Penyagina, I. M.; Zakharov, L. N.; Bochkarev, M. N.; Razuvaev, G. A.; Grishin, Yu. K.; Ustynyuk, Yu. A.; Struchkov, Yu. T. (J. Organomet. Chem. **335** [1987] 313/22).
[24] Bochkarev, M. N.; Silkin, V. B.; Maiorova, L. P.; Razuvaev, G. A.; Semchikov, Yu. D.; Sherstyanykh, V. I. (Metalloorg. Khim. **1** [1988] 196/200; Organomet. Chem. USSR [Engl. Transl.] **1** [1988] 108/11).
[25] Bochkarev, L. N.; Orlov, N. A.; Zhil'tsov, S. F. (Metalloorg. Khim. **2** [1989] 1431; Organomet. Chem. USSR [Engl. Transl.] **2** [1989] 759).
[26] Silkin, V. B.; Maiorova, L. P.; Bochkarev, M. N.; Semechikov, Yu. D.; Khvatova, N. L. (Vysokomol. Soedin. Ser. A **32** [1990] 2346/50; Polym. Sci. USSR [Engl. Transl.] **32** [1990] 2249/53).
[27] Pankratov, L. V.; Nevodchikov, V. I.; Cherkasov, V. K.; Bochkarev, M. N. (Metalloorg. Khim. **4** [1991] 516/20; Organomet. Chem. USSR [Engl. Transl.] **4** [1991] 247/9).
[28] Pankratov, L. V.; Nevodchikov, V. I.; Zakharov, L. N.; Bochkarev, M. N.; Zdanovich, I. V.; Latyaeva, V. N.; Lineva, A. N.; Batsanov, A. S.; Struchkov, Yu. T. (J. Organomet. Chem. **429** [1992] 13/26).
[29] Silkin, V. B.; Makarenko, N. P.; Bochkarev, M. N. (Metalloorg. Khim. **5** [1992] 621/4; Organomet. Chem. USSR [Engl. Transl.] **5** [1992] 299/301).

1.3.1.1.8.2 Other GeR_3H Compounds with R = Substituted Aryl, Naphthyl, or Furyl

The compounds in this section are listed in Table 21.

Table 21
Other GeR_3H Compounds with R = Substituted Aryl, or Furyl.
An asterisk indicates further information at the end of the table.
Explanations, abbreviations, and units are given on p. X.

No.	R group	preparation (yield) properties and remarks	Ref.
1	C_6H_4F-3	reacts with LiC_4H_9 in ether to give $Ge(C_6H_4F\text{-}3)_3Li$	[6]
2	C_6H_4F-4	reacts with Li in THF-HMPT to give $Ge(C_6H_4F\text{-}4)_3Li$	[6]
3	C_6H_4Cl-2	$Ge(C_6H_4Cl\text{-}2)_3Br + LiAlH_4$ like No. 12 b.p. 135 to 148°C/0.001 ^1H NMR (CCl_4): 5.51 (GeH) IR: ν(GeH) 2058(m), δ(GeH) 696(s)	[14]
4	C_6H_4Cl-3	$Ge(C_6H_4Cl\text{-}3)_3Br + LiAlH_4$; cf. No. 12 b.p. 160 to 168°C/0.002 ^1H NMR (CCl_4): 5.58 (GeH) IR: ν(GeH) 2043(m), δ(GeH) 715(s) or 687(s)	[14]
5	C_6H_4Cl-4	$Ge(C_6H_4Cl\text{-}4)_3Br + LiAlH_4$; cf. No. 12 b.p. 160 to 170°C/0.002 ^1H NMR ($CDCl_3$): 5.61 (GeH) IR: ν(GeH) 2042(m), δ(GeH) 735(s) or 695(s)	[14]

Table 21 (continued)

No.	R group	preparation (yield) properties and remarks	Ref.
6	$C_6H_4OCH_3$-2	$Ge(C_6H_4OCH_3\text{-}2)_3Br + LiAlH_4$; cf. No. 12 b.p. 160 to 180 °C/0.001 ^1H NMR ($CDCl_3$): 5.52 (GeH) IR: ν(GeH) 2033(m), δ(GeH) 694(s)	[14]
7	$C_6H_4OCH_3$-3	$Ge(C_6H_4OCH_3\text{-}3)_3Cl$ (obtained from $GeCl_4$ and 3-$CH_3OC_6H_4MgBr$ in THF) + $LiAlH_4$ (91% crude yield) m.p. 83.5 to 85 °C IR (Nujol): ν(GeH) 2041 with LiC_4H_9 in ether → $Ge(C_6H_4OCH_3\text{-}3)_3Li$	[6]
8	$C_6H_4OCH_3$-4	$Ge(C_6H_4OCH_3\text{-}4)_3Br + LiAlH_4$; cf. No. 12 $Ge(C_6H_4OCH_3\text{-}4)_3Cl$ (obtained from $GeCl_4$ and 4-$CH_3OC_6H_4MgBr$ in THF) + $LiAlH_4$ b.p. 170 to 180 °C/0.001 ^1H NMR ($CDCl_3$): 5.49 (GeH) IR: ν(GeH) 2037(m), δ(GeH) 731(s) or 689(s) [14]; ν(GeH) 2045(s) for the neat liquid [6] with Li in THF-HMPT → $Ge(C_6H_4OCH_3\text{-}4)_3Li$	[14] [6] [14] [6, 14] [6]
*9	$C_6H_4CH_2N(CH_3)_2$-2	$Ge(C_6H_4CH_2N(CH_3)_2\text{-}2)_3Cl + LiAlH_4$ in ether for 12 h (45%); the chloride was obtained by slowly adding 2-$(CH_3)_2NCH_2C_6H_4Li$ in ether to $GeCl_4$ in ether at 0 °C and reacting the mixture at 20 °C for 1 h m.p. 123 to 124 °C (from C_6H_6) ^1H NMR ($CDCl_3$): 1.95 (s, CH_3N), 3.45 (s, CH_2N), 6.05 (s, GeH), 7.1 to 7.4 (m, C_6H_4) IR (CCl_4): ν(GeH) 2080	[21]
*10	$C_6H_4CH_3$-2	$Ge(C_6H_4CH_3\text{-}2)_3Br + LiAlH_4$; cf. No. 12 fine white needles; m.p. 102 to 103 °C (from petroleum ether) [2], 102 to 105 °C [14] b.p. 120 to 150 °C/0.001 ^1H NMR (CCl_4): 5.60 (GeH) IR: ν(GeH) 2057(m) [4, 14] or 2050 [2], δ(GeH) 697(s) [4, 14]; other characteristic bands at 438(s), 403(m), 304(s), 291(s), and 245(w) [4]	[14] [2, 14] [2] [14] [2, 4, 14]
11	$C_6H_4CH_3$-3	$Ge(C_6H_4CH_3\text{-}3)_3Br + LiAlH_4$; cf. No. 12 first obtained (37%) by hydrolysis of $Ge(C_6H_4CH_3\text{-}3)_3MgBr$ (from 3-$CH_3C_6H_4Br$, excess Mg, and $GeCl_4$) b.p. 160 to 168 °C/0.001 [14] or 160 to 170 °C/0.001 [2]	[14] [2] [2, 14]

Table 21 (continued)

No.	R group	preparation (yield) properties and remarks	Ref.
		^{1}H NMR (CCl_4): 5.57 (GeH)	[14]
		IR: ν(GeH) 2037(m) [14], 2034(m) [2, 4], δ(GeH) 719(s) [4, 14]; other characteristic bands at 426(s), 379(m), 316(s), 308(s), and 270(w) [4]	[2, 4, 14]
		with Li in THF-HMPT → $Ge(C_6H_4CH_3\text{-}3)_3Li$	[6]
*12	$C_6H_4CH_3$-4	$Ge(C_6H_4CH_3\text{-}4)_3Br + LiAlH_4$ in refluxing ether (2 h), replacing the ether by petroleum ether and cooling the filtered and concentrated solution (90%)	[14]
		first obtained (28%) by hydrolysis of $Ge(C_6H_4CH_3\text{-}4)_3MgBr$ (from 4-$CH_3C_6H_4Br$, excess Mg, and $GeCl_4$)	[2]
		m.p. 81 °C (from petroleum ether) [2, 7], 87 to 88 °C (from petroleum ether) [14] (m.p. 124 to 125 °C [11] probably erroneous; confused with $Ge(C_6H_4CH_3\text{-}4)_3Br$?)	[2, 7, 14]
		b.p. 160 °C/0.001	[2]
		^{1}H NMR ($CDCl_3$): 5.52 (GeH)	[14]
		^{13}C NMR (THF): 21.35 (CH_3), 129.70 (C-3,5), 133.06 (C-1), 135.69 (C-2,6), 139.35 (C-4)	[18]
		IR: ν(GeH) 2034 [2], 2032(m) [14], 2031(m) [4], δ(GeH) 732(s) or 688(s) [4, 14]	[2, 4, 14]
13	$C_6H_4CF_3$-4	reacts with Li in THF-HMPT to give $Ge(C_6H_4CF_3\text{-}4)_3Li$	[6]
14	$C_6H_4C_6H_5$-4	$Ge(C_6H_4C_6H_5\text{-}4)_3Br + LiAlH_4$; cf. No. 12 b.p. 190 to 208 °C/0.001 ^{1}H NMR ($CDCl_3$): 5.63 (GeH) IR: ν(GeH) 2031(s), δ(GeH) 709(s) or 685(s)	[14]
*15	$C_6H_3(CH_3)_2$-3,4	$Ge(C_6H_3(CH_3)_2\text{-}3,4)_3Cl + LiAlH_4$ in ether (56%)	[16, 17]
		m.p. 76 °C (from CH_3OH)	[17]
		^{1}H NMR (CCl_4): 2.20 (CH_3-3,4), 5.53 (GeH), 7.13 (C_6H_3)	[16]
		IR: ν(GeH) 2042 (in C_7H_{16} or CCl_4 [15] and Nujol [16])	[15, 16]
*16	$C_6H_2(CH_3)_3$-2,4,6	$Ge(C_6H_2(CH_3)_3\text{-}2,4,6)_3Cl + LiAlH_4$ in ether for 3 h (73%) [11]; see also [10, 16, 24]	[11]
		colorless crystals	[3]
		m.p. 185 to 186 °C (from CH_3OH), 186 to 187 °C (from C_5H_{12} or C_6H_6-CH_3OH), 194 °C (from C_2H_5OH); see also [11, 29]	[17] [24] [3, 5]

References on p. 145

Table 21 (continued)

No.	R group	preparation (yield) properties and remarks	Ref.
*16 (continued)		b.p. 198°C/0.05	[17, 24]
		^{1}H NMR: 2.04 (s, CH_3-2,6), 2.17 (s, CH_3-4), 5.81 (s, GeH), 6.74 (s, C_6H_2) in CS_2;	[11]
		2.15, 2.23, 5.83, 6.75 in CCl_4; 2.10, 2.20, 5.80, 6.75 in THF-d_8	[16, 24, 27, 29]
		^{13}C NMR: 21.0 (CH_3-4), 23.5 (CH_3-2,6), 129.3 (C-3,5), 135.2 (C-1), 138.5 (C-4), 143.7 (C-2,6) in $N(C_2H_4OC_2H_4OCH_3)_3$;	[27]
		21.04, 23.70, 129.51, 135.46, 138.87, 144.07 in THF-d_8	[29]
		IR: ν(GeH) 2052 (in C_7H_{16} or CCl_4 [15] and in Nujol [16]), 2000 (in KBr) [11],	[11, 15, 16]
		2033(m), δ(GeH) 707(m) [4]; see also [3]	[3, 4]
17	$C_6H(CH_3)_4$-2,3,5,6	$GeBr_4$ + ($C_6H(CH_3)_4$-2,3,5,6)MgBr (1:5 mole ratio) in $C_6H_5CH_3$ at 110°C (12 h) followed by hydrolysis (35%) m.p. 211 to 212°C (from C_4H_9OH)	[5]
18	C_6Cl_5	$Ge(C_6Cl_5)_3Cl$ + $LiAlH_4$ in ether (3 h) followed by aqueous dilute HCl and extraction with $CHCl_3$ (88%) m.p. 248 to 249°C (from C_6H_6-C_6H_{14}) IR (KBr): 2180(w), 1515(m), 1340(s), 1325(s), 1300(s), 1220(w), 1200(w), 1165(m), 1085(m), 850(s), 765(s), 740(s), 700(w), and 675(s) UV ($CHCl_3$): $\lambda_{max}(\varepsilon)$ 285(1280,sh), 294(2290), 304(2600) refluxing with Br_2 in CCl_4 in the dark (1 d) → $Ge(C_6Cl_5)_3Br$ (93%)	[25]
*19	$C_{10}H_7$-1	$Ge(C_{10}H_7\text{-}1)_3Br$ + $LiAlH_4$ in refluxing ether for 2 h (82%); recrystallized from C_6H_6 and $CHCl_3$-petroleum ether colorless needles, m.p. 249 to 250°C	[1]
20		$Ge(C_4H_3O)_3Br$ + $LiAlH_4$ in C_6H_6 in the presence of $[N(C_2H_5)_3CH_2C_6H_5]Cl$ as the phase transfer catalyst at 25°C for 1 h (>95%); see also [26]	[20]

*Further information:

First-order rate constants for the loss of tritium from $Ge(C_6H_4X)_3T$ compounds during T-H exchange with CH_3OH-CH_3ONa have been measured at 20 to 40°C for Nos. 3 to 5 (X = Cl-2,3,4), 6 and 8 (X = OCH_3-2,4), 10 to 12 (X = CH_3-2,3,4), and No. 14 (X = C_6H_5-4). With respect to the rate constant k($Ge(C_6H_5)_3T$) = 1 (in CH_3OH containing 0.421 M CH_3ONa) at

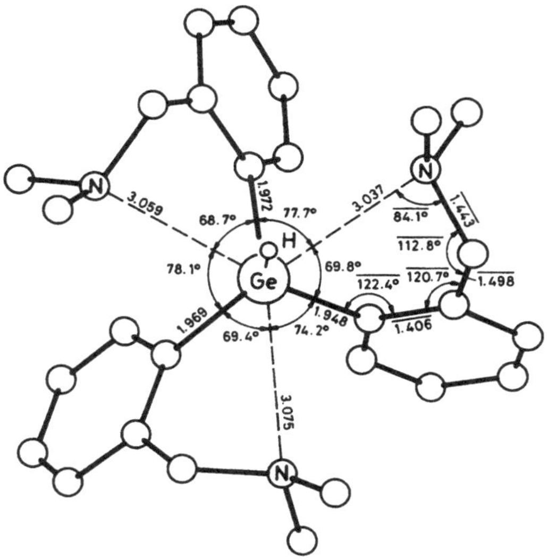

Fig. 1. Molecular structure of one of the two $Ge(C_6H_4CH_2N(CH_3)_2\text{-}2)_3H$ conformers [21].

30 °C, the relative rates k_{rel} in CH_3OH are as follows (molarity of CH_3ONa in parentheses) [14]: No. 3, 425 (0.0058 M); No. 4, 380 (0.001 M); No. 5, 71 (0.0058 M); No. 14, 1.03 (0.108 M); No. 11, 0.48 (0.108 M); Nos. 6 and 10, 0.081 (0.108 M); No. 12, 0.081 (0.078 M); and No. 8, 0.025 (0.108 M).

$Ge(C_6H_4CH_2N(CH_3)_2\text{-}2)_3H$ (Table **21**, No. **9**). An X-ray analysis showed that the solid compound exists in two slightly different conformations. Both molecules have a propeller shape with approximate C_3 geometry, the Ge-H bond being coincident with the axis of symmetry. The basic tetrahedral geometry is retained with the angles H-Ge-C = 112(2)° and C-Ge-C = 106.8(3)°. Formally the Ge atoms are seven-coordinated by three C atoms, three N atoms and the H atom, forming a distorted capped antiprism. Each N ligand lies opposite to a C-Ge bond with C-Ge-N angles from 172.5 to 176.8(3)° (mean 174.0°). The X-ray technique probably underestimates the Ge-H bond length (mean 1.58(6) Å). One of the conformers is illustrated in **Fig. 1** [21]. For a comparison with the structures of $Si(C_6H_4CH_2N(CH_3)_2\text{-}2)_3F$ and $Si(C_{10}H_6N(CH_3)_2\text{-}8)_3H$, see [32].

$Ge(C_6H_4CH_3\text{-}2)_3H$ (Table **21**, No. **10**) was first obtained with 12% yield by hydrolyzing $Ge(C_6H_4CH_3\text{-}2)_3MgBr$ which had been prepared from $GeCl_4$, $4\text{-}CH_3C_6H_4Br$, and an excess of Mg [2].

The compound crystallizes in the monoclinic system with a = 38.31, b = 5.25, c = 20.22 Å, and β = 121.0°, space group C2/c – C^6_{2h}; Z = 8 gives d_c = 1.32 g/cm³ while d_m = 1.31 g/cm³. The molecular structure is shown in **Fig. 2**, p. 144. The three ortho-CH_3 groups are all on the same side of the molecule, crowding close to the Ge-bonded H atom. Steric hindrance causes the C-Ge-C bond angle to vary between 98.7 and 110.3° and is also responsible for the angular distortions at C(2), C(4), and C(6) [9].

$Ge(C_6H_4CH_3\text{-}4)_3H$ (Table **21**, No. **12**). Isotopic exchange between $Ge(C_6H_5CH_3\text{-}4)_3H$ and $Ge(C_6H_5)_3D$ in the presence of hydrogermylation catalysts was studied with $Rh(P(C_6H_5)_3)_3Cl$ in C_6D_6 or cis-$Pt(P(C_6H_5)_3)_2Cl_2$ and H_2PtCl_6 in $CDCl_3$ at 20°. In the case of $Rh(P(C_6H_5)_3)_3Cl$ complete exchange occurred within less than 5 min [8].

References on p. 145

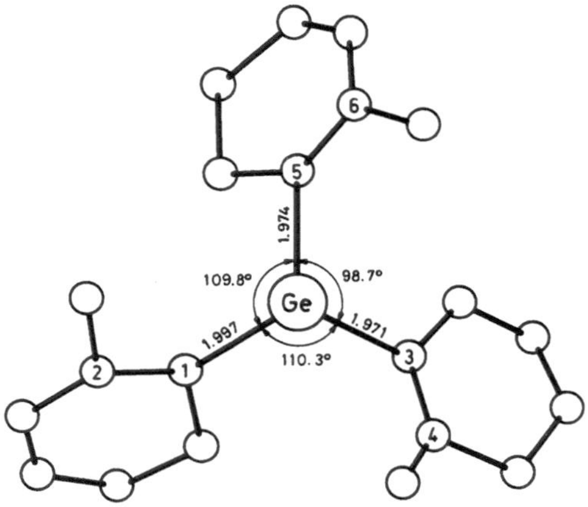

Fig. 2. Molecular structure of $Ge(C_6H_4CH_3\text{-}2)_3H$ [9]. The Ge-bonded H atom (Ge-H = 1.70 Å) is not shown.

Other bond angles:

C(1)-Ge-H	102.0°	C(3)-Ge-H	120.0°
C(5)-Ge-H	116.0° [9].		

The acidity of $Ge(C_6H_4CH_3\text{-}4)_3H$ in liquid NH_3 at +30 °C was determined from the equilibrium concentrations of No. 12 and its $[Ge(C_6H_4CH_3\text{-}4)_3]^-$ anion (determined by ^1H NMR) in the presence of 0.5 molar equivalents of an alkali metal: pK_a = 23.6 referred to fluorene pK_a = 20.6 [7].

Photochemically generated $t\text{-}C_4H_9O^{\bullet}$ reacted with $Ge(C_6H_4CH_3\text{-}4)_3H$ at −30° to yield the $Ge(C_6H_4CH_3\text{-}4)_3^{\bullet}$ radical [11]; cf. "Organogermanium Compounds" 3, 1990, p. 359.

Reactions with $Ru(CO)_2(P(C_6H_5)_3)_3$ in C_6H_6 or $Os(CO)_2(P(C_6H_5)_3)_3$ in C_6H_6 under irradiation were recently reported to yield 44% $Ru(CO)_2(P(C_6H_5)_3)_2)_2(H)Ge(C_6H_4CH_3\text{-}4)_3$ and 66% $Os(CO)_2(P(C_6H_5)_3)_2)_2(H)Ge(C_6H_4CH_3\text{-}4)_3$ [30].

$Ge(C_6H_3(CH_3)_2\text{-}3,4)_3H$ (Table **21**, No. **15**). The compound was part of studies on various organogermanium compounds dealing with the influence of the Ge substituents on the GeH bond frequency and their intensity [15].

$Ge(C_6H_2(CH_3)_3\text{-}2,4,6)_3H$ (Table **21**, No. **16**) was also obtained by adding RMgBr (R = $i\text{-}C_3H_7$, $t\text{-}C_4H_9$, C_6H_5, or C_9H_{11}) to $Ge(C_6H_2(CH_3)_3\text{-}2,4,6)Br$ (10:1 mole ratio) in $C_6H_5CH_3$ and heating to 110 °C for 30 h followed by hydrolysis (54, 30, 63, and 30% yield, respectively) [5]. It was formed along with polymers when GeI_2 and $2,4,6\text{-}(CH_3)_3C_6H_2MgBr$ were refluxed in THF for 8 h followed by hydrolysis [3]. UV irradiation or hydrolysis of a solution of $Ge(C_6H_2(CH_3)_3\text{-}2,4,6)_3HgCl$ in THF and hydrolysis of a $(Ge(C_6H_2(CH_3)_3\text{-}2,4,6)_3)_2Hg$ suspension in THF were also reported to yield No. 15 along with other products [27].

Treatment of $Ge(C_6H_2(CH_3)_3\text{-}2,4,6)_3Li$ with C_6H_5COCl in THF-ether and hydrolysis with aqueous $NaHCO_3$ yielded $Ge(C_6H_2(CH_3)_3\text{-}2,4,6)_3H$ as the main product, $Ge(C_6H_2(CH_3)_3\text{-}2,4,6)_3COC_6H_5$, $Ge(C_6H_2(CH_3)_3\text{-}2,4,6)_3Cl$, and $C_6H_5COCOC_6H_5$ [27]. Substantial amounts of No. 15 were also obtained in the reactions of $Ge(C_6H_2(CH_3)_3\text{-}2,4,6)_3Li$ with compound I in THF and with CH_3SSCH_3 or $Ge(C_6H_2(CH_3)_3\text{-}2,4,6)_3OCH_3$ in THF followed by methanolysis [28]. Small amounts of $Ge(C_6H_2(CH_3)_3\text{-}2,4,6)_3H$ were found in the product mixture obtained by irradiating or heating $Ge(C_6H_2(CH_3)_3\text{-}2,4,6)_3Cl$ and CH_3SSCH_3 in C_6H_6 in the presence of the electron-rich olefin II [29]. Variable amounts of $Ge(C_6H_2(CH_3)_3\text{-}2,4,6)_3H$ and $Ge(C_6H_2(CH_3)_3\text{-}2,4,6)_3Cl$ were formed via the $Ge(C_6H_2(CH_3)_3\text{-}2,4,6)_3^{\bullet}$ radical and germyl adducts when $Ge(C_6H_2(CH_3)_3\text{-}2,4,6)_3Li$ was reacted with compounds favoring single-electron transfer such as the galvinoxyl radical and compounds III or IV followed by treatment with aqueous 10% HCl [31].

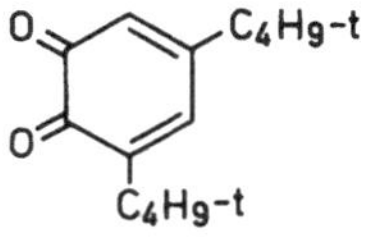

I

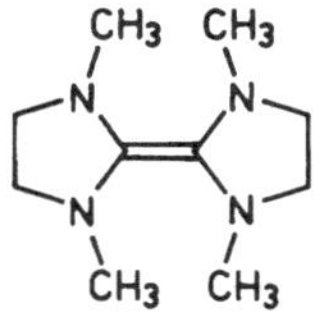

II

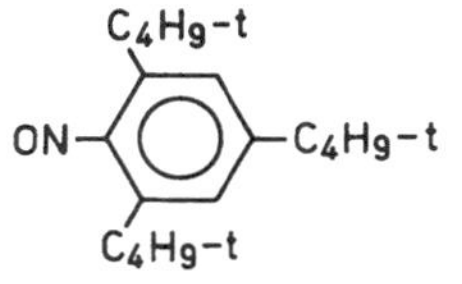

III

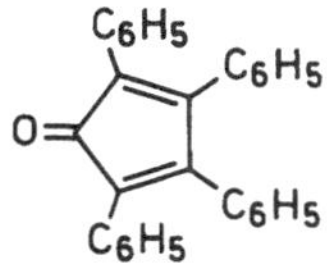

IV

The compound was part of studies on various organogermanium compounds dealing with the influence of the Ge substituents on the Ge-H bond frequency and their intensity [15].

The reaction of $Ge(C_6H_2(CH_3)_3\text{-}2,4,6)_3H$ with CCl_4 at 90 °C for 2 h in the presence of a trace AIBN gave $Ge(C_6H_2(CH_3)_3\text{-}2,4,6)_3Cl$ with 88% yield [16]; see also [10]. The chloride was obtained in the reaction with N-chlorosuccinimide [10]. $Ge(C_6H_2(CH_3)_3\text{-}2,4,6)_3H$ did not react with $C_{11}H_{23}I$ in C_6H_6 at 80 °C after being 8 h in the presence of AIBN [23]. Heating $Ge(C_6H_2(CH_3)_3\text{-}2,4,6)_3H$ and CH_3SSCH_3 in C_6H_6 in the presence of AIBN in a sealed tube at 100 °C for 2 h gave $Ge(C_6H_2(CH_3)_3\text{-}2,4,6)_3SCH_3$ with 61% yield [29].

Equimolar amounts of $Ge(C_6H_2(CH_3)_3\text{-}2,4,6)_3H$ and the oxaziridine $C_6H_5\overline{CHN(C_4H_9\text{-}t)O}$ in C_6D_6 at 80 °C for 15 h yielded $Ge(C_6H_2(CH_3)_3\text{-}2,4,6)_3OH$ (84%), $(Ge(C_6H_2(CH_3)_3\text{-}2,4,6)_3)_2O$ (traces), and $C_6H_5CH{=}NC_4H_9\text{-}t$ (95%) [13]. Treating the title compound with excess $LiC_4H_9\text{-}t$ in THF at 20 °C gave a 62% yield of $Ge(C_6H_2(CH_3)_3\text{-}2,4,6)_3Li$ [22, 27].

Photochemically generated $t\text{-}C_4H_9CO^{\bullet}$ reacted with $Ge(C_6H_2(CH_3)_3\text{-}2,4,6)_3H$ at 20 °C forming the $Ge(C_6H_2(CH_3)_3\text{-}2,4,6)_3^{\bullet}$ radical [11]; see also [10, 16] and "Organogermanium Compounds" 3, 1990, p. 360. An improved ESR signal of this radical resulted from UV irradiation of $Ge(C_6H_2(CH_3)_3\text{-}2,4,6)_3H$ and cyclohexanone in $C_6H_5CH_3$ at 0 °C [10]; see also [12, 16]. Irradiating $Ge(C_6H_2(CH_3)_3\text{-}2,4,6)_3H$ and $C_6H_5CH{=}N(C_4H_9\text{-}t)O$ in $C_6H_5CH_3$ gave $Ge(C_6H_2(CH_3)_3\text{-}2,4,6)CH(C_6H_5)N^{\bullet}(C_4H_9\text{-}t)O$, presumable via $Ge(C_6H_2(CH_3)_3\text{-}2,4,6)_3^{\bullet}$ [10]; see also [12]. The compound could not be used as a reagent for the reductive alkylation of 2-cyclohexen-1-one [23].

In $C_6H_5CH_3$ at 110 °C, the $Ge(C_6H_2(CH_3)_3\text{-}2,4,6)_3^{\bullet}$ radical abstracts hydrogen from the solvent yielding $Ge(C_6H_2(CH_3)_3\text{-}2,4,6)_3H$ and other products [19].

$Ge(C_{10}H_7\text{-}1)_3H$ (Table **21**, No. **19**) showed little reaction with cold basic solutions, but evolved H_2 when warmed gently with KOH in moist piperidine. A CCl_4 solution rapidly decolorized Br_2 in the cold to give HBr [1].

References:

[1] West, R. (J. Am. Chem. Soc. **74** [1952] 4363/5).
[2] Glockling, F.; Hooton, K. A. (J. Chem. Soc. **1962** 3509/12).
[3] Glockling, F.; Hooton, K. A. (J. Chem. Soc. **1963** 1849/54).
[4] Cross, R. J.; Glockling, F. (J. Organomet. Chem. **3** [1965] 146/55).
[5] Lapkin, I. I.; Dumler, V. A.; Ponosova, E. S. (Zh. Obshch. Khim. **39** [1969] 1455/9; J. Gen. Chem. USSR [Engl. Transl.] **39** [1969] 1426/9).

References on p. 145

[6] Steward, O. W.; Dziedzic, J. E.; Johnson, J. S. (J. Org. Chem. **36** [1971] 3475/80).
[7] Birchall, T.; Drummond, I. (Inorg. Chem. **11** [1972] 250/2).
[8] Corriu, R. J. P.; Moreau, J. J. E. (J. Organomet. Chem. **40** [1972] 55/72).
[9] Cameron, T. S.; Mannan, Kh. M.; Stobart, S. R. (Cryst. Struct. Commun. **4** [1975] 601/4).
[10] Gynane, M. J. S.; Lappert, M. F.; Rivière, P.; Rivière-Baudet, M. (J. Organomet. Chem. **142** [1977] C9/C11).

[11] Sakurai, H.; Mochida, K.; Kira, M. (J. Organomet. Chem. **124** [1977] 235/52).
[12] Rivière, P.; Richelme, S.; Rivière-Baudet, M.; Satgé, J.; Gynane, M. J. S.; Lappert, M. F. (J. Chem. Res. Synop. **1978** 218/9; J. Chem. Res. Miniprint **1978** 2801/16).
[13] Rivière, P.; Rivière-Baudet, M.; Richelme, S.; Satgé, J. (Bull. Soc. Chim. Fr. **1978** II 193/6).
[14] Eaborn, C.; Singh, B. (J. Organomet. Chem. **177** [1979] 333/48).
[15] Skobeleva, S. E.; Egorochkin, A. N.; Khorshev, S. Ya.; Ratushnaya, S. Kh.; Rivière, P.; Satgé, J.; Richelme, S.; Cazes, A. (J. Organomet. Chem. **182** [1979] 1/7).
[16] Gynane, M. J. S.; Lappert, M. F.; Riley, P. I.; Rivière, P.; Rivière-Baudet, M. (J. Organomet. Chem. **202** [1980] 5/12).
[17] Egorochkin, A. N.; Rivière, P. (unpublished results from [16]).
[18] Batchelor, R. J.; Birchall, T. (J. Am. Chem. Soc. **105** [1983] 3848/52).
[19] Neumann, W. P.; Schultz, K.-D.; Vieler, R. (J. Organomet. Chem. **264** [1984] 179/91).
[20] Gevorgyan, V. N.; Ignatovich, L. M.; Lukevics, E. (J. Organomet. Chem. **284** [1985] C31/C32).

[21] Brelière, C.; Carré, F.; Corriu, R. J. P.; Royo, G. (Organometallics **7** [1988] 1006/8).
[22] Castel, A.; Rivière, P.; Satgé, J.; Ko, Y.-H. (J. Organomet. Chem. **342** [1988] C1/C4).
[23] Pike, P.; Hershberger, S.; Hershberger, J. (Tetrahedron **44** [1988] 6295/304).
[24] Rivière, P.; Rivière-Baudet, M.; Satgé, J. (in King, R. B.; Eisch, J. J.; Organometallic Synthesis Vol. 4, New York 1988, pp. 545/8).
[25] Fajarí, L.; Juliá, L.; Riera, J.; Molins, E.; Miravitlles, C. (J. Organomet. Chem. **363** [1989] 31/7).
[26] Lukevics, E.; Gevorgyan, V. (Chem. Technol. Silicon Tin. Proc. Asian Network Anal. Inorg. Chem. 1st Int. Chem. Conf. Silicon Tin, Kuala Lumpur 1989 [1992], pp. 165/77).
[27] Castel, A.; Rivière, P.; Satgé, J.; Ko, Y. H.; Desor, D. (J. Organomet. Chem. **397** [1990] 7/15).
[28] Rivière, P.; Castel, A.; Ko, Y. H.; Desor, D. (J. Organomet. Chem. **386** [1990] 147/56).
[29] Rivière, P.; Rivière-Baudet, M.; Castel, A.; Desor, D.; Abdennadher, C. (Phosphorus Sulfur Silicon Relat. Elem. **61** [1991] 189/99).
[30] Clark, G. R.; Kevin, R. F.; Clifton, E. F. R.; Roper, W. R.; Salter, D. M.; Wright, L. J. (J. Organomet. Chem. **462** [1993] 331/41).

[31] Rivière, P.; Castel, A.; Desor, D.; Abdennadher, C. (J. Organomet. Chem. **443** [1993] 51/60).
[32] Brelière, C.; Carré, F.; Corriu, R. J. P.; Royo, G.; Man, M. W. C. (Organometallics **13** [1994] 307/14).

1.3.1.2 Triorganogermanium Hydrides of the $GeR_2(R')H$ Type

The $GeR_2(R')H$ compounds are subdivided into the following types:
1) $Ge(CH_3)_2(R')H$ (Table 22, p. 147)
2) $Ge(C_2H_5)_2(R')H$ (Table 23, p. 158)

3) $Ge(C_3H_7)_2(R')H$, $Ge(C_4H_9)_2(R')H$, and $GeR_2(R')H$ with R = larger alkyl and cycloalkyl (Table 24, p. 166)
4) $Ge(C_6H_5)_2(R')H$ (Table 25, p. 172)
5) other $GeR_2(R')H$ compounds where R represents various groups such as substituted alkyl, alkenyl, substituted aryl, and heterocyclic groups (Table 26, p. 180).

The compounds are listed according to the type of the R' group.

In the following tables, "$LiAlH_4$ reduction" stands for the frequently used preparation of the hydrides from the corresponding chloride, $GeR_2(R')Cl$, by treatment with $LiAlH_4$ in refluxing ether followed by hydrolysis. For typical examples of this method, see e.g. [5, 67, 79, 90, 129 to 131].

The references for Sections 1.3.1.2.1 to 1.3.1.2.5 are combined in a single list on p. 187.

1.3.1.2.1 $Ge(CH_3)_2(R')H$ Compounds

Table 22
$Ge(CH_3)_2(R')H$ Compounds.
An asterisk indicates further information at the end of the table.
Explanations, abbreviations, and units are given on p. X.

No.	R' group	preparation (yield) properties and remarks
*1	CF_3	preparation on p. 153 ^1H NMR: 0.52 (CH_3), 4.53 (GeH), J(H, H) = 3.37, J(F, H) = 6.85 ^{19}F NMR: −58.2 (CF_3, ^1J(C, F) = 335.6) MS: $[Ge(CH_3)(CF_3)H]^+$ (3), $[Ge(CH_3)CF_3]^+$ (2), $[GeCF_3]^+$ (2), $[Ge(CH_3)_2F]^+$ (30), $[Ge(CH_3)(H)F]^+$ (33), $[Ge(CH_3)_2H]^+$ (100), $[GeF]^+$ (20), $[GeCH_3]^+$ (45), $[GeH]^+$ (6), $[Ge]^+$ (8), $[CF_3]^+$ (3) [92]
*2	CH_2Cl	$LiAlH_4$ reduction (70%) [33] or 83% [71]; purified by preparative GLC [66] b.p. 80 to 90°C (crude product) [66], 99°C [71], 99.5°C/751 [33] d^{20} = 1.2698; n_D^{20} = 1.4490 [33] ^1H NMR (CCl_4): 0.31 (d, CH_3, ^3J = 3), 2.96 (d, CH_2Cl, ^3J = 2), 4.18 (m, GeH) [33, 71] IR (C_7H_{16}): ν(GeH) 2040 [49, 50], 2043 and 2064 [71]; see also p. 153
*3	$CHCl_2$	$LiAlH_4$ reduction of impure $Ge(CH_3)_2(CHCl_2)Cl$; contaminated with 10% $Ge(CH_3)_2(CH_2Cl)H$ [49] IR (C_7H_{16}): ν(GeH) 2060 [49, 50]
*4	CH_2Br	preparation on p. 153 IR (C_7H_{16}): ν(GeH) 2040 [49, 50]
*5	CH_2OH	$Ge(CH_3)_2(CH_2OOCCH_3)OOCCH_3$ + $LiAlH_4$ in refluxing ether (3 h) and at 20°C (15 h) followed by hydrolysis with 12.5% HCl (79%) [102]

Table 22 (continued)

No.	R′ group	preparation (yield) properties and remarks
*5 (continued)		colorless, sweet smelling liquid b.p. 52 °C/35 ^{1}H NMR ($CDCl_3$): 0.27 (d, CH_3Ge, J = 3.5), ca. 2.6 (br s, OH), 3.73 (d, CH_2Ge, J = 2.4), 3.91 (t-sept, GeH, J = 3.5 and 2.4) ^{13}C NMR ($CDCl_3$): −6.7 (CH_3Ge), 54.9 (CH_2Ge) IR (CCl_4): ν(GeH) 2030, ν(OH) 3620 (nonassociated) and 3320 (br, associated) MS: $[M]^+$ (1), $[M-H]^+$ (4), $[M-CH_3]^+$ (26), $[M-CH_2OH]^+$ (100), $[GeCH_3]^+$ (56) [102]
*6	$CH_2OSO_2CF_3$	no data published [102, footnote 9], [120]
7	$CH_2N(CH_3)C_2H_5$	$Ge(CH_3)_2(CH_2N(CH_3)COCH_3)Cl$ + $LiAlH_4$ in refluxing ether (2 h) and hydrolysis with aqueous NaOH-$CH_3COOC_2H_5$ (90%) [101] b.p. 80 °C ^{1}H NMR ($CDCl_3$): 0.27 (d, CH_3Ge, J = 3), 1.05 (t, CH_3C, J = 7), 2.24 (s, CH_3N), 2.26 (d, CH_2Ge, J = 3), 2.39 (q, CH_2C, J = 7), 3.9 to 4.1 (m, GeH) IR (neat): ν(GeH) 2020 [101]
*8	$CH_2N(CH_2CH{=}CH_2)C_2H_5$	$Ge(CH_3)_2(CH_2N(CH_2CH{=}CH_2)COCH_3)Cl$ + $LiAlH_4$ in refluxing ether and hydrolysis with aqueous NaOH-$CH_3COOC_2H_5$ (86%) [101] b.p. 100 °C/40 ^{1}H NMR ($CDCl_3$): 0.26 (d, CH_3Ge, J = 4), 1.03 (t, CH_3C, J = 7), 2.36 (d, CH_2Ge, J = 4), 2.51 (q, CH_2 of C_2H_5, J = 7), 3.08 (d, NCH_2CH, J = 6 Hz), 3.9 to 4.1 (m, GeH), 5.1 to 5.3 (m, $=CH_2$), 5.7 to 6.1 (m, =CH) IR (neat): ν(GeH) 2020 [101]
*9	$CH_2Si(CH_3)_3$	$LiAlH_4$ reduction [86] b.p. 121 °C n_D^{20} = 1.4440 ^{1}H NMR ($CDCl_3$): −0.18 (d, CH_2Ge, J = 4.0), 0 (s, CH_3Si), 0.19 (d, CH_3Ge, J = 4.0), 3.82 (m, GeH) IR (neat): ν(GeH) 2020, ν($GeCH_3$) 1225 [86]
*10	$CH_2Si(CH_3)_2H$	$Ge(CH_3)_2(CH_2Si(CH_3)_2Cl)Cl$ + $LiAlH_4$ (86%) [109] b.p. 97 °C/760 ^{1}H NMR (C_6H_6): −0.13 (t, CH_2, J = 4), 0.20 (d, CH_3Si), 0.30 (d, CH_3Ge), and 4.23 (m, SiH, GeH) IR: ν(SiH) 2100(s), ν(GeH) 2020(s) [109]
*11	$C(Si(CH_3)_3)_3$	$LiAlH_4$ reduction (70%); purified on SiO_2 with C_6H_{14} as eluent [70] m.p. 270 to 271 °C (with sublimation)

Table 22 (continued)

No.	R′ group	preparation (yield) properties and remarks
		^{1}H NMR (CCl_4): 0.20 (s, CH_3Si), 0.45 (d, CH_3Ge), 4.23 (m, GeH) IR: ν(GeH) 2020 [70]
*12	$CH_2C_6H_5$	formation on p. 154 no data reported [74, 117]
*13	C_2H_5	preparation on p. 154 b.p. 62 °C/755.5 [2] (similar in [15]), 62 °C/760 [36] d^{20} = 1.0077 [15], 1.0158 [2] n_D^{20} = 1.4078 [15], 1.4090 [2], 1.4093 [36] ^{1}H NMR: 0.17 (CH_3) [26], 3.93 (GeH) [9] IR (C_7H_{16}): ν(GeH) 2018 [49, 50]; spectrum displayed in [15] Raman: 2975(3, broad), 2925(2), 2904(8), 2871(4), 2021(5), 1468(1), 1422(1), 1240(4), 1220(3), 1106(0), 1021(1), 969(1), 847(0), 624(3), 586(5), 554(9), 286(3), 185(4), and 164(3) [2]; ν(GeH) 2022 [4]
*14	$CH_2CH_2P(H)C_6H_5$	preparation on p. 154 b.p. 74 to 76 °C/1 ^{1}H NMR (C_6D_6): 0.09 (d, CH_3, ^3J = 3.60), 0.90 to 1.33 (m, CH_2Ge), 1.50 to 2.07 (m, CH_2P), 3.93 (m, GeH), 4.13 (dt, PH, ^1J(P, H) = 205.9, ^3J(H, H) = 6.5) ^{31}P NMR: −42.4 IR (neat): ν(PH) 2302, ν(GeH) 2046 [77]
15	$CH_2CH_2C_6H_5$	$LiAlH_4$ reduction (76%) b.p. 81 °C/6 d^{20} = 1.1077; n_D^{20} = 1.5123 ^{1}H NMR (CCl_4): 0.2 (d, 6 H, J = 3.6), 1.0 to 1.4 (t, 2 H, J = 8), 2.75 (t, 2 H, J = 8), 3.9 (sept, 1 H, J = 3.6), 7.23 (s, 5 H) [54]
*16	C_3H_7-i	no preparation and data reported [44]
*17	$CH_2C(CH_3)_2CH_3$	$LiAlH_4$ reduction (83%) b.p. 122.0 to 122.5 °C n^{20} = 1.4302 ^{1}H NMR ($CDCl_3$): 0.21 (d, CH_3Ge, J = 4.0), 0.92 (d, CH_2Ge, J = 4.0), 0.97 (s, CH_3C), 3.89 (m, GeH) IR (neat): ν(GeH) 2020, ν($GeCH_3$) 1240 [86]
18	CH_2CH_2COOH	hydrolysis of No. 19 with alcoholic KOH followed by aqueous HCl almost colorless oil [52]
19	$CH_2CH_2COOCH_3$	$Ge(CH_3)_2H_2$ + CH_2=$CHCOOCH_3$ heated in a steel tube for 5 h; formed along with $Ge(CH_3)_2(CH_2CH_2COOCH_3)_2$

References on p. 187

Table 22 (continued)

No.	R′ group	preparation (yield) properties and remarks
19 (continued)		b.p. 72 to 75°C/34 [52] starting material for No. 18
*20	$CH_2CH_2CH_2P(H)C_6H_5$	preparation on p. 155 b.p. 88°C/0.7 ^{1}H NMR (C_6H_6): 0.13 (d, $Ge(CH_3)_2$, ^3J = 3.4), 3.96 (sept, GeH, ^3J = 3.4), 4.16 (td, PH, ^3J = 6.5, ^1J(P, H) = 201) ^{31}P NMR: 54.5 IR: ν(PH) 2320, ν(GeH) 2042 [58]
21	$CH_2CH_2CH_2C_6H_5$	$LiAlH_4$ reduction (72%) b.p. 84 to 85°C/3.5 d^{20} = 1.0717; n_D^{20} = 1.5115 ^{1}H NMR (CCl_4): 0.2 (d, 6 H, J = 3.6), 0.6 to 1.12 (m, 2 H), 1.45 to 2.04 (m, 2 H), 2.66 (t, 2 H, J = 8), 3.91 (sept, 1 H, J = 3.6), 7.25 (s, 5 H) [54]
*22	$CH_2CH_2CH(C_6H_5)_2$	$LiAlH_4$ reduction (77%) b.p. 114 to 115°C/0.16 d^{20} = 1.1083; n_D^{20} = 1.5603 ^{1}H NMR (CCl_4): 0.2 (d, 6 H, J = 3.6), 0.6 to 1.05 (m, 2 H), 1.95 to 2.4 (m, 2 H), 3.5 to 4.3 (m, 1 H), 7.3 (s, 10 H) [54]
*23	C_4H_9	formation on p. 155 no data reported [16]
*24	$CH_2(CH_2)_3C_6H_5$	$LiAlH_4$ reduction (64%) b.p. 115°C/9.5 d^{20} = 1.0592; n_D^{20} = 1.5055 ^{1}H NMR (CCl_4): 0.2 (d, 6 H, J = 3.6), 0.58 to 2.0 (m, 6 H), 2.45 to 2.85 (m, 2 H), 3.88 (sept, 1 H, J = 3.6), 7.22 (s, 5 H) [54]
25	$CH_2CH_2CH(CH_3)C_6H_5$	$LiAlH_4$ reduction (76%) b.p. 62°C/0.5 d^{20} = 1.0550; n_D^{20} = 1.5061 ^{1}H NMR (CCl_4): 0.16 (d, 6 H, J = 3.6), 0.35 to 0.97 (m, 2 H), 1.24 (d, 3 H, J = 6.8), 1.4 to 1.87 (m, 2 H), 2.61 (q, 1 H, J = 6.8), 3.73 (sept, 1 H, J = 3.6), 7.2 (s, 5 H) [54]
*26	$CH_2CH_2C(CH_3)_2C_6H_5$	$LiAlH_4$ reduction (78%) b.p. 123°C/7 d^{20} = 1.0575; n_D^{20} = 1.5097 ^{1}H NMR (CCl_4): 0.13 (d, 6 H, J = 3.6), 0.33 to 0.75 (m, 2 H), 1.3 (s, 6 H), 1.51 to 1.9 (m, 2 H), 3.77 (sept, 1 H, J = 3.6), 7.13 (s, 5 H) [54]

Table 22 (continued)

No.	R′ group	preparation (yield) properties and remarks
*27	$CH_2CH_2C(CH_3)_2$- $C_6H_4CH_3$-4	$LiAlH_4$ reduction (88%) b.p. 83 to 85 °C/2 d^{20} = 1.0458; n_D^{20} = 1.5115 ^{1}H NMR (CCl_4): 0.17 (d, 6 H, J = 3.6), 0.25 to 0.7 (m, 2 H), 1.23 (s, 6 H), 1.4 to 1.8 (m, 2 H), 2.3 (s, 3 H), 3.73 (sept, 1 H, J = 3.6), 7.08 (s, 4 H) [54]
28	$CH_2(CH_2)_4C_6H_5$	$LiAlH_4$ reduction (70%) b.p. 116 °C/5.5 d^{20} = 1.0404; n_D^{20} = 1.5049 ^{1}H NMR (CCl_4): 0.2 (d, 6 H, J = 3.6), 0.58 to 1.95 (m, 8 H), 2.4 to 2.8 (m, 8 H), 3.87 (sept, 1 H, J = 3.6), 7.24 (s, 5 H) [54]
*29	$C_{10}H_{15}$ = 1-adamantyl	$LiAlH_4$ reduction (82%) b.p. 100 to 102 °C/5 ^{1}H NMR (CCl_4): 0.13 to 0.2 (CH_3), 1.83 to 1.9 ($C_{10}H_{15}$), 3.53 to 3.73 (GeH) IR (mineral oil): ν(GeH) 2000 [89]
*30	$CH=CH_2$	preparation on p. 156 b.p. 42 °C/760 [77] ^{1}H NMR (C_6D_6): 0.30 (d, CH_3, ^3J = 3.4), 4.17 (d, GeH, ^3J = 3.6 and 3.4), 5.47 to 6.70 (m, $CH=CH_2$) [77] MS: m/e = 131 (100), 113 (5), 105 (40), 97 (8), 89 (10), 70 (8) [117]
*31	$CH_2CH_2CH=CH_2$	$LiAlH_4$ reduction (67%) b.p. 95 to 97 °C n_D^{20} = 1.4320 ^{1}H NMR (CCl_4): 0.20 (d, 6 H, J = 3.9), 0.57 to 1.33 (m, 4 H), 3.65 to 4.02 (m, 1 H), 4.73 to 5.14 (m, 2 H), 5.49 to 6.05 (m, 1 H) IR (neat): ν(GeH) 2050, ν(C=C) 1640, ν($GeCH_3$) 1240 [79]
32	$CH_2CH(OH)C(CH_3)=CH_2$	$LiAlH_4$ reduction (not given) no data reported [46]
*33	$C(OCH_3)=CHCH=CH_2$	preparation on p. 156 b.p. 67 to 69 °C/25 ^{1}H NMR (CCl_4): 0.37 (d, CH_3Ge, ^3J = 3.5), 3.56 (s, CH_3O), 4.43 (sept, GeH, ^3J = 3.5), 4.68 to 5.04 (m, 2 H), 5.67 to 6.66 (m, 2 H) IR (neat): ν(GeH) 2060, ν(C=C) 1620 and 1570 MS: m/e = 188(100), 173(32), 141(40), 135(80), 105(56), 89(40) [73]

References on p. 187

Table 22 (continued)

No.	R′ group	preparation (yield) properties and remarks
*34	$CH_2(CH_2)_2CH{=}CH_2$	$LiAlH_4$ reduction (33%) b.p. 70°C/90 n_D^{20} = 1.4450 ^1H NMR (CCl_4): 0.25 (d, 6 H, J = 3.8), 0.66 to 2.38 (m, 6 H), 3.74 to 4.00 (m, 1 H), 4.71 to 5.27 (m, 2 H), 5.40 to 6.00 (m, 1 H) IR (neat): ν(GeH) 2020, ν(C=C) 1640, ν($GeCH_3$) 1240 [79]
*35	$CH_2(CH_2)_3CH{=}CH_2$	$LiAlH_4$ reduction (31%) b.p. 83°C/75 n_D^{20} = 1.4511 ^1H NMR (CCl_4): 0.25 (d, 6 H, J = 3.8), 0.59 to 2.48 (m, 8 H), 3.68 to 4.00 (m, 1 H), 4.53 to 5.27 (m, 2 H), 5.37 to 6.20 (m, 1 H) IR (neat): ν(GeH) 2050, ν(C=C) 1640, ν($GeCH_3$) 1240 [79]
*36	$CH_2(CH_2)_4CH{=}CH_2$	$LiAlH_4$ reduction (52%) b.p. 84°C/25 n_D^{20} = 1.4506 ^1H NMR (CCl_4): 0.18 (d, 6 H, J = 3.5), 0.56 to 2.35 (m, 10 H), 3.74 to 4.00 (m, 1 H), 4.76 to 5.18 (m, 2 H), 5.48 to 6.15 (m, 1 H) IR (neat): ν(GeH) 2050, ν(C=C) 1640, ν($GeCH_3$) 1240 [79]
*37	C_6H_5	$LiAlH_4$ reduction [119] b.p. 65 to 66°C/20 [116], 80°C/50 [21, 77], 90°C/40 [126] d^{20} = 1.1583; n_D^{20} = 1.5170 [21] ^1H NMR: 4.50 (sept, GeH) (neat) [21]; 0.37 (d, CH_3, ^3J = 3.7 Hz), 4.50 (sept, GeH) [77] IR: ν(GeH) 2044 (in C_7H_{16} or CCl_4) [64]; 2025 (neat) [21, 77]
*38	$C_6H_2(CH_3)_3$-2,4,6	preparation on p. 157 ^1H NMR (neat): 0.44 (d, CH_3Ge), 2.13 (s, CH_3-4), 2.33 (s, CH_3-2,6), 4.77 (m, GeH, ^3J = 3.6), 6.60 (s, C_6H_2) [124]
*39	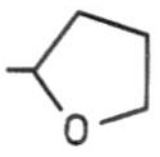	formation on p. 157 MS listed [97]
*40	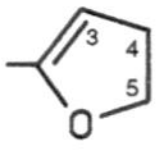	preparation on p. 157 ^1H NMR ($CDCl_3$): 0.36 (d, CH_3, J = 2.6), 2.42 to 2.71 (m, H-4), 4.16 (m, GeH), 4.27 (t, H-5, J = 9.3), 5.11 (t, H-3, J = 2.5) IR: ν(GeH) 2050, ν(C=C) 1602 MS listed in [106]
*41	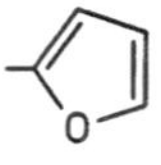	formation on p. 157 MS listed in [97]

Table 22 (continued)

No.	R′ group	preparation (yield) properties and remarks
*42	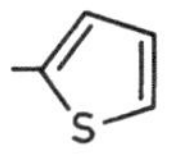	preparation on p. 157 b.p. 98 to 100°C/32 ^{1}H NMR: 0.53 (d, 6 H, CH_3), 4.64 (m, 1 H, GeH), 7.17, 7.22, and 7.58 (each 1 H of the ring) MS listed [99]

*Further information:

$Ge(CH_3)_2(CF_3)H$ (Table **22**, No. **1**) was obtained with 21% yield along with $Ge(CH_3)(CF_3)H_2$ and $Ge(CH_3)_3CF_3$ by condensing CH_3I and $Zn(CH_3)_2$ onto $Ge(CF_3)H_3$ (1:1:1 mole ratio). The products were separated by fractional condensation. Using CD_3I, $Zn(CH_3)_2$, and $Ge(CF_3)H_3$ yielded $Ge(CH_3)_2(CF_3)H$, **$Ge(CH_3)(CD_3)(CF_3)H$**, and **$Ge(CD_3)_2(CF_3)H$** in a 56:37:7 ratio along with other products. When CH_3I, $Zn(CH_3)_2$, and $Ge(CF_3)H_3$ were reacted in a 9:2:12 mole ratio, the product mixture contained 4.5% $Ge(CH_3)_2(CF_3)H$ [92].

$Ge(CH_3)_2(CH_2Cl)H$ (Table **22**, No. **2**). The doublet of the IR band ν(GeH) was interpreted as an intramolecular coordination between the Cl and Ge atoms; the less intense absorption at 2064 cm^{-1} was assigned to "free" molecules without such coordination. The type of solvent affects the shape of the ν(GeH) band [71]. But the appearance of two peaks was later explained by the presence of rotational conformers [78]. The compound was involved in several studies on the dependence of the Ge-H frequency [49, 50] and their integrated intensity [57] on the kind of substituents of various organogermanium compounds. It was also part of an investigation on the isolated Ge-H stretching frequency of various organogermanes [84].

Boiling No. 2 with CH_2=$CHCH_2Cl$ (2 h) in the presence of H_2PtCl_6 gave $Ge(CH_3)_2(CH_2Cl)CH_2CH_2CH_2Cl$ (56% yield) and $Ge(CH_3)_2(CH_2Cl)Cl$ (38%) [33]. Similar reactions with $Si(CH_3)_2(CH$=$CH_2)Cl$ [32, 34] or $Si(CH_3)_2(CH_2CH$=$CH_2)Cl$ [32] were reported to yield $Ge(CH_3)_2(CH_2Cl)CH_2CH_2Si(CH_3)_2Cl$ (78%) and $Ge(CH_3)_2(CH_2Cl)CH_2CH_2CH_2Si(CH_3)_2Cl$ (55%) [32, 34]. $Ge(CH_3)_2(CH_2Cl)CH_2CH_2COOSi(CH_3)_3$ resulted with 86% yield from the reaction of No. 2 with $Si(CH_3)_3OOCCH$=CH_2 at 170°C (1 h) in the presence of H_2PtCl_6 [34]. $Ge(CH_3)_2(CH_2Cl)H$ forms a complex with C_6H_5OH [111]. Photolysis of No. 2 in cyclopropane in the presence of t-$C_4H_9OOC_4H_9$-t and $Ge(C_2H_5)_3H$ gave the carbon-centered radical $Ge(CH_3)_2(H)CH_2^{\cdot}$, whereas $Ge^{\cdot}(CH_2Cl)(CH_3)_2$ was formed when $Ge(C_2H_5)_3H$ was absent [66].

$Ge(CH_3)_2(CHCl_2)H$ (Table **22**, No. **3**) was the subject of IR investigations on the Ge-H stretching frequency of various organogermanium compounds [49, 50, 57, 84] as described above for compound No. 2.

$Ge(CH_3)_2(CH_2Br)H$ (Table **22**, No. **4**) was prepared from $Ge(CH_3)_3Cl$ by bromination with a Cl_2-Br_2 mixture followed by $LiAlH_4$ reduction of the products. A sample of No. 4 contained 20% $Ge(CH_3)_2(CH_2Cl)H$ [49]. The compound was part of IR studies on the Ge-H bond [49, 50, 84] as described above for compound No. 2.

$Ge(CH_3)_2(CH_2OH)H$ (Table **22**, No. **5**). Reactions with acid anhydrides gave the corresponding esters. Treatment with CCl_4 at room temperature yielded $Ge(CH_3)_2(CH_2OH)Cl$ and $CHCl_3$ with ca. 60% conversion within 1 d [103].

References on p. 187

$Ge(CH_3)_2(CH_2OSO_2CF_3)H$ (Table **22**, No. **6**) rearranges under mild conditions to give $Ge(CH_3)_3OSO_2CF_3$ [102, p. 153, footnote 9], [120].

$Ge(CH_3)_2(CH_2N(CH_2CH{=}CH_2)C_2H_5)H$ (Table **22**, No. **8**). Heating in C_6H_6 at 75 °C (18 h) in the presence of AIBN yielded 53% of the cyclization product $\overline{Ge(CH_3)_2CH_2CH_2CH_2N(C_2H_5)C}H_2$ [101].

$Ge(CH_3)_2(CH_2Si(CH_3)_3)H$ (Table **22**, No. **9**). The attack of the photochemically generated $(CH_3)_3CO^{\bullet}$ radical on No. 9 led to the $Ge^{\bullet}(CH_3)_2CH_2Si(CH_3)_3$ radical [86].

$Ge(CH_3)_2(CH_2Si(CH_3)_2H)H$ (Table **22**, No. **10**). Irradiating the compound and $Fe(CO)_5$ (1 : 5 mole ratio) in C_5H_{12} at room temperature (6 h) yielded 62% of compound I (R = CH_3) [109].

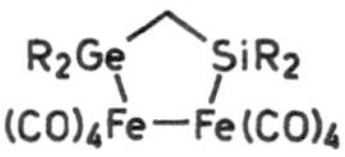

I

$Ge(CH_3)_2(C(Si(CH_3)_3)_3)H$ (Table **22**, No. **11**). Refluxing with I_2 in CCl_4 gave $Ge(CH_3)_2(C(Si(CH_3)_3)_3)I$ with 52% yield. Treatment with ICl in CCl_4 yielded always a ca. 3 : 2 (or 2 : 1 ?) mixture of $Ge(CH_3)_2(C(Si(CH_3)_3)_3)I$ and $Ge(CH_3)_2(C(Si(CH_3)_3)_3)Cl$, even with a deficient amount of ICl [70].

$Ge(CH_3)_2(CH_2C_6H_5)H$ (Table **22**, No. **12**) was formed (7% yield) in addition to other products upon irradiation of $Ge_2(CH_3)_4(CH_2C_6H_5)_2$ in C_6H_{12} at room temperature (3 to 4 h). Only traces were found, when $(CH_3)_3GeGe(CH_3)_2CH_2C_6H_5$ was irradiated under the same conditions [117].

Irradiation of $Ge(CH_3)_2(CH_2C_6H_5)H$ or $Ge(CH_3)_2(CHDC_6H_5)H$ in cyclopropane at low temperatures in the presence of t-$C_4H_9OOC_4H_9$-t generated the $Ge^{\bullet}(CH_3)_2CH_2C_6H_5$ and $Ge^{\bullet}(CH_3)_2CHDC_6H_5$ radicals, respectively [74].

$Ge(CH_3)_2(C_2H_5)H$ (Table **22**, No. **13**) was prepared by reducing a mixture of $Ge(CH_3)_2(C_2H_5)Cl$ and $Ge(CH_3)_2(C_2H_5)Br$ with LiH in dioxane [2]; see also [4]. It was obtained with 74% yield by treating $Ge(C_2H_5)(H)Cl_2$ with CH_3MgCl (or CH_3MgBr?) in ether [15]. A 23% yield of No. 13 was obtained via the $Ge(C_2H_5)Cl$ and $Ge(CH_3)_2(C_2H_5)Li$ intermediates by successively treating $Ge(C_2H_5)(H)Cl_2$ with $LiOCH_3$ and $LiCH_3$ followed by hydrolysis with acidic water [36].

The compound was included in IR studies on the Ge-H stretching frequency of various organogermanium compounds [4, 8, 49, 50, 83, 84] as described for No. 2.

$Ge(CH_3)_2(C_2H_5)D$ was similarly prepared from a mixture of $Ge(CH_3)_2(C_2H_5)Cl$ and $Ge(CH_3)_2(C_2H_5)Br$ using LiD in $O(C_4H_9)_2$ [2]; see also [4]. The compound boils at 60 °C/737.4 Torr; d^{20} = 1.0262 g/cm^3, n_D^{20} = 1.4083 [2]. The Raman spectrum shows bands at 2976(3), 2915(2), 2908(8), 2870(5), 1455(7, br), 1405(1), 1242(4), 1220(3), 1112(2), 1024(2), 968(1), 833(0), 704(0), 599(6, br), 568(10), 493(4, br), 282(2), 198(1), and 170(1) cm^{-1} [2]; ν(GeD) 1458 cm^{-1} [4].

$Ge(CH_3)_2(CH_2CH_2P(H)C_6H_5)H$ (Table **22**, No. **14**) was prepared by $LiAlH_4$ reduction in ether of either compound III (R = CH_3) or a mixture of compounds II and IV (R = CH_3); 74% and 78% yield, respectively [77].

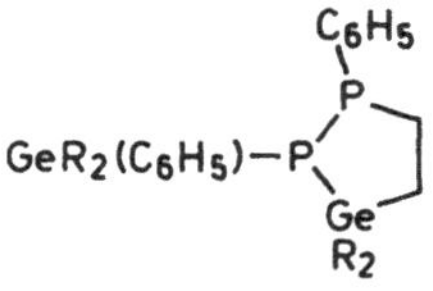

II

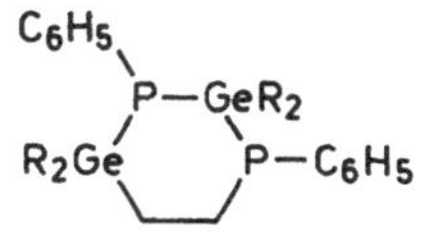

III

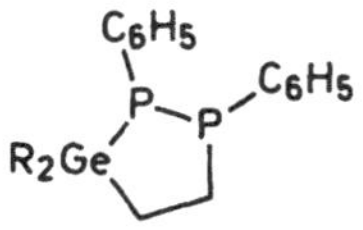

IV

$Ge(CH_3)_2(C_3H_7\text{-}i)H$ (Table **22**, No. **16**), on account of its symmetric and asymmetric conformations, was included in force-field calculations of conformational equilibria of several Si-, Ge-, and Sn-containing organometallic compounds [44].

$Ge(CH_3)_2(CH_2C_4H_9\text{-}t)H$ (Table **22**, No. **17**) reacted with photochemically generated $t\text{-}C_4H_9O^{\cdot}$ to give the $Ge^{\cdot}(CH_3)_2CH_2C_4H_9\text{-}t$ radical [86].

$Ge(CH_3)_2(CH_2CH_2CH_2P(H)C_6H_5)H$ (Table **22**, No. **20**) was prepared with 71% yield by ring cleavage of $\overline{Ge(CH_3)_2CH_2CH_2CH_2P}(C_6H_5)$ with $LiAlH_4$ in refluxing THF after 3 h [58].

$Ge(CH_3)_2(CH_2CH_2CH(C_6H_5)_2)H$ (Table **22**, No. **22**) rearranged on heating at 320 °C in an ampule (1 d) to give $Ge(CH_3)_2(CH_2CH_2CH_2C_6H_5)C_6H_5$ with 61 to 65% yield. In the presence of $t\text{-}C_4H_9OOC_4H_9\text{-}t$ in C_6H_6 at 135 °C this rearrangement occurred to only a small extent [54].

$Ge(CH_3)_2(C_4H_9)H$ (Table **22**, No. **23**) was obtained by $LiAlH_4$ reduction of a mixture of trialkylchlorogermanes formed from $Ge(CH_3)_4$ and $Ge(C_4H_9)_4$ in the presence of excess $AlCl_3$ at 220 °C. Chromatographic characteristics of No. 23 in GLC were described together with those of other alkyl derivatives of Si and Ge [16].

$Ge(CH_3)_2(CH_2(CH_2)_3C_6H_5)H$ (Table **22**, No. **24**) was a minor product along with $Ge(CH_3)_2(C_4H_9)C_6H_5$, when $Ge(CH_3)_2(CH_2CH_2CH_2CH_2Cl)C_6H_5$ was heated in C_6H_6 with $Sn(C_4H_9)_3H$ at 135 °C (20 h) in the presence of $t\text{-}C_4H_9OOC_4H_9\text{-}t$ [54].

$Ge(CH_3)_2(CH_2CH_2C(CH_3)_2C_6H_5)H$ (Table **22**, No. **26**). Heating the compound at 320 °C in an ampule (1 d) yielded 32% $Ge(CH_3)_2(CH_2CH_2CH(CH_3)_2)C_6H_5$; this rearrangement occurred to a small extent at 135 °C in C_6H_6 in the presence of $t\text{-}C_4H_9OOC_4H_9\text{-}t$ [54].

$Ge(CH_3)_2(CH_2CH_2C(CH_3)_2C_6H_4CH_3\text{-}4)H$ (Table **22**, No. **27**) rearranged like No. 26 at 320 °C to give $Ge(CH_3)_2(CH_2CH_2CH(CH_3)_2)C_6H_4CH_3\text{-}4$ (25% yield) [54].

$Ge(CH_3)_2(1\text{-}C_{10}H_{15})H$ ($1\text{-}C_{10}H_{15}$ = 1-adamantyl; Table **22**, No. **29**). Equimolar amounts of No. 29 and $SnCl_4$ reacted above 50 °C yielding $Ge(1\text{-}C_{10}H_{15})Cl_3$ but no $Ge(1\text{-}C_{10}H_{15})(H)Cl_2$ [88]. Attempts to achieve hydrogermylation by No. 29 failed with $CH_2{=}CHCH_2NCO$, $CH_2{=}CHCH_2NHSi(CH_3)_3$, and $CH_2{=}CHCH_2COOSi(CH_3)_3$ [88]. Other hydrogermylation reactions (all obviously in the presence of H_2PtCl_6 as the catalyst) are summarized below [88, 89]:

reactant	conditions products (yield)	Ref.
$CH_2{=}CHCOOSi(CH_3)_3$	120 to 140 °C, 4 h → $Ge(CH_3)_2(1\text{-}C_{10}H_{15})CH(CH_3)COOSi(CH_3)_3$ (50%)	[88]
$CH_2{=}CHCOOCH_3$	100 to 120 °C, 3.5 h → $Ge(CH_3)_2(1\text{-}C_{10}H_{15})CH_2CH_2COOCH_3$ (78%)	[89]
$CH_2{=}CHCOOH$	160 to 165 °C, 1 h → $Ge(CH_3)_2(1\text{-}C_{10}H_{15})CH(CH_3)COOH$,	[88]

References on p. 187

reactant	conditions products (yield)	Ref.
	$Ge(CH_3)_2(1\text{-}C_{10}H_{15})CH_2CH_2COOH$, and $Ge(CH_3)_2(1\text{-}C_{10}H_{15})OOCCH{=}CH_2$	
$CH_2{=}C(CH_3)COOCH_3$	126 to 135°C, 22 h → $Ge(CH_3)_2(1\text{-}C_{10}H_{15})CH_2CH(CH_3)COOCH_3$ (28%)	[88]
$CH_2{=}C(CH_3)COOC_4H_9$	126 to 135°C, 22 h → $Ge(CH_3)_2(1\text{-}C_{10}H_{15})CH_2CH(CH_3)COOC_4H_9$ (42%)	[88]

$Ge(CH_3)_2(CH{=}CH_2)H$ (Table **22**, No. **30**) was prepared with 90% yield from $Ge(CH_3)_2(H)Cl$ and $CH_2{=}CHMgBr$ in THF at room temperature (2 h) [77]. Irradiation of $(CH_3)_3Ge\text{-}Ge(CH_3)_2CH{=}CH_2$ in C_6H_{12} at room temperature (10 h) gave only trace amounts of No. 30 among several other products [117].

Bromination by N-bromosuccinimide in THF at 0°C yielded 77% of $Ge(CH_3)_2(CH{=}CH_2)Br$ [77].

$Ge(CH_3)_2(CH_2CH_2CH{=}CH_2)H$ (Table **22**, No. **31**). Ring closure to $\overline{Ge(CH_3)_2(CH_2)_3C}H_2$ resulted with low yield, when No. 31 was heated in C_6H_6 in the presence of a radical initiator such as AIBN at 42°C or $C_6H_5CO\text{-}OO\text{-}OCC_6H_5$ at 80°C [79].

(Z)-$Ge(CH_3)_2(C(OCH_3){=}CHCH{=}CH_2)H$ (Table **22**, No. **33**) was prepared with 66% yield by reacting $Ge(CH_3)_2(H)Cl$ with Z-$LiC(OCH_3){=}CHCH{=}CH_2$ in C_5H_{12} at −78°C (the Li reactant was obtained from trans-$CH_2{=}CHCH{=}CHOCH_3$ and LiC_4H_9-t) [73].

Heating No. 33 and H_2PtCl_6 at 160°C for 15 min yielded 22% $\overline{Ge(CH_3)_2CH_2\text{-}CH_2CH{=}C}OCH_3$ [73].

$Ge(CH_3)_2(CH_2(CH_2)_nCH{=}CH_2)H$ (n = 2, 3, and 4; Table **22**, Nos. **34** to **36**). Ring closure with almost exclusive formation of the corresponding compounds $\overline{Ge(CH_3)_2CH_2(CH_2)_n\text{-}CH_2C}H_2$ was achieved with generally low yields by radical initiators as described for No. 31 [79].

$Ge(CH_3)_2(C_6H_5)H$ (Table **22**, No. **37**) was obtained by $LiAlH_4$ cleavage of $(C_6H_5)_3Ge\text{-}Ge(CH_3)_2C_6H_5$ (1 d) or $Ge_2(CH_3)_4(C_6H_5)_2$ (2 h) in THF at room temperature; ca. 100% and 72% yield, respectively [63]. It was also formed from compound II (R = CH_3) by refluxing with $LiAlH_4$ in ether; 52% yield [77].

Other formations of $Ge(CH_3)_2(C_6H_5)H$ involved the following reactions: photolysis of $(CH_3)_3GeGe(CH_3)_2C_6H_5$ in C_6H_{12} or C_6H_6 at room temperature; 6 to 8% yield along with several other products [100] (see also [118]); photolysis of $(CH_3)_3SiGe(CH_3)_2C_6H_5$ in C_6H_{12} (trace amounts of No. 37) [118]; UV irradiation of $Ge(C_6H_5)H_3$ and excess CH_2N_2 in ether (6 h), 14% yield of No. 37 along with $Ge(CH_3)(C_6H_5)H_2$ [21]. Photolysis of $Ge_2(CH_3)_4(C_6H_5)_2$ in C_6H_{12} at room temperature (2 h) yielded $Ge(CH_3)_2(C_6H_5)H$ (11%) along with $Ge(CH_3)_2(C_6H_5)_2$ (17%) and $(Ge(CH_3)_2)_4$ (4%); the process in forming the hydrogermane remained unclear, since the photolysis in C_6D_{12} also gave $Ge(CH_3)_2(C_6H_5)H$ but no $Ge(CH_3)_2(C_6H_5)D$. Photolysis of $Ge_2(CH_3)_4(C_6H_5)_2$ in C_6H_{12} at room temperature in the presence of $CH_2{=}C(CH_3)C(CH_3){=}CH_2$ yielded $Ge(CH_3)_2(C_6H_5)H$ (21%), $Ge(CH_3)_2(C_6H_5)_2$ (5%), and $(Ge(CH_3)_2)_4$ (4%) [116]. Electrolysis of $Ge_2(CH_3)_4(C_6H_5)_2$ or $Ge(CH_3)_2(Ge(CH_3)_2C_6H_5)_2$ in CH_3CN ($[N(C_4H_9)_4][BF_4]$ or $[N(C_4H_9)_4][ClO_4]$ electrolyte) gave $Ge(CH_3)_2(C_6H_5)H$ along with $(Ge(CH_3)_2C_6H_5)_2O$ in the catholyte compartment [127].

The title compound was formed with 16% yield along with $Ge(CH_3)_3C_6H_5$ by the reaction of $Ge(C_6H_5)Cl$ with $LiCH_3$ (1:2 mole ratio) [36] (see also [53]) and with 15% yield along with $(Ge(CH_3)C_6H_5)_n$ by the reaction of $Ge(C_6H_5)Cl$ with CH_3MgI (1:2 mole ratio) [43] (see also [53]). It was also generated from $Ge(CH_3)C_6H_5$ and CH_3MgX [42], following hydrolysis in all cases.

$Ge(CH_3)_2(C_6H_5)H$ was involved in IR studies of the ν(GeH) frequency and its dependence on the nature of substituents of various organogermanium compounds [64]; cf. No. 2.

Photochemically generated t-$C_4H_9O^{\bullet}$ reacted with $Ge(CH_3)_2(C_6H_5)H$ to yield the $Ge^{\bullet}(CH_3)_2C_6H_5$ radical [56, 60]. Heating equimolar amounts of $Ge(CH_3)_2(C_6H_5)H$ and $Ge(CH_3)_2(C_6H_5)CH_2CH{=}CH_2$ at 200°C for 10 h in the presence of AIBN yielded 26% $Ge(CH_3)_2(C_6H_5)CH_2CH_2CH_2Ge(CH_3)_2C_6H_5$ [126].

$Ge(CH_3)_2(C_6H_2(CH_3)_3$-2,4,6)H (Table **22**, No. **38**) was obtained with 35% yield via $Ge(C_6H_2(CH_3)_3$-2,4,6)(H)Li_2, when $Ge(C_6H_2(CH_3)_3$-2,4,6)H_3 was reacted with LiC_4H_9-t (1:2 mole ratio) in THF-C_5H_{12} at −60°C followed by treatment with CH_3I [124].

$Ge(CH_3)_2(C_4H_5O)H$ (Table **22**, No. **40**) was obtained quantitatively by reducing $Ge(CH_3)_2$-$(C_4H_5O)_2$ with $LiAlH_4$ in ether at 25°C (180 h). In THF as the solvent, the formation of $Ge(CH_3)_2H_2$ was favored [106, 107]. Therefore, No. 40 did not react with $LiAlH_4$ in ether [107].

$Ge(CH_3)_2(C_4H_7O)H$ and **$Ge(CH_3)_2(C_4H_3O)H$** (Table **22**, Nos. **39** and **41**) were present (in the order of 10 to 20% yield) along with compounds VI, THF, and a trigermane in the product mixture resulting from the hydrogenation of compound V at 70°C (under 0.1 to 1.9 MPa pressure for 3 h) using several catalysts (Raney Ni, Pd/C, or Rh black) and solvents (C_2H_5OH, i-C_3H_7OH, or c-C_6H_{10}) [97].

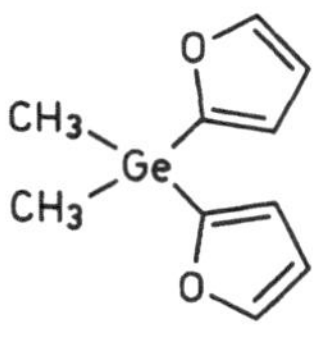

V

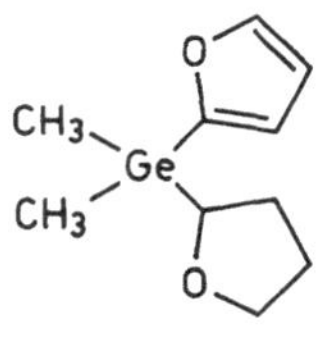

VI

$Ge(CH_3)_2(C_4H_3S)H$ (Table **22**, No. **42**) was prepared by $LiAlH_4$ reduction of a mixture of $Ge(CH_3)_2(C_4H_3S)Cl$ and $Ge(CH_3)_2(C_4H_3S)Br$ (2:11 ratio) in ether or in C_6H_6 in the presence of the phase transfer catalyst $[N(C_4H_9)_3CH_2C_6H_5]Br$ with almost quantitative yield [99]. Its hydrogermylation reactions in the presence of the H_2PtCl_6 catalyst are summarized below (C_4H_3S = 2-thienyl, $C_{10}H_{15}$ = 1-adamantyl) [99]:

reactant	conditions products (yield)
$CH{\equiv}CCH_2OCH_3$	50°C, 6 h → trans-$Ge(CH_3)_2(C_4H_3S)CH{=}CHCH_2OCH_3$ and $Ge(CH_3)_2(C_4H_3S)C(CH_2OCH_3){=}CH_2$ in a 84:16 ratio (68% total)
$CH{\equiv}CCOOCH_3$	100°C, 6 h → trans-$Ge(CH_3)_2(C_4H_3S)CH{=}CHCOOCH_3$, cis-$Ge(CH_3)_2(C_4H_3S)CH{=}CHCOOCH_3$ and $Ge(CH_3)_2(C_4H_3S)C(COOCH_3){=}CH_2$ in a 56:10:34 ratio (59% total)

References on p. 187

reactant	conditions products (yield)
$CH{\equiv}CC_4H_9$-t	100 °C, 6 h → trans-$Ge(CH_3)_2(C_4H_3S)CH{=}CHC_4H_9$-t and $Ge(CH_3)_2(C_4H_3S)C(C_4H_9\text{-t}){=}CH_2$ in a 95:5 ratio (60% total)
$CH{\equiv}CC_{10}H_{15}$	100 °C, 6 h → trans-$Ge(CH_3)_2(C_4H_3S)CH{=}CHC_{10}H_{15}$ (50%) (sole product)
$CH{\equiv}CC_6H_5$	100 °C, 6 h → trans-$Ge(CH_3)_2(C_4H_3S)CH{=}CHC_6H_5$ and $Ge(CH_3)_2(C_4H_3S)C(C_6H_5){=}CH_2$ in a 98:2 ratio (82% total)

1.3.1.2.2 $Ge(C_2H_5)_2(R')H$ Compounds

Table 23
$Ge(C_2H_5)_2(R')H$ Compounds.
An asterisk indicates further information at the end of the table.
Explanations, abbreviations, and units are given on p. X.

No.	R′ group	preparation (yield) properties and remarks
1	CF_3	$Ge(CF_3)H_3$ + $Zn(C_2H_5)_2$-C_2H_5I (slight excess) (20%); formed along with $Ge(C_2H_5)(CF_3)H_2$ and $Ge(CF_3)I_3$ ^{19}F NMR: −53.8 ppm (CF_3), $^3J(H, F)$ = 6.8 [92]
*2	CH_2Cl	$Ge(C_2H_5)_2(CHCl_2)H$ + $Sn(C_4H_9)_3H$, refluxing (1 h) and distillation in the presence of catalytic amounts of galvinoxyl and $C_6H_4(OH)_2$-1,4 (78%) b.p. 45 °C/20 n_D^{20} = 1.3930 ^1H NMR (C_6H_6): 0.65 to 1.20 (m, C_2H_5), 2.83 (d, CH_2Cl, ^3J = 2.5), 4.10 (m, GeH) IR (C_7H_{16}): ν(GeH) 2051 and 2027 [71]
*3	$CHCl_2$	$Ge(C_2H_5)_2(H)Cl$ + $LiCHCl_2$ (freshly prepared) at −90 °C (2 h), warming to room temperature, hydrolysis, and C_5H_{12} extraction (44%) [71] b.p. 78 to 80 °C/15 [61], 84 °C/20 [71] ^1H NMR: 1.20 (m, C_2H_5), 4.39 (m, GeH), 5.54 (d, $CHCl_2$) in C_6D_6 [61]; 1 to 1.40 (m, C_2H_5), 4.40 (m, GeH), 5.53 (d, $CHCl_2$, ^3J = 2.5) in CCl_4 [71] IR: ν(GeH) 2060 (neat) [61]; 2065(sh) and 2046 in C_7H_{16} [71]
*4	$CH_2C_6H_5$	$Ge(C_2H_5)_2H_2$ + $C_6H_5CH_2MgCl$ (excess) in refluxing THF for 5 h (50%) [49] colorless liquid [94]; b.p. 110 °C/12 [49] n_D^{20} = 1.5250 [49]

Table 23 (continued)

No.	R′ group	preparation (yield) properties and remarks
		^1H NMR (C_6D_6): 0.72 to 1.17 (m, C_2H_5), 2.20 (d, CH_2, J = 3), 3.90 to 4.20 (m, GeH), 6.87 to 7.14 (m, C_6H_5) [94] IR: ν(GeH) 2020 in C_7H_{16} or CCl_4 [64]; 2014 in C_7H_{16} [49, 50]; 2001 (neat) [94]
*5	CH_2CH_2OH	$Ge(C_2H_5)_2(CH_2COOR'')Cl + LiAlH_4$ (R″ not given) b.p. 86°C/12 d^{20} = 1.1661; n_D^{20} = 1.4700 [41] IR: ν(GeH) 2010 [69]
*6	CH_2CH_2SH	$Ge(C_2H_5)_2(CH_2CH_2SCOCH_3)Cl + LiAlH_4$ $Ge(C_2H_5)_2(CH_2CH_2SCOCH_3)Br$ + excess $LiAlH_4$ in refluxing ether (60%) b.p. 128°C/15 ^1H NMR (C_6D_6): 1.53 (m, CH_2Ge), 2.40 (m, CH_2S), 3.73 (sept, GeH) IR: ν(GeH) 2010 [69]
*7	$CH_2CH_2P(H)C_6H_5$	preparation given on p. 164 b.p. 105 to 108°C/0.5 ^1H NMR (C_6D_6): 0.72 to 1.42 (m, C_2H_5 and CH_2Ge), 1.62 to 2.12 (m, CH_2P), 3.87 (m, GeH), 4.03 (dt, PH, ^1J(P, H) = 204.8, ^3J(H, H) = 6.8) ^{31}P NMR: −42.0 ppm [77]
*8	C_3H_7	preparation given on p. 164 IR: ν(GeH) 2003 [20]
9	$CH_2CH_2CH_2Cl$	$LiAlH_4$ reduction (88%) b.p. 96°C/30 ^1H NMR (C_6H_6): 3.18 (t, CH_2Cl), 3.78 (m, GeH) IR (C_7H_{16} or CCl_4): ν(GeH) 2019; studies of the integrated intensity of ν(GeH) [64]; cf. No. 4
*10	$CH_2CH_2CH_2OH$	$\overline{Ge(C_2H_5)_2CH_2CH_2CH_2O}$ or $Ge(C_2H_5)_2(CH_2CH_2CH_2OH)Cl$ or $Ge(C_2H_5)_2(CH_2CH_2CH_2OOCCH_3)Cl + LiAlH_4$ in ether (72, 82, and 85%, respectively) b.p. 110°C/15 to 16 d^{20} = 1.1001; n_D^{20} = 1.4670 [37, 51]
*11	$CH_2CH(OH)CH_3$	$Ge(C_2H_5)_2(CH_2COCH_3)Cl + LiAlH_4$ b.p. 90°C/17 d^{20} = 1.1104; n_D^{20} = 1.4675 [41]
*12	$CH_2CH(CH_3)CH_2OH$	$Ge(C_2H_5)_2(CH_2CH(CH_3)CH_2OH)Cl$ or $Ge(C_2H_5)_2(CH_2CH(CH_3)CH_2OOCCH_3)Cl + LiAlH_4$ in ether (80% and 69%, respectively) b.p. 114°C/21 d^{20} = 1.0934; n_D^{20} = 1.4708 [51]

References on p. 187

Table 23 (continued)

No.	R′ group	preparation (yield) properties and remarks
*13	$CH_2CH(CH_3)CHO$	possibly formed by hydrogermylation of $CH_2{=}C(CH_3)CHO$; see $CH_2{=}CHCHO$, p. 29 no data reported [30, 38]
*14	CH_2CH_2COOH	$Ge(C_2H_5)_2(CH_2CH_2COOCH_3)H$ + alcoholic KOH followed by aqueous HCl b.p. 97 to 105 °C/1.5 [52]
*15	$CH_2CH_2COOCH_3$	$Ge(C_2H_5)_2H_2 + CH_2{=}CHCOOCH_3$ heated in a steel tube for 5 h; $Ge(C_2H_5)_2(CH_2CH_2COOCH_3)_2$ was the other product b.p. 67 to 69 °C/3.0 [52] starting material for No. 14
16	$CH_2CH_2CH_2Si(CH_3)H_2$	$Ge(C_2H_5)_2(CH_2CH_2CH_2Si(CH_3)Cl_2)H + LiAlH_4$ b.p. 87 °C/22 $d^{20} = 0.9793$; $n_D^{20} = 1.4570$ ^{1}H NMR (CCl_4): 0.12 (t, CH_3Si, J = 4), 3.5 to 3.8 (superimposed m's, 2(?) H, SiH, GeH) IR: ν(SiH) 2140, ν(GeH) 2010 [22]
17	$CH_2CH_2CH_2Si(CH_3)Cl_2$	$\overline{Ge(C_2H_5)_2CH_2CH_2CH_2}$ + $Si(CH_3)(H)Cl_2$ (1:5 mole ratio) in the presence of H_2PtCl_6 b.p. 125 °C/25 $d^{20} = 1.1623$; $n_D^{20} = 1.4682$ ^{1}H NMR (CCl_4): 0.75 (s, CH_3Si), 3.70 (sept, GeH, J = 2.5) IR: ν(GeH) 2010 with $LiAlH_4$ → No. 16 [22]
18	$CH_2CH_2CH_2$- $Si(CH_3)(C_6H_5)_2$	$\overline{Ge(C_2H_5)_2CH_2CH_2CH_2}$ + $Si(CH_3)(C_6H_5)_2H$ like No. 17 b.p. 148 °C/0.18 $d^{20} = 1.0752$; $n_D^{20} = 1.5481$ [22]
*19	C_4H_9	preparation given on p. 164 no data reported [16]
*20	$CH_2CH(CH_3)CH(CH_3)_2$	preparation given on p. 164 $d^{20} = 0.9870$; $n_D^{20} = 1.4540$ [29]
*21	C_4H_9-t	preparation not reported b.p. 126 to 127 °C/723 $n_D^{20} = 1.4442$ [108]
*22	$CH_2CH_2CH(OH)CH_3$	$LiAlH_4$ reduction of $\overline{Ge(C_2H_5)_2CH_2CH_2CH(CH_3)O}$ (90%), $(Ge(C_2H_5)_2CH_2CH_2COCH_3)_2O$ (80%) [29], or $Ge(C_2H_5)_2(CH_2CH_2CH(OH)CH_3)Cl$ (82%) [51] b.p. 105 °C/16 [29], 115 °C/22 [51] $d^{20} = 1.0795$ [29], 1.0873 [51] $n_D^{20} = 1.4638$ [29], 1.4673 [51]

Table 23 (continued)

No.	R′ group	preparation (yield) properties and remarks
*23	$CH_2CH(CH_3)CH(OH)CH_3$	$LiAlH_4$ reduction of $Ge(C_2H_5)_2(CH_2CH(CH_3)CH(CH_3)OOCCH_3)Cl$ (67%), $Ge(C_2H_5)_2(CH_2CH(CH_3)CH(OH)CH_3)Cl$ (77%) [51], or $Ge(C_2H_5)_2CH_2CH(CH_3)CH(CH_3)O$ (88%) [29]; also prepared from $(Ge(C_2H_5)_2CH_2CH(CH_3)COCH_3)_2O$ [29] b.p. 105 °C/11 [29], 117 °C/21 [51] d^{20} = 1.0712 [29], 1.0786 [51] n_D^{20} = 1.4684 [29], 1.4716 [51]
*24	$CH_2CH(OH)C_4H_9$-t	$Ge(C_2H_5)_2(CH_2COC_4H_9\text{-t})Cl + LiAlH_4$ (?) b.p. 107 °C/15 d^{20} = 1.0490; n_D^{20} = 1.4660 [41]
*25	$CH_2CH_2COCH_3$	no data reported [30, 38]
*26	$CH_2CH(CH_3)COCH_3$	no data reported [30, 38]
27	$C_4H_8NH_2$	$Ge(C_2H_5)_2(CH_2CH_2CH_2CN)Cl + LiAlH_4$ (70%) b.p. 95 °C/13 d^{20} = 1.0443; n_D^{20} = 1.4678 IR: ν(GeH) 2005 [10]
*28	C_5H_{11}-i	preparation given on p. 165 colorless liquid; vapor pressure: ca. 2 Torr at room temperature [1]
*29	$CHCH(OH)CH_2CH_2CH_2$	$Ge(C_2H_5)_2(CHCOCH_2CH_2CH_2)Cl + LiAlH_4$ b.p. 117 °C/13 d^{20} = 1.1474; n_D^{20} = 1.4922 [41]
*30	$CH{=}CH_2$	$Ge(C_2H_5)_2(H)Cl + CH_2{=}CHMgBr$ in THF for 1 h (23%) b.p. 43 °C/52 ^1H NMR (C_6D_6): 4.10 (GeH), 5.30 to 6.50 ($CH{=}CH_2$) IR: ν(GeH) 2040 [69]
*31	$CH{=}CHCH_2OH$	$LiAlH_4$ reduction (73%) b.p. 103 °C/26 d^{20} = 1.1429; n_D^{20} = 1.4851 [51]
32	$CH_2CH_2CH{=}CH_2$	$LiAlH_4$ reduction b.p. 160 °C/760 d^{20} = 1.0088; n_D^{20} = 1.4562 [28]
*33	$CH_2CH{=}CHCH_3$	prepared by hydrogermylation; see p. 165 b.p. 30 to 31 °C/10 n_D^{20} = 1.4260 ^1H NMR (CCl_4): 1.00 (m, C_2H_5), 1.54 (d, CH_2Ge, J = 6), 1.67 (d, CH_3, J = 6), 3.74 (m, GeH), 5.35 (m, CH=CH) IR (neat): 3030, 2030, 1650, 1460, 1380, 1250, 1150, 1020, 1000, 980, 780, 740, 700 [91]

References on p. 187

Table 23 (continued)

No.	R′ group	preparation (yield) properties and remarks
34	$CH_2CH(CH_3)CH{=}CH_2$	$LiAlH_4$ reduction of $Ge(C_2H_5)_2(CH_2CH(CH_3)CH{=}CH_2)X$ (with X = F, Cl, Br, I, OC_2H_5, $OOCCH_3$, or $OOCCHCl_2$) b.p. 68 °C/20 $d^{20} = 0.9906$; $n_D^{20} = 1.4548$ [28]
35	$CH_2C(CH_3){=}CHCH_3$	$Ge(C_2H_5)_2H_2 + CH_2{=}C(CH_3)CH{=}CH_2$ (22%) along with $Ge(C_2H_5)_2(CH_2C(CH_3){=}CHCH_3)_2$; conditions as for No. 33; see p. 165 b.p. 52 to 55 °C/5 $n_D^{20} = 1.4600$ ^1H NMR (CCl_4): 1.08 (m, C_2H_5), 1.48 (s, CH_2Ge), 1.56 (d, CH_3CH, J = 6), 1.66 (s, CH_3C), 3.84 (m, GeH), 5.00 (q, CH, J = 6) IR (neat): 3045, 2020, 1670, 1480, 1390, 1240, 1170, 1020, 980, 860, 770 [91]
*36	$CH_2C(CH_3){=}C(CH_3)_2$	$LiAlH_4$ reduction (65%) $d^{20} = 1.0099$; $n_D^{20} = 1.4750$ [29]
*37	$CH{=}CHC(OH)(CH_3)_2$	$LiAlH_4$ reduction (75%) b.p. 98 °C/21 $d^{20} = 1.0670$; $n_D^{20} = 1.4747$ [51]
38	$CH_2CH{=}CHCH_2$- $CH(CH_3)CH{=}CH_2$	$Ge(C_2H_5)_2H_2 + CH_2{=}CHCH{=}CHCH(CH_3)CH{=}CH_2$ (22%) along with $Ge(C_2H_5)_2(CH_2CH{=}CHCH_2CH(CH_3)CH{=}CH_2)_2$; conditions as for No. 33; see p. 165 b.p. 58 to 60 °C/2 $n_D^{20} = 1.4810$ ^1H NMR (CCl_4): 1.00 (d, CH_3, J = 6), 1.05 (m, C_2H_5), 1.59 (d, CH_2Ge, J = 6), 1.95 (m, CH_2CH), 3.75 (m, GeH), and 4.65 to 6.00 (m, $CH_2{=}CH$ and $CH{=}CH$) IR (neat): 3080, 3030, 2020, 1640, 1460, 1380, 1240, 1150, 1000, 980, 920, 850, 735, 710 [91]
39	$CH_2C(CH_3){=}CHCH_2$- $CH_2CH_2C(CH_3){=}CH_2$	$Ge(C_2H_5)_2H_2 + CH_2{=}C(CH_3)CH{=}CHCH_2CH_2C(CH_3){=}CH_2$ (30%) along with $Ge(C_2H_5)_2(CH_2C(CH_3){=}CHCH_2CH_2CH_2C(CH_3){=}CH_2)_2$; conditions the same as those for No. 33; see p. 165 b.p. 60 to 65 °C/2 $n_D^{20} = 1.4940$ ^1H NMR (CCl_4): 1.00 (m, C_2H_5), 1.45 (s, CH_2Ge), 1.66 (s, CH_3), 1.80 to 2.00 (m, $(CH_2)_3$), 4.70 (s, $CH_2{=}$), 4.90 (t, CH=, J = 6); δ(GeH) not listed IR (neat): 3040, 2020, 1640, 1245, 1010, 850, 775 [91]
*40	C_6H_5	$Ge(C_6H_5)(H)Cl_2 + C_2H_5MgBr$ (1:2 mole ratio) in refluxing ether (1 h) [75] $Ge(C_2H_5)_2(C_6H_5)Cl$ and $LiAlH_4$ in refluxing THF [75] $Ge(C_2H_5)_2(C_6H_5)Br + LiAlH_4$ in refluxing ether (83%) [93] b.p. 100 to 102 °C/21 [93]

Table 23 (continued)

No.	R′ group	preparation (yield) properties and remarks
		^{1}H NMR: 1.03 to 1.16 (m, C_2H_5), 4.40 to 4.60 (m, GeH), 7.20 to 7.60 (m, C_6H_5) in C_6D_6 [77]; 0.75 to 1.28 (m, C_2H_5), 4.40 to 4.63 (m, GeH), 7.03 to 7.60 (m, C_6H_5) in $CDCl_3$ or C_6D_6 [93] ^{13}C NMR: 4.90 (CH_3), 9.96 (CH_2), 128.17 (C-3,5), 128.55 (C-4), 132.95 (C-2,6), 137.65 (C-1) [108] IR (neat): ν(GeH) 2007 [93]

*Further information:

$Ge(C_2H_5)_2(CH_2Cl)H$ (Table **23**, No. **2**). The doublet character of the IR band ν(GeH) can be explained by intramolecular coordination between the Cl and the Ge atoms; the weaker peak at 2051 cm^{-1} was assigned to "free" molecules without such coordination. The influence of several solvents on the shape of the ν(GeH) band was considered [71]. But the appearence of two peaks was later attributed to the presence of rotational conformers [78].

An interaction of No. 2 with C_6H_5OH (in CCl_4) is indicated by a shift of the ν(OH) frequency [111].

$Ge(C_2H_5)_2(CHCl_2)H$ (Table **23**, No. **3**) was obtained with 19% yield by successively treating the germylene $Ge(C_2H_5)CHCl_2$ (obtained from $Ge(C_2H_5)(CHCl_2)(OCH_3)H$) with C_2H_5Br at 80 °C and $LiAlH_4$ at −40 °C [61].

For the interpretation of the ν(GeH) band doublet [71, 78], see No. 2; the weaker peak at 2065 cm^{-1} was assigned to molecules without coordination [71]. No. 2 interacts with C_6H_5OH (in CCl_4) as evidenced by a shift of the ν(OH) frequency [111].

Reducing No. 3 with $Sn(C_4H_9)_3H$ gave $Ge(C_2H_5)_2(CH_2Cl)H$ (No. 2) [71].

$Ge(C_2H_5)_2(CH_2C_6H_5)H$ (Table **23**, No. **4**). Small amounts of No. 4 were found among the products of irradiation of $Ge(C_2H_5)_2(CH_2C_6H_5)_2$ in C_6H_{14} for 3 min [94]. Treatment of $Ge(C_2H_5)_2(CH_2C_6H_5)CH_2NHCH_2C_6H_5$ with LiC_4H_9 (3:6.4 mole ratio) in THF-C_6H_{14} gave No. 4 with 83% yield along with $NH(C_5H_{11})CH_2C_6H_5$ after hydrolysis [121].

The compound was involved in IR studies on the dependence of the Ge-H frequency [49, 50] and their integrated intensity [57, 64] on the kind of substituents of various organogermanium compounds. It was included in an investigation of the isolated Ge-H stretching frequency of various organogermanes [84].

$Ge(C_2H_5)_2(CH_2CH_2OH)H$ (Table **23**, No. **5**) in ether was converted on Raney nickel at 20 °C into $\overline{Ge(C_2H_5)_2CH_2CH_2O}$ and compound VII (R = H), being the main products along with $(Ge(C_2H_5)_2O)_n$ and C_2H_4 [41].

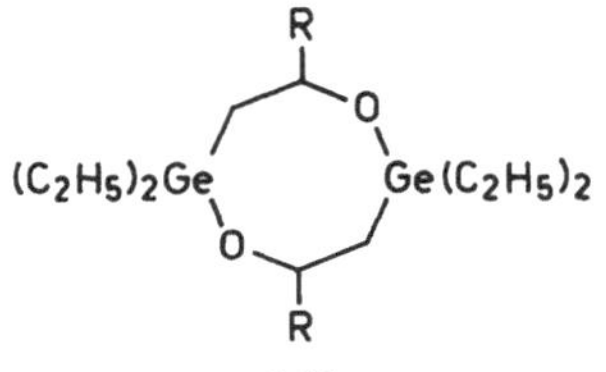

VII

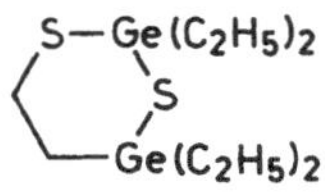

VIII

References on p. 187

$Ge(C_2H_5)_2(CH_2CH_2SH)H$ (Table **23** No. **6**) decomposes on heating (distillation) to give $(Ge(C_2H_5)_2S)_3$, compound VIII, and H_2. Treatment at room temperature with Raney nickel in C_6H_6 or with $Rh(P(C_6H_5)_3)_3Cl$ in C_6D_6 gave mixtures of $\overline{Ge(C_2H_5)_2CH_2CH_2S}$, $(Ge(C_2H_5)_2S)_3$, and compound VIII. Reacting with Raney nickel in the presence of excess $Ge(C_2H_5)_3SCH_3$ yielded $Ge(C_2H_5)_2(SCH_3)_2$, $(Ge(C_2H_5)_3)_2S$, $(Ge(C_2H_5)_2S)_3$, and compound VIII [69].

$Ge(C_2H_5)_2(CH_2CH_2P(H)C_6H_5)H$ (Table **23**, No. **7**) was prepared in ether by $LiAlH_4$ reduction of either compound III, p. 155 (R = C_2H_5) with 69% yield or a mixture of compounds II and IV, p. 155 (R = C_2H_5) with 73% yield [77].

$Ge(C_2H_5)_2(C_3H_7)H$ (Table **23**, No. **8**) was obtained by $LiAlH_4$ reduction of a mixture of trialkylchlorogermanes previously formed from $Ge(C_2H_5)_4$ and $Ge(C_3H_7)_4$ at 220 °C in the presence of an excess of $AlCl_3$. Chromatographic characteristics of $Ge(C_2H_5)_2(C_3H_7)H$ in GLC were described along with those of other alkyl derivatives of Si and Ge in [16].

$Ge(C_2H_5)_2(CH_2CH_2CH_2OH)H$ (Table **23**, No. **10**) was formed with 10% yield along with minor amounts of $Ge(C_2H_5)_3CH_2CH_2CH_2OH$ by successive treatment of the germylene $Ge(CH_2CH_2CH_2OH)H$ (prepared from $Ge(CH_2CH_2CH_2OH)H_3$ and $HgCl_2$ via $Ge(CH_2CH_2CH_2OH)H_2Cl$) with C_2H_5Br and C_2H_5MgBr. Only a 5% yield of No. 10 was obtained when the germylene $Ge(CH_2CH_2CH_2OH)Cl$ (prepared from $Ge(CH_2CH_2CH_2OH)(H)Cl_2$ and $Ge(C_4H_9)_3OCH_3$ via $Ge(CH_2CH_2CH_2OH)(H)(Cl)OCH_3$) was reacted with LiC_2H_5 [45].

Treatment with Raney nickel at room temperature [37, 38, 51] or with reduced Cu at 80 to 90 °C [38] gave $\overline{Ge(C_2H_5)_2CH_2CH_2CH_2O}$ [37, 38, 51] with 95 to 97% yield [38].

$Ge(C_2H_5)_2(CH_2CH(OH)CH_3)H$ (Table **23**, No. **11**) reacted in the presence of Raney nickel in ether at 20 °C to yield 60% of compound VII, p. 163 (R = CH_3) [41].

$Ge(C_2H_5)_2(CH_2CH(CH_3)CH_2OH)H$ (Table **23**, No. **12**) reacted in the presence of Raney nickel at 20 °C (1 d) to give $\overline{Ge(C_2H_5)_2CH_2CH(CH_3)CH_2O}$ with 95% yield [51].

$Ge(C_2H_5)_2(CH_2CH(CH_3)CHO)H$ (Table **23**, No. **13**) was mentioned as an intermediate in the two-step preparation of $\overline{Ge(C_2H_5)_2CH_2CH(CH_3)CH_2O}$ by hydrogermylation of $CH_2{=}C(CH_3)CHO$ with $Ge(C_2H_5)_2H_2$ (H_2PtCl_6 catalyst) followed by cyclization in the presence of AIBN at 80 °C [30, 38].

$Ge(C_2H_5)_2(CH_2CH_2COOH)H$ (Table **23**, No. **14**). Attempts to purify by distillation at 1.5 Torr led to $\overline{Ge(C_2H_5)_2CH_2CH_2COO}$ via elimination of H_2 from the carboxyl group and the GeH bond; the product condensed in the column, as the pot temperature was increased above 150 °C [52].

$Ge(C_2H_5)_2(CH_2CH_2COOCH_3)H$ (Table **23**, No. **15**). Additional heating of No. 15 with $CH_2{=}CHCOOCH_3$ gave $Ge(C_2H_5)_2(CH_2CH_2COOCH_3)_2$ [52].

$Ge(C_2H_5)_2(C_4H_9)H$ (Table **23**, No. **19**) was obtained by $LiAlH_4$ reduction of a mixture of trialkylchlorogermanes formed from $Ge(C_2H_5)_4$ and $Ge(C_4H_9)_4$ at 220 °C in the presence of excess $AlCl_3$. The chromatographic characteristics of $Ge(C_2H_5)_2(C_4H_9)H$ in GLC were described; cf. No. 8 [16].

$Ge(C_2H_5)_2(CH_2CH(CH_3)C_3H_7\text{-}i)H$ (Table **23**, No. **20**) was obtained along with $Ge(C_2H_5)_2(CH_2C(CH_3){=}C(CH_3)_2)H$ by treating $\overline{Ge(C_2H_5)_2CH_2C(CH_3)(OH)CH(CH_3)CH_2}$ with Raney nickel at 180 °C (20 h) and reducing the reaction mixture with $LiAlH_4$. It was also formed by ring cleavage of $\overline{Ge(C_2H_5)_2CH_2C(CH_3){=}C(CH_3)CH_2}$ with HCl gas (to yield $Ge(C_2H_5)_2(CH_2C(CH_3){=}C(CH_3)_2)Cl$ and $Ge(C_2H_5)_2(CH_2C(CH_3)(Cl)CH(CH_3)_2)Cl$) followed by $LiAlH_4$ reduction. The mixture of No. 20 and $Ge(C_2H_5)_2(CH_2C(CH_3){=}C(CH_3)_2)H$ boils at 92 to 97 °C/14 Torr [29].

$Ge(C_2H_5)_2(C_4H_9\text{-}t)H$ (Table **23**, No. **21**). The reaction of equimolar amounts of $Ge(C_2H_5)_2(C_4H_9\text{-}t)H$ and $Ge(C_2H_5)_3Li$ in HMPTA at 25°C (3 d) followed by addition of $Si(CH_3)_3Cl$ gave $Ge(C_2H_5)_3H$, $Ge(C_2H_5)_3Si(CH_3)_3$, $Ge(C_2H_5)_2(C_4H_9\text{-}t)Si(CH_3)_3$ and unreacted No. 21. The pK_a of No. 21 was calculated from the mole ratio of the components (28:17:26:29) and shown to differ only slightly from that of $Ge(C_2H_5)_3H$: $\Delta pK_a = 0.2$ [95, 108] (error of sign in [95]?), corresponding to $pK_a = 39.9$ for No. 21 [104].

$Ge(C_2H_5)_2(CH_2CH_2CH(OH)CH_3)H$ (Table **23**, No. **22**). Heating with Cu powder for 10 min resulted in ring closure, yielding 60% $\overline{Ge(C_2H_5)_2CH_2CH_2CH(CH_3)O}$ by elimination of H_2O [29]. Treatment with Raney nickel (2 h) gave the same compound with 89% yield [38]; see also [29, 51].

$Ge(C_2H_5)_2(CH_2CH(CH_3)CH(OH)CH_3)H$ (Table **23**, No. **23**). Treatment with Cu powder gave a 1:1 mixture of cis- and trans-$\overline{Ge(C_2H_5)_2CH_2CH(CH_3)CH(CH_3)O}$ with 87% total yield [29]. Raney nickel converted No. 23 at 20°C (1 d) into $\overline{Ge(C_2H_5)_2CH_2CH(CH_3)CH(CH_3)O}$ with 90% yield [51]; see also [29].

$Ge(C_2H_5)_2(CH_2CH(OH)C_4H_9\text{-}t)H$ (Table **23**, No. **24**) gave compound VII (R = t-C_4H_9) with 80% yield when treated with Raney nickel in ether at 20°C [41].

$Ge(C_2H_5)_2(CH_2CH(R'')COCH_3)H$ (R″ = H and CH_3, Table **23**, Nos. **25** and **26**) were only mentioned as intermediates in the two-step preparation of $\overline{Ge(C_2H_5)_2CH_2CH(R')CH(CH_3)O}$ compounds from $Ge(C_2H_5)_2H_2$ and $CH_2{=}C(R'')COCH_3$ [30, 38]; cf. No. 13.

$Ge(C_2H_5)_2(C_5H_{11}\text{-}i)H$ (Table **23**, No. **28**) was obtained with 87% yield by treating $Ge(C_2H_5)H_3$ with Li in $NH_2C_2H_5$ (→ $Ge(C_2H_5)H_2Li$), followed by reaction with i-$C_5H_{11}Br$ (→ $Ge(C_2H_5)(C_5H_{11}\text{-}i)H_2$), Li (→ $Ge(C_2H_5)(C_5H_{11}\text{-}i)(H)Li$), and finally C_2H_5I [1].

$Ge(C_2H_5)_2(\overline{CHCH(OH)CH_2CH_2CH_2})H$ (Table **23**, No. **29**) gave compound IX with 60% yield upon treatment with Raney nickel in ether at 20°C [41].

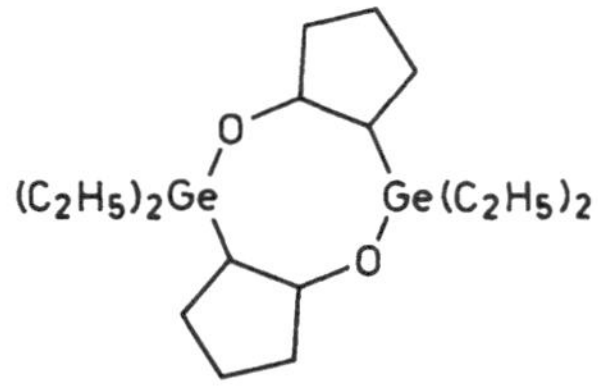

IX

$Ge(C_2H_5)_2(CH{=}CH_2)H$ (Table **23**, No. **30**) was mentioned as a possible product of the pyrolysis of $Ge(C_2H_5)_3CH_2CH_2OOCCH_3$ via $Ge(C_2H_5)_3CH{=}CH_2$ [76].

Bromination with N-bromosuccinimide in C_5H_{12} gave $Ge(C_2H_5)_2(CH{=}CH_2)Br$ with 81% yield [69].

cis-$Ge(C_2H_5)_2(CH{=}CHCH_2OH)H$ (Table **23**, No. **31**). Treatment with Raney nickel at 20°C (1 d) gave $\overline{Ge(C_2H_5)_2CH{=}CHCH_2O}$ with 76% yield; a small amount of $\overline{Ge(C_2H_5)_2CH_2CH_2CH_2O}$ was also formed [51].

$Ge(C_2H_5)_2(CH_2CH{=}CHCH_3)H$ (Table **23**, No. **33**) was obtained with 32% yield along with $Ge(C_2H_5)_2(CH_2CH{=}CHCH_3)_2$ by hydrogermylation of $CH_2{=}CHCH{=}CH_2$ with $Ge(C_2H_5)_2H_2$ (2:1 mole ratio) at 60°C (4 h) in an autoclave in the presence of a catalyst system consisting of $Ni(acac)_2$, $P(C_6H_5)_3$, and $Al(i\text{-}C_4H_9)_3$ (1:2:2 mole ratio) [91].

References on p. 187

$Ge(C_2H_5)_2(CH_2C(CH_3)=C(CH_3)_2)H$ (Table **23**, No. **36**) was also formed along with $Ge(C_2H_5)_2(CH_2CH(CH_3)C_3H_7\text{-}i)H$ from $Ge(C_2H_5)_2CH_2C(CH_3)(OH)CH(CH_3)CH_2$ on Raney nickel at 180 °C (20 h) followed by $LiAlH_4$ reduction of the reaction mixture. The mixture of No. 36 and $Ge(C_2H_5)_2(CH_2CH(CH_3)C_3H_7\text{-}i)H$ boils at 92 to 97 °C/14 Torr [29].

cis-$Ge(C_2H_5)_2(CH=CHC(OH)(CH_3)_2)H$ (Table **23**, No. **37**). Treatment with Raney nickel at 20 °C (1 d) gave $Ge(C_2H_5)_2CH=CHC(CH_3)_2O$ with 70% yield and a small amount of $Ge(C_2H_5)_2CH_2CH_2C(CH_3)_2O$ [51].

$Ge(C_2H_5)_2(C_6H_5)H$ (Table **23**, No. **40**) was also formed with 79% yield by refluxing compound II (p. 155, R = C_2H_5) and a large excess of $LiAlH_4$ in ether [77]. Treating $Ge(C_2H_5)_2(C_6H_5)CH_2NHCH_2C_6H_5$ with LiC_4H_9 (1.3:2.9 mole ratio) in THF-C_6H_{14}, followed by hydrolysis, gave No. 40 with 68% yield along with $NH(C_5H_{11})CH_2C_6H_5$ [121]. Irradiation ($\lambda >$ 250 nm) of $Ge(C_2H_5)_2(C_6H_5)_2$ in C_6H_{14} (30 min) under Ar yielded No. 40 along with several other products. With C_6D_{12} as the solvent, **$Ge(C_2H_5)_2(C_6H_5)D$** was formed under the same conditions [93].

Chlorination with CCl_4 in the presence of catalytic amounts of AIBN in a sealed tube at 80 °C (2 h) yielded $Ge(C_2H_5)_2(C_6H_5)Cl$ [75]. Refluxing with an equimolar amount of $Ge(C_2H_5)_2(CH_2CH=CH_2)C_6H_5$ (3 h) in the presence of AIBN gave $Ge(C_2H_5)_2(C_6H_5)CH_2$-$CH_2CH_2Ge(C_2H_5)_2C_6H_5$ with 33% yield [31].

Equimolar amounts of $Ge(C_2H_5)_2(C_6H_5)H$ and $Ge(C_2H_5)_3Li$ in HMPT and adding $Si(CH_3)_3Cl$ produced $Ge(C_2H_5)_3H$, $Ge(C_2H_5)_3Si(CH_3)_3$, $Ge(C_2H_5)_2(C_6H_5)Si(CH_3)_3$, and unreacted No. 40 in a 37:1:37:2 mole ratio [95, 108]. The pK_a was calculated to be 36.9 [104]; this is appreciably lower ($\Delta pK_a = -2.8$) than that of $Ge(C_2H_5)_3H$ [95, 108]; cf. No. 21.

1.3.1.2.3 $GeR_2(R')H$ Compounds with R = C_3H_7, C_4H_9, Larger Alkyl, and Cycloalkyl

Table 24
$GeR_2(R')H$ Compounds with R = C_3H_7, C_4H_9, Larger Alkyl, and Cycloalkyl.
An asterisk indicates further information at the end of the table.
Explanations, abbreviations, and units are given on p. X.

No.	R′ group	preparation (yield) properties and remarks
R = C_3H_7		
*1	C_2H_5	preparation given on p. 171 no data reported [16]
*2	CH_2CH_2OH	$Ge(C_3H_7)_2(CH_2COOCH_3)H + LiAlH_4$ (80%) [11] or $Ge(C_3H_7)_2(CH_2COOCH_3)Cl + LiAlH_4$ (60%) [17], both in ether b.p. 75 to 76 °C/1.5 [11], 98 to 99 °C/7.5 [17] d^{20} = 1.0885 [11], 1.0866 [17] n_D^{20} = 1.4670 [11, 17] IR: ν(GeH) ca. 2014 [49, 50]; see also [11]
*3	CH_2COOCH_3	$Ge(C_3H_7)_2H_2 + Hg(CH_2COOCH_3)_2$ (59%) [11] or $Ge(C_3H_7)_2(H)I$ and $Hg(CH_2COOCH_3)_2$ (1:1 mole ratio) in C_6H_6-petroleum ether (74%) [17]

Table 24 (continued)

No.	R′ group	preparation (yield) properties and remarks
		$Ge(C_3H_7)_2(H)Cl + Sn(C_4H_9)_3CH_2COOCH_3$ (70%) [25] b.p. 54 to 54.5°C/1.5 [11], 64 to 65.5°C/2 [17], 110 to 112°C/10 [25] d^{20} = 1.0903 [25], 1.0980 [11], 1.1006 [17] n_D^{20} = 1.4532 [11], 1.4539 [17], 1.4545 [25] ^1H NMR (34°C): 1.95 (CH_2, J = 2.0), 3.55 (OCH_3), 4.05 (GeH) [24] IR: ν(GeH) 2036 to 2038 [11] (wrong literature data cited in [25]) starting material for No. 2
4	CH_2COCH_3	$Ge(C_3H_7)_2(H)I + Sn(C_4H_9)_3CH_2COCH_3$ (78%) b.p. 69 to 70°C/1.5 d^{20} = 1.0991; n_D^{20} = 1.4642 [25]
5	CH_2CH_2COOH	$Ge(C_3H_7)_2H_2 + CH_2{=}CHCOOH$ (1:2 mole ratio) at 140°C for 18 h (30%) m.p. 31°C; b.p. 140 to 142°C/0.5 [5]
6	CH_2CH_2CN	$Ge(C_3H_7)_2H_2 + CH_2{=}CHCN$ (1:2 mole ratio) at 150°C (20 h) in the presence of $C_6H_4(OH)_2$-1,4 in a sealed tube (21%); obtained along with $Ge(C_3H_7)_2(CH_2CH_2CN)_2$ b.p. 75°C/0.4 d^{20} = 1.0632, n_D^{20} = 1.4621 IR: ν(GeH) 2015 [5]
R = i-C_3H_7		
*7	CH_3	formation given on p. 171 IR (CCl_4 or C_7H_{16}): ν(GeH) 2003(s) [19]
R = C_4H_9		
*8	CH_3	formation given on p. 171 no data reported [16]
*9	$CH_2C_6H_5$	formation given on p. 171 IR (neat): ν(GeH) 2009 [3, 5, 8, 39]
*10	C_2H_5	formation given on p. 171 no data reported [16]
*11	CH_2CH_2OH	$Ge(C_4H_9)_2(CH_2COOCH_3)H + LiAlH_4$ (72%) b.p. 82 to 84°C/1 d^{20} = 1.0446; n_D^{20} = 1.4679 [11] IR: ν(GeH) ca. 2014 [49, 50]; see also [11]
*12	CH_2COOCH_3	$Ge(C_4H_9)_2H_2 + Hg(CH_2COOCH_3)_2$ (1:1 mole ratio) in refluxing THF for 30 min (62%) [11]

References on p. 187

Table 24 (continued)

No.	R′ group	preparation (yield) properties and remarks
*12 (continued)		$Ge(C_4H_9)_2(H)I$ and $Hg(CH_2COOCH_3)_2$ (71%) [17] b.p. 73 to 74.5°C/1 [11], 106 to 109°C/6 [17] d^{20} = 1.0613 [11], 1.0661 [17] n_D^{20} = 1.4552 [11], 1.4560 [17] IR: ν(GeH) 2036 to 2038 [11] starting material for No. 11
13	$CH_2COOC_2H_5$	$Ge(C_4H_9)_2H_2$ + $CH(N_2)COOC_2H_5$ (1:3 mole ratio) in C_6H_6 at 100°C (8 h) in the presence of Cu powder (28%) b.p. 132 to 134°C/12 d^{20} = 1.0521; n_D^{20} = 1.4540 [5]
14	C_3H_7	$\overline{Ge(C_4H_9)_2CH_2CH_2CH_2}$ + $LiAlH_4$ in refluxing ether for 2 d (40%) b.p. 96°C/13 d^{20} = 0.9774; n_D^{20} = 1.4569 IR spectrum: ν(GeH) 2002 [20]
*15	$CH_2CH_2CH_2OH$	$Ge(C_4H_9)_2(CH_2CH_2COOCH_3)Cl$ + $LiAlH_4$ b.p. 143°C/12 d^{20} = 1.0229; n_D^{20} = 1.4701 IR: ν(OH) 3300(s), ν(GeH) 2005 [13]
16	$CH_2CH(CH_3)CH_2OH$	$Ge(C_4H_9)_2(CH_2CH(CH_3)COOCH_3)Cl$ + $LiAlH_4$ b.p. 151°C/14 d^{20} = 1.0221; n_D^{20} = 1.4674 IR: ν(OH) 3300(vs), ν(GeH) 2005 [13]
17	$CH_2CH_2COOCH_3$	$Ge(C_4H_9)_2H_2$ + CH_2=$CHCOOCH_3$ (1:2 mole ratio) at 100°C for 7 h (30%); formed along with $Ge(C_4H_9)_2(CH_2CH_2$-$COOCH_3)_2$ b.p. 112 to 113°C/1 d^{20} = 1.0545; n_D^{20} = 1.4568 [5]
18	$CH_2CH_2CH_2NH_2$	$Ge(C_4H_9)_2(CH_2CH_2CN)Cl$ + $LiAlH_4$ in ether b.p. 134°C/13 d^{20} = 0.9984; n_D^{20} = 1.4668 IR: ν(GeH) 2005 [10]
*19	CH_2CH_2CN	preparation not reported IR: ν(GeH) 2015 (neat) [8, 39]; 2018 (in C_7H_{16}) [8]
20	$CH_2CH_2CH_2Si(CH_3)H_2$	$Ge(C_4H_9)_2(CH_2CH_2CH_2Si(CH_3)Cl_2)H$ + $LiAlH_4$ b.p. 130°C/20 d^{20} = 0.9453; n_D^{20} = 1.4621 ^1H NMR (CCl_4): 0.12 (t, 3 H, $SiCH_3$, J = 4), 3.5 to 3.8 (m's, 2(?) H, (SiH, GeH)) IR: ν(SiH) 2140, ν(GeH) 2010 [22]

Table 24 (continued)

No.	R′ group	preparation (yield) properties and remarks
21	$CH_2CH_2CH_2Si(CH_3)Cl_2$	$\overline{Ge(C_4H_9)_2CH_2CH_2C}H_2$ + $Si(CH_3)(H)Cl_2$ (1:5 mole ratio) in the presence of H_2PtCl_6 [22, 23] b.p. 160 °C/18 [22] d^{20} = 1.0912; n_D^{20} = 1.4699 [22] ^1H NMR (CCl_4): 0.73 (s, $SiCH_3$), 3.70 (sept, GeH, J = 2.5) [22]; see also [23] IR: ν(GeH) 2010 [22] starting material for No. 20
22	$CH_2CH_2CH_2$-$Si(CH_3)(C_4H_9)_2$	$\overline{Ge(C_4H_9)_2CH_2CH_2C}H_2$ + $Si(CH_3)(C_4H_9)_2H$ (1:5 mole ratio), in the presence of H_2PtCl_6 [22, 23] b.p. 152 °C/0.75 d^{20} = 0.9246; n_D^{20} = 1.4650 [22] for ^1H NMR data, see [23]
23	$CH_2CH_2CH_2$-$Si(CH_3)(C_6H_5)_2$	$\overline{Ge(C_4H_9)_2CH_2CH_2C}H_2$ + $Si(CH_3)(C_6H_5)_2H$ like No. 22 [22, 23] b.p. 156 °C/0.1 d^{20} = 1.0261; n_D^{20} = 1.5353 [22] for ^1H NMR data, see [23]
24	$CH_2CH_2CH_2Si(C_2H_5)_3$	$\overline{Ge(C_4H_9)_2CH_2CH_2C}H_2$ + $Si(C_2H_5)_3H$ like No. 22 [22, 23] b.p. 143 °C/1.5 d^{20} = 0.9406; n_D^{20} = 1.4670 [22] for ^1H NMR data, see [23]
25	$CH_2CH_2CH_2Si(C_4H_9)_2H$	$Ge(C_4H_9)_2(CH_2CH_2CH_2Si(C_4H_9)_2Cl)H$ + $LiAlH_4$ [22] b.p. 148 °C/1.3 d^{20} = 0.9233; n_D^{20} = 1.4660 ^1H NMR (CCl_4): 3.5 to 3.8 (m's, SiH, GeH) IR: ν(SiH) 2110, ν(GeH) 2010 [22]
26	$CH_2CH_2CH_2Si(C_4H_9)_2Cl$	$\overline{Ge(C_4H_9)_2CH_2CH_2C}H_2$ + $Si(C_4H_9)_2(H)Cl$ like No. 22 [22, 23] b.p. 165 °C/1.3 d^{20} = 0.9814; n_D^{20} = 1.4702 [22] ^1H NMR (CCl_4): 3.70 (sept, GeH, J = 2.5) [22]; see also [23] IR: ν(GeH) 2010 [22] starting material for No. 25
*27	$CH_2CH_2COCH_3$	$Ge(C_4H_9)_2H_2$ + CH_2=$CHCOCH_3$ (1:2 mole ratio) refluxing (10 h) in the presence of $C_6H_4(OH)_2$-1,4 (30%); along with $Ge(C_4H_9)_2(CH_2CH_2COCH_3)_2$ [5] b.p. 121 °C/12 d^{20} = 1.0383; n_D^{20} = 1.4621 [5] IR: ν(GeH) 2003 (neat) [5, 8]; 2008 (C_7H_{16}) [8]
28	CH=CHC_6H_5	$Ge(C_4H_9)_2H_2$ + $CH{\equiv}CC_6H_5$ at 100 °C polymerized upon prolonged heating [5]

References on p. 187

Table 24 (continued)

No.	R′ group	preparation (yield) properties and remarks
*29	$CH_2CH{=}CH_2$	$Ge(C_4H_9)_2H_2$ + $CH_2{=}CHCH_2MgBr$ in THF in the presence of CuCl (35%) [5] b.p. 36°C/0.2 [85], 95 to 96°C/13 [5] d^{20} = 0.9702; n_D^{20} = 1.4608 [5] IR (neat): ν(GeH) 2012 [3, 5], 2013 [8], ν(C=C) 1630 [5]
*30	$CH_2C(CH_3){=}CH_2$	$Ge(C_4H_9)_2(H)Cl$ + $CH_2{=}C(CH_3)CH_2MgCl$ in refluxing THF-ether (5:1) for 3 h (73%); purified by fractional distillation b.p. 38.5 to 39.5°C/0.04 n_D^{20} = 1.4689 ^1H NMR ($CDCl_3$): 0.88 to 1.73 (m, 23 H), 3.80 (m, 1 H), 4.52 (s, 2 H) IR (neat): ν(GeH) 2050 [85]
*31	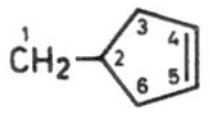	$Ge(C_4H_9)_2(CH_2C_5H_7)Cl$ + excess $LiAlH_4$ in ether (12 h) followed by hydrolysis with H_2O-NaOH (80%); purified by molecular distillation at 100°C/0.2 Torr unstable, decomposes on standing and readily in contact with air ^1H NMR ($CDCl_3$): 0.5 to 2.8 (m, 25 H), 3.6 to 3.8 (complex m, GeH), 5.6 (s, CH=CH) ^{13}C NMR ($CDCl_3$; C-α to C-δ = C_4H_9): 12.2 (C-α), 13.6 (C-δ), 20.2 (C-1), 26.0 and 28.5 (C-β,γ), 35.7 (C-2), 40.8 (C-3,6), 129.9 (C-4,5) IR (neat): ν(GeH) 2000 [113]
*32	C_6H_5	$Ge(C_6H_5)(H)Cl_2$ + C_4H_9MgBr (1:2 mole ratio) in refluxing ether for 1 h (82%) $Ge(C_4H_9)_2(C_6H_5)Cl$ + $LiAlH_4$ in refluxing THF [75] b.p. 145°C/10 ^1H NMR: 4.56 (quint, GeH) [75] IR (neat): ν(GeH) 2015 [3, 5, 7, 8]
R = t-C_4H_9		
33	C_6H_5	$Ge(C_6H_5)H_3$ + LiC_4H_9-t in THF-C_5H_{12} at −78°C; minor product along with $Ge(C_6H_5)H_2Li$ and $Ge(C_4H_9\text{-}t)(C_6H_5)H_2$ no data reported [124]
R = C_5H_{11}		
34	$CH(CH_3)CH_2CHO$	$Ge(C_5H_{11})_2H_2$ + $CH_3CH{=}CHCHO$ (1:2 mole ratio) at 100°C for 18 h in the presence of Pt asbestos (44%) b.p. 139 to 140°C/10 d^{20} = 0.9910, n_D^{20} = 1.4585 [5]

Table 24 (continued)

No.	R′ group	preparation (yield) properties and remarks
R = c-C_6H_{11}		
*35	C_6H_5	preparation not reported IR: ν(GeH) 2003 [39]

*Further information:

$Ge(C_3H_7)_2(C_2H_5)H$ (Table **24**, No. **1**) was obtained by $LiAlH_4$ reduction of a mixture of trialkylchlorogermanes (previously formed from $Ge(C_2H_5)_4$ and $Ge(C_3H_7)_4$ at 220 °C in the presence of excess $AlCl_3$). The chromatographic characteristics of No.1 in GLC were described together with those of other alkyl derivatives of Si and Ge [16].

$Ge(C_3H_7)_2(CH_2CH_2OH)H$ (Table **24**, No. **2**) was part of IR studies on various organogermanium compounds dealing with the dependence of the ν(GeH) frequency on the nature of the substituents [49, 50] and the frequency of an isolated GeH bond [84].

Reacting the compound with HgI_2 in ether yielded $Ge(C_3H_7)_2(H)I$ and $Ge(C_3H_7)_2(CH_2$-$CH_2OH)I$ with 32 and 52% yield, respectively [11].

$Ge(C_3H_7)_2(CH_2COOCH_3)H$ (Table **24**, No. **3**) reacted with $HgCl_2$ (2.1 : 1 mole ratio) in refluxing ether (30 min) to give $Ge(C_3H_7)_2(H)Cl$ with 35% yield and $Ge(C_3H_7)_2(CH_2COOCH_3)Cl$ with 40% yield [17].

$Ge(C_3H_7\text{-}i)_2(CH_3)H$ (Table **24**, No. **7**). Small quantities of the compound were probably present in a mixture of products obtained by reacting $GeCl_4$ with an i-C_3H_7MgBr-CH_3MgI mixture followed by hydrolysis with dilute HCl. The compound could not be separated from $Ge_2(CH_3)_6$ by GLC [19].

$Ge(C_4H_9)_2(CH_3)H$ (Table **24**, No. **8**) was obtained by $LiAlH_4$ reduction of a mixture of trialkylchlorogermanes (previously formed from $Ge(CH_3)_4$ and $Ge(C_4H_9)_4$ at 220 °C in the presence of excess $AlCl_3$). The chromatographic characteristics of $Ge(C_4H_9)_2(CH_3)H$ in GLC were described along with those of other alkyl derivatives of Si and Ge [16].

$Ge(C_4H_9)_2(CH_2C_6H_5)H$ (Table **24**, No. **9**). Refluxing $Ge(C_4H_9)_2H_2$ and $C_6H_5CH_2MgCl$ in THF (2 d) gave No. 9 (b.p. 155 °C/14 Torr) contaminated with $C_6H_5CH_2CH_2C_6H_5$ [5]. When $Ge(CH_2C_6H_5)_3H$ was treated with LiC_4H_9 in ether at −10 °C, many products were formed via H-Li exchange and Ge-C cleavage, probably including No. 9 and $Ge(C_4H_9)(CH_2C_6H_5)_2H$ (by ^1H NMR and IR) [12].

The compound was involved in IR studies on the ν(GeH) frequency of organogermanium compounds [8, 39, 49, 50, 84] as mentioned for No. 2.

$Ge(C_4H_9)_2(C_2H_5)H$ (Table **24**, No. **10**) was obtained by $LiAlH_4$ reduction of a mixture of trialkylchlorogermanes formed from $Ge(C_2H_5)_4$ and $Ge(C_4H_9)_4$ and studied by GLC as described for No. 1 [16].

$Ge(C_4H_9)_2(CH_2CH_2OH)H$ (Table **24**, No. **11**) was part of IR studies on the ν(GeH) frequency of organogermanium compounds [49, 50, 84]; see No. 2.

$Ge(C_4H_9)_2(CH_2COOCH_3)H$ (Table **24**, No. **12**) reacted with $Hg(CH_2COOCH_3)_2$ to yield 78% of $Ge(C_4H_9)_2(CH_2COOCH_3)_2$ [11].

References on p. 187

$Ge(C_4H_9)_2(CH_2CH_2CH_2OH)H$ (Table **24**, No. **15**). Small amounts of No. 15 were isolated (5% yield) when the germylene $Ge(CH_2CH_2CH_2OH)Cl$ (prepared from $Ge(CH_2CH_2$-$CH_2OH)(H)Cl_2$ and $Ge(C_4H_9)_3OCH_3$ via the $Ge(CH_2CH_2CH_2OH)(H)(Cl)OCH_3$ intermediate) was reacted with LiC_4H_9 followed by hydrolysis [45].

$Ge(C_4H_9)_2(CH_2CH_2CN)H$ (Table **24**, No. **19**) was involved in IR studies on the ν(GeH) frequency of organogermanium compounds [8, 39]; see also No. 2.

$Ge(C_4H_9)_2(CH_2CH_2COCH_3)H$ (Table **24**, No. **27**) was involved in IR studies on the ν(GeH) frequency of organogermanium compounds [8]; see also No. 2.

$Ge(C_4H_9)_2(CH_2CH{=}CH_2)H$ (Table **24**, No. **29**) was involved in IR studies on the ν(GeH) frequency of organogermanium compounds [8, 14, 49, 50, 84]; see also No. 2.

The compound polymerized on heating to 100 °C [5]. Heating in C_6H_6 at 80 °C (15 h) in a sealed tube in the presence of C_6H_5CO-OO-COC_6H_5 yielded $Ge(C_4H_9)_2H_2$ and $Ge(C_4H_9)_2(CH_2CH{=}CH_2)_2$. The reaction mechanism was discussed [85].

$Ge(C_4H_9)_2(CH_2C(CH_3){=}CH_2)H$ (Table **24**, No. **30**) was converted into $Ge(C_4H_9)_2H_2$ and $Ge(C_4H_9)_2(CH_2C(CH_3){=}CH_2)_2$ when heated in C_6H_6 in the presence of C_6H_5CO-OO-COC_6H_5 [85]; see No. 29.

$Ge(C_4H_9)_2(CH_2C_5H_7)H$ (Table **24**, No. **31**). Heating at 175 to 200 °C (3 d) in the presence of a catalytic amount of H_2PtCl_6 gave a complex mixture containing compound X [113].

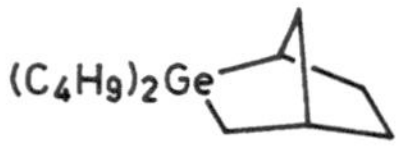

X

$Ge(C_4H_9)_2(C_6H_5)H$ (Table **24**, No. **32**). Refluxing $Ge(C_4H_9)_2H_2$ and C_6H_5MgBr in THF yielded 20% $Ge(C_4H_9)_2(C_6H_5)H$ (b.p. 140 °C/9 Torr) contaminated with $C_6H_5C_6H_5$ [5]. The compound was also generated by reacting the germylene $Ge(C_4H_9)C_6H_5$ with C_4H_9MgX [42]. It was a minor product of the reaction of $Ge(C_6H_5)H_3$ with LiC_4H_9 (4% yield) [53].

The compound was part of IR studies on the ν(GeH) frequency behavior of various organogermanium compounds [8, 39, 55]; see also No. 2.

Chlorination with CCl_4 in the presence of catalytic amounts of AIBN in a sealed tube at 80 °C for 2 h gave $Ge(C_4H_9)_2(C_6H_5)Cl$ [75].

$Ge(C_6H_{11}$-$c)_2(C_6H_5)H$ (Table **24**, No. **35**) was included in ν(GeH) frequency studies on various organogermanium compounds [39, 49, 55].

1.3.1.2.4 $Ge(C_6H_5)_2(R')H$ Compounds

Table 25
$Ge(C_6H_5)_2(R')H$ Compounds.
An asterisk indicates further information at the end of the table.
Explanations, abbreviations, and units are given on p. X.

No.	R′ group	preparation (yield) properties and remarks
*1	CH_3	$Ge(C_6H_5)_2(CH_3)Cl + LiAlH_4$ [119] $(Ge(C_6H_5)_2CH_3)_2 + LiAlH_4$ in THF at 20 °C for 3 d (77%) [63]

Table 25 (continued)

No.	R′ group	preparation (yield) properties and remarks
		b.p. 90°C/0.5 [21], 105°C/6 [60] d^{20} = 1.2128, n_D^{20} = 1.5830 [21] ^{1}H NMR (CCl_4): 5.05 (q, GeH) [21] IR: ν(GeH) 2047 in C_7H_{16} or CCl_4 [64], 2035(s) [18], 2030 (neat) [21]; $δ_s(CH_3)$ 1239(w), $ρ(CH_3)$ 797(s), δ(GeH) 707(s), ν(GeC) 594(m); other typical GeC_6H_5 bands at 1093, 309(s), 292(m), 254(w) [18]
*2	CH_2OH	$Ge(C_6H_5)_2(CH_2OOCCH_3)OOCCH_3$ + $LiAlH_4$ in refluxing ether for 6 h (90%); purified by distillation at 100 to 150°C/0.01 [102] colorless liquid ^{1}H NMR (CD_3CN): ca. 2.8 (br s, OH), 4.17 (d, CH_2Ge, J = 2.6), 5.12 (t, GeH, J = 2.6), 7.3 to 7.6 (m, C_6H_5) ^{13}C NMR (CD_3CN): 53.9 (CH_2Ge), 129.3 (C-3,5), 130.1 (C-4), 135.9 (C-2,6), 136.7 (C-1) IR (CCl_4): ν(OH) 3610 (nonassociated) and 3400 (br, associated), ν(GeH) 2045 MS: $[M]^+$ (<1), $[M - CH_2OH]^+$ (100), $[GeC_6H_5]^+$ (57) [102]
3	CH_2OOCCH_3	preparation not reported thermally stable at 160°C for 70 h [102, footnote 9], [120]; $Si(C_6H_5)_2(CH_2OOCCH_3)H$ rearranges at 135°C to give $Si(CH_3)(C_6H_5)_2OOCCH_3$ [102, footnote 9]
4	$CH_2OSO_2CF_3$	preparation not reported thermal rearrangement yields $Ge(CH_3)(C_6H_5)_2OSO_2CF_3$ [120]
*5	$CH(OH)C_6H_5$	$Ge(C_6H_5)_2(H)Li$ + C_6H_5CHO (1:1 mole ratio) in THF-ether at −30°C to +20°C (87%) along with $C_6H_5CH_2OH$ and traces of $Ge(C_6H_5)_2H_2$ [110] ^{1}H NMR (C_6D_6): 5.10 (d, CH), 5.33 (d, GeH, ^3J = 1.25) IR: ν(OH) 3320, ν(GeH) 2060 [80]
*6	$C(OH)(C_6H_5)_2$	preparation given on p. 177 ^{1}H NMR (C_6D_6): 2.15 (s, OH), 5.70 (s, GeH), 6.80 to 7.15, 7.25 to 7.50 (m's, C_6H_5) IR (KBr, Nujol): ν(OH) 3362, ν(GeH) 2033 [110]
*7	COC_6H_5	preparation given on p. 177 ^{1}H NMR (C_6D_6): 5.90 (s, GeH), 7.00 to 7.10, 7.25 to 7.65, 7.65 to 7.90 (m's, C_6H_5) IR (neat, KBr): ν(GeH) 2052, ν(C=O) 1686 [110]
*8	$COC_6H_2(CH_3)_3$-2,4,6	formation given on p. 177 ^{1}H NMR (C_6D_6): 2.00 (s, CH_3-4), 2.10 (s, CH_3-2,6), 5.75 (s, GeH), 6.48 (s, C_6H_2), 6.90 to 7.60 (m, C_6H_5) IR (neat, KBr): ν(GeH) 2048, ν(C=O) 1637 [110]

References on p. 187

Table 25 (continued)

No.	R′ group	preparation (yield) properties and remarks
*9	C_2H_5	$Ge(C_2H_5)(C_6H_5)_2Br$ + excess $LiAlH_4$ in ether (98%) [67] b.p. 97 to 103°C/0.5 [67], 110 to 115°C/0.8 [90] n_D^{25} = 1.5912 ^1H NMR ($CDCl_3$): 4.85 (GeH) IR: ν(GeH) 2033(ms), δ(GeH) 707(s) [67]
*10	CH_2CH_2OH	preparation given on p. 178 b.p. 130°C/0.02 n_D^{20} = 1.5848 [62, 68] ^1H NMR (C_6D_6): 1.48 (td, CH_2Ge, ^3J = 7 and 3), 3.63 (t, CH_2O, ^3J = 7), 5.15 (t, GeH, ^3J = 3) [62]; 1.53, 3.67, 5.20 [68] IR: ν(OH) 3320, ν(GeH) 2060 [62, 68]
*11	CH_2CH_2CHO	$Ge(C_6H_5)_2H_2$ + CH_2=CHCHO at 100°C (14 h) in the presence of AIBN in a sealed tube (51%) along with $\overline{Ge(C_6H_5)_2CH_2CH_2CH_2O}$ ^1H NMR: 1.42 (td, CH_2Ge, ^3J = 7.5, ^4J = 3), 2.44 (td, CCH_2C, ^3J = 7.5 and 1), 5.0 (t, GeH, ^4J = 3), 9.62 (t, CHO, ^3J = 1) IR: ν(GeH) 2040, ν(C=O) 1725 [47]
12	CH_2CH_2CN	$Ge(C_6H_5)_2H_2$ + CH_2=CHCN at 95°C (2.5 d) in the presence of AIBN in a sealed tube (33%) along with $Ge(C_6H_5)_2(CH_2CH_2CN)_2$ and polymers b.p. 136°C/0.01 d^{20} = 1.2472; n_D^{20} = 1.5862 [27]
13	$CH_2CH_2CH_2CN$	$Ge(C_6H_5)_2H_2$ + CH_2=$CHCH_2CN$ at 135°C (2 d) like No. 12 (35%) along with $Ge(C_6H_5)_2(CH_2CH_2CH_2CN)_2$ and polymers b.p. 163°C/0.05 d^{20} = 1.2105; n_D^{20} = 1.5796 [27]
*14	C_3H_7-i	preparation given on p. 178 b.p. 100 to 104°C/0.4 n_D^{20} = 1.5718 ^1H NMR ($CDCl_3$): 0.65 to 2.10 (m, 1 H), 0.91 to 1.40 (d, 6 H, J = 6.8), 4.80 to 5.00 (d, 1 H, J = 4.0), 6.85 to 7.75 (m, 10 H) IR (neat): ν(GeH) 2025 [90]
*15	C_4H_9	$Ge(C_6H_5)_2H_2$ + LiC_4H_9 (1:2 mole ratio) in ether-C_5H_{12} at −10°C (36%) [18] b.p. 107 to 109°C/0.08 [90], 120°C/0.01 [18] IR: ν(GeH) 2028(s) [18], 2025 [40b], δ(GeH) 707(s), $ν(GeC_4H_9)$trans 635(m), ν(GeC) (including $ν(GeC_4H_9)$gauche) 568(w); other bands at 1093, 881 (complex,w), 314(s), 301(m) [18]

Table 25 (continued)

No.	R′ group	preparation (yield) properties and remarks
*16	$CH_2CH_2CH(OH)CH_3$	$(Ge(C_6H_5)_2CH_2CH_2COCH_3)_2O$ + $LiAlH_4$ (71%) $Ge(C_6H_5)_2CH_2CH_2CH(CH_3)O$ + $LiAlH_4$ [40a] b.p. 135 °C/0.05 n_D^{20} = 1.5801 ^1H NMR: 1.05 (d, CH_3), 3.0 (OH), 3.6 (m, CH), 5.0 (t, GeH), 7 (C_6H_5) IR: ν(OH) 3360, ν(GeH) 2025 [46]
*17	$CH_2CH_2COCH_3$	$Ge(C_6H_5)_2H_2$ + CH_2=$CHCOCH_3$ (1:1) at 110 °C for 16 h (23%) along with $Ge(C_6H_5)_2CH_2CH_2CH(CH_3)O$, $Ge(C_6H_5)_2(CH_2CH_2COCH_3)_2$, and polymers b.p. 114 °C/0.03 n_D^{20} = 1.5792 [35]
*18	CH=CHCH=CH_2	formation given on p. 178 ^1H NMR: 5.0 to 6.8 (CH=CHCH=CH_2), 5.6 (GeH), 7.3 (C_6H_5) IR: ν(GeH) 2060, ν(C=C) 1560 to 1570 [65]
*19	$CH_2CH_2CH_2CH_2OH$	formation given on p. 178 ^1H NMR: 0.8 to 2.1 ($GeCH_2CH_2CH_2$), 3.4 (CH_2O), 5.0 (GeH), 7.3 (C_6H_5) IR: ν(OH) 3350, ν(GeH) 2020 [65]
*20	$CH_2CH_2CH_2$- $C(CH_3)OCH_2CH_2O$	$Ge(C_6H_5)_2(H)Cl$ + R′MgCl in THF at 20 °C for 3 d (60%) ^1H NMR (CCl_4): 1.15 (CH_3), 3.7 (CH_2O), 5.36 (GeH, ^3J = 2.6) IR (neat): ν(GeH) 2020 [59]
*21	$CH_2CH_2CH_2CH_2COCH_3$	$Ge(C_6H_5)_2H_2$ + CH_2=$CHCH_2CH_2COCH_3$ at 120 °C for 2 d (72%) along with $Ge(C_6H_5)_2(CH_2CH_2CH_2CH_2COCH_3)_2$ b.p. 148 °C/0.1 n_D^{20} = 1.565 ^1H NMR: 1.90 (s, CH_3), 5.0 (t, GeH) IR: ν(GeH) 2060, ν(C=O) 1730 [35]
*22	C_6H_{13}	preparation not reported IR: ν(GeH) 2039 (C_7H_{16} or CCl_4) [64], 2025 [40b]
23	C_8H_{17}	$Ge(C_6H_5)_2(C_8H_{17})Br$ + $LiAlH_4$ in refluxing ether (38%) $Ge(C_6H_5)_2K_2$ + excess $C_8H_{17}Br$ in THF-HMPT at 20 °C (30 min) and hydrolysis; minor product along with $Ge(C_8H_{17})_2(C_6H_5)_2$ b.p. 150 to 158 °C/0.7 n_D^{20} = 1.5429 ^1H NMR ($CDCl_3$): 0.35 to 1.95 (m, C_8H_{17}), 4.90 to 5.10 (m, GeH), 6.85 to 7.95 (m, C_6H_5) IR (neat): ν(GeH) 2030 [90]

References on p. 187

Table 25 (continued)

No.	R′ group	preparation (yield) properties and remarks
*24	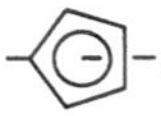 $-Re(P(C_6H_5)_3)(NO)H$	yellow powder m.p. 139 to 142°C (from CH_2Cl_2-C_6H_{14}) ^1H NMR (CD_2Cl_2): −9.66 (d, ReH, J(H, P) = 29.2), 4.62, 4.74, 4.90, 5.14 (br m, C_5H_4), 5.81 (s, GeH), 6.66 to 7.93 (m, C_6H_5) ^{13}C NMR (CD_2Cl_2): 86.8, 87.3, 88.8, 89.0, 91.6 (s's, C_5H_4), 127.0 (s, C-3,5/GeC_6H_5), 127.2 (d, C-3,5/PC_6H_5, J = 11.0), 127.5 (s, C-4/GeC_6H_5), 128.9 (s, C-4/PC_6H_5), 132.6 (d, C-2,6/PC_6H_5, J = 10.7), 134.0 (s, C-2,6/GeC_6H_5), 137.1 (d, C-1/PC_6H_5, J = 52.6), 145.5 (s, C-1/GeC_6H_5) ^{31}P NMR (CD_2Cl_2): 26.8 (s) IR (KBr): ν(GeH) and ν(ReH) 1962(m), ν(NO) 1642(vs) [115]
*25	C_6H_4F-4	$Ge(C_6H_5)_2(C_6H_4F\text{-}4)Br + LiAlH_4$ m.p. <30°C (from C_6H_{14}) ^1H NMR ($CDCl_3$): 5.59 (GeH) IR spectrum: ν(GeH) 2034(s), δ(GeH) 708(s) [67]
*26	C_6H_4Cl-3	$Ge(C_6H_5)_2(C_6H_4Cl\text{-}3)Br + LiAlH_4$ (?) b.p. 145 to 155°C/0.001 ^1H NMR (CCl_4): 5.61 (GeH) IR: ν(GeH) 2040(ms), δ(GeH) 706(s) [67]
*27	C_6H_4CN-4	$Ge(C_6H_5)_2(C_6H_4CN\text{-}4)Br + NaBH_4$ in THF-H_2O; purified by GLC on SiO_2 with $C_6H_5CH_3$-C_6H_{14} eluent m.p. <30°C ^1H NMR ($CDCl_3$): 5.44 (GeH) IR: ν(GeH) 2032(s), δ(GeH) 710(s) [67]
*28	$C_6H_4NO_2$-4	$Ge(C_6H_5)_2(C_6H_4NO_2\text{-}4)Br + NaBH_4$ (excess) in THF at 10°C b.p. 96 to 98°C ^1H NMR ($CDCl_3$): 5.68 (GeH) IR: ν(GeH) 2038(s), δ(GeH) 704(s) [67]
29	$C_6H_2R''_3$-2,4,6 R″ = $CH(Si(CH_3)_3)_2$	$Ge(C_6H_2(R'')_3\text{-}2,4,6)Cl_3 + C_6H_5MgBr$ (1 : 1.5 mole ratio) in THF at −78°C, followed by $LiAlH_4$ reduction under reflux (58%) along with $Ge(C_6H_5)(C_6H_2(R'')_3\text{-}2,4,6)H_2$ (15%) no data reported [122]

*Further information:

$Ge(C_6H_5)_2(CH_3)H$ (Table **25**, No. **1**) was also obtained from $Ge(C_6H_5)_2H_2$ and excess CH_2N_2 either by UV irradiation in ether (6 h) or by refluxing (10 h) in the presence of Cu powder; 43 and 25% yield, respectively [21]; see also [6]; from $Ge(C_6H_5)_2(H)Li$ and excess CH_3I [98] or $(CH_3)_2SO_4$ [110] in THF at 20°C with 68 [98] and 95% yield [110], respectively;

and from [Li(12-crown-4)][Ge(C_6H_5)$_2$H] and excess CH_3I at 20 °C; 82% yield [110]. Photolysis of $Ge_2(C_6H_5)_4(CH_3)_2$ in C_6H_{12} at room temperature (2 h) yielded 20% Ge(C_6H_5)$_2$(CH_3)H along with 15% Ge(CH_3)(C_6H_5)$_3$. The formation process of the hydrogermane is unclear [116].

The compound was involved in IR studies on various organogermanium compounds dealing with the integrated intensity of the ν(GeH) band and its dependence on the nature of the substituents [57, 64].

Constant-current electrolysis of Ge(C_6H_5)$_2$(CH_3)H in CH_3CN ([N(C_4H_9)$_4$][ClO_4] as supporting electrolyte) in a two-compartment cell gave Ge(C_6H_5)$_2$(CH_3)Cl and (Ge(C_6H_5)$_2$-CH_3)$_2$O in the cathode compartment. Ge(C_6H_5)$_2$(CH_3)F and (Ge(C_6H_5)$_2$$CH_3$)$_2$O were formed with [N($C_4H_9$)$_4$][$BF_4$] as the electrolyte. The halogenated products were readily hydrolyzed [119].

Photochemically generated t-$C_4H_9O^{\bullet}$ reacted with Ge(C_6H_5)$_2$(CH_3)H to yield the $Ge^{\bullet}$(C_6H_5)$_2$(CH_3) radical [56, 60].

The reaction of Ge(C_6H_5)$_2$(CH_3)H with $Os_3(CO)_{11}$(CH_3CN) in $C_6H_5CH_3$ at room temperature (12 h) gave $Os_3(CO)_{11}$(H)Ge(CH_3)(C_6H_5)$_2$ with 70 to 80% yield [87].

Ge(C_6H_5)$_2$(CH_2OH)H (Table **25**, No. **2**). The reactions with acid anhydrides gave the corresponding esters. Treatment with CCl_4 at room temperature yielded Ge(C_6H_5)$_2$(CH_2OH)Cl and $CHCl_3$ (ca. 60% conversion within 1 d) [103].

Ge(C_6H_5)$_2$(CH(OH)C_6H_5)H (Table **25**, No. **5**) was also formed by $LiAlH_4$ reduction of a reaction mixture containing compound XI obtained from equimolar amounts of Ge(C_6H_5)$_2$(H)Cl and C_6H_5CHO in THF at 60 °C in the presence of N(C_2H_5)$_3$ [80].

The compound is thermally unstable and decomposes partially above 100 °C [110]. Attempts to isolate No. 5 by distillation resulted in quantitative decomposition to Ge(C_6H_5)$_2H_2$ and C_6H_5CHO [80].

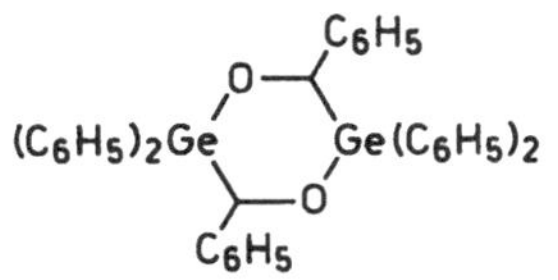

XI

Ge(C_6H_5)$_2$(C(OH)(C_6H_5)$_2$)H (Table **25**, No. **6**) was prepared with 80% yield via Ge(C_6H_5)$_2$-(C(OLi)(C_6H_5)$_2$)H by treating Ge(C_6H_5)$_2$(H)Li with a small excess of (C_6H_5)$_2$CO in THF-ether at room temperature (4 h) [110].

Ge(C_6H_5)$_2$(COC_6H_5)H (Table **25**, No. **7**) was prepared from Ge(C_6H_5)$_2$(H)Li and a small excess of C_6H_5COCl in THF-ether at −78 °C to +20 °C (1 h); 55% yield along with (Ge(C_6H_5)$_2$)$_n$ [110].

Ge(C_6H_5)$_2$(COC_6H_2(CH_3)$_3$-2,4,6)H (Table **25**, No. **8**) was a minor product (15% yield) along with (Ge(C_6H_5)$_2$)$_n$, Ge(C_6H_5)$_2$(COC_6H_2(CH_3)$_3$-2,4,6)$_2$, and Ge(C_6H_5)$_2H_2$ when Ge(C_6H_5)$_2$(H)Li was reacted with a small excess of C_6H_2(COCl)(CH_3)$_3$-2,4,6 in THF-ether at −78 °C to +20 °C (3 h) [110].

Ge(C_6H_5)$_2$(C_2H_5)H (Table **25**, No. **9**) was obtained along with Ge(C_2H_5)$_2$(C_6H_5)$_2$ by reacting Ge(C_6H_5)$_2K_2$ with excess C_2H_5X (X = Cl, Br, or I) in THF-HMPT at room temperature (30 min) [90]. Treatment of Ge(C_2H_5)(C_6H_5)$_2$$CH_2NHCH_2C_6H_5$ with LiC_4H_9 (1.3 : 1.5 or 1.3 : 2.9

References on p. 187

mole ratio) in THF-C_6H_{14} gave No. 9 with 63 or 91% yield, respectively, along with $NH(C_5H_{11})CH_2C_6H_5$ [121].

$Ge(C_6H_5)_2(C_2H_5)T$ was obtained by base-catalyzed tritiation of $Ge(C_6H_5)_2(C_2H_5)H$ with tritiated water in refluxing CH_3OH-DMSO containing CH_3ONa [67].

The first-order rate constant for the T-H exchange of $Ge(C_6H_5)_2(C_2H_5)T$ with CH_3OH-CH_3ONa was determined at 30°C: k_{rel} = 0.06 in CH_3OH (0.421 M CH_3ONa) relative to $k(Ge(C_6H_5)_3T)$ = 1 in CH_3OH (0.421 M CH_3ONa) [67].

$Ge(C_6H_5)_2(CH_2CH_2OH)H$ (Table **25**, No. **10**) was prepared by reducing compound XII [62] or $Ge(C_6H_5)_2(CH_2CH_2OOCCH_3)Cl$ [68] with $LiAlH_4$ in ether with 58% yield [68].

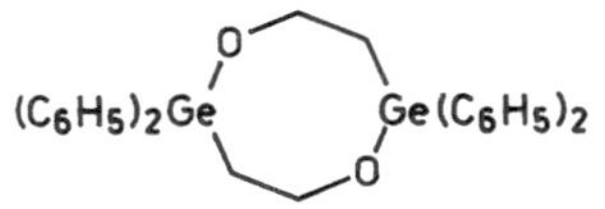

XII

Heating with CCl_4 in a sealed tube at 100°C (1 d) in the presence of AIBN gave $Ge(C_6H_5)_2(CH_2CH_2OH)Cl$ with 98% yield [68].

$Ge(C_6H_5)_2(CH_2CH_2CHO)H$ (Table **25**, No. **11**). A 68:32 mixture of No. 11 and $\overline{Ge(C_6H_5)_2CH_2CH_2CH_2O}$ boils at 140 to 145°C/0.05 Torr. Further heating of this mixture at 120°C (12 h) in the presence of AIBN increased the proportion of $\overline{Ge(C_6H_5)_2CH_2CH_2CH_2O}$ [47].

$Ge(C_6H_5)_2(C_3H_7\text{-}i)H$ (Table **25**, No. **14**) was prepared with 41% yield by adding $Ge(C_6H_5)_2Br_2$ in ether to i-C_3H_7MgCl in ether-C_6H_6, replacing the ether by $C_6H_5CH_3$, and refluxing for 7 h. It was also obtained from $Ge(C_6H_5)_2K_2$ and excess i-C_3H_7Cl in THF-HMPT at room temperature (30 min) [90].

$Ge(C_6H_5)_2(C_4H_9)H$ (Table **25**, No. **15**) was a minor product of the reaction of $Ge(C_6H_5)_2K_2$ with excess C_4H_9Br in THF-HMPA at room temperature (30 min) along with $Ge(C_4H_9)_2(C_6H_5)_2$ [90].

The compound was part of IR studies on the ν(GeH) frequency of organogermanium compounds and its dependence on the nature of the substituents [55].

$Ge(C_6H_5)_2(C_4H_9)D$ was prepared with 30% yield by deuterolysis of the reaction mixture obtained from $Ge(C_6H_5)_2H_2$ and LiC_4H_9. IR spectrum: ν(GeD) 1465(s) and δ(GeD) 525(s) cm^{-1} [18].

$Ge(C_6H_5)_2(CH_2CH_2CH(OH)CH_3)H$ (Table **25**, No. **16**) reacted with Raney nickel at 120°C to give $\overline{Ge(C_6H_5)_2CH_2CH_2CH(CH_3)O}$ with 64% yield [46].

$Ge(C_6H_5)_2(CH_2CH_2COCH_3)H$ (Table **25**, No. **17**). When the compound was heated at 100°C (or 120°C ?) for 16 h in the presence of AIBN, $\overline{Ge(C_6H_5)_2CH_2CH_2CH(CH_3)O}$ was the main product along with $(\text{-}Ge(C_6H_5)_2CH_2CH_2CH(CH_3)O\text{-})_n$. Thermal cyclization is strongly inhibited in the presence of $C_6H_4(OH)_2$-1,4 [35].

$Ge(C_6H_5)_2(CH{=}CHCH{=}CH_2)H$ (Table **25**, No. **18**) was obtained from a mixture of $\overline{Ge(C_6H_5)_2CH_2CH{=}CHCH(Br)}$ and $\overline{Ge(C_6H_5)_2CH_2CH(Br)CH{=}CH}$ by ring cleavage with $LiAlH_4$ in ether along with $\overline{Ge(C_6H_5)_2CH_2CH{=}CHCH_2}$ and $\overline{Ge(C_6H_5)_2CH_2CH_2CH{=}CH}$ [65].

$Ge(C_6H_5)_2(CH_2CH_2CH_2CH_2OH)H$ (Table **25**, No. **19**) was formed from $\overline{Ge(C_6H_5)_2\text{-}CH_2CH_2CH_2CH(OH)}$ and Na at 90°C (2 h) along with $\overline{Ge(C_6H_5)_2CH_2CH_2CH_2CH_2O}$; additional quantities of No. 19 were obtained by cleaving the latter cyclic product with $LiAlH_4$ in ether [65].

Heating with Raney nickel to 100°C (4 h) gave $Ge(C_6H_5)_2CH_2CH_2CH_2CH_2O$ [65].

$Ge(C_6H_5)_2(CH_2CH_2CH_2C(CH_3)OCH_2CH_2O)H$ (Table 25, No. **20**) reacted with excess $Sn(CH_3)_2(C_6H_5)N(C_2H_5)_2$ at 90°C (3 h) to give $Ge(C_6H_5)_2(R')Sn(CH_3)_2C_6H_5$ with 59% yield [59].

$Ge(C_6H_5)_2(CH_2CH_2CH_2CH_2COCH_3)H$ (Table **25**, No. **21**) When the compound was heated at 100°C for 16 h in the presence of AIBN, $Ge(C_6H_5)_2(CH_2)_4CH(CH_3)O$ was formed as the main product along with $(\text{-}Ge(C_6H_5)_2(CH_2)_4CH(CH_3)O\text{-})_n$ [35].

$Ge(C_6H_5)_2(C_6H_{13})H$ (Table **25**, No. **22**) was included in IR studies on the ν(GeH) behavior of various organogermanium compounds [55, 64].

$Ge(C_6H_5)_2(C_5H_4Re(P(C_6H_5)_3)(NO)H)H$ (Table **25**, No. **24**) was prepared with 46% yield by treating compound XIII with LiC_4H_9 in THF at −78°C (intermediate XIV), keeping the mixture at −15°C for 2 h (intermediate XV), and protonating with $H[BF_4] \cdot O(C_2H_5)_2$ at −78°C [115].

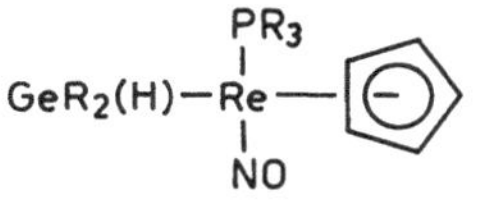

XIII

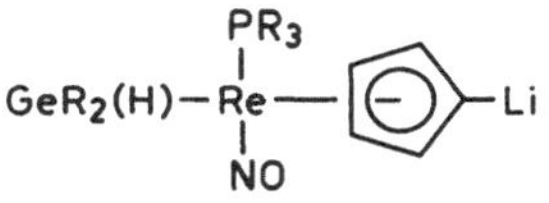

XIV

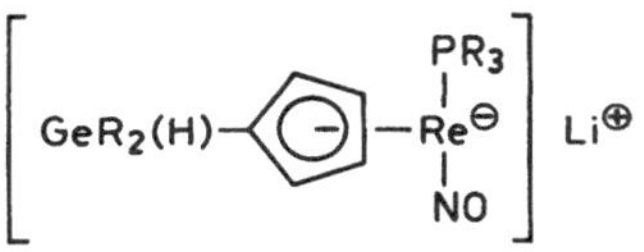

XV

$Ge(C_6H_5)_2(C_6H_4X)H$ (X = F-4, Cl-3, CN-4, NO_2-4, Table **25**, Nos. **25** to **28**). The tritiated derivatives of Nos. 25, 27, and 28 were also prepared: **$Ge(C_6H_5)_2(C_6H_4F\text{-}4)T$** by converting $Ge(C_6H_5)_2(C_6H_4F\text{-}4)H$ with LiC_4H_9 in ether into $Ge(C_6H_5)_2(C_6H_4F\text{-}4)Li$ followed by hydrolysis with tritiated water, and **$Ge(C_6H_5)_2(C_6H_4CN\text{-}4)T$** and **$Ge(C_6H_5)_2(C_6H_4NO_2\text{-}4)T$** by base-catalyzed tritiation of the corresponding $Ge(C_6H_5)_2(C_6H_4X)H$ with tritiated water in refluxing CH_3OH-DMSO containing CH_3ONa [67].

First-order rate constants for the loss of tritium from $Ge(C_6H_5)_2(C_6H_4X)T$ during T/H exchange with CH_3OH in the presence of CH_3ONa were measured at 30°C relative to the rate constant $k(Ge(C_6H_5)_3T) = 1$ in CH_3OH (0.421 M in CH_3ONa): k_{rel} (molarity of CH_3ONa) = 1.13 (0.420 M) for No. 25, 10 (0.0421 M) for No. 26, 680 (0.00490 M) for No. 27, and 1280 (0.00441 M) for No. 28. The results suggest a substancial delocalization of charge from the anionic Ge center into the aromatic rings of the $[Ge(C_6H_5)_2R']^-$ anion [67].

1.3.1.2.5 Other $GeR_2(R')H$ Compounds

Table 26 contains the remaining $GeR_2(R')H$ compounds in which R represents various substituted alkyl or aryl groups and a few heterocycles. The following ionic compound has been reported:

$[Ge(C_8H_7NS)_2(CH_3)H][CF_3SO_3]_2$ (C_8H_7NS = Formula XVI) was prepared with 71% yield by reacting $Ge(C_7H_4NS)_2(CH_3)H$ (Table 26, No. 23) with $CF_3SO_3CH_3$ (ca. 1:20 mole ratio) in c-$C_6H_{11}CH_3$ at 0°C (3 d) and then at room temperature (4 d). It was crystallized from CH_3NO_2-C_6H_6 [82].

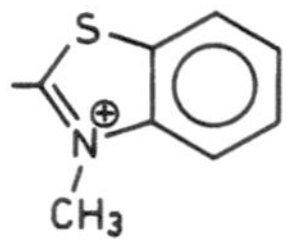

XVI

The salt melts at 155 to 158 °C with decomposition [82].

^1H NMR spectrum (in CD_3NO_2): δ(ppm) = 1.89 (d, CH_3Ge, J = 3.00 Hz), 4.70 (s, CH_3N), 7.02 (q, GeH), and 7.9 to 8.6 (m, aromatic H). IR spectrum (in CH_3NO_2): ν(GeH) 2183 cm^{-1} [82].

Table 26
Other $GeR_2(R')H$ Compounds.
An asterisk indicates further information at the end of the table.
Explanations, abbreviations, and units are given on p. X.

No.	R group R′ group	preparation (yield) properties and remarks
*1	CF_3 CH_3	preparation given on p. 184 ^1H NMR: 0.79 (CH_3), 5.08 (GeH), J(H, H) = 3.4, J(F, H) = 6.8 ^{19}F NMR spectrum: −54.4 ppm (CF_3), ^1J(C, F) = 331.6, ^4J(F, F) = 4.3 [92] IR and Raman on p. 185 MS: $[Ge(CH_3)(CF_3)(CF_2)H]^+$ (1), $[Ge(CH_3)(CF_3)F]^+$ (6), $[Ge(CH_3)(CF_3)H]^+$ (55), $[Ge(CH_3)F_2]^+$ (16), $[Ge(CH_3)(H)F]^+$ (100), $[GeF]^+$ (35), $[GeCH_3]^+$ (30), $[Ge]^+$ (4), $[CF_2H]^+$ (10) [92]
*2	CH_2OCH_3 C_6H_5	preparation given on p. 184 ^1H NMR (C_6D_6): 3.16 (s, CH_3O), 3.66 (d, CH_2Ge), 4.85 (quint, GeH, ^3J = 3), 7.00 to 7.80 (m, C_6H_5) IR (C_6D_6): ν(GeH) 2043 [124]
*3	$CH_2Si(CH_3)_3$ CH_3	$Ge(CH_2Si(CH_3)_3)_2(CH_3)Cl$ + $LiAlH_4$ in refluxing ether (69%) [86] b.p. 89 °C/30 n_D^{25} = 1.4627 ^1H NMR ($CDCl_3$): −0.16 (d, CH_2Ge, J = 4.0), 0 (s, CH_3Si), 0.22 (d, CH_3Ge, J = 4.0), 3.91 (m, GeH) IR (neat): ν(GeH) 2020, ν($GeCH_3$) and ν($SiCH_3$) 1250 [86]
4	$CH(Si(CH_3)_3)_2$ 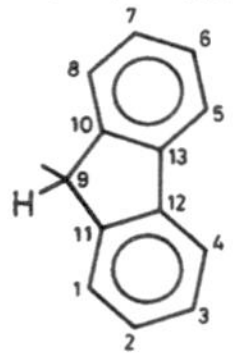	$Ge(CH(Si(CH_3)_3)_2)_2(CHC_{12}H_8)F$ (in ether) + LiC_4H_9-t in C_5H_{12} (1 : 2.7 mole ratio) at −78 °C to +20 °C followed by adding excess CH_3OH (85%) [114] white crystals; m.p. 118 to 119 °C (from C_5H_{12}) ^1H NMR ($CDCl_3$): −0.21 (d, GeCHSi, ^3J = 3.0), 0.01 (s, CH_3Si), 0.22 (s, CH_3Si), 4.52 (s, GeCH=), 4.75 (t, GeH, ^3J = 3.0), 7.18 to 7.55 (m, $C_{12}H_8$)

Table 26 (continued)

No.	R group R′ group	preparation (yield) properties and remarks
		^{13}C NMR ($CDCl_3$): 1.08 (GeCHSi), 2.95 and 3.40 (CH_3Si), 42.58 (C-9), 119.68 (C-4,5) 125.20, 125.59, 126.36 (C-1,2,3,6,7,8), 140.60 (C-12,13), 146.63 (C-10,11) IR (KBr): ν(GeH) 2076, 2068 MS: $[M - 1]^+$ (2), $[M - 15]^+$ (10), $[Ge(CH(Si(CH_3)_3)_2)_2H]^+$ (100), $[CHC_{12}H_8]^+$ (25) [114]
5	$CH(Si(CH_3)_3)_2$ 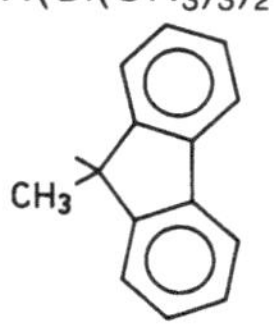	$Ge(CH(Si(CH_3)_3)_2)_2(CHC_{12}H_8)F$ in ether + LiC_4H_9-t in C_5H_{12} (1 : 2 mole ratio) at −78 °C to +20 °C followed by adding excess CH_3I (60%) [114] white crystals; m.p. 160 to 161 °C (from C_5H_{12}) ^1H NMR ($CDCl_3$): −0.1 and +0.14 (s's, CH_3Si), 1.79 (s, CH_3C), 4.96 (t, GeH, ^3J = 1.5), 7.30 to 7.84 (m, $C_{12}H_8$) ^{13}C NMR ($CDCl_3$): 3.51 and 3.72 (CH_3Si), 4.51 (GeCHSi), 25.48 (CH_3C), 47.94 (C-9), 120.24 (C-4,5), 124.13, 126.26, 126.74 (C-1,2, 3,6,7,8), 139.97, 151.63 (C-10,11,12,13) IR (KBr): ν(GeH) 2056 MS: $[M - 15]^+$ (2), $[Ge(CH(Si(CH_3)_3)_2)_2H]^+$ (35), $[C(CH_3)=C_{12}H_8]^+$ (59), $[Si(CH_3)_3]^+$ (100) [114]
*6	$CH_2C_6H_5$ CH_3	see further information on p. 184
*7	$CH_2C_6H_5$ C_2H_5	$Ge(CH_2C_6H_5)_2(H)Cl$ + excess C_2H_5MgBr in ether at 35 °C (2 h) followed by hydrolysis, extraction with ether-C_6H_6, and distillation (70%) [64]; similarly in refluxing C_6H_6-ether [72] b.p. 138 to 142 °C/0.05 [64, 72] ^1H NMR (C_6D_6): 0.60 to 1.00 (m, C_2H_5), 2.18 (d, CH_2), 4.18 (m, GeH) [64]; see also [72] IR (C_7H_{16} or CCl_4): ν(GeH) 2029 [64]
*8	$CH_2C_6H_5$ C_4H_9	formation given on p. 184 liquid IR (neat): ν(GeH) 2015 [39]
9	$CH_2C_6H_5$ C_7H_{15}	$Ge(C_7H_{15})H_3$ + $C_6H_5CH_2MgCl$ (1 : 2 mole ratio) in refluxing THF (15%) b.p. 180 °C/1 d^{20} = 1.079, n_D^{20} = 1.514 [5] IR: ν(GeH) 2014 [5]; 2015 (neat) [8]; included in studies of the ν(GeH) like No. 8 [8, 49, 50, 84]
10	$CH_2CH_2OC_4H_9$ C_7H_{15}	$Ge(C_7H_{15})H_3$ + CH_2=$CHOC_4H_9$ (1 : 3 mole ratio) at 130 to 140 °C (minor product) along with $Ge(C_7H_{15})(CH_2CH_2$-$OC_4H_9)_3$ and $Ge(C_7H_{15})(CH_2CH_2OC_4H_9)H_2$ not isolated; no data reported [5]

References on p. 187

Table 26 (continued)

No.	R group R′ group	preparation (yield) properties and remarks
11	$CH_2CH_2CH_2Cl$ C_2H_5	$Ge(CH_2CH_2CH_2Cl)_2(C_2H_5)Cl + LiAlH_4$ in ether (87%) ^1H NMR (CCl_4): 0.78 to 2.08 (m, C_2H_5), 3.47 (t, CH_2Cl), 3.80 (m, GeH) IR (C_7H_{16} or CCl_4): ν(GeH) 2023; integrated intensity of ν(GeH) studied [64]; cf. No. 8
*12	$CH_2C_4H_9$-t CH_3	$Ge(CH_2C_4H_9\text{-t})_2(CH_3)Cl + LiAlH_4$ in refluxing ether (94%) b.p. 83.0 to 84.0°C/19 n_D^{20} = 1.4420 ^1H NMR ($CDCl_3$): 0.25 (d, CH_3Ge, J = 4.0), 0.94 (d, CH_2Ge, J = 4.0), 0.97 (s, CH_3C), 4.02 (m, GeH) IR (neat): ν(GeH) 2040, ν($GeCH_3$) 1240 [86]
*13	C_6F_5 C_2H_5	$Ge(C_6F_5)_2(H)Br + C_2H_5MgBr$ in refluxing C_6H_6 for 10 h (24%) b.p. 120 to 124°C/5 d^{20} = 1.663; n_D^{20} = 1.4741 IR: ν(GeH) 2120; other bands at 1650, 1530, 1480, 1390, 1290, 1090, 980, 820, 625 (all C_6F_5), and 780 [48]
*14	$C_6H_2(CH_3)_3$-2,4,6 CH_3	preparation given on p. 185 white solid [110]; m.p. 90 to 94°C [98, 110] ^1H NMR (C_6D_6): 0.73 (d, CH_3Ge), 2.13 and 2.33 (CH_3-2,4,6), 5.49 (q, GeH), 6.69 (s, C_6H_2) [98]; 0.80 (d, CH_3Ge), 2.10 (s, CH_3-4), 2.30 (s, CH_3-2,6), 5.50 (q, GeH, ^3J = 4), 6.70 (s, C_6H_2) [110] IR (Nujol, KBr): ν(GeH) 2040 [98, 110]
*15	$C_6H_2(CH_3)_3$-2,4,6 $CH(OH)C_6H_5$	$Ge(C_6H_2(CH_3)_3\text{-2,4,6})_2(H)Li$ + excess C_6H_5CHO in THF-ether at −20°C for 30 min (62%) white powder; m.p. 78 to 80°C ^1H NMR (C_6D_6): 2.10 (s, CH_3-4), 2.23, 2.25 (s's, CH_3-2,6), 5.20 (d, GeCH), 5.30 (d, GeH, ^3J = 3), 6.60, 6.63 (s's, C_6H_2), 6.95 (s, C_6H_5) IR (KBr, Nujol): ν(OH) 3300, ν(GeH) 2060 [110]
*16	$C_6H_2(CH_3)_3$-2,4,6 $CH(OSi(CH_3)_3)C_6H_5$	preparation given on p. 185 b.p. 95°C/0.04 ^1H NMR: 0.05 (s, $SiCH_3$), 2.05 (s, CH_3-4), 2.30 (s, CH_3-2,6), 5.50 (s, GeH and GeCH), 6.65 (s, C_6H_2), 6.90 to 7.15 (m, C_6H_5) in C_6D_6; 5.20 (d, GeCH), 5.35 (d, GeH, ^3J = 3) in $CDCl_3$ IR (neat, KBr): ν(GeH) 2050 [110]
17	$C_6H_2(CH_3)_3$-2,4,6 COC_6H_5	$Ge(C_6H_2(CH_3)_3\text{-2,4,6})(H)Li$ + excess C_6H_5COCl in THF-ether at −78°C to +20°C (58%) [110]. yellow solid; m.p. 148 to 154°C ^1H NMR (C_6D_6): 2.10 (s, CH_3-4), 2.40 (s, CH_3-2,6), 6.50 (s, GeH), 6.70 (s, C_6H_2), 6.90 to 7.20, 7.90 to 8.20 (m's, C_6H_5) IR (Nujol, KBr): ν(GeH) 2048, ν(C=O) 1643 [110]

Table 26 (continued)

No.	R group R′ group	preparation (yield) properties and remarks
18	$C_6H_2(CH_3)_3$-2,4,6 $CH_2CH_2C_4H_9$-t	$Ge(C_6H_2(CH_3)_3$-2,4,6$)_2$=$CHCH_2C_4H_9$-t + $LiAlH_4$ (1:2 mole ratio) [125] waxy material, very difficult to crystallize ^1H NMR ($CDCl_3$): 0.87 (s, CH_3/R′), 1.28 (br s, CH_2CH_2), 2.26 (s, CH_3-4), 2.34 (s, CH_3-2,6), 5.25 (t, GeH, ^3J = 1.9), 6.81 (s, C_6H_2) ^{13}C NMR ($CDCl_3$): 13.34 (CH_2Ge), 21.04 (CH_3-4), 23.74 (CH_3-2,6), 28.92 (CH_3 of R′), 32.02 (γ-C of R′), 41.40 (β-CH_2 of R′), 128.60 (C-3,5), 134.25 (C-1), 138.11 (C-4), 143.46 (C-2,6) IR ($CDCl_3$): ν(GeH) 2062.5 MS: $[M]^+$ (1), $[M - R']^+$ (39), $[M - C_6H_3(CH_3)_3]^+$ (14), $[M - C_6H_3(CH_3)_3 - C_4H_9\text{-t}]^+$ (7), $[GeC_6H_2(CH_3)_3]^+$ (35), $[C_6H_2(CH_3)_3]^+$ (28), $[R']^+$ (7), $[C_4H_9\text{-t}]^+$ (80), $[CH(CH_3)_2]^+$ (100) [125]
*19	$C_6H_2(CH_3)_3$-2,4,6 $CHC_{12}H_8$ (fluorenyl); cf. No. 4	preparation given on p. 186 white crystals; m.p. 110 to 111 °C ^1H NMR (C_6D_6): 2.10 (s, CH_3-4), 2.20 (s, CH_3-2,6), 4.80 (d, GeCH, ^3J = 5.0), 5.20 (d, GeH, ^3J = 5.0), 6.73 (s, C_6H_2), 6.93 to 7.93 (m, $C_{12}H_8$) IR (Nujol): ν(GeH) 2095 and 2070 (the two bands probably due to Fermi resonance with a band at 1040) [96]
*20	$C_6H_2(C_3H_7\text{-i})_3$-2,4,6 CH_3	preparation given on p. 186 ^1H NMR (C_6D_6, 70 °C): 1.0 (d, $GeCH_3$, J = 4.1), 1.15, 1.16, 1.20 (d's, J = 6.8, 6.8, and 6.9, CH_3C), 2.78 (sept, CH/C_3H_7-4, J = 6.9), 3.47 (sept, CH/C_3H_7-2,6, J = 6.9), 5.78 (q, GeH, J = 4.1), 7.08 (s, C_6H_2) ^{13}C NMR (C_6D_6): 3.34 (CH_3Ge), 24.18, 24.64, 24.85 (CH_3/C_3H_7), 34.13 (CH/C_3H_7-2,6), 34.64 (CH/C_3H_7-4), 121.64 (C-3,5/C_6H_2), 134.95, 149.77, 154.1 (C-1,2,4,6/C_6H_2) IR (neat): ν(GeH) 2033(s) MS: $[M]^+$ (38), $[M - H]^+$ (100), and m/e = 479 (34), 289 (50), 275 (62), 231 (14), 189 (24), 91 (45) [123]
21	$CHC_{12}H_8$ $C_6H_2(CH_3)_3$-2,4,6	$LiAlH_4$ reduction of the chloride (impurity in $Ge(C_6H_2(CH_3)_3$-2,4,6$)(CHC_{12}H_8)Cl_2$) white crystals; m.p. 187 °C ^1H NMR ($CDCl_3$): 1.92 (CH_3-2,6), 2.17 (CH_3-4), 4.39 (t, HGe, $^3J(H,H)$ = 4.0), 4.76 (d, CH/$CHC_{12}H_8$, $^3J(H,H)$ = 4.0), 6.63 (s, C_6H_2), 7.13 to 7.89 (m, $C_{12}H_8$) [128]
*22	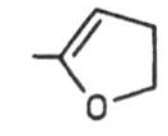 CH_3	see further information on p. 186

References on p. 187

Table 26 (continued)

No.	R group R' group	preparation (yield) properties and remarks
*23	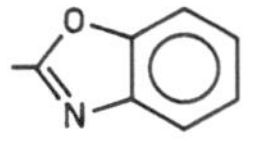 CH_3	preparation given on p. 186 m.p. 105°C ^1H NMR: 0.85 (d, CH_3Ge, J = 3.27), 5.48 (q, GeH) in C_6H_6; 7.1 to 7.9 (m, aromatic H) in CCl_4 IR: ν(GeH) 2120 [82]
*24	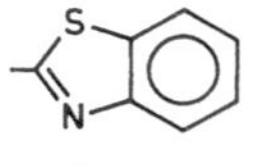 CH_3	preparation given on p. 187 m.p. 115°C ^1H NMR: 0.93 (d, CH_3Ge, J = 3.27), 5.72 (q, GeH) in C_6H_6; 7.3 to 8.3 (m, aromatic H) in CCl_4 IR: ν(GeH) 2090 [82]

*Further information:

$Ge(CF_3)_2(CH_3)H$ (Table **26**, No. **1**) was obtained with 85% yield by adding an aqueous solution of $NaBH_4$ to an ice-cooled solution of $Ge(CF_3)_2(CH_3)I$ in 30% H_3PO_4 (using a vacuum line; ca. 600 mbar). Condensation of CD_3I and $Zn(CH_3)_2$ onto $Ge(CF_3)_2H_2$ gave $Ge(CF_3)_2(CH_3)H$ and **$Ge(CF_3)_2(CD_3)H$** in a 68:32 ratio along with $Ge(CH_3)_2(CF_3)_2$, $Ge(CH_3)(CD_3)(CF_3)_2$, and $Ge(CD_3)_2(CF_3)_2$ [92]. The fundamental vibrations of the IR and Raman spectra are listed in Table 27.

$Ge(CH_2OCH_3)_2(C_6H_5)H$ (Table **26**, No. **2**) was prepared with 43% yield via $Ge(CH_2OCH_3)(C_6H_5)H_2$ and $Ge(CH_2OCH_3)(C_6H_5)(H)Li$ by successively treating $Ge(C_6H_5)H_2Li$ with CH_3OCH_2Cl, LiC_4H_9-t, and CH_3OCH_2Cl in THF at −78°C [124].

$Ge(CH_2Si(CH_3)_3)_2(CH_3)H$ (Table **26**, No. **3**). The reaction of photochemically generated t-$C_4H_9O^\bullet$ with $Ge(CH_2Si(CH_3)_3)_2(CH_3)H$ gave the $Ge^\bullet(CH_2Si(CH_3)_3)_2CH_3$ radical [86].

$Ge(CH_2C_6H_5)_2(CH_3)H$ (Table **26**, No. **6**) was probably a minor product along with $Ge(CH_3)_2(CH_2C_6H_5)_2$ and $Ge(CH_3)(CH_2C_6H_5)_3$, when $Ge(CH_2C_6H_5)_3H$ was treated with Li in DME at 0°C and subsequently with an excess of CH_3I. It was identified only by the strong ν(GeH) at 2034 cm^{-1} and by the fact that its volatility is slightly higher than that of $Ge(CH_3)_2(CH_2C_6H_5)_2$ [12].

$Ge(CH_2C_6H_5)_2(C_2H_5)H$ (Table **26**, No. **7**) was included in IR studies of the ν(GeH) integrated intensity of various organogermanium compounds and its correlation with the nature of the substituents [64].

The electronic absorption of the charge transfer complex formed between No. 7 and $(NC)_2C{=}C(CN)_2$ was used for calculating the σ_P value of the $CH_2Ge(C_2H_5)(CH_2C_6H_5)H$ substituent [72].

$Ge(CH_2C_6H_5)_2(C_4H_9)H$ (Table **26**, No. **8**). A mixture of $Ge(CH_2C_6H_5)_2(C_4H_9)H$ and $Ge(C_4H_9)_2(CH_2C_6H_5)H$ (identified by ^1H NMR and IR) was assumed to be formed (along with other products) by the reaction of $Ge(CH_2C_6H_5)_3H$ with LiC_4H_9 followed by CH_3I in ether [12].

No. 8 was included in IR studies of the ν(GeH) frequency of various organogermanium compounds, its correlation with the nature of the substituents [39, 49, 50], and behavior as an isolated GeH stretching frequency [84].

Table 27
Vibrational Spectra of $Ge(CF_3)_2(CH_3)H$ [92].
Wavenumbers in cm^{-1}.

IR gas	Raman liquid	assignment
3011(w)	3011(w)	$\nu_{as}(CH_3)$
2940(w)	2940(m,p)	$\nu_s(CH_3)$
2133(ms)	2133(s,p)	$\nu(GeH)$
1423(m)	1423(vw)	$\delta_{as}(CH_3)$
1267(w)	1267(w,p)	$\delta_s(CH_3)$
1199(vs)	1199(w,p)	$\nu_s(CF_3)$
1173(vvs)	1173(w)	$\nu_s(CF_3)$
1136(vvs)	1136(w,b)	$\nu_{as}(CF_3)$
1098(s)	1098(w,b)	$\nu_{as}(CF_3)$
858(ms)	858(w)	$\rho(CH_3)$
820(s)	820(vw)	$\rho(CH_3)$
727(m)	727(s,p)	$\delta_s(CF_3)$
684(s)	662(m)	$\rho(GeH)$
619(s)	619(m,p)	$\nu(GeCH_3)$
518(w)	518(w)	$\delta_{as}(CF_3)$
(510)(vw)		$\delta_{as}(CF_3)$
323(s)	323(w)	$\nu_{as}(GeC_2)$
308(m)	308(m,p)	$\rho(CF_3)$
276(w)	276(w,p)	$\rho(CF_3)$
	247(s,p)	$\nu_s(GeC_2)$
235(vw)	235(w)	$\rho(CF_3)$
200(vw)		$\rho(CF_3)$
	129(m)	$\delta(CGeCH_3)$
	78(wm)	$\delta(GeC_2)$

$Ge(CH_2C_4H_9\text{-}t)_2(CH_3)H$ (Table **26**, No. **12**). The reaction with photochemically generated $t\text{-}C_4H_9O^{\bullet}$ led to the $Ge^{\bullet}(CH_2C_4H_9\text{-}t)_2CH_3$ radical [86].

$Ge(C_6F_5)_2(C_2H_5)H$ (Table **26**, No. **13**) was formed with 17% yield among several products by treating $(Ge(C_6F_5)_2TlC_2H_5)_n$ with HCl in C_6H_6 [81].

$Ge(C_6H_2(CH_3)_3\text{-}2,4,6)_2(CH_3)H$ (Table **26**, No. **14**) was prepared by reacting $Ge(C_6H_2\text{-}(CH_3)_3\text{-}2,4,6)_2(H)Li$ with excess CH_3I [98] or $(CH_3)_2SO_4$ [110] in THF at 20 °C with 78 [98] or 95% yield [110], respectively, and from $[Li(12\text{-crown-}4)][Ge(C_6H_2(CH_3)_3\text{-}2,4,6)_2H]$ and excess CH_3I in C_7H_{16} at 20 °C with 98% yield [110].

$Ge(C_6H_2(CH_3)_3\text{-}2,4,6)_2(CH(OH)C_6H_5)H$ (Table **26**, No. **15**) decomposes partially in C_6D_6 above 100 °C to give $Ge(C_6H_2(CH_3)_3\text{-}2,4,6)_2H_2$ and C_6H_5CHO [110]. For the reaction with $Si(CH_3)_3Cl$, see No. 16.

$Ge(C_6H_2(CH_3)_3\text{-}2,4,6)_2(CH(OSi(CH_3)_3)C_6H_5)H$ (Table **26**, No. **16**) was prepared with 44% yield by treating No. 15 with $Si(CH_3)_3Cl$ and an excess of DBU (Formula XVII) in C_6H_6 at 20 °C for 15 h [110].

References on p. 187

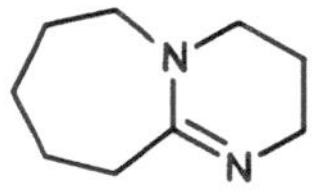

XVII

Ge(C$_6$H$_2$(CH$_3$)$_3$-2,4,6)$_2$(CHC$_{12}$H$_8$)H (Table **26**, No. **19**) was prepared by reduction of the germene Ge(C$_6$H$_2$(CH$_3$)$_3$-2,4,6)$_2$=CC$_{12}$H$_8$ [105] or its ether adduct [96] with LiAlH$_4$. Treating Ge(C$_6$H$_2$(CH$_3$)$_3$-2,4,6)$_2$(CHC$_{12}$H$_8$)F with LiC$_4$H$_9$-t in C$_5$H$_{12}$ gave an inseparable mixture of the starting compound, No. 19, and the germene above [105].

The compound crystallizes in the triclinic system with a = 10.405(5), b = 11.893(3), c = 12.091(5) Å, α = 94.00(4), β = 110.60(4), and γ = 110.13(4)°; space group $P\bar{1}-C_i^1$ and Z = 2. The molecular structure is shown in **Fig. 3**. Due to steric hindrance, the Ge-C(1) bond is unusually long [105].

Ge(C$_6$H$_2$(C$_3$H$_7$-i)$_3$-2,4,6)$_2$(CH$_3$)H (Table **26**, No. **20**) was prepared with 69% yield by reacting Ge(C$_6$H$_2$(C$_3$H$_7$-i)$_3$-2,4,6)$_2$H$_2$ with LiC$_4$H$_9$-t (1:1.4 mole ratio) in THF-C$_5$H$_{12}$ at −23°C (1 h), slowly adding the reaction mixture to an excess of (CH$_3$)$_2$SO$_4$ in THF at room temperature, stirring for 4 h, and hydrolysis with 6 M HCl [123].

Ge(C$_4$H$_5$O)$_2$(CH$_3$)H (Table **26**, No. **22**) was mentioned as an intermediate in the reaction between Ge(C$_4$H$_5$O)$_3$CH$_3$ and LiAlH$_4$ in THF or ether which finally led quantitatively to Ge(CH$_3$)H$_3$ [107]; see also [112].

Ge(C$_7$H$_4$NO)$_2$(CH$_3$)H (Table **26**, No. **23**) was obtained with 8% yield by treating Ge(CH$_3$)Cl$_3$ with Sn(CH$_3$)$_3$(C$_7$H$_4$NO) (ca. 1:1.8 mole ratio) in C$_6$H$_6$ at 0°C (5 h) followed by reduction of

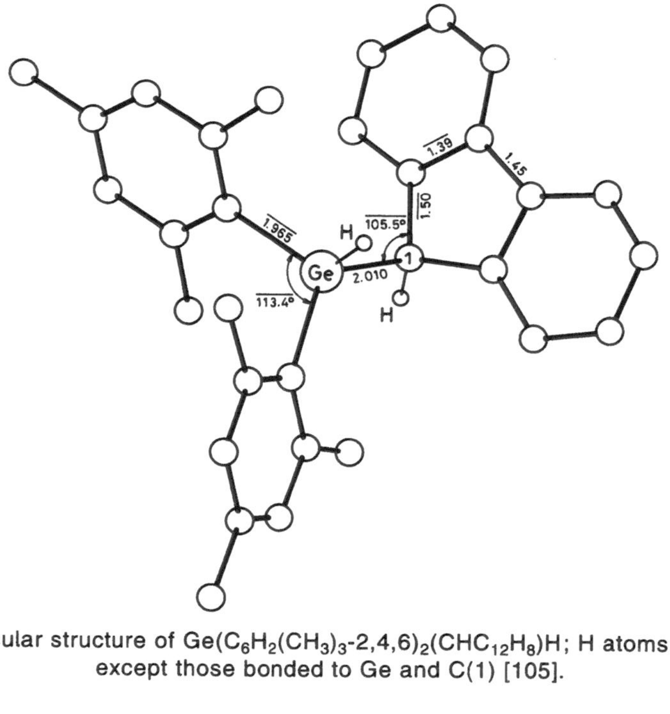

Fig. 3. Molecular structure of Ge(C$_6$H$_2$(CH$_3$)$_3$-2,4,6)$_2$(CHC$_{12}$H$_8$)H; H atoms are omitted except those bonded to Ge and C(1) [105].

the inseparable reaction mixture (containing $Ge(CH_3)(C_7H_4NO)_3$, $Ge(CH_3)(C_7H_4NO)_2Cl$, and $Ge(CH_3)(C_7H_4NO)Cl_2$ in a ca. 3:4:3 ratio) with $Sn(CH_3)_3H$ in C_6H_6 (3 h). It was purified by fractional crystallization from c-$C_6H_{11}CH_3$-C_6H_6 and recrystallization from CH_3NO_2 [82].

$Ge(C_7H_4NS)_2(CH_3)H$ (Table **26**, No. **24**) was prepared with 6% yield like No. 23 using $Ge(CH_3)Cl_3$ and $Sn(CH_3)_3(C_7H_4NS)$ (reaction mixture containing $Ge(CH_3)(C_7H_4NS)_3$, $Ge(CH_3)(C_7H_4NS)_2Cl$, and $Ge(CH_3)(C_7H_4NS)Cl_2$ in a ca. 1:4:1 ratio), reducing with $Sn(CH_3)_3H$, and purifying by fractional crystallization from c-$C_6H_{11}CH_3$-C_6H_6 and recrystallization from c-$C_6H_{11}CH_3$ [82].

The compound gradually decomposed in C_6H_6 solution to yield the germylene $(Ge(C_7H_4NS)CH_3)_n$ and compound XVIII. Equimolar amounts of $Ge(C_7H_4NS)_2(CH_3)H$ and $CF_3SO_3CH_3$ in a c-$C_6H_{11}CH_3$ suspension yielded within several days an unstable precipitate (probably XX), which immediately decomposed on addition of CD_3NO_2 to give $(Ge(C_7H_4NS)CH_3)_n$ and compound XIX. With a twentyfold amount of $CF_3SO_3CH_3$, the stable $[Ge(C_8H_7NS)_2(CH_3)H][CF_3SO_3]_2$ (see p. 179) was obtained under the same conditions [82].

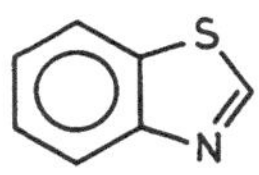

XVIII

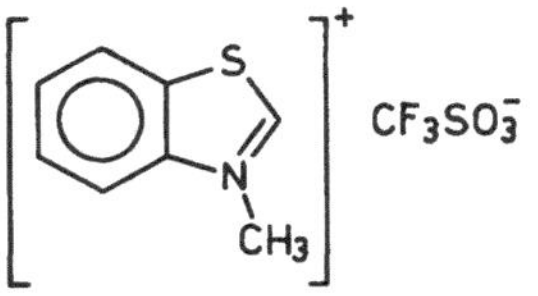

XIX

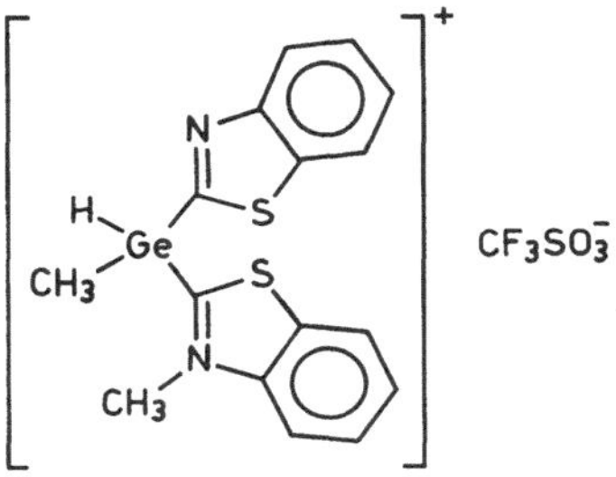

XX

References:

[1] Glarum, S. N.; Kraus, C. A. (J. Am. Chem. Soc. **72** [1950] 5398/401).
[2] Ponomarenko, V. A.; Vzenkova, G. Ia.; Egorov, Iu. P. (Dokl. Akad. Nauk SSSR **122** [1958] 405/8; Dokl. Chem. [Engl. Transl.] **118/123** [1958] 703/7).
[3] Mathis-Noël, R.; Mathis, F.; Satgé, J. (Bull. Soc. Chim. Fr. **1961** 676).
[4] Ponomarenko, V. A.; Zueva, G. Ya.; Andreev, N. S. (Izv. Akad. Nauk SSSR Ser. Khim. **1961** 1758/62; Bull. Acad. Sci. USSR Div. Chem. Sci. [Engl. Transl.] **1961** 1639/43).
[5] Satgé, J. (Ann. Chim. [Paris] [13] **6** [1961] 519/73).
[6] Kramer, K.; Wright, A. (Angew. Chem. **74** [1962] 468/9; Angew. Chem. Int. Ed. Engl. **1** [1962] 402/3).
[7] Mathis, R.; Mazerolles, P. (Bull. Soc. Chim. Fr. **1962** 1913/4).
[8] Mathis, R.; Satgé, J.; Mathis, F. (Spectrochim. Acta **18** [1962] 1463/72).
[9] Egorochkin, A. N.; Khidekel', M. L.; Ponomarenko, V. A.; Zueva, G. Ya.; Svirezheva, S. S.; Razuvaev, G. A. (Izv. Akad. Nauk SSSR Ser. Khim. **1963** 1865/8; Bull. Acad. Sci. USSR Div. Chem. Sci. [Engl. Transl.] **1963** 1717/9).
[10] Lesbre, M.; Satgé, J.; Massol, M. (C. R. Hebd. Seances Acad. Sci. **257** [1963] 2665/8).

[11] Baukov, Yu. I.; Lutsenko, I. F. (Zh. Obshch. Khim. **34** [1964] 3453/6; J. Gen. Chem. USSR [Engl. Transl.] **34** [1964] 3495/7).
[12] Cross, R. J.; Glockling, F. (J. Chem. Soc. **1964** 4125/33).

[13] Lesbre, M.; Satgé, J.; Massol, M. (C. R. Hebd. Seances Acad. Sci. **258** [1964] 2842/5).
[14] Mathis, R.; Constant, M.; Satgé, J.; Mathis, F. (Spectrochim. Acta **20** [1964] 515/21).
[15] Mironov, V. F.; Kravchenko, A. L. (Dokl. Akad. Nauk SSSR **158** [1964] 656/9; Dokl. Chem. [Engl. Transl.] **154/159** [1964] 949/52).
[16] Semlyen, J. A.; Walker, G. R.; Blofeld, R. E.; Phillips, C. S. G. (J. Chem. Soc. **1964** 4948/53).
[17] Baukov, Yu. I.; Belavin, I. Yu.; Lutsenko, I. F. (Zh. Obshch. Khim. **35** [1965] 1092/4; J. Gen. Chem. USSR [Engl. Transl.] **35** [1965] 1096/8).
[18] Cross, R. J.; Glockling, F. (J. Organomet. Chem. **3** [1965] 146/55).
[19] Semlyen, J. A.; Walker, G. R.; Phillips, C. S. G. (J. Chem. Soc. **1965** 1197/203).
[20] Mazerolles, P.; Dubac, J.; Lesbre, M. (J. Organomet. Chem. **5** [1966] 35/47).

[21] Satgé, J.; Rivière, P. (Bull. Soc. Chim. Fr. **1966** 1773/4).
[22] Mazerolles, P.; Dubac, J. (C. R. Seances Acad. Sci. C **265** [1967] 403/6).
[23] Mazerolles, P.; Dubac, J.; Lesbre, M. (Tetrahedron Lett. **1967** 255/8).
[24] Petrovskaya, L. I.; Belavin, I. Yu.; Burlachenko, G. S.; Fedin, É. I.; Baukov, Yu. I.; Lutsenko, I. F. (Zh. Strukt. Khim. **8** [1967] 168/9; J. Struct. Chem. [Engl. Transl.] **8** [1967] 141/2).
[25] Baukov, Yu. I.; Burlachenko, G. S.; Belavin, I. Yu.; Lutsenko, I. F. (Zh. Obshch. Khim. **38** [1968] 1899/900; J. Gen. Chem. USSR [Engl. Transl.] **38** [1968] 1846).
[26] Egorochkin, A. N.; Burov, A. I.; Mironov, V. F.; Gar, T. K.; Vyazankin, N. S. (Dokl. Akad. Nauk SSSR **180** [1968] 861/4; Dokl. Chem. [Engl. Transl.] **178/183** [1968] 500/3).
[27] Mazerolles, P.; Lesbre, M.; Lavergne, J.-P. (C. R. Seances Acad. Sci. C **266** [1968] 639/41).
[28] Mazerolles, P.; Manuel, G. (C. R. Seances Acad. Sci. C **267** [1968] 1158/61).
[29] Manuel, G.; Mazerolles, P. (J. Organomet. Chem. **19** [1969] 43/51).
[30] Massol, M.; Satgé, J.; Barrau, J. (C. R. Seances Acad. Sci. C **268** [1969] 1710/3).

[31] Mazerolles, P.; Lesbre, M.; Joanny, M. (J. Organomet. Chem. **16** [1969] 227/33).
[32] Mironov, V. F.; Gar, T. K.; Mikhailyants, S. A. (Dokl. Akad. Nauk SSSR **188** [1969] 120/3; Dokl. Chem. [Engl. Transl.] **184/189** [1969] 718/21).
[33] Mironov, V. F.; Mikhailyants, S. A.; Gar, T. K. (Zh. Obshch. Khim. **39** [1969] 397/400; J. Gen. Chem. USSR [Engl. Transl.] **39** [1969] 375/7).
[34] Mironov, V. F.; Mikhailyants, S. A.; Gar, T. K. (Zh. Obshch. Khim. **39** [1969] 2281/4; J. Gen. Chem. USSR [Engl. Transl.] **39** [1969] 2223/5).
[35] Satgé, J.; Rivière, P. (J. Organomet. Chem. **16** [1969] 71/82).
[36] Massol, M.; Satgé, J.; Rivière, P.; Barrau, J. (J. Organomet. Chem. **22** [1970] 599/610).
[37] Massol, M.; Barrau, J.; Satgé, J. (J. Organomet. Chem. **25** [1970] 81/90).
[38] Massol, M.; Barrau, J.; Satgé, J. (J. Heterocycl. Chem. **7** [1970] 783/90).
[39] Mathis, R.; Barthelat, M.; Mathis, F. (Spectrochim. Acta A **26** [1970] 1993/2000).
[40] Rivière, P. (Diss. Toulouse 1971; a. from [46], b. from [55]).

[41] Massol, M.; Mesnard, D.; Barrau, J.; Satgé, J. (C. R. Seances Acad. Sci. C **272** [1971] 2081/4).
[42] Rivière, P.; Satgé, J. (C. R. Seances Acad. Sci. C **272** [1971] 413/6).
[43] Rivière, P.; Satgé, J. (Synth. Inorg. Met.-Org. Chem. **1** [1971] 13/20).
[44] Quellette, R. J. (J. Am. Chem. Soc. **94** [1972] 7674/9).
[45] Barrau, J.; Satgé, J.; Massol, M. (Helv. Chim. Acta **56** [1973] 1638/46).
[46] Manuel, G.; Mazerolles, P.; Lesbre, M.; Pradel, J.-P. (J. Organomet. Chem. **61** [1973] 147/65).

[47] Rivière, P.; Satgé, J. (J. Organomet. Chem. **49** [1973] 173/89).
[48] Bochkarev, M. N.; Maiorova, L. P.; Korneva, S. P.; Bochkarev, L. N.; Vyazankin, N. S. (J. Organomet. Chem. **73** [1974] 229/36).
[49] Egorochkin, A. N.; Khorshev, S. Ya.; Ostasheva, N. S.; Satgé, J.; Rivière, P.; Barrau, J.; Massol, M. (J. Organomet. Chem. **76** [1974] 29/36).
[50] Egorochkin, A. N.; Khorshev, S. Ya.; Ostasheva, N. S.; Satgé, J.; Rivière, P.; Barrau, J.; Massol, M. (Dokl. Akad. Nauk SSSR **215** [1974] 858/60; Dokl. Chem. [Engl. Transl.] **214/219** [1974] 204/6).

[51] Massol, M.; Barrau, J.; Satgé, J.; Bouyssieres, B. (J. Organomet. Chem. **80** [1974] 47/69).
[52] Rice, L. M.; Wheeler, J. W.; Geschickter, C. F. (J. Heterocycl. Chem. **11** [1974] 1041/7).
[53] Rivière, P.; Satgé, J.; Soula, D. (J. Organomet. Chem. **72** [1974] 329/38).
[54] Sakurai, H.; Nozue, I.; Hosomi, A. (J. Organomet. Chem. **80** [1974] 71/8).
[55] Egorochkin, A. N.; Khorshev, S. Ya.; Satgé, J.; Rivière, P.; Barrau, J. (J. Organomet. Chem. **99** [1975] 239/49).
[56] Sakurai, H.; Mochida, K.; Kira, M. (J. Am. Chem. Soc. **97** [1975] 929/31).
[57] Egorochkin, A. N.; Khorshev, S. Ya.; Ostasheva, N. S.; Sevastyanova, E. I.; Satgé, J.; Rivière, P.; Barrau, J. (J. Organomet. Chem. **105** [1976] 311/20).
[58] Couret, C.; Escudié, J.; Satgé, J.; Redoulès, G. (Synth. React. Inorg. Met.-Org. Chem. **7** [1977] 99/110).
[59] Gielen, M.; Simon, S. (Bull. Soc. Chim. Belg. **86** [1977] 589/94).
[60] Sakurai, H.; Mochida, K.; Kira, M. (J. Organomet. Chem. **124** [1977] 235/52).

[61] Barrau, J.; Satgé, J. (J. Organomet. Chem. **148** [1978] C9/C12).
[62] Castel, A.; Rivière, P.; Satgé, J.; Cazes, A. (C. R. Seances Acad. Sci. C **287** [1978] 205/8).
[63] Duffaut, N.; Dunogues, J.; Calas, R.; Rivière, P.; Satgé, J.; Cazes, A. (J. Organomet. Chem. **149** [1978] 57/63).
[64] Egorochkin, A. N.; Sevast'yanova, E. I.; Khorshev, S. Ya.; Ratushnaya, S. Kh.; Satgé, J.; Rivière, P.; Barrau, J.; Richelme, S. (J. Organomet. Chem. **162** [1978] 25/35).
[65] Manuel, G.; Bertrand, G.; Mazerolles, P. (J. Organomet. Chem. **146** [1978] 7/16).
[66] Sakurai, H; Mochida, K. (J. Organomet. Chem. **154** [1978] 353/68).
[67] Eaborn, C.; Singh, B. (J. Organomet. Chem. **177** [1979] 333/48).
[68] Rivière-Baudet, M.; Rivière, P.; Satgé, J.; Lacrampe, G. (Recl. Trav. Chim. Pays-Bas **98** [1979] 42/52).
[69] Barrau, J.; Bouchaut, M.; Lavayssière, H.; Dousse, G.; Satgé, J. (Synth. React. Inorg. Met.-Org. Chem. **10** [1980] 515/29).
[70] Eaborn, C.; Siew, N. P. Y. (J. Organomet. Chem. **202** [1980] 157/62).

[71] Egorochkin, A. N.; Sevastyanova, E. I.; Khorshev, S. Ya.; Ratushnaya, S. Kh.; Richelme, S.; Satgé, J.; Rivière, P. (J. Organomet. Chem. **188** [1980] 73/83).
[72] Sennikov, P. G.; Skobeleva, S. E.; Kuznetsov, V. A.; Egorochkin, A. N.; Rivière, P.; Satgé, J.; Richelme, S. (J. Organomet. Chem. **201** [1980] 213/9).
[73] Soderquist, J. A.; Hassner, A. (J. Org. Chem. **45** [1980] 541/3).
[74] Mochida, K.; Kira, M.; Sakurai, H. (Chem. Lett. **1981** 645/8).
[75] Rivière, P.; Castel, A.; Satgé, J.; Cazes, A. (Synth. React. Inorg. Met.-Org. Chem. **11** [1981] 443/53).
[76] Eaborn, C.; Mahmoud, F. M. S.; Taylor, R. (J. Chem. Soc. Perkin Trans. II **1982** 1313/9).
[77] Escudié, J.; Couret, C.; Satgé, J.; Andriamizaka, J. D. (Organometallics **1** [1982] 1261/5).

[78] McKean, D. C.; Torto, I.; Morrisson, A. R. (J. Organomet. Chem. **226** [1982] C47/C51).
[79] Mochida, K.; Asami, K. (J. Organomet. Chem. **232** [1982] 13/9).
[80] Rivière, P.; Castel, A.; Satgé, J. (J. Organomet. Chem. **232** [1982] 123/35).

[81] Bochkarev, M. N.; Basalgina, T. A.; Kalinina, G. S.; Razuvaev, G. A. (J. Organomet. Chem. **243** [1983] 405/10).
[82] Jutzi, P.; Gilge, U. (J. Organomet. Chem. **244** [1983] 355/62).
[83] McKean, D. C.; Torto, I.; Mackenzie, M. W.; Morrisson, A. R. (Spectrochim. Acta A **39** [1983] 387/98).
[84] McKean, D. C.; Torto, I.; Mackenzie, M. W. (Spectrochim. Acta A **39** [1983] 399/408).
[85] Mochida, K.; Miyagawa, I. (Bull. Chem. Soc. Jpn. **56** [1983] 1875/6).
[86] Mochida, K. (Bull. Chem. Soc. Jpn. **57** [1984] 796/801).
[87] Burgess, K.; Guerin, C.; Johnson, B. F. G.; Lewis, J. (J. Organomet. Chem. **295** [1985] C3/C6).
[88] Chernysheva, O. N.; Gar, T. K.; Kisin, A. V.; Mironov, V. F. (Zh. Obshch. Khim. **55** [1985] 2333/8; J. Gen. Chem. USSR [Engl. Transl.] **55** [1985] 2073/7).
[89] Gar, T. K.; Chernysheva, O. N.; Kisin, A. V.; Mironov, V. F. (Zh. Obshch. Khim. **55** [1985] 1057/63; J. Gen. Chem. USSR [Engl. Transl.] **55** [1985] 942/7).
[90] Mochida, K.; Matsushige, N.; Hamashima, M. (Bull. Chem. Soc. Jpn. **58** [1985] 1443/7).

[91] Salimgareeva, I. M.; Bogatova, N. G.; Panasenko, A. A.; Monakov, Yu. B.; Yur'ev, V. P. (Izv. Akad. Nauk SSSR Ser. Khim. **1985** 1391/4 ; Bull. Acad. Sci. USSR Div. Chem. Sci. [Engl. Transl.] **1985** 1272/5).
[92] Eujen, R.; Mellies, R.; Petrauskas, E. (J. Organomet. Chem. **299** [1986] 29/40).
[93] Kobayashi, Mi.; Kobayashi, Ma. (Bull. Chem. Soc. Jpn. **59** [1986] 2807/10).
[94] Kobayashi, Mi.; Yoshida, M.; Kobayashi, Ma. (Bull. Chem. Soc. Jpn. **59** [1986] 3169/73).
[95] Bravo-Zhivotovskii, D. A.; Pigarev, S. D.; Vyazankina, O. A.; Vyazankin, N. S. (Zh. Obshch. Khim. **57** [1987] 2735/8; J. Gen. Chem. USSR [Engl. Transl.] **57** [1987] 2440/2).
[96] Couret, C.; Escudié, J.; Satgé, J.; Lazraq, M. (J. Am. Chem. Soc. **109** [1987] 4411/2).
[97] Lukevits, É.; Ignatovich, L. M.; Yuskovets, Zh. G.; Golender, L. O.; Shimanskaya, M. V. (Zh. Obshch. Khim. **57** [1987] 1294/9; J. Gen. Chem. USSR [Engl. Transl.] **57** [1987] 1158/63).
[98] Castel, A.; Rivière, P.; Satgé, J.; Ko, Y.-H. (J. Organomet. Chem. **342** [1988] C1/C4).
[99] Lukevits, É.; Strukovich, R. Ya.; Pudova, O. A. (Zh. Obshch. Khim. **58** [1988] 815/8; J. Gen. Chem. USSR [Engl. Transl.] **58** [1988] 721/3).
[100] Mochida, K.; Kikkawa, H.; Nakadaira, Y. (Chem. Lett. **1988** 1089/92).

[101] Shitara, K.; Sato, Y. (J. Organomet. Chem. **346** [1988] 1/6).
[102] Tacke, R.; Becker, B. (J. Organomet. Chem. **354** [1988] 147/53).
[103] Tacke, R.; Becker, B. (unpublished results from [102]).
[104] Bravo-Zhivotovskii, D. A.; Pigarev, S. D.; Voronkov, M. G.; Vyazankin, N. S. (Zh. Obshch. Khim. **59** [1989] 863/5; J. Gen. Chem. USSR [Engl. Transl.] **59** [1989] 761/3).
[105] Lazraq, M.; Escudié, J.; Couret, C.; Satgé, J.; Dräger, M.; Dammel, R. (Angew. Chem. **100** [1988] 885/7; Angew. Chem. Int. Ed. Engl. **27** [1988] 828).
[106] Gevorgyan, V.; Borisova, L.; Lukevics, É. (J. Organomet. Chem. **368** [1989] 19/21).
[107] Lukevics, É.; Gevorgyan, V. (Chem. Technol. Silicon Tin Proc. Asian Network Anal. Inorg. Chem. 1st Int. Chem. Conf. Silicon Tin, Kuala Lumpur 1989 [1992], pp. 165/77).

[108] Pigarev, S. D.; Bravo-Zhivotovskii, D. A.; Kalikhman, I. D.; Vyazankin, N. S.; Voronkov, M. G. (J. Organomet. Chem. **369** [1989] 29/41).
[109] Barrau, J.; Ben Hamida, N.; Satgé, J. (J. Organomet. Chem. **387** [1990] 65/76).
[110] Castel, A.; Rivière, P.; Satgé, J.; Ko, H. Y. (Organometallics **9** [1990] 205/10).

[111] Egorochkin, A. N.; Skobeleva, S. E.; Tsvetkova, V. L. (Metalloorg. Khim. **3** [1990] 656/61; Organometal. Chem. [Engl. Transl.] **3** [1990] 334/7).
[112] Gevorgyan, V.; Borisova, L.; Lukevics, É. (J. Organomet. Chem. **393** [1990] 57/67).
[113] Sommese, A. G.; Cremer, S. E.; Campbell, J. A.; Thompson, M. R. (Organometallics **9** [1990] 1784/92).
[114] Lazraq, M.; Couret, C.; Escudié, J.; Satgé, J.; Soufiaoui, M. (Polyhedron **10** [1991] 1153/61).
[115] Lee, K. E.; Arif, A. M.; Gladysz, J. A. (Organometallics **10** [1991] 751/60).
[116] Mochida, K.; Wakasa, M.; Sakaguchi, Y.; Hayashi, H. (Bull. Chem. Soc. Jpn. **64** [1991] 1889/95).
[117] Mochida, K.; Kikkawa, H.; Nakadaira, Y. (Bull. Chem. Soc. Jpn. **64** [1991] 2772/7).
[118] Mochida, K.; Kikkawa, H.; Nakadaira, Y. (J. Organomet. Chem. **412** [1991] 9/19).
[119] Okano, M.; Mochida, K. (Bull. Chem. Soc. Jpn. **64** [1991] 1381/2).
[120] Tacke, R.; Becker, B. (unpublished results); Becker, B. (Diss. TU Braunschweig 1989 from Tacke, R.; Wiesenberger, F.; Z. Naturforsch. **46b** [1991] 275/9).

[121] Terunuma, D.; Kizaki, H.; Sato, T.; Masuo, K.; Nohira, H. (Chem. Lett. **1991** 97/100).
[122] Tokitoh, N.; Takahashi, M.; Matsumoto, T.; Suzuki, H.; Matsuhashi, Y.; Okazaki, R. (Phosphorus Sulfur Silicon Relat. Elem. **59** [1991] 455/8).
[123] Baines, K. M.; Groh, R. J.; Joseph, B.; Parshotam, U. R. (Organometallics **11** [1992] 2176/80).
[124] Castel, A.; Rivière, P.; Satgé, J.; Desor, D. (J. Organomet. Chem. **433** [1992] 49/61).
[125] Couret, C.; Escudié, J.; Delpon-Lacaze, G.; Satgé, J. (Organometallics **11** [1992] 3176/7).
[126] Aoyagi, S.; Tanaka, K.; Zicmane, I.; Takeuchi, Y. (J. Chem. Soc. Perkin Trans. II **1992** 2217/20).
[127] Okano, M.; Mochida, K. (Denki Kagaku Oyobi Kogyo Butsuri Kagaku **61** [1993] 772/3).
[128] Chaubon-Deredempt, M. A.; Escudié, J.; Couret, C. (J. Organomet. Chem. **467** [1994] 37/46).
[129] Finholt, A. E.; Bond, A. C.; Wilzbach, K. E.; Schlesinger, H. I. (J. Am. Chem. Soc. **69** [1947] 2692/6).
[130] Anderson, H. H. (J. Am. Chem. Soc. **79** [1957] 326/8).

[131] Clark, K. B.; Griller, D. (Organometallics **10** [1991] 746/50).

1.3.1.3 Ge(R)(R′)(R″)H Compounds

The compounds in this section listed in Table 28 are arranged in the order of the types $Ge(CH_3)(C_2H_5)(R'')H$ (Nos. 1 and 2), $Ge(CH_3)(R')(C_6H_5)H$ (R′ = substituted alkyl, Nos. 3 to 9), and $Ge(R)(C_6H_5)(C_{10}H_7\text{-}1)H$ (Nos. 10 to 12). The compound $Ge(CH_3)(CD_3)(CF_3)H$ is described together with $Ge(CH_3)_2(CF_3)H$ in Section 1.3.1.2.1, p. 147.

Compounds No. 1 and 2 were part of an IR study on the integrated intensity of the ν(GeH) frequency and its dependence on the nature of Ge substituents of various organogermanium compounds [37].

Table 28
Ge(R)(R′)(R″)H Compounds.
An asterisk indicates further information at the end of the table.
Explanations, abbreviations, and units are given on p. X.

No.	R R′ groups R″ group	preparation (yield) properties and remarks
1	CH_3 C_2H_5 $CH_2CH_2CH_2Cl$	$Ge(CH_3)(C_2H_5)(CH_2CH_2CH_2Cl)Cl$ + $LiAlH_4$ in ether (85%) b.p. 62°C/12 ^{1}H NMR (C_6H_6): 0.01 (d, CH_3), 0.40 to 0.75 (m, C_2H_5), 3.13 (t, CH_2Cl), 3.82 (m, GeH) IR (C_7H_{16} or CCl_4): ν(GeH) 2021 [37]
2	CH_3 C_2H_5 C_6H_5	$Ge(CH_3)(C_6H_5)(H)Cl$ + excess C_2H_5MgBr in ether (78%) b.p. 110°C/55 n_D^{20} = 1.5180 ^{1}H NMR (CCl_4): 4.47 (sext, GeH, ^3J = 3.0) IR (C_7H_{16} or CCl_4): ν(GeH) 2038 [37]
*3	CH_3 C_6H_5 CH_2Cl	No. 4 + $Sn(C_4H_9)_3H$ under UV irradiation for 15 min (68%) b.p. 65 to 70°C/0.2 ^{1}H NMR: 0.48 (d, CH_3), 3.0 (d, CH_2Ge, ^3J = 2.5), 4.72 (m, GeH) IR (C_7H_{16}): ν(GeH) 2066, 2044 [39]; see further information on p. 194
4	CH_3 C_6H_5 $CHCl_2$	$Ge(CH_3)(C_6H_5)(Cl)H$ in ether slowly added to $LiCHCl_2$ in THF at −90°C, then stirring 2 h at −90°C (42%) b.p. 60 to 66°C/0.02 ^{1}H NMR (C_6D_6): 0.48 (d, CH_3), 4.80 (m, GeH), 5.24 (d, CHGe, ^3J = 2.5) starting material for the preparation of No. 3 [39]
5	CH_3 C_6H_5 CH_2OH	$Ge(CH_3)(C_6H_5)(CH_2OOCCH_3)OOCCH_3$ + $LiAlH_4$ in refluxing ether (3 h) followed by aqueous HCl (82%) colorless liquid, b.p. 60°C/0.02 ^{1}H NMR (C_6D_6): 0.22 (d, CH_3Ge, J = 3.5), 1.5 (br s, position not constant, OH), 3.46, 3.50 (A and B of an ABX system, $GeCH_2O$, J(A, B) = 12.0, J(A, X) = 2.5, J(B, X) = 2.4), 4.47 (m, X, GeH), 7.0 to 7.3 (m, C_6H_5) ^{13}C NMR (C_6D_6): −7.7 (CH_3Ge), 54.4 ($GeCH_2$); 128.4 (C-3,5), 129.0 (C-4), 134.6 (C-2,6), 137.2 (C-1), all of C_6H_5 IR (CCl_4): ν(OH) 3500, 3320 (non-associated and associated), ν(GeH) 2030 MS: $[M - CH_3]^+$ (10), $[M - CH_2OH]^+$ (100), $[GeC_6H_5]^+$ (28), $[GeCH_3]^+$ (30) starting material for the preparation of No. 6 [52]
6	CH_3 C_6H_5 CH_2OOCCH_3	No. 5 in C_5H_{12} + $(CH_3CO)_2O$ in ether in the presence of $N(C_2H_5)_3$, 6 h at 60°C (89%) colorless liquid with sweet floral odor b.p. 60°C/0.01

Table 28 (continued)

No.	R R′ groups R″ group	preparation (yield) properties and remarks
		^{1}H NMR (C_6D_6): 0.38 (d, CH_3Ge, J = 3.4), 1.60 (s, CH_3CO), 4.13, 4.17 (A and B of an ABX system, CH_2Ge, J(A, B) = 12.3, J(A, X) = 2.7, J(B, X) = 2.3), 4.70 (m, X, GeH), 7.1 to 7.4 (m, C_6H_5) ^{13}C NMR (C_6D_6): −6.8 (CH_3Ge), 20.1 (CH_3 of CH_3CO), 56.2 (CH_2Ge); 128.5 (C-3,5), 129.2 (C-4), 134.5 (C-2,6), 136.4 (C-1), all of C_6H_5; 170.5 (C=O) IR (neat): ν(GeH) 2040, ν(C=O) 1740 MS: $[M - H]^+$ (14), $[M - CH_3]^+$ (34), 165 (100), $[GeC_6H_5]^+$ (62), $[COCH_3]^+$ (51) thermally stable in C_6D_6 at 160°C over 2 d [52]
7	CH_3 C_6H_5 $CH_2C_6H_5$	$Ge(CH_3)(C_6H_5)(CH_2C_6H_5)CH_2NHCH_2C_6H_5$ + LiC_4H_9 in THF followed by quenching with H_2O [53]
8	CH_3 C_6H_5 CH_2CH_2OH	$Ge(CH_3)C_6H_5 \cdot N(C_2H_5)_3$ (obtained from $Ge(CH_3)(C_6H_5)(H)Cl$ and $N(C_2H_5)_3$ in C_5H_{12}) + $\overline{CH_2CH_2O}$ at 150°C (15 h) in a sealed tube, followed by $LiAlH_4$ reduction of the mixture containing 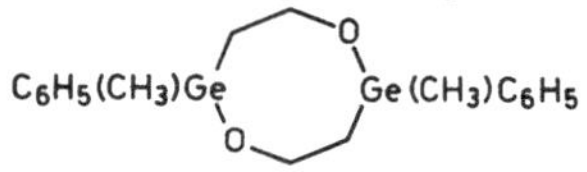 (47%) b.p. 90°C/0.04 ^{1}H NMR (C_6D_6): 0.52 (CH_3), 3.55 (m, CH_2O), 4.10 (m, HGe) [47]
9	CH_3 C_6H_5 $CH_2CH(CH_3)CH_2$- NHC_6H_5	$LiAlH_4$ reduction of 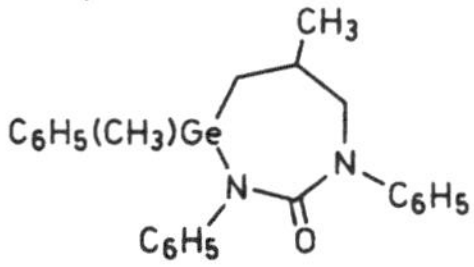 in refluxing THF-ether (3 h) (28%) b.p. 130 to 132°C/0.2 ^{1}H NMR(C_6D_6): 0.36 (d, CH_3Ge, ^3J = 3.5), 0.84 (d, CH_3C, ^3J = 6.5), 0.98 (m, CH_2Ge), 1.73 (m, CH), 4.27 (d, CH_2N, J = 7), 4.70 (sext, HGe) IR: ν(NH) 3380, ν(GeH) 2020 [38]
*10	CH_3 C_6H_5 $C_{10}H_7$-1	*racemate:* m.p. 50 to 51°C, b.p. 155°C/0.5 [2] MS: $[M]^+$, $[Ge(C_{10}H_7)C_6H_5]^+$, $[Ge(CH_3)(C_{10}H_7)H]^+$, $[Ge(CH_3)C_{10}H_7]^+$, $[GeC_{10}H_7]^+$, $[Ge(CH_3)C_6H_5]^+$, $[GeC_6H_5]^+$ (base peak), and m/e = 99, 89, 75, 74 [36] *(+)-isomer* (R configuration [2]): m.p. 74 to 75°C (from C_2H_5OH) $[\alpha]_D^{25}$ = +26.7° (c 10.6 in C_6H_{12}) [2]; racemizes in the presence of $LiAlH_4$ [5] *(−)-isomer* (S configuration [2]): m.p. 74 to 75°C $[\alpha]_D^{25}$ = −25.5° (c 11.3 in C_6H_{12}) [2]

References on p. 199

Table 28 (continued)

No.	R R′ groups R″ group	preparation (yield) properties and remarks
*11	C_2H_5 C_6H_5 $C_{10}H_7$-1	*racemate:* oil, b.p. 140 to 141 °C/0.01 n_D^{25} = 1.6270 [6] *(+)-isomer* (R configuration [12]): crystals [6]; m.p. 31 to 32.5 °C (from C_2H_5OH) [6], 32 to 33 °C [23] $[\alpha]_D^{25}$ = +23.6° (c 5.5 in C_6H_6) [6]; possibly optically not pure [23] fluoresces strongly in UV light [6] racemizes readily (without significant decomposition) when heated to 100 °C [11]; see also [22]; racemizes in the presence of $LiAlH_4$ [24] *(−)-isomer* (S configuration [12]): oil n_D^{25} = 1.6226 $[\alpha]_D^{20}$ = −10.9° (c 13.3 in C_6H_6); optically impure [6]
*12	C_3H_7-i C_6H_5 $C_{10}H_7$-1	*racemate:* b.p. 149 to 152 °C/0.10 to 0.15 [16] ^{1}H NMR: 1.18, 1.28 (d's, CH_3, J = 6.7), 1.85 (m, GeCH), 5.22 (d, HGe, J = 3.0), 7.1 to 8.0 (m, C_6H_5 and $C_{10}H_7$) [16] IR: ν(GeH) 2005; other bands at 1500 ($C_{10}H_7$) and 1380, 1362 (C_3H_7) [16] *(+)-isomer* (R configuration [16]): oil $[\alpha]_D^{25}$ = +1.6° (in C_6H_6) [16] or +1.71° [31] *(−)-isomer:* $[\alpha]_D^{25}$ = −0.52° (in C_6H_{12}) [30], maximum $[\alpha]_D$ = −1.8° [31] IR: ν(GeH) 2005 [30]

*Further information:

Explanations. In the following text “Reduction” refers to reactions of Ge(R)(R′)(R″)X compounds with $LiAlH_4$. The abbreviations “ret.” and “inv.” indicate retention and inversion of the configuration at the Ge center. The formula $OC_{10}H_{19}$ stands for the menthoxy group.

$Ge(CH_3)(C_6H_5)(CH_2Cl)H$ (Table **28**, No. **3**). The ν(GeH) doublet band was explained by intramolecular coordination between the Cl and Ge atoms. The weaker peak at 2066 cm^{-1} was assigned to “free” molecules without such coordination [39]. But the appearance of two peaks was later attributed to the presence of rotational conformers [46]. The influence of several solvents on the shape of the ν(GeH) band was considered [39].

$Ge(CH_3)(C_6H_5)(C_{10}H_7\text{-}1)H$ (Table **28**, No. **10**) was first prepared with 70% yield by treating $Ge(CH_3)(C_6H_5)Br_2$ with one equivalent 1-$C_{10}H_7MgBr$ in ether-C_6H_6 and then adding the reaction mixture to excess $LiAlH_4$ in ether [2].

The (+)-$Ge(CH_3)(C_6H_5)(C_{10}H_7\text{-}1)H$ enantiomer was obtained with 96% yield by reducing the less levorotatory diastereomer of $Ge(CH_3)(C_6H_5)(C_{10}H_7\text{-}1)OC_{10}H_{19}$ in ether [2]. The influence of the solvent and the presence of LiCl on this reaction was studied in detail. Predominant formation of (+)-$Ge(CH_3)(C_6H_5)(C_{10}H_7\text{-}1)H$ (with ret.) was observed in solutions of ether or $O(C_3H_7\text{-}i)_2$ and in THF containing large amounts of LiCl; predominant formation of the

(−)-enantiomer (with inv.) occurred in THF or pyridine [17]. (+)-$Ge(CH_3)(C_6H_5)(C_{10}H_7\text{-}1)H$ was also obtained with 99% yield by reducing (+)-$Ge(CH_3)(C_6H_5)(C_{10}H_7\text{-}1)Cl$ (with inv.) [3].

(−)-$Ge(CH_3)(C_6H_5)(C_{10}H_7\text{-}1)H$ was prepared with 85% yield by reducing the more levorotatory $Ge(CH_3)(C_6H_5)(C_{10}H_7\text{-}1)OC_{10}H_{19}$ diastereomer [2] or with 97% yield by reducing (−)-$Ge(CH_3)(C_6H_5)(C_{10}H_7\text{-}1)OCH_3$ (with ret.) [3].

The title compound was obtained by reducing various compounds containing a Ge-transition metal bond [40, 44, 45]: from optically active $Ge(CH_3)(C_6H_5)(C_{10}H_7\text{-}1)Co(CO)_3L$ (L = CO, $P(OC_6H_5)_3$, $P(C_6H_5)_3$ [40, 45], $P(OC_2H_5)_3$, or $P(C_4H_9\text{-}t)_3$ [45]) with 60 to 87% inv. [40, 45] (see also [44]); from (S)-$Ge(CH_3)(C_6H_5)(C_{10}H_7\text{-}1)W(CO)_4NO$ to give (R)-(+)-$Ge(CH_3)(C_6H_5)(C_{10}H_7\text{-}1)H$ with 67% ret.; and from (S)-$Ge(CH_3)(C_6H_5)(C_{10}H_7\text{-}1)\text{-}Co(CO)_3C(OC_2H_5)C_4H_9$, (R)-(+)-$Ge(CH_3)(C_6H_5)(C_{10}H_7\text{-}1)Mn(CO)_5$, cis-(S)-(−)-$Ge(CH_3)\text{-}(C_6H_5)(C_{10}H_7\text{-}1)Mn(CO)_4C(OC_2H_5)CH_3$, or cis-(S)-(−)-$Ge(CH_3)(C_6H_5)(C_{10}H_7\text{-}1)Re(CO)_4C\text{-}(OC_2H_5)CH_3$ (all with predominant ret.) [44]; see also [45].

Reduction of optically active $Ge(CH_3)(C_6H_5)(C_{10}H_7\text{-}1)Co(CO)_3D$ (D = $P(C_6H_{11}\text{-}c)_3$ or $P(C_4H_9)_3$) in ether yielded $Ge(CH_3)(C_6H_5)(C_{10}H_7\text{-}1)H$ as a racemate or with 58% ret., respectively. When the same process was carried out in DME, the product was obtained with 81 or 82% inversion of the configuration [45]. Reducing $Ge(CH_3)(C_6H_5)(C_{10}H_7\text{-}1)Mo(CO)\text{-}(C_5H_5)(CH_3)NO$ in ether gave 25% (−)-$Ge(CH_3)(C_6H_5)(C_{10}H_7\text{-}1)H$ with 80% ret. [50].

(+)-$Ge(CH_3)(C_6H_5)(C_{10}H_7\text{-}1)H$ was also formed from $[N(C_2H_5)_4][(S)\text{-}(-)\text{-}Ge(CH_3)\text{-}(C_6H_5)(C_{10}H_7\text{-}1)Mo(CO)_5]$ in THF-ether at room temperature (67% yield) [43] and from (S)-(−)-$Ge(CH_3)(C_6H_5)(C_{10}H_7\text{-}1)Mn(CO)_2(C_5H_4CH_3)H$ and $P(C_6H_5)_3$ in $C_6H_5CH_3$ at room temperature (with ret.) [49].

(+)-$Ge(CH_3)(C_6H_5)(C_{10}H_7\text{-}1)H$ crystallizes in the orthorhombic system with a = 8.82, b = 20.09, and c = 7.93 Å; space group $P2_12_12_1-D_2^4$ [8].

Rotatory dispersion curves (in C_6H_{12} at ca. 25 °C) are similar for (+)-$Ge(C_2H_5)\text{-}(C_6H_5)(C_{10}H_7\text{-}1)H$ and the analogous C and Si compounds indicating the same R configuration for all three (+)-enantiomers. Comparison of the $[\alpha]_D^{23}$ values of theses compounds evidences that the size of the central atom, and hence steric crowding, has far less to do with the specific rotation than has its electronegativity [4].

Treating (−)-$Ge(CH_3)(C_6H_5)(C_{10}H_7\text{-}1)H$ with Cl_2 gave 100% (+)-$Ge(CH_3)(C_6H_5)(C_{10}H_7\text{-}1)\text{-}Cl$ with ret. [3]. Bromination of $Ge(CH_3)(C_6H_5)(C_{10}H_7\text{-}1)H$ with one equivalent of N-bromosuccinimide in refluxing CCl_4 yielded 82% $Ge(CH_3)(C_6H_5)(C_{10}H_7\text{-}1)Br$. Reacting (+)-$Ge(CH_3)(C_6H_5)(C_{10}H_7\text{-}1)H$ with either N-bromosuccinimide or Br_2 in CCl_4 at 0 °C gave $Ge(CH_3)(C_6H_5)(C_{10}H_7\text{-}1)Br$ with no observable optical activity and led to considerable cleavage of the Ge-$C_{10}H_7$ bond by Br_2 [2]. Refluxing $Ge(CH_3)(C_6H_5)(C_{10}H_7\text{-}1)H$ in excess CH_2Br_2 or $CHBr_3$ in the presence of a catalytic amount of t-$C_4H_9OOC_4H_9$-t yielded $Ge(CH_3)\text{-}(C_6H_5)(C_{10}H_7\text{-}1)Br$ quantitatively [27].

The rate of hydrogen abstraction by photochemically generated t-$C_4H_9O^\bullet$ was determined from the rise of the transient absorption of the $Ge^\bullet(CH_3)(C_6H_5)(C_{10}H_7\text{-}1)$ radical in C_6H_6 at 28.5 °C after N_2-laser excitation: $k = (1.2 \pm 0.1) \times 10^8\ L \cdot mol^{-1} \cdot s^{-1}$ [48]. The $Ge(CH_3)\text{-}(C_6H_5)(C_{10}H_7\text{-}1)H$ enantiomers reacted with CCl_4 in the presence of ca. 10 mol% $(C_6H_5\text{-}COO)_2$ at 80 °C to give the enantiomers of $Ge(CH_3)(C_6H_5)(C_{10}H_7\text{-}1)Cl$ while retaining their configuration, demonstrating the nonplanar structure of the intermediate germyl radical [19]. But progressive dilution of CCl_4 with C_6H_{12} showed that the chiral germyl radical also can undergo inversion [51]. Irradiating $Ge(CH_3)(C_6H_5)(C_{10}H_7\text{-}1)H$ and a small amount of CCl_4 in C_6H_{12} with a mercury lamp produced $Ge(CH_3)(C_6H_5)(C_{10}H_7\text{-}1)Cl$ and $CHCl_3$ [48].

References on p. 199

The platinum-catalyzed hydrogermylation reaction between (+)-$Ge(CH_3)(C_6H_5)(C_{10}H_7$-1)H and $C_6H_5C{\equiv}CC_6H_5$ gave a noncrystallizable oil which on the basis of spectral evidence appeared to be a mixture of cis- and trans-(+)-$Ge(CH_3)(C_6H_5)(C_{10}H_7$-1)$C(C_6H_5)$=$CHC_6H_5$, but a pure crystalline material could not be isolated [9]. Other hydrogermylation reactions of $Ge(CH_3)(C_6H_5)$-($C_{10}H_7$-1)H are summarized in Table 29. The reactions proceed with predominant retention of configuration at the Ge atom [18, 20].

Table 29
Hydrogermylation with $Ge(CH_3)(C_6H_5)(C_{10}H_7$-1)H and $Ge(C_3H_7)(C_6H_5)(C_{10}H_7$-1)H.
The Ge(R)(R′)(R″) unit is abbreviated GE; Rh and Pt stand for the $Rh(P(C_6H_5)_3)_3Cl$ and cis-$Pt(P(C_6H_5)_3)_2Cl_2$ catalysts, respectively.

reactant	conditions → products (yield)	Ref.
$Ge(CH_3)(C_6H_5)(C_{10}H_7$-1)H		
CH_2=CHC_5H_{11}	50°C, 8 h; Rh or Pt → GE-$CH_2CH_2C_5H_{11}$	[20]
CH_2=CHC_6H_5	50°C, 4 h; Rh → GE-$CH_2CH_2C_6H_5$ and GE-$CH(CH_3)C_6H_5$ in a 91:9 ratio (72% total)	[20]
	50°C, 8 h; Pt → GE-$CH_2CH_2C_6H_5$ and GE-$CH(CH_3)C_6H_5$ in a 90:10 ratio (53% total)	[20]
$C_6H_5C{\equiv}CH$	50°C, 192 h; no catalyst → trans-GE-CH=CHC_6H_5, cis-GE-CH=CHC_6H_5 and GE-$C(C_6H_5)$=CH_2 in a 56:25:19 ratio, which varied considerably in several runs (12% total)	[21]; see also [18]
	20°C, 2.5 h; Rh → trans-GE-CH=CHC_6H_5, cis-GE-CH=CHC_6H_5, and GE-$C(C_6H_5)$=CH_2 in a 89:8:3 ratio (82% total)	[18, 21]
	50°C, 10 h; Pt → trans-GE-CH=CHC_6H_5, cis-GE-CH=CHC_6H_5, and GE-$C(C_6H_5)$=CH_2 in a 84:9:7 ratio (71% total)	[18, 21]
	20°C, 0.5 h, H_2PtCl_6 cat. → trans-GE-CH=CHC_6H_5, cis-GE-CH=CHC_6H_5, and GE-$C(C_6H_5)$=CH_2 in a 71:2:27 ratio (68% total)	[18, 21]
$Ge(C_3H_7$-i)$(C_6H_5)(C_{10}H_7$-1)H		
CH_2=CHC_6H_5	50°C, 48 h; Rh → GE-$CH_2CH_2C_6H_5$ and GE-$CH(CH_3)C_6H_5$ in a 87:13 ratio (49% total)	[20]
	50°C, 96 h; Pt → GE-$CH_2CH_2C_6H_5$ and GE-$CH(CH_3)C_6H_5$ in a 79:21 ratio (39% total)	[20]
$C_6H_5C{\equiv}CH$	20°C, 24 h; Rh → trans-GE-CH=CHC_6H_5, cis-GE-CH=CHC_6H_5, and GE-$C(C_6H_5)$=CH_2 in a 70:24:6 ratio (70% total)	[18, 21]
	50°C, 48 h; Pt → trans-GE-CH=CHC_6H_5, cis-GE-CH=CHC_6H_5, and GE-$C(C_6H_5)$=CH_2 in a 53:28:19 ratio (51% total)	[18, 21]
	20°C, 4 to 6 h, H_2PtCl_6 cat. → trans-GE-CH=CHC_6H_5, cis-GE-CH=CHC_6H_5, and GE-$C(C_6H_5)$=CH_2 in a 69:ca. 0:31 ratio (59% total)	[18, 21]

Metalation of (−)-$Ge(CH_3)(C_6H_5)(C_{10}H_7)H$ with one equivalent LiC_4H_9 in ether at room temperature (0.5 h) gave a solution of $Ge(CH_3)(C_6H_5)(C_{10}H_7)Li$ while retaining the configuration [3, 15] (similar for the (+)-enantiomer; see [31, 43, 50]). Successive treatment of (−)-$Ge(CH_3)(C_6H_5)(C_{10}H_7)H$ with LiC_4H_9 and C_2H_5I yielded (+)-$Ge(CH_3)(C_2H_5)(C_6H_5)C_{10}H_7$ [12].

Reacting $Ge(CH_3)(C_6H_5)(C_{10}H_7)H$ with RMgBr (R = C_2H_5, CH_2=$CHCH_2$, or C_6H_5) in the presence of several Ni-containing catalysts yielded $Ge(CH_3)(C_6H_5)(C_{10}H_7)R$ while retaining the configuration. The influence of the catalyst type ($NiCl_2$, $NiBr_2$, $Ni(P(C_6H_5)_2C_2H_4P(C_6H_5)_2)Cl_2$, or $Ni(P(C_6H_5)_3)_2Cl_2$) and the reaction time on the product yield (up to 94%) was studied [33]. A similar reaction of (+)-$Ge(CH_3)(C_6H_5)(C_{10}H_7)H$ with $C_6H_5CH_2MgCl$ in the presence of $NiCl_2$ or $Ni(P(C_6H_5)_3)_2Cl_2$ gave (+)-$Ge(CH_3)(C_6H_5)(C_{10}H_7)CH_2C_6H_5$ and the digermane (−)-$Ge_2(CH_3)_2(C_6H_5)_2(C_{10}H_7)_2$ [32]. Reaction mechanisms were discussed [32, 33].

The reaction of $Ge(CH_3)(C_6H_5)(C_{10}H_7)H$ with $Sn(CH_3)_2(C_6H_5)N(C_2H_5)_2$ at 90 °C (1 d) yielded 70% $Ge(CH_3)(C_6H_5)(C_{10}H_7)Sn(CH_3)_2C_6H_5$ [36]; see also [7].

(−)-$Ge(CH_3)(C_6H_5)(C_{10}H_7)H$ and $Hg(C_6H_5)CBr_3$ reacted in refluxing C_6H_6 (2 h) to form (−)-$Ge(CH_3)(C_6H_5)(C_{10}H_7)CHBr_2$ with 62% yield [15]; see also [14]. (+)-$Ge(CH_3)(C_6H_5)(C_{10}H_7)H$ and $Hg(C_4H_9\text{-t})_2$ in C_6H_{14} at 80 °C (8 h) yielded 63% (−)-$Hg(Ge(CH_3)(C_6H_5)C_{10}H_7)_2$ [29].

The following reactions with metal carbonyls were reported: (−)-$Ge(CH_3)(C_6H_5)(C_{10}H_7)H$ + $Mn_2(CO)_{10}$ at 150 °C → (R)-(+)-$Ge(CH_3)(C_6H_5)(C_{10}H_7)Mn(CO)_5$ [44] (see also [45]); (+)-$Ge(CH_3)(C_6H_5)(C_{10}H_7)H$ + $Co_2(CO)_8$ (2:1 mole ratio) → (S)-(+)-$Ge(CH_3)(C_6H_5)(C_{10}H_7)Co(CO)_4$ with 86% yield [35] (see also [40, 45]); and $Ge(CH_3)(C_6H_5)(C_{10}H_7)H$ + $Co_2(CO)_6D_2$ (D = $P(C_4H_9)_3$, $P(C_4H_9\text{-t})_3$, $P(C_6H_{11}\text{-c})_3$ [45], $P(C_6H_5)_3$ [40, 45], $P(OC_2H_5)_3$ [45], or $P(OC_6H_5)_3$ [40, 45]) → $Ge(CH_3)(C_6H_5)(C_{10}H_7)Co(CO)_3D$ and H_2 [40, 45]. Although (+)- and (−)-$Ge(CH_3)(C_6H_5)(C_{10}H_7)H$ are known to racemize thermally [42], they reacted with a slight excess of $HCCo_3(CO)_9$ in refluxing $C_6H_5CH_3$ to yield 30% optically active $Ge(CH_3)(C_6H_5)(C_{10}H_7)CCo_3(CO)_9$ [41].

Some optical cycles (Walden cycles) starting from (+)- and (−)-$Ge(CH_3)(C_6H_5)(C_{10}H_7)H$ are listed below using the abbreviation GE for the $Ge(CH_3)(C_6H_5)(C_{10}H_7)$ unit:

(+)-GE-H → (R)-GE-Li → (S)-[GE-Mo(CO)$_5$]Li →
(S)-(−)-[GE-Mo(CO)$_5$][N(C_2H_5)$_4$] → (+)-GE-H, 81% ret. [43];
(+)-GE-H → (R)-GE-Li → (S)-(−)-[GE-Mn(CO)$_2$($C_5H_4CH_3$)]Li →
(S)-(−)-GE-Mn(CO)$_2$($C_5H_4CH_3$)H → (+)-GE-H, 91% ret. [49];
(−)-GE-H → GE-Li → (−)-GE-H, 86% ret. [3];
(−)-GE-H → Ge-Li → (+)-GE-COOH → (+)-GE-COOCH$_3$ → (−)-GE-OCH$_3$ →
(−)-GE-H, 94% ret. [3]; and
(−)-GE-H → (+)-GE-Cl → (+)-GE-H, inv. [3].

$Ge(C_2H_5)(C_6H_5)(C_{10}H_7\text{-1})H$ (Table **28**, No. **11**) was obtained by reducing racemic $Ge(C_2H_5)(C_6H_5)(C_{10}H_7)OC_{10}H_{19}$ [6] (see also [1, 10]) or $Ge(C_2H_5)(C_6H_5)(C_{10}H_7)OCH(CH_3)C_6H_5$ [1, 6]. Small amounts of the racemate were formed along with the main product (±)-$Ge(C_2H_5)(C_6H_5)(C_{10}H_7)CH_2OH$ by reducing (−)-$Ge(C_2H_5)(C_6H_5)(C_{10}H_7)COOH$ [25].

The (+)-$Ge(C_2H_5)(C_6H_5)(C_{10}H_7)H$ enantiomer was prepared from the more levorotatory $Ge(C_2H_5)(C_6H_5)(C_{10}H_7)OC_{10}H_{19}$ diastereoisomer and $LiAlH_4$ in refluxing ether after 6 h (81% yield) [6] or 2 h (95% yield) [23]. High yields were also obtained by reducing (+)-$Ge(C_2H_5)(C_6H_5)(C_{10}H_7)OCH_3$ (98%), (−)-$Ge(C_2H_5)(C_6H_5)(C_{10}H_7)NC_4H_4$ (90%), or (+)-$Ge(C_2H_5)(C_6H_5)(C_{10}H_7)SC_6H_5$ (90%) [23]. For the formation by hydrolysis of (R)-$Ge(C_2H_5)(C_6H_5)(C_{10}H_7)Li$, see [6, 13, 25].

References on p. 199

An optically impure sample of the (−)-$Ge(C_2H_5)(C_6H_5)(C_{10}H_7)H$ enantiomer was obtained by reducing the less pentane-soluble $Ge(C_2H_5)(C_6H_5)(C_{10}H_7)OCH(CH_3)C_6H_5$ diastereoisomer in refluxing ether-$O(C_4H_9)_2$ [6]. The (−)-enantiomer was also formed from (−)-$Ge(C_2H_5)(C_6H_5)(C_{10}H_7)Cl$ [6, 23] (see also [1]), (−)-$Ge(C_2H_5)(C_6H_5)(C_{10}H_7)OCH_3$ [6, 23], or (−)-$Ge(C_2H_5)(C_6H_5)(C_{10}H_7)OOCGe(C_2H_5)(C_6H_5)C_{10}H_7$ and $LiAlH_4$ [25, 28].

In the reduction of optically active $Ge(C_2H_5)(C_6H_5)(C_{10}H_7)X$ compounds by $LiAlH_4$ to give $Ge(C_2H_5)(C_6H_5)(C_{10}H_7)H$, the predominant stereochemistry is inversion in the case of relatively good leaving groups such as X = Cl, SC_6H_5, NC_6H_4 [10, 23], and OOCGe-$(C_2H_5)(C_6H_5)C_{10}H_7$ [10], but retention in the case of poorer leaving groups such as X = OCH_3, $OC_{10}H_{19}$ [10, 23], and $OGe(C_2H_5)(C_6H_5)C_{10}H_7$ [10].

(+)-$Ge(C_2H_5)(C_6H_5)(C_{10}H_7)H$ crystallizes in the orthorhombic system with a = 8.94, b = 19.32, and c = 8.80 Å; space group $P2_12_12_1-D_2^4$. Z = 4 gives d_c = 1.306 g/cm³ [23].

An influence of the solvent (C_6H_6 or C_6H_{12}) on the specific rotation of (+)-Ge-$(C_2H_5)(C_6H_5)(C_{10}H_7)H$ was noted, and the optical rotatory dispersion was measured in the range 589 to 320 nm [23].

Treating (+)-$Ge(C_2H_5)(C_6H_5)(C_{10}H_7)H$ with Cl_2 in CCl_4 gave (−)-$Ge(C_2H_5)(C_6H_5)$-$(C_{10}H_7)Cl$ with ret. [6, 26] (see also [1]), whereas a similar treatment with Br_2 yielded racemic $Ge(C_2H_5)(C_6H_5)(C_{10}H_7\text{-}1)Br$ [6] (see also [23]). Reacting it with LiC_4H_9 in ether gave $Ge(C_2H_5)(C_6H_5)(C_{10}H_7)Li$ [6] while retaining the R configuration [13, 25]; successive treatment with LiC_4H_9 and CH_3I yielded (+)-$Ge(CH_3)(C_2H_5)(C_6H_5)C_{10}H_7$ [12].

(+)-$Ge(C_2H_5)(C_6H_5)(C_{10}H_7\text{-}1)H$ reacted with trans-$Pt(P(CH_3)_2C_6H_5)_2(H)Cl$ at 70 °C (8 h) in a slow flow of N_2 to give (+)-trans-$Pt(P(CH_3)_2C_6H_5)_2(Cl)Ge(C_2H_5)(C_6H_5)C_{10}H_7$ with 75% yield [22].

Some optical cycles (Walden cycles) starting from (+)-$Ge(C_2H_5)(C_6H_5)(C_{10}H_7)H$ are listed below, where GE stands for the $Ge(C_2H_5)(C_6H_5)(C_{10}H_7)$ unit:

(+)-GE-H → (−)-GE-Cl → (+)-GE-SC_6H_5 → (+)-GE-H, 68% ret. [10, 23];
(+)-GE-H → (−)-GE-Cl → (−)-GE-NC_4H_4 → (+)-GE-H, 72% ret. [10, 23];
(+)-GE-H → (−)-GE-Cl → (+)-GE-S-GE → (+)-GE-H, 71% ret. [23];
(+)-GE-H → (R)-GE-Li → (+)-GE-H, 89% ret. [25], 90% ret. [13];
(+)-GE-H → (R)-GE-Li → (−)-GE-COOH → (R)-GE-Li → (+)-GeE-H, 90% [13], 91% [25] ret.;
(+)-GE-H → (R)-GE-Li → (−)-GE-COOH → (−)-GE-$COOCH_3$ → (+)-GE-OCH_3 → (+)-GE-H, 70% ret. [25];
(+)-GE-H → (−)-GE-Cl → (−)-GE-H, 90% [10], 93% [23] inv.; and
(+)-GE-H → (−)-GE-Cl → (−)-GE-OCH_3 → (−)-GE-H, 67% inv. [23].

$Ge(C_3H_7\text{-}i)(C_6H_5)(C_{10}H_7\text{-}1)H$ (Table **28**, No. **12**) was first prepared with 75% yield from $Ge(C_3H_7\text{-}i)(C_6H_5)Br_2$ and one equivalent $C_{10}H_7MgBr$ in ether-$C_6H_5CH_3$ and adding the reaction mixture to excess $LiAlH_4$ in ether [16]. It was formed by Ge-Si cleavage of racemic $Ge(C_3H_7\text{-}i)(C_6H_5)(C_{10}H_7)Si(C_6H_5)_3$ with LiR (R = CH_3, C_3H_7) or CH_3ONa followed by hydrolysis [34].

The (+)-$Ge(C_3H_7\text{-}i)(C_6H_5)(C_{10}H_7)H$ enantiomer was prepared with 65% yield by reducing the less pentane-soluble $Ge(C_3H_7\text{-}i)(C_6H_5)(C_{10}H_7)OC_{10}H_{19}$ diastereomer in ether (6 h) [16].

Reducing (−)-$Ge(C_3H_7\text{-}i)(C_6H_5)(C_{10}H_7)OCH_3$ in C_5H_{12}-ether yielded 86% of the (−)-$Ge(C_3H_7\text{-}i)(C_6H_5)(C_{10}H_7)H$ enantiomer [30]. It was a minor product (22% yield) along with (+)-$Ge_2(C_3H_7\text{-}i)_2(C_6H_5)_2(C_{10}H_7)_2$ when (−)-$Ge(C_3H_7\text{-}i)(C_6H_5)(C_{10}H_7)OC_{10}H_{19}$ was reacted with $Si(C_6H_5)_3Li$ followed by hydrolysis (via $Ge(C_3H_7\text{-}i)(C_6H_5)(C_{10}H_7)Li$) [34].

Chlorination of (+)-$Ge(C_3H_7\text{-}i)(C_6H_5)(C_{10}H_7)H$ with Cl_2 in CCl_4 at −15°C gave (−)-$Ge(C_3H_7\text{-}i)(C_6H_5)(C_{10}H_7)Cl$ [16], retaining the configuration [30, 31]. Bromination with N-bromosuccinimide in CCl_4 yielded 90% $Ge(C_3H_7\text{-}i)(C_6H_5)(C_{10}H_7)Br$ [16].

Optically active $Ge(C_3H_7\text{-}i)(C_6H_5)(C_{10}H_7)H$ reacted with LiC_4H_9 to give $Ge(C_3H_7\text{-}i)(C_6H_5)(C_{10}H_7)Li$, retaining the configuration [31]. $Ge(C_3H_7\text{-}i)(C_6H_5)(C_{10}H_7)H$ and RMgBr (R = CH_3, C_2H_5, or CH_2=$CHCH_2$) in the presence of several Ni-containing catalysts yielded $Ge(C_3H_7\text{-}i)(C_6H_5)(C_{10}H_7)R$, retaining the configuration. The dependence of the product yield (<78%) on the catalyst type ($NiCl_2$, $Ni(P(C_6H_5)_3)_2Cl_2$, or $Ni(P(C_6H_5)_2C_2H_4P(C_6H_5)_2)Cl_2$) and the reaction time was studied and a reaction mechanism discussed [33]. For a similar reaction with $C_6H_5CH_2MgCl\text{-}NiCl_2$, see [32].

Hydrogermylation reactions of $Ge(C_3H_7\text{-}i)(C_6H_5)(C_{10}H_7)H$ are summarized in Table 29; they proceed predominantly by retaining the configuration at the Ge atom [18].

References:

[1] Bott, R. W.; Eaborn, C.; Varma, I. D. (Chem. Ind. [London] **1963** 614).
[2] Brook, A. G.; Peddle, G. J. D. (J. Am. Chem. Soc. **85** [1963] 1869/70).
[3] Brook, A. G.; Peddle, G. J. D. (J. Am. Chem. Soc. **85** [1963] 2338/9).
[4] Brook, A. G. (J. Am. Chem. Soc. **85** [1963] 3051/2).
[5] Peddle, G. J. D. (Diss. Univ. Toronto 1963 from [23]).
[6] Eaborn, C.; Simpson, P.; Varma, I. D. (J. Chem. Soc. A **1966** 1133/6).
[7] Creemers, H. M. J. C. (Diss. Utrecht 1967 from Gielen, M.; Simon, S.; Tondeur, Y.; van de Steen, M.; Hoogzand, C.; Bull. Soc. Chim. Belg. **83** [1974] 337/8).
[8] Okaya, Y. (Rep. Meet. Am. Crystallogr. Assoc., Minnesota 1967 p. 60 from [23]).
[9] Brook, A. G.; Pannell, K. H.; Anderson, D. G. (J. Am. Chem. Soc. **90** [1968] 4374/7).
[10] Eaborn, C.; Hill, R. E. E.; Simpson, P. (Chem. Commun. **1968** 1077/8).

[11] Eaborn, C.; Simpson, P.; Walsingham, R. A. (unpublished work from [10]).
[12] Eaborn, C.; Hill, R. E. E.; Simpson, P.; Brook, A. G.; MacRae, D. (J. Organomet. Chem. **15** [1968] 241/43).
[13] Eaborn, C.; Hill, R. E. E.; Simpson, P. (J. Organomet. Chem. **15** [1968] P1/P3).
[14] Brook, A. G.; Duff, J. M.; Anderson, D. G. (Can. J. Chem. **48** [1970] 561/9).
[15] Brook, A. G.; Duff, J. M.; Anderson, D. G. (J. Am. Chem. Soc. **92** [1970] 7567/72).
[16] Carré, F.; Corriu, R. (J. Organomet. Chem. **25** [1970] 395/402).
[17] Jean, A.; Lequan, M. (Tetrahedron Lett. **1970** 1517/9).
[18] Corriu, R. J. P.; Moreau, J. J. E. (J. Chem. Soc. Chem. Commun. **1971** 812/3).
[19] Sakurai, H.; Mochida, K. (J. Chem. Soc. Chem. Commun. **1971** 1581).
[20] Corriu, R. J. P.; Moreau, J. J. E. (J. Organomet. Chem. **40** [1972] 55/72).

[21] Corriu, R. J. P.; Moreau, J. J. E. (J. Organomet. Chem. **40** [1972] 73/96).
[22] Eaborn, C.; Kapoor, P. N.; Tune, D. J.; Turpin, C. L.; Walton, D. R. M. (J. Organomet. Chem. **34** [1972] 153/4).
[23] Eaborn, C.; Hill, R. E. E.; Simpson, P. (J. Organomet. Chem. **37** [1972] 251/65).
[24] Walsingham, R. A. (personal communication from [23]).
[25] Eaborn, C.; Hill, R. E. E.; Simpson, P. (J. Organomet. Chem. **37** [1972] 267/74).
[26] Eaborn, C.; Hill, R. E. E.; Simpson, P. (J. Organomet. Chem. **37** [1972] 275/9).
[27] Sakurai, H.; Mochida, K.; Hosomi, A.; Mita, F. (J. Organomet. Chem. **38** [1972] 275/80).
[28] Eaborn, C.; Hill, R. E. E.; Simpson, P. (J. Organomet. Chem. **51** [1973] 147/51).
[29] Vodolazskaya, V. M.; Fedot'ev, B. V.; Baukov, Yu. I.; Kruglaya, O. A.; Vyazankin, N. S. (J. Organomet. Chem. **63** [1973] C5/C6).
[30] Carré , F.; Corriu, R. (J. Organomet. Chem. **65** [1974] 343/8).

[31] Carré, F.; Corriu, R. (J. Organomet. Chem. **65** [1974] 349/59).
[32] Carré, F.; Corriu, R. (J. Organomet. Chem. **73** [1974] C49/C52).
[33] Carré, F. H.; Corriu, R. J. P. (J. Organomet. Chem. **74** [1974] 49/58).
[34] Carré, F. H.; Corriu, R. J. P. (J. Organomet. Chem. **86** [1975] C47/C49).
[35] Colomer, E.; Corriu, R. J. P. (J. Organomet. Chem. **133** [1977] 159/68).
[36] Gielen, M.; Simon, S. (Bull. Soc. Chim. Belg. **86** [1977] 589/94).
[37] Egorochkin, A. N.; Sevast'yanova, E. I.; Khorshev, S. Ya.; Ratushnaya, S. Kh.; Satgé, J.; Rivière, P.; Barrau, J.; Richelme, S. (J. Organomet. Chem. **162** [1978] 25/35).
[38] Rivière-Baudet, M.; Rivière, P.; Satgé, J.; Lacrampe, G. (Recl. Trav. Chim. Pays-Bas **98** [1979] 42/52).
[39] Egorochkin, A. N.; Sevastyanova, E. I.; Khorshev, S. Ya.; Ratushnaya, S. Kh.; Richelme, S.; Satgé, J.; Rivière, P. (J. Organomet. Chem. **188** [1980] 73/83).
[40] Cerveau, G.; Colomer, E.; Corriu, R. J. P. (Angew. Chem. **93** [1981] 489/90; Angew. Chem. Int. Ed. Engl. **20** [1981] 478).
[41] Combes, C. E. J.; Corriu, R. J. P.; Henner, B. J. L. (J. Organomet. Chem. **221** [1981] 257/69).
[42] Carré, F. H. (personal communication from [41]).
[43] Carré, F.; Cerveau, G.; Colomer, E.; Corriu, R. J. P. (J. Organomet. Chem. **229** [1982] 257/73).
[44] Cerveau, G.; Colomer, E.; Corriu, R. J. P. (Organometallics **1** [1982] 867/9).
[45] Cerveau, G.; Colomer, E.; Corriu, R. J. P. (J. Organomet. Chem. **236** [1982] 33/52).
[46] McKean, D. C.; Torto, I.; Morrisson, A. R. (J. Organomet. Chem. **226** [1982] C47/C51).
[47] Barrau, J.; Bouchaut, M.; Lavayssière, H.; Dousse, G.; Satgé, J. (J. Organomet. Chem. **243** [1983] 281/90).
[48] Hayashi, H.; Mochida, K. (Chem. Phys. Lett. **101** [1983] 307/11).
[49] Carré, F.; Colomer, E.; Corriu, R. J. P.; Vioux, A. (Organometallics **3** [1984] 1272/8).
[50] Cerveau, G.; Colomer, E.; Corriu, R. J. P.; Vioux, A. (J. Organomet. Chem. **321** [1987] 327/34).
[51] Mochida, K.; Yamauchi, T.; Sakurai, H. (Bull. Chem. Soc. Jpn. **62** [1989] 1982/5).
[52] Tacke, R.; Wiesenberger, F. (Z. Naturforsch. **46b** [1991] 275/9).
[53] Terunuma, D.; Kizaki, H.; Nagao, T.; Nohira, H. (59th Natl. Meet. Chem. Soc. Jpn., Tokyo 1990, Vol. 1, Abstr. No. 3D47 from Terunuma, D.; Kizaki, H.; Sato, T.; Masuo, K.; Nohira, H.; Chem. Lett. **1991** 97/100).

1.3.1.4 Triorganogermanium Hydrides of the Ge-R(R')H Type

The Ge-R(R')H compounds are listed in Table 30. The iron complex I (R = C_6H_5) was already described in "Organoiron Compounds" B 7, 1981, pp. 130 and 133.

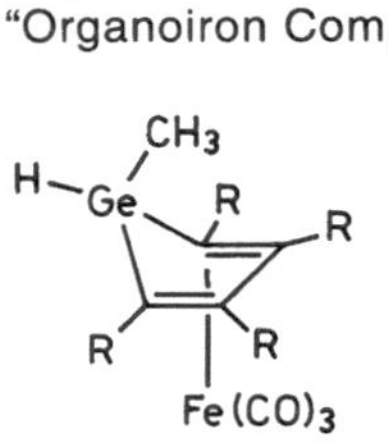

I

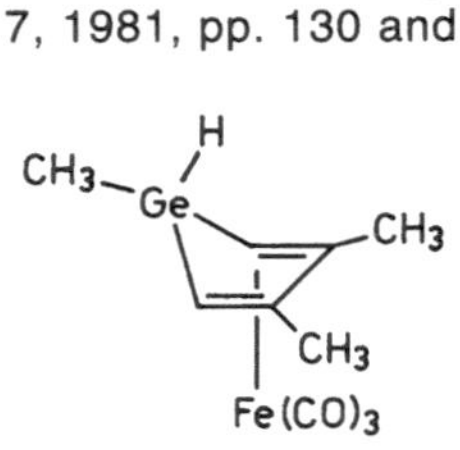

II

The related complex **GeCH=C(CH$_3$)C(CH$_3$)=CH(Fe(CO)$_3$)(CH$_3$)H** (Formula II) was later identified by ^{1}H NMR spectroscopy in a crude mixture obtained by treating

GeCH=C(CH_3)C(CH_3)=CH(Fe(CO)$_3$)(CH_3)Cl with Al(C_4H_9-i)$_2$H in C_6H_{14} at −50°C. ^{1}H NMR spectrum (in C_6D_6): δ = 6.03 ppm (q, GeH, J ≈ 4 Hz); the CH_3Ge group appears as a doublet. During column chromatography on neutral Al_2O_3 with C_6H_{14} as the eluent, the title compound was presumably converted into GeCH=C(CH_3)C(CH_3)=CH(Fe(CO)$_3$)(CH_3)OH which decomposed upon attempted purification [26].

In order to study the conformation and estimate the conformational energy of group R′ bonded to Ge, molecular mechanics calculations were performed on the germacyclopentane No. 5 [37] and the germacyclohexanes No. 16 [29, 31], 17 [35], 20, 23 [29], 24, 26 [35], and 27 [28]. MO calculations were carried out for the germacyclohexanes No. 17 [35], 19 [30], and 26 [35].

Table 30
Ge-R(R′)H Compounds.
An asterisk indicates further information at the end of the table.
Explanations, abbreviations, and units are given on p. X.

No.	compound	preparation (yield) properties and remarks
4-membered rings		
*1	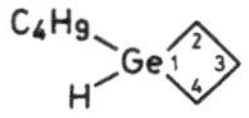	Ge($CH_2CH_2CH_2$Cl)(C_4H_9)(H)Cl + Na in $C_6H_5CH_3$ (45%) $LiAlH_4$ reduction of GeCH$_2$CH$_2$CH$_2$(C_4H_9)X (X = Cl, Br, or I) in refluxing ether (along with the minor product Ge(C_3H_7)(C_4H_9)H_2) b.p. 70°C/25 d^{20} = 1.0879; n_D^{20} = 1.4781 ^{1}H NMR (C_6H_6): 1.4, 2.2 (H-2,4 and H-3), 5 (HGe) IR: ν(GeH) 2020; other band at 1110 to 1120 (characteristic for germacyclobutanes) unstable at 150°C [7]
*2		Ge(CH_2CH(CH_3)CH_2Cl)(C_4H_9)(H)Cl + Na in $C_6H_5CH_3$ (20%) b.p. 61°C/15 d^{20} = 1.0361; n_D^{20} = 1.4701 ^{1}H NMR (C_6H_6): 1.4, 2.6 (H-2,4 and H-3), 5 (HGe) IR: ν(GeH) 2020; other band at 1110 to 1120 (characteristic for germacyclobutanes) unstable at 150°C [7]
3	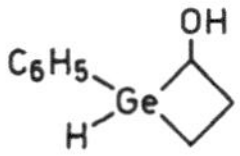	$LiAlH_4$ reduction of Ge(CH(OH)CH=CH_2)(C_6H_5)Cl_2 in a large amount of ether (10%) along with polymers b.p. 96°C/0.04 [10, 12] ^{1}H NMR (CCl_4): 1.0 to 1.80 (m, H-3,4), 3.45 (m, H-2), 4.37 (m, HGe) [10, 12], 7.2 to 7.8 (m, C_6H_5) [12] IR (neat): ν(OH) 3340, ν(GeH) 2040 [10, 12]
4	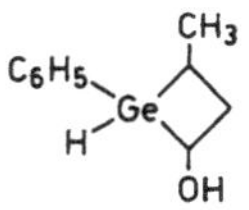	UV irradiation of Ge(CH(OH)CH=CHCH$_3$)(C_6H_5)H_2 in ether for 2 h (12%) [12]; see also [10] b.p. 98 to 103°C/0.05 [10], 98 to 105°C/0.05 [12]

References on p. 215

Table 30 (continued)

No.	compound	preparation (yield) properties and remarks
4 (continued)		^{1}H NMR (CCl_4): 0.90 to 1.30 (m, CH_3) [10, 12], 1.50 to 2.0 (m, H-2,3) [12], 3.45 (m, H-4), 4.38 (m, HGe) [10, 12] IR (neat): ν(OH) 3300, ν(GeH) 2030 [10, 12]
5-membered rings		
5	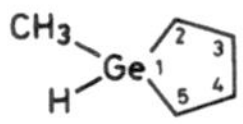	$GeCH_2CH_2CH_2CH_2(CH_3)C_6H_5$ + Br_2 and $LiAlH_4$ reduction of the intermediate $GeCH_2CH_2CH_2CH_2(CH_3)Br$ ^{13}C NMR ($CDCl_3$): −5.20 (CH_3), 11.86 (C-2,5), 28.60 (C-3,4) ^{73}Ge NMR ($CDCl_3$): −27.4 [37]
6	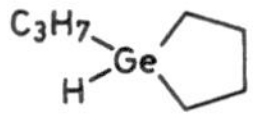	$LiAlH_4$ reduction of $GeCH_2CH_2CH_2CH_2(CH_2CH_2CH_2Br)Br$ in refluxing ether for 2 d (90%) b.p. 89 °C/65 d^{20} = 1.0995; n_D^{20} = 1.4754 IR: ν(GeH) 2023 [4] MS: $[M]^+$, $[M - CH_2{=}CH_2]^+$, $[Ge^{\cdot}]^+$ (100); complete spectrum displayed [6]
*7	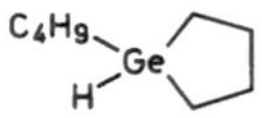	$LiAlH_4$ reduction of $GeCH_2CH_2CH_2CH_2(C_4H_9)Br$ (90%) b.p. 179 °C/745 d^{20} = 1.0694; n_D^{20} = 1.4756 [3] MS of $GeCH_2CD_2CD_2CH_2(C_4H_9)H$: $[M]^+$, $[M - CH_2{=}CD_2]^+$, $[Ge^{\cdot}]^+$ (100); spectrum displayed [6]
8	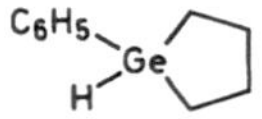	$LiAlH_4$ reduction of $GeCH_2CH_2CH_2CH_2(C_6H_5)Br$ in refluxing ether for 20 h (90%) b.p. 127 °C/30 d^{20} = 1.2215; n_D^{20} = 1.5601 [2] IR (neat): ν(GeH) 2035(s) [1, 2] MS (70 eV) of $GeCH_2CD_2CD_2CH_2(C_6H_5)H$: $[M]^+$, $[M - CH_2{=}CD_2]^+$, $[GeC_6H_5]^+$ (100), $[M - C_6H_6]^+$, $[Ge^{\cdot}]^+$ (16); spectrum displayed [6]
*9	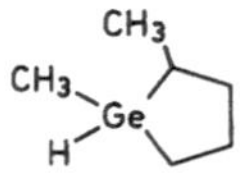 cis and trans	preparation given on p. 211 b.p. 78 to 82 °C/200 configurationally stable isomers; readily separable by fractional distillation or GC ^{1}H NMR (C_6H_6): 0.15 (d, CH_3Ge), 4.2 (m, HGe) for the cis isomer; 0.22 (d, CH_3Ge), 4.0 (m, HGe) for the trans isomer ^{13}C NMR (C_6D_6): −8.6 (CH_3Ge), 12.7 (C-5), 16.4 (CH_3-2), 19.4 (C-2), 26.7 (C-4), 38.5 (C-3) for the cis isomer; −5.4 (CH_3Ge), 12.7 (C-5), 18.6 (CH_3-2), 22.7 (C-2), 26.9 (C-4), 38.5 (C-3) for the trans isomer [14, 15]

Table 30 (continued)

No.	compound	preparation (yield) properties and remarks
10	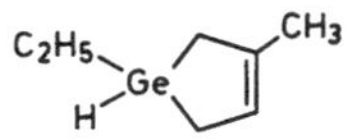	$Ge(C_2H_5)H$ (from $Ge(C_2H_5)(OCH_3)H_2$) + CH_2=CHC(CH_3)=CH_2 at 120 °C in a sealed tube (15%) $LiAlH_4$ reduction of $GeCH_2CH$=C(CH_3)CH_2(C_2H_5)Cl in ether b.p. 58 °C/16 d^{20} = 1.1253; n_D^{20} = 1.4870 ^{1}H NMR (CCl_4): 1.53 (H-2), 1.75 (CH_3-3), 4.33 (HGe), 5.54 (H-4) IR (neat): ν(GeH) 2030, ν(C=C) 1635 [9]
11	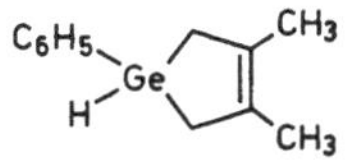	$Ge(C_6H_5)H$ (from $Ge(C_6H_5)(OCH_3)H_2$ and RO^-) + CH_2=C(CH_3)C(CH_3)=CH_2 at 120 °C? (17%) [9] $Ge(C_6H_5)H \cdot N(C_2H_5)_3$ (from $Ge(C_6H_5)H_2Cl$ and $N(C_2H_5)_3$) + CH_2=C(CH_3)C(CH_3)=CH_2 in C_6H_6 at 130 °C (6 h) in a sealed tube (8%) [24] $LiAlH_4$ reduction of $GeCH_2C$(CH_3)=C(CH_3)CH_2(C_6H_5)X (X = Cl or OCH_3) in ether (high yield) [9] b.p. 94 °C/1.5 n_D^{20} = 1.5605 ^{1}H NMR (CCl_4): 1.74 (CH_3-3,4), 1.82 (H-2,5), 4.78 (HGe) IR (neat): ν(GeH) 2045, ν(C=C) 1640 [9]
12	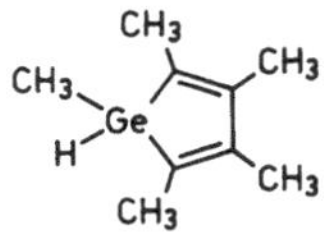	GeC(CH_3)=C(CH_3)C(CH_3)=CCH_3Cl_2 + $LiCH_3$ in THF-ether from −78 to +20 °C followed by $LiAlH_4$ reduction for 5 h (5%); obtained along with GeC(CH_3)=C(CH_3)C(CH_3)=$CCH_3(CH_3)_2$ and GeC(CH_3)=C(CH_3)C(CH_3)=CCH_3H_2 as the major products MS: $[M]^+$ (35), $[M - CH_3]^+$ (74), $[M - CH_3 - C_2(CH_3)_2]^+$ (50), $[GeCH_3]^+$ (100) [33]
*13	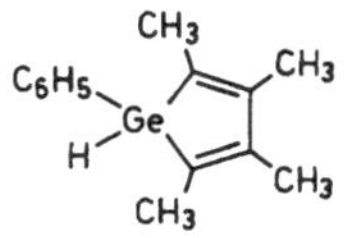	GeC(CH_3)=C(CH_3)C(CH_3)=CCH_3Cl_2 + C_6H_5MgBr in ether at −70 to +20 °C (2 h) followed by $LiAlH_4$ reduction in ether (80%) [32]; see also [33] GeC(CH_3)=C(CH_3)C(CH_3)=CCH_3Cl_2 + LiC_6H_5 followed by $LiAlH_4$ reduction $Ge(C_6H_5)Cl_3$ + ZrC(CH_3)=C(CH_3)C(CH_3)=$CCH_3(C_5H_5)_2$ in THF at 50 °C followed by $LiAlH_4$ reduction (20%) [33] yellow liquid [33]; b.p. 102 °C/0.52 [32, 33] ^{1}H NMR (C_6D_6): 1.80, 1.98 (s's, CH_3), 5.82 (s, HGe), 7.21, 7.55 (m's, C_6H_5) [32, 33] ^{13}C NMR: 128.2 (C-2,5), 148.9 (C-3,4), 129.6, 130.0, 130.2, 135.8 (C-3′,5′, C-1′, C-4′, and C-2′,6′ of C_6H_5, respectively), in THF-d_8 [32]; 14, 15 (CH_3), 134, 148 (C_4Ge), 128, 129, 130, 135 (C_6H_5) in CD_3COCD_3 [33] IR: ν(GeH) 2031 [32, 33] MS: $[M]^+$ (24), $[M - C_6H_5]^+$ (13), $[GeC_6H_5]^+$ (100), $[M - C_6H_5 - C_2(CH_3)_2]^+$ (24) [32]

References on p. 215

Table 30 (continued)

No.	compound	preparation (yield) properties and remarks
*14	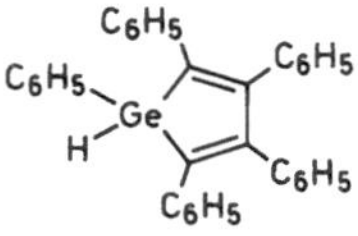	LiAlH$_4$ reduction of GeC(C$_6$H$_5$)=C(C$_6$H$_5$)C(C$_6$H$_5$)=CC$_6$H$_5$(C$_6$H$_5$)Cl (>70%) [5] m.p. 187 to 188°C IR: ν(GeH) 2056 [5, 8]
*15	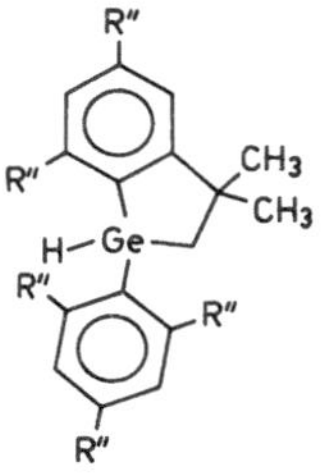 R″ = C$_4$H$_9$-t	formation described on p. 214 ^{1}H NMR: 1.00 (9 H), 1.04 (9 H), 1.23 (3 H), 1.33 (9 H), 1.34 (9 H), 1.65 (9 H), 1.74 to 1.75 (s, 3 H + AB part of ABX system, 2 H), 6.2 (pseudo t, HGe), 7.37, 7.39, 7.41, 7.46, 7.51 (two overlapping AB systems, 4 H) ^{13}C NMR: 29.64, 31.36, 31.49, 32.37, 33.56, 33.88, 34.50, 34.83, 37.51, 38.95, 43.4, 118.7, 122.5, 123.4, 123.7, 148.8, 152.2, 154.8, 157.9, 159.5, 160.4 IR: ν(GeH) 2080 MS (70 eV, 50°C): [M]$^+$ (14), [M − 245]$^+$ (75) [27]
6-membered rings		
*16	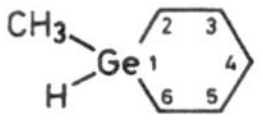	LiAlH$_4$ reduction of GeCH$_2$CH$_2$CH$_2$CH$_2$CH$_2$(CH$_3$)Br in refluxing ether for 3 h (41%) Ge(CH$_3$)(H)Cl$_2$ + BrMgCH$_2$CH$_2$CH$_2$CH$_2$CH$_2$MgBr (6%) oil; b.p. 43 to 46°C/18 ^{1}H NMR (CDCl$_3$): 0.23 (d, CH$_3$Ge, J = 3), 3.79 (m, HGe) [29] ^{13}C NMR (CDCl$_3$): −7.01 (CH$_3$Ge), 12.10 (C-2,6), 25.58 (C-3,5), 29.61 (C-4) ^{73}Ge NMR (CDCl$_3$): −65.3 (J(Ge, H) = 90.8) [25, 28, 30] IR (neat): ν(GeH) 2030 MS: [M]$^+$, [M − 15]$^+$, [M − 28]$^+$ [29] for the conformation, see p. 214
17	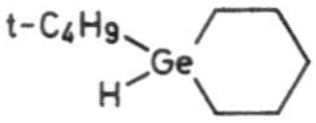	LiAlH$_4$ reduction of GeCH$_2$CH$_2$CH$_2$CH$_2$CH$_2$(C$_4$H$_9$-t)Cl in ether (50%) b.p. 98 to 102°C/20 ^{1}H NMR (CDCl$_3$): 1.00 (s, CH$_3$), 3.58 (m, HGe) ^{13}C NMR (CDCl$_3$): 9.44 (C-2,6), 20.79 (t-CGe), 26.60 (C-3,5), 28.68 (CH$_3$), 30.34 (C-4) ^{73}Ge NMR (CDCl$_3$): −27.8 IR (neat): ν(GeH) 2042 polymerizes rapidly in air [35]
*18	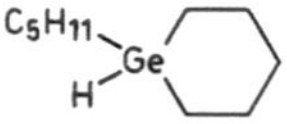	identified only by IR [2]; see p. 214
19	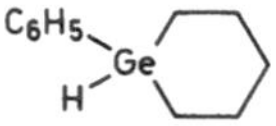	LiAlH$_4$ reduction of GeCH$_2$CH$_2$CH$_2$CH$_2$CH$_2$(C$_6$H$_5$)Br in refluxing ether for 20 h (82%) [2]; see also [30] b.p. 128°C/14 [2, 30]

Table 30 (continued)

No.	compound	preparation (yield) properties and remarks
		d^{20} = 1.1989; n_D^{20} = 1.5550 [2] ^{1}H NMR ($CDCl_3$): 4.36 (m, HGe) ^{13}C NMR ($CDCl_3$): 11.96 (C-2,6), 26.13 (C-3,5), 30.12 (C-4), 134.27, 138.00 (C-2′,6′ and C1′ of C_6H_5, respectively) ^{73}Ge NMR ($CDCl_3$): −69.4 [30] IR (neat): ν(GeH) 2023(s) [1, 2] 1:1 to 2:3 mixture of axial and equatorial conformers; the conformational energy of the C_6H_5 group is essentially zero in contrast to that in phenylcyclohexanes (based on NMR results and MNDO calculations) [30]
20	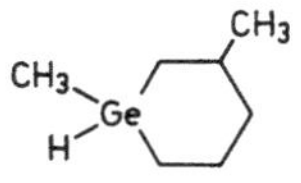	$LiAlH_4$ reduction of $GeCH_2CH(CH_3)CH_2CH_2CH_2(CH_3)Br$ (43%) cis,trans mixture in a 34:66 ratio b.p. 75 to 78 °C/27 ^{1}H NMR ($CDCl_3$): 0.23 (d, CH_3Ge, J = 3) for the cis isomer; 0.21 (d, CH_3Ge, J = 3) for the trans isomer [29] ^{13}C NMR ($CDCl_3$): −5.28 (q, CH_3Ge), 11.95 (t, C-6), 21.75 (t, C-2), 26.12 (t, C-5), 27.60 (q, CH_3-3), 31.97 (d, C-3), 38.43 (t, C-4) for the cis isomer; −7.58 (q, CH_3Ge), 11.00 (t, C-6), 20.66 (t, C-2), 24.61 (t, C-5), 27.29 (q, CH_3-3), 31.97 (d, C-3), 38.26 (t, C-4) for the trans isomer [28, 29, 30] ^{73}Ge NMR ($CDCl_3$): −60.3 [29, 30], −60.6 [28] for the cis isomer; −70.9 for the trans isomer [28, 29, 30] IR (neat): ν(GeH) 2040 MS: $[M]^+$, $[M - 15]^+$, $[M - 28]^+$ [29] predominant conformations: 1,3-equatorial CH_3 groups in the cis isomer; 1-axial,3-equatorial CH_3 groups in the trans isomer [29, 30]
21	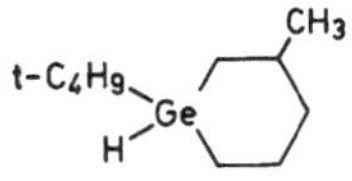	$LiAlH_4$ reduction of $GeCH_2CH(CH_3)CH_2CH_2CH_2(C_4H_9\text{-t})Cl$ (30%) cis,trans isomers in a 1:1 ratio b.p. 86 °C/14 ^{1}H NMR ($CDCl_3$): 0.96 (d, CH_3-3, J = 7), 1.10 (s, CH_3/C_4H_9-t), 3.56 (m, HGe) ^{13}C NMR ($CDCl_3$): 8.27 (C-6), 18.15 (C-2), 21.16 (t-CGe), 26.16 (C-5), 27.96 (CH_3-3), 28.58 (CH_3/C_4H_9-t), 33.57 (C-3), 38.72 (C-4) for the cis isomer; 8.45 (C-6), 17.76 (C-2), 21.16 (t-CGe), 24.63 (C-5), 26.68 (CH_3-3), 29.26 (CH_3/C_4H_9-t), 32.12 (C-3), 37.79 (C-4) for the trans isomer ^{73}Ge NMR ($CDCl_3$): −37.5 (unresolved) IR (neat): ν(GeH) 2040 [35]

References on p. 215

Table 30 (continued)

No.	compound	preparation (yield) properties and remarks
22	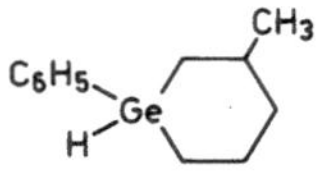	$LiAlH_4$ reduction of $GeCH_2CH(CH_3)CH_2CH_2CH_2(C_6H_5)Br$ (?) (53%) mixture of isomers; b.p. 113°C/11 ^1H NMR ($CDCl_3$): 0.99 (d, CH_3-3, J = 6.0), 4.41 (m, HGe) ^{13}C NMR ($CDCl_3$): 11.34 (C-6), 19.81 (C-2), 25.87 (C-5), 27.61 (CH_3-3), 33.37 (C-3), 38.27 (C-4), 134.35, 141.59(?) (C-2′,6′ and C-1′ of C_6H_5, respectively) for the cis isomer; 10.36 (C-6), 20.80 (C-2), 24.66 (C-5), 27.17 (CH_3-3), 31.90 (C-3), 37.92 (C-4), 134.57, 141.59 (C-2′,6′ and C-1′ of C_6H_5, respectively) for the trans isomer ^{73}Ge NMR: extensive broadening did not allow the signal to be observed predominant conformations: equatorial C_6H_5 and CH_3 in the cis isomer; axial C_6H_5 and equatorial CH_3 in the trans isomer [30]
23	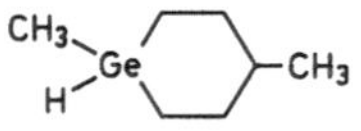	$LiAlH_4$ reduction of cis,trans-$GeCH_2CH_2CH(CH_3)CH_2CH_2(CH_3)Br$ (70:30 isomer ratio) (38%); cis,trans mixture in a 58:42 ratio b.p. 46 to 48°C/10 ^1H NMR ($CDCl_3$): 0.21 and 0.23 (d's, CH_3Ge, J = 3) for the cis and trans isomers, respectively [29] ^{13}C NMR ($CDCl_3$): −7.83 (CH_3Ge), 10.84 (C-2,6), 23.70 (CH_3-4), 33.51 (C-3,5), 35.19 (or 35.27?, C-4) for the cis isomer; −5.77 (CH_3Ge), 11.76 (C-2,6), 23.70 (CH_3-4), 34.81 (C-3,5), 35.27 (or 35.19?, C-4) for the trans isomer ^{73}Ge NMR ($CDCl_3$): −73.4 for the cis isomer; −61.5 for the trans isomer [25, 30] IR (neat): ν(GeH) 2040 MS: $[M]^+$, $[M-15]^+$, $[M-28]^+$ [29] predominant conformation: 1-axial, 4-equatorial CH_3 in the cis isomer; 1-equatorial, 4-equatorial CH_3 in trans-isomer [25, 29, 30]
24	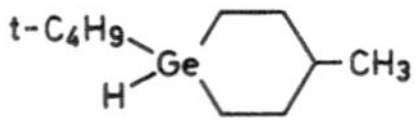	$LiAlH_4$ reduction of $GeCH_2CH_2CH(CH_3)CH_2CH_2(C_4H_9\text{-t})Cl$ (47%); cis,trans mixture in a 7:3 ratio b.p. 88°C/9 ^1H NMR ($CDCl_3$): 1.03 (d, CH_3-4, J = 6), 1.06 (s, CH_3/C_4H_9-t), 3.46 (m, HGe) for the cis isomer ^{13}C NMR ($CDCl_3$): 7.26 (C-2,6), 21.08 (t-CGe), 22.05 (CH_3-4), 28.88 (CH_3/C_4H_9-t), 33.69 (C-3,5), 34.16 (C-4) for the cis isomer; 8.38 (C-2,6), 20.17 (t-CGe), 23.74 (CH_3-4), 28.30 (CH_3/C_4H_9-t), 34.96 (C-3,5), 35.84 (C-4) for the trans isomer

Table 30 (continued)

No.	compound	preparation (yield) properties and remarks
		^{73}Ge NMR ($CDCl_3$): −41.2 (unresolved) IR (neat): ν(GeH) 2038 [35]
25		$LiAlH_4$ reduction of $GeCH_2CH_2CH(CH_3)CH_2CH_2(C_6H_5)Br$ (?) (73%) mixture of isomers; b.p. 118 °C/6 ^{1}H NMR ($CDCl_3$): 0.86 (d, CH_3-4, J = 6.0), 4.33 (m, HGe) ^{13}C NMR ($CDCl_3$): 10.19 (C-2,6), 23.32 (CH_3-4), 33.63 (C-3,5), 35.02 (C-4), 134.27 (C-2,6/C_6H_5), 141.46 (C-1/C_6H_5) for the cis isomer; 11.09 (C-2,6), 23.49 (CH_3-4), 34.63 (C-3,5), 35.23 (C-4), 134.40 (C-2,6/C_6H_5), 141.46(? C-1/C_6H_5) for the trans isomer ^{73}Ge NMR ($CDCl_3$): −74.9 (the signal could not be resolved into two peaks) predominant conformation: axial C_6H_5 and equatorial CH_3 in the cis isomer; equatorial C_6H_5 and CH_3 in the trans isomer [30]
26	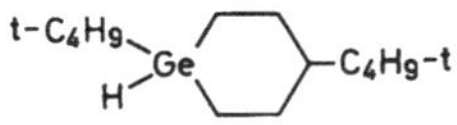	$LiAlH_4$ reduction of $GeCH_2CH_2CH(C_4H_9\text{-t})CH_2CH_2(C_4H_9\text{-t})Cl$ (12%); cis,trans isomers in a 3:1 ratio b.p. 106 °C/12 ^{1}H NMR ($CDCl_3$): 0.83, 1.07 (s's, C_4H_9-t), 3.56 (m, HGe) for the cis isomer ^{13}C NMR ($CDCl_3$): 10.18 (C-2,6), 21.65 (t-CGe), 27.74 (C-3,5), 27.99 (CH_3/C_4H_9-4), 29.44 (CH_3/C_4H_9Ge), 33.76 (t-C/C_4H_9-4), 51.96 (C-4) for the cis isomer; 10.01 (C-2,6), 28.14 (CH_3/C_4H_9-4), 28.38 (C-3,5), 28.72 (CH_3/C_4H_9Ge), 52.54 (C-4) (t-C unidentified) for the trans isomer ^{73}Ge NMR: no signal recorded in spite of a reasonably high concentration IR (neat): ν(GeH) 2035 [35]
27	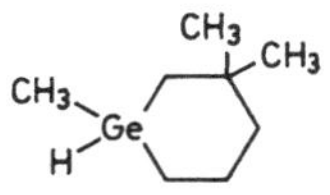	$Ge(CH_3)(H)Cl_2$ + $BrMgCH_2C(CH_3)_2CH_2CH_2CH_2MgBr$ (6%) b.p. 90 to 93 °C/13 ^{1}H NMR ($CDCl_3$): 0.21 (d, CH_3Ge, J = 3), 0.98 (s, CH_3-3,3′), 3.88 (m, HGe) ^{13}C NMR ($CDCl_3$ or C_6D_6): 11.31 (C-6), 21.33(C-5), 26.48 (C-2), 30.07 (CH_3-3), 32.46 (C-3), 34.15 (CH_3-3′), 42.42 (C-4); δ(CH_3Ge) not given ^{73}Ge NMR ($CDCl_3$): −70.6 IR (neat): ν(GeH) 2050 MS: $[M]^+$, $[M-15]^+$, $[M-28]^+$ predominant conformation: equatorial CH_3Ge [28]

References on p. 215

Table 30 (continued)

No.	compound	preparation (yield) properties and remarks
28	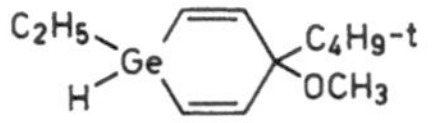	LiAlH$_4$ reduction of GeCH=CHC(C$_4$H$_9$-t)(OCH$_3$)CH=CH(C$_2$H$_5$)Cl (ca. 1:2 mole ratio) in refluxing ether for 3 h (71%) colorless liquid; mixture of cis,trans isomers b.p. 75 to 85 °C/0.01 ^{1}H NMR (CDCl$_3$): 0.90 to 1.20 (m, C$_2$H$_5$), 0.93 (s, CH$_3$/C$_4$H$_9$-t), 3.10 (s, CH$_3$O), 4.30, 4.35 (t's, HGe), 6.03, 6.10, 6.28, 6.38 (H-2,6), 6.41, 6.66 (H-3,5, ^{3}J(H, H = 14) IR (neat): ν(CH/OCH$_3$) 2835(w), ν(GeH) 2040(s), ν(C=C) 1610(m), ν(COC) 1080(s), δ(cis-CH=CH) 705(m) [18] thermolysis at 600 °C/15 to 20 Torr → Nos. 35 and 36 in a 76:24 ratio [21]
29	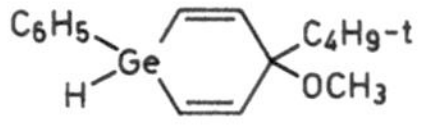	LiAlH$_4$ reduction of GeCH=CHC(C$_4$H$_9$-t)(OCH$_3$)CH=CH(C$_6$H$_5$)Cl (ca. 1:2 mole ratio) in refluxing ether for 3 h (73%) yellowish green liquid; mixture of cis,trans isomers b.p. 110 to 120 °C/0.01, 115 to 125 °C/0.01 ^{1}H NMR (CDCl$_3$): 0.93, 0.98 (s's, C$_4$H$_9$-t), 3.12, 3.20 (s's, CH$_3$O), 4.74, 4.84 (s's, HGe), 6.14, 6.22, 6.37, 6.47 (H-2,6), 6.52, 6.75 (H-3,5), 7.05 to 7.55 (m, C$_6$H$_5$); ^{3}J(H-2,3) = ^{3}J(H-5,6) = 13.8 IR (neat): ν(CH/OCH$_3$) 2830(w), ν(GeH) 2040(s), ν(C=C) 1600(m), ν(COC) 1070(s), δ(cis-CH=CH) 695(s) [18]
30	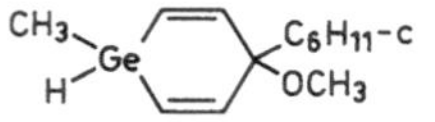	LiAlH$_4$ reduction of GeCH=CHC(C$_6$H$_{11}$-c)(OCH$_3$)CH=CH(CH$_3$)Cl (5:12 mole ratio) in refluxing ether for 3 h (47%) colorless liquid; mixture of cis,trans isomers b.p. 130 to 140 °C (bath)/0.01 ^{1}H NMR (CDCl$_3$): 0.24 to 0.40 (m (or two d's with ^{3}J(H, H) = 6?), CH$_3$Ge), 0.66 to 2.10 (m, C$_6$H$_{11}$-c), 3.08 (s, CH$_3$O), 4.25, 4.32 (q's, HGe), 6.20, 6.28 (s's, H-2,3,5,6) IR (neat): ν(CH/OCH$_3$) 2830(m), ν(GeH) 2060(s), ν(C=C) 1610(m), ν(COC) 1080(s), δ(cis-CH=CH) 705 [18]
31	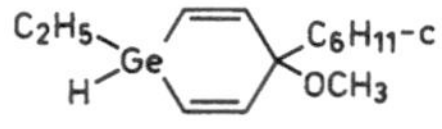	LiAlH$_4$ reduction of GeCH=CHC(C$_6$H$_{11}$-c)(OCH$_3$)CH=CH(C$_2$H$_5$)Cl (ca. 1:2 mole ratio) in refluxing ether for 3 h (61%) colorless liquid; mixture of cis,trans isomers b.p. 135 to 145 °C/0.01 ^{1}H NMR (CDCl$_3$): 0.60 to 2.10 (m, C$_2$H$_5$ and C$_6$H$_{11}$-c), 3.09 (s, CH$_3$O), 4.30, 4.36 (t's, HGe), 6.20, 6.29 (s's, H-2,3,5,6) IR (neat): ν(CH/OCH$_3$) 2820(w), ν(GeH) 2020(s), ν(C=C) 1610(m), ν(COC) 1065(s), δ(cis-CH=CH) 690(m)

Table 30 (continued)

No.	compound	preparation (yield) properties and remarks
		MS: $[M]^+$ (<1), $[M - CH_3]^+$ (15), $[M - C_6H_{11}]^+$ (70), $[M - C_6H_{11} - CH_3OH]^+$ (20), $[GeOCH_3]^+$ (100); fragmentation pathway displayed [18]
32		$LiAlH_4$ reduction of $GeCH{=}CHC(C_6H_{11}\text{-c})(OCH_3)CH{=}CH(C_6H_5)Cl$ (ca. 1:2 mole ratio) in refluxing ether for 3 h (87%) light yellowish liquid; mixture of cis,trans isomers b.p. 150 to 155 °C/0.01 ^{1}H NMR ($CDCl_3$): 0.60 to 2.10 (m, C_6H_{11}-c), 3.13, 3.20 (s's, CH_3O), 4.74, 4.83 (s's, HGe), 6.38, 6.40 (s's, H-2,3,5,6), 7.10 to 7.60 (m, C_6H_5) IR (neat): ν(CH/OCH_3) 2810(m), ν(GeH) 2025(s), ν(C=C) 1600(m), ν(COC) 1060(s), δ(cis-CH=CH) 685(m) MS: $[M]^+$ (2), $[M - CH_3]^+$ (35), $[M - C_6H_{11}]^+$ (100), $[M - C_6H_{11} - CH_3OH]^+$ (22), $[GeOCH_3]^+$ (55); fragmentation pathway displayed [18]
33	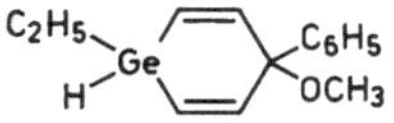	slow addition of $GeCH{=}CHC(C_6H_5)(OCH_3)CH{=}CH(C_2H_5)Cl$ to a refluxing suspension of $LiAlH_4$ in ether (cf. preparation of compound No. 37) no data published [18]
34		$LiAlH_4$ reduction of trans-$GeCH{=}CHC(C_6H_5)(OCH_3)CH{=}CH(C_6H_5)Cl$ (ca. 1:2 mole ratio) in refluxing ether for 3 h (74%) colorless crystals; only one isomer (probably the trans isomer) m.p. 95 to 96 °C (from petroleum ether) ^{1}H NMR ($CDCl_3$): 3.40 (s, CH_3O), 4.93 (s, HGe), 6.13, 6.36 (H-2,6), 6.39, 6.62 (H-3,5, ^3J(H, H) = 13.8), 7.00 to 7.66 (m, C_6H_5Ge and C_6H_5C) IR (KBr): ν(CH/OCH_3) 2830(m), ν(GeH) 2040(s), ν(C=C) 1605(m), 1595(m), ν(COC) 1080(s), 1070(s), δ(cis-CH=CH) 680(m), 670(m) [18]
35	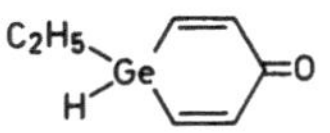	thermolysis of compound No. 28 (39%) along with No. 36, from which it was chromatographically separated on SiO_2 followed by distillation yellowish liquid with a stinging odor b.p. 107 to 112 °C/15 ^{1}H NMR ($CDCl_3$): 0.93 to 1.27 (m, C_2H_5), 4.53 (m, HGe), 6.91 (H-2,6), 7.24 (H-3,5), ^3J(H-2,3) = ^3J(H-5,6) = 13.83, ^4J(H-2,6) = ^4J(H-3,5) = 2.27, ^5J(H-2,5) = ^5J(H-3,6) = 0.14; spectrum displayed IR (neat): ν(GeH) 2080, ν(CO) 1635, ν(C=C) 1580

References on p. 215

Table 30 (continued)

No.	compound	preparation (yield) properties and remarks
35 (continued)		UV: $\lambda_{max}(\varepsilon)$ = 231(14800) in C_6H_{14}; bathochromic shift to 246(50000) or 249(10600) in the presence of small amounts of $N(C_2H_5)_3$ or CF_3COOH MS: $[M]^+$ (14), $[M-H]^+$ (11), $[M-CH_3]^+$ (18), $[M-C_2H_5]^+$ (100), $[M-C_2H_5-C_2H_2]^+$ (41), $[M-C_2H_5-C_2H_2-CO]^+$ (48) [21]
36	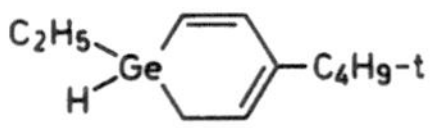	thermolysis of No. 28 (10%) along with No. 35; see above No. 28 + Na in liquid NH_3 at −70°C (38%) colorless liquid; b.p. 60 to 65°C/0.01 ^{1}H NMR ($CDCl_3$): 0.73 to 1.17 (m, C_2H_5), 1.05 (s, C_4H_9-t), 1.70 (m, H-6), 4.03 (m, HGe), 5.83 (t, H-5, J(H-5,6) ≈ 6), 6.06 to 6.30 (H-2), 6.73 to 6.97 (H-3, J(H-2,3) = 14) [21]
37		$LiAlH_4$ reduction of $GeCH{=}CHC(C_6H_5)(OCH_3)CH{=}CH(C_2H_5)Cl$ (ca. 1:2 mole ratio) in ether at 20°C (3 h) and refluxing for 2 h (95%); see also preparation of No. 33 colorless liquid; mixture of enantiomers b.p. 95 to 105°C (bath)/ 0.01 ^{1}H NMR ($CDCl_3$): 0.74 to 1.15 (C_2H_5), 1.82, 2.01 (H-6,6′), 4.13 (complex m, HGe), 6.15 (H-5), 6.33 (H-2), 6.95 (H-3); ^{2}J(H-6,6′) = −18.25, ^{3}J(H-2,3) = 13.33, ^{3}J(H-5,6), ^{3}J(H-5,6′) = 5.75 and 5.52, ^{3}J(HGe, H-6,6′) = 4.50 and 4.25, ^{4}J(H-3,5) = 1.33, ^{4}J(HGe, H-5) = 1.15, ^{5}J(H-2,5) = 1.14 [18]
38	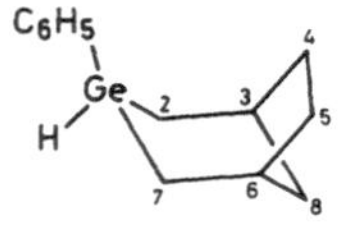 Ge-R = GeC_7H_{12}	$LiAlH_4$ reduction of $GeC_7H_{12}(C_6H_5)Br$ (55:45 isomeric mixture) in refluxing ether for 14 h (95%) liquid; 65:35 mixture of endo,exo isomers ^{1}H NMR ($CDCl_3$): 0.8 to 2.3 (m, 10 H), 4.0 to 4.4 (m, 3 H), 7.2 to 7.8 (m, 5 H) ^{13}C NMR ($CDCl_3$): 21.0 (C-2,7), 31.9 (C-4,5), 34.4 (C-3,6), 41.4 (C-8) for the H_{endo} isomer; 22.1 (C-2,7), 32.3 (C-4,5), 35.0 (C-3,6), 41.8 (C-8) for the H_{exo} isomer; 127.9, 128.4, 133.6, 134.0 (C_6H_5) [34]

11- and 12-membered rings

No.	compound	preparation (yield) properties and remarks
39		$LiAlH_4$ reduction of $GeCH_2(CH_2)_8CH_2(CH_3)Cl$ in ether (92%) treatment of $GeCH_2(CH_2)_3CO(CH_2)_4CH_2(CH_3)Cl$ with Zn-HCl at 40°C; obtained as a 3:2 mixture with compound VIII, p. 214 b.p. 114°C/12 d^{20} = 1.0778; n_D^{20} = 1.4964 [11]
*40	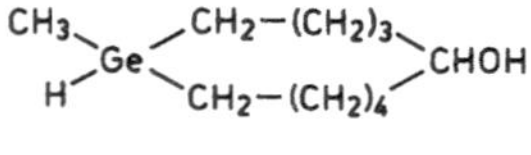	$LiAlH_4$ reduction of the digermoxane $(GeCH_2(CH_2)_3CO(CH_2)_4CH_2(CH_3))_2O$ in refluxing ether for 5 h (78%)

Table 30 (continued)

No.	compound	preparation (yield) properties and remarks
		$LiAlH_4$ reduction of $GeCH_2(CH_2)_3CO(CH_2)_4CH_2(CH_3)Cl$ in ether (difficult to isolate) b.p. 90 °C/0.01 d^{20} = 1.1543; n_D^{20} = 1.5122 ^1H NMR (CCl_4): 0.18 (d, CH_3Ge, ^3J = 4), 0.70 to 1.30 (m, CCH_2Ge), 1.20 to 1.90 (m, CCH_2C), 3.05 (s, OH), 3.50 to 3.95 (m, HGe and OCH); also monitored in the presence of $Eu(fod)_3$ shift reagent IR: ν(OH) 3500 to 3100, ν(GeH) 2000 (neat?); ν(OH) 3610 in CCl_4 [11]
*41	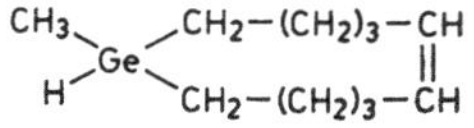	not isolated; see p. 214 ^1H NMR: 3.5 to 3.9 (HGe), 5.3 (CH=) IR: ν(GeH) 2010; ν(C=C) detected [13]
42	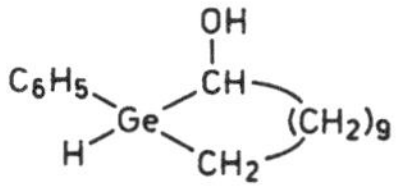	heating $Ge(C_6H_5)(CH(OH)(CH_2)_8CH{=}CH_2)H_2$ in C_6H_{14} in a sealed tube at 100 °C (12 h) in the presence of AIBN (70%) [12]; see also [10] b.p. 150 to 160 °C/0.06 (dec.) [10, 12], 150 to 170 °C/0.06 (dec.) [12] ^1H NMR (CCl_4): 3.48 (m, H-2), 4.40 (m, HGe) IR (neat): ν(OH) 3340, ν(GeH) 2045 [10, 12]

*Further information:

$GeCH_2CH_2CH_2(C_4H_9)H$ (Table **30**, No. **1**) reacted with $CH_2{=}CHC_4H_9$ or $CH{\equiv}CC_6H_{13}$ at 150 °C in the absence of any catalyst to yield 50% $GeCH_2CH_2CH_2(C_4H_9)C_6H_{13}$ or 66% $GeCH_2CH_2CH_2(C_4H_9)CH{=}CHC_6H_{13}$ (cis : trans = 9 : 1), respectively. Refluxing with CCl_4, $C_6H_{13}Br$, or CH_3I gave $GeCH_2CH_2CH_2(C_4H_9)X$ compounds (X = Cl, Br, or I) [7].

$GeCH_2CH(CH_3)CH_2(C_4H_9)H$ (Table **30**, No. **2**) reacted with CH_3I under reflux to give $GeCH_2CH(CH_3)CH_2(C_4H_9)I$ [7].

$GeCH_2CH_2CH_2CH_2(C_4H_9)H$ (Table **30**, No. **7**) reacted with $Ge(C_2H_5)_3CH{=}CH_2$ to give compound III with 70% yield [3].

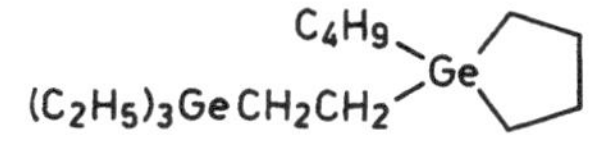

III

$GeCH(CH_3)CH_2CH_2CH_2(CH_3)H$ (Table **30**, No. **9**) was prepared by reducing several $GeCH(CH_3)CH_2CH_2CH_2(CH_3)X$ compounds. The stereochemistry of these reduction processes depends on the electronic character of the nucleophile and the nature of the leaving group [14, 15, 20, 22, 23]; reactions of $GeCH(CH_3)CH_2CH_2CH_2(CH_3)X$ compounds at room temperature are listed below (inv. and ret. stand for inversion and retention of configuration at the Ge atom):

starting material X group	Z/E ratio	reducing agent	No. 9 Z/E ratio	predominant stereochemistry	Ref.
Cl	50/50	$LiAlH_4$-ether	52/48 (80% yield)		[15]
	70/30	$LiAlH_4$-ether	65/35	inv. (87%)	[15, 20, 22]
		$LiAlH_4$-THF	68/32	inv. (94%)	[20, 22]
		$LiBH_4$-ether	68/32	inv. (94%)	[20, 22]
Br	55/45	$LiAlH_4$-ether	55/45 (90% yield)	inv.	[14, 15]
OCH_3	30/70	$LiAlH_4$-ether	70/30	ret. (100%)	[20, 22]
		$LiAlH_4$-THF	42/58	inv. (70%)	[20, 22]
		$LiBH_4$-ether	30/70	inv. (100%)	[20, 22]
OC_2H_5	30/70	$LiAlH_4$-ether	68/32	ret. (94%)	[20, 22]
		$LiAlH_4$-THF	40/60	inv. (75%)	[20, 22]
		$LiBH_4$-ether	38/62	inv. (80%)	[20, 22]
OC_3H_7-i	20/80	$LiAlH_4$-ether	70/30	ret. (85%)	[20, 22]
		$LiAlH_4$-THF	22/78	inv. (92%)	[20, 22]
		$LiBH_4$-ether	45/55	inv. (58%)	[20, 22]
		$Al(C_4H_9\text{-}i)_2H$-ether, C_6H_{14}	70/30	ret. (85%)	[22]
		$LiAlH_4$ + LiBr-THF	62/38	ret. (70%)	[22]
SC_2H_5	32/68	$LiAlH_4$-ether	32/68	inv. (100%)	[20, 22]
		$LiAlH_4$-THF	35/65	inv. (92%)	[20, 22]
		$LiBH_4$-ether	32/68	inv. (100%)	[20, 22]
		$Al(C_4H_9\text{-}i)_2H$-ether, C_6H_{14}	45/55	inv. (64%)	[22]
		$Al(C_4H_9\text{-}i)_2H$ + LiBr-ether, C_6H_{14}	58/42	ret. (70%)	[22]
SC_4H_9-t	20/80	$LiAlH_4$-ether	20/80	inv. (100%)	[20, 22]
		$LiAlH_4$-THF	20/80	inv. (100%)	[20, 22]
		$LiBH_4$-ether	20/80	inv. (100%)	[20, 22]
		$Al(C_4H_9\text{-}i)_2H$-ether, C_6H_{14}	30/70	inv. (83%)	[22]
		$Al(C_4H_9\text{-}i)_2H$ + LiBr-ether, C_6H_{14}	45/55	inv. (58%)	[22]
$N(C_2H_5)_2$	20/80	$LiAlH_4$-ether	75/25	ret. (90%)	[20, 22]
		$LiAlH_4$-THF	80/20	ret. (100%)	[20, 22]
		$LiBH_4$-ether	25/75	inv. (90%)	[20, 22]
$P(C_2H_5)_2$	30/70	$LiAlH_4$-ether	70/30	ret. (100%)	[20, 22, 23]
	60/40	$LiAlH_4$-ether	42/58	ret. (90%)	[23]
	30/70	$LiAlH_4$-THF	62/38	ret. (80%)	[20, 22, 23]
	60/40	$LiAlH_4$-THF	45/55	ret. (75%)	[23]
	30/70	$LiBH_4$-ether	30/70	inv. (100%)	[20, 22, 23]
	60/40	$LiBH_4$-ether	60/40	inv. (100%)	[23]

Halogenations of isomer-enriched samples of $\overline{GeCH(CH_3)CH_2CH_2C}H_2(CH_3)H$ yielded $\overline{GeCH(CH_3)CH_2CH_2C}H_2(CH_3)X$ compounds (X = Cl or Br) [15]; these reactions are listed below:

Z/E ratio of No. 9	reactant, solvent (t in °C)	Z/E ratio of the product	predominant stereochemistry
20/80	Cl_2, CCl_4 (−20)	70/30	retention
70/30	Cl_2, CCl_4 (−20)	33/67	retention
20/80	CCl_4, − (+20)	72/28	retention
70/30	CCl_4, − (+20)	35/65	retention
20/80	SO_2Cl_2, C_6H_6 (+20)	55/45	retention
70/30	SO_2Cl_2, C_6H_6 (+20)	40/60	retention
70/30	CH_3OCH_2Cl + $AlCl_3$ cat. (+20)	42/58	retention
20/80	Br_2, CCl_4 (−20)	53/47	epimerization
70/30	Br_2, CCl_4 (−20)	55/45	epimerization
20/80	$CHBr_3$ (+20)	55/45	epimerization
70/30	$CHBr_3$ (+20)	55/45	epimerization

Hydrogermylation reactions of isomer-enriched samples of the title compound (Z/E = 30/70 or 80/20) were performed with CH_2=CH_2 or CH≡CH at room temperature in C_6H_6 in the presence of H_2PtCl_6 as catalyst and with CH_2=CHC_4H_9 under the same conditions or without catalyst by heating (100°C for 20 h, no solvent). In all cases, the $\overline{GeCH(CH_3)CH_2CH_2C}H_2(CH_3)R$ product (R = C_2H_5, CH=CH_2, or C_6H_{13}, respectively) was formed while retaining the configuration at the Ge atom [14, 16].

No. 9 reacted with $LiP(C_2H_5)_2$ in C_6H_{14} at 60°C (20 h) in a sealed tube to yield 51% $\overline{GeCH(CH_3)CH_2CH_2C}H_2(CH_3)P(C_2H_5)_2$, predominantly retaining the configuration at the Ge atom [17].

$\overline{GeC(CH_3)=C(CH_3)C(CH_3)=C}CH_3(C_6H_5)H$ (Table **30**, No. **13**) reacted with an equimolar amount of LiC_4H_9 in THF-C_6H_{14} at −70°C (1 h) giving $\overline{GeC(CH_3)=C(CH_3)C(CH_3)=C}CH_3(C_6H_5)Li$ [32]. With CCl_4 at 90°C (4 h) in a sealed tube $\overline{GeC(CH_3)=C(CH_3)C(CH_3)=C}CH_3(C_6H_5)Cl$ was obtained [33]. Compound IV resulted with 65% yield by treating No. 13 with $Co_2(CO)_8$ (2:1 mole ratio) in C_5H_{12} at 35°C (2 d) [36].

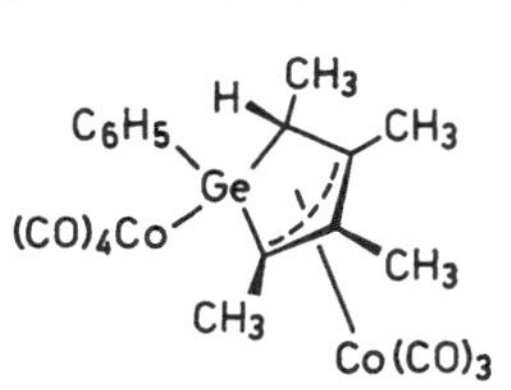

IV

$\overline{GeC(CH_3)=C(CH_3)C(CH_3)=C}CH_3(C_6H_5)D$ was obtained by deuteriolysis of $\overline{GeC(CH_3)=C(CH_3)C(CH_3)=C}CH_3(C_6H_5)Li$ in THF. ^1H NMR spectrum (in C_6D_6): δ(ppm) = 1.87, 2.04 (s's, CH_3), 7.27 and 7.58 (C_6H_5). IR spectrum: ν(GeD) band at 1459 cm^{-1}. The mass spectrum shows the ions $[M]^+$ (72), $[M - C_6H_5]^+$ (17), $[GeC_6H_5]^+$ (100), and $[M - C_6H_5 - C_2(CH_3)_2]^+$ (19) [32].

References on p. 215

$\overline{GeC(C_6H_5)=C(C_6H_5)C(C_6H_5)=C}C_6H_5(C_6H_5)H$ (Table **30**, No. **14**) was also obtained by treating $\overline{GeC(C_6H_5)=C(C_6H_5)C(C_6H_5)=C}C_6H_5(C_6H_5)Li$ (see below) with H_2O or with $FeCl_2$ in THF in the absence of water. In the latter case a series of reactions was assumed to occur involving hydrogen abstraction from the solvent [8].

Treating No. 14 with LiC_4H_9 in THF at −78 °C produced a bright red color which was attributed to the formation of $\overline{GeC(C_6H_5)=C(C_6H_5)C(C_6H_5)=C}C_6H_5(C_6H_5)Li$ [5, 8]. The same product was obtained (along with $Ge(C_6H_5)_3H$) using an equimolar amount of $Ge(C_6H_5)_3Li$. Equilibrium studies of this reaction indicated that No. 14 is at least 10^6-times more acidic than $Ge(C_6H_5)_3H$ [8].

The reaction with an equimolar amount of $Fe_2(CO)_9$ in refluxing ether (1.5 h) gave compound V (R = C_6H_5) with a nearly quantitative yield [19].

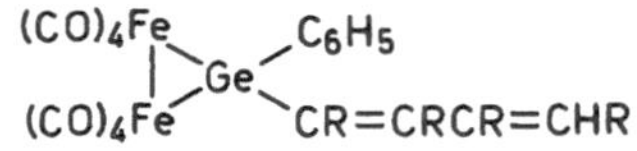

V

$\overline{GeCH_2C(CH_3)_2C}_6H_2(C_4H_9\text{-t})_2(C_6H_2(C_4H_9\text{-t})_3\text{-2,4,6})H$ (Table **30**, No. **15**) was formed by rearrangement at 20 °C of the germylene $Ge(C_6H_2(C_4H_9\text{-t})_3\text{-2,4,6})_2$ which had been prepared from $GeCl_2 \cdot C_4H_8O_2$ and $LiC_6H_2(C_4H_9\text{-t})_3$-2,4,6 at −70 °C [27].

$\overline{GeCH_2CH_2CH_2CH_2C}H_2(CH_3)H$ (Table **30**, No. **16**). Consistent with molecular mechanics calculations, the compound was shown by ^{13}C and ^{73}Ge NMR to be a ca. 3:2 mixture of axial and equatorial conformers [29]; see also [31]. Attempts to observe the two conformers by freezing the inversion process on the NMR time scale failed because of the very low barrier for inversion. The folding of the ring is reduced around the Ge atom of No. 16 compared to that of c-$C_6H_{11}CH_3$ and $\overline{SiCH_2CH_2CH_2CH_2C}H_2(CH_3)H$ [29]; see also [38].

$\overline{GeCH_2CH_2CH_2CH_2C}H_2(C_5H_{11})H$ (Table **30**, No. **18**) was probably formed along with compounds VI and VII by reacting $GeCl_4$ with $BrMg(CH_2)_5MgBr$, followed by $LiAlH_4$ reduction of the reaction mixture. It was detected by a ν(GeH) band in the fraction of compound VII (b.p. 124 °C/30 Torr) and was eliminated by treating this fraction with concentrated H_2SO_4 [2]. The ν(GeH) band of $\overline{GeCH_2CH_2CH_2CH_2C}H_2(R')H$ compounds (R′ = unnamed aliphatic group) was reported to be at 2013 cm^{-1} [1].

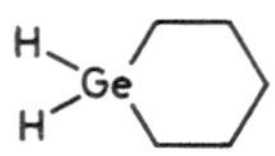

VI

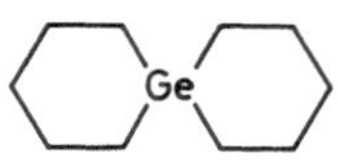

VII

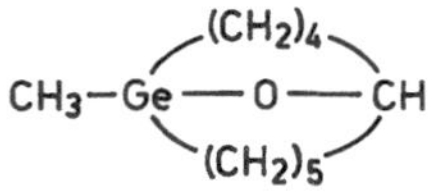

VIII

$\overline{GeCH_2(CH_2)_3CH(OH)(CH_2)_4C}H_2(CH_3)H$ (Table **30**, No. **40**) was also obtained by $LiAlH_4$ reduction of compound VIII in ether. On heating, the compound slowly lost hydrogen. In the presence of Raney nickel, evolution of hydrogen occurred at 20 °C, regenerating compound VIII within a few days; small amounts of $Ge(CH_3)H_3$ and $CH_3(CH_2)_3CH(OH)(CH_2)_4CH_3$ were also formed by hydrogenation [11].

$\overline{GeCH_2(CH_2)_3CH=CH(CH_2)_3C}H_2(CH_3)H$ (Table **30**, No. **41**) was obtained along with its isomer IX in a 3:7 ratio by $LiAlH_4$ reduction of $\overline{GeCH_2(CH_2)_3CH=CH(CH_2)_3C}H_2(CH_3)Cl$ or the digermoxane $(\overline{GeCH_2(CH_2)_3CH=CH(CH_2)_3C}H_2(CH_3)\text{-})_2O$. The isomer mixture could not be separated by distillation, but on heating in the presence of AIBN to 80 °C, No. 41 polymerized and IX was isolated unchanged [13].

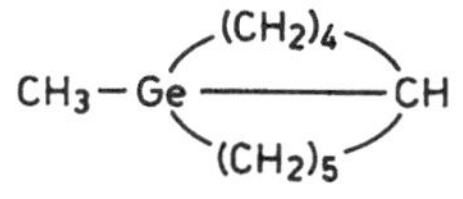

IX

References:

[1] Mathis, R.; Mazerolles, P. (Bull. Soc. Chim. Fr. **1962** 1913/4).
[2] Mazerolles, P. (Bull. Soc. Chim. Fr. **1962** 1907/13).
[3] Mazerolles, P.; Dubac, J. (C. R. Hebd. Seances Acad. Sci. **257** [1963] 1103/6).
[4] Mazerolles, P.; Dubac, J.; Lesbre, M. (J. Organomet. Chem. **5** [1966] 35/47).
[5] Curtis, M. D. (J. Am. Chem. Soc. **89** [1967] 4241/2).
[6] Duffield, A. M.; Djerassi, C.; Mazerolles, P.; Dubac, J.; Manuel, G. (J. Organomet. Chem. **12** [1968] 123/32).
[7] Mazerolles, P.; Dubac, J.; Lesbre, M. (C. R. Seances Acad. Sci. C **266** [1968] 1794/6).
[8] Curtis, M. D. (J. Am. Chem. Soc. **91** [1969] 6011/8).
[9] Massol, M.; Rivière, P.; Barrau, J.; Satgé, J. (C. R. Seances Acad. Sci. C **270** [1970] 237/9).
[10] Rivière, P.; Satgé, J. (Angew. Chem. **83** [1971] 286/7; Angew. Chem. Int. Ed. Engl. **10** [1971] 267/8).
[11] Mazerolles, P.; Faucher, A. (J. Organomet. Chem. **63** [1973] 195/203).
[12] Rivière, P.; Satgé, J. (J. Organomet. Chem. **49** [1973] 173/89).
[13] Mazerolles, P.; Faucher, A. (J. Organomet. Chem. **85** [1975] 159/63).
[14] Dubac, J.; Mazerolles, P.; Joly, M.; Piau, F. (J. Organomet. Chem. **127** [1977] C69/C74).
[15] Dubac, J.; Mazerolles, P.; Joly, M.; Cavezzan, J. (J. Organomet. Chem. **165** [1979] 163/73).
[16] Dubac, J.; Mazerolles, P.; Cavezzan, J.; Joly, M. (J. Organomet. Chem. **165** [1979] 175/85).
[17] Dubac, J.; Cavezzan, J.; Mazerolles, P.; Escudié, J.; Couret, C.; Satgé, J. (J. Organomet. Chem. **174** [1979] 263/74).
[18] Märkl, G.; Rudnick, D. (J. Organomet. Chem. **181** [1979] 305/28).
[19] Curtis, M. D.; Butler, W. M.; Scibelli, J. (J. Organomet. Chem. **192** [1980] 209/18).
[20] Dubac, J.; Cavezzan, J.; Laporterie, A.; Mazerolles, P. (J. Organomet. Chem. **197** [1980] C15/C18).
[21] Märkl, G.; Hofmeister, P.; Rudnick, D.; Schiessl, R. (J. Organomet. Chem. **193** [1980] 175/89).
[22] Dubac, J.; Cavezzan, J.; Laporterie, A.; Mazerolles, P. (J. Organomet. Chem. **209** [1981] 25/36).
[23] Dubac, J.; Escudié, J.; Couret, C.; Cavezzan, J.; Satgé, J.; Mazerolles, P. (Tetrahedron **37** [1981] 1141/51).
[24] Rivière, P.; Castel, A.; Satgé, J. (J. Organomet. Chem. **232** [1982] 123/35).
[25] Takeuchi, Y.; Shimoda, M.; Tomoda, S. (Magn. Reson. Chem. **23** [1985] 580/1).
[26] Burns, G. T.; Colomer, E.; Corriu, R. J. P.; Lheureux, M.; Dubac, J.; Laporterie, A.; Iloughmane, H. (Organometallics **6** [1987] 1398/406).
[27] Lange, L.; Meyer, B.; du Mont, W. W. (J. Organomet. Chem. **329** [1987] C17/C20).
[28] Takeuchi, Y.; Ichikawa, Y.; Tanaka, K.; Kakimoto, N. (Bull. Chem. Soc. Jpn. **61** [1988] 2875/80).
[29] Takeuchi, Y.; Shimoda, M.; Tanaka, K.; Tomoda, S.; Ogawa, K.; Suzuki, H. (J. Chem. Soc. Perkin Trans. II **1988** 7/13).

[30] Takeuchi, Y.; Tanaka, K.; Harazono, T.; Ogawa, K.; Yoshimura, S. (Tetrahedron **44** [1988] 7531/9).

[31] Allinger, N. L.; Quinn, M. I.; Chen, K.; Thompson, B.; Frierson, M. R. (J. Mol. Struct. **194** [1989] 1/18).

[32] Dufour, P.; Dubac, J.; Dartiguenave, M.; Dartiguenave, Y. (Organometallics **9** [1990] 3001/3).

[33] Dufour, P.; Dartiguenave, M.; Dartiguenave, Y.; Dubac, J. (J. Organomet. Chem. **384** [1990] 61/9).

[34] Sommese, A. G.; Cremer, S. E.; Campbell, J. A.; Thompson, M. R. (Organometallics **9** [1990] 1784/92).

[35] Takeuchi, Y.; Tanaka, K.; Harazono, T.; Yoshimura, S. (Bull. Chem. Soc. Jpn. **63** [1990] 708/15).

[36] Dufour, P.; Menu, M.-J.; Dartiguenave, M.; Dartiguenave, Y.; Dubac, J. (Organometallics **10** [1991] 1645/7).

[37] Takeuchi, Y.; Tanaka, K.; Harazono, T. (Bull. Chem. Soc. Jpn. **64** [1991] 91/8).

[38] Takeuchi, Y.; Zicmane, I.; Manuel, G.; Boukherroub, R. (Bull. Chem. Soc. Jpn. **66** [1993] 1732/7).

1.3.2 Diorganogermanium Dihydrides

1.3.2.1 GeR_2H_2 Compounds

1.3.2.1.1 Dimethylgermane, Dimethylgermanium Dihydride, $Ge(CH_3)_2H_2$

Preparation and Formation. $Ge(CH_3)_2H_2$ was prepared by reducing $Ge(CH_3)_2Br_2$ with LiH in dioxane (99% yield) [2] or $Ge(CH_3)_2Cl_2$ with NaH in mineral oil at 110 °C (3 h) in the presence of $B(C_6H_5)_3$ [18]. Treating $Ge(CH_3)_2Br_2$ with aqueous $NaBH_4$ in 1 M HBr at 35 to 50 °C and 400 to 10 Torr followed by condensation into a trap at −150 °C, yielded 94% of the title compound [7]; see also [14]. The reduction of $Ge(CH_3)_2Cl_2$ was also carried out with $LiAlH_4$ in $O(C_4H_9)_2$ at 0 °C (95% yield) [35] (see also [28]) or 40 °C [23], and with a 100% excess of $Li[AlH(OC_4H_9\text{-}t)_3]$ in refluxing dioxane (60 to 70% yield) [12] and subsequent trap-to-trap fractionation. $Ge(CH_3)_2H_2$ was also formed quantitatively (according to GLC) from $Ge(CH_3)_2(\overline{C{=}CHCH_2CH_2O})_2$ and $LiAlH_4$ (1 : 2 mole ratio) in THF at 25 °C (72 h) [52]; see also [57].

Adding $(CH_3)_2\overline{GeP(C_6H_5)Ge(CH_3)_2P(C_6H_5)CH_2C}H_2$ to excess $LiAlH_4$ in ether, refluxing for 2 h, extraction with ether after hydrolysis, and distillation yielded 67% $Ge(CH_3)_2H_2$ (trapped at −78 °C) along with $Ge(CH_3)_2(CH_2CH_2P(C_6H_5)H)H$; using $(Ge(CH_3)_2PC_6H_5)_n$ as the starting material gave 64% yield [39]. The redistribution reaction of $Ge(CH_3)_2(H)PH_2$ at room temperature yielded $Ge(CH_3)_2H_2$ and $Ge(CH_3)_2(PH_2)_2$ and reached equilibrium after 50 h; at 84 °C $Ge(CH_3)_2H_2$ and PH_3 were obtained after 17 d (monitored by ^1H NMR) [31].

$Ge_2(CH_3)_4H_2$ decomposes near 70 °C yielding $Ge(CH_3)_2H_2$ and the germylene $Ge(CH_3)_2$ [59]. $Ge(CH_3)_2H_2$, $Ge_2(CH_3)_2H_4$, and $Ge(CH_3)_3H$ were the products of the pyrolysis of $Ge(CH_3)H_3$ at 420 °C [19]; see also p. 272. $Ge(CH_3)_2H_2$ was identified (by ^1H NMR) in the products formed by UV-irradiating a mixture of $Ge(CH_3)_2(N_3)_2$ and excess $(Ge(CH_3)_2O)_n$ in C_6H_{12} and subsequently treating it with $LiAlH_4$ in ether [54]. Small amounts of the compound were found after adding $Ge(CH_3)_2S$ in C_2H_5OH to amalgamated Zn dust followed by treatment with aqueous 12 M HCl [1].

$Ge(CH_3)_2D_2$ and **$Ge(CD_3)_2H_2$** were prepared from $Ge(CH_3)_2Br_2$ and LiD in dioxane (67% yield) [2] and $Ge(CD_3)_2Br_2$ and aqueous $NaBH_4$ in HBr (99% yield), respectively [7].

The heat of formation of $Ge(CH_3)_2H_2$ and the ionization potential were derived from AM1 calculations: ΔH_f = 1.4 kcal/mol and IP = 10.98 eV [55].

The Molecule, Spectra, and Physical Properties. The experimental structure parameters of $Ge(CH_3)_2H_2$, r(Ge-C) = 1.950±0.003 Å and <(C-Ge-C) = 110.0±0.5°, are based on the microwave spectrum, assuming the CH_3 group dimensions r(C-H) = 1.083 Å and <(H-C-H) = 108.5°. No tilt of the CH_3 groups was detected [20]. Structural parameters for many germanium compounds were also calculated in the context of an MNDO parametrization for germanium (for $Ge(CH_3)_2H_2$: r(Ge-H) = 1.486 and r(Ge-C) = 1.932 Å, <(H-Ge-H) = 109.0° and <(C-Ge-C) = 111.0°) [50]. MM2 force field calculations led to good agreement between the observed structures and those calculated for the staggered conformations of methylgermanes including $Ge(CH_3)_2H_2$ [56].

The rotational spectra of the five Ge-isotopic species were investigated in the range 8 to 35 GHz. Rotational constants for the A_1A_1 ground torsional state of the ^{74}Ge species are A = 12522.67, B = 5587.68, and C = 4264.57 MHz. No hyperfine structure due to the quadrupole moment of ^{73}Ge was resolved. All identified transitions were split into quartets by the internal rotation of the CH_3 groups. This gave a barrier to internal CH_3 rotation of 1.182 kcal/mol [20], which is similar to the barrier in $Ge(CH_3)H_3$ and is in contrast to the often large barrier differences between organic molecules with two equivalent CH_3 groups and their corresponding one-top analogs [25].

The dipole moment μ = 0.616±0.006 D was determined from the Stark effect on transitions of the ^{74}Ge A_1A_1 species [20]; μ = 0.758 D resulted from measurements of the dielectric constant in the gas phase [12]. Calculated values are 0.63 D (based on a Ge-H bond moment of 0 and a Ge-C bond moment of 0.55 D) [56] and 0.59 D (from the evaluation of formal charge distributions) [29]; for an inconsistent value from AM1 calculations, see [55].

The derivative of the bond polarizability of some organometallic compounds including $Ge(CH_3)_2H_2$ was calculated by different methods. The data showed an increasing E-C polarizability in the order E = Si < Ge < Sn [38].

Ionization energies (in eV) from He(I) and He(II) photoelectron spectra (displayed) are listed below along with the proposed assignments based on a comparison with spectra of related species and semi-empirical CNDO/2 calculations:

vertical IP	10.74	11.04	11.50	13.60	13.80	14.30	14.90	16.90
assignment	$3b_1$	$4a_1$	$3b_2$	$1a_2$	$2b_1$	$3a_1$	$2b_2$	$2a_1$

The first and third band represent predominantly Ge-C and Ge-H bonding, respectively [37]; see also [47].

The core-electron binding energies from the X-ray photoelectron spectrum are 36.68 eV for Ge 3d (relative to Ne 2s = 48.47 eV) and 290.04 eV for C 1s (relative to Ar $2p_{3/2}$ = 248.63 eV). For $Ge(CH_3)_nH_{4-n}$ (n = 0 to 3) the binding energy increases stepwise as additional CH_3 groups are displaced by H atoms. Calculations show that the binding energies are related to the partial charge at the corresponding element [36].

^1H NMR spectrum: δ(ppm) = 0.29 (t, CH_3, $^1J(C, H)$ = 128.5 Hz) and 3.72 (sept, HGe, $^3J(H, H)$ = 3.95 Hz) in CCl_4 at 35°C [9, 28]; −0.14 (t, CH_3, $^1J(C, H)$ = 126.8 Hz) and 3.34 (sept, HGe, $^3J(H, H)$ = 3.8 Hz) in $CHCl_3$ [39] or for the neat liquid [10, 12]; 0.22 and 3.72 for the neat liquid [21]; see also [8]. For $Ge(CH_3)_nH_{4-n}$ compounds (n = 0 to 4) the chemical shifts $\delta(CH_3)$ and δ(HGe) and the $^3J(H, H)$ coupling constant depend linearly on the degree of substitution n [9, 10]. Linear relations were also found between $^3J(H, H)$ and the lengths of the involved bonds H-C-Ge-H [9] and between the chemical shifts $\delta(CH_3)$ [11] or δ(HGe) [8]

References on p. 221

and the sum of the Taft polar constants of the other substituents [8, 11] or the ν(GeH) frequencies [8]. For a correlation between the ^{1}H NMR data and the Ge-X π(d-p) interaction in a series of methyl and ethyl derivatives of Si and Ge, see also [15]. Nuclear magnetic double-resonance experiments revealed that the geminal H, H and H, C coupling constants of group-IV hydrides increase from C to Pb, consistent with a qualitative argument based on MO theory. Geminal H, H coupling constants were determined using deuterium-substituted compounds; the following values were reported for $Ge(CH_3)_2H_2$ and $Ge(CH_3)_2(H)D$ (the sign relative to J(H, C) > 0): ^3J(H, D) = 12.4 ± 0.3, ^1J(C, H) = 7.6 ± 0.5, and ^2J(C, H) = 3.8 ± 0.1 Hz [16, 21]. The ^{1}H NMR spectra of some hydrides of the Main Groups IV and V including $Ge(CH_3)_2H_2$ were used to determine the relative acidities of these compounds in liquid ammonia (at 31 °C): δ(ppm) = 0.12 (CH_3) and 3.43 (HGe), ^3J(H, H) = 3.82 Hz [14].

^{13}C NMR spectrum: δ = −7.0 ppm [49].

^{73}Ge NMR spectrum (neat): δ = −127.6 ± 0.1 ppm; ^1J(Ge, H) = 92.3 ± 0.5 and ^2J(Ge, H) = 3.42 ± 0.2 Hz [49]; see also [45]. The dependence of δ(^{73}Ge) and J(Ge, H) on the number of H atoms in $Ge(CH_3)_nH_{4-n}$ compounds (n = 0 to 4) was demonstrated graphically. The nuclear relaxation and correlation times were listed for $Ge(CH_3)_2H_2$ in mixtures of methylgermanes, and the ^{73}Ge nuclear quadrupole coupling constant were estimated [49]. A correlation between the chemical shifts for ^{73}Ge versus ^{29}Si (29 species studied) and ^{119}Sn (26 species studied) was given [45]; see also $Ge(CH_3)_3H$, p. 4. For recent ab initio calculations of ^{73}Ge chemical shifts, see [62].

The vibrational spectra of $Ge(CH_3)_2H_2$ (also displayed) are listed in Table 31 [13] (some data also reported in [1]). The fundamental skeletal frequencies observed were compared with those calculated using the force constants derived from $Ge(CH_3)_4$ and GeH_4 [13]. Force constants of $Ge(CH_3)_2H_2$ were estimated based on the force constants of $E(CH_3)_2H_2$ compounds (E = C, Si, Sn) and used to predict the IR frequencies of $Ge(CH_3)_2H_2$ and $Ge(CD_3)_2D_2$ [32]. GeH stretch-stretch interaction force constants of a series of compounds were found to be very small and positive (except for GeX_2H_2 with X = halogen or CF_3), reflecting the virtual decoupling of GeH stretching motions in GeH_2 or GeH_3 groups where the bonds differ only slightly in strength ("local mode behavior") [43]. Force field calculations were also carried out to determine the conformational equilibria of some Group IV organometallic compounds including $Ge(CH_3)_2H_2$ (staggered and eclipsed conformations) [24].

For correlations between the ν(GeH) frequency and the substituents at the Ge atom, see [6, 44]; ν(GeD) 1480 cm^{-1} for liquid $Ge(CH_3)_2D_2$ [4].

Treating the GeH bond as a diatomic molecule, the dissociation energy was deduced from the fundamental and first-overtone GeH stretching bands to be D° = 74.6 kcal/mol (at 0 K) for $Ge(CH_3)_2(H)D$. Studies on a series of related compounds indicated that CH_3 substitution weakens the E-H bond [40]. Isolated CH stretching frequencies were observed in the IR spectra of partially deuterated methylgermanes such as $Ge(CHD_2)_2H_2$ (gas-phase spectrum illustrated) were used to predict C-H bond lengths, H-C-H angles, and C-H dissociation energies, revealing a quite large difference in strength within a single CH_3 group [41, 46]. Attempts to evaluate a π(d-p) bonding contribution from the ν(GeH) data failed [44].

$Ge(CH_3)_2H_2$ is a colorless liquid boiling at 3.0 °C [35], 6.5 °C/744 Torr [2]; melting point: −144 °C [35] or −144.3 ± 0.2 °C [7]. $Ge(CH_3)_2D_2$ boils at 6.5 °C/745 Torr [2]. $Ge(CH_3)_2H_2$ gas may be stored at room temperature [35].

Equilibrium vapor pressures can be expressed by log (p/Torr) = 5.4643 − 0.005872 T + 1.75 log T − 1445.2/T and give a boiling point of 3.0 °C, a heat of vaporization of ΔH_v = 5525 cal/mol, and a Trouton constant of 20.0 cal · mol^{-1} · K^{-1} [7]. Somewhat different data

Table 31
Vibrational Spectra of $Ge(CH_3)_2H_2$ [13].
Wavenumbers in cm^{-1} (relative intensity and degree of depolarization).

IR gas	fundamental vibration	Raman neat liquid	fundamental vibration	assignment
2987(s)	ν_1,ν_{15},ν_{21}	2987(2, 0.88)	$\nu_1,\nu_{10},\nu_{15},\nu_{21}$	$\nu_{as}(CH_3)$
2936(s)	ν_2,ν_{22}	2919(4, p, 0.09)	ν_2,ν_{22}	$\nu_s(CH_3)$
2080(vs)	ν_{18}			$\nu_{as}(GeH)$
2062(vs)	ν_7	2056(10, p, 0.23)	ν_7,ν_{18}	$\nu_s(GeH)$
1446(w)	ν_{16}			$\delta_{as}(CH_3)$
1419(w)	ν_3,ν_{23}	1409(0)		$\delta_{as}(CH_3)$
1260(w) 1250	ν_4,ν_{24}	1247(2, p, 0.21)		$\delta_s(CH_3)$
880(s) 899	ν_9	886(2, 0.80)		$\delta(GeH_2)$
843(vs)	ν_5,ν_{17},ν_{25}	836(0)	$\nu_5,\nu_{12},\nu_{17},\nu_{25}$	$\rho(CH_3)$
670 662 (w) 654	ν_{27}	662(1, 0.88)	ν_{13},ν_{27}	$\rho(GeH_2)$ in-plane
612 604 (vs) 596	ν_{26}	590(10, p, 0.24)	ν_6,ν_{26}	$\nu_{as}(GeC)$
430(w)	ν_{19}	452(0)	ν_{19}	$\rho(GeH_2)$ out-of-plane
		188(2, 0.80)	ν_8	$\delta(GeC_2)$

were obtained from the equation $\log(p/Torr) = 7.798 - 1340.4/T$ [5]. Selected vapor pressures from [7]:

T in K	189.2	209.4	227.7	237.3	249.9	258.2
p in Torr	4.9	24.7	80.6	137.0	256.4	372.7

Mass Spectrum and Chemical Behavior. The ionic decomposition processes of $E(CH_3)_2H_2$ compounds (E = Si, Ge, Sn) were studied by mass spectrometry. The mass spectrum of $Ge(CH_3)_2H_2$ (displayed) shows a very weak molecular ion $[M]^+$. The fragments $[M-1]^+$ and $[M-2]^+$ are formed in competitive reactions by cleavage of the Ge-H bonds. The $[Ge]^+$ ion is very intense [47]. The ionization efficiency curves for ions formed from $[Ge(CH_3)_2H_2]^+$ are displayed. An analysis of the second derivative of the total ion current showed qualitative agreement between ionic states corresponding to the first PE band and those formed by electron impact in the same energy region [47].

Dissociative chemisorption of $Ge(CH_3)_2H_2$ on Si(100) at room temperature was found by low-energy electron diffraction (LEED) and UV photoelectron spectroscopy (UPS). $H(CH_3)_2Ge\text{-}SiSi\text{-}H$ surface complexes appear to be formed. Photochemical decomposition ($h\nu < 60$ eV) at room temperature results in desorption of ligands, but some of H atoms and CH_3 groups are left on the surface [61].

References on p. 221

Absolute rate constants for the reaction of $Ge(CH_3)_nH_{4-n}$ compounds (n = 1 to 3) with H atoms (formed by mercury photosensitation of H_2-substrate mixtures at 32 °C) were determined in experiments involving competitive reactions of H atoms with Si_2H_6. Arrhenius parameters and Ge-H bond dissociation energies were estimated (E_a = 0.86 kcal/mol and D(Ge-H) = 79 kcal/mol for $Ge(CH_3)_2H_2$) showed a continuous lowering of the activation energy with successive replacement of the first three H atoms in GeH_4 by CH_3 groups [34].

An excess of $Ge(CH_3)_2H_2$ reacted with I_2 at −78 °C to give after 1 h $Ge(CH_3)_2(H)I$ and small amounts of $Ge(CH_3)_2I_2$ [28, 35]. Treatment with HCl in the vapor phase in the presence of $AlCl_3$ yielded $Ge(CH_3)_2Cl_2$ (main product) and $Ge(CH_3)_2(H)Cl$ [5]. The latter product was also obtained with high yield when $Ge(CH_3)_2H_2$ was reacted with a deficient amount (according to [3]) of $HgCl_2$ [26] or BCl_3 at −196 to +20 °C [28, 35], or with $SnCl_4$ (1 : 1 mole ratio) in ether at −25 °C [22].

The relative rates of GeH_2 insertion into Ge-H bonds on a per bond basis were determined from competitive reactions of GeH_2 with Ge_2H_6 and methylgermanes at 280 °C giving the order $Ge(CH_3)_3H > Ge(CH_3)_2H_2 > Ge_2H_6 > Ge(CH_3)H_3$. $H_3GeGe(CH_3)_2H$ was the insertion product formed from $Ge(CH_3)_2H_2$. The relations between the relative rate and the bond energy and the negative charge of the H atom in the bond undergoing insertion was discussed. Insertion of SiH_2 at 350 °C gave $Ge(CH_3)_2(H)SiH_3$ [27].

Heating an excess of $Ge(CH_3)_2H_2$ with 2,3-$C_2B_4H_8$ at 300 °C (1 h) in a sealed evacuated tube gave $GeC_2B_5H_n(CH_3)_{7-n}$ (n = 1 to 5; 10% estimated yield) and a large number of methylcarboranes (mainly $C_2B_5H_n(CH_3)_{7-n}$ with n = 2 to 6) in addition to a polymeric solid with the composition $(GeCH_3)_n$. In contrast, reacting with 2,4-$C_2B_5H_7$ at 270 °C (2 h) yielded $Ge(CH_3)_3H$, $Ge(CH_3)_4$, $B(CH_3)_3$, and unidentified products [23].

An excess of $Ge(CH_3)_2H_2$ reacted with $Fe_3(CO)_{12}$ at 65 °C for 5 h to give $Fe_2(CO)_6$-$(\mu$-$Ge(CH_3)_2)_3$ [17] and with $Co_2(CO)_8$ in $C_6H_5CH_3$ at room temperature for 3 h to give Co_2-$(CO)_6(\mu$-$Ge(CH_3)_2)_2$ [33]. A large excess of $Ge(CH_3)_2H_2$ reacted with μ_4-$Ge(Co_2(CO)_7)_2$ in C_6H_{14} at room temperature in a sealed tube (20 weeks in the dark) to yield the disubstituted product $[\mu_4$-$Ge(Co_2(CO)_6(\mu$-$Ge(CH_3)_2))_2]$, whereas lower reactant ratios (2 to 2.5-fold excess of $Ge(CH_3)_2H_2$) and shorter reaction times (ca. 8 weeks) gave the mono-substituted species $[\mu_4$-$Ge(Co_2(CO)_6(\mu$-$Ge(CH_3)_2))(Co_2(CO)_7)]$ [58]; see also [42]. Reacting $Co_4(CO)_{11}(\mu_4$-$SiCH_3)_2$ (Formula I) with an excess of $Ge(CH_3)_2H_2$ in C_5H_{12} at 30 °C (7 d) in a sealed tube yielded 99% $Co_4(CO)_{10}(\mu_4$-$SiCH_3)_2(\mu$-$Ge(CH_3)_2)$ (Formula II) along with CO and H_2 [60].

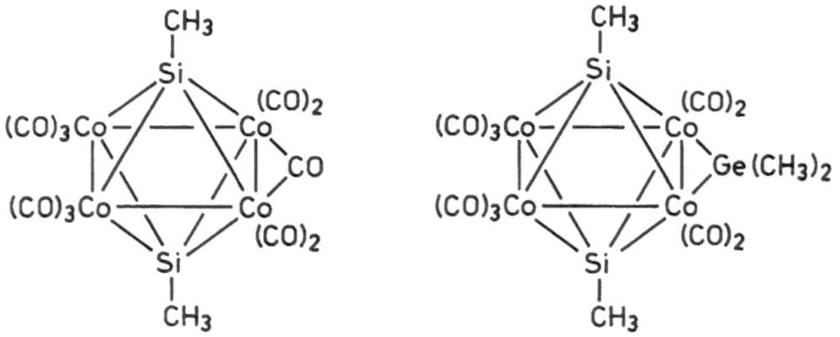

I II

Hydrogermylation of CH_2=$CHCOOCH_3$ with $Ge(CH_3)_2H_2$ was carried out in a heated steel cylinder (5 h) to yield $Ge(CH_3)_2(CH_2CH_2COOCH_3)H$ and $Ge(CH_3)_2(CH_2CH_2COOCH_3)_2$ [30].

Methylgermanes including $Ge(CH_3)_2H_2$ were detected in natural waters at the parts-per-trillion level via hydride generation, graphite furnace atomization, and atomic absorption spectrometry [48].

$Ge(CH_3)_2H_2$ was used for producing epitaxial Ge films achieving a fine-scale pattern on a Si substrate [51]. Based on the results of surface dynamic processes, such as $Ge(CH_3)_2H_2$ adsorption and CH_3 group photodesorption, a model of stepwise photochemical, Ge growth was proposed [52].

References:

[1] West, R. (J. Am. Chem. Soc. **75** [1953] 6080/1).
[2] Ponomarenko, V. A.; Vzenkova, G. Ya.; Egorov, Yu. P. (Dokl. Akad. Nauk SSSR **122** [1958] 405/8; C.A. **1959** 112).
[3] Anderson, H. H. (J. Am. Chem. Soc. **82** [1960] 3016/8).
[4] Ponomarenko, V. A.; Zueva, G. Ya.; Andreev, N. S. (Izv. Akad. Nauk SSSR Ser. Khim. **1961** 1758/62; Bull. Acad. Sci. USSR Div. Chem. Sci. [Engl. Transl.] **1961** 1639/43).
[5] Amberger, E.; Boeters, H. (Angew. Chem. **73** [1961] 114).
[6] Mathis, R.; Satgé, J.; Mathis, F. (Spectrochim. Acta **18** [1962] 1463/72).
[7] Griffiths, J. E. (Inorg. Chem. **2** [1963] 375/7).
[8] Egorochkin, A. N.; Khidekel', M. L.; Ponomarenko, V. A.; Zueva, G. Ya.; Svirezheva, S. S.; Razuvaev, G. A. (Izv. Akad. Nauk SSSR Ser. Khim. **1963** 1865/8; Bull. Acad. Sci. USSR Div. Chem. Sci. [Engl. Transl.] **1963** 1717/9).
[9] Schmidbaur, H. (Chem. Ber. **97** [1964] 1639/48).
[10] Van der Kelen, G. P.; Verdonck, L.; van de Vondel, D. (Bull. Soc. Chim. Belg. **73** [1964] 733/40).

[11] Egorochkin, A. N.; Khidekel', M. L.; Ponomarenko, V. A.; Zueva, G. Ya.; Razuvaev, G. A. (Izv. Akad. Nauk SSSR Ser. Khim. **1964** 373/5; Bull. Acad. Sci. USSR Div. Chem. Sci. [Engl. Transl.] **1964** 347/8).
[12] Van de Vondel, D. F. (J. Organomet. Chem. **3** [1965] 400/5).
[13] Van de Vondel, D. F.; van der Kelen, G. P. (Bull. Soc. Chim. Belg. **74** [1965] 467/78).
[14] Birchall, T.; Jolly, W. L. (Inorg. Chem. **5** [1966] 2177/80).
[15] Egorochkin, A. N.; Burov, A. I.; Mironov, V. F.; Gar, T. K.; Vyazankin, N. S. (Dokl. Akad. Nauk SSSR **180** [1968] 681/4; Dokl. Chem. [Engl. Transl.] **178/183** [1968] 500/3).
[16] Dreeskamp, H.; Schumann, C. (Chem. Phys. Lett. **1** [1968] 555/6).
[17] Brooks, E. H.; Elder, M.; Graham, W. A. G.; Hall, D. (J. Am. Chem. Soc. **90** [1968] 3587/8).
[18] Berger, A. (U.S. 3401183 [1965/1968]; C.A. **70** [1969] No. 29058).
[19] Kohanek, J. J.; Estacio, P.; Ring, M. A. (Inorg. Chem. **8** [1969] 2516/7).
[20] Thomas, E. C.; Laurie, V. W. (J. Chem. Phys. **50** [1969] 3512/5).

[21] Schumann, C.; Dreeskamp, H. (J. Magn. Reson. **3** [1970] 204/17).
[22] Kuz'min, O. V.; Nametkin, N. S.; Chernysheva, T. I.; Gar, T. K.; Lapetukhina, N. A.; Mironov, V. F. (Dokl. Akad. Nauk SSSR **193** [1970] 826/7; Dokl. Chem. [Engl. Transl.] **190/195** [1970] 550/2).
[23] Ledoux, W. A.; Grimes, R. N. (J. Organomet. Chem. **28** [1971] 37/48).
[24] Ouellette, R. J. (J. Am. Chem. Soc. **94** [1972] 7674/9).
[25] Ingham, K. C. (J. Phys. Chem. **76** [1972] 551/3).
[26] Job, R. C.; Curtis, M. D. (Inorg. Nucl. Chem. Lett. **8** [1972] 251/5).
[27] Sefcik, M. D.; Ring, M. A. (J. Organomet. Chem. **59** [1973] 167/73).
[28] Barker, G. K.; Drake, J. E.; Hemmings, R. T. (Can. J. Chem. **52** [1974] 2622/33).
[29] Ramalingam, S. K.; Soundararajan, S. (J. Organomet. Chem. **72** [1974] 59/63).
[30] Rice, L. M.; Wheeler, J. W.; Geschickter, C. F. (J. Heterocycl. Chem. **11** [1974] 1041/7).

[31] Dahl, A. R.; Heil, C. A.; Norman, A. D. (Inorg. Chem. **14** [1975] 1095/8).
[32] Bozhko, N. V.; Timoshinin, V. S.; Sharkov, V. I. (Zh. Prikl. Spectrosk. **23** [1975] 736/8; J. Appl. Spectrosc. [Engl. Transl.] **23** [1975] 1408/10).
[33] Adams, R. D.; Cotton, F. A.; Cullen, W. R.; Hunter, D. L.; Mihichuk, L. (Inorg. Chem. **14** [1975] 1395/9).
[34] Austin, E. R.; Lampe, F. W. (J. Phys. Chem. **81** [1977] 1546/9).
[35] Drake, J. E.; Glavincevski, B. M.; Hemmings, R. T.; Henderson, H. E. (Inorg. Synth. **18** [1978] 154/61).
[36] Drake, J. E.; Riddle, C.; Glavincevski, B.; Gorzelska, K.; Henderson, H. E. (Inorg. Chem. **17** [1978] 2333/6).
[37] Drake, J. E.; Glavincevski, B. M.; Gorzelska, K. (Can. J. Chem. **57** [1979] 2278/84).
[38] Gupta, S. L.; Verma, U. P.; Rastogi, V. K. (Curr. Sci. **49** [1980] 190/2).
[39] Escudié, J.; Couret, C.; Satgé, J.; Andriamizaka, J. D. (Organometallics **1** [1982] 1261/5).
[40] McKean, D. C.; Torto, I.; Morrison, A. R. (J. Phys. Chem. **86** [1982] 307/9).

[41] McKean, D. C.; Mackenzie, M. W.; Torto, I. (Spectrochim. Acta A **38** [1982] 113/8).
[42] Foster, S. P.; Mackay, K. M. (J. Organomet. Chem. **238** [1982] C46/C48).
[43] McKean, D. C.; Torto, I.; Mackenzie, M. W.; Morrison, A. R. (Spectrochim. Acta A **39** [1983] 387/98).
[44] McKean, D. C.; Torto, I.; Mackenzie, M. W. (Spectrochim. Acta A **39** [1983] 399/408).
[45] Watkinson, P. J.; Mackay, K. M. (J. Organomet. Chem. **275** [1984] 39/42).
[46] McKean, D. C.; Mackenzie, M. W.; Morrison, A. R. (J. Mol. Struct. **116** [1984] 331/44).
[47] Lango, J.; Szepes, L.; Csaszar, P.; Innorta, G. (J. Organomet. Chem. **269** [1984] 133/45).
[48] Hambrick, G. A.; Froelich, P. N.; Andreae, M. O.; Lewis, B. L. (Anal. Chem. **56** [1984] 421/4).
[49] Wilkins, A. L.; Watkinson, P. J.; Mackay, K. M. (J. Chem. Soc. Dalton Trans. **1987** 2365/72).
[50] Dewar, M. J. S.; Grady, G. L.; Healy, E. F. (Organometallics **6** [1987] 186/9).

[51] Fujinaga, K.; Takahashi, T.; Ishii, H.; Nippon Telegraph and Telephone Public Corp. (Japan Kokai Tokkyo Koho 01-28914 [89-28914] [1987/89]; C.A. **111** [1989] No. 48673).
[52] Ishii, H.; Takahashi, Y.; Fujinaga, K. (Proc. 1st Int. Conf. Electron. Mater.: New Mater. New Phys. Phenom. Electron. 21st Century, Tokyo 1988 [1989], pp. 137/40).
[53] Gevorgyan, V.; Borisova, L.; Lukevics, E. (J. Organomet. Chem. **368** [1989] 19/21).
[54] Barrau, J.; Bean, D. L.; Welsh, K. M.; West, R.; Michl, J. (Organometallics **8** [1989] 2606/8).
[55] Dewar, M. J. S.; Jie, C. (Organometallics **8** [1989] 1544/7).
[56] Allinger, N. L.; Quinn, M. I.; Chen, K.; Thompson, B.; Frierson, M. R. (J. Mol. Struct. **194** [1989] 1/18).
[57] Lukevics, E.; Gevorgyan, V. (Chem. Technol. Silicon Tin. Proc. Asian Network Anal. Inorg. Chem. 1st Int. Chem. Conf. Silicon Tin, Kuala Lumpur 1989 [1992], pp. 165/77).
[58] Lee, S. K.; Mackay, K. M.; Nicholson, B. K.; Service, M. (J. Chem. Soc. Dalton Trans. **1992** 1709/16).
[59] Barrau, J.; Rima, G.; Cassano, V.; Satgé, J. (Inorg. Chim. Acta **198/200** [1992] 461/7).
[60] Anema, S. G.; Lee, S. K.; Mackay, K. M.; Nicholson, B. K. (J. Organomet. Chem. **444** [1993] 211/8).

[61] Namba, H.; Yamaguchi, T.; Kuroda, H. (Surface Sci. **283** [1993] 132/6).
[62] Nakatsuji, H.; Nakao, T. (Intern. J. Quantum Chem. **49** [1994] 279/90).

1.3.2.1.2 Diethylgermane, Diethylgermanium Dihydride, $Ge(C_2H_5)_2H_2$

Preparation and Formation. $Ge(C_2H_5)_2H_2$ was prepared by reducing $Ge(C_2H_5)_2Cl_2$ with excess $LiAlH_4$ in refluxing ether (30 min) [41] (see also [2, 11]) or in $O(C_4H_9)_2$ (ca. 90% yield) [3]; see also [33]. It was purified by trap-to-trap vacuum distillation [41]. The compound was also obtained from a $Ge(C_2H_5)_2Cl_2$-$Ge(C_2H_5)_2Br_2$ mixture and LiH in $O(C_4H_9)_2$ [1] or from $Ge(C_2H_5)_2X_2$ (X = halogen) and $NaBH_4$ in refluxing THF after 3 to 4 h (ca. 70% yield) [3] or with aqueous $NaBH_4$ in acidic solution [12] (according to [7]); cf. $Ge(CH_3)_2H_2$ on p. 216.

Reacting $Ge(C_2H_5)_2$=O, (obtained from $Ge(C_2H_5)_2 \cdot N(C_2H_5)_3$ and $\overline{CH_2CH_2O}$ at 150 °C in a sealed tube) with $LiAlH_4$ in ether yielded 32% $Ge(C_2H_5)_2H_2$ along with 13% $Ge_2(C_2H_5)_4H_2$ (analyzed by GC) [37]; see also [32]. $\overline{Ge(C_2H_5)_2P(C_6H_5)Ge(C_2H_5)_2P(C_6H_5)CH_2CH_2}$ and $(Ge(C_2H_5)_2PC_6H_5)_n$ were reduced with $LiAlH_4$ as described for $Ge(CH_3)_2H_2$ (see p. 216) to give $Ge(C_2H_5)_2H_2$ with ca. 68% yield [34].

Hydrolysis of $Ge(C_2H_5)_2Li_2$ (formed from $Ge(C_2H_5)_2(Si(CH_3)_3)Li$ and $Si(CH_3)_3Li$ in HMPT at 25 °C) yielded 68% $Ge(C_2H_5)_2H_2$ [43, 48]. The reaction of $Ge(C_2H_5)_3Li$ with $Ge(C_2H_5)_2$-$(Si(CH_3)_3)Li$ in HMPT at 25 °C (94 h) followed by hydrolysis and extraction with ether gave $Ge(C_2H_5)_2H_2$ (26%), $Ge(C_2H_5)_3H$ (28%), and $(C_2H_5)_3GeGe(C_2H_5)_2H$ (27%) as the main products along with smaller amounts of mixed Ge-Si compounds [44, 48]. For the formation from $GeLi_4$ and C_2H_5Br, see also [29].

$Ge(C_2H_5)_2H_2$ was the main product in addition to di- and trigermanes when Ge_2H_6 was reacted with CH_2=CH_2 at 160 °C in a sealed tube (80 h) [14]. ^{60}Co-γ irradiation of the same reactants at 0 °C (ca. 20 min) yielded a complex mixture of alkylgermanes including $Ge(C_2H_5)_2H_2$ [23]. Similar irradiation of GeH_4 and CH_2=CH_2 followed by trap-to-trap distillation in vacuum, also yielded small amounts of $Ge(C_2H_5)_2H_2$ [24].

$Ge(C_2H_5)_2H_2$ and $Ge(C_2H_5)_2F_2$ resulted from the disproportionation of $Ge(C_2H_5)_2(H)F$ after standing for several days [13]. The decomposition of $Ge(C_2H_5)_2(H)N(C_4H_9)_2$ at 20 °C under daylight yielded ca. 10% $Ge(C_2H_5)_2H_2$ as indicated by the ^1H NMR spectrum [26]. ^1H NMR and IR spectra also proved the presence of $Ge(C_2H_5)_2H_2$ in a reaction mixture obtained from $Ge(C_2H_5)_2(H)P(C_2H_5)_2$ and $(CHO)_2$ in C_6H_6 at 35 °C [21].

Small amounts (<5% yield) of $Ge(C_2H_5)_2H_2$ were formed in the photolysis of $Ge(C_2H_5)_2(H)Fe(CO)_2(C_5H_5)$ in C_6D_6 for 3.5 h [53].

$Ge(C_2H_5)_2D_2$ was prepared by reducing a mixture of $Ge(C_2H_5)_2Cl_2$ and $Ge(C_2H_5)_2Br_2$ with LiD in $O(C_4H_9)_2$ [1].

Spectra. ^1H NMR spectrum: δ(ppm) = 0.57 to 1.23 (m, C_2H_5), 3.76 (quint, HGe, $^3J(H, H)$ = 2.7 Hz) in C_6D_6 [34]; 0.78 (CH_2), 1.03 (CH_3, $^3J(H, H)$ = 7.8 Hz), 3.76 (50% solution), 3.83 (10% solution) (HGe, $^3J(H, H)$ = 2.7 Hz) in C_6H_6 [11]; 0.87 (CH_2) [9, 15] and 3.70 (HGe) [8] for the neat liquid. (^1H NMR spectrum displayed in [11]).

The ^1H NMR spectra of $Ge(C_2H_5)_nH_{4-n}$ compounds (n = 0 to 3) were discussed with respect to variations of the chemical shift of the HGe, CH_2, and CH_3 protons with n [11]. Relations between δ(HGe) and the ν(GeH) frequency, δ(HGe) or δ(CH_2) and the sum of Taft σ^* constants of the substituents were pointed out in [8, 9] for series of Ge compounds including $Ge(C_2H_5)_2H_2$. For the effect of π(d-p) interaction on δ(CH_2) in ethyl derivatives of Si and Ge, see also [15]. The ^1H NMR spectra of some hydrides of groups IV and V were used to determine the relative acidities of these compounds in liquid NH_3 (at 31 °C): δ(ppm) = 0.89, 0.95 (C_2H_5), 3.47 (HGe, $^3J(H, H)$ = 2.73 and $^4J(H, H) \approx$ 0.5 Hz for $Ge(C_2H_5)_2H_2$ [12].

^{13}C NMR spectrum: δ = 0.0 and 8.3 ppm [45].

References on p. 226

^{73}Ge NMR spectrum (neat): δ = −88.4 ± 1 ppm (J(Ge, H) = 88.7 ± 2.0 Hz) [45]; see also [38]. The dependence of the ^{73}Ge chemical shift and the coupling constant on the number of hydrogen atoms in $Ge(C_2H_5)_nH_{4-n}$ compounds (n = 0 to 4) was studied [45]. A close correlation was found between the chemical shift δ(^{73}Ge) versus δ(^{29}Si) and δ(^{73}Ge) versus δ(^{119}Sn) (29 and 26 compounds were compared, respectively) [38].

The IR spectrum (illustrated in [2]) is listed in Table 32 [11] along with selected unassigned Raman frequencies reported earlier in [1]. The following data were also given for $Ge(C_2H_5)_2H_2$: ν(GeH) 2032 in CCl_4 [6, 10]; 2036 [6], 2037 [10], 2042 [30] in C_7H_{16}; 2036 for the neat liquid [6]; 2049.1, 2042.4 cm^{-1} for the gas [35]. For $Ge(C_2H_5)_2(H)D$ as a gas: ν(GeH) 2048.7, 2042.6 cm^{-1} (from 2100 to 2000 cm^{-1} displayed) [35].

The only slight effect of various solvents on the ν(GeH) frequency (e.g. 2037 cm^{-1} in C_6H_{14} and 2028 cm^{-1} in dioxane) is attributed to the polarity of the Ge^+-H^- bond which is unable to form a true hydrogen bond with proton acceptors [17]. The ν(GeH) frequency [6, 18] and its integrated intensity (A) [10, 30] were studied on a series of Ge hydrides including $Ge(C_2H_5)_2H_2$. There is an approximately linear relation between ν(GeH) [6, 36] or A [10, 30] and the sum of Taft σ* constants of the substituents, providing an appropriate treatment of the conformational effects is carried out [36]. Relations between the ν(GeH) or ν(GeD) frequency and the inductive effect of the substituents were also discussed for a variety of Ge hydrides and deuterides [4]. $Ge(C_2H_5)_2H_2$ was part of a study on the π(d-p) interaction between Ge and various ligands and its effect on the ν(GeH) frequency [28].

Conformational effects of gauche and trans CH_3 groups on the isolated ν(GeH) of $Ge(C_2H_5)_2(H)D$ result in a ν(GeH) doublet spaced by 6 cm^{-1}, which expectedly is equal to

Table 32
IR Spectrum of $Ge(C_2H_5)_2H_2$ (presumably gas phase) [11] and Selected Raman Lines of $Ge(C_2H_5)_2H_2$ and $Ge(C_2H_5)_2D_2$ [1].
Wavenumbers in cm^{-1}.

IR [11] $Ge(C_2H_5)_2H_2$	assignment [11]	Raman [1] $Ge(C_2H_5)_2H_2$	$Ge(C_2H_5)_2D_2$
2085(sh)	ν(GeH)		
2044(vs)		2032(10)	
1468(m)	$\delta_{as}(CH_3)$	1460(1)	1461(10)
1430(sh)	$\delta_{as}(CH_2)$		
1380(w)	$\delta_s(CH_3)$		
1228(w)	$\delta_s(CH_2)$	1221(4)	1222(7)
1020(m)	ν(CC)	1023(1)	1023(3)
968(m)		966(1)	968(3)
872(s)	} δ(GeH) + ρ(GeC_2H_5)	870(3)	
865(sh)			
830(w)			
763(vs)			
688(w)			
630(w,br)	} ν(GeC) + ρ(GeH_2)		624(3)
567(m)		545(5)	578(8)
448(m)			474(4)
		287(2)	278(3)
		170(1)	172(1)

the doublet separation of $Ge(C_2H_5)(H)D_2$ [35]. Force field calculations were carried out to determine the conformational equilibria of some group IV organometallic compounds, including four conformations of $Ge(C_2H_5)_2H_2$ [22].

The Raman spectra of $Ge(C_2H_5)_2H_2$ and $Ge(C_2H_5)_2D_2$ are completely listed in [1] without giving any assignment; cf. Table 32.

The He(I) photoelectron spectrum shows four bands with the following vertical ionization potentials (orbital type for C_{2v} symmetry): 9.8 (b_2, GeC_2), 10.5 (a_1, H_2GeC_2), 11.1 (b_1, GeH_2), and ca. 12.3 eV ($1e_g$ and $3a_{1g}$, CC and CH of C_2H_5). The ionization potentials of the Ge-C and C-H bonds in $Ge(C_2H_5)_nH_{4-n}$ compounds (n = 0 to 4) smoothly decrease with increasing n (spectra of these compounds displayed) [33].

Physical properties. Boiling point 72.5°C/740.5 Torr [1], 74°C/760 Torr [2, 3, 34], 77°C [33]; density: d^{20} = 1.0378 [1], 1.0390 g/cm^3 [2, 3]; and refractive index: n_D^{20} = 1.4202 [1], 1.4219 [2, 3]. Respective data for $Ge(C_2H_5)_2D_2$: 71.5°C/743.5 Torr, d^{20} = 1.0525 g/cm^3, and n_D^{20} = 1.4200 [1].

Mass Spectrum and Chemical Behavior. The complete mass spectrum of naturally abundant $Ge(C_2H_5)_2H_2$ was reported (the region around the parent ion is displayed). The peaks may be grouped into the following envelopes: m/e 128 to 136 ($[GeC_4H_n]^+$, n = 0 to 12), 113 to 121 ($[GeC_3H_n]^+$, n = 0 to 9), 94 to 109 ($[GeC_2H_n]^+$, n = 0 to 7; m/e 104 base peak), 82 to 93 ($[GeCH_n]^+$, n = 0 to 5), 70 to 79 ($[GeH_n]^+$, n = 0 to 3), 35 to 39 ($[GeH_n]^{++}$, n = 0 to 3), 24 to 29 ($[C_2H_n]^+$, n = 0 to 5), and 12 to 15 ($[CH_n]^+$, n = 0 to 3). $[Ge(C_2H_5)_2H_n]^+ \rightarrow [Ge(C_2H_5)H_n]^+ \rightarrow [GeH_n]^+$ is given as a general fragmentation route [24].

Refluxing $Ge(C_2H_5)_2H_2$ for 15 h under Ar yielded $H(C_2H_5)_2GeGe(C_2H_5)_2H$ (17%), $H(C_2H_5)_2GeGe(C_2H_5)_2Ge(C_2H_5)_2H$ (41%), and $H(C_2H_5)_2Ge(Ge(C_2H_5)_2)_2Ge(C_2H_5)_2H$ (19%) [31].

The oxidation of $Ge(C_2H_5)_2H_2$ and other organometallic compounds (in $C_6H_5CH_3$ at 330 K) by air proceeds through the formation of peroxides as indicated by the observation of kinetic chemiluminescence maxima [39, 42]. Interaction with XeF_2 (in $C_6H_5CH_3$ at 300 K) is accompanied by chemiluminescence [39].

$Ge(C_2H_5)_2H_2$ reacted with excess alkali metal in HMPT-THF to give the monoanion $[Ge(C_2H_5)_2H]^-$ as shown by the formation of $Ge(C_2H_5)_3H$ after treating the reaction mixture with C_2H_5Br and H_2O [48]. $Ge(C_2H_5)_2(H)Cl$ (80% yield) was obtained in the reaction with $HgCl_2$ [13].

$Ge(C_2H_5)_2H_2$ reacted with CH_3OCH_2Cl under reflux (15 min) to yield 80% $Ge(C_2H_5)_2(H)Cl$ and CH_3OCH_3 [5]; see also [3, 13]. Reactions of $Ge(C_2H_5)_2H_2$ with C_4H_9Br or C_3H_7I (ca. 1:0.7 mole ratio) under reflux gave $Ge(C_2H_5)_2(H)Br$ (45%) and $Ge(C_2H_5)_2(H)I$ (40%, along with $Ge(C_2H_5)_2I_2$), respectively [3, p. 569]. Treating $Ge(C_2H_5)_2H_2$ with CH_3OH at 20°C in the presence of Raney Ni yielded $Ge(C_2H_5)_2(H)OCH_3$ [16, 19].

Addition of $Ge(C_2H_5)_2H_2$ to vinyl ketones CH$_2$=CR'COR" (R' = H or CH_3, R" = CH_3 or H; R' = R" = CH_3) in the presence of H_2PtCl_6 yielded after refluxing for several hours $Ge(C_2H_5)_2(H)CH_2CH(R')COR''$ which isomerized in the presence of AIBN to $\overline{Ge(C_2H_5)_2CH_2CHR'CHR''O}$ compounds [20]; see also [16]. The addition of $Ge(C_2H_5)_2H_2$ to CH$_2$=CHCOOCH$_3$ (1:2 mole ratio) at 100°C (5 h) in a pressure vessel was used to prepare $Ge(C_2H_5)_2(H)CH_2CH_2COOCH_3$ and $Ge(C_2H_5)_2(CH_2CH_2COOCH_3)_2$ [27]. Refluxing $Ge(C_2H_5)_2H_2$ with CH$_2$=CHCOOCH$_3$, CH$_2$=C(CH$_3$)COOCH$_3$, or CH$_3$CH=CHCOOC$_2$H$_5$ in the presence of AIBN or under UV irradiation also yielded $Ge(C_2H_5)_2(CHR'CHR'COOR'')_2$ compounds (R' = H or CH_3; R" = CH_3 or C_2H_5). Reactions with CH$_2$=CHCOOCH$_3$ and CH$_3$CH=CHCOOC$_2$H$_5$ gave $Ge(C_2H_5)_2(CH_2CH_2COOCH_3)CH(CH_3)CH_2COOC_2H_5$ [25].

References on p. 226

$Ge(C_2H_5)_2(H)R'$ and $Ge(C_2H_5)_2R'_2$ compounds (R' = $CH_2CH=CHCH_3$, $CH_2C(CH_3)=CHCH_3$, $CH_2CH=CHCH_2CH(CH_3)CH=CH_2$, $H_2C(CH_3)=CHCH_2CH_2CH_2C(CH_3)=CH_2$) were obtained by treating $Ge(C_2H_5)_2H_2$ with $CH_2=CHCH=CH_2$, $CH_2=C(CH_3)CH=CH_2$, $CH_2=CHCH(CH_3)CH=CHCH=CH_2$ at 60 °C (4 h) or with $CH_2=C(CH_3)CH_2CH_2CH=CH\text{-}C(CH_3)=CH_2$ at 80 °C (3 h) in the presence of $Ni(acac)_2$, $P(C_6H_5)_3$, and $Al(C_4H_9\text{-}i)_3$ (1:2:2 ratio). Cyclic products such as $\overline{Ge(C_2H_5)_2CH_2CH_2CH(CH_3)CH_2CH=CHCH_2}$ and $\overline{Ge(C_2H_5)_2CH_2CH(CH_3)CH_2CH_2CH_2CH=C(CH_3)CH_2}$ were also formed in the reactions with the heptatriene and octatriene, especially at 120 °C. Similar treatment of cyclo-C_8H_8 with $Ge(C_2H_5)_2H_2$ at 100 °C (3 h) yielded compound I as the only product [40].

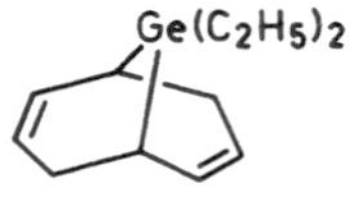

I

The metallation of $Ge(C_2H_5)_2H_2$ with $Ge(C_2H_5)_3Li$ (1:3 mole ratio) in HMPT followed by treatment with C_2H_5Br or $Si(CH_3)_3Cl$ is described in [48]. UV irradiation of $Ge(C_2H_5)_2H_2$ and $Fe(CO)_5$ in C_5H_{12} at room temperature (4 h) yielded $Fe(CO)_4(\mu\text{-}Ge(C_2H_5)_2)_2Fe(CO)_4$ [49]; cf. "Eisen-Organische Verbindungen" C 1, p. 42.

$Ge(C_2H_5)_2H_2$ was tested as a precursor for chemical vapor deposition of germanium films. Photolytic CO_2-laser-induced deposition was first reported in [41]. A controlled Ge epitaxial growth layer by layer was achieved by chemisorbing $Ge(C_2H_5)_2H_2$ (0.001 to 0.1 Torr) on a Ge surface at 220 °C (formation of a $Ge(C_2H_5)_2$ monolayer suggested), evacuating, desorbing the C_2H_5 groups above 400 °C ($\rightarrow CH_2=CH_2 + H_2$), and repeating this process several times [50]. Carbon-free Ge films could also be formed by photolysis with an ArF laser of a solution of $Ge(C_2H_5)_2H_2$ in $C_{12}H_{26}$; C_2H_4 was formed as gaseous product [52].

The preparation of Ge films from $Ge(C_2H_5)_2H_2$ on Si substrates was claimed in [46, 47]. $Ge(C_2H_5)_2H_2$ was shown to deposit Ge on Si(111) by adsorption and thermal decomposition. Desorption of $CH_2=CH_2$ at ca. 700 K (consistent with a β-hydride elimination mechanism) and H_2 at ca. 800 K was observed by laser-induced thermal and temperature-programmed desorption techniques. These reactions occur with negligible carbon deposition [51, 55]. Transmission Fourier transform IR spectra showed that $Ge(C_2H_5)_2H_2$ dissociatively adsorbs on porous silicon surfaces at 200 K to form SiH, GeH, and SiC_2H_5 surface species. No spectral features were observed for GeC_2H_5 surface species. On thermal annealing, the GeH species transferred hydrogen to the silicon surface and formed SiH species between 300 and 640 K [56]. The decomposition kinetics of the SiC_2H_5 species was investigated in detail and compared with the decomposition following exposure to $Si(C_2H_5)_2H_2$ and $Si(C_2H_5)H_3$ [54]. Germanium chemical vapor deposition on silicon using $Ge(C_2H_5)_2H_2$ was also discussed in [57, 58].

References:

[1] Ponomarenko, V. A.; Vzenkova, G. Ya.; Egorov, Yu. P. (Dokl. Akad. Nauk SSSR **122** [1958] 405/8; C.A. **53** [1959] 112).

[2] Satgé, J.; Mathis-Noël, R.; Lesbre, M. (C.R. Hebd. Seances Acad. Sci. **249** [1959] 131/3).

[3] Satgé, J. (Ann. Chim. [Paris] **1961** 519/73).

[4] Ponomarenko, V. A..; Zueva, G. Ya.; Andreev, N. S. (Izv. Akad. Nauk SSSR Ser. Khim. **1961** 1758/62; Bull. Acad. Sci. USSR Div. Chem. Sci. [Engl. Transl.] **1961** 1639/43).
[5] Lesbre, M.; Satgé, J. (C.R. Hebd. Seances Acad. Sci. **252** [1961] 1971/8).
[6] Mathis, R.; Satgé, J.; Mathis, F. (Spectrochim. Acta **18** [1962] 1463/72).
[7] Griffiths, J. E. (Inorg. Chem. **2** [1963] 375/7).
[8] Egorochkin, A. N.; Khidekel', M. L.; Ponomarenko, V. A.; Zueva, G. Ya.; Svirezheva, S. S.; Razuvaev, G. A. (Izv. Akad. Nauk SSSR Ser. Khim. **1963** 1865/8; Bull. Acad. Sci. USSR Div. Chem. Sci. [Engl. Transl.] **1963** 1717/9).
[9] Egorochkin, A. N.; Khidekel', M. L.; Ponomarenko, V. A.; Zueva, G. Ya.; Razuvaev, G. A. (Izv. Akad. Nauk SSSR Ser. Khim. **1964** 373/5; Bull. Acad. Sci. USSR Div. Chem. Sci. [Engl. Transl.] **1964** 347/8).
[10] Mathis, R.; Constant, M.; Satgé, J.; Mathis, F. (Spectrochim. Acta **20** [1964] 515/21).

[11] Mackay, K. M.; Watt, R. (J. Organomet. Chem. **6** [1966] 336/51).
[12] Birchall, T.; Jolly, W. L. (Inorg. Chem. **5** [1966] 2177/80).
[13] Massol, M.; Satgé, J. (Bull. Soc. Chim. Fr. **1966** 2737/43).
[14] Mackay, K. M.; Watt, R. (J. Organomet. Chem. **14** [1968] 123/9).
[15] Egorochkin, A. N.; Burov, A. I.; Mironov, V. F.; Gar, T. K.; Vyazankin, N. S.; (Dokl. Akad. Nauk SSSR **180** [1968] 861/4; Dokl. Chem. [Engl. Transl.] **180** [1968] 500/3).
[16] Massol, M.; Satgé, J.; Barrau, J. (C.R. Hebd. Seances Acad. Sci. C **268** [1969] 1710/3).
[17] Mathis, R.; Berthelat, M.; Mathis, F. (Spectrochim. Acta A **26** [1970] 2001/5).
[18] Mathis, R.; Berthelat, M.; Mathis, F. (Spectrochim. Acta A **26** [1970] 1993/2000).
[19] Massol, M.; Satgé, J.; Rivière, P.; Barrau, J. (J. Organomet. Chem. **22** [1970] 599/610).
[20] Massol, M.; Barrau, J.; Satgé, J. (J. Heterocycl. Chem. **7** [1970] 783/90).

[21] Couret, C.; Satgé, J.; Couret, F. (Inorg. Chem. **11** [1972] 2274/7).
[22] Ouellette, R. J. (J. Am. Chem. Soc. **94** [1972] 7674/9).
[23] Khandelwal, J. K.; Pinson, J. W. (Inorg. Nucl. Chem. Lett. **9** [1973] 393/7).
[24] Pinson, J. W.; Khandelwal, J. K. (Spectrosc. Lett. **6** [1973] 745/61).
[25] Nederlandse Centrale Organisatie Voor Toegepast-Natuurwetenschappelijk Onderzoek (Neth. Appl. 73-08463 [1973/74] 1/6; C.A. **83** [1975] No. 10402).
[26] Rivière, P.; Rivière-Baudet, M.; Couret, C.; Satgé, J. (Synth. React. Inorg. Met.-Org. Chem. **4** [1974] 295/307).
[27] Rice, L. M.; Wheeler, J. W.; Geschickter, C. F. (J. Heterocycl. Chem. **11** [1974] 1041/7).
[28] Egorochkin, A. N.; Khorshev, S. Ya.; Satgé, J.; Rivière, P.; Barrau, J. (J. Organomet. Chem. **99** [1975] 239/49).
[29] Morrison, J. A.; Lagow, R. J. (Inorg. Chem. **16** [1977] 2972/4).
[30] Egorochkin, A. N.; Khorshev, S. Ya.; Sevastyanova, E. I.; Ratushnaya, S. Kh.; Satgé, J.; Rivière, P.; Barrau, J.; Richelme, S. (J. Organomet. Chem. **155** [1978] 175/83).

[31] Marchand, A.; Gerval, P.; Rivière, P.; Satgé J. (J. Organomet. Chem. **162** [1978] 365/87).
[32] Barrau, J.; Bouchaut, M.; Lavayssière, H.; Dousse, G.; Satgé, J. (Helv. Chim. Acta **62** [1979] 152/4).
[33] Beltram, G.; Fehlner, T. P.; Mochida, K.; Kochi, J. K. (J. Electron Spectrosc. Rel. Phenom. **18** [1980] 153/9).
[34] Escudié, J.; Couret, C.; Satgé, J.; Andriamizaka, J. D. (Organometallics **1** [1982] 1261/5).
[35] McKean, D. C.; Torto, I.; Mackenzie, M. W.; Morrison, A. R. (Spectrochim. Acta A **39** [1983] 387/98).

[36] McKean, D. C.; Torto, I.; Mackenzie, M. W. (Spectrochim. Acta A **39** [1983] 399/408).
[37] Barrau, J.; Bouchaut, M.; Lavayssière, H.; Dousse, G.; Satgé, J. (J. Organomet. Chem. **243** [1983] 281/90).
[38] Watkinson, P. J.; Mackay, K. M. (J. Organomet. Chem. **275** [1984] 39/42).
[39] Bulgakov, R. G.; Maistrenko, G. Ya.; Yakovlev, V. N.; Kuleshov, S. P.; Tolstikov, G. A.; Kazakov, V. P. (Dokl. Akad. Nauk SSSR **282** [1985] 1385/9; Dokl. Chem. [Engl. Transl.] **280/285** [1985] 173/6).
[40] Salimgareeva, I. M.; Bogatova, N. G.; Panasenko, A. A.; Monakov, Yu. B.; Yur'ev, V. P. (Izv. Akad. Nauk SSSR Ser. Khim. **1985** 1391/4; Bull. Acad. Sci. USSR Div. Chem. Sci. [Engl. Transl.] **1985** 1272/5).

[41] Stanley, A. E.; Johnson, R. A.; Turner, J. B.; Roberts, A. H. (Appl. Spectrosc. **40** [1986] 374/8).
[42] Bulgakov, R. G.; Kuleshov, S. P.; Yakovlev, V. N.; Maistrenko, G. Ya.; Tolstikov, G. A.; Kazakov, V. P. (Izv. Akad. Nauk SSSR Ser. Khim. **1986** 2216/8; Bull. Acad. Sci. USSR Div. Chem. Sci. [Engl. Transl.] **1986** 2022/4).
[43] Bravo-Zhivotovskii, D. A.; Pigarev, S. D.; Vyazankina, O. A.; Vyazankin, N. S. (Zh. Obshch. Khim. **57** [1987] 2644/5; J. Gen. Chem. USSR [Engl. Transl.] **57** [1987] 2356/7).
[44] Bravo-Zhivotovskii, D. A.; Pigarev, S. D.; Vyazankina, O. A.; Vyazankin, N. S. (Zh. Obshch. Khim. **57** [1987] 2735/8; J. Gen. Chem. USSR [Engl. Transl.] **57** [1987] 2440/2).
[45] Wilkins, A. L.; Watkinson, P. J.; Mackay, K. M. (J. Chem. Soc. Dalton Trans. **1987** 2365/72).
[46] Takahashi, T.; Ishii, H.; Fujinaga, K.; Nippon Telegraph and Telephone Public Corp. (Jpn. Kokai Tokkyo Koho 01-05011 [89-05011] 1987/89; C.A. **110** [1989] No. 223253).
[47] Fujinaga, K.; Takahashi, T.; Ishii, H.; Nippon Telegraph and Telephone Public Corp. (Jpn. Kokai Tokkyo Koho 01-28914 [89-28914] 1987/89; C.A. **111** [1989] No. 48673).
[48] Pigarev, S. D.; Bravo-Zhivotovskii, D. A.; Kalikhman, I. D.; Vyazankin, N. S.; Voronkov, M. G. (J. Organomet. Chem. **369** [1989] 29/41).
[49] Barrau, J.; Ben Hamida, N.; Agrebi, A.; Satgé, J. (Organometallics **8** [1989] 1585/93).
[50] Takahashi, Y.; Ishii, H.; Fujinaga, K. (J. Electrochem. Soc. **136** [1989] 1826/7).

[51] Coon, P. A.; Wise, M. L.; Walker, Z. H.; George, S. M.; Roberts, D. A. (Appl. Phys. Lett. **60** [1992] 2002/4).
[52] Pola, J.; Parsons, J. P.; Taylor, R. (J. Mater. Chem. **2** [1992] 1289/92).
[53] Castel, A.; Rivière, P.; Ahbala, M.; Satgé, J.; Soufiaoui, M.; Knouzi, N. (J. Organomet. Chem. **447** [1993] 123/30).
[54] Coon, P. A.; Wise, M. L.; Walker, Z. H.; George, S. M. (Surf. Sci. **291** [1993] 337/48).
[55] Coon, P. A.; Wise, M. L.; George, S. M. (J. Chem. Phys. **98** [1993] 7485/95).
[56] Dillon, A. C.; Robinson, M. B.; George, S. M.; Roberts, D. A. (Surf. Sci. **286** [1993] L535/L541).
[57] Coon, P. A.; Wise, M. L.; Dillon, A. C.; George, S. M. (Mater. Res. Soc. Symp. Proc. **282** [1993] 413/9 from C.A. **120** [1994] No. 39041).
[58] Greenlief, C. M.; Klug, D. A.; Du, W.; Keeling, L. A. (Mater. Res. Soc. Symp. Proc. **282** [1993] 427/32 from C.A. **120** [1994] No. 39043).

1.3.2.1.3 GeR_2H_2 Compounds with R Larger than Ethyl

The compounds of this section are listed in Table 33.

Table 33
GeR_2H_2 Compounds with R Groups Larger than Ethyl.
An asterisk indicates further information at the end of the table.
Explanations, abbreviations, and units are given on p. X.

No.	R group	formation (yield) properties and remarks
1	C_3H_7	$Ge(C_3H_7)_2X_2$ (X = halogen) + excess $LiAlH_4$ in ether or $O(C_4H_9)_2$ [2] (ca. 90% [5]) $Ge(C_3H_7)_2X_2$ (X = halogen) + $NaBH_4$ in refluxing THF (3 to 4 h) followed by addition of H_2O at 20 °C (ca. 70%) [5] b.p. 126 to 127 °C/760 d^{20} = 1.0030; n_D^{20} = 1.4340 [2, 5] IR (C_7H_{16}): ν(GeH) 2042 [25] with $HgCl_2$ [16] or HgI_2 in $CHCl_3$-THF at 50 °C [14] → $Ge(C_3H_7)_2(H)X$ (X = Cl, I) with CH_2=CHCOOH at reflux (18 h) → $Ge(C_3H_7)_2(CH_2CH_2COOH)H$ [5] with CH_2=$CHCOOCH_3$ under reflux in the presence of AIBN or UV irradiation → $Ge(C_3H_7)_2(CH_2CH_2COOCH_3)_2$ [21] with CH_2=CHCN at 15 °C (20 h) in the presence of hydroquinone in a sealed tube → $Ge(C_3H_7)_2(CH_2CH_2CN)H$ and $Ge(C_3H_7)_2(CH_2CH_2CN)_2$ [5] with $Hg(CH_2COOCH_3)_2$ (1:1 or 1:2 mole ratio) in refluxing THF → $Ge(C_3H_7)_2(CH_2COOCH_3)H$ or $Ge(C_3H_7)_2(CH_2COOCH_3)_2$, respectively [14]
*2	C_3H_7-i	$Ge(C_3H_7\text{-i})_2I_2$ + $LiAlH_4$ in refluxing ether (2 h) followed by adding H_2O and distilling the ether layer after treatment with dilute H_2SO_4 (54% crude) [1] b.p. 110 to 111 °C/760 d^{20} = 0.982; n_D^{20} = 1.432 [1, 5] ^{1}H NMR: 3.68 (HGe) [13] IR: ν(GeH) 2026 for $Ge(C_3H_7\text{-i})_2H_2$ [13]; 2033.6 for $Ge(C_3H_7\text{-i})_2(H)D$ as a gas (also illustrated) [27] slowly oxidized in air → $(Ge(C_3H_7\text{-i})_2O)_3$ with excess $AgOOCCH_3$ under reflux (1 h) → $Ge(C_3H_7\text{-i})_2(OOCCH_3)_2$ and $(Ge(C_3H_7\text{-i})_2O)_3$ [1]
*3	C_4H_9	$Ge(C_4H_9)_2X_2$ (X = halogen) + $LiAlH_4$ in ether [2] $Ge(C_4H_9)_2I_2$ + excess $NaBH_4$ in refluxing THF (4 h) followed by H_2O at 15 °C, extraction of the aqueous layer with ether, solvent removal, and distillation (70%) [5] b.p. 37 to 40 °C/2 [14], 76.2 to 76.7 °C/24 [3], 86 °C/46 [2, 5], 173 °C/760 [3] d^{20} = 0.977 [3], 0.9782 [2, 5] n_D^{20} = 1.4420 [14], 1.4423 [3], 1.4428 [2, 5] ^{1}H NMR (neat): 0.9 (C_4H_9), 3.74 (HGe) [22] IR: ν(GeH) 2031 in CCl_4 [9, 15]; 2032 for the neat liquid [9]; 2035 [9, 15] or 2042 [25] in C_7H_{16} (spectrum illustrated) [5]

References on p. 232

Table 33 (continued)

No.	R group	formation (yield) properties and remarks
*3 (continued)		with H_2O at 100 °C in the presence of Cu powder $\rightarrow Ge(C_4H_9)_2O$ with excess H_2SO_4 at 20 °C $\rightarrow Ge(C_4H_9)_2SO_4$; formula not given [5] with a deficient amount of $HgCl_2$ [3, 17], $HgBr_2$ [3], $Hg(SCN)_2$ and heating [6], or $Hg(CN)_2$ under reflux (90 min) [6] $\rightarrow Ge(C_4H_9)_2(H)X$ (X = Cl, Br, NCS, or CN) with a deficient amount of CH_3OCH_2Cl, C_4H_9Br, or C_3H_7I under reflux $\rightarrow Ge(C_4H_9)_2(H)X$ (X = Cl, Br, or I) [5] with $C(C_6H_5)_3Cl \rightarrow Ge(C_4H_9)_2Cl_2$ [5] with C_6H_5I in C_6H_6 at 175 °C $\rightarrow Ge(C_4H_9)_2I_2$ [7] with $C_7H_{15}OH$ at 150 °C $\rightarrow Ge(C_4H_9)_2(H)OC_7H_{15}$ with c-$C_6H_{11}OH$ at 160 °C $\rightarrow Ge(C_4H_9)_2(OC_6H_{11}\text{-c})_2$ with $HO(CH_2)_4OH$ at 140 °C $\rightarrow (C_4H_9)_2\overline{GeOCH_2CH_2CH_2CH_2O}$ with $C_6H_4(OH)_2$-1,4 at ca. 150 °C $\rightarrow$ chain polymer [10] with $CH_2{=}CHCOCH_3$ (1:2 mole ratio), refluxing (in the presence of hydroquinone) $\rightarrow Ge(C_4H_9)_2(CH_2CH_2COCH_3)H$ and $Ge(C_4H_9)_2(CH_2CH_2COCH_3)_2$ [5] with $CH_2{=}CHCOOCH_3$ (1:2 mole ratio) at 100 °C (7 h) $\rightarrow Ge(C_4H_9)_2(CH_2CH_2COOCH_3)H$ and $Ge(C_4H_9)_2(CH_2CH_2COOCH_3)_2$ [5]; refluxing the same reactants in the presence of AIBN or UV-irradiation $\rightarrow Ge(C_4H_9)_2(CH_2CH_2COOCH_3)_2$ [21] with excess $CH(=N_2)COOC_2H_5$ in C_6H_6 at 100 °C (8 h) in the presence of Cu powder $\rightarrow Ge(C_4H_9)_2(CH_2COOC_2H_5)H$ and $Ge(C_4H_9)_2(CH_2COOC_2H_5)_2$ [5] with $C_6H_5C{\equiv}CH$ at 100 °C in the presence of H_2PtCl_6 $\rightarrow Ge(C_4H_9)_2(CH{=}CHC_6H_5)H$ [4] with $CH_2{=}CHCH_2MgCl$ in refluxing THF (48 h) in the presence of $CuCl_2$ $\rightarrow Ge(C_4H_9)_2(CH_2CH{=}CH_2)H$ [5] with $Hg(CH_2COOCH_3)_2$ (1:1 or 1:2 mole ratio) in refluxing THF (ca. 1 h) $\rightarrow Ge(C_4H_9)_2(CH_2COOCH_3)H$ or $Ge(C_4H_9)_2(CH_2COOCH_3)_2$, respectively [14]
4	C_4H_9-t	$GeCl_4$ + LiC_4H_9-t (1:3 mole ratio) in petroleum ether-C_5H_{12} at −78 to +20 °C (overnight), refluxing (2 h), solvent removal from the filtrate, distillation, and GC (ca. 6%); other products were $Ge(C_4H_9\text{-t})_2Cl_2$, $Ge(C_4H_9\text{-t})_3Cl$, and $Ge_2(C_4H_9\text{-t})_4H_2$ b.p. 57 °C/24 ^1H NMR (CH_2Cl_2): 1.15 (s, CH_3), 3.67 (s, br, HGe) IR: ν_{as}(GeH) 2015 MS: $[M]^+$, superimposed by $[M - H]^+$ and $[M - 2H]^+$; $[C_4H_9]^+$ (base peak); $[M - C_4H_9]^+$ and $[M - C_4H_9 - C_4H_8]^+$ (both strong) [26]
5	C_5H_{11}	$Ge(C_5H_{11})_2X_2$ + $LiAlH_4$ like No. 1 (ca. 90%) [5]; see also [2] $Ge(C_5H_{11})_2X_2$ + $NaBH_4$ like No. 1 (ca. 70%) [5] b.p. 92 °C/12 d^{20} = 0.9595; n_D^{20} = 1.4478 [2, 5]

Table 33 (continued)

No.	R group	formation (yield) properties and remarks
		IR (liquid): ν(GeH) 2032 [9] with C_3H_7I at 100 °C (1 h) → $Ge(C_5H_{11})_2I_2$ [5] with $CH_3CH{=}CHCHO$ (1:2 mole ratio) at 100 °C (18 h) in the presence of Pt asbestos → $Ge(C_5H_{11})_2(CH(CH_3)CH_2CHO)H$ and $Ge(C_5H_{11})_2(CH(CH_3)CH_2CHO)_2$ [5] with $CH_2{=}CHCOOCH_3$, refluxing in the presence of AIBN or UV irradiation → $Ge(C_5H_{11})_2(CH_2CH_2COOCH_3)_2$ [21]
6	C_6H_{13}	$Ge(C_6H_{13})_2X_2 + LiAlH_4$ like No. 1 (ca. 90%) [5]; see also [2] $Ge(C_6H_{13})_2X_2 + NaBH_4$ like No. 1 (ca. 70%) [5] b.p. 113 to 114 °C/8, 245 °C/750 without decomposition [5] d^{20} = 0.9484; n_D^{20} = 1.4522 [2, 5] IR (liquid): ν(GeH) 2032 [5, 8, 9] (illustrated [2])
7	C_7H_{15}	$Ge(C_7H_{15})_2X_2 + LiAlH_4$ like No. 1 (ca. 90%) [5]; see also [2] $Ge(C_7H_{15})_2X_2 + NaBH_4$ like No. 1 (ca. 70%) [5] b.p. 148 °C/10 d^{20} = 0.9348; n_D^{20} = 1.4543 [2, 5] IR (neat liquid): ν(GeH) 2032 [5, 9]
8	C_8H_{17}	$Ge(C_8H_{17})_2X_2 + LiAlH_4$ like No. 1 (ca. 90%) [5]; see also [2] $Ge(C_8H_{17})_2X_2 + NaBH_4$ like No. 1 (ca. 70%) [5] b.p. 164 to 165 °C/9 d^{20} = 0.9274; n_D^{20} = 1.4568 [2, 5] IR (neat liquid): ν(GeH) 2032 [8, 9]

*Further information:

General Remarks. The IR spectra of the compounds in this section in addition to other alkylgermanes have been the subject of discussions on the dependence of the ν(GeH) frequency and intensity on the kind of the alkyl substituent [9, 15, 18, 23, 25, 27, 28].

$Ge(C_3H_7\text{-}i)_2H_2$ (Table **33**, No. **2**). A relation between the ^1H NMR shifts of the H_2Ge protons and the ν(GeH) frequencies or the sum of the Taft polar constants σ^* of the substituents was pointed out in [13] for a series of Ge hydrides including No. 2. The fact that only one ν(GeH) band was present in the gas-phase IR spectrum of $Ge(C_3H_7\text{-}i)_2(H)D$ (2033.6 cm^{-1}) was taken as evidence for the predominant presence of the conformer lacking any CH_3-CH_3 steric interaction [27].

ArF-laser (λ = 193 nm) photolysis of a $C_{12}H_{26}$ solution of $Ge(C_3H_7\text{-}i)_2H_2$ caused carbon-free Ge films to deposit on a quartz window of the photolysis cell [30].

$Ge(C_4H_9)_2H_2$ (Table **33**, No. **3**) was also obtained by $LiAlH_4$ reduction of a higher boiling mixture formed by the reaction of $GeCl_4$ with C_4H_9MgCl [3] or by hydrolytic workup of a reaction mixture resulting from the treatment of $Ge(C_4H_9)_2Cl_2$ with four equivalents of potassium in HMPT followed by distillation (22% yield) [20]. $Ge(C_4H_9)_2(H)F$ disproportionated when left standing for several days yielding No. 3 and $Ge(C_4H_9)_2F_2$ [17]. Heating $Ge(C_4H_9)_2(CH_2CR'{=}CH_2)H$ (R' = H, CH_3) in C_6H_6 at 80 °C (15 h) in the presence of C_6H_5CO-

References on p. 232

OO-OCC$_6$H$_5$ in a sealed tube yielded No. 3 in addition to Ge(C$_4$H$_9$)$_2$(CH$_2$CR'=CH$_2$)$_2$. The mechanism of its formation was discussed [29].

^{1}H NMR studies on the organohydrides of Si, Ge, and Sn including No. 3 revealed that these compounds are weakly associated in pure form; in aprotic donor solvents they form labile complexes with variable composition, corresponding to 5- and 6-coordinate central atoms [22].

The rather small effect of various solvents on the ν(GeH) frequency (e.g. 2036 cm^{-1} in C$_6$H$_{14}$ and 2027 cm^{-1} in HMPT) is attributed to the polarity of the Ge$^{(+)}$-H$^{(-)}$ bond which is unable to form a true hydrogen bond with proton acceptors [19].

The reaction of No. 3 with a slight excess of Br$_2$ in C$_2$H$_5$Br (2 h) quantitatively yielded Ge(C$_4$H$_9$)$_2$Br$_2$ [6]. It rapidly reacted with a deficient amount of I$_2$ at 25°C to give Ge(C$_4$H$_9$)$_2$(H)I [3]. However, gradually adding a slight excess of powdered I$_2$ over a period of 2 h followed by refluxing with a Cu wire led to quantitative conversion to Ge(C$_4$H$_9$)$_2$I$_2$ [6].

The influence of various donor solvents such as C$_5$H$_5$N on the rate constants of the reactions of No. 3 with CH$_3$COOH, CH$_2$ClCOOH, C$_6$H$_5$COOH, or 3-NO$_2$C$_6$H$_4$COOH at 110°C in the presence of Cu (dehydrocondensation with replacement of hydrogen by the corresponding anion) showed that the overall reaction rate depends not only on the electron-donor properties of the solvent but also on its steric feature [24].

Refluxing Ge(C$_4$H$_9$)$_2$H$_2$ and Ge(C$_2$H$_5$)$_2$(CH=CH$_2$)$_2$ for 2 h in the presence of C$_6$H$_5$CO-OO-OCC$_6$H$_5$ gave (C$_2$H$_5$)$_2$GeCH$_2$CH$_2$Ge(C$_4$H$_9$)$_2$CH$_2$CH$_2$ and (-CH$_2$CH$_2$Ge(C$_2$H$_5$)$_2$CH$_2$CH$_2$-Ge(C$_4$H$_9$)$_2$-)$_n$ (major product) [11]. Heating No. 3 with 1,5-hexadiyne (1:1 mole ratio) at 120°C (7 d) in the presence of H$_2$PtCl$_6$ in a closed tube yielded a low-molecular-weight (-Ge(C$_4$H$_9$)$_2$CH=CHCH$_2$CH$_2$CH=CH-)$_n$ polymer [12].

References:

[1] Anderson, H. H. (J. Am. Chem. Soc. **78** [1956] 1692/4).
[2] Satgé, J.; Mathis-Noël, R.; Lesbre, M. (C.R. Hebd. Seances Acad. Sci. **249** [1959] 131/3).
[3] Anderson, H. H. (J. Am. Chem. Soc. **82** [1960] 3016/8).
[4] Lesbre, M.; Satgé, J. (C.R. Hebd. Seances Acad. Sci. **250** [1960] 2220/2).
[5] Satgé, J. (Ann. Chim. [Paris] [13] **6** [1961] 519/73).
[6] Anderson, H. H. (J. Am. Chem. Soc. **83** [1961] 547/8).
[7] Lesbre, M.; Satgé, J. (C.R. Hebd. Seances Acad. Sci. **252** [1961] 1976/8).
[8] Mathis-Noël, R.; Mathis, F.; Satgé, J. (Bull. Soc. Chim. Fr. **1961** 676).
[9] Mathis, R.; Satgé, J.; Mathis, F. (Spectrochim. Acta **18** [1962] 1463/72).
[10] Lesbre, M.; Satgé, J. (C.R. Hebd. Seances Acad. Sci. **254** [1962] 4051/3).

[11] Mazerolles, P. (Bull. Soc. Chim. Fr. **1962** 1907/13).
[12] Noltes, J. G.; van der Kerk, G. J. M. (Recl. Trav. Chim. Pays-Bas **81** [1962] 41/8).
[13] Egorochkin, A. N.; Khidekel', M. L.; Ponomarenko, V. A.; Zueva, G. Ya.; Svirezheva, S. S.; Razuvaev, G. A. (Izv. Akad. Nauk SSSR Ser. Khim. **1963** 1865/8; Bull. Acad. Sci. USSR Div. Chem. Sci. [Engl. Transl.] **1963** 1717/9).
[14] Baukov, Yu. I.; Lutsenko, I. F.; Lomonosov, M. V. (Zh. Obshch. Khim. **34** [1964] 3453/6; J. Gen. Chem. USSR [Engl. Transl.] **34** [1964] 3495/7).
[15] Mathis, R.; Constant, M.; Satgé, J.; Mathis, F. (Spectrochim. Acta **20** [1964] 515/21).
[16] Baukov, Yu. I.; Belavin, I. Yu.; Lutsenko, I. F. (Zh. Obshch. Khim. **35** [1965] 1092/4; J. Gen. Chem. USSR [Engl. Transl.] **35** [1965] 1096/8).
[17] Massol, M.; Satgé, J. (Bull. Soc. Chim. Fr. **1966** 2737/43).

[18] Mathis, R.; Berthelat, M.; Mathis, F. (Spectrochim. Acta A **26** [1970] 1993/2000).
[19] Mathis, R.; Berthelat, M.; Mathis, F. (Spectrochim. Acta A **26** [1970] 2001/5).
[20] Bulten, E. J.; Noltes, J. G. (J. Organomet. Chem. **29** [1971] 397/407).

[21] Nederlandse Centrale Organisatie Voor Toegepast-Natuurwetenschappelijk Onderzoek (Neth. Appl. 73-08463 [1973/74] 1/6; C.A. **83** [1975] No. 10402).
[22] Ivanov, V. A.; Reikhsfel'd, V. O.; Saratov, I. E. (Zh. Obshch. Khim. **45** [1975] 2036/40; J. Gen. Chem. USSR [Engl. Transl.] **45** [1975] 1999/2002).
[23] Egorochkin, A. N.; Khorshev, S. Ya.; Satgé, J.; Rivière, P.; Barrau, J. (J. Organomet. Chem. **99** [1975] 239/49).
[24] Saratov, L. E.; Ivanov, V. A.; Reikhsfel'd, V. O. (Zh. Obshch. Khim. **46** [1976] 1052/7; J. Gen. Chem. USSR [Engl. Transl.] **46** [1976] 1048/52).
[25] Egorochkin, A. N.; Khorshev, S. Ya.; Sevastyanova, E. I.; Ratushnaya, S. Kh.; Satgé, J.; Rivière, P.; Barrau, J.; Richelme, S. (J. Organomet. Chem. **155** [1978] 175/83).
[26] Wojnowska, M.; Noltemeyer, M.; Füllgrabe, H.-J.; Meller, A. (J. Organomet. Chem. **228** [1982] 229/38).
[27] McKean, D. C.; Torto, I.; Mackenzie, M. W.; Morrison, A. R. (Spectrochim. Acta A **39** [1983] 387/98).
[28] McKean, D. C.; Torto, I.; Mackenzie, M. W. (Spectrochim. Acta A **39** [1983] 399/408).
[29] Mochida, K.; Miyagawa, I. (Bull. Chem. Soc. Jpn. **56** [1983] 1875/6).
[30] Pola, J.; Parsons, J. P.; Taylor, R. (J. Mater. Chem. **2** [1992] 1289/92).

1.3.2.1.4 GeR_2H_2 Compounds with R = Substituted Alkyl, Alkenyl, and Cycloalkenyl

The compounds in this section are summarized in Table 34.

Table 34
GeR_2H_2 Compounds with R = Substituted Alkyl, Alkenyl, and Cycloalkenyl.
An asterisk indicates further information at the end of the table.
Explanations, abbreviations, and units are given on p. X.

No.	R group	formation (yield) properties and remarks
*1	CF_3	$Ge(CF_3)_2I_2$ + aqueous $NaBH_4$ in 30% H_3PO_4 at 0°C in a vacuum followed by fractional condensation (>90%) [16] or in 1 M aqueous HBr followed by trap-to-trap fractionation (85%) [8] b.p. 20.5°C [16] ^{1}H NMR: 4.55 [8] or 5.05 [16] (HGe, ^3J(H, F) = 7.8) [8, 16]; ^3J(D, F) = 1.20 , ^2J(H, D) = 3.0 for $Ge(CF_3)_2(H)D$ [16] ^{13}C NMR (C_6D_6): 129.1 (CF_3, ^2J(C, H) = 12.3, ^3J(C, F) = 5.3) [16] ^{19}F NMR: −50.3 (CF_3, ^1J(C, F) = 330.7, (^4J(F, F) = 4.72) [16]; see also [13]; −26.2 (CF_3) for the neat compound relative to CF_3COOH [8] IR (gas): 2155(s), 1168(vs), 1137(vvs), 722(s) [8]; see also p. 235
*2	CH_2SiH_3	$Ge(CH_2SiCl_3)_2Cl_2$ + excess $LiAlH_4$ in tetralin at 100°C (10 h) in the presence of $[N(C_2H_5)_3CH_2C_6H_5]Cl$ followed by vacuum

Table 34 (continued)

No.	R group	formation (yield) properties and remarks
*2 (continued)		condensation of the volatiles at 50 °C into a trap at −196 °C and distillation (71%) [18] b.p. 78 °C [18] ^{1}H NMR: −0.25 (sext, CH_2), 3.71 (t, H_3Si, $^3J(H, H)$ = 4.6, $^1J(Si, H)$ = 196.5), 3.91 (quint, HGe, $^3J(H, H)$ = 3.7) in C_6D_6 [18]; 0.02 (CH_2), 3.69 (HGe, J(H, H) = 3.7), 3.98 (H_3Si, $^3J(H, H)$ = 4.6) in C_6H_{12} [4, 6] ^{13}C NMR (C_6D_6): −15.76 (t, CH_2, $^1J(C, H)$ = 123.3 [18] ^{29}Si NMR (neat): −59.45 (q, $^1J(Si, H)$ = 196.7, $^2J(Si, H)$ = 8.3, $^3J(Si, H)$ = 4.2) [18] ^{73}Ge NMR (neat at 0°C): −95.66 (t, $^1J(Ge, H)$ = 96.1) [18] IR (KBr): ν(CH) 2973, 2938, 2873, 2802, ν(SiH) 2154, ν(GeH) 2048, δ(CH) 1047, δ(SiH_3) 929, δ(GeH) 876, 800, ν(SiC) 752, 712, ν(GeC) 658 [18]; for the gas-phase spectrum [6], see p. 236
*3	$CH_2C_6H_5$	$Ge(CH_2C_6H_5)_4$ + Li shot in DME at 20 to 0 °C (16 h), hydrolysis with 10% aqueous monoglyme, extraction with C_6H_6 and H_2O, and distillation of the organic layer under vacuum (8%) along with $Ge(CH_2C_6H_5)_3H$ (21%) [1] b.p. 80 to 85 °C/0.001 ^{1}H NMR: 2.13 (t, CH_2), 4.02 (quint, HGe), 6.94 (C_6H_5) [1] IR: ν(GeH) 2037(s), δ(GeH) 864(s), (704(s)?), ν_{as}(GeC) 556(w), ν_s(GeC) 532(w) [2]; see also [1]; ν(GeH) 2057 in C_7H_{16} [9] for $Ge(CH_2C_6H_5)_2H_2$; ν(GeD) 1466(m), δ(GeD) 615(m), 495(m), ν(GeC) 564(w) for $Ge(CH_2C_6H_5)_2D_2$ [2]; see also [1] with $HgCl_2$ (1.2:1 mole ratio) at 20 °C (24 h) → $Ge(CH_2C_6H_5)_2(H)Cl$ [10]
4	$CH_2CH_2CH_2Cl$	IR (C_7H_{16}): ν(GeH) 2055 [9]
5	$CH_2(CH_2)_3NH_2$	$Ge(CH_2CH_2CH_2CN)_2Br_2$ + $LiAlH_4$ in ether b.p. 140 °C/20 d^{20} = 1.1078; n_D^{20} = 1.4988 IR: ν(GeH) 2020 [3]
*6	$CH{=}CH_2$	see p. 236
*7	C_5H_5-c	GeH_2Br_2 + excess KC_5H_5 at −20 °C (90 min), warming to +20 °C, and pumping through a trap at 0 °C [7]; see also [5] colorless volatile liquid [7] ^{1}H NMR (neat liquid): 3.61 (HGe), 5.61 (C_5H_5) at 35 °C to 40 °C [5, 7]; 3.61, 5.70 (br) at −20 °C; 3.60 (br), 5.90 (v br) at −50 °C [5] Raman (liquid): 3115(m), 3080(m), 2935(m,br), 2040(s,p), 1605(vw), 1550(vw), 1462(vs,p), 1380(m), 1115(s), 1090(w), 1020(m), 970(vs,p), 950(s), 910(m), 860(w,br), 830(m), 779(m), 716(w), 683(m), 668(m), 620(m,br), 480(m,sh), 359(vs,p), 300(w), 230(w,br), 190(w,br) [7]

*Further information:

General Remarks. IR studies on the behavior of the ν(GeH) vibration of various germanium hydrides included compounds No. 1, 3 and 4 [9, 14].

$Ge(CF_3)_2H_2$ (Table **34**, No. **1**) was formed in small amounts along with $Ge(CF_3)_3H$ (main product) by reducing $Ge(CF_3)_3I$ with $NaBH_4$ in H_3PO_4 [16]. The deuterated derivatives, $Ge(CF_3)_2(H)D$ [11, 16] and $Ge(CF_3)_2D_2$, were prepared from $Ge(CF_3)_2I_2$, $NaBD_4$, and H_3PO_4 or D_3PO_4, respectively, and purified by GC [11].

A complete analysis of the vibrational spectra of $Ge(CF_3)_2H_2$, $Ge(CF_3)_2(H)D$, and $Ge(CF_3)_2D_2$ was reported [11]. Frequencies and assignments for $Ge(CF_3)_2H_2$ are listed in Table 35 (Raman polarization spectrum of $Ge(CF_3)_2H_2$ illustrated). A normal-coordinate analysis was carried out giving the stretching force constants f(Ge-C) = 2.25, f(Ge-H) = 2.743, and f(C-F) = 5.75 N/cm, all being slightly lower than the corresponding force constants of $Ge(CF_3)_3H$. The bonding parameters were discussed with respect to those of related molecules [11]. For Ge-H stretch-stretch interaction force constants, see $Ge(CH_3)_2H_2$ [15], p. 218.

Table 35
Vibrational Spectra of $Ge(CF_3)_2H_2$ [11].
Wavenumbers in cm^{-1}.

IR gas	Raman liquid	assignment
2161(s)	2170(w)	ν_{as}(GeH)
2143(m), 2147(m)	2150(s,p)	ν_s(GeH)
1196(s)	1195(w,p)	$\nu_s(CF_3)$
1168(vs)	1168(w)	
1136(vs)	1120(vw,br)	$\nu_{as}(CF_3)$
1111(ms)		
1106(s)		
846(ms)	843(m,p)	$\delta(GeH_2)$
	728(s,p)	$\delta_s(CF_3)$
720(s)	720(w)	δ(CGeH)
644(vw)	652(m)	
548(w)	550(w)	$\delta_{as}(CF_3)$
530(sh)	530(vw)	
516(w)		
508(sh)	510(w)	
454(m)	458(w)	δ(CGeH)
324(s)	323(sh)	$\nu_{as}(GeC_2)$
	315(s,p)	$\rho(CF_3)$
248(vw)	250(vs,p)	$\nu_s(GeC_2)$
	206(vw)	$\rho(CF_3)$
203(w)		
75(w,br)	79(m)	$\delta(GeC_2)$

He(I) and He(II) photoelectron spectra (illustrated) of $Ge(CF_3)_nH_{4-n}$ compounds (n = 1 to 3) were studied and compared. The experimental vertical ionization energies in the range

References on p. 237

10 to 27 eV, the values for the orbitals calculated by the CNDO/2 method, and the contributions to the molecular orbitals were listed. The following ionization energies were measured: 12.40, 12.75 (both mainly Ge-C and F), and 13.50 eV (mainly Ge-H). Core-level binding energies: 38.3 (Ge 3d, relative to Ne 2s at 48.47 eV), 130.75 (Ge $3p_{3/2}$), 134.80 (Ge $3p_{1/2}$), 297.90 (C 1s), and 694.40 eV (F 1s) (all relative to Ar $2p_{2/3}$ at 248.63 eV). The energies increase with the number of CF_3 groups along the the series $Ge(CF_3)H_3$, $Ge(CF_3)_2H_2$, $Ge(CF_3)_3H$ [12].

The vapor pressure is expressed by log (p/mbar) = 7.855 − 1424/T; ΔH_v = 27.26 kJ/mol and ΔS_v = 92.8 kJ · mol^{-1} · K^{-1} [16].

Mass spectrum: $[Ge(CF_3)(CF_2)H_2]^+$ (10), $[Ge(CF_3)HF]^+$ (11), $Ge(CF_3)H_2]^+$ (95), $[GeCF_3]^+$ (25), $[GeF_2H]^+$ (25), $[GeFH_2]^+$ (55), $[GeF]^+$ (100), $[C_2F_3H]^+$ (?), $[GeH]^+$ (50), $[Ge]^+$ (30), $[CF_3]^+$ (8), $[CF_2H]^+$ (20) [16]; see also [8].

Condensation with CD_3I and $Zn(CH_3)_2$ gave $Ge(CH_3)(CF_3)_2H$ and $Ge(CD_3)(CF_3)_2H$ in a 68:32 ratio and $Ge(CH_3)_2(CF_3)_2$, $Ge(CH_3)(CD_3)(CF_3)_2$, and $Ge(CD_3)_2(CF_3)_2$ in a 37:46:17 ratio [16].

$Ge(CH_2SiH_3)_2H_2$ (Table **34**, No. **2**) was obtained in small amounts along with $Ge(CH_2\text{-}SiH_2Cl)H_3$, $Ge(CH_2Si(H)Cl_2)H_3$, and GeH_4 by reacting $Ge(CH_2SiH_3)H_3$ with HCl in the vapor phase at room temperature in the presence of $AlCl_3$ (24 h) followed by trap-to-trap distillation [4, 6].

The vapor pressure at 0 °C is 8.6 Torr [4, 6].

IR spectrum (vapor): ν(CH) 2942, 2910(d,vw), ν(SiH) 2160(s), ν(GeH) 2056(s), $\delta(CH_2)$ 1380(vw), 1055, 1049(d,m), $\delta(SiH_3)$ 940(vs), ν(SiC) 750(m), ν(GeC) 662(w) cm^{-1}; other bands at 885, 876(d,w), 805(m), and 718(m) cm^{-1} [6].

Mass spectrum: m/e = 166 to 155 $[GeSi_2C_2H_n]^+$, 136 to 125 $[GeSiC_2H_n]^+$, 123 to 110 $[GeSiCH_n]^+$, 107 to 98 $[GeSiH_n]^+$, 87 $[Si_2C_2H_7]^+$, 59 $[Si_2H]^+$, 43 $[SiCH_3]^+$, 28 $[Si]^+$ [18]; see also [6].

$Ge(CH_2C_6H_5)_2H_2$ (Table **34**, No. **3**). The compound forms a charge-transfer complex with $C_2(CN)_4$ in CH_2Cl_2, showing a CT absorption band at 19800 cm^{-1}. The linear relation between ν_{CT} and the σ_p^+-Brown-Okamoto constants was used to calculate the σ_p^+ constants of Ge- and Si-containing substituents in C_6H_5R' compounds (for No. 3, R' = $CH_2Ge(CH_2C_6H_5)H_2$). The resonance constants σ_R of the corresponding substituents were determined from the integrated intensity of the IR stretching vibrations of the aromatic ring (1600 cm^{-1} region) [10].

$Ge(CH_2C_6H_5)_2D_2$ was obtained in small amounts as described for No. 3 in Table 34 except hydrolysing with D_2O [1].

$Ge(CH{=}CH_2)_2H_2$ (Table **34**, No. **6**). The effect of bridging ER_2 groups (E = C, Si, Ge, Sn) on the symmetry of the π levels in planar conformers of $H_2C{=}CH\text{-}ER_2\text{-}CH{=}CH_2$ molecules including No. 6 was studied by the MNDO method [17].

$Ge(C_5H_5\text{-}c)_2H_2$ (Table **34**, No. **7**). The temperature-dependent ^1H NMR spectrum was compared with that of $Ge(C_5H_5\text{-}c)H_3$. Progressive broadening of the single C_5H_5 resonance on cooling indicates fluxional behavior [5, 7].

The mass spectrum showed $[M]^+$ which fragmented to give $[C_5H_5GeH]^+$. The $[C_5H_6]^+$ and $[C_5H_7]^+$ ions supposedly originate from transfer of Ge-bonded hydrogen to the C_5H_5 groups (all fragments listed) [7].

References:

[1] Cross, R. J.; Glockling, F. (J. Chem. Soc. **1964** 4125/33).
[2] Cross, R. J.; Glockling, F. (J. Organomet. Chem. **3** [1965] 146/55).
[3] Mazerolles, P.; Lesbre, M.; Lavergne, J.-P. (C.R. Seances Acad. Sci. C **266** [1968] 639/41).
[4] Gibbon, G. A.; Kifer, E. W.; van Dyke, C. H. (Inorg. Nucl. Chem. Lett. **6** [1970] 617/20).
[5] Stobart, S. R. (J. Organomet. Chem. **33** [1971] C11/C12).
[6] van Dyke, C. H.; Kifer, E. W.; Gibbon, G. A. (Inorg. Chem. **11** [1972] 408/12).
[7] Angus, P. C.; Stobart, S. R. (J. Chem. Soc. Dalton Trans. **1973** 2374/80).
[8] Lagow, R. J.; Eujen, R.; Gerchman, L. L.; Morrison, J.-A. (J. Am. Chem. Soc. **100** [1978] 1722/6).
[9] Egorochkin, A. N.; Khorshev, S. Ya.; Sevastyanova, E. I.; Ratushnaya, S. Kh.; Satgé, J.; Rivière, P.; Barrau, J.; Richelme, S. (J. Organomet. Chem. **155** [1978] 175/83).
[10] Sennikov, P. G.; Skobeleva, S. E.; Kuznetsov, V. A.; Egorochkin, A. N.; Rivière, P.; Satgé, J.; Richelme, S. (J. Organomet. Chem. **201** [1980] 213/9).

[11] Eujen, R.; Mellies, R. (Spectrochim. Acta A **38** [1982] 533/40).
[12] Drake, J. E.; Gorzelska, K.; Helbing, R.; Eujen, R. (J. Electron Spectrosc. Relat. Phenom. **26** [1982] 19/30).
[13] Eujen, R.; Mellies, R. (J. Fluorine Chem. **22** [1983] 263/80).
[14] McKean, D. C.; Torto, I.; Mackenzie, M. W. (Spectrochim. Acta A **39** [1983] 399/408).
[15] McKean, D. C.; Torto, I.; Mackenzie, M. W.; Morrison, A. R. (Spectrochim. Acta A **39** [1983] 387/98).
[16] Eujen, R.; Mellies, R.; Petrauskas, E. (J. Organomet. Chem. **299** [1986] 29/40).
[17] Shorygin, P. P.; Isaev, A. N. (Izv. Akad. Nauk SSSR Ser. Khim. **1989** 307/10; Bull. Acad. Sci. USSR Div. Chem. Sci. [Engl. Transl.] **1989** 260/2).
[18] Schmidbaur, H.; Rott, J. (Z. Naturforsch. **45b** [1990] 961/6).

1.3.2.1.5 Diphenylgermane, Diphenylgermanium Dihydride, $Ge(C_6H_5)_2H_2$

Preparation and Formation. $Ge(C_6H_5)_2H_2$ was prepared by reducing $Ge(C_6H_5)_2Cl_2$ with $LiAlH_4$ in ether [13]. Mixed phenylbromogermanes obtained from bromination of $Ge(C_6H_5)_4$ or $Ge_2(C_6H_5)_6$ were also reduced with $LiAlH_4$ in refluxing ether (2 h) followed by fractionation to yield $Ge(C_6H_5)_2H_2$ with 55 [1, 3] and 67% [1] yield, respectively.

$Ge(C_6H_5)_2H_2$ was formed when $Ge_2(C_6H_5)_4H_2$ was reacted with $LiAlH_4$ in THF at room temperature for 24 h (90 to 95% yield) [38], with Li [38] or Li-Hg for 30 h (31% yield), or with GeR_3Li [30], in the latter three cases along with $(Ge(C_6H_5)_2)_n$. The formation (20% yield) from $Ge(C_6H_5)_2(H)Li$ and $Ge(C_6H_5)H_2Cl$ in THF at −20 to +20°C (2 h) was also reported [52] (see also [49]); $H(C_6H_5)_2GeGe(C_6H_5)H_2$ was the other product [52]. The reaction of $Ge(C_6H_5)_2(H)Li$ with $Ge(C_6H_5)(H)Cl_2$ in THF-ether at −70°C (1 h) yielded $Ge(C_6H_5)_2H_2$ (43%) and $H(C_6H_5)_2GeGe(C_6H_5)H_2$ [52]. Reacting $Ge_2(C_6H_5)_4H_2$ with LiC_4H_9-t in a 2:1 mole ratio in C_5H_{12}-THF at −40°C followed by CH_3I gave $Ge(C_6H_5)_2H_2$ (15%), $CH_3(C_6H_5)_2GeGe(C_6H_5)_2H$ (13%), $(Ge(C_6H_5)_2)_n$ (20%), and 50% starting material [55]. $Ge(C_6H_5)_2H_2$ was identified by ^1H NMR spectroscopy as one of the products of the reaction of $Ge(C_6H_5)_2(H)Li$ with α,α'-dibromo-o-xylene at −78°C followed by hydrolysis [53].

The compound was formed along with $Ge(C_6H_5)_2(H)Cl$ by treating $Ge(C_6H_5)_2Cl_2$ with $Sn(C_4H_9)_3H$ in the presence of $CH(=N_2)COOC_2H_5$ at 20°C (33% yield after 24 h) [44]. Electrochemical reduction of $Ge(C_6H_5)_2Cl_2$ in DME (with $[N(C_4H_9)_4]ClO_4$ electrolyte) gave $Ge(C_6H_5)_2H_2$ [14].

$Ge(C_6H_5)_2H_2$ was obtained as a product of the decomposition of the following compounds: $Ge(C_6H_5)H_3$ at 150 °C in a sealed tube (12 h) or at room temperature in the presence of $AlCl_3$ (along with GeH_4) [17]; $Ge(C_6H_5)_3H$ at 300 °C (along with $Ge(C_6H_5)_4$) [2]; $Ge(C_6H_5)_2(H)Cl$ at 20 °C in a sealed tube or more rapidly in C_5H_{12} at 200 °C (4 h) or in the presence of AIBN or H_2PtCl_6 (along with $Ge(C_6H_5)_2Cl_2$ and $Ge(C_6H_5)_3Cl$) [17]; $Ge(C_6H_5)_2(H)SCH_3$ at 80 °C (24 h) or UV-irradiated (5 h) at 30 °C (along with $Ge(C_6H_5)_2(SCH_3)_2$ [31]; $Ge(C_6H_5)_2(H)N(C_4H_9)_2$ at 20 °C (48 h); $Ge(C_6H_5)_2(H)P(C_2H_5)_2$ at 20 °C (15 min) [32]; $Ge(C_6H_5)_2(CH(OH)C_6H_5)H$ at 100 °C [52]; and decomposition of $Ge_2(C_6H_5)_4H_2$ upon distillation at 0.02 Torr [27].

$Ge(C_6H_5)_2H_2$, digermanes, and polymers were formed when $Ge(C_6H_5)_2K_2$ was reacted with 4-$CH_3C_6H_4X$ (X = F or Cl) in HMPT-THF at room temperature followed by hydrolysis [46]. For the formation from $Ge(C_6H_5)_2(COC_6H_2(CH_3)_3\text{-}2,4,6)H$ by hydrolysis, see [52]. Photolysis of $Ge(C_6H_5)_2(H)Fe(CO)_2(C_5H_5)$ in C_6D_6 (3.5 h) yielded 18% $Ge(C_6H_5)_2H_2$ [54].

$Ge(C_6H_5)_2(H)D$ was prepared by reducing $Ge(C_6H_5)_2(H)Cl$ with excess $LiAlD_4$ in ether at 0 °C [52] followed by treatment with D_2O, extraction with ether, and distillation (67% yield) [38]. It was also formed from $Ge(C_6H_5)_2(H)Li$ by hydrolysis with D_2O in THF at −40 °C, extraction, and concentration under vacuum [52]. **$Ge(C_6H_5)_2D_2$** was prepared from $Ge(C_6H_5)_2Cl_2$ and $LiAlD_4$ [22] or $Ge(C_6H_5)_2K_2$ (formed from $Ge(C_6H_5)_2H_2$ and K in THF-HMPT) and D_2O (84 to 96% yield) [46]. $Ge(C_6H_5)_2(H)T$ was prepared by adding tritiated water to $Ge(C_6H_5)_2H_2$ in CH_3OH, containing CH_3ONa and DMSO, refluxing for 20 h, adding the mixture to $C_6H_5CH_3$, and neutralizing with HCl [40].

The enthalpy of combustion of liquid $Ge(C_6H_5)_2H_2$, $\Delta H_{comb} = -7214 \pm 3.3$ kJ/mol, was determined calorimetrically, yielding the thermodynamic parameters of formation: $\Delta H_f = 224 \pm 3.3$ kJ/mol, $\Delta G_f = 386 \pm 3.8$ kJ/mol, and $\Delta S_f = 545.2 \pm 0.4\ J \cdot mol^{-1} \cdot K^{-1}$ (at 298.15 K) [42].

Spectra and Physical Properties. ^1H NMR spectrum: δ(ppm) = 4.90 (HGe) [40] or 5.14 [10] in $CDCl_3$; 5.02 in CCl_4 [47]; 4.99 (s, HGe), 7.15 to 7.64 (m, C_6H_5) in THF-d_8 [46]; 5.00, 7.10 to 7.50 in C_5H_{12}-THF [52]; 5.09, 7.23 [9] or 5.17, 7.14 [33] were reported for neat $Ge(C_6H_5)_2H_2$; see also [6, 13, 32]. Data for $Ge(C_6H_5)_2(H)D$: δ(ppm) = 5.10 (t, HGe, J(H, D) = 0.9 Hz), 6.90 to 7.20 and 7.20 to 7.49 (m, C_6H_5) in C_6D_6 [52]; see also [38].

A correlation between the chemical shift of the GeH protons and the sum of the Taft polar constants of the other substituents at the Ge atom was presented in [6] for a series of Ge hydrides including $Ge(C_6H_5)_2H_2$. ^1H NMR spectra of phenyl-substituted hydrides of group IV and V elements were also evaluated in order to obtain the effective shielding constant of the C_6H_5 group which depends on the electronegativity of the element [10].

^{13}C NMR spectrum (in THF): δ(ppm) = 129.11 (C-3), 129.83 (C-4), 134.65 (C-1), 135.84 (C-2) [45]; 129.4, 130.1, 135.0, 136.1 [52]; 130.2, 130.9, 135.7, 137.7 (in THF-d_8) [46]. ^{13}C NMR spectra of $Ge(C_6H_5)_nH_{4-n}$ and $Ge(C_6H_5)_nH_{3-n}Na$ compounds (n = 1 to 3) were compared with those of analogous group V derivatives, indicating less delocalization of the negative charge of the germyl anions into the aromatic ring than found in the anions of C_6H_5-substituted phosphines and arsines [45].

IR spectrum: ν(GeH) 2051(s), δ(GeH) 866(s), (758(s) or 723(s)?), 511(m); characteristic GeC_6H_5 bands at 1094, 312(s), 292(m) cm^{-1} for the neat compound [11] (see also [34]); ν(GeH) 2055 to 2062 cm^{-1} in C_7H_{16} or CCl_4 [25, 39]; 2024 cm^{-1} in THF-C_5H_{12} [52]. Data for $Ge(C_6H_5)_2(H)D$: ν(GeH) 2065 and ν(GeD) 1455 cm^{-1} [38]. The IR and Raman spectra of $Ge(C_6H_5)_2H_2$ and $Ge(C_6H_5)_2D_2$ were fully displayed. The spectral region below 900 cm^{-1} was listed and interpreted in detail (C_6H_5 vibrations assigned assuming C_{2v} local ring symmetry) [22]. The Ge-H and Ge-D frequencies reported in [22] are given below (in cm^{-1}):

$Ge(C_6H_5)_2H_2$ IR	Raman	$Ge(C_6H_5)_2D_2$ IR	Raman	assignment
2053(vs)		1474(vs)		stretching
866(vs,br)	864(10)	613(vs)	614(17)	deformation
758(vs)	757(5)	542(vs)	543(6)	wagging
	724(3)	515(w,sh)	516(4)	twisting
513(vs)		355(m)		rocking

$Ge(C_6H_5)_2H_2$ was part of an earlier discussion on possible relations between the ν(GeH) frequency and the ^1H NMR chemical shift of a series of Ge hydrides [6]. The ν(GeH) frequency is related to the sum of Taft σ^* constants and steric effects of the substituents at the Ge atom [25]. Steric effects of phenyl, xylyl, and mesityl substituents on the frequency and integrated intensity of the ν(EH) band were studied for a series of Si and Ge hydrides including the title compound [41]. The integrated intensity of the ν(GeH) band was also discussed with respect to π(d-p) interaction between the substituents and the Ge atom [39].

The ν(GeH) and $\nu(GeC_6H_5)$ bands of the neat compound are shifted in CH_3COCH_3 (2046 → 2060 and 1092 → 1061 cm^{-1}, respectively) owing to association via Ge←O coordination [34].

Extended X-ray absorption-fine structure spectra of $Ge(C_6H_5)_nH_{4-n}$ (n = 2 to 4) and $Ge_2(C_6H_5)_6$ were studied to elucidate the structure of the germylene $Ge(C_6H_2(C_4H_9\text{-t})_3\text{-}2,4,6)_2$. The spectrum of the germylene is very similar to that of $Ge(C_6H_5)_2H_2$ [48].

$Ge(C_6H_5)_2H_2$ melts at 239.78±0.01 K; the enthalpy and entropy of melting were determined to be ΔH_m = 11.91±0.02 kJ/mol and ΔS_m = 49.74±0.08 J · mol^{-1} · K^{-1} [42]. Reported boiling points: 66 to 67°C/0.2 Torr [37], 70°C/0.02 Torr [27], 78°C/0.1 Torr [30], 91 to 94°C/1 Torr [40], or 95°C/1 Torr [3]. Refractive index n_D^{20} = 1.5935 [1, 30] and n_D^{25} = 1.5918 [1]. Respective values for $Ge(C_6H_5)_2(H)D$: 64°C/0.01 Torr and n_D^{20} = 1.5958 [38]; a boiling point of 140 to 143°C/0.7 Torr for $Ge(C_6H_5)_2D_2$ [46] is hardly consistent with the data above.

Measurements of the heat capacity from 5 to 330 K and the calculated thermodynamic functions are reported in [42].

Mass Spectrum and Chemical Behavior. $Ge(C_6H_5)_2D_2$ was briefly mentioned in an investigation of the mass spectra of organogermanes. Only the metastable-supported cleavage of two bonds, $[Ge(C_6H_5)D]^{+\bullet}$ → $[Ge]^{+\bullet}$ + C_6H_5D, was noted [19].

Phenylgermanes are weaker acids than GeH_4, supposedly resulting from the absence of a significant π(p-p) interaction between the C_6H_5 ring and the Ge atom [26]. ^1H NMR spectroscopy was used to determine the acidity of some aryl-substituted germanes in liquid NH_3. The acidity decreases with increasing aryl substitution. Factors such as solvation, ion pairing, and structural effects were discussed to explain the order of acidities [28].

$Ge(C_6H_5)_2H_2$ is fairly stable below 0°C [1, 3]. In the absence of air the compound remained unchanged after 8 h heating at 250°C, even when UV-irradiated. Decomposition occurred on heating at 280 to 340°C for 48 h in an evacuated tube to give $Ge(C_6H_5)_4$, Ge, C_6H_6, and H_2 [7]. $Ge(C_6H_5)_2(H)D$ formed $Ge(C_6H_5)_2H_2$ in ether solution at 20°C over a period of several days [52].

^1H NMR studies on organohydrides of Si, Ge, and Sn showed that these compounds are weakly associated in the pure state and form in aprotic donor solvents labile complexes of variable composition corresponding to a 5- or 6-coordinate central atom [33]. A charge trans-

References on p. 241

fer complex between $Ge(C_6H_5)_2H_2$ and $C_2(CN)_4$ in CH_2Cl_2 (ν(CT) absorption at 25600 cm^{-1}) was used to calculate the Hammett σ_p and the Brown-Okamoto σ_p^+ constants of the $Ge(C_6H_5)H_2$ group [35].

Base-catalyzed hydrogen exchange between tritiated Ge hydrides and CH_3OH containing CH_3ONa was studied at 20 to 40°C. For $Ge(C_6H_5)_2(H)T$ the rate relative to that of $Ge(C_6H_5)_3H$ was found to be k_{rel} = 0.11 at 30°C. The exchange rates decrease from $Ge(C_6H_5)_3T$ to $Ge(C_6H_5)H_2T$; this order of kinetic acidities [40] is opposite to the order reported for equilibrium acidities in liquid NH_3 [15].

Reactions of $Ge(C_6H_5)_2H_2$ with alkali metals in HMPT-THF gave $Ge(C_6H_5)_2M_2$ compounds (M = Li, Na, K) [46]; see also [49]. Adding $Ge(C_6H_5)_2H_2$ to a slight excess of Na or $NaNH_2$ in liquid NH_3 yielded $Ge(C_6H_5)_2(H)Na$ [45].

$Ge(C_6H_5)_2H_2$ was brominated (1:2 mole ratio) in $CHCl_3$ at ca. 0°C to give $Ge(C_6H_5)_2Br_2$ [1, 22], whereas treatment with two moles of I_2 in $CHCl_3$ at 70 to 100°C yielded GeI_4 [1]; see also [3]. For the reaction with tritiated H_2O [40], see the formation of $Ge(C_6H_5)_2(H)T$ above.

The reaction with HgX_2 (1:0.8 mole ratio) at 50°C (12 h for X = Cl) or at 70°C (24 h for X = Br) led to $Ge(C_6H_5)_2(H)X$ (main product), $Ge(C_6H_5)_2X_2$, and $Ge(C_6H_5)X_3$ (only for X = Cl, 2% yield) [17].

Dehydrocondensation between $Ge(C_6H_5)_2H_2$ and RCOOH (R = CH_3, CH_2Cl, C_6H_5, or $C_6H_4NO_2$-3) at 110°C in the presence of Cu (to yield $Ge(C_6H_5)_2(OOCR)_2$) was studied in various N-donor solvents such as C_5H_5N. The overall reaction rate depends not only on the electron donor properties of the solvent but also on its steric feature [36].

The following reactions of the hydrogermylation type were reported: with CH_2=CHCHO (1:10 mole ratio) at 100°C (14 h) → $Ge(C_6H_5)_2(CH_2CH_2CHO)H$, $Ge(C_6H_5)_2(CH_2CH_2CHO)_2$ (small amount), $(C_6H_5)_2\overline{GeOCH_2CH_2CH_2}$, and polymers [29]; with CH_2=$CHCOCH_3$ at 110°C (16 h) → $Ge(C_6H_5)_2(CH_2CH_2COCH_3)H$, $Ge(C_6H_5)_2(CH_2CH_2COCH_3)_2$, $(C_6H_5)_2\overline{GeCH_2CH_2CH(CH_3)O}$, and polymers [24]; with CH_2=$CHCH_2CH_2COCH_3$ at 120°C (48 h) → $Ge(C_6H_5)_2$-$(CH_2(CH_2)_3COCH_3)H$ and $Ge(C_6H_5)_2(CH_2(CH_2)_3COCH_3)_2$ [24]; with CH_2=$CH(CH_2)_nCN$ in the presence of AIBN in a sealed tube at 95°C/60 h (for n = 0) or 135°C/48 h (for n = 1) → $Ge(C_6H_5)_2((CH_2)_{n+2}CN)H$, $Ge(C_6H_5)_2((CH_2)_{n+2}CN)_2$, and polymers [21]; and with 1,5-hexadiyne (1:1 mole ratio) at 80°C (1 d) in the presence of H_2PtCl_6 in a sealed tube → (-Ge-$(C_6H_5)_2CH$=$CHCH_2CH_2CH$=CH-$)_n$ polymer of low molecular weight [5]. The thermodynamic parameters of the reaction with $E(C_6H_5)_2(C{\equiv}CH)_2$ (E = Si or Ge) to give the linear polymers (-$Ge(C_6H_5)_2CH$=$CHE(C_6H_5)_2CH$=CH-$)_n$ were calculated from available enthalpies of formation; this polyaddition is thermodynamically favored in the range 0 to 320 K [42].

$Ge(C_6H_5)_2(H)Cl$ and a small amount of $Ge(C_6H_5)_2Cl_2$ were obtained on treating $Ge(C_6H_5)_2H_2$ with CH_3OCH_2Cl (1:1.1 mole ratio) at 60°C (30 min) in the presence of $AlCl_3$ [17]; the proportion of $Ge(C_6H_5)_2Cl_2$ increased during distillation of the mixture, suggesting the equilibrium 2 $Ge(C_6H_5)_2(H)Cl$ ⇌ $Ge(C_6H_5)_2H_2$ + $Ge(C_6H_5)_2Cl_2$ [37]. Halogenation was also observed with N-chlorosuccinimide (1:1 mole ratio) in CH_3NO_2 at 100°C (24 h) to yield $Ge(C_6H_5)_2(H)Cl$ and with one mole N-bromosuccinimide (see also [13]) at 60°C (3 h) or N-iodosuccinimide at 45°C (2 h) in petroleum ether to give $Ge(C_6H_5)_2(H)X$ (72%/30% for X = Br/I) and $Ge(C_6H_5)_2X_2$ (8%/40% for X = Br/I) [17].

$Ge(C_6H_5)_2H_2$ reacted with cyclohexanone (1:2 mole ratio) at 120°C (24 h) in the presence of AIBN in a sealed tube to yield $Ge(C_6H_5)_2(OC_6H_{11}$-c$)_2$ [17].

$Ge(C_6H_5)_2H_2$ cleaved CH_3SSCH_3 (1:2 mole ratio) at 120°C (4 h) yielding $Ge(C_6H_5)_2$-$(SCH_3)_2$. $Ge(C_6H_5)_2H_2$ and $Ge(C_6H_5)_2(SCH_3)_2$ produced $Ge(C_6H_5)_2(H)SCH_3$ when heated at

80 °C (24 h) or UV-irradiated at 30 °C (5 h). A radical mechanism was proposed as galvinoxyl almost inhibited this reaction [31].

$Ge(C_6H_5)_2H_2$ reacted with excess CH_2N_2 to give $Ge(CH_3)(C_6H_5)_2H$ and $Ge(CH_3)_2(C_6H_5)_2$ [4]. These two products were also obtained by UV irradiation of the reactants in ether (6 h), but 52% of $Ge(C_6H_5)_2H_2$ remained unchanged; refluxing (10 h) in the presence of Cu powder gave only $Ge(CH_3)(C_6H_5)_2H$ and 75% $Ge(C_6H_5)_2H_2$ unchanged [13].

The reaction with LiC_4H_9-t (1:1 mole ratio) in $THF-C_5H_{12}$ at −20 °C gave $Ge(C_6H_5)_2(H)Li$ (characterized in solution) [49, 52]; when this reaction was carried out at +20 °C in the presence of $NH(C_2H_5)_2$ followed by hydrolysis, polygermanes $H(Ge(C_6H_5)_2)_nH$ were formed, where n = 2 to 4 (depending on the reaction time in the range of minutes to several hours) [52]. Using a 1:2 mole ratio of $Ge(C_6H_5)_2H_2$ and LiC_4H_9-t at −40 °C [49, 52] or LiC_4H_9-n in ether at −10 °C [9] (see also [11]) yielded $Ge(C_6H_5)_2(H)Li$ and $Ge(C_6H_5)_2Li_2$, the latter identified as $Ge(CH_3)_2(C_6H_5)_2$ after reaction with CH_3I [49]; see also [56].

$Ge(C_6H_5)_2H_2$ reacted with $SnR_3N(C_2H_5)_2$ (R = C_2H_5, C_6H_5; 1:2 mole ratio) to give $Ge(C_6H_5)_2(SnR_3)_2$. With $Sn(C_6H_5)_3N(C_2H_5)_2$ (1:1 mole ratio) in refluxing butyronitrile (1 h) $Ge(C_6H_5)_2(H)Sn(C_6H_5)_3$, $Ge(C_6H_5)_2(H)N(C_2H_5)_2$, and $Sn_2(C_6H_5)_6$ were formed. $Ge(C_6H_5)_2$-$(Sn(C_2H_5)_2Ge(C_6H_5)_3)_2$ was obtained from $Ge(C_6H_5)_2H_2$ and $Ge(C_6H_5)_3Sn(C_2H_5)_2N(C_2H_5)_2$ (1:2 ratio). $(\text{-}Ge(C_6H_5)_2SnR_2\text{-})_n$ polymers were formed from $Ge(C_6H_5)_2H_2$ and $SnR_2(N(C_2H_5)_2)_2$ (R = C_2H_5, C_6H_5) [18]. The formation of $(C_6H_5)_3PbGe(C_6H_5)_2Pb(C_6H_5)_3$ from $Ge(C_6H_5)_2H_2$ and $Pb(C_6H_5)_2(N(C_2H_5)_2)_2$ was also reported [16].

$Ge(C_6H_5)_2H_2$ was selectively dimerized to $Ge_2(C_6H_5)_4H_2$ by ca. 3 mol% $Ti(C_5H_5)_2(CH_3)_2$ in $C_6H_5CH_3$ at 50 °C. Reacting $Ge(C_6H_5)_2H_2$ with $Ti(C_5H_5)_2(CH_3)_2$ in a 1:1 mole ratio in $C_6H_5CH_3$ (at 65 to 70 °C or UV-irradiated) produced a purple product which effectively catalyzed the oligomerization of a $Ge(C_6H_5)_2H_2$-$Ge_2(C_6H_5)_4H_2$ mixture to $H(Ge(C_6H_5)_2)_4H$ as the main product [50]; see also [51].

$Ge(C_6H_5)_2Fe_2(CO)_8$ was among the products isolated from the reaction of $Ge(C_6H_5)_2H_2$ with $Fe_3(CO)_{12}$ [20], and $Ge(C_6H_5)_2Co_2(CO)_7$ was formed from $Ge(C_6H_5)_2H_2$ and $Co_2(CO)_8$ in $C_6H_5CH_3$ at room temperature [23].

Adding $Ge(C_6H_5)_2H_2$ to $Hg(C_2H_5)_2$ in refluxing xylene over several hours followed by separation from Hg, evaporation, and heating the residue at 190 °C (8 h) generated cyclo-$(Ge(C_6H_5)_2)_4$ along with polymeric $(Ge(C_6H_5)_2)_n$ [12]; see also [8]. With $Hg(C_4H_9)_2$ in C_6H_6 at 20 °C, $(Ge(C_6H_5)_2Hg)_n$ was formed which decomposed with loss of Hg yielding $(\text{-}Ge$-$(C_6H_5)_2Ge(C_6H_5)_2Ge(C_6H_5)_2Hg\text{-})_n$ ($n \geq 2$) [43].

References:

[1] Johnson, O. H.; Harries, D. M. (J. Am. Chem. Soc. **72** [1950] 5564/6).
[2] Johnson, O. H.; Harries, D. M. (J. Am. Chem. Soc. **72** [1950] 5566/8).
[3] Johnson, O. H.; Harries, D. M. (Inorg. Synth. **5** [1957] 74/6).
[4] Kramer, K.; Wright, A. (Angew. Chem. **74** [1962] 468/9; Angew. Chem. Int. Ed. Engl. **1** [1962] 402).
[5] Noltes, J. G.; van der Kerk, G. J. M. (Recl. Trav. Chim. Pays-Bas **81** [1962] 41/8).
[6] Egorochkin, A. N.; Khidekel', M. L.; Ponomarenko, V. A.; Zueva, G. Ya.; Svirezheva, S. S.; Razuvaev, G. A. (Izv. Akad. Nauk SSSR Ser. Khim. **1963** 1865/8; Bull. Acad. Sci. USSR Div. Chem. Sci. [Engl. Transl.] **1963** 1717/9).
[7] Glockling, F.; Hooton, K. A. (J. Chem. Soc. **1963** 1849/54).
[8] Neumann, W. P. (Angew. Chem. **75** [1963] 679/80; Angew. Chem. Int. Ed. Engl. **2** [1963] 555).

[9] Cross, R. J.; Glockling, F. (J. Chem. Soc. **1964** 4125/33).
[10] Ryan, M. T.; Lehn, W. L. (J. Organomet. Chem. **4** [1965] 455/60).

[11] Cross, R. J.; Glockling, F. (J. Organomet. Chem. **3** [1965] 146/55).
[12] Neumann, W. P.; Kühlein, K. (Justus Liebigs Ann. Chem. **683** [1965] 1/11).
[13] Satgé, J.; Rivière, P. (Bull. Soc. Chim. Fr. **1966** 1773/4).
[14] Dessy, R. E.; Kitching, W.; Chivers, T. (J. Am. Chem. Soc. **88** [1966] 453/9).
[15] Birchall, T.; Jolly, W. L. (Inorg. Chem. **5** [1966] 2177/80).
[16] Neumann, W. P.; Kühlein, K. (Tetrahedron Lett. **1966** 3419/21).
[17] Rivière, P.; Satgé, J. (Bull. Soc. Chim. Fr. **1967** 4039/46).
[18] Creemers, H. M. J. C.; Noltes, J. G. (J. Organomet. Chem. **7** [1967] 237/47).
[19] Glockling, F.; Light, J. R. C. (J. Chem. Soc. A **1968** 717/34).
[20] Brooks, E. H.; Elder, M.; Graham, W. A. G.; Hall, D. (J. Am. Chem. Soc. **90** [1968] 3587/8).

[21] Mazerolles, P.; Lesbre, M.; Lavergne, J. P. (C. R. Seances Acad. Sci. C **266** [1968] 639/41).
[22] Durig, J. R.; Turner, J. B.; Gibson, B. M.; Sink, C. W. (J. Mol. Struct. **4** [1969] 79/89).
[23] Fieldhouse, S. A.; Freeland, B. H.; O'Brien, R. J. (J. Chem. Soc. Chem. Commun. **1969** 1297/8).
[24] Satgé, J.; Rivière, P. (J. Organomet. Chem. **16** [1969] 71/82).
[25] Mathis, R.; Berthelat, M.; Mathis, F. (Spectrochim. Acta A **26** [1970] 1993/2000).
[26] Jolly, W. L. (Inorg. Chem. **10** [1971] 2364/5).
[27] Rivière, P.; Satgé, J. (Synth. Inorg. Met.-Org. Chem. **2** [1972] 57/63).
[28] Birchall, T.; Drummond, I. (Inorg. Chem. **11** [1972] 250/2).
[29] Rivière, P.; Satgé, J. (J. Organomet. Chem. **49** [1973] 173/89).
[30] Rivière, P.; Satgé, J.; Soula, D. (J. Organomet. Chem. **72** [1974] 329/38).

[31] Rivière, P.; Dousse, G.; Satgé, J. (Synth. React. Inorg. Met.-Org. Chem. **4** [1974] 281/93).
[32] Rivière, P.; Rivière-Baudet, M.; Couret, C.; Satgé, J. (Synth. React. Inorg. Met.-Org. Chem. **4** [1974] 295/307).
[33] Ivanov, V. A.; Reikhsfel'd, V. O.; Saratov, I. E. (Zh. Obshch. Khim. **45** [1975] 2036/40; J. Gen. Chem. USSR [Engl. Transl.] **45** [1975] 1999/2002).
[34] Reikhsfel'd, V. O.; Ivanov, V. A.; Saratov, I. E. (Zh. Obshch. Khim. **45** [1975] 2243/5; J. Gen. Chem. USSR [Engl. Transl.] **45** [1975] 2202/4).
[35] Kuznetsov, V. A.; Egorochkin, A. N.; Skobeleva, S. E.; Razuvaev, G. A.; Pritula, N. A.; Zueva, G. Ya. (Zh. Obshch. Khim. **45** [1975] 2439/44; J. Gen. Chem. USSR [Engl. Transl.] **45** [1975] 2396/400).
[36] Saratov, I. E.; Ivanov, V. A.; Reikhsfel'd, V. O. (Zh. Obshch. Khim. **46** [1976] 1052/7; J. Gen. Chem. USSR [Engl. Transl.] **46** [1976] 1048/52).
[37] Gielen, M.; Simon, S. (Bull. Soc. Chim. Belg. **86** [1977] 589/94).
[38] Duffaut, N.; Dunogues, J.; Calas, R.; Rivière, P.; Satgé, J.; Cazes, A. (J. Organomet. Chem. **149** [1978] 57/63).
[39] Egorochkin, A. N.; Khorshev, S. Ya.; Sevastyanova, E. I.; Ratushnaya, S. Kh.; Satgé, J.; Rivière, P.; Barrau, J.; Richelme, S. (J. Organomet. Chem. **155** [1978] 175/83).
[40] Eaborn, C.; Singh, B. (J. Organomet. Chem. **177** [1979] 333/48).

[41] Skobeleva, S. E.; Egorochkin, A. N.; Khorshev, S. Ya.; Ratushnaya, S. Kh.; Rivière, P.; Satgé, J.; Richelme, S.; Cazes, A. (J. Organomet. Chem. **182** [1979] 1/7).
[42] Lebedev, N. K.; Kiparisova, E. G.; Lebedev, B. V.; Sladkov, A. M.; Vasneva, N. A. (Izv. Akad. Nauk SSSR Ser. Khim. **1980** 555/8; Bull. Acad. Sci. USSR Div. Chem. Sci. [Engl. Transl.] **1980** 374/6).

[43] Rivière, P.; Castel, A.; Satgé, J. (J. Organomet. Chem. **212** [1981] 351/67).
[44] Castel, A.; Rivière, P.; Satgé, J. (J. Organomet. Chem. **232** [1982] 137/46).
[45] Batchelor, R. J.; Birchall, T. (J. Am. Chem. Soc. **105** [1983] 3848/52).
[46] Mochida, K.; Matsushige, N.; Hamashima, M. (Bull. Chem. Soc. Jpn. **58** [1985] 1443/7).
[47] Rivière, P.; Rivière-Baudet, M.; Castel, A.; Satgé, J.; Lavabre, A. (Synth. React. Inorg. Met.-Org. Chem. **17** [1987] 539/57).
[48] Mochida, K.; Fujii, A.; Tsuchiya, N.; Tohji, K.; Udagawa, Y. (Organometallics **6** [1987] 1811/2).
[49] Castel, A.; Rivière, P.; Satgé, J.; Ko, Y.-H. (J. Organomet. Chem. **342** [1988] C1/C4).
[50] Aitken, C.; Harrod, J. F.; Malek, A.; Samuel, E. (J. Organomet. Chem. **349** [1988] 285/91).

[51] Harrod, J. F. (ACS Symp. Ser. **360** [1988] 89/100).
[52] Castel, A.; Rivière, P.; Satgé, J.; Ko, H. Y. (Organometallics **9** [1990] 205/10).
[53] Castel, A.; Rivière, P.; Satgé, J. Desor, D. (J. Organomet. Chem. **433** [1992] 49/61).
[54] Castel, A.; Rivière, P.; Ahbala, M.; Satgé, J.; Soufiaoui, M.; Knouzi, N. (J. Organomet. Chem. **447** [1993] 123/30).
[55] Castel, A.; Rivière, P.; Satgé, J.; Desor, D.; Ahbala, M.; Abdennadher, C. (Inorg. Chim. Acta **212** [1993] 51/5).
[56] Castel, A.; Rivière, P.; Satgé, J. (J. Organomet. Chem. **462** [1993] 97/102).

1.3.2.1.6 GeR_2H_2 Compounds with R = Substituted Aryl and Heterocycle

The compounds in this section are listed in Table 36.

Table 36
GeR_2H_2 Compounds with R = Substituted Aryl and Heterocycle.
An asterisk indicates further information at the end of the table.
Explanations, abbreviations, and units are given on p. X.

No.	R group	formation (yield) properties and remarks
1	$C_6H_4CH_3$-4	$Ge(C_6H_4CH_3\text{-}4)_2Br_2$ + $LiAlH_4$ in refluxing ether (3 h) followed by hydrolysis, solvent removal from the organic layer, and fractionation (75%) b.p. 115°C/2; n_D^{20} = 1.5839 ^{1}H NMR (THF-d_8): 2.25 (s, CH_3), 4.91 (s, HGe), 6.99 to 7.50 (dd, C_6H_4) ^{13}C NMR (THF-d_8): 22.6 (CH_3), 131.0 (C-3), 132.4 (C-4), 137.0 (C-2), 140.7 (C-1) IR (neat): ν(GeH) 2050 with potassium in THF-HMPT → $Ge(C_6H_4CH_3\text{-}4)_2K_2$ [9]
2	$C_6H_3(C_2H_5)_2$-2,6	$Ge_2(C_6H_3(C_2H_5)_2\text{-}2,6)_4$ + excess Li naphthalenide in DME at −78 to +20°C (1 h), addition of CH_3OH by vacuum condensation at −78°C, warming to +20°C, and concentration (7%) along with $Ge_2R_4H_2$ (14%) and polymers ^{1}H NMR (C_6D_6): 5.83 (s) [16]

Table 36 (continued)

No.	R group	formation (yield) properties and remarks
3	$C_6H_3(C_3H_7\text{-i})_2\text{-}2,6$	$GeCl_4$ + $Mg(C_6H_3(C_3H_7\text{-i})_2\text{-}2,6)Br$ (1:2.1 mole ratio) in ether-THF at −78 to +20°C (24 h) followed by $LiAlH_4$ in ether at 20°C (5 h); workup by hydrolysis at 0°C, extraction with ether, concentration, and purification by flash chromatography using C_6H_{14} (72%) colorless oil ^{1}H NMR (C_6D_6): 1.12 (d, CH_3, J = 6.8), 3.45 (sept, CH/C_3H_7-i, J = 6.8), 5.55 (s, HGe), 7.08 (d, 4 H, J = 7.8), 7.23 (t, 2 H, J = 7.8) MS: $[M]^+$ cluster at m/e = 393 to 400 with refluxing CCl_4 in the presence of 10 mol% $C_6H_5CO\text{-}OO\text{-}COC_6H_5$ (3 h) → $Ge(C_6H_3(C_3H_7\text{-i})_2\text{-}2,6)_2Cl_2$ [16]
*4	$C_6H_2(CH_3)_3\text{-}2,4,6$	$Ge(C_6H_2(CH_3)_3\text{-}2,4,6)_2Cl_2$ (from $GeCl_4$ and RMgBr in THF) + $LiAlH_4$ in refluxing ether (6 h), hydrolysis, extraction with ether, and distillation (45%) along with $GeRH_3$ and GeR_3H (small amount) [14]; see also [12] $Ge(C_6H_2(CH_3)_3\text{-}2,4,6)_2(H)Li$ + $Ge(C_6H_5)H_2Cl$ in THF at −20 to +20°C (14 h) followed by hydrolysis, extraction, and concentration under vacuum (41%) along with $H(C_6H_2(CH_3)_3\text{-}2,4,6)_2GeGe(C_6H_5)H_2$ and $(Ge(C_6H_5)H)_n$ [18]; see also [13] m.p. 118 to 120°C [14, 21]; b.p. 140°C/0.05 [14] ^{1}H NMR: 2.23 (s, CH_3-4), 2.30 (s, CH_3-2,6), 5.05 (s, HGe), 6.75 (s, C_6H_2) in CCl_4 [14, 21]; 5.00 (HGe), 6.80 (C_6H_2) in C_5H_{12}-THF-d_8 [18] ^{13}C NMR: 21.46 (CH_3-4), 24.17 (CH_3-2,6), 129.46 (C-3), 131.90 (C-1), 139.43 (C-4), 144.37 (C-2) in THF-d_8 [21] or THF [18] IR: ν(GeH) 2074 presumably in CCl_4 [12, 14]; 2060 in Nujol; 2040 in THF-C_5H_{12} [18] stable under an N_2 atmosphere [14] with CCl_4 at 90°C in the presence of AIBN → $Ge(C_6H_2(CH_3)_3\text{-}2,4,6)_2Cl_2$ [12] with N-chlorosuccinimide (1:1 mole ratio) in THF at 80°C in a sealed tube (20 h) → $Ge(C_6H_2(CH_3)_3\text{-}2,4,6)_2(H)Cl$ [12]; with excess N-chlorosuccinimide in refluxing THF → $Ge(C_6H_2(CH_3)_3\text{-}2,4,6)_2Cl_2$ [22] with N-bromosuccinimide (1:1 mole ratio) in THF at 30°C (1.5 h) → $Ge(C_6H_2(CH_3)_3\text{-}2,4,6)_2(H)Br$ [12] with CH_3SSCH_3 in C_6H_6 at 100°C (2 h) in the presence of AIBN in a sealed tube → $Ge(C_6H_2(CH_3)_3\text{-}2,4,6)_2(SCH_3)_2$ [21]; see also [14] with LiC_4H_9-t in THF-C_5H_{12} at −40°C (or −20°C?) → $Ge(C_6H_2(CH_3)_3\text{-}2,4,6)_2(H)Li$ characterized in solution [13, 18]

Table 36 (continued)

No.	R group	formation (yield) properties and remarks
*5	$C_6H_2(C_3H_7\text{-i})_3$-2,4,6	$Ge(C_6H_2(C_3H_7\text{-i})_3\text{-2,4,6})_2Cl_2$ + $LiAlH_4$ in refluxing ether (11 h) followed by aqueous 6 M HCl, extraction with ether, and solvent removal (71%) colorless oil ^{1}H NMR (C_6D_6): 1.19 (d, CH_3/C_3H_7-2,6, J = 6.8), 1.20 (d, CH_3/C_3H_7-4, J = 6.9), 2.77 (sept, CH/C_3H_7-4, J = 6.9), 3.52 (sept, CH/C_3H_7-2,6, J = 6.8), 5.57 (s, HGe), 7.12 (s, C_6H_2) ^{13}C NMR (C_6D_6): 24.19 (CH_3/C_3H_7-4), 24.6 (CH_3/C_3H_7-2,6), 34.7 (CH/C_3H_7-4), 34.84 (CH/C_3H_7-2,6), 121.55 (C-3,5), 131.81, 150.16, 154.45 (C-2,4,6) IR (neat): ν(GeH) 2056(s) MS: $[M]^+$ (6), $[M - C_{15}H_{24}]^+$ (92), $[C_{15}H_{23}]^+$ (100), and m/e = 235 (15), 189 (22) with LiC_4H_9-t in THF-C_5H_{12} at −23°C (1 h) → $Ge(C_6H_2(C_3H_7\text{-i})_3\text{-2,4,6})_2(H)Li$ [24]
*6	C_6F_5	$Ge(C_6F_5)_2Br_2$ + $LiAlH_4$ in $C_6H_5CH_3$-ether at 0°C, hydrolysis, solvent removal from the organic layer, and vacuum distillation (86%) b.p. 103°C/5 d^{20} = 1.837; n_D^{20} = 1.4720 [1] IR: ν(GeH) 2140 in C_7H_{16} [3]; 2142 [1] with Br_2 at 40 to 60°C → $Ge(C_6F_5)_2(H)Br$ (21%) and $Ge(C_6F_5)_2Br_2$ (3 to 5%) [25]; with Br_2 in C_6H_6 high yields of $Ge(C_6F_5)_2Br_2$ [1] with S_n at 170 to 190°C → $(C_6F_5)_2GeSGe(C_6F_5)_2S$ [2] with $Sn(C_2H_5)_3N(C_2H_5)_2$ in C_6H_6 at 100°C (4 h) → $Ge(C_6F_5)_2(Sn(C_2H_5)_3)_2$ [1]; using a 100% excess of No. 6 at 0°C → $Ge(C_6F_5)_2(H)Sn(C_2H_5)_3$ [6] with $V(C_5H_5)_2$ in $C_6H_5CH_3$ at 85°C (1.5 h) in a sealed tube → $Ge(C_6F_5)_2(H)V(C_5H_5)_2$ [10] with $Ni(C_5H_5)_2$ in $C_6H_5CH_3$ at 70°C (3 h) in a sealed tube → $(Ge(C_6F_5)_2NiC_5H_5)_2$ [26] with $Pt(P(C_6H_5)_3)_3$ in C_6H_6 at 20°C (1 h) → $Ge(C_6F_5)_2(H)Pt(P(C_6H_5)_3)_2H$ [4] with $Hg(C_2H_5)_2$ at 150°C (3 to 4 h) → $Ge_2(C_6F_5)_4H_2$ and Hg [23]
7	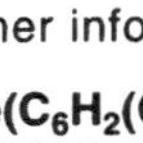	Ge(C=CHCH=CHO)$_2Br_2$ + $LiAlH_4$ (1:4 ratio) in the $LiAlH_4$-C_6H_6 two-phase system at 25°C (6 h) or at 80°C (1 h) in the presence of $[N(C_2H_5)_3CH_2C_6H_5]Cl$ as phase-transfer catalyst (>95%) [11, 17]

* Further information:

$Ge(C_6H_2(CH_3)_3\text{-2,4,6})_2H_2$ (Table **36**, No. **4**) was formed with 24% yield by reacting $GeR_2(H)Li$ (R = $C_6H_2(CH_3)_3$-2,4,6) with 3,5-di-t-butyl-ortho-quinone in $C_6H_5CH_3$ at 20°C

References on p. 247

(24 h) followed by hydrolysis, extraction with ether, concentration, and GC. Other Ge-containing products were the germadioxolane I and $(GeR_2)_n$ [19].

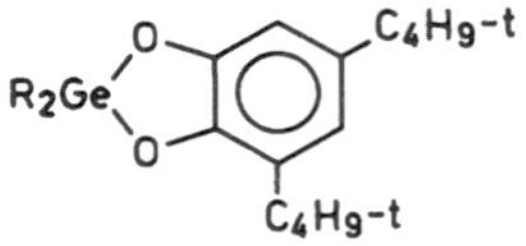

I

The formation of No. 4 by photolysis of $GeR_2(H)Li$ in the presence of anthracene was briefly mentioned in [19]. The compound was obtained along with $Ge(C_6H_5)_3H$ and $(C_6H_5)_3GeGeR_2H$ by the reaction of $Ge(C_6H_5)_3Li$ with $GeR_2(OCH_3)_2$ in THF-ether at 0 to 20 °C (12 h) followed by aqueous 6 M HCl and $LiAlH_4$ in ether [21]. It was a by-product in the synthesis of GeR_3H from GeR_3Cl and $LiAlH_4$ [5] and of $Ge_2R_4H_2$ from $GeR_2(H)Cl$ and Mg-Hg in refluxing THF [22]. Decomposition of $GeR_2(CH(OH)C_6H_5)H$ in C_6D_6 at 160 °C (4 h) in a sealed tube also yielded 25% No. 4 [18]. **$Ge(C_6H_2(CH_3)_3$-2,4,6$)_2D_2$** was prepared from GeR_2X_2 (X = halogen) and $LiAlD_4$ [24].

The influence of mesityl substituents on the IR frequency and integrated intensity of ν(EH) bands (E = Si, Ge) was shown for No. 4 and related compounds to be determined not only by the inductive effect and π(d-p) interaction, but also by the steric effects of these groups [5].

$Ge(C_6H_2(CH_3)_3$-2,4,6$)_2H_2$ was metalated with a slight excess of LiC_4H_9-t in THF at low temperature to yield $Ge(C_6H_2(CH_3)_3$-2,4,6$)_2(H)Li$; 28% $Ge(C_6H_2(CH_3)_3$-2,4,6$)_2(H)Li$ and 72% $Ge(C_6H_2(CH_3)_3$-2,4,6$)_2Li_2$ were obtained with a large excess of LiC_4H_9-t [27].

$Ge(C_6H_2(C_3H_7$-i$)_3$-2,4,6$)_2H_2$ (Table **36**, No. **5**) was formed with 9% yield along with $Ge(C_6H_2(C_3H_7$-i$)_3$-2,4,6$)_2(Si(C_6H_2(CH_3)_3$-2,4,6$)_2H)H$ by reacting $Ge(C_6H_2(C_3H_7$-i$)_3$-2,4,6$)_2$-$(Si(C_6H_2(CH_3)_3$-2,4,6$)_2Cl)H$ with excess LiC_4H_9-t in THF-C_5H_{12} at −23 to +20 °C (5.5 h) followed by CH_3OH, aqueous workup, and TLC on SiO_2 [24].

$Ge(C_6F_5)_2H_2$ (Table **36**, No. **6**) was obtained by heating $Ge(C_6F_5)_2Br_2$ and excess $Ge(C_2H_5)_3H$ at 120 °C (6 h) followed by GC (ca. 24% yield) or vacuum fractionation (8% yield). Other products were $Ge(C_6F_5)_2(H)Br$ and $Ge(C_2H_5)_3Br$ [1]. Treating $(C_5H_5)_2V$-$(Ge(C_6F_5)_2)_nV(C_5H_5)_2$ (n = 1 to 6) with HCl in DME in an evacuated sealed ampule yielded 57% of No. 6 along with $Ge(C_6F_5)_2(H)Cl$, both identified by GC [10].

$Ge(C_6F_5)_2H_2$ was part of an investigation of the integrated intensity of the ν(GeH) band in the IR spectra of a series of GeR_2H_2 compounds [3].

$Ge(C_6F_5)_2H_2$ reacted with excess $N(C_2H_5)_3$ in THF at 40 °C to form a network polymer with the composition $(Ge_6(C_6F_5)(C_6F_4)_{11})_n$. Side reactions led to loss of C_6F_5H. A general scheme for the formation of the polymer was proposed [15]; see also [20].

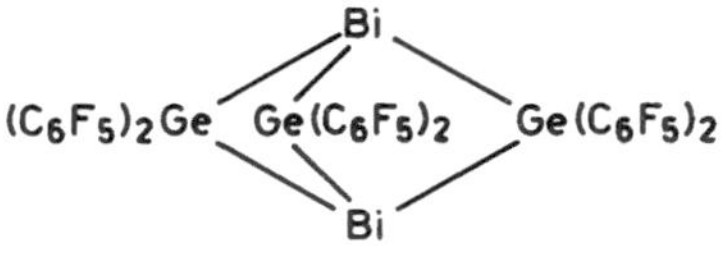

II

$Ge(C_6F_5)_2H_2$ and $Ge(C_6F_5)_2Br_2$ reached a dynamic equilibrium with $Ge(C_6F_5)_2(H)Br$ within ca. 3 h when heated at 150 °C. Below 140 °C no reaction took place [2]. Heating with an excess of the dihydride at 160 to 165 °C produced after 6 h $Ge(C_6F_5)_2(H)Br$, $Ge(C_6F_5)_2H_2$, and $Ge(C_6F_5)_2Br_2$ in a 74:22:4 ratio [6].

The reaction of $Ge(C_6F_5)_2H_2$ with $Bi(C_2H_5)_3$ in $C_6H_5CH_3$ at 110 °C (ca. 4 h) yielded $(Ge(C_6F_5)_2)_3Bi_2$ (Formula II) [7], which reacted further with an excess of No. 6 at 180 °C (3 h) to give $Ge_2(C_6F_5)_4H_2$ and Bi [23]. Adding $Tl(C_2H_5)_3$ to $Ge(C_6F_5)_2H_2$ in C_6H_{14} at 20 to 40 °C yielded $(Ge(C_6F_5)_2TlC_2H_5)_n$ which on heating with excess No. 6 at 100 °C (1 h) produced $Ge_2(C_6F_5)_4H_2$, $(Ge(C_6F_5)_2)_n$, and Tl [8]. $Ge_2(C_6F_5)_4H_2$ was also formed by reacting of No. 6 with $Ge(C_6F_5)_2(H)Br$ and excess Zn in THF at 60 °C (1 h) [23], with $Ge_2(C_6F_5)_6$ (along with $Ge(C_6F_5)_3Ge(C_6F_5)_2H$ and $Ge(C_6F_5)_3H$) [25], and with $Rh(P(C_6H_5)_3)_3Cl$ or $Rh(P(C_6H_5)_3)_2R'$ (R′ = 3,6-di-t-butyl-o-benzoquinone or 3,6-di-t-butyl-4,5-dimethoxy-o-benzoquinone) in $C_6H_5CH_3$ at 20 °C (2 to 3 h) [23].

References:

[1] Bochkarev, M. N.; Maiorova, L. P.; Korneva, S. P.; Bochkarev, L. N.; Vyazankin, N. S. (J. Organomet. Chem. **73** [1974] 229/36).

[2] Bochkarev, M. N.; Maiorova, L. P.; Vyazankin, N. S.; Razuvaev, G. A. (J. Organomet. Chem. **82** [1974] 65/71).

[3] Egorochkin, A. N.; Khorshev, S. Ya.; Sevastyanova, E. I.; Ratushnaya, S. Kh.; Satgé, J.; Rivière, P.; Barrau, J.; Richelme, S. (J. Organomet. Chem. **155** [1978] 175/83).

[4] Bochkarev, M. N.; Maiorova, L. P.; Skobeleva, S. E.; Razuvaev, G. A. (Izv. Akad. Nauk SSSR Ser. Khim. **1979** 1854/8; Bull. Acad. Sci. USSR Div. Chem. Sci. [Engl. Transl.] **1979** 1717/22).

[5] Skobeleva, S. E.; Egorochkin, A. N.; Khorshev, S. Ya.; Ratushnaya, S. Kh.; Rivière, P.; Satgé, J.; Richelme, S.; Cazes, A. (J. Organomet. Chem. **182** [1979] 1/7).

[6] Bochkarev, M. N.; Maiorova, L. P.; Razuvaev, G. A. (Zh. Obshch. Khim. **50** [1980] 903/7; J. Gen. Chem. USSR [Engl. Transl.] **50** [1980] 730/3).

[7] Bochkarev, M. N.; Razuvaev, G. A.; Zakharov, L. N.; Struchkov, Yu. T. (J. Organomet. Chem. **199** [1980] 205/16).

[8] Bochkarev, M. N.; Basalgina, T. A.; Kalinina, G. S.; Razuvaev, G. A. (J. Organomet. Chem. **243** [1983] 405/10).

[9] Mochida, K.; Matsushige, N.; Hamashima, M. (Bull. Chem. Soc. Jpn. **58** [1985] 1443/7).

[10] Pankratov, L. V.; Lineva, A. N.; Latyaeva, V. N.; Cherkasov, V. K.; Bochkarev, M. N.; Razuvaev, G. A. (Zh. Obshch. Khim. **55** [1985] 1767/71; J. Gen. Chem. USSR [Engl. Transl.] **55** [1985] 1568/71).

[11] Gevorgyan, V. N.; Ignatovich, L. M.; Lukevics, E. (J. Organomet. Chem. **284** [1985] C31/C32).

[12] Rivière, P.; Rivière-Baudet, M.; Castel, A.; Satgé, J.; Lavabre, A. (Synth. React. Inorg. Met.-Org. Chem. **17** [1987] 539/57).

[13] Castel, A.; Rivière, P.; Satgé, J.; Ko, Y.-H. (J. Organomet. Chem. **342** [1988] C1/C4).

[14] Rivière, P.; Rivière-Baudet, M.; Satgé, J. (in: Organometallic Synthesis, Vol. 4, New York 1988, pp. 545/8).

[15] Silkin, V. B.; Maiorova, L. P.; Bochkarev, M. N. (Metalloorg. Khim. **1** [1988] 1338/41; Organomet. Chem. USSR [Engl. Transl.] **1** [1988] 731/3).

[16] Park, J.; Batcheller, S. A.; Masamune, S. (J. Organomet. Chem. **367** [1989] 39/45).

[17] Lukevics, E.; Gevorgyan, V. (Chem. Technol. Silicon Tin. Proc. 1st Asian Network Anal. Inorg. Chem. Int. Chem. Conf. Silicon Tin, Kuala Lumpur 1989 [1992], pp. 165/77).

[18] Castel, A.; Rivière, P.; Satgé, J.; Ko, H. Y. (Organometallics **9** [1990] 205/10).
[19] Rivière, P.; Castel, A.; Ko, Y. H.; Desor, D. (J. Organomet. Chem. **386** [1990] 147/56).
[20] Silkin, V. B.; Maiorova, L. P.; Bochkarev, M. N.; Semechikov, Yu. D.; Khvatova, N. L. (Vysokomol. Soedin. A **32** [1990] 2346/50; Polym. Sci. USSR [Engl. Transl.] **32** [1990] 2249/53).
[21] Rivière, P.; Rivière-Baudet, M.; Castel, A.; Desor, D.; Abdennadher, C. (Phosphorus Sulfur Silicon Relat. Elem. **61** [1991] 189/99).
[22] Baines, K. M.; Cooke, J. A. (Organometallics **10** [1991] 3419/21).
[23] Pankratov, L. V.; Nevodchikov, V. I.; Cherkasov, V. K.; Bochkarev, M. N. (Metalloorg. Khim. **4** [1991] 516/20; Organomet. Chem. USSR [Engl. Transl.] **4** [1991] 247/9).
[24] Baines, K. M.; Groh, R. J.; Joseph, B.; Parshotam, U. R. (Organometallics **11** [1992] 2176/80).
[25] Silkin, V. B.; Makarenko, N. P.; Bochkarev, M. N. (Metalloorg. Khim. **5** [1992] 621/4; Organomet. Chem. USSR [Engl. Transl.] **5** [1992] 299/301).
[26] Pankratov, L. V.; Nevodchikov, V. I.; Zakharov, L. N.; Bochkarev, M. N.; Zdanovich, I. V.; Latyaeva, V. N.; Lineva, A. N.; Batsanov, A. S.; Struchkov, Y. I. (J. Organomet. Chem. **429** [1992] 13/26).
[27] Castel, A.; Rivière, P.; Satgé, J. (J. Organomet. Chem. **462** [1993] 97/102).

1.3.2.2 Ge(R)(R′)H_2 Compounds

1.3.2.2.1 Ge(R)(R′)H_2 Compounds with R = Alkyl

The compounds in this section are listed in Table 37.

Table 37
Ge(R)(R′)H_2 Compounds with R = Alkyl.
An asterisk indicates further information at the end of the table.
Explanations, abbreviations, and units are given on p. X.

No.	R′ group	formation (yield) properties and remarks
R = CH_3		
1	CF_3	Ge(CF_3)H_3 + CH_3I + Zn(CH_3)$_2$ (1 : 1 : 1 mole ratio) in a vacuum system at low temperature and warming to 20 °C; formed along with Ge(CH_3)$_2$(CF_3)H and Ge(CH_3)$_3$$CF_3$; products separated by fractional condensation ^{1}H NMR (neat?): 0.61 (CH_3), 4.39 (H_2Ge, ^3J(H, H) = 3.75, ^3J(F, H) = 7.75) ^{19}F NMR (neat?): −53.9 (CF_3); ^1J(C, F) = 333.3 MS: $[M]^+$ (4), $[M - H]^+$ (5), $[GeCF_3]^+$ (10), $[Ge(CH_3)HF]^+$ (42), $[GeF]^+$ (72), $[GeCH_3]^+$ (100), $[GeH]^+$ (34), $[Ge]^+$ (44), $[CF_3]^+$ (12) [19]
2	CH_2Cl	preparation not reported IR (C_7H_{16}): ν(GeH) 2085, 2064; ν(GeH) region illustrated [14]; see also General Remarks, p. 253

Table 37 (continued)

No.	R′ group	formation (yield) properties and remarks
*3	C_2H_5	sodium salt of $Ge(C_2H_5)H_3$ + CH_3Cl in liquid NH_3; no details reported [2] IR of $Ge(CH_3)(C_2H_5)(H)D$ (gas): ν(GeH) 2058.7, 2052.2 (illustrated) [16]
*4	C_3H_7-i	only mentioned in the context of force-field calculations [6]
*5	C_5H_5-c	$Ge(CH_3)H_2Cl$ + KC_5H_5 followed by trap-to-trap fractionation (25%) colorless, volatile liquid ^1H NMR (neat, 35°C): −0.20 (t, CH_3), 3.72 (q, H_2Ge, $^3J(HGeCH_3)$ = 3.5), 5.62 (br, C_5H_5) IR (gas): 3095(m), 3000(w), 2934(m), 2084(s,sh), 2063, 2054(vs), 1635(vw,br), 1472(m), 1390(vw,br), 1247(m), 1090(vw), 983(s), 950(m), 881(s), 867(m,sh), 830(vs), 773(m), 694(vs,sh), 675(vs), 666(vs,sh), and 600(m) Raman (neat liquid): ν(GeH) 2065(s,p), $\delta(GeH_2)$ 831(m,p), ν(GeC) 603(vs,p), $\nu(Ge\text{-}C_5H_5)$ 371(vs,p), $\delta(C\text{-}Ge\text{-}C_5H_5)$ 228(m,p), $\delta(Ge\text{-}C_5H_5)$ 194(m,dp) UV (gas): λ_{max} = 264 [7]
*6	C_6H_5	$Ge(C_6H_5)H_3$ + CH_2N_2 (50% excess) in ether under UV irradiation for 6 h followed by GC (59%) along with $Ge(CH_3)_2(C_6H_5)H$ $Ge(C_6H_5)H_2X$ (X = halogen) + CH_3MgI b.p. 70°C/35 d^{20} = 1.2024; n_D^{20} = 1.5292 ^1H NMR (neat): 4.40 (q, H_2Ge) [3] IR: ν(GeH) 2061 in C_7H_{16} [10], 2055 for the neat liquid [3]
7	$C_6H_2(CH_3)_3$-2,4,6	$Ge(C_6H_2(CH_3)_3\text{-}2,4,6)H_2Li$ + 100% excess CH_3I in THF at −60 to +20°C followed by hydrolysis and distillation of the concentrated extract (76%) $Ge(C_6H_2(CH_3)_3\text{-}2,4,6)H_3$ + LiC_4H_9-t (1:2 mole ratio) in THF-C_5H_{12} at −60°C followed by excess CH_3I (36%) along with $Ge(CH_3)_2(C_6H_2(CH_3)_3\text{-}2,4,6)H$ (35%) b.p. 45 to 46°C/0.07 ^1H NMR (neat): 0.40 (t, CH_3), 2.16 (s, CH_3-4), 2.36 (s, CH_3-2), 4.44 (q, H_2Ge, $^3J(H, H)$ = 4), 6.69 (s, C_6H_2) IR (neat): ν(GeH) 2062 [23]
8	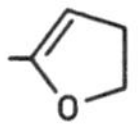	$Ge(\overline{C{=}CHCH_2CH_2O})_3CH_3$ + $LiAlH_4$ in THF (15 min) or ether (3 h) at 25°C or in C_6H_6 at 80°C (24 h) in the presence 18-crown-6; identified as an intermediate by GC/MS, the end-product being $Ge(CH_3)H_3$ [22]; see also [21] MS: $[M]^+$ (20) and m/e = 89 (100); other m/e values and relative intensities given [22]

References on p. 254

Table 37 (continued)

No.	R′ group	formation (yield) properties and remarks
9	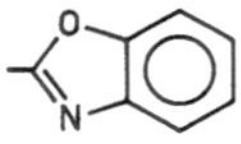	$Ge(CH_3)(R')Cl_2$, $Ge(CH_3)R'_2Cl$, and $Ge(CH_3)R'_3$ (R′ = benzoxazol-2-yl; prepared as an inseparable mixture from $Ge(CH_3)Cl_3$ and $Sn(CH_3)_3R'$) + $Sn(CH_3)_3H$ in C_6H_6 after 3 h; isolated by fractional crystallization and distillation (12%) along with $Ge(CH_3)R'_2H$ b.p. 78°C/0.15 ^{1}H NMR: 0.42 (t, CH_3, J = 3.67), 4.48 (q, H_2Ge) in C_6H_6; 7.2 to 8.0 (m, C_6H_4) in CCl_4 IR: ν(GeH) 2095 [17]
10	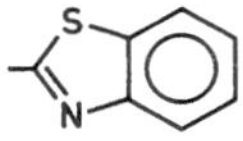	prepared like No. 9, R′ = benzthiazol-2-yl (11%) b.p. 81°C/0.05 ^{1}H NMR: 0.45 (t, CH_3, J = 3.73), 4.60 (q, H_2Ge) in C_6H_6; 7.5 to 8.4 (m, C_6H_4) in CCl_4 IR: ν(GeH) 2070 [17]
R = CHF_2		
*11	CF_3	$Ge(CF_3)I_3$ in 30% H_3PO_4 + aqueous $NaBH_4$ at 20°C in a vacuum line, followed by fractional condensation; purified by GC (11%) along with other Ge-containing products b.p. 41.9°C vapor pressure: log (p/mbar) = 8.355 − 1685/T; ΔH_v = 32.26 kJ/mol, ΔS_v = 102.4 kJ · mol^{-1} · K^{-1} ^{1}H NMR: 4.83 (H_2Ge), 6.48 (CH); ^3J(H, H) = 2.1, ^2J(F, H) = 46.0, $^3J(F_3CGeH)$ = 7.8, $^3J(F_2CGeH)$ = 9.9, ^4J(F, H) = 0.6 ^{19}F NMR: −125.3 (CHF_2), −49.9 (CF_3); ^4J(F, F) = 4.1 IR and Raman spectra on p. 254 MS: $[Ge(CHF_2)(CF_2)H_2]^+$ (0.7), $[Ge(CF_3)H_2]^+$ (98), $[Ge(CHF_2)H_2]^+$ (26), $[GeF]^+$ (100), $[C_2F_3H]^+$ (23), $[GeH]^+$ (30), $[Ge]^+$ (10), $[CF_3]^+$ (5), $[CHF_2]^+$ (35) [19]
R = C_2H_5		
12	CF_3	$Ge(CF_3)H_3$ + $Zn(C_2H_5)_2$ + slight excess of C_2H_5I (35%) along with $Ge(C_2H_5)_2(CF_3)H$ and $Ge(CF_3)I_3$ ^{19}F NMR: −51.3 (CF_3); ^3J(H, F) = 7.3 [19]
13	CH_2Cl	$Ge(C_2H_5)(CH_2Cl)Cl_2$ + $LiAlH_4$ in ether at 0°C followed by hydrolysis, extraction with ether, and distillation (84%) [12] b.p. 102°C ^{1}H NMR (C_6D_6): 2.80 (t, CH_2, ^3J(H, H) = 2.50), 4.33 (m, H_2Ge) [12] IR (C_7H_{16}): ν(GeH) 2078, 2054; ν(GeH) region illustrated [14] with CH_3OCH_2Cl (slowly added) in the presence of a catalytic amount of $AlCl_3$ → $Ge(C_2H_5)(CH_2Cl)(H)Cl$ → $Ge(C_2H_5)(CH_2Cl)Cl_2$ [12]

Table 37 (continued)

No.	R′ group	formation (yield) properties and remarks
14	$CHCl_2$	$Ge(C_2H_5)H_2Cl + LiCHCl_2$ in ether-THF at $-95\,°C$ for 2 h followed by hydrolysis at 20 °C, extraction with C_5H_{12}, and distillation (43%) [14]; see also [11] b.p. 33 to 35 °C/10 [11], 59 °C/40 [14] ^1H NMR: 1.57 (m, C_2H_5), 4.49 (m, H_2Ge), 5.60 (t, $CHCl_2$) for the neat liquid [11]; 4.52 (m, H_2Ge), 5.62 (t, $CHCl_2$) [14] IR: ν(GeH) 2095, 2074 in C_7H_{16} (illustrated) [14]; 2107 for the neat compound [11] with $CH_3OCH_2Cl \rightarrow Ge(C_2H_5)(CHCl_2)(H)Cl$ [11]
*15	$CH_2C_6H_5$	$Ge(C_2H_5)H_2Cl + C_6H_5CH_2MgCl$ in ether at 0 °C, then refluxed for 1 h, followed by hydrolysis and distillation of the concentrated extract (ca. 70%) b.p. 82 to 84 °C/14 ^1H NMR (C_6D_6): 0.70 to 1.20 (m, C_2H_5), 2.28 (t, CH_2), 3.94 (q, H_2Ge) [13] IR (C_7H_{16}): ν(GeH) 2049 [10]
*16	CH_2CH_2OH	$Ge(C_2H_5)Cl + BrCH_2COOCH_3$ followed by treatment with $LiAlH_4$ in refluxing ether for 1 h and usual workup (60%) $Ge(C_2H_5)(CH_2CH_2OOCCH_3)Cl_2 + LiAlH_4$ in ether at 0 °C (70%) b.p. 67 °C/14 $d^{20} = 1.2138$; $n_D^{20} = 1.4697$ ^1H NMR (CCl_4): 3.70 (m, H_2Ge) [5] IR: ν(GeH) 2054 in C_7H_{16} [10]; ν(OH) 3300, ν(GeH) 2040 [5]
17	$CH_2CH_2CH_2Cl$	preparation not reported IR (C_7H_{16}): ν(GeH) 2049 [10]
18	$CH_2CH(OH)CH_3$	$Ge(C_2H_5)Cl + ClCH_2COCH_3$ followed by treatment with excess $LiAlH_4$ in ether and hydrolysis (15%) $Ge(C_2H_5)(CH_2CH(CH_3)OOCCH_3)Cl_2 + LiAlH_4$ in ether (65%) b.p. 76.5 °C/18 $d^{20} = 1.1677$; $n_D^{20} = 1.4652$ ^1H NMR (C_6H_6): 1.20 (d, CH_3), 3.81 (quint, H_2Ge, $^3J(H, H) = 3$), 3.93 (sext, CH, J = 6) IR: ν(OH) 3350, ν(GeH) 2040 [5]
19	C_5H_{11}-i	$Ge(C_2H_5)H_3$ + Li in $N(C_2H_5)H_2$ at $-33\,°C$ followed by i-$C_5H_{11}Br$ (64%, based on i-$C_5H_{11}Br$) vapor pressure ca. 10 Torr at 20 °C with Li in $N(C_2H_5)H_2$ at $-33\,°C$ followed by $C_2H_5I \rightarrow Ge(C_2H_5)_2(C_5H_{11}\text{-i})H$ insoluble in and unaffected by H_2O; soluble in petroleum ether and C_6H_6 vapor causes severe headache [1]
20	$CH_2CH{=}CHCH_3$	$Ge(C_2H_5)Cl + BrCH_2CH{=}CHCH_3$, then treated with $LiAlH_4$ at 0 °C followed by usual workup (79%)

References on p. 254

Table 37 (continued)

No.	R′ group	formation (yield) properties and remarks
20 (continued)		b.p. 94 to 95 °C/760 d^{20} = 1.0216; n_D^{20} = 1.4522 ^{1}H NMR (C_6D_6): 1.58 (d, CH_3), 3.85 (quint, H_2Ge, $^3J(H, H)$ = 3), 5 to 5.8 (m, CH=CH) IR: ν(GeH) 2040 [5]
21	C_6H_5	$Ge(C_6H_5)Cl + Hg(C_2H_5)_2$ in C_6H_6 at 100 °C to give $Ge(C_2H_5)(C_6H_5)Cl_2$ and $Ge(C_2H_5)_2(C_6H_5)Cl$ followed by $LiAlH_4$ in refluxing THF and GC (88%) along with $Ge(C_2H_5)_2(C_6H_5)H$ (5%) $Ge(C_6H_5)H_2Cl + C_2H_5MgBr$ in refluxing ether for 1 h followed by hydrolysis and distillation of the ether extract (85%) b.p. 65 °C/10 ^{1}H NMR (C_6D_6): 4.50 (t, H_2Ge) [15] IR (C_7H_{16}): ν(GeH) 2056 [10] with CCl_4 at 80 °C in the presence of AIBN in a sealed tube → $Ge(C_2H_5)(C_6H_5)Cl_2$ [15]
R = C_3H_7		
22	C_4H_9	$X(C_4H_9)GeCH_2CH_2CH_2$ (X = halogen) + $LiAlH_4$ in refluxing ether along with $H(C_4H_9)GeCH_2CH_2CH_2$ in a 4:6 ratio [4]
R = C_4H_9		
23	CH_2Cl	No. 24 + $Sn(C_4H_9)_3H$ (exothermic) refluxed for 1 h, then distillation in the presence of catalytic amounts of galvinoxyl and hydroquinone (75%) b.p. 44 to 46 °C/11 ^{1}H NMR: 3.03 (t, CH_2), 4.08 (quint, H_2Ge) IR (C_7H_{16}): ν(GeH) 2076, 2053; ν(GeH) region illustrated [14]
24	$CHCl_2$	$Ge(C_4H_9)H_2Cl + LiCHCl_2$ like No. 14 (48%) b.p. 65 °C/15 ^{1}H NMR: 4.54 (q, H_2Ge), 5.60 (t, CH) IR (C_7H_{16}): ν(GeH) 2094, 2075; ν(GeH) region illustrated starting material for No. 23 [14]
25	C_6H_5	$Ge(C_6H_5)Cl + Hg(C_4H_9)_2$ like No. 21 (87%) along with $Ge(C_4H_9)_2(C_6H_5)H$ (12%) $Ge(C_6H_5)H_2Cl + C_4H_9MgBr$ like No. 21 (78%) b.p. 100 °C/10; n_D^{20} = 1.5162 ^{1}H NMR (C_6D_6): 4.49 (t, H_2Ge) with CCl_4 → $Ge(C_4H_9)(C_6H_5)Cl_2$ as described for No. 21 [15]
R = C_4H_9-t		
26	C_6H_5	for the formation, see further information on No. 6, p. 253 [23]

Table 37 (continued)

No.	R′ group	formation (yield) properties and remarks
27	$C_6H_2(C_4H_9\text{-t})_3\text{-}2,4,6$	$Ge(C_4H_9\text{-t})(C_6H_2(C_4H_9\text{-t})_3\text{-}2,4,6)(OCH_3)_2$ + $LiAlH_4$ in ether at 0 to 20 °C (2 h); recrystallized from C_5H_{12} at −30 °C (88%) white crystals; m.p. 55 to 57 °C ^1H NMR (C_6D_6): 1.17 (s, t-C_4H_9/R), 1.40 (s, t-C_4H_9-4/R′), 1.63 (s, t-C_4H_9-2,6/R′), 5.28 (s, HGe), 7.57 (s, C_6H_2) IR (film): ν(GeH) 2058 [24]

* Further information:

General Remarks. The effect of π(d-p) interaction between Ge and the corresponding substituents on the ν(GeH) frequency [9, 18] or its integrated intensity [10] was studied on a series of compounds including Nos. 3, 6, 15, 16, 17, 20, and 21. The doublet character of the ν(GeH) band of Nos. 2, 13, 14, 23, and 24 (see Table 37) was attributed to intramolecular coordination between the chlorine and Ge atoms. Thus, the two bands result from coordinated (at lower frequencies) and free molecules. The enthalpy of intramolecular coordination was determined by IR spectroscopy [14]. A linear correlation was found between the ΔH values [14], the ν(GeH) frequencies [18], or their integrated intensities [10] and the sum of Taft σ^* constants of the substituents.

$Ge(CH_3)(C_2H_5)H_2$ (Table **37**, No. **3**). The ^1H NMR spectra of some Group IV and V hydrides were used to determine the relative acidities of these compounds in liquid NH_3 (at 31 °C): δ(ppm) = 0.36 (CH_3), 1.12, 1.21 (C_2H_5), 3.75 (H_2Ge); $^3J(H, CH_3)$ = 3.85 Hz for No. 3 [2].

Molecular mechanics calculations using the MM2 force field revealed the staggered conformations to be more stable than the eclipsed one, but the barrier height is small because of the long Ge-C bond (1.25 kcal/mol) [20]. Force field calculations were also carried out to determine the conformational equilibria of some Group IV organometallic compounds; ΔG = −0.6 kcal/mol was estimated for the anti ↔ gauche equilibrium of No. 3 [6]. Conformational effects on the ν(GeH) frequencies of gauche and trans CH_3 groups were cited as evidence for conformer abundance in ethyl and isopropyl germanes [16].

$Ge(CH_3)(C_3H_7\text{-i})H_2$ (Table **37**, No. **4**). According to force field calculations the symmetric conformation (CH_3 in gauche position with respect to the two CH_3 groups of i-C_3H_7) is the more stable one by 0.19 kcal/mol [6].

$Ge(CH_3)(C_5H_5\text{-c})H_2$ (Table **37**, No. **5**). The ^1H NMR spectrum at variable temperatures (−60 to +100 °C) showed that the compound is fluxional at 35 °C [7].

The mass spectrum contains $[M]^+$ which fragments to give $[Ge(CH_3)H]^+$ and $[Ge(C_5H_5)H]^+$. The $[C_5H_6]^+$ and $[C_5H_7]^+$ ions are supposed to result from transfer of Ge-H hydrogen to the C_5H_5 group. All fragments observed were listed [7].

$Ge(CH_3)(C_6H_5)H_2$ (Table **37**, No. **6**) was formed with 2 to 4% yield by adding $Ge_2(C_6H_5)_2H_4$ in THF to Li/Hg and treating the mixture with CH_3I [8].

No. 6 and $Ge(C_4H_9\text{-t})(C_6H_5)H_2$ (No. 26) were obtained by the reaction of $Ge(C_6H_5)H_3$ with LiC_4H_9-t in THF-C_5H_{12} at −78 °C followed by excess CH_3I at 20 °C for 1 h. Hydrolysis, extraction with ether (?), and concentration gave a residue which consisted of No. 6, No. 26, $Ge(C_4H_9\text{-t})_2(C_6H_5)H$, and $Ge(C_6H_5)H_3$ in a 69:9:8:14 ratio as indicated by GC and ^1H NMR [23].

References on p. 254

Table 38
Vibrational Spectra of $Ge(CHF_2)(CF_3)H_2$ [19].
Wavenumbers in cm^{-1}.

IR gas	Raman liquid	assignment	IR gas	Raman liquid	assignment
2951(m)	2155(w)	ν_{as}(GeH)	642(m)	653(w)	twist(CGeH)
2131(s)	2137(s,p)	ν_s(GeH)	610(m)	614(m,p)	$\delta(CF_2)$
1338(sh)	1337(vw)	ρ(CH)	538(vw)	552(wm,p)	$\delta_{as}(CF_3)$
1311(s)	1312(w,p)	ρ(CH)	517(vw)	512(w)	$\delta_{as}(CF_3)$
1184(vs)	1182(w,p)	$\nu_s(CF_3)$	441(ms)	442(w,p)	δ(CGeH)
1118(vvs)	1090(w,br)	$\nu_{as}(CF_3)$	323(s)	320(m,p)	ν_{as}(GeC)
1093(sh)		$\nu(CF_2)$	305(sh)	303(wm,p)	$\rho(CF_3/CF_2)$
1055(vs)	1030(vw)	$\nu(CF_2)$	262(m)	265(vw)	$\rho(CF_3/CF_2)$
849(s)	843(m)	$\delta(GeH_2)$	250(sh)	253(s,p)	ν_s(GeC)
729(vs)	726(s,p)	$\delta_s(CF_3)$	207(w)	212(w,p)	$\rho(CF_3)$
691(ms)	700(w)	wag (CGeH)		82(w)	$\delta(GeC_2)$

$Ge(CHF_2)(CF_3)H_2$ (Table **37**, No. **11**). The IR and Raman spectra are given in Table 38.

$Ge(C_2H_5)(CH_2C_6H_5)H_2$ (Table **37**, No. **15**) forms a charge-transfer complex with $C_2(CN)_4$ in CH_2Cl_2 showing a CT absorption band at 19300 cm^{-1}. The linear relation between ν_{CT} and the σ^+-Brown-Okamoto constants was used to calculate the σ_p^+ constants of the Ge- and Si-containing substituents in C_6H_5R'' compounds ($R'' = CH_2Ge(C_2H_5)H_2$ for No. 15). The resonance constants σ_R^o of the corresponding substituents were determined from the integrated intensities of aromatic ring stretching modes in the 1600 cm^{-1} region of the IR spectrum [13].

$Ge(C_2H_5)(CH_2CH_2OH)H_2$ (Table **37**, No. **16**). The increase of the ν(GeH) intensity with increasing concentration in alcohol-substituted Ge compounds like No. 16 was attributed to the formation of self-associates such as $H_2(C_2H_5)GeCH_2CH_2O(H) \rightarrow Ge(C_2H_5)(CH_2$-$CH_2OH)H_2$ where the Ge atom forms an additional $(p \rightarrow d)\sigma$ bond [10].

References:

[1] Glarum, S. N.; Kraus, C. A. (J. Am. Chem. Soc. **72** [1950] 5398/401).
[2] Birchall, T.; Jolly, W. L. (Inorg. Chem. **5** [1966] 2177/80).
[3] Satgé, J.; Rivière, P. (Bull. Soc. Chim. Fr. **1966** 1773/4).
[4] Mazerolles, P.; Dubac, J.; Lesbre, M. (C. R. Seances Acad. Sci. C **266** [1968] 1794/6).
[5] Massol, M.; Barrau, J.; Rivière, P.; Satgé, J. (J. Organomet. Chem. **30** [1971] 27/41).
[6] Ouellette, R. J. (J. Am. Chem. Soc. **94** [1972] 7674/9).
[7] Angus, P. C.; Stobart, S. R. (J. Chem. Soc. Dalton Trans. **1973** 2374/80).
[8] Rivière, P.; Satgé, J.; Soula, D. (J. Organomet. Chem. **72** [1974] 329/38).
[9] Egorochkin, A. N.; Khorshev, S. Ya.; Satgé, J.; Rivière, P.; Barrau, J. (J. Organomet. Chem. **99** [1975] 339/49).
[10] Egorochkin, A. N.; Khorshev, S. Ya.; Sevastyanova, E. I.; Ratushnaya, S. Kh.; Satgé, J.; Rivière, P.; Barrau, J.; Richelme, S. (J. Organomet. Chem. **155** [1978] 175/83).

[11] Barrau, J.; Satgé, J. (J. Organomet. Chem. **148** [1978] C9/C12).
[12] Egorochkin, A. N.; Sevastyanova, E. I.; Khorshev, S. Ya.; Ratushnaya, S. Kh.; Richelme, S.; Satgé, J.; Rivière, P. (J. Organomet. Chem. **188** [1980] 73/83).

[13] Sennikov, P. G.; Skobeleva, S. E.; Kuznetsov, V. A.; Egorochkin, A. N.; Rivière, P.; Satgé, J.; Richelme, S. (J. Organomet. Chem. **201** [1980] 213/9).
[14] Egorochkin, A. N.; Sevastyanova, E. I.; Khorshev, S. Ya.; Richelme, S.; Satgé, J. (J. Organomet. Chem. **205** [1981] 311/8).
[15] Rivière, P.; Castel, A.; Satgé, J.; Cazes, A. (Synth. React. Inorg. Met. Org. Chem. **11** [1981] 443/53).
[16] McKean, D. C.; Torto, I.; Mackenzie, M. W.; Morrison, A. R. (Spectrochim. Acta A **39** [1983] 387/98).
[17] Jutzi, P.; Gilge, U. (J. Organomet. Chem. **244** [1983] 355/62).
[18] McKean, D. C.; Torto, I.; Mackenzie, M. W. (Spectrochim. Acta A **39** [1983] 399/408).
[19] Eujen, R.; Mellies, R., Petrauskas, E. (J. Organomet. Chem. **299** [1986] 29/40).
[20] Allinger, N. L.; Quinn, M. I.; Chen, K.; Thompson, B.; Frierson, M. R. (J. Mol. Struct. **194** [1989] 1/18).

[21] Lukevics, E.; Gevorgyan, V. (Chem. Technol. Silicon Tin Proc. Asian Network Anal. Inorg. Chem. 1st Int. Chem. Conf. Silicon and Tin, Kuala Lumpur 1989 [1992], pp. 165/77).
[22] Gevorgyan, V.; Borisova, L.; Lukevics, E. (J. Organomet. Chem. **393** [1990] 57/67).
[23] Castel, A.; Rivière, P.; Satgé, J.; Desor, D. (J. Organomet. Chem. **433** [1992] 49/61).
[24] Richelme, S.; Andrianarison, M.; Couret, C.; Escudié, J.; Satgé, J. (Main Group Met. Chem. **10** [1987] 69/76).

1.3.2.2.2 Ge(R)(R′)H_2 Compounds with R = Aryl

Most compounds in this section and listed in Table 39 have the composition $Ge(C_6H_5)(R')H_2$. Four compounds containing a trisubstituted phenyl group as the R ligand are at the end of the table.

Table 39
Ge(R)(R′)H_2 Compounds with R = Aryl.
An asterisk indicates further information at the end of the table.
Explanations, abbreviations, and units are given on p. X.

No.	R′ group	preparation and formation (yield) properties and remarks
R = C_6H_5		
1	CH_2OCH_3	$Ge(C_6H_5)H_2Li + CH_3OCH_2Cl$ in THF at −78 °C to +20 °C followed by hydrolysis and distillation of the extracts (73%) b.p. 72 to 78 °C/13 ^{1}H NMR (C_6D_6): 3.08 (s, CH_3O), 3.55 (t, CH_2), 4.62 (t, H_2Ge, $^3J(H, H) = 3$), 7.00 to 7.60 (m, C_6H_5) IR (C_6D_6): ν(GeH) 2052 MS: $[M]^+$ with LiC_4H_9-t in THF-C_5H_{12} at −78 °C followed by CH_3OCH_2Cl and hydrolysis → $Ge(C_6H_5)(CH_2OCH_3)_2H$ [22]

Table 39 (continued)

No.	R′ group	preparation and formation (yield) properties and remarks
2	$CH(C_6H_5)OH$	$Ge(C_6H_5)(CH(C_6H_5)OH)Cl_2$ (from $Ge(C_6H_5)(H)Cl_2$ and C_6H_5CHO at 80 °C) + excess $LiAlH_4$ in ether at 0 °C followed by hydrolysis and distillation (48%) b.p. 108 to 110 °C/4 ^1H NMR (C_6D_6): 4.66 (d, H_2Ge, $^3J(H, H)$ = 3), 4.83 (t, CH) IR (film): ν(OH) 3330, ν(GeH) 2075 [17]
3	$CH(C_6H_5)N(C_6H_5)H$	$Ge(C_6H_5)(CH(C_6H_5)N(C_6H_5)H)Cl_2$ + $LiAlH_4$ in THF at 45 °C; workup like No. 2 (59%) [15] $Cl(C_6H_5)\overline{GeCH(C_6H_5)N(C_6H_5)CONC_6H_5}$ + $LiAlH_4$ in ether [15, 16] b.p. 104 °C/0.03 ^1H NMR (C_6D_6): 4.65 (t, CH), 4.80 (d, H_2Ge, $^3J(H, H)$ = 2.5) IR (film): ν(NH) 3395, ν(GeH) 2080 [15]
4	$CH(C_6H_5)N(C_6H_5)OH$	$Ge(CH(C_6H_5)N(C_6H_5)OH)(C_6H_5)Cl_2$ + $LiAlH_4$ in ether at 0 °C followed by hydrolysis and concentration (90%) ^1H NMR (C_6D_6): 4.20 (d, H_2Ge, $^3J(H, H)$ = 2.5), 5.30 (t, CH) IR: ν(OH) 3490, ν(GeH) 2080 decomposes during distillation to yield $(Ge(C_6H_5)H_2)_2O$ and $C_6H_5CH{=}NC_6H_5$ [12]
*5	CH_2CH_2OH	$Ge(C_6H_5)(CH_2CH_2OOCCH_3)Cl_2$ + $LiAlH_4$ in ether followed by hydrolysis and distillation of the ether layer (74%) [7] b.p. 91 °C/0.06 [7, 13]; n_D^{20} = 1.5590 [7] ^1H NMR: 1.25 (m, CH_2Ge, $^3J(H, H)$ = 7.5), 3.55 (t, CH_2O), 4.48 (t, H_2Ge, $^3J(H, H)$ = 3) IR: ν(OH) 3400, ν(GeH) 2055 with excess CCl_4 (exothermic, 60 °C) → $Ge(CH_2CH_2OH)(C_6H_5)Cl_2$ [7]
6	$CH(OH)CH_3$	like No. 2 using $Ge(CH(OH)CH_3)(C_6H_5)Cl_2$ (40%) b.p. 130 °C/18 ^1H NMR (C_6D_6): 1.33 (d, CH_3), 4.00 (q of d, CH, $^3J(H, H)$ = 7), 4.53 (d, H_2Ge, $^3J(H, H)$ = 3) IR (film): ν(OH) 3320, ν(GeH) 2070 [17]
7	$CH(OH)CH(C_6H_5)_2$	$Ge(COCH(C_6H_5)_2)(C_6H_5)Cl_2$ + $LiAlH_4$ [1, 2] ^1H NMR: 4.03 (d, CH, $^3J(H, H)$ = 10), 4.35 (d, H_2Ge, $^3J(H, H)$ = 1.5), 4.60 (sext, CHGe) [1] IR: ν(OH) 3400, ν(GeH) 2050 [1, 2] rather unstable, decomposes at 120 to 140 °C [1, 2]
8	$CH_2CH_2CH_2OH$	$Ge(CH_2CH_2CH_2OH)(C_6H_5)Cl_2$ or $Cl(C_6H_5)\overline{GeCH_2CH_2CH_2O}$ with excess $LiAlH_4$ in ether and workup like No. 5 (90 and 93%, respectively) [7]

Table 39 (continued)

No.	R′ group	preparation and formation (yield) properties and remarks
		$Ge(C_6H_5)(H)Cl_2$ + CH_2=CHCHO UV-irradiated and subsequently treated with $LiAlH_4$ in ether followed by distillation (53%) [8] (-$Ge(C_6H_5)(X)OCH$=$CHCH_2$-$)_n$ (from $Ge(C_6H_5)X$ and CH_2=CHCHO) + $LiAlH_4$ (34% for X = F and 49% for X = Cl) [11] b.p. 88 °C/0.04; n_D^{20} = 1.5492 [4, 7, 8] ^1H NMR: 1.12 (m, CH_2Ge), 1.68 (quint, CH_2, $^3J(H, H)$ = 6.5), 3.44 (t, CH_2O), 4.38 (t, H_2Ge, $^3J(H, H)$ = 2.75) [7, 8] IR: ν(GeH) 2062 in C_7H_{16} [14]; ν(OH) 3380, ν(GeH) 2055 [7, 8]
9	$CH_2CH(OH)CH_3$	$Ge(C_6H_5)(CH_2CH(OOCCH_3)CH_3)Cl_2$ + $LiAlH_4$ in ether (60%) $Ge(C_6H_5)Cl$ + CH_3COCH_2Cl followed by treatment with $LiAlH_4$ in ether (15 to 20%) b.p. 140 °C/17 d^{20} = 1.2308; n_D^{20} = 1.5464 ^1H NMR (CD_3COCD_3): 1.22 (d, CH_3), 1.46 (dt, CH_2Ge), 4.05 (sext, CH, J = 6.5), 4.42 (t, H_2Ge, $^3J(H, H)$ = 3) [5] IR: ν(GeH) 2066 in C_7H_{16} [14]; ν(OH) 3350, ν(GeH) 2055 [5]
*10	$CH(OH)C_3H_7$-i	$Ge(C_6H_5)(CH(OH)C_3H_7\text{-i})Cl_2$ + LiH in THF-ether followed by hydrolysis and distillation of the organic phase (79%) b.p. 58 °C/0.06; n_D^{20} = 1.5400 ^1H NMR: 0.86 and 1.00 (d, CH_3, $^3J(H, H)$ = 7.0), 1.90 (m, CH), 3.74 (d of t, CHO, $^3J(H, H)$ = 6.5), 4.55 (d, H_2Ge, $^3J(H, H)$ = 2.5) [7] IR: ν(GeH) 2066 in C_7H_{16} [14]; ν(OH) 3380, ν(GeH) 2035 [7]
11	$CH(NHC_2H_5)C_3H_7$-i	$Ge(C_6H_5)(CH(NHC_2H_5)C_3H_7\text{-i})Cl_2$ (from $Ge(C_6H_5)(H)Cl_2$ and i-C_3H_7CH=NC_2H_5 in ether) + $LiAlH_4$ (64%) (-$Ge(C_6H_5)(X)N(C_2H_5)CH(C_3H_7\text{-i})$-$)_n$ (from $Ge(C_6H_5)X$ (X = F or Cl) and i-C_3H_7CH=NC_2H_5) + $LiAlH_4$ b.p. 115 °C/0.2 ^1H NMR (C_6D_6): 2.55 (d, CHN), 4.64 (d, H_2Ge, $^3J(H, H)$ = 2) IR: ν(NH) 3350, ν(GeH) 2050 [10]
12	$CH(CH_3)CH_2CH_2OH$	(-$Ge(C_6H_5)(X)OCH$=$CHCH(CH_3)$-)n (from $Ge(C_6H_5)X$ and CH_3CH=CHCHO, X = F or Cl) + $LiAlH_4$ (ca. 65%) b.p. 95 °C/0.3

References on p. 260

Table 39 (continued)

No.	R′ group	preparation and formation (yield) properties and remarks
12 (continued)		^{1}H NMR (C_6D_6): 1.10 (d, CH_3, $^3J(H, H) = 6.5$), 3.50 (t, CH_2O, $^3J(H, H) = 7$), 4.50 (d, H_2Ge, $^3J(H, H) = 2$) IR: ν(OH) 3350, ν(GeH) 2055 [11]
13	$CH_2(CH_2)_3CH(OH)CH_3$	$Ge(C_6H_5)(CH_2(CH_2)_3COCH_3)(H)Cl$ or $Cl(C_6H_5)\overline{GeCH_2(CH_2)_3CH(CH_3)O}$ + $LiAlH_4$ [2]
14	$CH_2(CH_2)_9CH_2OH$	$Ge(C_6H_5)(CH_2(CH_2)_9CHO)Cl_2$ + $LiAlH_4$ in ether (74%) b.p. 144 to 147 °C/0.04; n_D^{20} = 1.5120 ^{1}H NMR: 1.0 to 1.70 (m, CH_2, $^3J(H, H) = 6$), 3.50 (t, CH_2O), 4.37 (t, H_2Ge, $^3J(H, H) = 3$) IR: ν(OH) 3340, ν(GeH) 2060 [8]
*15	C_6H_{11}-c	preparation not reported IR: ν(GeH) 2037 for the neat liquid, 2039 in C_7H_{14}, 2040 in CCl_4 [3]
16	$CH_2CH{=}CH_2$	$Ge(C_6H_5)(CH_2CH{=}CH_2)Cl_2$ (from $Ge(C_6H_5)Cl$ and $CH_2{=}CHCH_2Cl$) + $LiAlH_4$ at −20 °C and workup like No. 2 (32%) along with $Ge(C_6H_5)H_3$ b.p. 85 °C/10; n_D^{20} = 1.5320 ^{1}H NMR: 2.00 (m, CH_2Ge), 4.43 (t, H_2Ge), 4.70 to 6.30 ($CH{=}CH_2$) IR: ν(GeH) 2060, ν(C=C) 1640 polymerizes at room temperature to give an insoluble solid [5]
17	$CH(OH)CH{=}CHCH_3$	like No. 5 except using $Ge(C_6H_5)(CH(OH)CH{=}CHCH_3)Cl_2$ (88%) [8]; see also [6] $(-Ge(C_6H_5)(X)OCH(CH{=}CHCH_3)-)_n$ (from $Ge(C_6H_5)X$ and $CH_3CH{=}CHCHO$, X = F or Cl) + $LiAlH_4$ [11] b.p. 78 °C/0.05; n_D^{20} = 1.5550 [8] ^{1}H NMR: 1.63 (m, CH_3), 4.43 (m, CHGe), 4.50 (d, H_2Ge, $^3J(H, H) = 2.75$), 5.45 (m, CH=CH) IR: ν(OH) 3350, ν(GeH) 2060, ν(C=C) 1650 [8] UV irradiation in ether for 2 h → $H(C_6H_5)\overline{GeCH(OH)CH_2C}HCH_3$ and a nondistillable polymer [8]; see also [6]
18	$CH(OH)(CH_2)_8CH{=}CH_2$	$Ge(C_6H_5)(CH(OH)(CH_2)_8CH{=}CH_2)Cl_2$ + $LiAlH_4$ in ether (67%) [8]; see also [6] b.p. 154 °C/0.05; n_D^{20} = 1.5135 ^{1}H NMR: 4.00 (m, CHO), 4.45 (d, H_2Ge, $^3J(H, H) = 2$), 4.55 to 6.10 (m, $CH{=}CH_2$) IR: ν(OH) 3360, ν(GeH) 2065, ν(C=C) 1635 [8] heating in C_6H_{14} at 100 °C for 12 h in the presence of AIBN in a sealed tube → $H(C_6H_5)\overline{Ge(CH_2)_{10}C}HOH$ [8]; see also [6]

Table 39 (continued)

No.	R′ group	preparation and formation (yield) properties and remarks
19	$C_6H_2R''_3$-2,4,6 R″ = $CH(Si(CH_3)_3)_2$	$Ge(C_6H_2(CH(Si(CH_3)_3)_2)_3$-2,4,6)$Cl_3$ + C_6H_5MgBr (1:1.5 mole ratio) in THF at −78 °C followed by treatment with $LiAlH_4$ under reflux (15%); along with $Ge(C_6H_5)_2(C_6H_2(CH(Si(CH_3)_3)_2)_3$-2,4,6)H (58%) [18] $Ge(C_6H_5)Cl_3$ + $Li(C_6H_2(CH(Si(CH_3)_3)_2)_3$-2,4,6) in THF at −78 °C and $LiAlH_4$ as above (9%) [19] white crystals; m.p. 134 to 136 °C ^{1}H NMR ($CDCl_3$): −0.03 (s, 36 H), 0.04 (s, 18 H), 1.32 (s, 1 H), 2.11 (s, 2 H), 5.14 (s, 2 H), 6.35 (br s, 2 H), 7.3 to 7.6 (m, 5 H) MS: $[M]^+$ (33), m/e = 615 (25), 552 (33), 73 (100) [19] sulfurization like with No. 22 → inseparable mixture of cyclic germapolysulfides [18]
R = $C_6H_2(CH_3)_3$-2,4,6		
20	$CH(OH)C_6H_5$	$Ge(C_6H_2(CH_3)_3$-2,4,6)H_2Li + C_6H_5CHO in THF at −40 to +20 °C (2 h) followed by hydrolysis and distillation of the concentrated extract (23%) b.p. 95 °C/0.05 ^{1}H NMR (C_6D_6): 2.00 (s, CH_3-4), 2.15 (s, CH_3-2,6), 4.70 (d, H_2Ge, J = 3), 4.83 (t, CH), 6.57 (s, C_6H_2), 6.96 (m, C_6H_5) IR (neat): ν(OH) 3400, ν(GeH) 2050 [22]
*21	$CH(OH)C_6H_2$- $(CH_3)_3$-2,4,6	like No. 20 using 2,4,6-$(CH_3)_3C_6H_2CHO$; workup by adding ether-C_5H_{12} to the ether extract and drying the precipitate under reduced pressure (53%) m.p. 130 to 132 °C ^{1}H NMR (C_6D_6): 2.08, 2.12 (s, CH_3-4), 2.26, 2.30 (s, CH_3-2,6), 4.89, 4.93 (d, H_2Ge), 5.42 (t, CH, ^3J(H, H) = 2.5), 6.71 (s, C_6H_2) IR (Nujol): ν(OH) 3500, ν(GeH) 2075 [22]
22	$C_6H_2R''_3$-2,4,6 R″ = $CH(Si(CH_3)_3)_2$	$Ge(C_6H_2(CH(Si(CH_3)_3)_2)_3$-2,4,6)$Cl_3$ + 2,4,6-$(CH_3)_3C_6H_2MgBr$ (1:2 mole ratio) like No. 19 (54%) [18]; alkylation also with 2,4,6-$(CH_3)_3C_6H_2Li$ [20] white crystals; m.p. 162 to 163 °C (dec.) ^{1}H NMR ($CDCl_3$; 500 MHz): −0.043 (s, 18 H), 0.001 (s, 18 H), 0.052 (s, 18 H), 1.31 (s, 1 H), 2.01 (s, 1 H), 2.22 (s, 1 H), 2.26 (s, 3 H), 2.42 (s, 6 H), 5.14 (s, 2 H), 6.29 (s, 1 H), 6.44 (s, 1 H), 6.83 (s, 2 H) ^{13}C NMR ($CDCl_3$): 0.59 (q), 0.79 (q), 0.93 (q), 21.01 (q), 24.69 (q), 28.93 (2 d's), 30.27 (d), 122.26 (d), 127.08 (d), 128.14 (s), 128.44 (d), 133.00 (s), 138.51 (s), 143.24 (s), 143.29 (s), 150.28 (s), 150.38 (s) IR (KBr): ν(GeH) 2070

References on p. 260

Table 39 (continued)

No.	R′ group	preparation and formation (yield) properties and remarks
22 (continued)		MS: $[M]^+$ (9), m/e = 731 (4), 552 (16), 105 (100), 73 (50) [21] heating in excess sulfur at 180 °C for 10 min → $Ge(R)(R')S_3S$; similarly during 10 h → $Ge(R)(R')S_3S$ and $Ge(R)(R')SSGe(R)(R')S_2S$ [18]; see also [20] with excess LiC_4H_9-t in THF-C_5H_{12} at −20 °C in the presence of HMPT (→ Ge(R)(R′)(H)Li) followed by Se at 20 °C → $Ge(R)(R')Se_3Se$; similarly with S → $Ge(R)(R')S_3S$ [23]
23	$CHC_{12}H_8$ (fluorenyl)	$LiAlH_4$ reduction of the chloride in ether at 0 °C (91%) white needles; m.p. 92 °C ^1H NMR ($CDCl_3$): 2.25 (s, CH_3-2,6), 2.30 (s, CH_3-4), 4.29 (t, CH/$CHC_{12}H_8$, $^3J(H, H)$ = 3.3), 4.52 (d, HGe, $^3J(H, H)$ = 3.3), 6.88 (s, C_6H_2), 7.19 to 7.90 (m, $C_{12}H_8$) IR: ν(GeH) 2061 and 2083 with N-chlorosuccinimide in THF at 95 °C in a sealed tube → $Ge(C_6H_2(CH_3)_3$-2,4,6)($CHC_{12}H_8$)(H)Cl [24]
R = $C_6H_2(C_3H_7$-i)$_3$-2,4,6		
24	$C_6H_2R''_3$-2,4,6 R″ = $CH(Si(CH_3)_3)_2$	with LiC_4H_9-t and Se like No. 22 → $Ge(R)(R')Se_3Se$ [23]

* Further information:

General Remarks. A linear relation between the integrated intensity of the ν(GeH) frequency and the sum of the Taft constants σ^* of the substituents at Ge was found for Nos. 8, 9, 10, and related compounds in which the substituents exhibit only an inductive effect [14].

$Ge(C_6H_5)(CH_2CH_2OH)H_2$ (Table **39**, No. **5**) was obtained by reducing with $LiAlH_4$ $C_6H_5(XCH_2CH_2O)GeCH_2CH_2OGe(OCH_2CH_2X)(C_6H_5)CH_2CH_2O$ (X = F, Cl) formed from $Ge(C_6H_5)X$ and excess CH_2CH_2O at 120 °C [13].

$Ge(C_6H_5)(CH(OH)C_3H_7$-i$)H_2$ (Table **39**, No. **10**) was also formed by reducing with $LiAlH_4$ the products resulting from the reaction of $Ge(C_6H_5)OCH_3$ or $Ge(C_6H_5)SCH_3$ with i-C_3H_7CHO. It was similarly obtained from phenylphosphino- or phenylaminogermylenes [9].

$Ge(C_6H_5)(C_6H_{11}$-c$)H_2$ (Table **39**, No. **15**) was only mentioned in the context of IR studies of the Ge-H vibration in a series of variably substituted organogermanes [3].

$Ge(C_6H_2(CH_3)_3$-2,4,6)($CH(OH)C_6H_2(CH_3)_3$-2,4,6)H_2 (Table **39**, No. **21**). The formation of a mixture of No. 21 (11%), $Ge(R)(COR)_2Cl$ (26%), $Ge_2(R)_2H_4$ (5%), and $Ge(R)H_3$ (10%) (R = $C_6H_2(CH_3)_3$-2,4,6) was indicated by ^1H NMR spectroscopy after the reaction of $Ge(R)H_2Li$ with RCOCl in THF at −78 to +20 °C followed by hydrolysis with aqueous $NaHCO_3$ [22].

References:

[1] Rivière, P.; Satgé, J. (C. R. Seances Acad. Sci. C **267** [1968] 267/9).
[2] Satgé, J.; Rivière, P. (J. Organomet. Chem. **16** [1969] 71/82).

[3] Mathis, R.; Berthelat, M.; Mathis, F. (Spectrochim. Acta A **26** [1970] 1993/2000).
[4] Rivière, P.; Satgé, J. (C. R. Seances Acad. Sci. C **272** [1971] 413/6).
[5] Massol, M.; Barrau, J.; Rivière, P.; Satgé, J. (J. Organomet. Chem. **30** [1971] 27/41).
[6] Rivière, P.; Satgé, J. (Angew. Chem. **83** [1971] 286/7; Angew. Chem. Int. Ed. Engl. **10** [1971] 267).
[7] Rivière, P.; Satgé, J. (J. Organomet. Chem. **49** [1973] 157/72).
[8] Rivière, P.; Satgé, J. (J. Organomet. Chem. **49** [1973] 173/89).
[9] Rivière, P.; Rivière-Baudet, M.; Satgé, J. (J. Organomet. Chem. **97** [1975] C37/C40).
[10] Rivière, P.; Satgé, J.; Castel, A. (C. R. Seances Acad. Sci. C **282** [1976] 971/4).

[11] Rivière, P.; Satgé, J.; Castel, A. (C. R. Seances Acad. Sci. C **284** [1977] 395/8).
[12] Rivière, P.; Richelme, S.; Rivière-Baudet, M.; Satgé, J. (J. Chem. Res. Synop. **1978** 220/1; J. Chem. Res. Miniprint **1978** 2817/28).
[13] Castel, A.; Rivière, P.; Satgé, J.; Cazes, A. (C. R. Seances Acad. Sci. C **287** [1978] 205/8).
[14] Egorochkin, A. N.; Khorshev, S. Ya.; Sevastyanova, E. I.; Ratushnaya, S. Kh.; Satgé, J.; Rivière, P.; Barrau, J.; Richelme, S. (J. Organomet. Chem. **155** [1978] 175/83).
[15] Rivière, P.; Rivière-Baudet, M.; Richelme, S.; Castel, A.; Satgé, J. (J. Organomet. Chem. **168** [1979] 43/52).
[16] Rivière-Baudet, M.; Rivière, P.; Satgé, J.; Lacrampe, G. (Recl. Trav. Chim. Pays-Bas **98** [1979] 42/52).
[17] Castel, A.; Rivière, P.; Satgé, J. (J. Organomet. Chem. **232** [1982] 137/46).
[18] Tokitoh, N.; Takahashi, M.; Matsumoto, T.; Suzuki, H.; Matsuhashi, Y.; Okazaki, R. (Phosphorus Sulfur Silicon Relat. Elem. **59** [1991] 161/4).
[19] Tokitoh, N.; Matsumoto, T.; Suzuki, H.; Okazaki, R. (Tetrahedron Lett. **32** [1991] 2049/52).
[20] Tokitoh, N.; Suzuki, H.; Matsumoto, T.; Matsuhashi, Y.; Okazaki, R. (J. Am. Chem. Soc. **113** [1991] 7047/9).

[21] Gmelin-Sonderdruck Nr. F30128; supplementary material of [20].
[22] Castel, A.; Rivière, P.; Satgé, J.; Desor, D. (J. Organomet. Chem. **433** [1992] 49/61).
[23] Tokitoh, N.; Matsumoto, T.; Okazaki, R. (Tetrahedron Lett. **33** [1992] 2531/4).
[24] Chaubon-Deredempt, M. A.; Escudié, J.; Couret, C. (J. Organomet. Chem. **467** [1994] 37/46).

1.3.2.3 Germacyclic Compounds of the $\overline{Ge\text{-}RH_2}$ Type

The compounds in this section are listed in Table 40. They consist of five- and six-membered saturated (Nos. 1 to 9) and unsaturated (Nos. 10 to 13) rings. Two three-membered derivatives are only mentioned in the context of theoretical studies described below.

$\overline{GeCH_2CH_2}H_2$. Ab initio quantum-mechanical methods were used to determine the strain energies of three-membered rings containing C, Si, or Ge and the energetics of their decomposition to give in this case GeH_2 and $CH_2{=}CH_2$. The decomposition enthalpy was predicted from a simple model using the strain energies along with single-bond dissociation energies, π-bond energies, and divalent state stabilization energies. $\overline{GeCH_2CH_2}H_2$ is the least stable species with respect to dissociation [20].

$\overline{GeCH{=}CH}H_2$. The geometry of $\overline{ECH{=}CH}H_2$ compounds (E = C, Si, Ge, Sn) was predicted from SCF wave functions. The analysis of the total electron density in terms of bent bond lengths revealed little or no π-complex character in the cyclopropene ring. Intrinsic vibrational frequencies and force constants were also computed and showed that the thermodynamic stability of these compounds decreases in the order E = C > Si > Ge > Sn. A calculation of the energetics of the reactions EH_2 + HC≡CH → ECH=CHH$_2$ indicated the absence of barriers to the formation of the metallacyclopropenes [16].

For a recent investigation of ^{13}C and ^{73}Ge NMR spectra of 3-germabicyclo[3.1.0]hexanes and their structures as predicted by MM calculations, see [22].

Table 40
Germacyclic Compounds, Ge-RH$_2$.
An asterisk indicates further information at the end of the table.
Explanations, abbreviations, and units are given on p. X.

No.	compound	preparation and formation (yield) properties and remarks
*1	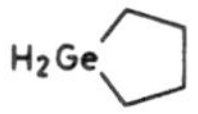	$GeCH_2CH_2CH_2CH_2Cl_2$ (from $GeCl_4$ + $BrMg(CH_2)_4MgBr$) + $LiAlH_4$ in refluxing ether (8 h) followed by hydrolysis and distillation (ca. 19% based on $GeCl_4$) [1]; $GeCH_2CH_2CH_2CH_2Br_2$ also used as the starting material (90%) [3] b.p. 91 to 92°C/760 [1, 3] d^{20} = 1.2261; n_D^{20} = 1.4838 [1] ^{13}C NMR ($CDCl_3$): 13.22 (C-2,5), 29.20 (C-3,4) ^{73}Ge NMR ($CDCl_3$, 30°C): −101.1 [19] IR: ν(GeH) 2052(s) [1, 3]; see also p. 265 rapidly oxidized in air → $(GeCH_2CH_2CH_2CH_2\text{-}O)_3$ [1] with I_2 (1:0.7 mole ratio) at 0°C → $GeCH_2CH_2CH_2CH_2(H)I$ [1] with $HgCl_2$ (1:1 mole ratio) under reflux (48 h) → $GeCH_2CH_2CH_2CH_2(H)Cl$ [3] with SO_2Cl_2 (1:2 mole ratio) → $GeCH_2CH_2CH_2CH_2Cl_2$ [3] refluxed with excess C_4H_9Br → $GeCH_2CH_2CH_2CH_2Br_2$ [1] heated with excess C_4H_9I for 72 h → $GeCH_2CH_2CH_2CH_2I_2$ [1]
2	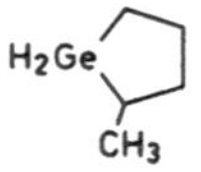	from $GeCH(CH_3)CH_2CH_2CH_2Cl_2$ like No. 1, but after 24 h (10% based on $GeCl_4$) [10, 11]; GC and ^{1}H NMR indicated contamination, probably with $Ge(C_5H_{11})H_3$ [11] b.p. 70°C/400 [10], 70 to 72°C/300 [11] ^{1}H NMR (C_6H_6): 3.9 and 4.0 (m, HGe) [10, 11]
*3	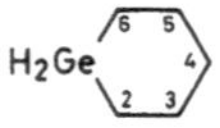	from $GeCH_2(CH_2)_3CH_2Cl_2$ like No. 1 (20% based on $GeCl_4$) [1]; see also [13] b.p. 65 to 67°C/150 [13], 119 to 120°C [1] d^{20} = 1.1846; n_D^{20} = 1.4835 [1] ^{1}H NMR ($CDCl_3$): 3.69 (m, HGe) [13] ^{13}C NMR ($CDCl_3$, 30°C): 9.32 (C-2,6), 26.78 (C-3,5), 29.89 (C-4) [12, 14] ^{73}Ge NMR ($CDCl_3$): −131.2 (J(Ge, H) = 92.8) [12], −130.3 [14] IR: ν(GeH) 2044(s) [1]; 2050 for the neat compound [13]

Table 40 (continued)

No.	compound	preparation and formation (yield) properties and remarks
		reacts with I_2, C_4H_9Br, and C_4H_9I in the same way as described for No. 1 [1] with $CH_2{=}CHCOOCH_3$ under reflux in the presence of AIBN or UV-irradiation → $\overline{GeCH_2(CH_2)_3C}H_2(CH_2CH_2COOCH_3)_2$ [9]
4	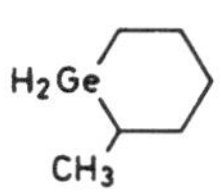	only mentioned in the context of molecular mechanics calculations [15]; see General Remarks, p. 264
5	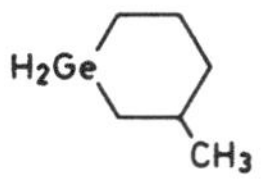	from $\overline{GeCH_2CH(CH_3)CH_2CH_2C}H_2Cl_2$ like No. 1 (11%) b.p. 72 to 74°C/55 ^{1}H NMR ($CDCl_3$): 3.65 (m, HGe) [13] ^{13}C NMR ($CDCl_3$ at 30°C): 8.10 (C-6), 17.55 (C-2), 25.71 (C-5), 27.09 (CH_3), 33.18 (C-3), 37.90 (C-4) [12, 14] ^{73}Ge NMR ($CDCl_3$): −131.2 [12, 14] IR (neat): ν(GeH) 2050 [13] MS: $[M]^+$, $[M-28]^+$ [13]
6	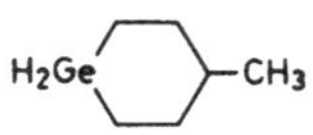	from $\overline{GeCH_2CH_2CH(CH_3)CH_2C}H_2Cl_2$ like No. 1 (22%) b.p. 72 to 75°C/20 ^{1}H NMR ($CDCl_3$): 3.62 (m, HGe) [13] ^{13}C NMR ($CDCl_3$): 8.02 (C-2,6), 23.45 (CH_3), 34.76 (C-3,5), 34.98 (C-4) [12] ^{73}Ge NMR ($CDCl_3$, 30°C): −134.3 (J(Ge, H) = 94.0) [12] IR (neat): ν(GeH) 2050 [13] MS: $[M]^+$, $[M-28]^+$ [13]
7	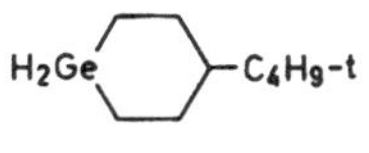	from $\overline{GeCH_2CH_2CH(C_4H_9\text{-}t)CH_2C}H_2Cl_2$ like No. 1, but after 1.5 h (51%) b.p. 125°C/23 ^{1}H NMR ($CDCl_3$): 0.83 (s, CH_3), 2.16 (m, H-4), 3.41 (m, HGe) ^{13}C NMR ($CDCl_3$): 9.53 (C-2,6), 27.84 (CH_3), 28.00 (C-3,5), 33.47 (q, C/C_4H_9), 51.52 (C-4) ^{73}Ge NMR ($CDCl_3$, 30°C): −131.6 IR (neat): ν(GeH) 2050 [17]
*8	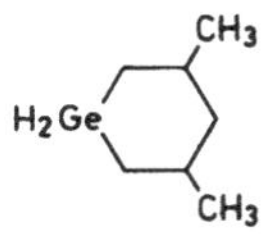	from $\overline{GeCH_2CH(CH_3)CH_2CH(CH_3)C}H_2Cl_2$ like No. 1 (8%); ca. 6:1 mixture of cis,trans isomers indicated by ^{13}C NMR b.p. 97 to 100°C/17 ^{1}H NMR ($CDCl_3$): 1.00 (d, CH_3, J = 7), 3.63 (m, HGe) ^{13}C NMR ($CDCl_3$): 15.62 (C-2), 23.91 (CH_3), 26.95 (C-3,5), 43.37 (C-4) for the trans isomer ^{73}Ge NMR ($CDCl_3$, 30°C): −145.0 for the trans isomer IR (neat): ν(GeH) 2045 [14]
9	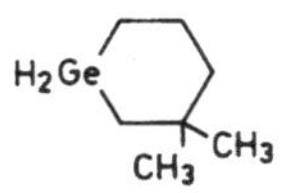	from $\overline{GeCH_2C(CH_3)_2CH_2CH_2C}H_2Cl_2$ like No. 1 (14%) b.p. 88 to 90°C/22 ^{1}H NMR ($CDCl_3$): 1.00 (s, CH_3), 3.60 (m, HGe)

References on p. 266

Table 40 (continued)

No.	compound	preparation and formation (yield) properties and remarks
9 (continued)		^{13}C NMR ($CDCl_3$, 30°C): 8.22 (C-6), 22.04 (C-5), 23.17 (C-2), 32.23 (CH_3), 32.60 (C-3), 42.67 (C-4) ^{73}Ge NMR ($CDCl_3$): −142.1 IR (neat): ν(GeH) 2040 [14]
10	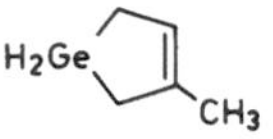	$GeCH_2C(CH_3){=}CHCH_2I_2$ + $LiAlH_4$ in refluxing ether (3 h) and workup like for No. 1 (50%) b.p. 117°C/760 d^{20} = 1.2751; n_D^{20} = 1.5212 IR: ν(GeH) 2040, ν(C=C) 1640 polymerizes when kept in an inert atmosphere [4]
11	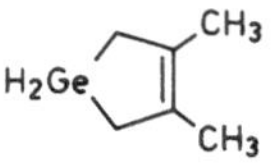	from $GeCH_2C(CH_3){=}C(CH_3)CH_2I_2$ like No. 10; purified by chromatography (27%) b.p. 142°C/760 d^{20} = 1.1765; n_D^{20} = 1.5025 IR: ν(GeH) 2060, ν(C=C) 1640 [4]
12	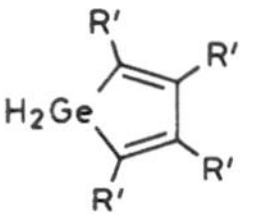 R′ = CH_3	$GeC(CH_3){=}C(CH_3)C(CH_3){=}C(CH_3)Cl_2$ in THF + $LiCH_3$ in ether at −78 to +20°C followed by $LiAlH_4$ reduction in ether at 0°C and hydrolysis after 5 h (30%); analyzed in solution by GC/MS; other products were $GeC_4(CH_3)_4(CH_3)H$ (5%) and $GeC_4(CH_3)_4(CH_3)_2$ (60%); similarly prepared with 20% yield using LiC_6H_5 (or C_6H_5MgBr) in C_6H_6-ether along with $GeC_4(CH_3)_4(C_6H_5)H$ (60%) and $GeC_4(CH_3)_4(C_6H_5)_2$ unstable; transformed into a viscous oil in the absence of a solvent MS: $[M]^+$ (54), $[M - H - C_2(CH_3)_2]^+$ (38), $[GeCH_3]^+$ (100) [18]
13	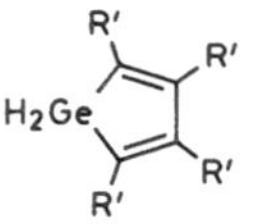 R′ = C_6H_5	$GeC(C_6H_5){=}C(C_6H_5)C(C_6H_5){=}C(C_6H_5)Cl_2$ + $LiAlH_4$ (70 to 90%) m.p. 192 to 193°C IR: ν(GeH) 2060 [5]

* Further information:

General Remarks. The optimized structures and steric energies of Nos. 3, 5, 6 [13], and the stereoisomers of Nos. 8 and 9 [14] were calculated by the MM2 (Molecular Mechanics) force field program. The calculations revealed a chair conformation flattened at the Ge atom [13]. They also showed that the axial conformation of Nos. 4, 5, and 6 is 1.41 kcal/mol less stable than the equatorial one if the CH_3 group is at C-4. The energy difference declines to 1.34 kcal/mol for CH_3 in 3-position and to 0.89 kcal/mol for CH_3 in 2-position [15].

A large upfield ^{73}Ge chemical shift is recorded, if the C-3 atom of germacyclohexanes has an axial CH_3 group. The effect of this CH_3 group upon other carbon shifts is similar to those of cyclohexanes [14].

$\overline{GeCH_2CH_2CH_2CH_2}H_2$ (Table **40**, No. **1**). The deuterated **$\overline{GeCH_2CH_2CH_2CH_2}D_2$** was obtained by reducing $\overline{GeCH_2CH_2CH_2CH_2}I_2$ with $LiAlD_4$ in refluxing ether (2 h) and isolating the compound by vapor phase chromatography (97% GeD_2 species) [2].

The dipole moment $\mu = 0.665 \pm 0.007$ D was determined from the Stark effect on some transitions of the ^{74}Ge microwave spectrum [7]; $\mu = 0.76$ D was calculated assigning a value of zero to the Ge-H bond moment and a value of $\mu = 0.55$ D to the Ge-C bond moment [15].

The microwave spectrum of four Ge isotopic species of No. 1 was investigated in the range 8 to 22 GHz. Rotational constants for the ground state of the ^{74}Ge species are: A = 5323.7, B = 2842.20, and C = 2041.52 MHz. From the excited states assigned to the lowest ring deformation mode, a lower limit of about 500 cm^{-1} was estimated for the barrier to pseudorotation. The molecule was found to be permanently twisted into a C_2 conformation with a twist angle of 18° and <(C-Ge-C) = 98°, <(Ge-C-C) = 106°, and <(C-C-C) = 115°. The nuclear quadrupole coupling constants for the ^{73}Ge species are: $\chi_{aa} = 15.89$, $\chi_{bb} = -10.29$, and $\chi_{cc} = -5.60$ MHz [7].

Table 41
Selected Fundamental Vibrations of $\overline{GeCH_2CH_2CH_2CH_2}(H,D)_2$ [8].
Ra denotes Raman lines; otherwise IR bands. Wavenumbers in cm^{-1}.

$\overline{GeCH_2CH_2CH_2CH_2}H_2$	$\overline{GeCH_2CH_2CH_2CH_2}D_2$	approximate description, fundamental vibration	
2978(m)	2978(s)	ν(CH)	ν_{21}
2946(s)	2944(s)		ν_{22}
2936(s)	2934(s)		ν_{23}
2932(83)/Ra	2933(89)/Ra		ν_1
2904(31)/Ra	2900(34)/Ra		ν_2
2865(m)	2864(s)		ν_4
2853(39)/Ra	2853(41)/Ra		ν_4
2066(vs)		ν(GeH)	ν_{25}
2054(100)/Ra			ν_5
	1485(vs)	ν(GeD)	ν_{25}
	1477(100)/Ra		ν_5
1455/1453(w)	1453(m)	$\delta(CH_2)$	ν_6, ν_{26}
1429/1419(w)	1425(w)/1415(8)/Ra		ν_7, ν_{27}
1324/1312(vw)	1323/1311(w)	CH_2 wagging	ν_8, ν_{28}
1078(m)	1077(m)	CH_2 rocking	ν_{11}
1025(m)	1028(?)	CH_2 twisting	ν_{12}
881(ms)		$\delta(GeH_2)$	ν_{13}
848(11)/Ra	847(12)/Ra	δ(ring)	ν_{14}
804(m)	789(s)	rocking CH_2	ν_{34}
762(w)	735(w)	δ(ring)	ν_{16}
	631(s)	$\delta(GeD_2)$	ν_{13}
635(vw)	617(33)/Ra	ring breathing	ν_{17}
543(vw)	427(vw)	twisting $Ge(H,D)_2$	ν_{18}
431(mw)		GeH_2 rocking	ν_{38}
	340(m)	GeD_2 rocking	ν_{38}
345(w)	340(m)	δ(ring)	ν_{19}

References on p. 266

The IR (vapor) and Raman (liquid) spectra of No. 1 and the deuterated derivative were recorded at high resolution and analyzed in detail (spectra displayed) [8]. The data (large number of polarized Raman lines) rule out a planar configuration and are consistent with a twisted C_2 molecular structure. Table 41 gives a selection of bands from a total of 39 fundamental modes which were assigned. The effect of hindered pseudo-rotation could be observed in the far-IR spectrum. The barriers to pseudorotation were calculated to be 5.9 ± 0.1 kcal/mol for No. 1 and 6.0 ± 0.3 kcal/mol for the deuterated compound [8].

MM2 force field calculations revealed the half-chair conformation to be more stable than the envelope by 3.32 kcal/mol and more stable than the planar form by 4.34 kcal/mol. The calculated moments of inertia were in good agreement with the experimental values [15]. MM2 calculations using other parameters [19] than in [15] indicated that the symmetric twist form is ca. 2 kcal/mol more stable than the envelope.

The mass spectrum (displayed) showed $[M]^+$ and the fragments $[M-1]^+$, $[M-2]^+$, $[M-C_2H_4]^+$, and $[Ge]^+$ as the most intense peaks [2, 6]. At low electron energy (10 eV) the $[M]^+$ ion was entirely suppressed [6]. In the spectrum of $\overline{GeCH_2CH_2CH_2C}H_2D_2$, $[M-1]^+$ almost completely became an $[M-2]^+$ fragment, and the original $[M-2]^+$ ion was divided evenly between $[M-2]^+$, $[M-3]^+$, and $[M-4]^+$. Another prominent ion was $[GeH]^+$ [2].

$\overline{GeCH_2CH_2CH_2CH_2C}H_2H_2$ (Table **40**, No. **3**). The molecular structure was investigated by electron diffraction. The chair conformation reproduces the experimental data very well, yielding r(C-H) = 1.108(6), r(Ge-H) = 1.530(32), r(C-C) = 1.535(3) (average), and r(Ge-C) = 1.956(4) Å and the angles C(2)-C(3)-C(4) = 114.4(38)° and Ge-C(2)-C(3) = 110.9(30)°. The chair at the Ge end is less puckered than the carbon end of the molecule [21]. The difference in the steric energy between the chair and a twist boat form is 3.83 or 5.44 kcal/mol, depending on the parameters used in the MM2 method [17].

Oxidation of No. 3 proceeds stepwise, the end product being compound I [1].

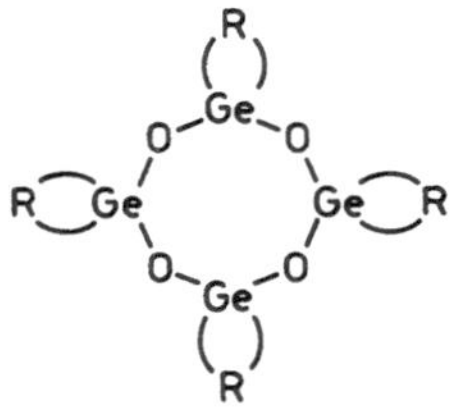

I

$\overline{GeCH_2CH(CH_3)CH_2CH(CH_3)C}H_2H_2$ (Table **40**, No. **8**). The ^{13}C NMR spectrum indicates that the major component of the reaction product was the trans isomer (axial and equatorial CH_3); the minor component was the eq,eq conformer of the cis forms, since the ax,ax conformer has a much higher steric energy [14].

References:

[1] Mazerolles, P. (Bull. Soc. Chim. Fr. **1962** 1907/13).
[2] Duffield, A. M.; Budzikiewicz, H.; Djerassi, C. (J. Am. Chem. Soc. **87** [1965] 2920/6).
[3] Mazerolles, P.; Dubac, J.; Lesbre, M. (J. Organomet. Chem. **5** [1966] 35/47).
[4] Mazerolles, P.; Manuel, G. (Bull. Soc. Chim. Fr. **1966** 327/31).
[5] Curtis, M. D. (J. Am. Chem. Soc. **89** [1967] 4241/2).
[6] Duffield, A. M.; Djerassi, C.; Mazerolles, P.; Dubac, J.; Manuel, G. (J. Organomet. Chem. **12** [1968] 123/32).

[7] Thomas, E. C.; Laurie, V. W. (J. Chem. Phys. **51** [1969] 4327/30).
[8] Durig, J. R.; Willis, J. N. (J. Chem. Phys. **52** [1970] 6108/19).
[9] Nederlandse Centrale Organisatie voor Toegepast-Natuurwetenschappelijk Onderzoek (Neth. Appl. 73-08463 [1973/74] 1/6; C.A. **83** [1975] No. 10402).
[10] Dubac, J.; Mazerolles, P.; Joly, M.; Piau, F. (J. Organomet. Chem. **127** [1977] C69/C74).
[11] Dubac, J.; Mazerolles, P.; Joly, M.; Cavezzan, J. (J. Organomet. Chem. **165** [1979] 163/73).
[12] Takeuchi, Y.; Shimoda, M.; Tomoda, S. (Magn. Reson. Chem. **23** [1985] 580/1).
[13] Takeuchi, Y.; Shimoda, M.; Tanaka, K.; Tomoda, S.; Ogawa, K.; Suzuki, H. (J. Chem. Soc. Perkin Trans. II **1988** 7/13).
[14] Takeuchi, Y.; Ichikawa, Y.; Tanaka, K.; Kakimoto, N. (Bull. Chem. Soc. Jpn. **61** [1988] 2875/80).
[15] Allinger, N. L.; Quinn, M. I.; Chen, K.; Thompson, B.; Frierson, M. R. (J. Mol. Struct. **194** [1989] 1/18).
[16] Boatz, J. A.; Gordon, M. S.; Sita, L. R. (J. Phys. Chem. **94** [1990] 5488/93).
[17] Takeuchi, Y.; Tanaka, K.; Harazono, T.; Yoshimura, S. (Bull. Chem. Soc. Jpn. **63** [1990] 708/15).
[18] Dufour, P.; Dartiguenave, M.; Dartiguenave, Y.; Dubac, J. (J. Organomet. Chem. **384** [1990] 61/9).
[19] Takeuchi, Y.; Tanaka, K.; Harazono, T. (Bull. Chem. Soc. Jpn. **64** [1991] 91/8).
[20] Horner, D. A.; Grev, R. S.; Schaefer, H. F. (J. Am. Chem. Soc. **114** [1992] 2093/8).
[21] Shen, Q.; Rhodes, S.; Takeuchi, Y.; Tanaka, K. (Organometallics **11** [1992] 1752/4).
[22] Takeuchi, Y.; Zismane, I.; Manuel, G.; Boukherroub, R. (Bull. Chem. Soc. Jpn. **66** [1993] 1732/7).

1.3.3 Monoorganogermanium Trihydrides

1.3.3.1 Methylgermane, Methylgermanium Trihydride, $Ge(CH_3)H_3$

Preparation and Formation. $Ge(CH_3)H_3$ was prepared by reducing $Ge(CH_3)X_3$ in xylene with 50% NaH in mineral oil at 110°C (exothermic) in the presence of $Al(CH_3)_2Cl$-$Al(CH_3)Cl_2$ (X = Cl, 100% yield) [24], with LiH in dioxane (X = Br, 50% yield) [7], with $LiAlH_4$ (X = Cl [30, 45, 61] or I [8]) or excess $Li[AlH(OC_4H_9\text{-}t)_3]$ in refluxing dioxane (X = Cl, 60 to 70% yield) [22], or with $NaBH_4$ in aqueous 1 M HBr at 35 to 45°C and ca. 400 Torr (X = Br, ca. 100% yield) [14].

$Ge(CH_3)H_3$ was also obtained by reacting GeH_4 with potassium in HMPT at 10°C in a vacuum system [30] or with KOH powder in DME at −196 to +20°C [32], followed by treatment with CH_3I (94% [30] and 80% yield [32]), or directly from GeH_3Na [2, 9] or GeH_3K [33] (see also [69]) and CH_3I (51% yield [9]) in liquid NH_3 [2] or DME at 20°C (70% yield) [33]. The mechanism of formation of $Ge(CH_3)H_3$ from GeH_3K and a series of halomethanes was studied [37]. $Ge(CH_3)H_3$ was the main product of the reaction of GeH_3Na with CH_2Br_2 in liquid NH_3 [2].

$Li[GeH_3AlH_3]$ in diglyme (prepared from GeH_4 and $LiAlH_4$ at 95°C) reacted with excess CH_3I at 25°C (1 h) to yield 4 to 5% $Ge(CH_3)H_3$ along with larger amounts of GeH_4, Ge_2H_6, and $H_3GeGe(CH_3)H_2$ [81]. The compound was the end-product (>95% yield) of the stepwise cleavage of dihydrofuryl groups in $Ge(CH_3)(\overline{C{=}CHCH_2CH_2O})_3$ on treatment with $LiAlH_4$ in THF or ether at 25°C or in C_6H_6 in the presence of 18-crown-6 at 80°C [91]; see also [90].

The preparation from GeH_3Cl and $LiCH_3$ [51] or by the Grignard reaction of GeH_3Br with CH_3I [52] is briefly mentioned. Usually the reaction mixtures were worked up by trap-to-trap distillation.

Ligand redistribution of $Ge(CH_3)(H)(PH_2)_2$ at room temperature led to small amounts of $Ge(CH_3)H_3$ along with $Ge(CH_3)(H)_2PH_2$ and $Ge(CH_3)(PH_2)_3$, reaching an equilibrium after 2 h (monitored by ^1H NMR) [54]. $Ge(CH_3)H_3$ was a by-product of the intramolecular dehydrocondensation of 6-methyl-6-germacycloundecan-1-ol at 20 °C in the presence of Raney Ni, giving an oxagermabicyclodecane as the main product [46]; cf. "Organogermanium Compounds" 5, 1993, p. 138. Cis-$Fe(CO)_4(Ge(CH_3)H_2)_2$ was slowly converted in the dark into $Ge(CH_3)H_3$ and $(Fe(CO)_4Ge(CH_3)H)_2$ [62].

$Ge(CH_3)D_3$ was prepared from $Ge(CH_3)X_3$ (X = Cl [45, 69, 89] or I [8]) and $LiAlD_4$ in $O(C_4H_9)_2$ [89] or LiD in dioxane (X = Br, 52% yield) [7], from GeD_3K and CH_3I in DME [38], or from GeD_3Br and CH_3MgBr [52].

$Ge(CH_3)(H)D_2$ was studied as an impurity species in $Ge(CH_3)D_3$ prepared from GeD_4 and CH_3I in a DMSO-KOH slurry. A sample prepared from $Ge(H)D_3$ contained in excess of 90% $Ge(CH_3)(H)D_2$ [76].

$Ge(CD_3)H_3$ was obtained from GeH_3Br and CD_3MgBr in $O(C_4H_9)_2$ [52, 89], from $Ge(CD_3)Br_3$ and aqueous $NaBH_4$ in HBr (ca. 96% yield) [14], from GeH_3Na and CD_3Br in liquid NH_3 [8], or from GeH_3K (GeH_4 and KOH in DME) and CD_3I [64].

$Ge(CHD_2)H_3$ was formed by reacting GeH_4 with CHD_2I in a DMSO-KOH slurry [76].

$Ge(CH_2D)H_3$ was obtained from GeH_3Br and CH_2DMgBr in $O(C_4H_9)_2$ [89] or from GeH_3Na and CH_2DBr in liquid NH_3 [8].

$Ge(^{13}CH_3)H_3$ was prepared from GeH_3Br and $^{13}CH_3MgBr$ in $O(C_4H_9)_2$ [89] or from GeH_3Na and $^{13}CH_3I$ in liquid NH_3 [8].

The heat of formation and the ionization potential derived from AM1 calculations are ΔH_f = 15.2 kcal/mol and IP = 11.05 eV [87]. $\Delta H_f^\circ = 4.4 \pm 2.0$ kcal/mol was estimated by an empirical equation [86].

The Molecule, Spectra, and Physical Properties. The following structural parameters were derived from the rotational constants of the microwave spectrum (involving 28 isotopic species): r(Ge-C) 1.9453 ± 0.0005, r(Ge-H) 1.529 ± 0.005, r(C-H) 1.083 ± 0.005 Å; <(H-Ge-H) = $109°15' \pm 30'$, <(H-C-H) = $108°25' \pm 30'$ [8] (r(Ge-C) = 1.946 ± 0.001 Å in [5]). The microwave results [8] together with data derived from the rotational structure of perpendicular-type bands in the IR spectra of $Ge(CHD_2)H_3$ and $Ge(CH_3)(H)D_2$ were used to obtain the ground-state geometry r(Ge-C) = 1.949_0, r(Ge-H) = 1.528_5, r(C-H) = 1.092_1 Å, <(H-Ge-H) = 108.7_8°, and <(H-C-H) = 108.8_4° [76].

The barrier to internal rotation about the Ge-C bond was determined from the splitting of the K = 1 transitions of the asymmetric rotors $Ge(CH_2D)H_3$, $Ge(CH_3)H_2D$, and $Ge(CH_3)(H)D_2$: 1.239 ± 0.025 kcal/mol [8] (585 cm^{-1} in [5]; see also [6]). An average value of 1.270 kcal/mol was derived from IR combination bands involving the torsional modes [16]. An analytic expression for the barrier height as a function of the experimental torsional frequencies and the geometry (using the self-consistent harmonic approximation) was tested on $Ge(CH_3)H_3$ and several other one-top molecules [56] (improved calculations in [57]). Barriers to internal rotation of $E(CH_3)H_3$ compounds (E = C, Si, Ge, Sn) were also computed using the pseudopotential theory and were compared with conventional self-consistent field results and experimental values [58, 59]. $Ge(CH_3)H_3$ was part of ab initio pseudopotential and all-electron calculations of barriers of all $H_3EE'H_3$ compounds (E,E' = C, Si, Ge, Sn, Pb). Stabilizing vicinal σ-EH → σ*-E′H and σ-E′H → σ*-EH interactions in the staggered conformation are responsible for the rotational barriers of the complete series [96]. Other earlier discussions of the barrier problem including $Ge(CH_3)H_3$ dealt with a semiempirical electrostatic model developed from the integral Hellmann-Feynman theorem [28], a comparison of rotational barriers in molecules with one and two CH_3 groups and the kind of nonbonded intramolecular forces [42] (cf. $Ge(CH_3)_2H_2$, p. 217), and comments on the effect of nonbonded electrons on the barrier height [29].

Molecular mechanics calculations (MM2 force field) gave a barrier (1.24 kcal/mol) consistent with the experimental value. Good agreement between the calculated staggered conformation of methylgermanes and the observed structure was obtained. The eclipsed forms show little structural change because of the long Ge-C bond [88]. Force field calculations were carried out to determine the conformational equilibria of several Group-IV organometallic compounds including $Ge(CH_3)H_3$ (staggered and eclipsed conformations) [43]. Structural parameters of $Ge(CH_3)H_3$ were also calculated by the ab initio LCAO-MO-SCF method yielding the staggered form as the most stable conformation [82]. Other ab initio calculations were used for obtaining the molecular geometry and the IR spectrum [100]. FSGO model potentials were given for Group IV atomic cores and applied to $Ge(CH_3)H_3$ and other Group-IV molecules [97]. For a comparison of observed geometrical parameters with those derived from AM1 calculations, see [87]. The valence electronic structure of $E(CH_3)H_3$ molecules (E = C, Si, Ge) was determined using the pseudopotential method within the self-consistent field approximation [59].

Within the generalization of an empirical model for bond dissociation energies, a value of D(Ge-C) = 82.6 kcal/mol was reported for $Ge(CH_3)H_3$ [101].

The analysis of the Stark effect of several isotopic species gave the dipole moment μ = 0.67 (±1.5%) [5], 0.635±0.006 [8], and 0.644±0.005 D [13]; μ = 0.64 D (at 25°C) resulted from measurements of the dielectric constant of the gas at different pressures [22]. The calculated value of 0.55 D is based on a Ge-H bond moment of 0 and a Ge-C bond moment of 0.55 D [88]; μ = 0.6363 D for the staggered conformation was obtained from LCAO-MO-SCF calculations [82]. For a dipole moment from AM1 calculations, see [87].

^1H NMR spectrum: δ(ppm) = 0.29 (CH_3, ^3J(H, H) = 4.22±0.03 Hz), 3.45 (HGe) in C_6H_{12} [18]; 0.35 (q, CH_3), 3.49 (q, HGe), ^3J(H, H) = 4.33 Hz, ^1J(C, H) = 129.0 Hz in CCl_4 [19] (see also [55]); −0.18 (q, CH_3), 3.00 (q, HGe), ^3J(H, H) = 4.2 Hz, ^1J(C, H) = 127.8 Hz for the neat liquid [20, 22]. Correlations between ^1H NMR data of $Ge(CH_3)_nH_{4-n}$ compounds and the degree of substitution [19], the bond length d(H-C-Ge-H) [19], the sum of the Taft σ* constants of the substituents [15, 21], and the ν(Ge-H) frequency [15] were reported.

The ^1H and ^1D NMR spectra of $Ge(CD_3)H_3$ and $Ge(CH_3)D_3$ dissolved under pressure in the nematic phase of $4\text{-}C_2H_5OC_6H_4CH{=}NC_6H_4C_4H_9\text{-}4$ were used to calculate several molecular parameters (spectra also displayed) [52]. The relative acidities of some hydrides of Groups IV and V were derived from the ^1H NMR spectra of these compounds in liquid ammonia (at 31°C) including $Ge(CH_3)H_3$: δ(ppm) = 0.19 (CH_3) and 3.22 (HGe); ^3J(H,H) = 4.22 Hz [26].

^{13}C NMR spectrum: δ = −1.5 ppm [84].

^{73}Ge NMR spectrum (in $O(C_4H_9)_2$): δ = −209.2±0.1 ppm [77, 78, 84]; ^1J(Ge,H) = 94.5±0.5 and ^2J(Ge,H) = 3.48±0.1 Hz [78, 84]. The variation of the ^{73}Ge chemical shift or the coupling constant with the number of hydrogen atoms in $Ge(CH_3)_nH_{4-n}$ compounds (n = 0 to 4) was studied. The ^{73}Ge nuclear quadrupole coupling constant was derived from the relaxation time measured at 27.0°C to be 4.5 MHz (rough estimate) [84]. A close correlation

References on p. 274

was found between the chemical shifts of ^{73}Ge and ^{29}Si (29 species) or ^{119}Sn (26 species) [77]. The application of proton polarization transfer to a high-spin nucleus such as ^{73}Ge was investigated for several Ge compounds including $Ge(CH_3)H_3$ [78]. A theoretical study on germanium chemical shifts of $Ge(CH_3)_nH_{4-n}$ compounds (n = 0 to 4) was recently published [105].

The fundamental vibrations of $Ge(CH_3)H_3$ (see Table 42) and $Ge(CD_3)H_3$ were obtained from gas-phase IR and liquid-phase Raman spectra and were assigned and compared [73] with previous IR [16, 23, 64] and Raman [23] data. IR spectra were also recorded for polycrystalline $Ge(CH_3)D_3$ and $Ge(CD_3)H_3$ at 77 K and for $Ge(CH_3)H_3$ and $Ge(CD_3)H_3$ in argon matrices at 8 K [64]. Fundamental vibrations of $Ge(CHD_2)H_3$ from IR (gas and polycrystalline) and Raman spectra (liquid) were listed and assigned (some bands illustrated at 0.06 cm^{-1} resolution) [76], and the IR spectrum of $Ge(CH_3)(H)D_2$ (impurity in $Ge(CH_3)D_3$) was briefly described [76] (IR spectra of $Ge(CH_3)H_3$ [16, 23], $Ge(CH_3)D_3$, and $Ge(CD_3)H_3$ [16] and the Raman spectrum of $Ge(CH_3)H_3$ [23] displayed).

A normal coordinate analysis for $Ge(CH_3)H_3$, $Ge(CD_3)H_3$, $Ge(CH_3)D_3$, and $Ge(CD_3)D_3$ yielded the fundamental frequencies, Coriolis-coupling coefficients, and ground-state centrifugal-distortion constants [27] which were compared with experimental data from [16] (not available for $Ge(CD_3)D_3$); see also [31] for a report on other calculations of frequencies and Coriolis constants and [41] for the fundamental vibrations of the same species resulting from calculations with the modified Urey-Bradley force field. A normal coordinate analysis for $Ge(CH_3)H_3$ adopting the hybrid orbital force field gave the force constants f(Ge-C) = 2.700, f(Ge-H) = 2.570, and f(C-H) = 4.873 mdyn/Å [25]; for force constants obtained from Urey-Bradley force field calculations, see [41]. Force field calculations for $Ge(R)(H,D)_3$ compounds (R = CH_3, C≡CH, C≡N) were briefly discussed in an abstract [47].

The analysis of the rotational structures of some IR bands gave values of rotational and Coriolis coupling constants [16]. Q-branch maxima from the five naturally occurring Ge iso-

Table 42
Vibrational Spectra of $Ge(CH_3)H_3$; Fundamental Vibrations.
Wavenumbers in cm^{-1}.

IR gas [73]	polycrystalline [a)] [64]	Raman liquid [73]	assignment	
2998.7	2987.9	2991	$\nu_{as}(CH_3)$	ν_7
2939.0	2920.3	2931	$\nu_s(CH_3)$	ν_1
2086.4	2064.1	2075	$\nu_s(GeH)$	ν_2
2084.7	2069.8	2075	$\nu_{as}(GeH)$	ν_8
b)	1422.8	1426	$\delta_{as}(CH_3)$	ν_9
1254.9	1249.1	1249	$\delta_s(CH_3)$	ν_3
900.4	898.6, 895.9	896	$\delta_{as}(CH_3)$	ν_{10}
c)	823.0	c)	$\rho(CH_3)$	ν_{11}
843.3	830.3(br), 820.3	840	$\delta_s(GeH_3)$	ν_4
603.9	605.3	598	$\nu(GeC)$	ν_5
505.8	507.3, 499.8	510	$\rho(GeH_3)$	ν_{12}

a) Recorded at 77 K. – b) $\delta_{as}(CH_3)$ is suspected to be in Fermi resonance with at least $\nu_{10} + \nu_{12}$. – c) $\rho(CH_3)$ lies under the very intense $\delta_s(GeH_3)$.

topes have been resolved in the gas-phase IR spectra of $Ge(CH_3)H_3$ and $Ge(CD_3)H_3$ yielding new values of rotational and Coriolis coupling constants for several fundamental vibrations [73].

The isolated ν_{is}(GeH) frequencies in the IR spectra of alkyl germanes correlate well with the sum of Taft σ^* coefficients of the ligands at Ge, provided an appropriate treatment of the conformational effects is carried out [75]. A linear relation was also found between the ν(GeH) or ν(GeD) frequency in the Raman spectra and the inductive effect of the substituents at Ge for a series of germanes including $Ge(CH_3)H_3$ and $Ge(CH_3)D_3$ [12].

Isolated CH stretching frequencies obtained from the IR spectra of partially deuterated methyl germanes including $Ge(CHD_2)H_3$ (gas phase spectrum illustrated) were used to predict CH bond lengths, HCH angles, and CH dissociation energies [74]. Treating the GeH bond as a diatomic molecule, the dissociation energy could be deduced from the fundamental and first overtone GeH stretching bands, giving D° = 76.5 kcal/mol (at 0 K) for $Ge(CH_3)HD_2$ [70]. Studies on a series of related compounds indicated that methyl substitution weakens the E-H bond [70]; see also [80]. A linear relation between r_0(Ge-H) and ν_{is}(GeH) was pointed out for GeH_4, $Ge(CH_3)H_3$, GeH_3F, and $Ge(CH_3)_3H$ [76, 80].

The isotopically isolated stretching frequencies of $Ge(CH_3)H_2D$ and $Ge(CH_3)(H)D_2$ in the $Ge(CH_3)H_3$ and $Ge(CH_3)D_3$ matrices indicate that the crystalline four compounds have C_s symmetry [64].

The symmetrical EH_3 deformation frequency of EH_3X compounds (E = Si, Ge) increases with the electronegativity of X (X including CH_3, halogen, and various other groups) which may be explained in terms of repulsions between the E-H bonding electrons and the electrons in the valence orbitals of the X atom [17]. Hartree-Fock-Roothaan calculations for CH_3 group vibrations of $Ge(CH_3)H_3$ and $Ti(CH_3)H_3$ were mentioned in a study dealing with a possible H-Ti interaction in $Ti(CH_3)Cl_3$ [85].

The He(I) and He(II) photoelectron spectra of $Ge(CH_3)H_3$ gave the following vertical ionization energies (orbital symmetry and principal contributions to the molecular orbitals): 11.0 ($3a_1$, GeC), 11.50 (2e, GeH_3), 13.90 (1e, CH_3), and 17.60 eV ($2a_1$, Ge s orbital). The assignments are based on a comparison with spectra of GeH_4 and $Ge(CH_3)_4$ and semiempirical CNDO/2 calculations. Features corresponding to the carbon 2s orbital are apparent at 21 to 22 eV. The spectra were displayed and compared with those of $Ge(CH_3)F_3$ and $Ge(CH_3)Cl_3$, and a linear correlation between the ionization energies and the CNDO/2 eigenvalues was demonstrated [65].

The core-electron binding energies from the X-ray photoelectron spectrum are 36.98 eV for Ge 3d (relative to Ne 2s at 48.47 eV) [63], 128.78 eV for Ge $3p_{3/2}$ relative to Ar $2p_{3/2}$ at 248.45 eV) [51], and 290.26 eV for C 1s (relative to Ar $2p_{3/2}$ at 248.63 eV) [51, 63]. A comparison of calculated and experimental energies for the Ge 3d and C 1s levels in the $Ge(CH_3)_nH_{4-n}$ series (n = 0 to 3) is based on CNDO/2 calculations and the electronegativity-equalization method involving s-orbital participation. The binding energies increase as the CH_3 groups are replaced by H atoms [63]. The energies of analogous C, Si, and Ge compounds were correlated by the electrostatic potential equation using charge distributions from extented Huckel theory, CNDO/2, and an electronegativity-equalization method. The data could be rationalized without considering any π(p-d) bonding in the Si and Ge compounds [51]; see also [44]. Relaxation energy differences (relative to GeH_4) were estimated from Auger and core binding energies of several Ge compounds including $Ge(CH_3)H_3$ [48]. Studies on the Ge 3 $p_{3/2}$ level by the semiempirical SCC-MO method indicate that the relaxation energy plays a decisive role in determining shifts along the series $GeH_4 \rightarrow Ge(CH_3)H_3 \rightarrow Ge(CH_3)_4$ [67]. A report on a linear four-parameter equation for predicting core-electron

References on p. 274

binding energies also mentioned $Ge(CH_3)H_3$ along with various types of organic compounds [60].

Gaseous $Ge(CH_3)H_3$ has a pungent odor characteristic of volatile Ge compounds [1]. Melting point: -153.7 ± 0.3 °C [9], -154.5 ± 0.2 °C [14], ca. −158 °C [2, 11]. Boiling point: −23.5 °C/745 Torr [7]; see also [2, 10]. $Ge(CH_3)D_3$ boils at −23.5 °C/752 Torr [7].

The vapor pressure of $Ge(CH_3)H_3$ can be expressed by the following equations:

log (p/Torr) = 3.9624 − 0.003034 T + 1.75 log T − 1080.3/T; this yields the b.p. −34.1 °C, ΔH_v = 4981 cal/mol, and the ΔS_v = 20.8 cal · mol^{-1} · K^{-1} [14] (see also [32]);

log (p/Torr) = 7.578 − 1118.2/T (−109 to −76 °C range) gives the b.p. −35.1 °C and ΔH_v = 5114 cal/mol [11].

Other boiling points reported are close to −23 °C [2, 7, 9] and obviously due to impurities [14]. The vapor pressure data listed below were obtained on a sample contaminated with 0.1% $Ge(CH_3)_2H_2$ [14]:

t in °C	−111.8	−95.0	−78.5	−63.8	−53.2	−45.6
p in Torr	4.25	19.55	67.7	169.4	303.5	446.7

For the application of Egloff's boiling point equation [1], see [4].

Molar thermodynamic properties such as heat content, free energy, entropy, and heat capacity were evaluated in the range 100 to 1200 K and with a rigid rotor, harmonic oscillator approximation for the ideal gaseous state at one atmosphere pressure [25]. For a comparison of thermodynamic properties of $GeRH_3$ compounds (R = CH_3, C_2H_5, C_3H_7), see [53].

Mass Spectrum and Chemical Behavior. The mass spectrum consisted of peaks due to the ions $[Ge(C)H_{6-n}]^+$, $[GeH_{3-n}]^+$, and $[CH_{3-n}]^+$ [34]. The relative abundances of significant ions in the ion trap mass spectra of $Ge(CH_3)H_3$ (4.0×10^{-7} Torr) at 60 °C and 7.0×10^{-5} Torr (total) He pressure at reaction times ranging from 0 to 100 ms are listed in a table. The most important ions are $[GeH_n]^+$ (n = 0 to 3), $[GeCH_n]^+$ (n = 1 to 5), $[GeC_2H_n]^+$ (n = 6, 7), $[Ge_2CH_n]^+$ (n = 4, 5), $[Ge_2C_2H_n]^+$ (n = 6, 7), and $[Ge_3C_3H_n]^+$ (n = 9, 12). $[GeCH_5]^+$ was the most abundant ion at any time studied [102].

The thermodynamic stability of $M(CH_3)H_3$ compounds with respect to loss of CH_4 was calculated by ab initio methods and was found to decrease in the series M = Si > Ge > Pb [100]. The geometry of the $[Ge(CH_3)H_4]^+$ cation was also calculated by the ab initio LCAO-MO-SCF method. This yielded a proton affinity of 7.34 eV for $Ge(CH_3)H_3$ which is little larger than that of GeH_4 [82]. The gas-phase acidity of $Ge(CH_3)H_3$ was found to be smaller than that of GeH_4 by ca. 35 kJ/mol; this resulted from proton transfer reactions to various reference acids monitored by Fourier transform ion cyclotron resonance spectrometry [104].

The pyrolysis of $Ge(CH_3)H_3$ at 420 °C for 24 h in a flow system yielded H_2, CH_4, and $Ge(CH_3)_2H_2$ as the major products and small amounts of $Ge(CH_3)_3H$, $Ge_2(CH_3)_2H_4$; 65% of the germanium was deposited in polymeric solids (pyrolysis compared with that of $Si(CH_3)H_3$ at 520 °C) [34]. The homogeneous gas-phase decomposition was studied by the comparative rate method (cyclopropane as the comparative standard) using a single-pulse shock tube at 3100 Torr total pressure between 1050 and 1250 K. It indicated three primary processes: $Ge(CH_3)H_3 \rightarrow Ge(CH_3)H + H_2$, $Ge(CH_3)H_3 \rightarrow GeH_2 + CH_4$, and $Ge(CH_3)D_3 \rightarrow CH_2{=}GeD_2$ + HD. The overall decomposition rate is expressed by log (k/s^{-1}) = 13.34 − (50420 ± 3700) cal/Θ and it comprises ca. 40% of the first reaction and ca. 30% each of the second and third reactions. Rate constants for the primary processes were obtained by RRKM calculations [69].

Absolute rate constants of the reaction of H atoms (formed by Hg photosensitization of H_2-substrate mixtures at 32 °C) with $Ge(CH_3)_nH_{4-n}$ compounds (n = 1 to 3) were determined by experiments involving competitive reaction of H atoms with Si_2H_6. The Arrhenius parameters $A = 2.0 \times 10^{-11}$ cm^3/s and $E_a = 1.1$ kcal/mol were calculated for $Ge(CH_3)H_3$, and the dissociation energy D(Ge-H) = 83 kcal/mol estimated by the BEBO method. The activation energy falls off slowly as the Ge-H bonds in GeH_4 are successively replaced by $Ge\text{-}CH_3$ bonds [61]; cf. $Ge(CH_3)_3H$, p. 7, and $Ge(CH_3)_2H_2$, p. 220.

Gas-phase ion-molecule reactions in the systems $Ge(CH_3)H_3\text{-}O_2$ and $Ge(CH_3)H_3\text{-}NH_3$ were studied by Fourier transform and high-pressure mass spectrometry. The self-condensation processes in pure $Ge(CH_3)H_3$ and the effects of the reagent concentration and total pressure on the formation of Ge-O and Ge-N bonds were investigated. The results suggest a very low reactivity of $Ge(CH_3)H_3$ towards oxygen. Many N-containing ions were detected in the $Ge(CH_3)H_3\text{-}NH_3$ system ($[GeNH_4]^+$ most abundant) under the conditions of chemical ionization (at 0.5 Torr). The reactivity of $Ge(CH_3)H_3$ was compared with that of GeH_4 [98].

$Ge(CH_3)H_3$ reacted with Br_2 (ca. 1 : 0.7 mole ratio) at −196 to +20 °C to give $Ge(CH_3)H_2Br$, $Ge(CH_3)(H)Br_2$, and $Ge(CH_3)Br_3$ in a 10 : 3 : 1 ratio (based on the ^1H NMR spectrum) [35]; see also [39, 89]. A similar treatment with I_2 at −78 °C (1 h) yielded $Ge(CH_3)H_2I$ and traces of $Ge(CH_3)(H)I_2$ and $Ge(CH_3)I_3$ [35]; see also [40]. $Ge(CH_3)(D)I_2$ and $Ge(CH_3)I_3$ were obtained from $Ge(CH_3)D_3$ and I_2 (ca. 1 : 2 mole ratio) at −196 to +20 °C (1 h) [38].

Excess $Ge(CH_3)H_3$ reacted with Na in liquid NH_3 at −33 °C to liberate H_2 and to form a viscous, yellow liquid at room temperature after evaporation of NH_3. H_2 was also produced upon treatment with Li in $NH_2C_2H_5$. In both cases, $Ge(CH_3)H_3$ was regenerated on addition of NH_4Br to the reaction mixtures [3].

$Ge(CH_3)H_3$ reacted with HCl in the gas phase at 20 °C (36 h) [11, 35], with HBr at −78 °C (1 h) [35], or with HI (all 1 : 1 mole ratio) at 20 °C (15 min) [35] (see also [68]) in the presence of AlX_3 (X = Cl, Br, or I, respectively) to give $Ge(CH_3)H_2X$, $Ge(CH_3)(H)X_2$, and small amounts of $Ge(CH_3)X_3$. According to [9] the reaction of $Ge(CH_3)H_3$ with HCl (1.3 : 1 mole ratio) at 100 °C (4 h) in the presence of $AlCl_3$ generated $Ge(CH_3)(H)Cl_2$, whereas the corresponding reaction with HBr (1 : 1 mole ratio) in the presence of $AlBr_3$ gave $Ge(CH_3)H_2Br$. $Ge(CH_3)H_2X$ (X = Cl or Br) was also obtained by reacting $Ge(CH_3)H_3$ with excess HX in a sealed tube at several atmospheres pressure or below 1 atm in the presence of traces of Hg [55].

Ion trap mass spectroscopy was used to identify the ionic species formed in $Ge(CH_3)H_3\text{-}SiH_4$ mixtures and to determine their relative abundances as a function of the reaction time. The results were compared with those of $Si(CH_3)H_3$ and of a $GeH_4\text{-}SiH_4$ mixture for their relevance in photovoltaic technology [103].

The relative rate of GeH_2 insertion into the Ge-H bonds on a per bond basis was determined by competitive reactions of GeH_2 with Ge_2H_6 and methylgermanes at 280 °C giving the rate order $Ge(CH_3)_3H > Ge(CH_3)_2H_2 > Ge_2H_6 > Ge(CH_3)H_3$. $H_3GeGe(CH_3)H_2$ resulted from insertion into $Ge(CH_3)H_3$. Relations between the relative rates and the bond energy and negative charge of the H atom in the bond undergoing insertion were discussed. Pyrolysis of Ge_2H_6 in the presence of $Ge(CH_3)D_3$ at 295 °C was assumed to produce $DH_2GeGe(CH_3)D_2$ which decomposes to give GeH_2D_2 among other products [45]. The pyrolysis of $Ge(CH_3)H_3$ and Si_2H_6 at 350 °C (to give $Ge(CH_3)H_2SiH_3$) was also discussed in the context of relative insertion rates of SiH_2 into several methylsilanes [49].

All $Ge(CH_3)H_nX_{3-n}$ compounds were formed from $Ge(CH_3)H_3$ and BX_3 (X = Cl, Br, I) at −78 °C (1 h); increasing amounts of BX_3 gave markedly improved yields of $Ge(CH_3)X_3$ [35]. The reaction of $Ge(CH_3)D_3$ with BX_3 (X = Cl, Br) to give $Ge(CH_3)(D)X_2$ was also mentioned [38].

References on p. 274

$Ge(CH_3)H_3$ and CCl_4 (1:15 mole ratio) yielded after 4 h $Ge(CH_3)Cl_3$ as the main product along with $Ge(CH_3)H_2Cl$ and $Ge(CH_3)(H)Cl_2$. In contrast, only $Ge(CH_3)H_2Cl$ was detected with ca. 2% yield (by ^1H NMR) after three weeks in $GeCl_4$ [55]. Treatment with $SnCl_4$ (1:1 mole ratio) in $O(C_4H_9)_2$ at room temperature (5 to 6 h) yielded 60% $Ge(CH_3)H_2Cl$ [36]. $Ge(CH_3)H_3$ did not react with excess PCl_3 and showed only ca. 5% conversion into $Ge(CH_3)H_2Br$ with PBr_3 after three months [55]. Passing $Ge(CH_3)H_3$ over AgCl at 90°C or over AgBr at 125°C gave only the $Ge(CH_3)H_2X$ compounds [40].

Gas-phase ion-molecule reactions were studied by Fourier-transform mass spectrometry and high-pressure mass spectrometry (0.01- to 0.5-Torr range) on mixtures of $Ge(CH_3)H_3$ and unsaturated hydrocarbons such as $CH_2{=}CH_2$, $CH_3CH{=}CH_2$, $CH_2{=}C{=}CH_2$, and $CH_3C{\equiv}CH$. The formation of ionic species with new Ge-C bonds was discussed in conjunction with the formation of amorphous Ge carbides for photovoltaic applications [98].

The reaction of excess $Ge(CH_3)H_3$ with $Fe_3(CO)_{12}$ in C_5H_{12} at 50°C (24 h) in an evacuated, sealed tube yielded the trigonal-bipyramidal cluster $Fe_3(CO)_9(\mu_3\text{-}Ge(CH_3)_2)$ together with complexes tentatively identified as $Fe_2(CO)_7(\mu\text{-}Ge(CH_3)H)_2$ and $Fe_2(CO)_6(\mu\text{-}Ge(CH_3)H)_3$. Heating $Ge(CH_3)H_3$ with excess $Fe(CO)_5$ in petroleum ether at 135°C (3 h) also yielded the cluster along with $Ge(Fe_2(CO)_8)_2$ [92].

$Ge(CH_3)H_3$ reacted with $Co_2(CO)_8$ (1:1 mole ratio) in C_6H_{14} at 10 to 20°C (12 h) to yield $Co_2(CO)_7(\mu\text{-}Ge(CH_3)Co(CO)_4)$. The same product in addition to $Co_4(CO)_{11}(\mu_4\text{-}Ge(CH_3)_2)$ was formed using a 1:2 ratio (in a sealed tube at 20°C for 6 months), whereas $Co_2(CO)_6(\mu\text{-}Ge(CH_3)Co(CO)_4)_2$ and $Co_2(CO)_n(Ge(CH_3)H)$ (n = 7 or 8) were obtained using 2:1 ratio [94]; see also [66]. Excess $Ge(CH_3)H_3$ reacted with $Co_4(CO)_{12}$ in C_6H_{14} at 20°C (13 d) or at 30°C (7 d) in a sealed tube to give $Co_4(CO)_{11}(\mu_4\text{-}GeCH_3)_2$ with a high yield [95]. According to [71] the same product also resulted from $Ge(CH_3)H_3$ and $\mu_4\text{-}Ge(Co_2(CO)_7)_2$ (1:1 mole ratio) after 6 months in C_6H_{14} in a sealed tube, but later studies identified $\mu_4\text{-}Ge(Co_2(CO)_6(\mu\text{-}Ge(CH_3)H))(Co_2(CO)_7)$ as the product [99]. At 35 to 40°C, $\mu_4\text{-}Ge(Co(CO)_4)Co_3(CO)_9$ was obtained in addition, whereas a large excess of $Ge(CH_3)H_3$ gave only $\mu_4\text{-}Ge(Co_2(\mu\text{-}Ge(CH_3)H)(CO)_6)_2$. Treating $Ge(CH_3)H_3$ with $\mu_4\text{-}Ge(Co(CO)_4)Co_3(CO)_9$ (3:1 mole ratio) in C_6H_{14} at 20°C (20 weeks) in a sealed tube yielded $\mu_4\text{-}Ge(Co_2(CO)_6(\mu\text{-}Ge(CH_3)H))(Co_2(CO)_7)$ [99].

$Ge(CH_3)H_3$ reacted with trans-$Rh(CO)(P(C_2H_5)_3)_2X$ (X = Cl or I) at −75°C to give the octahedral complexes $Rh(CO)(P(C_2H_5)_3)_2(Ge(CH_3)H_2)(H)X$ which dissociate reversibly. Thermodynamic parameters for the dissociation were determined from NMR spectra [72].

The equilibrium distribution of microimpurities of $Ge(CH_3)H_3$, hydrocarbons, H_2S, and AsH_3 in GeH_4 between the liquid and vapor was investigated by the radioactive tracer method (^{14}C, ^{35}S, ^{76}As isotopes) [50].

Methylgermanes including $Ge(CH_3)H_3$ were detected in natural waters at the parts-per-trillion level via hydride generation, graphite furnace atomization, and atomic absorption spectrometry. The major Ge species in seawater was found to be $Ge(CH_3)H_3$ [79].

$Ge(CH_3)H_3$ was mentioned in connection with the catalytic properties of a metal-containing acidic zeolite used for hydrocarbon conversion processes. Treating the zeolite with a volatile organometallic compound improved the catalyst by converting the metal to an intermetallic compound [83, 93].

References:

[1] Egloff, G.; Sherman, J.; Dull, R. B. (J. Phys. Chem. **44** [1940] 730/45).
[2] Teal, G. K.; Kraus, C. A. (J. Am. Chem. Soc. **72** [1950] 4706/9).

[3] Glarum, S. N.; Kraus, C. A. (J. Am. Chem. Soc. **72** [1950] 5398/401).
[4] English, W. D. (J. Am. Chem. Soc. **74** [1952] 2927/8).
[5] Barchukov, A. I.; Prokhorov, A. M. (Opt. Spektrosk. **4** [1958] 799; C.A. **1958** 16875).
[6] Barchukov, A. I.; Prokhorov, A. M. (Opt. Spektrosk. **5** [1958] 530/4).
[7] Ponomarenko, V. A.; Vzenkova, G. Ya.; Egorov, Yu. P. (Dokl. Akad. Nauk SSSR **122** [1958] 405/8; C.A. **1959** 112).
[8] Laurie, V. W. (J. Chem. Phys. **30** [1959] 1210/4).
[9] Griffiths, J. E.; Onyszchuk, M. (Can. J. Chem. **39** [1961] 339/47).
[10] Satgé, J. (Ann. Chim. [Paris] [13] **6** [1961] 519/73).

[11] Amberger, E.; Boeters, H. (Angew. Chem. **73** [1961] 114).
[12] Ponomarenko, V. A.; Zueva, G. Ya.; Andreev, N. S. (Izv. Akad. Nauk SSSR Ser. Khim. **1961** 1758/62; Bull. Acad. Sci. USSR Div. Chem. Sci. [Engl. Transl.] **1961** 1639/43).
[13] Barchukov, A. I.; Petrov, Yu. N. (Opt. Spektrosk. **11** [1961] 129; Opt. Spectrosc. [Engl. Transl.] **11** [1961] 67).
[14] Griffiths, J. E. (Inorg. Chem. **2** [1963] 375/7).
[15] Egorochkin, A. N.; Khidekel', M. L.; Ponomarenko, V. A.; Zueva, G. Ya.; Svirezheva, S. S.; Razuvaev, G. A. (Izv. Akad. Nauk SSSR Ser. Khim. **1963** 1865/8; Bull. Acad. Sci. USSR Div. Chem. Sci. [Engl. Transl.] **1963** 1717/9).
[16] Griffiths, J. E. (J. Chem. Phys. **38** [1963] 2879/91).
[17] Jolly, W. L. (J. Am. Chem. Soc. **85** [1963] 3083/5).
[18] Ebsworth, E. A. V.; Frankiss, S. G.; Robiette, A. G. (J. Mol. Spectrosc. **12** [1964] 299/300).
[19] Schmidbaur, H. (Chem. Ber. **97** [1964] 1639/48).
[20] Van der Kelen, G. P.; Verdonck, L.; Van de Vondel, D. (Bull. Soc. Chim. Belg. **73** [1964] 733/40).

[21] Egorochkin, A. N.; Khidekel', M. L.; Ponomarenko, V. A.; Zueva, G. Ya.; Razuvaev, G. A. (Izv. Akad. Nauk SSSR Ser. Khim. **1964** 373/5; Bull. Acad. Sci. USSR Div. Chem. Sci. [Engl. Transl.] **1964** 347/8).
[22] Van de Vondel, D. F. (J. Organomet. Chem. **3** [1965] 400/5).
[23] Van de Vondel, D. F.; Van der Kelen, G. P. (Bull. Soc. Chim. Belg. **74** [1965] 467/78).
[24] Berger, A.; Compagnie Francaise Thomson-Houston (Fr. 1429930 [1966]; C.A. **65** [1966] 18620).
[25] Galasso, V.; Bigotto, A.; de Alti, G. (Z. Phys. Chem. [Munich] **50** [1966] 38/45).
[26] Birchall, T.; Jolly, W. L. (Inorg. Chem. **5** [1966] 2177/80).
[27] Clark, E. A.; Weber, A. (J. Chem. Phys. **45** [1966] 1759/66).
[28] Lowe, J. P.; Parr, R. G. (J. Chem. Phys. **44** [1966] 3001/9).
[29] Harvey, A. B. (J. Phys. Chem. **70** [1966] 3370/1).
[30] Cradock, S.; Gibbon, G. A.; Van Dyke, C. H. (Inorg. Chem. **6** [1967] 1751/2).

[31] Ponomarev, Yu. I.; Kovalev, I. F.; Orlov, V. A. (Opt. Spektrosk. **23** [1967] 483/5; Opt. Spectrosc. [Engl. Transl.] **23** [1967] 258/9).
[32] Rustad, D. S.; Birchall, T.; Jolly, W. L. (Inorg. Synth. **11** [1968] 128/30).
[33] Amberger, E.; Mühlhofer, E. (J. Organomet. Chem. **12** [1968] 55/62).
[34] Kohanek, J. J.; Estacio, P.; Ring, M. A. (Inorg. Chem. **8** [1969] 2516/7).
[35] Drake, J. E.; Hemmings, R. T.; Riddle, C. (J. Chem. Soc. A **1970** 3359/62).
[36] Kuz'min, O. V.; Nametkin, N. S.; Chernysheva, T. I.; Gar, T. K.; Lepetukhina, N. A.; Mironov, V. F. (Dokl. Akad. Nauk SSSR **193** [1970] 826/7; Dokl. Chem. [Engl. Transl.] **190/195** [1970] 550/2).
[37] Dreyfuss, R. M.; Jolly, W. L. (Inorg. Chem. **10** [1971] 2567/71).

[38] Barker, G. K.; Drake, J. E.; Hemmings, R. T.; Rapp, B. (J. Chem. Soc. A **1971** 3291/6).
[39] Van Dyke, C. H.; Bulkowski, J. E.; Viswanathan, N. (Inorg. Nucl. Chem. Lett. **7** [1971] 1057/61).
[40] Bellama, J. M.; McCormick, C. J. (Inorg. Nucl. Chem. Lett. **7** [1971] 533/6).

[41] Ohno, K.; Hayashi, M.; Murata, H. (J. Sci. Hiroshima Univ. A **36** [1972] 121/39).
[42] Ingham, K. C. (J. Phys. Chem. **76** [1972] 551/3).
[43] Ouellette; R. J. (J. Am. Chem. Soc. **94** [1972] 7674/9).
[44] Perry, W. B.; Jolly, W. L. (Chem. Phys. Lett. **17** [1972] 611/3).
[45] Sefcik, M. D.; Ring, M. A. (J. Organomet. Chem. **59** [1973] 167/73).
[46] Mazerolles, P.; Faucher, A. (J. Organomet. Chem. **63** [1973] 195/203).
[47] Yarandina, V. N.; Sverdlov, L. M. (Zh. Fiz. Khim. **47** [1973] 267; Russ. J. Phys. Chem. [Engl. Transl.] **47** [1973] 153).
[48] Perry, W. B.; Jolly, W. L. (Chem. Phys. Lett. **23** [1973] 529/32).
[49] Sefcik, M. D.; Ring, M. A. (J. Am. Chem. Soc. **95** [1973] 5168/73).
[50] Efremov, A. A.; Falaleev, V. A.; Petrik, A. G.; Zel'venskii, Ya. D.; Rostunova, R. P. (Zh. Fiz. Khim. **47** [1973] 1036; Russ. J. Phys. Chem. [Engl. Transl.] **47** [1973] 589).

[51] Perry, W. B.; Jolly, W. L. (Inorg. Chem. **13** [1974] 1211/7).
[52] Ader, R.; Loewenstein, A. (J. Am. Chem. Soc. **96** [1974] 5336/40).
[53] Zaitsev, N. M.; Maslov, P. G.; Ivolgin, V. I. (Zh. Prikl. Khim. [Leningrad] **47** [1974] 2020/3; J. Appl. Chem. USSR [Engl. Transl.] **47** [1974] 2075/8).
[54] Dahl, A. R.; Heil, C.A.; Norman, A. D. (Inorg. Chem. **14** [1975] 1095/8).
[55] Graham, B. W. L.; Mackay, K. M.; Stobart, S. R. (J. Chem. Soc. Dalton Trans. **1975** 475/80).
[56] Dellepiane, G.; Piseri, L. (J. Mol. Spectrosc. **59** [1976] 209/15).
[57] Dellepiane, G.; Piseri, L.; Bosi, P. (Spectrosc. Lett. **9** [1976] 881/4).
[58] Ewig, C. S.; Van Wazer, J. R. (J. Chem. Phys. **65** [1976] 2035/6).
[59] Nicolas, G.; Barthelat, J. C.; Durand, Ph. (J. Am. Chem. Soc. **98** [1976] 1346/50).
[60] Jolly, W. L.; Bakke, A. A. (J. Am. Chem. Soc. **98** [1976] 6500/4).

[61] Austin, E. R.; Lampe, F. W. (J. Phys. Chem. **81** [1977] 1546/9).
[62] Bonny, A.; Mackay, K. M. (J. Chem. Soc. Dalton Trans. **1978** 506/11).
[63] Drake, J. E.; Riddle, C.; Glavincevski, B. M.; Gorzelska, K.; Henderson, H. E. (Inorg. Chem. **17** [1978] 2333/6).
[64] Oxton, I. A. (J. Mol. Struct. **56** [1979] 57/68).
[65] Drake, J. E.; Glavincevski, B. M.; Gorzelska, K. (J. Electron Spectrosc. Relat. Phenom. **17** [1979] 73/80).
[66] Gerlach, R. F.; Graham, B. W. L.; Mackay, K. M. (J. Organomet. Chem. **182** [1979] 285/98).
[67] Maksic, Z. B.; Rupnik, K. (Croat. Chem. Acta **53** [1980] 413/8).
[68] Hasegawa, A.; Uchimura, S.; Hayashi, M. (J. Chem. Soc. Perkin Trans. II **1980** 1690/5).
[69] Dzarnoski, J.; O'Neal, H. E.; Ring, M. A. (J. Am. Chem. Soc. **103** [1981] 5740/6).
[70] McKean, D. C.; Torto, I.; Morrisson, A. R. (J. Phys. Chem. **86** [1982] 307/9).

[71] Foster, S. P.; Mackay, K. M.; Nicholson, B. K. (J. Chem. Soc. Chem. Commun. **1982** 1156/7).
[72] Ebsworth, E. A. V.; de Ojeda, M. R.; Rankin, D. W. H. (J. Chem. Soc. Dalton Trans. **1982** 1513/9).
[73] Mackenzie, M. W. (Spectrochim. Acta A **38** [1982] 1083/7).
[74] McKean, D. C.; Mackenzie, M. W.; Torto, I. (Spectrochim. Acta A **38** [1982] 113/8).

[75] McKean, D. C.; Torto, I.; Mackenzie, M. W. (Spectrochim. Acta A **39** [1983] 399/408).
[76] McKean, D. C.; Mackenzie, M. W.; Morrisson, A. R. (J. Mol. Struct. **116** [1984] 331/44).
[77] Watkinson, P. J.; Mackay, K. M. (J. Organomet. Chem. **275** [1984] 39/42).
[78] Mackay, K. M.; Watkinson, P. J.; Wilkins, A. L. (J. Chem. Soc. Dalton Trans. **1984** 133/9).
[79] Hambrick, G. A.; Froelich, P. N.; Andreae, M. O.; Lewis, B. L. (Anal. Chem. **56** [1984] 421/4).
[80] McKean, D. C. (J. Mol. Struct. **113** [1984] 251/66).

[81] Wingleth, D. C.; Norman, A. D. (Inorg. Chim. Acta **114** [1986] 191/6).
[82] Kohda-Sudoh, S.; Katagiri, S.; Ikuta, S.; Nomura, O. (J. Mol. Struct. **138** [1986] 113/5).
[83] Chen, N. Y.; Degnan, T. F.; Weisz, P. B.; Mobil Oil Corp. (U.S. 4803186 [1986/89]).
[84] Wilkins, A. L.; Watkinson, P. J.; Mackay, K. M. (J. Chem. Soc. Dalton Trans. **1987** 2365/72).
[85] Williams, R. L.; Hall, M. D. (J. Am. Chem. Soc. **110** [1988] 4428/9).
[86] Luo, Y.-R.; Benson, S. W. (J. Am. Chem. Soc. **111** [1989] 2480/2).
[87] Dewar, M. J. S.; Jie, C. (Organometallics **8** [1989] 1544/7).
[88] Allinger, N. L.; Quinn, M. I.; Chen, K.; Thompson, B.; Frierson, M. R. (J. Mol. Struct. **194** [1989] 1/18).
[89] Hayashi, M.; Kaminaka, S.; Fujitake, M.; Miyazaki, S. (J. Mol. Spectrosc. **135** [1989] 289/304).
[90] Lukevics, E.; Gevorgyan, V. (Chem. Technol. Silicon Tin Proc. Asian Network Anal. Inorg. Chem. 1st Int. Chem. Conf. Silicon Tin, Kuala Lumpur 1989 [1992], pp. 165/77).

[91] Gevorgyan, V.; Borisova, L.; Lukevics, E. (J. Organomet. Chem. **393** [1990] 57/67).
[92] Anema, S. G.; Mackay, K. M.; Nicholson, B. K.; Van Tiel, M. (Organometallics **9** [1990] 2436/42).
[93] Chen, N. Y.; Degman, T. F., Jr.; Weisz, P. B.; Mobil Oil Corp. (Eur. Appl. 0-363 531 A1 [1990]).
[94] Anema, S. G.; Lee, S. K.; Mackay, K. M.; Nicholson, B. K.; Service, M. (J. Chem. Soc. Dalton Trans. **1991** 1201/8).
[95] Anema, S. G.; Lee, S. K.; Mackay, K. M.; McLeod, L. C.; Nicholson, B. K.; Service, M. (J. Chem. Soc. Dalton Trans. **1991** 1209/17).
[96] Schleyer, P. v. R.; Kaupp, M.; Hampel, F.; Bremer, M.; Mislow, K. (J. Am. Chem. Soc. **114** [1992] 6791/7).
[97] Walther, P.; Gruendler, W. (Collect. Czech. Chem. Commun. **57** [1992] 997/1004).
[98] Operti, L.; Splendore, M.; Vaglio, G. A.; Volpe, P.; Speranza, M.; Occhiucci, G. (J. Organomet. Chem. **433** [1992] 35/48).
[99] Lee, S. K.; Mackay, K. M.; Nicholson, B. K.; Service, M. (J. Chem. Soc. Dalton Trans. **1992** 1709/16).
[100] Hein, T. A.; Thiel, W.; Lee, T. J. (J. Phys. Chem. **97** [1993] 4381/5).

[101] Luo, Y.-R.; Pacey, P. D. (Can. J. Chem. **71** [1993] 572/7).
[102] Operti, L.; Splendore, M.; Vaglio, G. A.; Volpe, P. (Organometallics **12** [1993] 4509/15).
[103] Operti, L.; Splendore, M.; Vaglio, G. A.; Volpe, P. (Organometallics **12** [1993] 4516/22).
[104] Decouzon, M.; Gal, J.-F.; Gayraud, J.; Maria, P.-C.; Vaglio, G.-A.; Volpe, P. (J. Am. Soc. Mass Spectrom. **4** [1993] 54/7).
[105] Nakatsuji, H.; Nakao, T. (Intern. J. Quantum Chem. **49** [1994] 279/90).

1.3.3.2 Ethylgermane, Ethylgermanium Trihydride, $Ge(C_2H_5)H_3$

Preparation and Formation. $Ge(C_2H_5)H_3$ was prepared by reducing $Ge(C_2H_5)Cl_3$ with LiH in dioxane (54% yield) [5], with $NaBH_4$ in acidic solution [7], or with $LiAlH_4$ in $O(C_4H_9)_2$ at room temperature [22] (see also [15]), in diglyme (71% yield) [10], in ether or tetraglyme [8], or in refluxing ether [30]. $Ge(C_2H_5)H_3$ was also obtained by reacting GeH_4 with K in HMPT at 10 °C in a vacuum system followed by treatment with C_2H_5Br (65% yield) [11] or directly from GeH_3Na and C_2H_5Br in liquid NH_3 [3]. The reaction mixtures were usually worked up by trap-to-trap distillation. In one case the compound was purified by low-temperature sublimation [22].

$Ge(C_2H_5)H_3$ was formed in addition to di- and trigermanes by the reaction of Ge_2H_6 with $CH_2{=}CH_2$ (0.4 : 1 mole ratio) at 154 °C (65 h) in a sealed tube and was separated by GC [12]. Irradiating the same reactants (0.6 : 1 mole ratio) with ^{60}Co γ rays at 0 °C (ca. 20 min) resulted in a complex mixture of alkylgermane species including $Ge(C_2H_5)H_3$; but very low concentrations of $CH_2{=}CH_2$ (ca. 760 : 1 mole ratio) favored the formation of only $Ge(C_2H_5)H_3$ and $H_3GeGe(C_2H_5)H_2$ [20]. $Ge(C_2H_5)H_3$ was also generated by irradiating a mixture of GeH_4 (600 Torr) and $CH_2{=}CH_2$ (100 Torr) at 50 °C (1 h) with ^{60}Co γ rays [18]. The disproportionation of $Ge(C_2H_5)H_2F$ yielded $Ge(C_2H_5)H_3$ and $Ge(C_2H_5)(H)F_2$ and was catalyzed by small amounts of H_2O and/or HF [16].

$Ge(C_2H_5)D_3$ was prepared from $Ge(C_2H_5)Cl_3$ and LiD in dioxane (56% yield) [5] or $LiAlD_4$ in $O(C_4H_9)_2$ [22] or diglyme (90% yield) [10].

$Ge(CH_2CD_3)H_3$ was obtained from GeH_3K and CD_3CH_2Br [22].

The Molecule, Spectra, and Physical Properties. The microwave spectra of $Ge(C_2H_5)H_3$, $Ge(C_2H_5)D_3$, and $Ge(CH_2CD_3)H_3$ were recorded in the range 18000 to 40000 MHz and were assigned to the three most abundant naturally occurring isotopes of germanium (^{70}Ge, ^{72}Ge, and ^{74}Ge). R- and Q-branch, A-type transitions were identified. A least-squares fit of 27 moments of inertia gave the following structural parameters (estimated error limits ±0.01 Å and ±0.75°): r(Ge-C) = 1.949, r(Ge-H) = 1.522, r(C-C) = 1.545, r(C-H/CH_2) = 1.093, and r(C-H/CH_3) = 1.091 Å; <(Ge-C-C) = 112.16°, <(H-Ge-H) = 108.59°, <(H-Ge-C) = 109.74°, <(H-C-H/CH_2) = 106.43°, <(Ge-C-H) = 111.64°, <(C-C-H/CH_2) = 107.87°, <(C-C-H/CH_3) = 110.86°, <(H-C-H/CH_3) = 108.04° [22]. For a comparison with geometrical parameters derived from AM1 calculations, see [32]. Barriers to internal rotation of 1.41 ± 0.03 kcal/mol (489 ± 10 cm^{-1}) for the germyl top and 2.85 ± 0.03 kcal/mol (998 ± 10 cm^{-1}) for the methyl top were derived from an analysis of the microwave splitting pattern of the coupled internal rotors of $Ge(C_2H_5)H_3$ [22]. MM2 calculations gave 1.33 and 2.37 kcal/mol, respectively [33]. Force field calculations were carried out to determine the conformational equilibrium of several Group-IV organometallic compounds and included three conformations of $Ge(C_2H_5)H_3$. A barrier of 1.11 kcal/mol was obtained for the rotation about the Ge-C bond [19]. The structure of $Ge(C_2H_5)H_3$ was critically discussed in conjunction with the IR spectra of deuterated germanes. It appears to be premature to claim that the values for the ground-state skeleton distances Ge-H, Ge-C, and C-C are reliable [28].

The dipole moment was determined from the Stark effect to be μ = 0.76 ± 0.02 D; the dipole vector is at an angle of 23.9° to the A rotational axis, indicating that the GeH_3-CH_2 moiety accounts for most of the dipole of $Ge(C_2H_5)H_3$ [22].

The bond dissociation energy D(Ge-C) = 78.7 kcal/mol was estimated on the basis of an empirical model for homolytic bond dissociation energies [37].

^1H NMR spectrum (in C_6H_6): δ(ppm) = 0.74 (CH_2), 0.97 (CH_3, ^3J(H, H) = 7.8 Hz), 3.55 (HGe, ^3J(H,H) = 3.0 Hz) [8]. The ^1H NMR spectra of $Ge(C_2H_5)_nH_{4-n}$ compounds (n = 1 to 4) were discussed with respect to the variation of the chemical shifts of the HGe, CH_2, and CH_3 protons with n [8]. The ^{1}H NMR spectra of some Groups-IV and -V hydrides including $Ge(C_2H_5)H_3$ were used to determine the relative acidities of these compounds in liquid NH_3 (at 31 °C): δ(ppm) = 0.93 (C_2H_5), 3.26 (HGe, $^3J(H,H)$ = 2.8 and $^4J(H,H)$ = 0.5 Hz) [7].

^{13}C NMR spectrum: δ = 1.5 and 12.2 ppm [31].

^{73}Ge NMR spectrum (neat): δ = −186.4 ± 0.6 ppm; J(Ge,H) = 92.4 ± 0.5 Hz. The relation between the ^{73}Ge chemical shift or the coupling constant and the number of hydrogen atoms in $Ge(C_2H_5)_nH_{4-n}$ compounds (n = 0 to 4) was studied [31]. A correlation between the chemical shifts of ^{73}Ge vs. ^{29}Si (based on 29 species) and vs. ^{119}Sn (based on 26 species) was reported [27].

Gas-phase and solid-film IR spectra of $Ge(C_2H_5)H_3$ and $Ge(C_2H_5)D_3$ were recorded (displayed between 3000 and 400 cm^{-1}) and the fundamental vibrations assigned [10]; see also [8]. The data for the gas-phase IR spectrum [10] are listed in Table 43, p. 280, along with the Raman spectra of the liquid [10] and polycrystalline solid obtained after annealing at 75 K [22]. Raman spectra of gaseous and solid $Ge(C_2H_5)H_3$, $Ge(C_2H_5)D_3$, and $Ge(CH_2CD_3)H_3$ and IR spectra of gaseous and solid $Ge(CH_2CD_3)H_3$ are completely listed (all spectra displayed) [22].

The gas-phase spectrum of $Ge(C_2H_5)H_3$ shows a ν(GeH) doublet (2080.7 and 2074.7 cm^{-1}) which is attributed to two conformers with one Ge-H bond trans and two Ge-H bonds gauche to the CH_3 group, respectively. Almost identical frequencies of the ν(GeH) doublet of the isolated Ge-H bond in $Ge(C_2H_5)(H)D_2$ demonstrate "local mode" behavior, even with only a 6-cm^{-1} difference in the Ge-H bond strength [26]; see also [29].

A correlation between the ν(GeH) or ν(GeD) frequency and the inductive effect of the substituents at the Ge atom was earlier found for a series of organogermanes [6]. The isolated ν_{is}(GeH) frequency in the IR spectra of gaseous alkylgermanes correlates well with the sum of Taft σ^* constants of the substituents, when ν_{is}(GeH) values were chosen for equal numbers of CH_3 groups trans and gauche to the Ge-H bond [25, 29].

Low-frequency vibrations (according to [15] incorrectly assigned in [10]) were recorded in the IR spectrum (solid at −190 °C) and Raman spectrum (liquid at −180 °C) and were assigned as follows (IR/Raman): δ(GeCC) 241(s)/232(s), torsion(CH_3) 189(w)/−, torsion(GeH_3) 113/116(m) (low-frequency range displayed). The torsion modes gave barriers to internal rotation of 1.37 kcal/mol for the GeH_3 group and 2.33 kcal/mol for the CH_3 group (compared with barriers of other ethane- and propane-type molecules) [15]. Low-frequency Raman spectra at 17 K were also reported for $Ge(C_2H_5)H_3$, $Ge(C_2H_5)D_3$, and $Ge(CH_2CD_3)H_3$ and their assignments discussed [22].

A normal coordinate analysis was carried out for $Ge(C_2H_5)H_3$, $Ge(C_2H_5)D_3$, and $Ge(CH_2CD_3)H_3$ (calculated frequencies and potential energy distribution listed), yielding the force constants f(Ge-C) = 2.668, f(Ge-H) = 2.535, and f(C-C) = 4.454 mdyn/Å [22].

Isolated ν(CH) frequencies in the gas-phase IR spectra of partially deuterated ethylgermanes, such as $Ge(CD_2CH_3)H_3$, $Ge(CD_2CHD_2)D_3$, and $Ge(CHDCD_3)D_3$ (spectra displayed from 3000 to 2900 cm^{-1}), were used to predict CH bond lengths, HCH angles, and CH dissociation energies [24].

Ab initio molecular orbital theory was used to study the stabilization of $[E(CH_2CH_2^+)H_3]$ ions (E = Group IV element) by E. The magnitude of this β effect was predicted to increase in the order C < Si < Ge < Sn [35].

The UV photoelectron spectrum of $Ge(C_2H_5)H_3$ (displayed for all $Ge(C_2H_5)_nH_{4-n}$, where n = 1 to 4) shows three bands at the following vertical ionization potentials (in eV; orbitals

References on p. 282

Table 43
Vibrational Spectra of $Ge(C_2H_5)H_3$; Fundamental Modes (C_s Symmetry).
Wavenumbers in cm^{-1}.

IR gas [10]		Raman liquid [10]	Raman solid [22] [a]	assignment [10]	
2970.0(s)	Q	2960(3)	2952(54)	$\nu_{as}(CH_3, CH_2)$	ν_{17}, ν_{18}
2957.2(s)		2925(2) [c]	2931(108)	$\nu_{as}(CH_3)$	ν_1
2936(m,sh)		2925(6)	2910(104)	$\nu_s(CH_2)$	ν_2
2925(m,sh)					
2893.4(m)	R				
2885.8(m)	Q	2880(4)	2890(ca. 20)	$\nu_s(CH_3)$	ν_3
2876.2(m)	P		2867(50)		
2085.5(vs)					
2079.5(vs)	Q	2070(10)	2052(vvs)	$\nu_{s,as}(GeH_3)$, wag(CH_2)	ν_4, ν_5, ν_{19}
2072.0(vs)					
1470.6(mw)	Q	1465(0)	1451(12)	$\delta_{as}(CH_3)$	ν_{20}
1464.1(mw)		not observed	1462(10)	$\delta_{as}(CH_3)$	ν_6
1456.0(mw,sh)					
1413.5(m) [b]			1413(23)	scis(CH_2)	ν_7
1393.9(vw)			1383(3)	$\delta_s(CH_3)$	ν_8
1379(vw,sh)					
1231.1(w,br)		1230(2)	1229(13)	wag(CH_2), twist(CH_2)	ν_9, ν_{21}
			1215(15)		
1037.5(mw)	R				
1030.6(mw)	Q	1025(1) [c]	1025(18)	$\nu(CC)$	ν_{10}
1021.1(mw)	P				
979.1(mw)	Q	974(1) [c]	975(36)	$\rho(CH_3)$	ν_{11}, ν_{22}
892.6(m)	R?				
884.2(m)	Q	885(3)	880(67)	$\delta_{as}(GeH_3)$	ν_{12}, ν_{23}
880(sh)			873(117)		
842.7(vs)	R				
834.9(vs)	Q	830	826(45)	$\delta_s(GeH_3)$	ν_{13}
828.9(vs)	P				
738.6(m)	Q	not observed	733(0.5)	$\rho(CH_2)$	ν_{24}
623.9(m)	R	619(4)	610(76)	$\nu(GeC)$	ν_{14}
612.6(m)	P				
528.5(m)	R				
522.4(m)	Q	525(5)	520(105)	$\rho(GeH_3)$	ν_{15}
513.9(m)	P				
492.8(m)	Q	not observed	497(8)	$\rho(GeH_3)$	ν_{25}
		230(0)	240(11)	$\delta(GeCC)$	ν_{16}

[a] Selected data from a polycrystalline sample at 20 K. – [b] IR of the solid. – [c] Values from [5]; presumably solid sample.

according to C_{3v} symmetry): 10.4 (a_1, GeC), 11.6 (e, GeH_3), and ca. 12.3 ($1e_g$ and $3a_{1g}$, CC and CH). The ionization potentials of the Ge-C and Ge-H bands of $Ge(C_2H_5)_nH_{4-n}$ compounds smoothly decrease with the number of C_2H_5 groups [23].

$Ge(C_2H_5)H_3$ has a pungent odor characteristic of volatile Ge compounds [3]; boiling point ca. 9.2°C [3], 11.5°C/743.5 Torr [5]; a boiling point of 11.3°C/748.5 Torr was given for $Ge(C_2H_5)D_3$ [5]. $Ge(C_2H_5)H_3$ solidifies at liquid N_2 but not liquid O_2 temperature [3].

The following vapor pressure equations, vapor pressures at 0°C, extrapolated (?) boiling points, enthalpies of vaporization (in kcal/mol), and Trouton's constants (in cal · mol^{-1} · K^{-1}) were reported [10]:

log (p/Torr) = 7.496 − 1300/T, p = 534.0 Torr, b.p. 8.5°C, ΔH_v = 5.95, and ΔS_v = 21.1 for $Ge(C_2H_5)H_3$;

log (p/Torr) = 7.596 − 1326/T, p = 548.8 Torr, b.p. 8.2°C, ΔH_v = 6.07, and ΔS_v = 21.5 for $Ge(C_2H_5)D_3$.

For the application of Egloff's boiling point equation [1], see [4].

The specific heat C_p^0, enthalpy $[H_T^0 - H_0^0]$, and entropy S_T^0 were calculated for $Ge(C_2H_5)H_3$ and $Ge(C_2H_5)D_3$ in the ideal gaseous state using the method of statistical thermodynamics for a rigid rotor, harmonic oscillator model and taking into account the potential barrier of internal rotation. Self-consistent formulas for the explicit dependence of the thermodynamic properties on temperature and pressure were obtained by an approximation method. C_p^0 and S_T^0 values were compared with those of $Ge(CH_3)H_3$ and $Ge(C_3H_7)H_3$ [21].

Mass Spectrum and Chemical Behavior. The complete 70-eV mass spectrum (m/e values and relative intensities) of naturally abundant $Ge(C_2H_5)H_3$ was reported. The Ge-containing ions may be grouped into the envelopes $[GeC_2H_n]^+$ (n = 0 to 8), $[GeCH_n]^+$ (n = 0 to 5), $[GeH_n]^+$ (n = 0 to 3), $[GeC_2H_n]^{2+}$ (n = 0 to 8). The base peak at m/e = 74 has ca. 300-times the intensity of the highest peak in the molecular ion region (displayed). Metastable transitions of the types $[Ge(C_2H_5)H_n]^+ \rightarrow [GeH_{n-1}]^+ + C_2H_5 + H$ and $[Ge(C_2H_5)H_n]^+ \rightarrow [GeH_n]^+ + C_2H_5$ were also observed [18].

The acidity constant pK_a = 23.9 in liquid NH_3 at 30°C (referenced to a value of pK_a = 20.6 for fluorene) was derived for $Ge(C_2H_5)H_3$ by determining the concentrations of the parent acid and the $[Ge(C_2H_5)H_2]^-$ anion by ^1H NMR. The much lower acidity of $Ge(C_2H_5)H_3$ compared to that of $Ge(C_6H_5)H_3$ can be attributed to the opposite inductive effects of the C_2H_5 and C_6H_5 groups along with the acidity-reducing solvation effects [17].

$Ge(C_2H_5)H_3$ reacted with Br_2 at low temperatures to give $Ge(C_2H_5)H_2Br$ [16]. $Ge(C_2H_5)H_2Br$ and small amounts of $Ge(C_2H_5)(H)Br_2$ (contaminated with $Ge(C_2H_5)Br_3$) resulted from the bromination with N-bromosuccinimide in C_5H_{12} at room temperature (15 h). $Ge(C_2H_5)H_2I$ and the unstable $Ge(C_2H_5)(H)I_2$ were similarly obtained from $Ge(C_2H_5)H_3$ and N-iodosuccinimide [9]. $Ge(C_2H_5)H_3$ and an equimolar amount of $SnCl_4$ in ether at −25°C [13] or $HgCl_2$ (1.5:1 mole ratio) at room temperature (24 h) [9] yielded $Ge(C_2H_5)H_2Cl$.

Excess $Ge(C_2H_5)H_3$ reacted with Na in liquid NH_3 at −33°C with evolution of H_2 to form a glassy material after evaporation of NH_3; addition of excess NH_4Br regenerated $Ge(C_2H_5)H_3$. Considerable amounts of C_2H_6 were additionally evolved during the reaction with Li in $NH_2C_2H_5$. A subsequent treatment with i-C_5H_{11}Br yielded $Ge(C_2H_5)(C_5H_{11}\text{-i})H_2$ [2].

$Ge_2(C_2H_5)_2H_4$ was obtained from $Ge(C_2H_5)H_3$ and $Ge(C_2H_5)H_2OCH_3$ at 120°C (2 h) in the presence of CH_3ONa-CH_3OH. The reaction was reported to proceed via insertion of $Ge(C_2H_5)H$ formed from the unstable $Ge(C_2H_5)H_2OCH_3$ [14].

The reaction of excess $Ge(C_2H_5)H_3$ with $Fe_3(CO)_{12}$ in C_5H_{12} at 50°C (48 h) in an evacuated, sealed tube gave a mixture of $Fe_3(CO)_9(\mu_3\text{-}GeC_2H_5)_2$, $Fe_2(CO)_7(\mu\text{-}Ge(C_2H_5)H)_2$, and $Fe_2(CO)_6(\mu\text{-}Ge(C_2H_5)H)_3$ as indicated by IR spectroscopy [34]. Long-time reactions of (Co_2-

References on p. 282

$(CO)_7)_2(\mu_4\text{-}Ge)$ (Formula V) with excess $Ge(C_2H_5)H_3$ in C_6H_{14} at 20 to 30°C gave $Co_4(CO)_{13}(\mu_4\text{-}Ge)(\mu\text{-}Ge(C_2H_5)H)$ (Formula VI, R = C_2H_5) and $Co_4(CO)_{12}(\mu_4\text{-}Ge)(\mu\text{-}Ge(C_2H_5)H)_2$ (Formula VII, R = R′ = C_2H_5) along with H_2 and CO. Increasing amounts of $Ge(C_2H_5)H_3$ and slightly elevated temperatures favor the formation of compound VII [36].

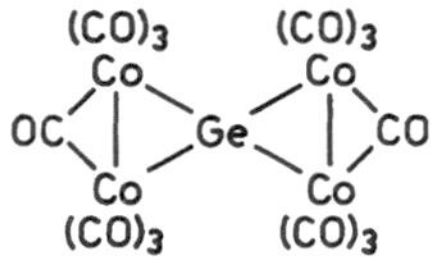

V

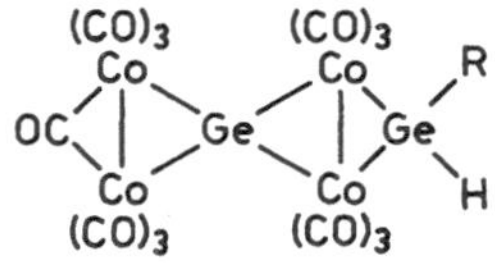

VI

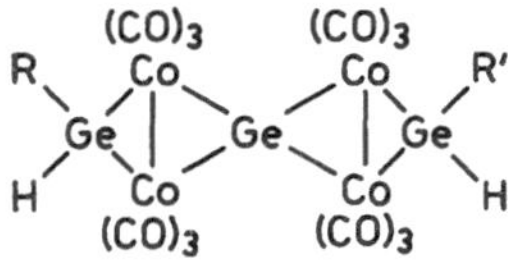

VII

Germanium films were produced from $Ge(C_2H_5)H_3$ at 80 Torr by chemical vapor deposition induced by a CO_2 laser; other products were H_2 and $CH_2{=}CH_2$ [30].

References:

[1] Egloff, G.; Sherman, J.; Dull, R. B. (J. Phys. Chem. **44** [1940] 730/45).
[2] Glarum, S. N.; Kraus, C. A. (J. Am. Chem. Soc. **72** [1950] 5398/401).
[3] Teal, G. K.; Kraus, C. A. (J. Am. Chem. Soc. **72** [1950] 4706/9).
[4] English, W. D. (J. Am. Chem. Soc. **74** [1952] 2927/8).
[5] Ponomarenko, V. A.; Vzenkova, G. Ya.; Egorov, Yu. P. (Dokl. Akad. Nauk SSSR **122** [1958] 405/8; C.A. **53** [1959] 112).
[6] Ponomarenko, V. A.; Zueva, G. Ya.; Andreev, N. S. (Izv. Akad. Nauk SSSR Ser. Khim. **1961** 1758/62; Bull. Acad. Sci. USSR Div. Chem. Sci. [Engl. Transl.] **1961** 1639/43).
[7] Birchall, T.; Jolly, W. L. (Inorg. Chem. **5** [1966] 2177/80).
[8] Mackay, K. M.; Watt, R. (J. Organomet. Chem. **6** [1966] 336/51).
[9] Massol, M.; Satgé, J. (Bull. Soc. Chim. Fr. **1966** 2737/43).
[10] Mackay, K. M.; Watt, R. (Spectrochim. Acta A **23** [1967] 2761/78).

[11] Cradock, S.; Gibbon, G. A.; Van Dyke, C. H. (Inorg. Chem. **6** [1967] 1751/2).
[12] Mackay, K. M.; Watt, R. (J. Organomet. Chem. **14** [1968] 123/9).
[13] Kuz'min, O. V.; Nametkin, N. S.; Chernysheva, T. I.; Gar, T. K.; Lepetukhina, N. A.; Mironov, V. F. (Dokl. Akad. Nauk SSSR **193** [1970] 826/7; Dokl. Chem. [Engl. Transl.] **190/195** [1970] 550/2).
[14] Massol, M.; Satgé, J.; Rivière, P.; Barrau, J. (J. Organomet. Chem. **22** [1970] 599/610).
[15] Durig, J. R.; Hawley, C. W. (J. Phys. Chem. **75** [1971] 3993/4000).
[16] Van Dyke, C. H.; Bulkowski, J. E.; Viswanathan, N. (Inorg. Nucl. Chem. Lett. **7** [1971] 1057/61).
[17] Birchall, T.; Drummond, I. (Inorg. Chem. **11** [1972] 250/2).
[18] Foster, A. G.; Khandelwal, J. K.; Pinson, J. W. (Spectrosc. Lett. **5** [1972] 263/70).
[19] Ouelette, R. J. (J. Am. Chem. Soc. **94** [1972] 7674/9).
[20] Khandelwal, J. K.; Pinson, J. W. (Inorg. Nucl. Chem. Lett. **9** [1973] 393/7).

[21] Zaitsev, N. M.; Maslov, P. G.; Ivolgin, V. I. (Zh. Prikl. Khim. [Leningrad] **47** [1974] 2020/3; J. Appl. Chem. USSR [Engl. Transl.] **47** [1974] 2075/8).
[22] Durig, J. R.; Lopata, A. D.; Groner, P. (J. Chem. Phys. **66** [1977] 1888/900).
[23] Beltram, G.; Fehlner, T. P.; Mochida, K.; Kochi, J. K. (J. Electron Spectrosc. Relat. Phenom. **18** [1980] 153/9).
[24] McKean, D. C.; Mackenzie, M. W.; Torto, I. (Spectrochim. Acta A **38** [1982] 113/8).
[25] McKean, D. C.; Torto, I.; Mackenzie, M. W. (Spectrochim. Acta A **39** [1983] 399/408).

[26] McKean, D. C.; Torto, I.; Mackenzie, M. W.; Morrisson, A. R. (Spectrochim. Acta A **39** [1983] 387/98).
[27] Watkinson, P. J.; Mackay, K. M. (J. Organomet. Chem. **275** [1984] 39/42).
[28] McKean, D. C.; Mackenzie, M. W.; Morrisson, A. R. (J. Mol. Struct. **116** [1984] 331/44).
[29] McKean, D. C. (J. Mol. Struct. **113** [1984] 251/66).
[30] Stanley, A. E.; Johnson, R. A.; Turner, J. B.; Roberts, A. H. (Appl. Spectrosc. **40** [1986] 374/8).

[31] Wilkins, A. L.; Watkinson, P. J.; Mackay, K. M. (J. Chem. Soc. Dalton Trans. **1987** 2365/72).
[32] Dewar, M. J. S.; Jie, C. (Organometallics **8** [1989] 1544/7).
[33] Allinger, N. L.; Quinn, M. I.; Chen, K.; Thompson, B.; Frierson, M. R. (J. Mol. Struct. **194** [1989] 1/18).
[34] Anema, S. G.; Mackay, K. M.; Nicholson, B. K.; Van Tiel, M. (Organometallics **9** [1990] 2436/42).
[35] Nguyen, K. A.; Gordon, M. S.; Wang, G.; Lambert, J. B. (Organometallics **10** [1991] 2798/803).
[36] Lee, S. K.; Mackay, K. M.; Nicholson, B. K. (J. Chem. Soc. Dalton Trans. **1993** 715/22).
[37] Luo, Y.-R.; Pacey, P. D. (Can. J. Chem. **71** [1993] 572/7).

1.3.3.3 $GeRH_3$ Compounds with R Larger than Ethyl

The compounds in this section are listed in Table 44.

Table 44
$GeRH_3$ Compounds with R Larger than Ethyl.
An asterisk indicates further information at the end of the table.
Explanations, abbreviations, and units are given on p. X

No.	R group	formation (yield) properties and remarks
*1	C_3H_7	$Ge(C_3H_7)Cl_3 + LiAlH_4$ in refluxing $O(C_3H_7\text{-i})_2$ and distillation (85%) [4] $Ge(C_3H_7)Cl_3 + LiH$ in dioxane (ca. 35%) [9] $GeH_3Na + C_3H_7Br$ in liquid NH_3 followed by vapor distillation with NH_3, distillation over $CaCl_2$, and condensation at −193°C [2] b.p. 30°C [4], 41 to 42.5°C/757; d^{20} = 1.0391; n_D^{20} = 1.4130 for $Ge(C_3H_7)H_3$ [9] b.p. 41.5°C/754; d^{20} = 1.0508; n_D^{20} = 1.4055 for $Ge(C_3H_7)D_3$ [9] ^{1}H NMR: 3.50 (HGe) [13] IR: ν(GeH) 2067 [13] Raman (liquid): ν(GeH) 2067 for $Ge(C_3H_7)H_3$; ν(GeD) 1486 for $Ge(C_3H_7)D_3$ [9] with Br_2 in $BrCH_2CH_2Br \rightarrow Ge(C_3H_7)Br_3$ [4]
*2	C_3H_7-i	prepared by standard methods; details not given [22]
*3	C_4H_9	$Ge(C_4H_9)Cl_3 + LiAlH_4$ in ether or $O(C_4H_9)_2$ [6, 7] followed by distillation (99%) [7] $Ge(C_4H_9)X_3$ (X = halogen) + $NaBH_4$ in THF [10]

Table 44 (continued)

No.	R group	formation (yield) properties and remarks
*3 (continued)		b.p. 74°C/760 [6, 10], 75.6°C [7] d^{20} = 1.033 [7], 1.0220 [6, 10] n_D^{20} = 1.4207 [7], 1.4200 [6, 10] ^{1}H NMR (neat): 0.89 (C_4H_9), 3.52 (HGe) [20] IR: ν(GeH) 2065 in C_7H_{16} [12, 14]; 2063 for the neat liquid [12, 16]; 2062 in CCl_4 [12, 14] (spectrum displayed [10])
4	C_5H_{11}	from $Ge(C_5H_{11})X_3$ like No. 3 [6, 10] b.p. 104 to 105°C/760 d^{20} = 1.0138; n_D^{20} = 1.4302 [6, 10] IR (liquid): ν(GeH) 2063 [12] with excess $C_5H_{11}MgBr$ in refluxing THF (72 h) → $Ge(C_5H_{11})_2H_2$ [10]
5	C_5H_{11}-i	GeH_3Na + i-$C_5H_{11}Br$ in NH_3 with Li in $NH_2C_2H_5$ at −33°C, followed by NH_4Br → $Ge(C_2H_5)(C_5H_{11}\text{-i})H_2$ [3]
6	C_6H_{13}	from $Ge(C_6H_{13})X_3$ like No. 3 [6, 10] b.p. 128 to 129°C/760 d^{20} = 0.9972; n_D^{20} = 1.4350 [6, 10] IR: spectrum illustrated [6] with excess KOH in C_2H_5OH at 80°C for 15 h → $(Ge(C_6H_{13})O)_2O$ + H_2 with organic bromides (not defined) at 85°C → $Ge(C_6H_{13})Br_3$ (100%) with C_3H_7I at 60°C (exothermic) → $Ge(C_6H_{13})I_3$ (96%) with CCl_3COOH at 120°C → $Ge(C_6H_{13})Cl_3$ (10%) [10]
7	C_7H_{15}	from $Ge(C_7H_{15})X_3$ like No. 3 [6, 10] b.p. 85°C/74 [6, 10], 156°C/750 [10] d^{20} = 0.9819; n_D^{20} = 1.4390 [6, 10] IR (liquid): ν(GeH) 2063 [12] (spectrum illustrated [6]) with $C_6H_5CH_2MgCl$ (1:2 mole ratio) in refluxing THF (48 h) → $Ge(C_7H_{15})(CH_2C_6H_5)_2H$ [10] with CH_2=$CHOC_4H_9$ (1:3 mole ratio) at 130 to 140°C (5 h) → $Ge(C_7H_{15})(CH_2CH_2OC_4H_9)_3$ [10] with CH≡C$(CH_2)_4CH_3$ (1:3 mole ratio) at 200°C in the presence of H_2PtCl_6 → $Ge(C_7H_{15})(CH{=}CH(CH_2)_4CH_3)_3$ [8, 10]
*8	C_8H_{17}	from $Ge(C_8H_{17})X_3$ like No. 3 [6, 10] b.p. 80°C/31 d^{20} = 0.9717; n_D^{20} = 1.4422 [6, 10] IR: ν(GeH) 2067 in C_6H_{14}; 2064 for the neat liquid; 2063 in CCl_4; 2062 in C_5H_5N [17] with O_2 → $(Ge(C_8H_{17})O)_2O$ [10]
supplement		
9	C_4H_9-t	$Ge(C_4H_9\text{-t})Cl_3$ or $Ge(C_4H_9\text{-t})(OC_2H_5)_3$ + $LiAlH_4$ in $O(C_4H_9)_2$ followed by vacuum distillation at low temperature (98%)

Table 44 (continued)

No.	R group	formation (yield) properties and remarks
		colorless liquid ^{1}H NMR (C_6D_6): 1.04 (s, CH_3), 3.72 (s, HGe) ^{13}C NMR (C_6D_6): 20.8 (CH_3), 30.2 (C/C_4H_9) IR (neat): ν(GeH) 2060(s) MS: $[M]^+$ (100), $[M - C_4H_9]^+$ (23) irradiation with $Fe(C_5H_5)(CO)_2Si(CH_3)_3$ (1:2 mole ratio) in C_5H_{12} at 0°C (4 h) $\rightarrow$ $Fe_2(C_5H_5)_2(CO)_2(\mu$-CO)(μ-Ge(C_4H_9-t)H) (40%) and $Fe_2(C_5H_5)_2(CO)_2(\mu$-Ge(C_4H_9-t)H$)_2$ (11%) [27]

*Further information:

General Remarks. The ν(GeH) frequencies in the IR spectra (liquid) of compounds No. 1, 3, 4, 6, 7, and 8 were observed at 2063 to 2065 cm^{-1} [10, 11]. For general remarks on the complete IR spectra of $GeRH_3$ (Nos. 4, 6, 7, and 8), GeR_2H_2, and GeR_3H compounds, see [6]. A linear correlation between the ν(GeH) frequencies and the sum of Taft σ^* constants of the ligands at Ge was found for a series of compounds including Nos. 2 [23, 24], 3 [12, 16], 4, 7, and 8 [12] provided that an appropriate treatment of the conformational effects was carried out [22].

Ge-C bond dissociation energies of 75.7 kcal/mol for $Ge(C_3H_7\text{-i})H_3$ (No. 2) and 73.1 kcal/mol for $Ge(C_4H_9\text{-t})H_3$ were recently derived from a generalized empirical model for bond dissociation energies [26].

$Ge(C_3H_7)H_3$ (Table **44**, No. **1**) was formed with 19% yield by reacting $GeCl_4$ with LiC_3H_7 (formation of propylchlorogermanes) followed by $LiAlH_4$ reduction in $O(C_3H_7\text{-i})_2$ [4]. $Ge(C_3H_7)D_3$ was obtained from $Ge(C_3H_7)Cl_3$ and LiD in dioxane (29% yield) [9].

$Ge(C_3H_7)H_3$ is a liquid with a pungent odor resembling that of other volatile germanes [2]. Egloff's boiling point equation [1] gave a boiling point of 311 K [5].

$Ge(C_3H_7)H_3$ was part of studies on correlations between the ^{1}H NMR shift δ(HGe) and the IR frequency ν(GeH) or the sum of Taft σ^* constants of the other substituents at Ge [13]. For a correlation between the ν(GeH) or ν(GeD) frequencies and the inductive effect of the substituents at Ge, see also [9].

Force field calculations were carried out to determine the conformational equilibria of several Group-IV organometallic compounds, including four conformations of $Ge(C_3H_7)H_3$ [18]. Molecular mechanics calculations using the MM2 force field revealed the gauche conformation to be slightly more stable with respect to the anti form (620 cal/mol). Otherwise, the torsional profile looked much like that for butane [25].

Four calculated values of C_p^0 have been listed in the range 298 to 1000 K and compared with data for $Ge(CH_3)H_3$ and $Ge(C_2H_5)H_3$ [19].

$Ge(C_3H_7\text{-i})H_3$ (Table **44**, No. **2**). $Ge(C_3H_7\text{-i})(H)D_2$ was obtained as an impurity in the fully deuterated compound [22].

Pairs of IR bands appearing in the ν(GeH) region of $Ge(C_3H_7\text{-i})(H)D_2$ (2073.2 and 2066.7 cm^{-1}, also illustrated [22]) were explained by the presence of conformers with trans and gauche CH_3 groups [22, 24].

References on p. 286

$Ge(C_4H_9)H_3$ (Table **44**, No. **3**). ^{1}H NMR studies on organohydrides of Si, Ge, and Sn including $Ge(C_4H_9)H_3$ revealed that these compounds are weakly associated as pure liquids and form in aprotic donor solvents labile complexes with various compositions corresponding to 5- and 6-coordinate central atoms [20]. Correlations between the ^{1}H NMR shifts $\delta(CH_3)$ and $\delta(EH_3)$ of $E(C_4H_9)H_3$ compounds (E = Si, Ge, Sn) and the diamagnetic susceptibilities of the solvents C_6H_{14}, C_6H_{12}, CCl_4, and $CHCl_3$ gave an estimate of the hydride character of the EH_3 hydrogen atom [21].

An almost linear correlation was found between the integrated intensity of the ν(GeH) frequency and the sum of Taft σ^* constants of the ligands at Ge for $Ge(C_4H_9)H_3$ and related compounds [14].

Gradual addition of deficient amounts of $HgCl_2$ (0 to 25 °C for ca. 2 h, then slight heating), $HgBr_2$ (31 °C for 5 h, then gentle reflux), or powdered I_2 (0 °C for 2 h) converted $Ge(C_4H_9)H_3$ into partially substituted $Ge(C_4H_9)H_2X$ compounds (X = Cl, Br, I). Treatment with excess I_2 (at 0 to 30 °C, then slight heating) resulted in the formation of $Ge(C_4H_9)I_3$ [7]. The reaction with CH_3OCH_2Cl in the presence of $AlCl_3$ gave $Ge(C_4H_9)H_2Cl$ [15].

$Ge(C_8H_{17})H_3$ (Table **44**, No. **8**). A slight effect of various solvents on the ν(GeH) frequency (e.g. 2067 cm^{-1} in C_6H_{14} and 2061 cm^{-1} in HMPT) was attributed to the polarity of the Ge^+-H^- bond which is unable to form a true hydrogen bond with proton acceptors [17].

References:

[1] Egloff, G.; Sherman, J.; Dull, R. B. (J. Phys. Chem. **44** [1940] 730/45).
[2] Teal,' G. K.; Kraus, C. A. (J. Am. Chem. Soc. **72** [1950] 4706/9).
[3] Glarum, S. N.; Kraus, C. A. (J. Am. Chem. Soc. **72** [1950] 5398/401).
[4] Johnson, O. H.; Jones, L. V. (J. Org. Chem. **17** [1952] 1172/6).
[5] English, W. D. (J. Am. Chem. Soc. **74** [1952] 2927/8).
[6] Satgé, J.; Mathis-Noël, R.; Lesbre, M. (C.R. Hebd. Seances Acad. Sci. **249** [1959] 131/3).
[7] Anderson, H. H. (J. Am. Chem. Soc. **82** [1960] 3016/8).
[8] Lesbre, M.; Satgé, J. (C.R. Hebd. Seances Acad. Sci. **250** [1960] 2220/2).
[9] Ponomarenko, V. A.; Zueva, G. Ya.; Andreev, N. S. (Izv. Akad. Nauk SSSR Ser. Khim. **1961** 1758/62; Bull. Acad. Sci. USSR Div. Chem. Sci. [Engl. Transl.] **1961** 1639/43).
[10] Satgé, J. (Ann. Chim. [Paris] [13] **6** [1961] 519/73).

[11] Mathis-Noël, R.; Mathis, F.; Satgé, J. (Bull. Soc. Chim. Fr. **1961** 676).
[12] Mathis, R.; Satgé, J.; Mathis, F. (Spectrochim. Acta **18** [1962] 1463/72).
[13] Egorochkin, A. N.; Khidekel', M. L.; Ponomarenko, V. A.; Zueva, G. Ya.; Svirezheva, S. S.; Razuvaev, G. A. (Izv. Akad. Nauk SSSR Ser. Khim. **1963** 1865/8; Bull. Acad. Sci. USSR Div. Chem. Sci. [Engl. Transl.] **1963** 1717/9).
[14] Mathis, R.; Constant, M.; Satgé, J.; Mathis, F. (Spectrochim. Acta **20** [1964] 515/21).
[15] Massol, M.; Satgé, J. (Bull. Soc. Chim. Fr. **1966** 2737/43).
[16] Mathis, R.; Barthelat, M.; Mathis, F. (Spectrochim. Acta A **26** [1970] 1993/2000).
[17] Mathis, R.; Barthelat, M.; Mathis, F. (Spectrochim. Acta A **26** [1970] 2001/5).
[18] Ouelette, R. J. (J. Am. Chem. Soc. **94** [1972] 7674/9).
[19] Zaitsev, N. M.; Maslov, P. G.; Ivolgin, V. I. (Zh. Prikl. Khim. [Leningrad] **47** [1974] 2020/3; J. Appl. Chem. USSR [Engl. Transl.] **47** [1974] 2075/8).
[20] Ivanov, V. A.; Reikhsfel'd, V. O.; Saratov, I. E. (Zh. Obshch. Khim. **45** [1975] 2036/40; J. Gen. Chem. USSR [Engl. Transl.] **45** [1975] 1999/2002).

[21] Saratov, I. S.; Reikhsfel'd; V. O.; Ivanov, V. A. (Zh. Obshch. Khim. **47** [1977] 1776/81; J. Gen. Chem. USSR [Engl. Transl.] **47** [1977] 1625/9).

[22] McKean, D. C.; Torto, I.; Mackenzie, M. W.; Morrisson, A. R. (Spectrochim. Acta A **39** [1983] 387/98).
[23] McKean, D. C.; Torto, I.; Mackenzie, M. W. (Spectrochim. Acta A **39** [1983] 399/408).
[24] McKean, D. C. (J. Mol. Struct. **113** [1984] 251/66).
[25] Allinger, N. L.; Quinn, M. I.; Chen, K.; Thompson, B.; Frierson, M. R. (J. Mol. Struct. **194** [1989] 1/18).
[26] Luo, Y.-R.; Pacey, P. D. (Can. J. Chem. **71** [1993] 572/7).
[27] Kawano, Y.; Sugawara, K.; Tobita, H.; Ogino, H. (Chem. Lett. **1994** 293/6).

1.3.3.4 $GeRH_3$ Compounds with R = Substituted Alkyl

The compounds described in this chapter are listed in Table 45.

Table 45
$GeRH_3$ Compounds with R = Substituted Alkyl.
An asterisk indicates further information at the end of the table.
Explanations, abbreviations, and units are given on p. X.

No.	R group	formation (yield) properties and remarks
*1	CH_2F	$GeH_3K + CH_2FBr$ in HMPT, followed by trap-to-trap fractionation (35%) [17] IR (vapor): $\nu_{as}(GeH_2)$ 2109.4 (H trans to H/CH_2F), ν(GeH) 2089.8 (H trans to F) for rotational conformers of $Ge(CH_2F)H_3$; ν(GeH) 2107.4 (H trans to H/CH_2F), 2090.1 (H trans to F) for rotational conformers of $Ge(CH_2F)(H)D_2$ which was present as an impurity in $Ge(CH_2F)D_3$; ν(GeH) region illustrated [32]
*2	CF_3	$Ge(CF_3)I_3$ in 30% H_3PO_4 + aqueous $NaBH_4$ (ice cooling) for 1 h in vacuum followed by fractional condensation (>90%) [31, 40, 43] b.p. −22.1 °C [40] ^{1}H NMR: 4.13 (q, HGe, ^3J(H, F) = 8.7) [18]; 4.27 (HGe, ^3J(H, F) = 8.7, ^3J(D, F) = 1.35, ^2J(H, D) = 2.1) for partly deuterated species [40] ^{13}C NMR (C_6D_6): 131.0 (^1J(F, C) = 331.7, ^2J(C, H) = 9.1) [40] ^{19}F NMR: −49.2 (^1J(F, C) = 331.6, ^3J(F, H) = 8.7) [37]; see also [40]
*3	CH_2Cl	$GeCl_4 + CH_2N_2$ in ether at −60 °C (→ $Ge(CH_2Cl)Cl_3$) and $LiAlH_4$ reduction in ether [10] or $O(C_4H_9)_2$ [21]; $Ge(CH_2Cl)D_3$ similarly prepared with $LiAlD_4$ [21] m.p. −95.4 °C; b.p. 44.0 °C (extrapolated) [10] ^{1}H NMR (C_6H_{12}): 3.06 (CH), 3.97 (HGe); ^3J(H, H) = 3.3 ± 0.1, ^1J(C, H) = 150.3 ± 0.5 [10] IR (vapor): $\nu_{as}(GeH_2)$ 2112.58 (H trans to H/CH_2Cl), ν(GeH) 2084.74 (H trans to Cl) for rotational conformers of $Ge(CH_2Cl)H_3$; ν(GeH) 2110.2 (H trans to H/CH_2Cl), 2084.74 (H trans to Cl) for rotational conformers of $Ge(CH_2Cl)(H)D_2$

Table 45 (continued)

No.	R group	formation (yield) properties and remarks
*3 (continued)		which was present as an impurity in $Ge(CH_2Cl)D_3$; ν(GeH) region displayed [32]; for the complete spectra, see p. 298 MS: $[GeH_nCCl]^+$, $[GeH_nC]^+$, $[GeH_n]^+$, and rearranged $[GeH_nCl]^+$ fragments stable towards air and light with NaI in refluxing CH_3COCH_3 → $Ge(CH_2I)H_3$ (No. 5) no reaction with Hg [10]
*4	CH_2Br	like No. 3 using $GeBr_4$ [10, 14]; $Ge(CH_2Br)D_3$ similarly prepared using $LiAlD_4$ in $O(C_4H_9)_2$ [14] m.p. −94.3°C; b.p. 75.7°C (extrapolated) [10] ^1H NMR (C_6H_{12}): 2.63 (CH), 4.16 (HGe); $^3J(H, H) = 3.4 \pm 0.1$, $^1J(C, H) = 150.1 \pm 0.5$ [10] IR (vapor): ν_{as}(GeH) 2112(vs), ν_s(GeH) 2081(vs), $\delta_{as}(GeH_3)$ 887(s), 883(s), $\delta_s(GeH_3)$ 830(vs), ν(CBr) 660(s), ν(GeC) 634(s) [14]; see also p. 297 MS: fragments like those of No. 3 stable towards air and light no reaction with Hg [10]
5	CH_2I	No. 3 + NaI in refluxing CH_3COCH_3 m.p. −96.5°C; b.p. 115.1°C (extrapolated) ^1H NMR (C_6H_{12}): 2.10 (CH), 4.59 (HGe); $^3J(H, H) = 3.6 \pm 0.1$, $^1J(C, H) = 149.5 \pm 0.5$ IR: 2980, 2960, 2112, 2102, 2081, 1386, 1188, 1105, 884, 860 to 830, 745, 666, 640, 610, 438 MS: fragments like those of No. 3 light-sensitive; rearranges slowly at room temperature to give $Ge(CH_3)H_2I$ reacts with Hg [10]
6	CH_2OH	only mentioned in connection with studies on the electronic effects of EH_3 substituents; see General Remarks, p. 295
*7	CH_2OCH_3	$GeHCl_3$ + CH_3OCH_2Cl at 20°C (0.5 h) followed by $LiAlH_4$ in $O(C_4H_9)_2$ (85%); purified by vacuum distillation at −85 to −68°C [20] GeH_4 + K in HMPT at 10°C, then treatment with CH_3OCH_2Cl and workup by vacuum fractionation (41%); lower yields obtained with Li (9%) or Na (19%); product analyzed by GC [3] GeH_3Na + CH_3OCH_2Cl (small amount NH_3 present) at −78 to 0°C (1 h) followed by trap-to-trap distillation and purification by GC (ca. 11%) [4] m.p. -121.6 ± 0.3°C [4] ^1H NMR (C_6H_6): 3.11 (CH_3O, $^1J(C, H) = 140.5$), 3.46 (CH_2O), 3.61 (HGe, $^3J(H, H) = 3.10$) [4, 20] IR (vapor): ν(CH) 3030(sh), 2915(w), 2865(sh), ν(GeH)

Table 45 (continued)

No.	R group	formation (yield) properties and remarks
		2092(s), $\delta_s(CH_3)$ 1445(vw), ν(COC) 1119, 1107(s), $\delta_s(GeH_3)$ 824(vs); other bands at 1269(vw), 1188, 1175(w); 932(w); 707(w) [4] spectrum displayed in [4] thermally quite stable MS: $[^{76}Ge(CH_2OCH_3)H_3]^+$ and $[HCO]^+$ (base peak) [4]
*8	COOH	formed by acidification of aqueous $Ge(COOK)H_3$ [6]
*9	$COOCH_3$	$Ge(COOK)H_3 + [O(CH_3)_3]BF_4$ in ether or triglyme at 0°C followed by fractional condensation or GC vapor pressure (in Torr): 31.0 at 0°C, 79.5 at 22°C ^1H NMR (neat): 3.57 (s, CH_3), 4.0 (s, HGe) IR: ν(C-O) 1140(vs); other bands at 2950(m), 2115(s), 1720(s), 1430(w), 1180(s), 945(m), 810(vs), 675(w), 565(m) MS: $[CH_3]^+$, $[CH_3O]^+$, $[GeH_n]^+$, $[GeH_nCO]^+$, $[GeH_nCO_2]^+$, $[GeH_nCOOCH_3]^+$ decomposes at 260°C (1 h) in a sealed tube to yield GeH_4, CH_3OH, CO, and a Ge mirror [19]
*10	$COOC_2H_5$	like No. 9 using $[O(C_2H_5)_3]BF_4$ vapor pressure (in Torr): 22.0 at 0°C, 41.5 at 22°C ^1H NMR (neat): 0.92 (t, CH_3, J = 7.0), 3.82 (s, HGe), 3.84 (q, CH_2, J = 7.0) IR: ν(C-O) 1145(vs); other bands at 3010(m), 2110(s), 1720(s), 1415(w), 1370(w), 1030(m), 810(vs), 675(w), 570(m) MS: $[CH_3]^+$, $[C_2H_5]^+$, $[C_2H_3O]^+$, $[GeH_n]^+$, $[GeH_nCO]^+$, $[GeH_nCO_2]^+$, $[GeH_nCOOC_2H_5]^+$ decomposes at 160°C (8 h) in a sealed tube to give GeH_4, C_2H_5OH, CO, and a Ge mirror; above 300°C (0.5 h) → CO, H_2, CH_4, and C_2H_4 [19]
11	CH_2NH_2	see No. 6
*12	CH_2SiH_3	$GeH_3Na + Si(CH_2Cl)H_3$ at 20°C (15 min) in vacuum; purified by low-temperature vacuum fractionation (35%) [9, 13] $Ge(CH_2SiH_3)Cl_3$ + excess $LiAlH_4$ in tetralin at 100°C (8 h) in the presence of $[N(C_2H_5)_3CH_2C_6H_5]Cl$ followed by vacuum condensation of the volatiles at 60°C into a trap at −196°C (29%) [44] m.p. −134.8±0.2°C [9, 13]; b.p. 29.5°C [44], 29.9°C (extrapolated) [13] ^1H NMR: −0.01 (CH_2), 3.63 (H_3Ge, $^3J(H, H) = 4.0$, $^4J(H, H) = 0.5$), 3.71 (H_3Si, $^3J(H, H) = 4.5$, $^1J(Si, H) = 198.7$) in C_6H_{12} [9, 13] (displayed in [13]); −0.43 (sept, CH_2), 3.56 (t, H_3Ge, $^3J(H, H) = 3.9$), 3.71 (t, H_3Si, $^3J(H, H) = 4.4$), $^1J(Si, H) = 196.8$) in C_6D_6 [44]

References on p. 301

Table 45 (continued)

No.	R group	formation (yield) properties and remarks
*12 (continued)		^{13}C NMR (C_6D_6): −20.56 ("t", CH_2, $^1J(C, H)$ = 124.2, ^2J(C, H/Ge) = 4.9, ^2J(C, H/Si) = 6.1) [44] ^{73}Ge NMR (tetralin at 0°C): −188.95 (q, ^1J(Ge, H) = 99.0) [44] IR: ν(CH) 2940, ν(SiH) 2153, ν(GeH) 2074, δ(CH_2) 1047, δ(SiH_3) 946, δ(GeH_3) 853 in KBr [44]; ν(SiH) 2153(s), ν(GeH) 2080(s), δ(SiH_3) 941(vs); other bands at 2955, 2920(d,vw), 1368(vw), 1048(m), 880(vw), 850(s), 840(sh), 775(sh), 752(m), 745(sh), 720(m), 710(sh), 540(vw,br), 475(vw,br) for the vapor [13] thermally rather stable [9] MS: $[M]^+$ (low intensity) [13], $[GeCH_nSi]^+$, $[GeH_nSi]^+$, $[GeCH_n]^+$, $[GeH_n]^+$, $[CH_5Si]^+$, $[Si]^+$ [44] starting material for Nos. 13 and 14
*13	CH_2SiH_2Cl	No. 12 + HCl (2.5:1 mole ratio) in the gas phase at 20°C (24 h) in the presence of $AlCl_3$ followed by trap-to-trap distillation (ca. 65% [9, 13]) m.p. −129.2 ± 0.2°C [9, 13]; b.p. 104°C [9] ^1H NMR (C_6H_{12}): 0.40 (CH_2), 3.65 (HGe, ^3J(H, H) = 3.9), 4.78 (HSi, ^3J(H, H) = 3.9, ^1J(Si, H) = 234.3) [9, 13] IR (vapor): ν(SiH) 2180(s), ν(GeH) 2088(s), ν(SiCl) 544(w); other bands at 2954, 2915(d,vw), 1362(w), 1051(m), 1008(w), 955(m), 880, 870(d,vs), 841(vs), 758(s), 736(m), 673(w), 647(w), 510, 502(d,vw), 483(w) [13] with H_2O at 20°C (15 min) or passing through an HgO-sand mixture → $(GeH_3CH_2SiH_2\text{-})_2O$ [13]
14	$CH_2Si(H)Cl_2$	like No. 13 but with a 1.25:1 mole ratio for 12 h (36%) [13]; see also [9] vapor pressure: 3.7 Torr at 0°C [9, 13] ^1H NMR (C_6H_{12}): 0.64 (CH_2), 3.70 (HGe, ^3J(H, H) = 3.9), 5.53 (HSi, ^3J(H, H) = 1.9, ^1J(Si, H) = 285.0) [9, 13] IR (vapor): ν(SiH) 2210(s), ν(GeH) 2095(s), ν(SiCl) 570(m); other bands at 2942, 2910(d,vw), 1382, 1365(d,w), 1055(m), 1005(w), 960(vw), 863(m), 812(vs), 762(s), 701, 694(d,w), 672(w), 630(w), 570(m), 480(w) [13] MS showed the expected pattern; no details given [13]
*15	$C(Si(CH_3)_3)_3$	only mentioned in connection with the determination of cone angles for M-E$(Si(CH_3)_3)_3$ fragments; see p. 300
*16	$CH_2C_6H_5$	only mentioned in connection with a quantum-chemical study on the electronic structure of $E(CH_2C_6H_5)H_3$ compounds; see p. 300
17	CH_2CH_2F	see No. 6
18	CH_2CH_2OH	$GeCl_2$ + $CH_2BrCOOCH_3$ followed by $LiAlH_4$ reduction; isolated by fractionation (15%)

Table 45 (continued)

No.	R group	formation (yield) properties and remarks
		$Ge(CH_2OOCCH_3)Cl_3$ + $LiAlH_4$ and distillation; obtained as a mixture with No. 19 (25%) b.p. 57 to 58 °C/40 d^{20} = 1.2571; n_D^{20} = 1.4518 ^{1}H NMR (C_6H_6): 1.0 to 1.45 (m, CH_2Ge), 3.10 to 3.75 (HGe and CH_2O) IR: ν(GeH) 2080 [11]
19	$CH(OH)CH_3$	formed as a mixture with No. 18 [11] ^{1}H NMR: 1.36 (d, CH_3), 3.75 (HGe), 4.0 (q, CH, J = 7.50) IR (mixture with No. 18): ν(OH) 3300, ν(GeH) 2100 [11]
*20	$CH_2CH_2OCH_3$	condensation of $CH_3OCH_2CH_2Cl$ onto GeH_3K (from GeH_4 and K in HMPT) followed by warming to 20 °C and trap-to-trap fractionation; purified by low-temperature distillation (31%) m.p. −157.1 °C ^{1}H NMR (C_6H_{12}): 1.34 (CH_2Ge), 3.23 (CH_3O, ^1J(C, H) = 139.9), 3.44 (CH_2O, ^3J(H, H) = 7.5), 3.46 (HGe, ^3J(H, H) = 3.5) IR (vapor): major bands at 2975, 2942, 2912, 2886, 2842, 2814(m), 2066(s), 1381(d,m), 1132(s), 844, 836(s) stable for days at room temperature MS: $[M]^+$ (10); m/e = 105 (100); m/e values and intensities of other peaks given [20]
21	$CH_2CH_2NH_2$	see No. 6
*22	$CH_2CH_2SiH_3$	$Ge(CH_2CH_2SiCl_3)Cl_3$ + $LiAlH_4$ like No. 12 followed by distillation of the condensate at normal pressure (79%) b.p. 42 °C ^{1}H NMR (C_6D_6): 0.58 ("tq", CH_2Si, ^3J(HC, CH) = 8.2), 0.78 ("tq", CH_2Ge), 3.55 (t, HSi, ^3J(H, H) = 3.7, ^1J(Si, H) = 193.4), 3.56 (t, HGe, ^3J(H, H) = 3.3) ^{13}C NMR (neat at 10 °C): 4.90 (CGe), 4.93 (CSi, ^1J(C, H) = 125.8, ^1J(C, Si) = 52.8) ^{29}Si NMR (neat at 10 °C): −55.95 ("qtt", ^1J(Si, H) = 193.9, ^2J(Si, H) = 7.4, ^3J(Si, H) = 4.4) ^{73}Ge NMR (neat at 10 °C): −182.0 (q, ^1J(Ge, H) = 92.9) IR (KBr): ν(CH) 2936, 2907, ν(SiH) 2154, ν(GeH) 2066, δ(SiH_3) 926, δ(GeH_3) 877, 818, ν(SiC) 705, ν(GeH) 603, 549 [44] MS: $[GeH_3CH_2CH_2]^+$ (17.2), $[GeH_3]^+$ (12), $[SiH_3CH_2CH_2]^+$ (100), $[SiH_3]^+$ (52.1) [42]; see also [44]
23	$CH(OH)CH(C_6H_5)_2$	$Ge(COCH(C_6H_5)_2)Cl_3$ + $LiAlH_4$ ^{1}H NMR: 3.50 (d, HGe, ^3J(H, H) = 2), 3.90 (d, CHC_6H_5, ^3J(H, H) = 10), 4.54 (oct, CHGe)

References on p. 301

Table 45 (continued)

No.	R group	formation (yield) properties and remarks
23 (continued)		IR: ν(OH) 3400, ν(GeH) 2060 rather unstable, decomposes at 120 to 140 °C [7]
24	$C(OH)(CH_3)_2$	$(\text{-}GeF_2OC(CH_3)_2\text{-})_n$ (from GeF_2 and CH_3COCH_3) + $LiAlH_4$ [22]
25	$CH_2CH_2CH_2F$	see No. 6
*26	$CH_2CH_2CF_3$	$Ge(CH_2CH_2CF_3)Cl_3$ + LiH in dioxane (25%) b.p. 46 °C/750.5 d^{20} = 1.3362; n_D^{20} = 1.3530 [1] ^1H NMR: 3.65 (HGe) [2] IR: ν(GeH) 2079 [1, 2] Raman (liquid): ν(GeH) 2079 [1]
27	$CH_2CH_2CH_2Cl$	$Ge(CH_2CH_2CH_2Cl)Cl_3$ + $LiAlH_4$ (77%) b.p. 52 °C/80 d^{20} = 1.2929; n_D^{20} = 1.4644 with $SnCl_4$ (1 : 1 mole ratio) in ether at 25 to 30 °C (5 to 6 h) → $Ge(CH_2CH_2CH_2Cl)H_2Cl$ [8]
28	$CH_2CH_2CH_2OH$	$Ge(CH_2CH_2COOH)Cl_3$ + $LiAlH_4$ in ether at −30 to +20 °C followed by hydrolysis, extraction with ether, and distillation of the organic phase (72%) [15] $Ge(CH_2CH_2COOH)I_3$ in $CHCl_3$-C_5H_{12} + $LiAlH_4$ in refluxing ether (2 h); workup as above (46%) [5] b.p. 90 °C/89 [15], 143 °C/758 [5] d^{20} = 1.258(5) [5], 1.2651 [15] n_D^{20} = 1.462(8) [5], 1.4689 [15] ^1H NMR (C_6D_6): 0.5 to 1.15 (m, CH_2Ge), 1.15 to 1.82 (m, CCH_2C), 3.34 (t, CH_2O, $^3J(H, H)$ = 6.4), 3.56 (t, HGe, $^3J(H, H)$ = 3.6) [15] IR: ν(GeH) 2050 [5], 2060 [15] with $HgCl_2$ (1 : 1 mole ratio) in THF at 20 °C (2 h) → $Ge(CH_2CH_2CH_2OH)H_2Cl$ (85%); similarly, except using a 1 : 2 mole ratio (24 h) → $Ge(CH_2CH_2CH_2OH)(H)Cl_2$ (95%); similarly, except using a 1 : 3 mole ratio (120 h) → $Ge(CH_2CH_2CH_2OH)Cl_3$ (30%) [15] with a stoichiometric amount of CCl_4 at 20 °C (30 h) → $Ge(CH_2CH_2CH_2OH)H_2Cl$ (75%) and $Ge(CH_2CH_2CH_2OH)(H)Cl_2$ (5%) [15] starting material for No. 32 [15]
29	$CH(OH)C_3H_7\text{-i}$	$(\text{-}GeF_2OCH(C_3H_7\text{-i})\text{-})_n$ (from GeF_2 and i-C_3H_7CHO) + $LiAlH_4$ b.p. 80 °C/60 ^1H NMR (C_6D_6): 1.10 (s, CH_3), 3.20 (s, HGe) IR: ν(OH) 3330, ν(GeH) 2070 [22]
30	$CH(CH_3)CH_2CH_2OH$	$Ge(CH(CH_3)CH_2CHO)I_3$ in C_6H_{14} + $LiAlH_4$ in refluxing ether (3 h); workup as for No. 28 (37%) [5]

Table 45 (continued)

No.	R group	formation (yield) properties and remarks
		(-$GeF_2OCH{=}CHCH(CH_3)$-)$_n$ (from GeF_2 and $CH_3CH{=}CHCHO$) + $LiAlH_4$ (11%) [23] b.p. 68 °C/18 [5], 70 to 72 °C/10 [23] d^{20} = 1.197(6); n_D^{20} = 1.468(2) [5] ^1H NMR (C_6D_6): 3.65 (d, HGe, ^3J(H, H) = 2), 3.74 (t, CH_2O, ^3J(H, H) = 7) [23] IR: ν(OH) 3400 to 3200, ν(GeH) 2060 [5], 2070 [23]
*31	$(CH_2)_3OCH_3$	like No. 20 except using $CH_3OCH_2CH_2CH_2Cl$ (40%) m.p. −97.7 °C ^1H NMR (C_6H_{12}): 1.04 (CH_2Ge), 3.21 (CH_3O, ^1J(C, H) = 141.6), 3.27 (CH_2O, ^3J(H, H) = 6.3), 3.47 (HGe, ^3J(H, H) = 3.5) IR (vapor): 2978, 2920, 2856, 2821(m), 2069(vs), 1131(s), 836(vs) MS: m/e = 41 (13), 43 (12), 45 (100), 75 (10), 77 (10), 101 (14), 103 (26), 104 (15), 105 (35), 106 (18), 107 (26), 108 (17), 109 (11) [20]
32	$(CH_2)_3OCH_2OCH_3$	No. 28 + CH_3OCH_2Cl (ca. 1 : 1 mole ratio) followed by fractionation (43%) along with $Ge(CH_2CH_2CH_2OH)H_2Cl$ and $Ge(CH_2CH_2CH_2OCH_2OCH_3)H_2Cl$ b.p. 60 °C/30 d^{20} = 1.1758; n_D^{20} = 1.4458 ^1H NMR (C_6D_6): 0.6 to 1.08 (m, CH_2Ge), 1.3 to 1.95 (m, CCH_2C), 3.17 (s, CH_3O), 3.35 (t, CCH_2O, ^3J(H, H) = 6), 3.55 (t, HGe, ^3J(H, H) = 3.2), 4.45 (s, OCH_2O) IR: ν(GeH) 2070 [15]
*33	CH_2CH_2COOH	$Ge(CH_2CH_2COOH)Cl_3$ + $NaBH_4$ in the presence of $N(C_2H_5)_3$ for 20 h (80%) [26] $(GeCH_2CH_2COOH)_2O_3$ in aqueous KOH + $NaBH_4$ (30 min) followed by acidification with CH_3COOH; workup by extraction with $CH_3COOC_2H_5$, solvent removal, and vacuum distillation of the residual oil [45] b.p. 68 °C/3 ^1H NMR ($CDCl_3$): 1.28 (m, CH_2Ge), 2.53 (t, CH_2CO), 3.56 (t, HGe), 10.80 (s, OH) IR (KBr): ν(GeH) 2070, ν(C=O) 1790 [45] with $C_6H_5CH_2Br$ at 80 °C (90 min) → $C_6H_5CH_3$ [26]
*34	$CH(CH_3)CH_2COOH$	$Ge(CH(CH_3)CH_2COOH)Cl_3$ in aqueous KOH + $NaBH_4$ (30 min) followed by acidification with CH_3COOH; workup as for No. 33 (75%) b.p. 71 °C/2 ^1H NMR ($CDCl_3$): 1.23 (d, CH_3), 1.87 (m, CHGe), 2.51 (d, CH_2CO), 3.63 (d, HGe), 11.69 (s, OH) IR (KBr): ν(GeH) 2060, ν(C=O) 1705 [45]

References on p. 301

Table 45 (continued)

No.	R group	formation (yield) properties and remarks
*35	$CH_2CH(CH_3)COOH$	from $(GeCH_2CH(CH_3)COOH)_2O_3$ like No. 33 or from $Ge(CH_2CH(CH_3)COOH)Cl_3$ like No. 34 [45]
*36	$CH(CH_3)CH(CH_3)$-COOH	from $(Ge(CH(CH_3)CH(CH_3COOH)_2O_3$ like No. 33 or from $Ge(CH(CH_3)CH(CH_3)COOH)Cl_3$ like No. 34 [45]
*37	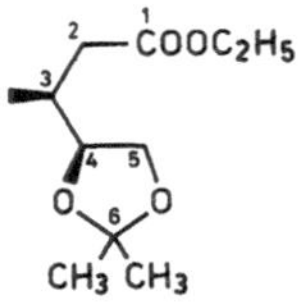 threo	preparation given on p. 300 $[\alpha]_D = -1.4°$ (c = 1.22) in CH_3OH at 20°C ^1H NMR (C_6D_6): 1.224 and 1.370 (s's, $(CH_3)_2C$), 1.948 (m, H-3), 2.314 (dd, H-2′, ^2J = 16.5, J(H-2′,3) = 7.7), 2.456 (dd, H-2, ^2J = 16.49, J(H-2,3) = 7.33), 3.366 (dd, H-5′, ^2J = 8.1, J(H-4,5′) = 7.33), 3.744 (dd, H-5, ^2J = 8.06, J(H-4,5) = 6.23), 3.745 (d, HGe, J = 2.56), 4.050 (dt, H-4, J(H-3,4) = 6.6, J(H-4,5) = 6.23) ^{13}C NMR (C_6D_6): 14.2 (CH_3/C_2H_5), 24.2 (C-3), 25.5 and 26.6 (CH_3-6), 35.6 (C-2), 60.4 (CH_2/C_2H_5), 68.8 (C-5), 78.1 (C-4), 109.0 (C-6), 172.1 (C-1) IR (film): ν(GeH) 2090(s), ν(C=O) 1735(s) MS (CI): $[M - 1]^+$ [47]
*38	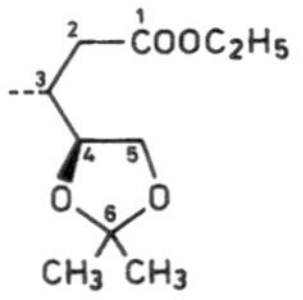 erythro	preparation given on p. 300 $[\alpha]_D = -4.1°$ (c = 1.29) in CH_3OH at 20°C ^1H NMR (C_6D_6): 1.258 and 1.354 (s's, $(CH_3)_2C$), 2.674 (dd, H-2, ^2J = 19, J(H-2,3) = 5.15), 2.740 (dd, H-2′, ^2J = 16.9, J(H-2′,3) = 8.43), 3.369 (t, H-5′, J = 7.69), 3.639 (d, HGe, J = 2.56), 3.797 (dd, H-5, ^2J = 8.06, J(H-4,5) = 6.23), 4.045 (ddd, H-4, J(H-3,4) = 8.8, J(H-4,5) = 7.33, J(H-4,5′) = 5.86) ^{13}C NMR (C_6D_6): 14.2 (CH_3/C_2H_5), 24.6 (C-3), 25.8 and 27.0 (CH_3-6), 35.1 (C-2), 60.3 (CH_2/C_2H_5), 68.9 (C-5), 78.0 (C-4), 109.1 (C-6), 172.3 (C-1) IR (film): ν(GeH) 2100(s), ν(C=O) 1730(br,s) MS (CI): $[M - 1]^+$ [47]
39	$CH_2CH_2CH_2NH_2$	see No. 6
40	$CH(NHC_2H_5)$-C_3H_7-i	$(-GeF_2N(C_2H_5)CH(C_3H_7\text{-}i)-)_n$ (from GeF_2 and i-$C_3H_7CH{=}NC_2H_5$) + $LiAlH_4$ b.p. 120°C/25 ^1H NMR (C_6D_6): 2.50 (d, CHN), 3.71 (d, HGe); ^3J(H, H) = 2 IR: ν(NH) 3375, ν(GeH) 2060 [22]
*41	$CH_2CH_2CH_2SiH_3$	$Ge(CH_2CH_2CH_2SiH_3)Cl_3$ + $LiAlH_4$ in $O(C_4H_9)_2$; purified by distillation [42] MS: $[GeH_3(CH_2)_3]^+$ (19.7), $[GeH_3(CH_2)_2]^+$ (14), $[SiH_3(CH_2)_3]^+$ (70), $[SiH_3CH_2]^+$ (100) [42]
*42	$CH(OH)CH_2CH_2CH_3$	$(-GeF_2OCH(CH_2CH_2CH_3)-)_n$ (from GeF_2 and C_3H_7CHO) + $LiAlH_4$ [22]

Table 45 (continued)

No.	R group	formation (yield) properties and remarks
		b.p. 70 to 75°C/40 [34], 85°C/60 [22] ^{1}H NMR (C_6D_6): 0.93 (CH_3), 1.18 to 1.72 (m, CH_2, J = 6), 3.75 (d, HGe, J = 2), 3.93 (m, CHO) IR (neat): ν(OH) 3440, ν(GeH) 2075 [34]; see also [22]
43	$(CH_2)_4OCH_3$	like No. 20 except using $CH_3O(CH_2)_4Br$ (70%, based on $CH_3O(CH_2)_4Br$) m.p. −109.8°C ^{1}H NMR (C_6H_{12}): 1.00 (CH_2Ge), 3.23 (CH_3O, J(C, H) = 140.7), 3.30 (CH_2O, ^{3}J(H-3,4) = 6.4), 3.47 (HGe, ^{3}J(H, H) = 3.4), IR (vapor): 2918, 2847, 2813(m), 2057(s), 1132(s), 835(vs) MS: m/e = 15 (50), 27 (26), 28 (17), 29 (37), 41 (19), 45 (100), 55 (21), 58 (17) [20]
44	$(CH_2)_6OCH_3$	like No. 20 except using $CH_3O(CH_2)_6Cl$ (3%, based on $CH_3O(CH_2)_6Cl$) m.p. −72.9°C vapor pressure: 1.6 Torr at 30°C ^{1}H NMR (C_6H_{12}): 1.02 (CH_2Ge), 3.21 (CH_3O, ^{1}J(C, H) = 139.7), 3.28 (CH_2O), 3.45 (HGe, ^{3}J(H, H) = 3.4) IR (neat): 2976(m), 2923, 2855(s), 2057(s), 1451(m), 1120(s), 878(m), 829(vs) MS: m/e = 41 (13), 45 (100), 54 (14), 55 (27), 82 (28), 105 (12), 118 (11) [20]

*Further information:

General Remarks. An almost linear relation was found between the isolated ν(GeH) frequency and the sum of Taft σ^* parameters of the substituents at Ge in $Ge(CH_2X)H_3$ compounds (Nos. 1, 3, and 4), but not for $Ge(CF_3)H_3$. Possible reasons for the different behavior were discussed [35].

CNDO/2 calculations were used to study the effect of EH_3 substituents on the gas-phase basicity of $E(CH_2X)H_3$, $E(CH_2CH_2X)H_3$, and $E(CH_2CH_2CH_2X)H_3$ compounds (E = C, Si, Ge) such as Nos. 1, 17, 25 (X = F), 6, 18, 28 (X = OH), and 11, 21, 39 (X = NH_2). The protonation energies, calculated for various energetically preferred conformations, revealed that the electronic effects of GeH_3 and SiH_3 are crucially influenced by the molecular conformation [27]; see also [38] for Nos. 6, 11, 18, and 21. The same conclusion was drawn from calculations of the potential curves of internal rotation around the C-X bond of $E(CH_2X)H_3$ and $E(CH_2CH_2X)H_3$ compounds (X = F, OH, NH_2) [28, 29, 30].

The base strength of $E((CH_2)_nOCH_3)H_3$ compounds (E = C, Si, Ge; n = 1 to 6) including Nos. 7, 20, 31, 43, and 44 were determined by measuring the IR shift $\Delta\nu$(OH) of CH_3OH (in CCl_4 or C_6H_{12}) upon complexation with the oxygen atom of the ethers. The $\Delta\nu$(OH) values indicated that some factors other than inductive effects significantly influence the electron-donating ability of oxygen in the Si and Ge hydrides when n = 1 and 2, whereas for n = 3 to

References on p. 301

6, the basicities converge toward those of the carbon analogs for which only inductive effects are operative [20].

$Ge(CH_2F)H_3$ (Table **45**, No. **1**). The microwave spectra of three isotopic species (^{70}Ge, ^{72}Ge, and ^{74}Ge) were measured in the range 8000 to 30000 MHz; rotational constants for $^{74}Ge(CH_2F)H_3$: A = 26166.35, B = 3515.274, and C = 3288.729 MHz. The height of the potential barrier hindering the internal rotation of the GeH_3 group was determined to be 1.390 ± 0.40 kcal/mol from A-E frequency differences using the principal axis method. The electric dipole moment was obtained from Stark effect measurements on four low J transitions: $\mu = 1.64 \pm 0.03$ D (nearly parallel to the C-F bond) [17]. For comments on the IR frequency ν(GeH), see Nos. 3 and 4.

$Ge(CF_3)H_3$ (Table **45**, No. **2**) was formed with less than 5% yield by vacuum distillation of excess CF_3Br onto a solution of GeH_3K in HMPT at low temperature, warming to 0 °C, and trap-to-trap fractionation [18]. Reacting $Ge(CF_3)I_3$ with $NaBD_4$ in H_3PO_4 yielded a sample containing 35% $Ge(CF_3)H_3$, 50% $Ge(CF_3)H_2D$, and 15% $Ge(CF_3)HD_2$; using $NaBH_4$-D_3PO_4 gave 16% $Ge(CF_3)H_2D$, 72% $Ge(CF_3)HD_2$, and 12% $Ge(CF_3)D_3$ (determined by NMR analysis) [31, 40]. $Ge(CF_3)D_3$ was obtained from $Ge(CF_3)I_3$ in D_3PO_4 and $NaBD_4$ in D_2O [31, 43].

The microwave spectra of five isotopic species of $Ge(CF_3)H_3$ were investigated in the range 15000 to 31000 MHz. The height of the potential barrier hindering the internal rotation around the Ge-C bond was determined to be 1.280 ± 0.15 kcal/mol using the frequency method for rotational satellite transitions [18]. The microwave spectra of five isotopic species of $Ge(CF_3)D_3$ (from 18000 to 39000 MHz at room temperature) and three isotopic species of $Ge(CF_3)H_2D$ and $Ge(CF_3)(H)D_2$ (from 26500 to 39000 MHz at −70 °C) were assigned for the ground vibrational state and the excited states of torsion. In addition, the spectra of $Ge(CF_3)D_3$ were assigned for the excited states of CF_3 rocking modes. From a diagnostic least-squares adjustment to fit fifteen rotational constants for the ground vibrational state, the following r_0 molecular parameters were obtained: r(Ge-H) = 1.499 ± 0.004, r(Ge-C) = 1.997 ± 0.017, and r(C-F) = 1.352 ± 0.008 Å; $\angle$(C-Ge-H) = $107.1 \pm 0.2°$ and $\angle$(Ge-C-F) = $111.6 \pm 0.9°$. The rotational barrier of the CF_3 moiety was estimated to be 1.30 ± 0.10 kcal/mol. The data were compared with those of $Si(CF_3)H_3$ and CF_3CH_3 [43].

The complete gas-phase IR and liquid phase Raman spectra of $Ge(CF_3)H_3$ (Raman polarization spectrum illustrated), $Ge(CF_3)H_2D$, $Ge(CF_3)HD_2$, and $Ge(CF_3)D_3$ were reported in [31]. A normal coordinate analysis yielded the force constants f(GeH) = 2.70, f(GeC) = 2.21, and f(CF) = 5.69 N/cm. The fundamental vibrations of $Ge(CF_3)H_3$ are listed in Table 46. The observed rotational fine structure of several perpendicular bands of $Ge(CF_3)H_3$ was in agreement with the Coriolis constants computed for a rigid symmetric top molecule and Ge isotopic shifts [31].

He(I) and He(II) photoelectron spectra (illustrated) of $Ge(CF_3)_nH_{4-n}$ (n = 1 to 3) were studied and compared. The experimental vertical ionization energies over the range 10 to 27 eV, the values of the orbitals calculated by the CNDO/2 method, and the contributions to the molecular orbitals were listed. For orbitals involved with the Ge bonding in $Ge(CF_3)H_3$, the following ionization energies were measured: 12.10 ($5a_1$, GeC) and 12.9 eV (5e, GeH). The core-level binding energies (in eV) are 37.8 for Ge 3d (relative to Ne 2s at 48.47), 130.40 for Ge $3p_{3/2}$, 134.50 for Ge $3p_{1/2}$, 297.50 for C 1s, and 694.15 for F 1s (relative to $Ar2p_{2/3}$ at 48.63); the energies increase with the number of CF_3 groups in the series $Ge(CF_3)H_3$, $Ge(CF_3)_2H_2$, and $Ge(CF_3)_3H$ [33].

The evaporation parameters ΔH_v = 19.78 kJ/mol and ΔS_v = 78.8 kJ · mol^{-1} · K^{-1} were deduced from the equation log(p/mbar) = 7.120 − 1033/T [40].

Table 46
Vibrational Spectra of $Ge(CF_3)H_3$; Fundamental Modes (C_{3v} Symmetry) [31].
Wavenumbers in cm^{-1}; PQR indicates a_1 fundamentals with PQR structure.

IR gas	Raman liquid	assignment	
2138 (s)	2140 (w)	ν_{as}(GeH)	ν_9
2129 (s) (PQR)	2124.5 (vs, p)	ν_s(GeH)	ν_4
1185 (vs) (PQR)	1180 (m, p)	ν_s(CF)	ν_1
1117 (vvs)	1100 (vw, br)	ν_{as}(CF)	ν_6
874 (s)	872 (m)	$\delta_{as}(GeH_3)$	ν_{10}
809 (vs) (PQR)	805 (m, p)	$\delta_s(GeH_3)$	ν_5
727 (w) (PQR)	723.3 (s, p)	$\delta_s(CF_3)$	ν_2
590 (s)	592 (m)	$\rho(GeH_3)$	ν_{11}
494 (ms)	496 (w)	$\delta_{as}(CF_3)$	ν_7
306 (m) (PQR)	302.0 (s, p)	ν(GeC)	ν_3
212 (w)		$\rho(CF_3)$	ν_8

Mass spectrum: $[Ge(CF_3)H_2]^+$ (10), $[Ge(CF_3)H]^+$ (15), $[GeCF_3]^+$ (20), $[GeF]^+$ (80), $[GeH_3]^+$ (100), $[GeH]^+$ (25), $[Ge]^+$ (20), $[CF_3]^+$ (25), $[CF_2H]^+$ (20) [40].

Cocondensation of $Ge(CF_3)H_3$ with CH_3I and $Zn(CH_3)_2$ (1 : 1 : 1 mole ratio) at low temperatures and warming to room temperature gave $Ge(CH_3)(CF_3)H_2$, $Ge(CH_3)_2(CF_3)H$, and $Ge(CH_3)_3CF_3$. With a 6 : 4.5 : 1 mole ratio the following iodogermanes were additionally obtained: $Ge(CF_3)H_2I$, $Ge(CH_3)(CF_3)(H)I$, $Ge(CH_3)(CF_3)I_2$, $Ge(CH_3)_2(CF_3)I$, $Ge(CH_3)(CF_3)I_2$, and $Ge(CF_3)I_3$. The reaction with $Zn(C_2H_5)_2$ and a slight excess of C_2H_5I gave $Ge(CF_3)(C_2H_5)H_2$, $Ge(CF_3)(C_2H_5)_2H$, and $Ge(CF_3)I_3$. With $Zn(C_2H_5)_2$-CH_3I methyl- and ethyl-containing species were formed, whereas $Zn(CH_3)_2$-C_2H_5I yielded only methylated germanes [40].

$Ge(CH_2Cl)H_3$ and **$Ge(CH_2Br)H_3$** (Table **45**, Nos. **3** and **4**). The microwave spectra of five isotopic species of $Ge(CH_2Cl)H_3$ and $Ge(CH_2Cl)D_3$ were studied in the range 8000 to 34000 MHz. From splittings of the spectra due to the first excited GeH_3 torsional state, the barrier to internal rotation was determined to be 1.740 ± 0.030 kcal/mol. The quadrupole coupling constants were determined for the $^{74}Ge(CH_2{}^{35}Cl)H_3$ and $^{74}Ge(CH_2{}^{37}Cl)H_3$ species. The following structural parameters were calculated for $Ge(CH_2Cl)H_3$: r(Ge-C) = 1.961 and r(Ge-H) = 1.517 Å; <(H-Ge-C) = 107°43′ and <(Ge-C-Cl) = 100°11′ [21]; see also [16]. The geometry of $Ge(CH_2Cl)H_3$ was examined critically [39] by comparison with the data from [21].

The frequencies observed in the IR spectra of $Ge(CH_2X)H_3$ and $Ge(CH_2X)D_3$ (X = Cl [12], Br [14]) for the vapor and the solid in the range 4000 to 250 cm^{-1} were listed and assigned. All the fundamental vibrations, except the skeletal GeCX bending (ν_{11}) and $Ge(H, D)_3$ torsional vibrations (ν_{18}), were observed [12, 14]; the fundamental vibrations of $Ge(CH_2Cl)H_3$ and $Ge(CH_2Cl)D_3$ are listed in Table 47, p. 298 [12]. Somewhat different data for Nos. 3 and 4 were reported in [10]. A normal vibration calculation was carried out using a modified Urey-Bradley force field. The transferability of force constants of similar molecules was studied [12, 14]. The IR spectra of $Ge(CH_2X)H_3$ compounds (X = F, Cl, Br) show a marked differences between the gauche and trans effects of the halogen atom, resulting in two ν(GeH) bands; the virtual identity of the ν(GeH) regions of $Ge(CH_2X)H_3$ and $Ge(CH_2X)(H)D_2$ (for X = F and Cl illustrated) demonstrates "local mode" behavior in the fundamental region [36].

References on p. 301

Table 47
Vapor Phase IR Spectra of $Ge(CH_2Cl)H_3$ and $Ge(CH_2Cl)D_3$;
Fundamental Modes (C_s Symmetry) [12]. Wavenumbers in cm^{-1}.

$Ge(CH_2Cl)H_3$	$Ge(CH_2Cl)D_3$	assignment	
3001(w)	3004(w)	$\nu_{as}(CH_2)$	ν_{12}
2957(m)	2962(m)	$\nu_s(CH_2)$	ν_1
2115(vs)	1523(vs), 1504(vs)	ν_{as}(GeH, GeD)	ν_2, ν_{13}
2086(vs)	1497(vs)	ν_s(GeH, GeD)	ν_3
1408(w)	1408(w)	scissor (CH_2)	ν_4
1177(w)	1178(m)	wagging (CH_2)	ν_5
1116(w)	1112(w)	twist (CH_2)	ν_{14}
879(s)	624(s)	$\delta_{as}(GeH_3, GeD_3)$	ν_{15}
881(s)	607(s)	$\delta_{as}(GeH_3, GeD_3)$	ν_6
826(vs)	591(vs)	$\delta_s(GeH_3, GeD_3)$	ν_7
787(s)	759(s)	$\rho(CH_2)$	ν_{16}
743(m)	743(m)	ν(CCl)	ν_8
641(m)	638(s)	ν(GeC)	ν_9
492(s)	393(s)	$\rho(GeH_3, GeD_3)$	ν_{10}
(492)	379(m)	$\rho(GeH_3, GeD_3)$	ν_{17}

$Ge(CH_2OCH_3)H_3$ (Table **45**, No. **7**). The ^1H NMR spectrum in C_6H_{12} or $Si(CH_3)_4$ consists of two unresolved peaks at δ = 3.23 and 3.58 ppm in a 3:5 ratio, apparently owing to a chemical shift coincidence of the GeH_3 and $GeCH_2$ protons. The GeH_3CH_2 portion of the spectrum was monitored in C_6H_6 solution at high resolution (displayed) [4].

The vapor pressures in the range −64.0 to 0 °C can be expressed by log(p/Torr) = 8.1674 − 1678.6/T. This gives an extrapolated boiling point of 44.4 °C, ΔH_v = 7.679 kcal/mol, and ΔS_v = 24.2 cal · mol^{-1} · K^{-1} [4]. Selected experimental vapor pressures:

t in °C	−64.0	−45.8	−31.1	−21.8	0
p in Torr	1.3	6.2	17.2	31.3	103.2

The Lewis basicity was determined from the $\Delta\nu$(OH) shift (126 cm^{-1}) of 0.01 M CH_3OH in CCl_4 (see also General Remarks, p. 295): No. 7 is a weaker Lewis base than $C_2H_5OCH_3$ and has nearly the base strength of $O(CH_3)_2$ [4].

$Ge(COOH)H_3$ (Table **45**, No. **8**). The pK_a of the acid at ca. 25 °C was determined to be 3.47 ± 0.15 by measuring the pH at the midpoint of rapidly repeated back-and-forth titrations of $Ge(COOK)H_3$ with 0.1 M HCl and NaOH [6].

$Ge(COOH)H_3$ decomposes in dilute aqueous acidic solutions to form CO, an orange-yellow solid of the approximate composition $GeH_{0.6}$, and small amounts of GeH_4 (see also [6]). The reaction is first-order in both [H^+] and [$Ge(COOH)H_3$] with k = (5.59 ± 0.14) × 10^{-4} M^{-1} · s^{-1} at 22.5 °C. Rate measurements between 0 and 39.5 °C gave E_a = 16.9 kcal/mol. In strongly acidic solutions CO was evolved quantitatively, but no solid hydride or GeH_4 was formed. The resulting solution contained the $[GeH_3]^+$ cation, presumably stabilized as $[GeH_3OH_2]^+$ [25].

The salt **$Ge(COOK)H_3$** was obtained by shaking a solution of GeH_3K in DME at 0 °C under approximately 1 atm CO_2 for ca. 2 h and filtration of the precipitated product [24, 25]. Condensation of excess CO_2 onto GeH_3K in DME at −196°C, warming to room temperature, and removal of the volatiles yielded the salt quantitatively [6].

The white solid was recrystallized from C_2H_5OH-ether. An X-ray powder pattern of this sample (d values listed) could not be indexed properly for either a cubic or tetragonal lattice system. The compound hydrolyzes slowly in moist air [6].

^1H NMR spectrum (in D_2O): single sharp resonance at δ = 4.30 ppm. IR spectrum (in Nujol): ν(GeH) 2060, $\nu(COO^-)$ 1540, 1325, $\delta(GeH_3)$ 873, 825, 800, $\rho(GeH_3)$ 678, and ν_{as}(GeC) 557 cm^{-1} (spectrum illustrated). The electronic spectrum of the $[Ge(COO)H_3]^-$ ion (in H_2O) consists of a band at 239 nm (ε < 500 $M^{-1} \cdot cm^{-1}$) as a shoulder on an intense end absorption (presumably n → π^* transition) [6].

The effects of π(p-d) back-bonding from the carboxyl group to the vacant d orbital of Ge on the physical and chemical properties of the $[Ge(COO)H_3]^-$ ion were discussed [6].

The thermal decomposition of $Ge(COOK)H_3$ in a sealed tube at 480°C for 3 h may be represented by the equation 2 $Ge(COOK)H_3$ → 3 H_2 + CO + 2 Ge + K_2CO_3 [6]. Product analysis after heating at 510°C over a period of 24 h was used to estimate the purity of the samples [24, 25].

Various solutions of $Ge(COOK)H_3$ were electrolyzed at 0°C (at Pt, 10 to 30 V) in a vacuum system with the aim of obtaining Ge_2H_6 from the GeH_3 radicals expected to be generated. But only minor amounts of Ge_2H_6 were observed: ca. 5% yield in aqueous 1 M $KHSO_4$, traces in H_2O, and no Ge_2H_6 in ca. 80% CH_3OH [6].

Aqueous solutions of $Ge(COOK)H_3$ very slowly hydrolyse at room temperature; essentially complete hydrolysis was observed at 85°C within two days yielding GeH_4 and CO_2. The rate of hydrolysis in 1 M $KHSO_4$ (pH ≈ 2) in D_2O was measured by ^1H NMR giving a half-life of ca. 30 min. Acidification of aqueous $Ge(COOK)H_3$ yielded $Ge(COOH)H_3$ which rapidly decomposed [6]. The reaction with 8 M HCl was followed by ^1H NMR as a function of temperature and time [25]. The salt decomposed in neutral or alkaline solutions (pH 6 to 13) to give GeH_4 and HCO_3^- (or CO_3^{2-}). The rate of this decarboxylation was pH independent and first-order in the germaacetate ion: rate constant k = 1.62×10^{-5} s^{-1} at 60°C (E_a = 28.6 kcal/mol). At pH ≥ 13 another decomposition reaction competes with the decarboxylation producing H_2 and an insoluble brown polymer, $H_2Ge_2O_3 \cdot n\,H_2O$. This reaction is first-order in $[OH^-]$ and has the rate constant k = 2.0×10^{-5} $M^{-1} \cdot s^{-1}$ at 60°C (E_a = 20.8 kcal/mol). Mechanisms of these reactions were discussed [24].

$Ge(COOK)H_3$ was used as the starting material for Nos. 8, 9, and 10 [19].

$Ge(COOCH_3)H_3$ and **$Ge(COOC_2H_5)H_3$** (Table **45**, Nos. **9** and **10**). The mass spectra show a cluster of peaks centered at 91 mass units, suggesting either decarbonylation followed by $[GeH_nOR]^+$ → $[GeH_nO]^+$ + R or the metastable process $[GeH_nCO_2]^+$ → $[GeH_nCO]^+$ + O [19].

The compounds hardly react with D_2O at room temperature, but at 80°C over a period of days the formation of $Ge(COOD)H_3$ was indicated by NMR. The base-catalyzed hydrolysis (aqueous KOH) of $Ge(COOCH_3)H_3$ at 0°C gave mainly CH_3OH along with GeH_4, a polymeric Ge-containing ether, and CO (ca. 10%), whereas the acid-catalyzed hydrolysis (aqueous H_2SO_4) at 0°C yielded mainly CO (up to 93%). $Ge(COOC_2H_5)H_3$ behaves similarly [19].

Adding GeH_3K in triglyme to $Ge(COOC_2H_5)H_3$, acidification with H_2SO_4, and heating at 80°C overnight produced GeH_4, Ge_2H_6, and H_2 [19].

$Ge(CH_2SiH_3)H_3$ (Table **45**, No. **12**). The vapor pressures in the range −64 to −31°C fit the equation log(p/Torr) = 7.9811 − 1545.7/T; the extrapolated boiling point is 29.9°C; selected vapor pressures [13] are as follows:

References on p. 301

t in °C	−64.0	−50.0	−44.3	−36.4	−31.2
p in Torr	3.8	11.5	16.9	28.3	38.5

By-products of the reactions with HCl (see Table 45; Nos. 13 and 14) were GeH_4 and $Ge(CH_2SiH_3)_2H_2$, presumably resulting from disproportionation of $Ge(CH_2SiH_3)H_3$ [9, 13].

$Ge(CH_2SiH_2Cl)H_3$ (Table **45**, No. **13**). The vapor pressures (measured in the ranges −15 to +15°C and −196°C to +4°C) can be expressed by the equation log(p/Torr) = 6.9293 − 1526.2/T; the extrapolated boiling point is 104.3°C; selected vapor pressures [13] are as follows:

t in °C	−22.9	−11.4	0.0	10.3	15.4
p in Torr	6.9	12.3	21.7	34.2	43.0

The mass spectrum showed the expected fragmentation pattern (details not reported) [13].

$Ge(C(Si(CH_3)_3)_3)H_3$ (Table **45**, No. **15**). Cone angles were determined for M-E$(Si(CH_3)_3)_3$ fragments (M = C, Si, Ge, Sn; E = C, Si) of energy-minimized M(E$(Si(CH_3)_3)_3)H_3$ structures (determined by MM calculations) to gauge the steric requirements of E$(Si(CH_3)_3)_3$ ligands [48].

$Ge(CH_2C_6H_5)H_3$ (Table **45**, No. **16**). The proton affinity of 3.4 and 0.5 kcal/mol (referred to zero of $C_6H_5C_2H_5$) for the gauche and planar conformers, respectively, and the barrier to internal rotation around the $C_6H_5CH_2$ bond (2.8 kcal/mol) were reported in connection with a quantum-chemical study of the electronic structure of $E(CH_2C_6H_5)H_3$ compounds (E = Si, Ge, Sn, Pb) and their protonated derivatives [41].

Carbon-free Ge films were deposited by ArF laser photolysis of a solution of $Ge(CH_2C_6H_5)H_3$ in $C_{12}H_{26}$. GeH_4 and $C_6H_5CH_3$ were formed in the gaseous and liquid phase, respectively [49].

$Ge(CH_2CH_2OCH_3)H_3$ (Table **45**, No. **20**). Over the range −32 to +24°C the vapor pressure data fit the equation ln(p/Torr) = 18.095 + 4038.6/T; this gives an extrapolated boiling point of 79.1°C, ΔH_v = 8.025 kcal/mol, and a Trouton constant of 22.8 cal · mol^{-1} · K^{-1} [20].

$Ge(CH_2CH_2SiH_3)H_3$ and **$Ge(CH_2CH_2CH_2SiH_3)H_3$** (Table **45**, Nos. **22** and **41**) were used to prepare Ge-Si alloys by chemical vapor deposition at 425, 450, and 475°C and atmospheric pressure. During the pyrolysis, $CH_2{=}CH_2$, CH≡CH (from No. 22) and $CH_3CH{=}CH_2$ (from No. 41) were evolved and a solid phase was deposited composed of Ge and Si [42].

$Ge(CH_2CH_2CF_3)H_3$ (Table **45**, No. **26**) was part of investigations of relations between δ(HGe) shifts, ν(GeH) frequencies, Taft σ^* parameters [2], and the inductive effect of the substituents at the Ge atom [1].

$Ge(CH_2CH_2CH_2OCH_3)H_3$ (Table **45**, No. **31**). Over the temperature range −16 to +26°C the vapor pressures fit the equation ln(p/Torr) = 18.818 + 4567.6/T; this gives an extrapolated boiling point of 101.7°C, ΔH_v = 9.076 kcal/mol, and a Trouton constant of 24.2 cal · mol^{-1} · K^{-1} [20].

$Ge(CHR'CHR'COOH)H_3$ (R' = H or CH_3, Table **45**, Nos. **33** to **36**) reacted with Br_2 in $CHCl_3$ at −77°C to +20°C to give the corresponding $Ge(CHR'CHR'COOH)(H)Br_2$ compounds [45].

$Ge(CHR'CH_2COOC_2H_5)H_3$ (R' = $CHOC(CH_3)_2OCH_2$), threo and erythro, Table **45**, Nos. **37** and **38**) were prepared from $GeHCl_3$ and Z-$OC(CH_3)_2OCH_2CHCH{=}CHCOOC_2H_5$ (ca. 1.7:1 mole ratio) in ether at 5°C followed by adding excess t-C_4H_9OK-C_2H_5OH and excess KBH_4;

after 18 h at room temperature, the mixture was quenched with CH_3COOH, concentrated under vacuum, and neutralized in $CH_3COOC_2H_5$ solution. Column chromatography over SiO_2 in C_6H_{14}-$CH_3COOC_2H_5$ (5:1) yielded 25% threo and 26% erythro compound as an oil [47]; see also [46]. A similar treatment of E-$\overline{OC(CH_3)_2OCH_2C}HCH=CHCOOC_2H_5$ with $GeHCl_3$ gave a 2:1 ratio of the threo and erythro isomers (ca. 20% total yield) [47].

Heating No. 37 or 38 with Amberlyst 15 in 4:1 dioxane-H_2O at 80°C for 2 h yielded threo- and erythro-Ge($\overline{CHCH_2COOC}HCH_2OH)H_3$ (cf. Section 1.3.3.8, p. 317) [47]; see also [46].

Ge(CH(OH)CH₂CH₂CH₃)H₃ (Table **45**, No. **42**) was obtained with 41% yield by reacting $GeCl_2 \cdot \overline{N(C_2H_5)CH_2C}H_2$ with $CH_3CH_2CH_2CHO$ at 80°C (14 h) in a sealed tube and reducing the mixture with $LiAlH_4$ followed by hydrolysis, concentration, and distillation [34].

References:

[1] Ponomarenko, V. A.; Zueva, G. Ya.; Andreev, N. S. (Izv. Akad. Nauk SSSR Ser. Khim. **1961** 1758/62; Bull. Acad. Sci. USSR Div. Chem. Sci. [Engl. Transl.] **1961** 1639/43).
[2] Egorochkin, A. N.; Khidekel', M. L.; Ponomarenko, V. A.; Zueva, G. Ya.; Svirezheva, S. S.; Razuvaev, G. A. (Izv. Akad. Nauk SSSR Ser. Khim. **1963** 1865/8; Bull. Acad. Sci. USSR Div. Chem. Sci. [Engl. Transl.] **1963** 1717/9).
[3] Cradock, S.; Gibbon, G. A.; Van Dyke, C. H. (Inorg. Chem. **6** [1967] 1751/2).
[4] Gibbon, G. A.; Wang, J. T.; Van Dyke, C. H. (Inorg. Chem. **6** [1967] 1989/94).
[5] Mazerolles, P.; Manuel, G. (Bull. Soc. Chim. Fr. **1967** 2511/5).
[6] Kuznesof, P. M.; Jolly, W. L. (Inorg. Chem. **7** [1968] 2574/7).
[7] Rivière, P.; Satgé, J. (C.R. Seances Acad. Sci. C **267** [1968] 267/9).
[8] Kuz'min, O. V.; Nametkin, N. S.; Chernysheva, T. I.; Gar, T. K.; Lepetukhina, N. A.; Mironov, V. F. (Dokl. Akad. Nauk SSSR **193** [1970] 826/7; Dokl. Chem. [Engl. Transl.] **190/195** [1970] 550/2).
[9] Gibbon, G. A.; Kifer, E. W.; Van Dyke, C. H. (Inorg. Nucl. Chem. Lett. **6** [1970] 617/20).
[10] Bellama, J. M.; McCormick, C. J. (Inorg. Nucl. Chem. Lett. **7** [1971] 533/6).

[11] Massol, M.; Barrau, J.; Rivière, P.; Satgé, J. (J. Organomet. Chem. **30** [1971] 27/41).
[12] Ohno, K.; Murata, H. (Bull. Chem. Soc. Jpn. **45** [1972] 3333/42).
[13] Van Dyke, C. H.; Kifer, E. W.; Gibbon, G. A. (Inorg. Chem. **11** [1972] 408/12).
[14] Ohno, K.; Murakami, M.; Hayashi, M.; Murata, H. (J. Sci. Hiroshima Univ. Phys. Chem. A **37** [1973] 345/56).
[15] Barrau, J.; Satgé, J.; Massol, M. (Helv. Chim. Acta **56** [1973] 1638/46).
[16] Nakagawa, J.; Hayashi, M. (Chem. Lett. **1974** 1379/80).
[17] Krisher, L. C.; Watson, W. A.; Morrison, J. A. (J. Chem. Phys. **60** [1974] 3417/20).
[18] Krisher, L. C.; Watson, W. A.; Morrison, J. A. (J. Chem. Phys. **61** [1974] 3429/33).
[19] Strom, K. A.; Jolly, W. L. (J. Organomet. Chem. **69** [1974] 201/4).
[20] Bellama, J. M.; Gerchman, L. L. (Inorg. Chem. **14** [1975] 1618/21).

[21] Nakagawa, J.; Hayashi, M. (Bull. Chem. Soc. Jpn. **49** [1976] 3441/8).
[22] Rivière, P.; Satgé, J.; Castel, A. (C.R. Seances Acad. Sci. C **282** [1976] 971/4).
[23] Rivière, P.; Satgé, J.; Castel. A. (C.R. Seances Acad. Sci. C **284** [1977] 395/8).
[24] Yang, D. J.; Jolly, W. L. (Inorg. Chem. **16** [1977] 2834/7).
[25] Yang, D. J.; Jolly, W. L. (Inorg. Chem. **17** [1978] 621/4).
[26] Ikegami, S.; Suzuki, Y. (Jpn. Kokai Tokkyo Koho 80-35007 [1978/80]; C.A. **93** [1980] No. 168411).
[27] Ponec, R.; Dejmek, L.; Chvalovsky, V. (J. Organomet. Chem. **197** [1980] 31/7).
[28] Ponec, R.; Dejmek, L.; Chvalovsky, V. (Collect. Czech. Chem. Commun. **5** [1980] 2895/902).

[29] Dejmek, L.; Ponec, R.; Chvalovsky, V. (Collect. Czech. Chem. Commun. **45** [1980] 3518/24).
[30] Dejmek, L.; Ponec, R.; Chvalovsky, V. (Collect. Czech. Chem. Commun. **45** [1980] 3510/7).
[31] Eujen, R.; Bürger, H. (Spectrochim. Acta A **37** [1981] 1029/34).
[32] McKean, D. C.; Torto, I.; Morrisson, A. R. (J. Organomet. Chem. **226** [1982] C47/C51).
[33] Drake, J. E.; Gorzelska, K.; Helbing, R.; Eujen, R. (J. Electron Spectrosc. Relat. Phenom. **26** [1982] 19/30).
[34] Barrau, J.; Bouchaut, M.; Lavayssière, H.; Dousse, G.; Satgé, J. (J. Organomet. Chem. **243** [1983] 281/90).
[35] McKean, D. C.; Torto, I.; Mackenzie, M. W. (Spectrochim. Acta A **39** [1983] 399/408).
[36] McKean, D. C.; Torto, I.; Mackenzie, M. W.; Morrisson, A. R. (Spectrochim. Acta A **39** [1983] 387/98).
[37] Eujen, R.; Mellies, R. (J. Fluorine Chem. **22** [1983] 263/80).
[38] Ponec, R.; Chvalovsky, V.; Voronkov, M. G. (J. Organomet. Chem. **264** [1984] 163/8).
[39] McKean, D. C.; Mackenzie, M. W.; Morrisson, A. R. (J. Mol. Struct. **116** [1984] 331/44).
[40] Eujen, R.; Mellies, R.; Petrauskas, E. (J. Organomet. Chem. **299** [1986] 29/40).
[41] Burshtein, K. Ya.; Shorygin, P. P. (Dokl. Akad. Nauk SSSR **296** [1987] 903/6; Dokl. Phys. Chem. [Engl. Transl.] **292/297** [1987] 953/6).
[42] Reynes, A.; Dufor, C.; Mazerolles, P.; Morancho, R. (J. Phys. Colloq. [Paris] **1989** C5-757/C5-764).
[43] Sullivan, J. F.; Whang, C. M.; Durig, J. R.; Bürger, H.; Eujen, R.; Cradock, S. (J. Mol. Struct. **223** [1990] 457/70).
[44] Schmidbaur, H.; Rott, J. (Z. Naturforsch. **45b** [1990] 961/6).
[45] Kakimoto, N.; Yoshihara, T.; Asai Germanium Research Institute (Ger. Offen. 40 02 651 A 1 [1990]; C.A. **114** [1991] No. 62357).
[46] Takahashi, Y.; Kakimoto, N.; Asai Germanium Research Institute (Jpn. Kokai Tokkyo Koho 02-131490 [90-131490] [1988/90]; C.A. **113** [1990] No. 212348).
[47] Takahashi, Y.; Kakimoto, N. (J. Organomet. Chem. **399** [1990] 47/51).
[48] Aggarwal, M.; Geanangel, R. A.; Ghuman, M. A. (Main Group Met. Chem. **14** No. 5 [1991] 263/9).
[49] Pola, J.; Parsons, J. P.; Taylor, R. (J. Mater. Chem. **2** [1992] 1289/92).

1.3.3.5 $GeRH_3$ Compounds with R = Cycloalkyl

The compounds in this section are listed in Table 48. Further information on each compound is given at the end of the Table.

Table 48 '
$GeRH_3$ Compounds with R = Cycloalkyl.
Explanations, abbreviations, and units are given on p. X.

No.	R group	formation (yield) properties and remarks
1	C_3H_5-c	$Ge(C_3H_5\text{-c})Cl_3$ (from $GeCl_4$ + c-C_3H_5Li [12]) + $LiAlH_4$ in $O(C_4H_9)_2$ at 20 to 50 °C (4 h) followed by fractionation (14%) [2]; see also [3]; $Ge(C_3H_5\text{-c})D_3$ similarly prepared with $LiAlD_4$ [3, 13] and partially

Table 48 (continued)

No.	R group	formation (yield) properties and remarks
		deuterated species prepared from a mixture of $LiAlH_4$ and $LiAlD_4$ [13] b.p. 42.3 to 42.5°C/710 [2, 3], 44°C/760 [12] n_D^{20} = 1.4363 [2] IR (gas): ν(CH) 3080 to 3017, ν(GeH) 2070, ν(GeC) 600 MS: $[M]^+$ (1), $[M-H]^+$ (3), $[M-H_2]^+$ (22), $[GeCH_3]^+$ (55), $[Ge]^+$ (100), $[C_3H_5]^+$ (14) [2]
2	C_4H_7-c	$Ge(C_4H_7\text{-c})Cl_3$ + $LiAlH_4$ in ether (45%) [2]; see also [5]; purified on a low-temperature vacuum sublimation column [9] b.p. 73 to 74°C/715 [2, 5] n_D^{20} = 1.4673 [2] IR (gas): ν(CH) 2980 to 2880, ν(GeH) 2065, ν(GeC) 565; see also p. 306 MS: $[M]^+$ (8), $[M-H]^+$ (3), $[M-H_2]^+$ (14), $[GeC_2H_5]^+$ (24), $[GeC_2H_4]^+$ (25), $[Ge]^+$ (58), $[C_4H_7]^+$ (100) [2]
3	C_5H_9-c	$Ge(C_5H_9\text{-c})Cl_3$ + $LiAlH_4$ in refluxing ether (4 h) and at 20°C (14 h) followed by fractionation (51%); purified by GC [2]; $Ge(C_5H_9\text{-c})D_3$ similarly prepared with $LiAlD_4$ [3] b.p. 105 to 105.4°C/706 n_D^{20} = 1.4666 [2] IR (neat): ν(CH) 2950 to 2865, ν(GeH) 2055, ν(GeC) 555 [2]; see also p. 305 MS: $[M]^+$ (6), $[M-H]^+$ (3), $[M-H_2]^+$ (7), $[GeC_3H_8]^+$ (7), $[GeC_3H_6]^+$ (4), $[GeCH_3]^+$ (8), $[Ge]^+$ (21), $[C_5H_9]^+$ (100); m/e = 68 (58), 41 (37) [2]
4	C_6H_{11}-c	only mentioned in connection with force field calculations [1]; see p. 306
5	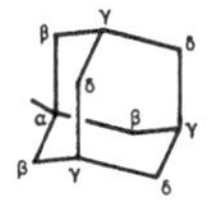	$Ge(C_{10}H_{15}\text{-}1)Cl_3$ + $LiAlH_4$ in refluxing ether (4 h) followed by hydrolysis with aqueous HCl (61%) b.p. 76 to 77°C/5 n_D^{20} = 1.5240 ^1H NMR (CCl_4): 1.8 to 1.9 ($C_{10}H_{15}$), 3.6 (HGe) ^{13}C NMR (C_6D_6): 27.2 (C-α), 29.6 (C-γ), 37.8 (C-β), 42.6 (C-δ) IR (suspension in mineral oil): ν(GeH) 2050 MS: m/e = 212 $[M]^+$, 135 $[C_{10}H_{15}]^+$ [6] with excess I_2 in o-xylene at 0 to 20°C (ca. 10 h) → $Ge(C_{10}H_{15}\text{-}1)I_3$ [6] with $SnCl_2$ in ether at ca. 20°C (ca. 6 h) → $Ge(C_{10}H_{15}\text{-}1)H_2Cl$ [7]

*Further information:

$Ge(C_3H_5\text{-c})H_3$ (Table **48**, No. **1**). For the preparation of cyclopropyl-deuterated derivatives the corresponding deuterated bromocyclopropane had to be prepared first which then was metallated with Li, reacted with $GeCl_4$, and reduced by $LiAlH_4$ [13].

References on p. 306

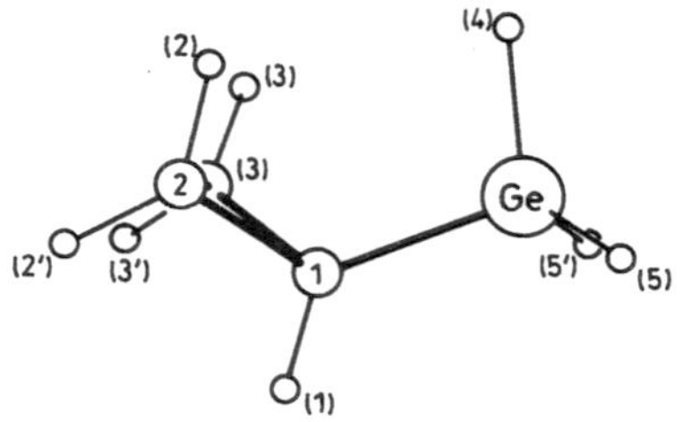

Fig. 4. Perspective view of $Ge(C_3H_5\text{-}c)H_3$ [13].

The gas-phase molecular structure of No. 1 was investigated by electron diffraction giving the following parameters (for the labeling of atoms, see Fig. 4): bond lengths Ge-C(1) = 1.924(2), C(1)-C(2,3) = 1.521(7), C(2)-C(3) = 1.502(9), Ge-H = 1.530, C-H = 1.091(3) Å, and angles C(1)-Ge-H(4) = 108.8(1.2)°, C(1)-Ge-H(5,5′) = 113.9° (calculated), and H-C-H = 118.2(2.3)°; the angles with the ring plane R are C(1)-Ge-R = 55.5(1.6)° and C(1)-H(1)-R = 57.3(1.9)°. The agreement with data obtained from ab initio MO calculations was fairly good. Both studies revealed the GeH_3 group to be tilted toward the ring plane. This was interpreted as a result of hyperconjugative interaction. The structural results assess the strong π-donor character of the cyclopropyl system and demonstrate that the π-acceptor ability of the GeH_3 group is less pronounced than that of the SiH_3 group [12].

The molecular structure of $Ge(C_3H_5\text{-}c)H_3$ was also derived from the microwave spectra of 41 isotopomers. The two vicinal bonds C(1)-C(2,3) are longer (1.520 Å) than the distal C(2)-C(3) bond (1.504 Å). The germyl group is tilted by 3.7° towards the ring, and the inward Ge-H(4) bond (1.538 Å) is longer than the two equivalent outward Ge-H(5,5′) bonds (1.531 Å). The C(2,3)-H(2,3) bonds are not only longer (1.096 Å) than the C(2,3)-H(2′,3′) bonds (1.085 Å), but are also bent towards the GeH_3 group; see **Fig. 4.** This structure was obtained with no assumptions other than molecular C_s symmetry. The rotational constants A, B, and C of 40 spectra of $Ge(C_3H_5\text{-}c)H_3$ isotopomers are listed in [13].

From several doublet transitions of the first excited state of germyl torsion, the barrier to internal rotation was calculated: V_3 = 5.58(20) kJ/mol = 1.334(50) kcal/mol [13]; ab initio calculations gave V_3 = 1.370 kcal/mol [12]. V_3 = 473.9 cm^{-1} or 1.360 kcal/mol resulted from combination band spectrum in the range 1530 to 1600 cm^{-1} (depicted) arising from a combination of the internal rotation with the Ge-H stretching mode; this barrier is only slightly higher than that of $Ge(CH_3)H_3$ in contrast to the Si analog [10].

The electric dipole moment, μ = 0.732(9) D, was obtained from Stark-effect measurements. It is primarily directed along the Ge-C bond [13].

The IR spectrum of $Ge(C_3H_5\text{-}c)(H)D_2$ vapor (impurity in $Ge(C_3H_5\text{-}c)D_3$) showed two ν(GeH) bands (2079.5 and 2073.8 cm^{-1}; illustrated) indicating bond strength differences within the germyl group. The higher-frequency band was attributed to the Ge-H(5,5′) bonds and the other band to the Ge-H(4) bond [3, 4]. The Taft σ^* parameter of the cyclopropyl group was deduced to be −0.08 using a correlation equation between ν(GeH) and the sum of the Taft σ^* parameters; this value reflects the smaller +I effect of cyclopropyl compared with conventional values of −0.19 for i-C_3H_7 and −0.10 for C_2H_5 [3].

$Ge(C_4H_7\text{-}c)H_3$ (Table **48**, No. **2**). The molecular structure and conformation of $Ge(C_4H_7\text{-}c)H_3$ was determined by electron diffraction (see **Fig. 5**). The predominance (77:23 ratio) of the quasi-equatorial conformer (ΔG = 3.1(1) kJ/mol), the near equality of the skeleton C-C bond lengths and the values of the puckering angles for the equatorial (25.3(3.1)°)

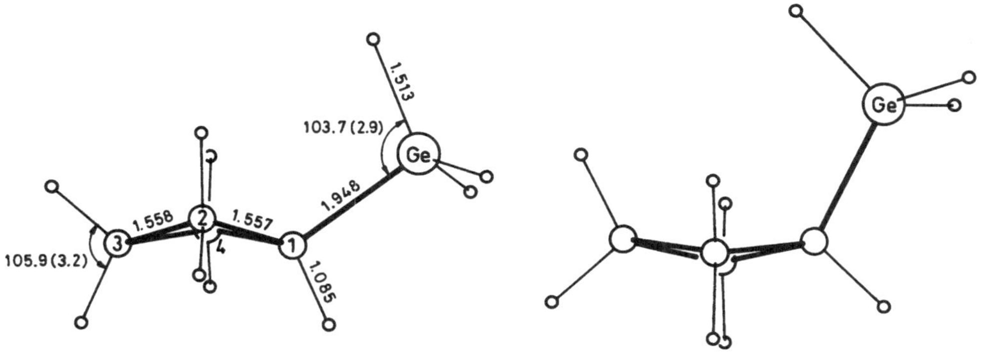

a b

Fig. 5. Molecular models for (a) quasi-equatorial and (b) quasi-axial conformers of $Ge(C_4H_7\text{-}c)H_3$; the angle between Ge-C(1) and the C(1), C(2), C(4) plane is 131.2(1.1)° [5].

and axial form (20.4(3.6)°) were seen as evidence for a correlation between structural parameters in four-membered rings and the electronegativity of substituents. A comparison of structural factors in $Ge(C_4H_7\text{-}c)H_3$, $Si(C_4H_7\text{-}c)H_3$, and other monosubstituted cyclobutanes indicates that the Ge atom is more electronegative than the Si atom [5]. A series of Q branches obtained from the low-frequency Raman spectrum of the vapor was assigned to the ring-puckering vibration of the two conformers and was fitted to an asymmetric potential function which is consistent with a more stable equatorial conformation by 191 cm^{-1} (546 cal/mol), an equatorial-to-axial barrier of 432 cm^{-1} (1240 cal/mol), and ring-puckering angles of 18° (equatorial) and 14° (axial) [11].

The barriers to internal rotation of the GeH_3 group, $V = 440 \pm 2$ cm^{-1} (1.26 kcal/mol) and 415 ± 10 cm^{-1} (1.19 kcal/mol) for the equatorial and axial conformers, respectively, were calculated from the frequencies of the germyl torsional modes which were obtained from sum and difference bands on the GeH_3 stretching modes [11].

An analysis of the quadratic Stark effect indicated an electric dipole moment along the a axis with a small μ_c component for both conformers [9].

The IR spectra (gas and solid) and Raman spectra (gas, liquid, and solid) were all listed (illustrated) and assigned [11]. Some fundamental vibrations are given in Table 49, p. 306. A comparison of the spectra of the liquid with that of the solid showed that two stable conformers exist in the liquid at ambient temperature with the equatorial conformer the predominant form in the gas and liquid and the only conformer present in the solid [11].

The microwave spectrum (26.5 to 39.5 GHz; displayed) of No. 2 was investigated, and the spectra of three isotopic species of the equatorial and axial conformers were identified and assigned for the ground state. Rotational constants were determined for all six species. There is a difference of approximately 6° in the puckering angle of the four-membered ring between the equatorial and axial conformers, with the axial being less puckered [9].

$Ge(C_5H_9\text{-}c)H_3$ (Table **48**, No. **3**). The IR spectrum of gaseous $Ge(C_5H_9\text{-}c)(H)D_2$ (impurity in $Ge(C_5H_9\text{-}c)D_3$) showed two ν(GeH) bands at 2072.3 and 2067.3 cm^{-1} (displayed) assigned as No. 1, indicating bond strength differences within the germyl group [3, 4]. For the σ^* parameter of the germyl group, see No. 1 [3].

References on p. 306

Table 49
Vibrational Spectra of $Ge(C_4H_7\text{-}c)H_3$ [11]; Selected Fundamental Vibrations. Wavenumbers in cm^{-1}.

IR gas	Raman liquid	solid	assignment*)	
		3006(w,p)	$\nu_{as}(CH/CH_2\text{-}\gamma)$	$\nu_{1'}$
2980(s)	2970(sh,p)	2969(s)	$\nu_{as}(CH/CH_2\text{-}\gamma)$	ν_1
2968(s)	2954(s,p)	2964(s)	$\nu_{as}(CH/CH_2\text{-}\beta)$	ν_2
2927(sh)	2911(s,p)	2915(s)	$\nu_s(CH/CH_2\text{-}\gamma)$	ν_3
2073(sh)	2859(s,p)	2851(s)	$\nu_s(CH/CH_2\text{-}\beta)$	ν_3
2072(vs)		2060(s)	$\nu_{as}(GeH)$	ν_6, ν_{26}
2068(vs)	2059(vs,p)	2050(vs)	$\nu_s(GeH)$	ν_7
1071(w)	1065(w,p)	1063(m)	ring deformation	ν_{13}
1036(w)	1032(w,p)			$\nu_{13'}$
883(s)		877(m)	$\delta_{as}(GeH)$	ν_{15}
876(s)	878(m,dp)	873(m)	$\delta_{as}(GeH)$	ν_{34}
828(vs)	824(m)	828(m)	$\delta_s(GeH)$	ν_{17}
675(w)	665(w,p)		$\rho(GeH)$	$\nu_{20'}$
588(m)			$\rho(GeH)$	ν_{20}
566(m)	563(w,dp,br)	562(m)	$\rho(GeH)$	ν_{37}
484(w)	486(m,p)		$\nu(GeC)$	$\nu_{21'}$
440(w)	441(m,p)	447(m)	$\nu(GeC)$	ν_{21}
262(m,br)	264(m,p,br)	265(w)	ring-GeH_3 bend	ν_{22}
186(vw)		212(vw)	ring-GeH_3 bend	ν_{38}
135(w,br)		157(vw)	ring puckering	ν_{23}
		130(vw)	GeH_3 torsion	ν_{39}

*) ν' refers to the assignments made for the axial conformer.

$Ge(C_6H_{11}\text{-}c)H_3$ (Table **48**, No. **4**). Force field calculations were carried out to determine the conformational equilibria of several Group-IV organometallic compounds including No. 4. The difference of the conformational energies for the equatorial and axial conformations of $Ge(C_6H_{11}\text{-}c)H_3$ (1.72 and 2.92 kcal/mol, respectively; ΔE = 1.20 kcal/mol) is significantly lower than that of CH_3 in methylcyclohexane (ΔE = 1.7 kcal/mol) [1].

$Ge(C_{10}H_{15}\text{-}1)H_3$ (Table **48**, No. **5**). The higher dipole moments of (1-adamantyl)germanes, $Ge(C_{10}H_{15}\text{-}1)X_3$, ($\mu$ = 0.88 D for No. 5 at 25°C) compared with those of the corresponding alkyl germanes suggest electron-acceptor properties of the adamantyl substituent with respect to the Ge atom [8].

References:

[1] Ouelette, R. J. (J. Am. Chem. Soc. **94** [1972] 7674/9).
[2] Dakkouri, M.; Kehrer, H. (Chem. Ber. **116** [1983] 2041/3).
[3] McKean, D. C.; Morrison, A. R.; Dakkouri, M. (Spectrochim. Acta A **40** [1984] 771/4).
[4] McKean, D. C. (J. Mol. Struct. **113** [1984] 251/66).
[5] Dakkouri, M. (J. Mol. Struct. **130** [1985] 289/300).
[6] Gar, T. K.; Chernysheva, O. N.; Kisin, A. V.; Mironov, V. F. (Zh. Obshch. Khim. **55** [1985] 1057/63; J. Gen. Chem. USSR [Engl. Transl.] **55** [1985] 942/7).

[7] Chernysheva, O. N.; Gar, T. K.; Kisin, A. V.; Mironov, V. F. (Zh. Obshch. Khim. **55** [1985] 2333/8; J. Gen. Chem. USSR [Engl. Transl.] **55** [1985] 2073/7).
[8] Kartsev, G. N.; Akin'shina, G. A.; Gar, T. K.; Chernysheva, O. N. (Zh. Obshch. Khim. **57** [1987] 1576/80; J. Gen. Chem. USSR [Engl. Transl.] **57** [1987] 1404/7).
[9] Durig, J. R.; Geyer, T. J.; Groner, P.; Dakkouri, M. (Chem. Phys. **125** [1988] 299/305).
[10] Kelly, M. B.; Laane, J.; Dakkouri, M. (J. Mol. Spectrosc. **137** [1989] 82/6).
[11] Durig, J. R.; Little, T. S.; Geyer, T. J.; Dakkouri, M. (J. Phys. Chem. **93** [1989] 6296/303).
[12] Dakkouri, M. (J. Am. Chem. Soc. **113** [1991] 7109/14).
[13] Epple, K. J.; Rudolph, H. D. (J. Mol. Spectrosc. **152** [1992] 355/76).

1.3.3.6 $GeRH_3$ Compounds with R = Alkenyl and Alkynyl

The compounds in this section are listed in Table 50. The formation of Ge(CH(OH)-CH=CHR')H_3 compounds (R' = H or CH_3) as intermediates during the reduction of polymeric (-F_2GeOCH(CH=CHR')-$)_n$ (from GeF_2 and R'CH=CHCHO) with $LiAlH_4$ was briefly mentioned. The end products were the cyclic compounds I and polymers [15].

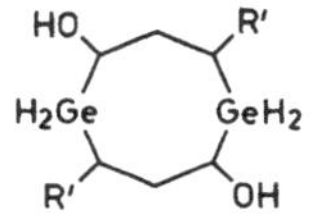

I

Table 50
$GeRH_3$ Compounds with R = Alkenyl and Alkynyl.
An asterisk indicates further information at the end of the table.
Explanations, abbreviations, and units are given on p. X.

No.	R group	formation (yield) properties and remarks
*1	$CH=CH_2$	Ge($CH=CH_2$)Cl_3 + $LiAlH_4$ in diglyme at ca. 100 °C (30 min) followed by trap-to-trap distillation under high vacuum and GC (83%); 58% yield with O$(C_4H_9)_2$ as the solvent [1]; see also [10]; Ge($CH=CH_2$)D_3 similarly prepared except using $LiAlD_4$ [10] b.p. −3.5 °C/760 [1, 2]
*2	$CH_2CH=CH_2$	Ge($CH_2CH=CH_2$)Cl_3 + LiH in dioxane (28%); Ge($CH_2CH=CH_2$)D_3 similarly prepared except using LiD (24%) b.p. 37 °C/752.7; d^{20} = 1.0797; n_D^{20} = 1.4315 for Ge($CH_2CH=CH_2$)H_3 b.p. 37 °C/744; d^{20} = 1.1146; n_D^{20} = 1.4320 for Ge($CH_2CH=CH_2$)D_3 Raman (liquid): ν(GeH) 2069 for Ge($CH_2CH=CH_2$)H_3; ν(GeD) 1480 for Ge($CH_2CH=CH_2$)D_3 [3]
3	$CH_2CH=CHCH_3$	pyrolysis of Ge_2H_6 and CH_2=CHCH=CH_2 (1:10 mole ratio) at 230 °C (1 h) in a flow system (6%)

Table 50 (continued)

No.	R group	formation (yield) properties and remarks
3 (continued)		^1H NMR ($CDCl_3$): 1.65 (complex, CH_3 and CH_2), 3.54 (t, HGe), 5.40 (complex, CH=CH) IR (vapor): 3015(m), 2920(s), 2850(w), 2060(vs), 1410(m), 965(m), 882(s), 833(vs), 698(m), 622(m), 570(w), 500(m) [14]
4	$CH(CH_2CH_2OH)$-$CH{=}CHCH_3$	$Ge(CH(CH_2COOH)CH{=}CHCH_3)I_3$ + $LiAlH_4$ in refluxing ether for 4 h (47%); purified by GC b.p. 89 °C/13 d^{20} = 1.152(2); n_D^{20} = 1.490(6) ^1H NMR (CCl_4): 1.30 (d, CH_3), 1.7 and 2.2 (m's, CH_2CHGe), 3.50 (t, CH_2O), 3.65 (d, HGe), 4 (s, OH), 5.0 to 5.9 (m, CH=CH) IR: ν(OH) 3300, ν(GeH) 2060, ν(C=C) 1640 [5]
*5	C≡CH	GeH_3Br + CH≡CMgBr in THF at −195 to +20 °C (2 h) in a high-vacuum system followed by trap-to-trap distillation and GC (ca. 25%); $Ge(C{\equiv}CH)D_3$ similarly prepared except using GeD_3Br m.p. −86 °C [6]
*6	C≡CCl	GeH_3Cl + LiC≡CCl (from $LiCH_3$ + ClCH=CHCl in ether) followed by trap-to-trap distillation (ca. 55%) m.p. 117 K; b.p. 329 K vapor pressure: log (p/Torr) = 8.232 − 1761/T ^1H NMR ($Si(CH_3)_4$): 4.00 (s, HGe) IR (neat): ν(C≡C) 2172, ν_{as}(GeH) 2117.5, ν_s(GeH) 2116, $\delta_{as}(GeH_3)$ 884.1, $\delta_s(GeH_3)$ 833, $\rho(GeH_3)$ 624.3 [16]
*7	$C{\equiv}CCF_3$	GeH_3Cl + $CF_3C{\equiv}CMgI$ in DME followed by trap-to-trap distillation (40%) ^1H NMR ($CFCl_3$-$Si(CH_3)_4$): 4.11 (q, H_3Ge, ^5J(F, H) = 1.8) MS: $[GeH_nCCCF_3]^+$, $[GeH_nCCCF_2]^+$, $[GeH_nCCCF]^+$, $[GeH_nCC]^+$ (n = 0 to 3); $[CCCF_m]^+$, $[CF_m]^+$ (m = 1 to 3) [21]

* Further information:

$Ge(CH{=}CH_2)H_3$ (Table **50**, No. **1**). The heat of formation, ΔH_f° = −18.9 ± 2 kcal/mol, was estimated based on a linear relation between ΔH_f° values of C_6H_5X and $CH_2{=}CHX$ compounds [29].

The following structural parameters were calculated from 25 rotational constants and assuming a structure for the vinyl group (except the C=C bond distance): r(Ge-C) = 1.926, r(Ge-H) = 1.521, and r(C-C) = 1.347 Å; angles <(C-C-Ge) = 122°54′ and <(C-Ge-H) = 110°42′ [13]. For a comparison of observed geometrical parameters with those derived from AM1 calculations, see [25]. The barrier to internal rotation of the GeH_3 group around the Ge-C bond was determined from the ground- and excited-state splittings of the rotational transitions: V = 1.238 ± 0.057 kcal/mol [13]. Ab initio MO calculations gave a value of 1.440 kcal/mol [27].

The dipole moment μ = 0.50 ± 0.03 D was obtained from measurements of the quadratic Stark effect of two ground-state transitions [13].

The IR spectra of gaseous $Ge(CH{=}CH_2)H_3$ (see also [1]) and $Ge(CH{=}CH_2)D_3$ were recorded and assigned from 4000 to 33 cm^{-1}. The fundamental vibrations of the IR and Raman spectra are listed in Table 51 (all spectra displayed). The rotational fine structure of the best resolvable band at 950 cm^{-1} was discussed [10].

Table 51
Vibrational Spectra of $Ge(CH{=}CH_2)H_3$; Fundamental Modes (C_s Symmetry) [10].
Wavenumbers in cm^{-1}.

IR gas	Raman*)	assignment	
3075 (R)			
3066 (Q) (m)	3062 (55; 0.50)	$\nu_{as}(CH_2)$	ν_1
3059 (P)			
3011 (R)			
3002 (Q) (m)	2992 (66; 0.06)	$\nu_s(CH_2)$	ν_2
2993 (P)			
2969 (R)			
2959 (Q) (w)	2950 (16; 0.08)	$\nu(CH)$	ν_3
2950 (P)			
2096 (R)			
2089 (centered) (vs)	2078 (100; 0.01)	$\nu(GeH)$	ν_4,ν_5,ν_{15}
2082 (P)			
	1595 (5;?)	$\nu(C{=}C)$	ν_6
1406 (R)			
1398 (Q) (ms)	1396 (40; 0.15)	$\delta(CH_2)$	ν_7
1390 (P)			
1268 (m)	1255 (5; 0.05)	δ(CH in-plane)	ν_8
1007 (R)			
998 (Q) (m)	1006 (1;?)	twist(CH_2)	ν_{16}
950 (Q) (m)		wag(CH_2)	ν_{18}
894 (R)			
887 (centered) (mw)	879 (4; 0.68)	$\delta_{as}(GeH_3)$	ν_9, ν_{17}
880 (P)			
876 (Q) (m)		$\rho(CH_2)$	ν_{10}
843 (R)			
834 (Q) (vs)	829 (4; 0.23)	$\delta_s(GeH_3)$	ν_{11}
828 (P)			
645 (R)			
639 (Q) (vs)	633 (2; 0.06)	$\nu(GeC)$	ν_{12}
633 (P)			
550 (R)			
543 (Q) (s)	524 (2; 0.11)	$\rho(GeH_3)$	ν_{13}
535 (P)			
420 (w)	426 (1; dp)	$\rho(GeH_3)$	ν_{20}
	267 (1.5; 0.15)	$\delta(GeC)$	ν_{14}

*) relative intensity and depolarization ratio in parentheses.

References on p. 312

The microwave spectra of ten isotopic species of $Ge(CH{=}CH_2)H_3$ and $Ge(CH{=}CH_2)D_3$ were recorded in the range 12.4 to 40 MHz [13].

The vapor pressure of $Ge(CH{=}CH_2)H_3$ can be expressed by the equation log (p/Torr) = 7.564 − 1263/T; this gives an extrapolated boiling point of −3.5 °C and a Trouton constant of 21.4 cal · mol^{-1} · K^{-1}. Selected experimental vapor pressures are as follows [1]:

t in °C	−77	−63.3	−51.6	−45.1	−36.6
p in Torr	13.4	35.3	73.3	106.4	177.0

$Ge(CH{=}CH_2)H_3$ polymerized in day light, especially in the presence of Hg, to yield a white solid of the composition $(GeC_2H_6)_n$ [1].

$Ge(CH_2CH{=}CH_2)H_3$ (Table **50**, No. **2**) and $Ge(CH_2CH{=}CH_2)D_3$ were part of an investigation dealing with a relation between the ν(GeH) or ν(GeD) frequencies in the Raman spectra and the inductive effect of substituents at Ge for a series of Ge hydrides and deuterides [3].

$Ge(CH_2CH{=}CH_2)H_3$ was included in quantum-chemical studies on the problem of conjugation effects between a double bond and a group containing a heavy atom. Various aspects of σ-π interaction in $E(CH_2CH{=}CH_2)H_3$ compounds (E = C, Si, Ge, Sn, Pb) in the ground and excited states were discussed [23, 24]. MNDO calculations showed that the interaction energy between the π(C=C) and σ(E-C,E-H) orbitals does not depend on the nature of E and is similar to the π-π interaction energy between occupied π orbitals of butadiene [26].

$Ge(CH_2CH{=}CH_2)H_3$ polymerized on standing or heating to form a solid of the composition $(Ge(CH_2CH_2CH_2)H_2)_n$ [3].

$Ge(C{\equiv}CH)H_3$ (Table **50**, No. **5**). The following structural parameters were calculated from the rotational constants assuming r(C-H) = 1.056 Å: r(Ge-C) = 1.896, r(Ge-H) = 1.521, r(C≡C) = 1.208 Å, and <(H-Ge-H) = 109.9° [4].

The dipole moment μ = 0.136 D was calculated from second-order Stark effect measurement; the very low value could not be explained [4]. It can possibly be attributed to a mesomeric effect, as the evaluation of the formal charge distribution yielded a higher moment of 0.84 D [12].

The IR spectra of $Ge(C{\equiv}CH)H_3$ and $Ge(C{\equiv}CH)D_3$ are listed in Table 52 (all spectra displayed) [6, 28]; see also [9]. The Q-branch structure was resolved for the ν(GeH), $\delta_{as}(GeH_3)$, and $\rho(GeH_3)$ bands of $Ge(C{\equiv}CH)H_3$, and values for all five Coriolis coupling constants were obtained from analysis or from the zeta sum rule [6]. A vibrational analysis was carried out for $Ge(C{\equiv}CH)H_3$ and $Ge(C{\equiv}CH)D_3$ using Wilson's F-G matrix technique; the resulting valence force constants (see also [7]), Coriolis coupling and centrifugal distortion constants (see also [18]) were listed [22]; for $Ge(C{\equiv}CH)H_3$, see also [28]. Force constants and Coriolis coupling constants were calculated and compared for the analogous $E(C{\equiv}CH)H_3$ molecules with E = C, Si, Ge) and the importance of π(d-p) bonding discussed [9]. Other methods of force field calculations for $Ge(C{\equiv}CH)H_3$ were also reported [8], and force field calculations for Ge(R)-$(H,D)_3$ compounds (R = CH_3, C≡CH, C≡N) were briefly discussed in an abstract [11]. $Ge(C{\equiv}CH)H_3$ was included in studies on the isolated ν(GeH) frequency [19] and its relation with the sum of Taft σ* values [20].

The microwave spectrum of $Ge(C{\equiv}CH)H_3$ was investigated in the range 7 to 36 GHz. Rotational constants and moments of inertia were listed for nine isotopic species. The hyperfine components of $^{73}Ge(C{\equiv}CH)H_3$ transitions led to a nuclear quadrupole coupling constant of eQq = +32.5 MHz [4].

Table 52
IR Spectra of $Ge(C{\equiv}CH)H_3$ and $Ge(C{\equiv}CH)D_3$; Fundamental Modes (C_{3v} Symmetry). Wavenumbers in cm^{-1}.

$Ge(C{\equiv}CH)H_3$ gas [6]	in KBr [28]	$Ge(C{\equiv}CH)D_3$ gas [6]	assignment	
3320 (R)		3324 (R)		
3314.5 (Q)		3317.5 (Q)		
3313.5 (Q′)	3312(s)	3316.5 (Q′)	ν(CH)	ν_1
3312.0 (Q′)		3315.5 (Q′)		
3305 (P)		3308 (P)		
2127 (R)				
2120 (Q)	2122	1525 (Q)	ν_s(GeH, GeD)	ν_2
2112 (P)				
2117.2	2118(w)	1525	ν_{as}(GeH, GeD)	ν_6
2065 (R)		2064 (R)		
2060 (Q)	2055(vs)	2059.8 (Q)	ν(C≡C)	ν_3
2040 (P)		2047 (P)		
886.0	889(vs)	643.2	$\delta_{as}(GeH_3, GeD_3)$	ν_7
850 (R)		614 (R)		
843.8 (Q)	845	608 (Q)	$\delta_s(GeH_3, GeD_3)$	ν_4
836 (P)		602 (P)		
673	675(vs)	673	δ(C≡C-H)	ν_8
643.8	639	484	$\rho(GeH_3, GeD_3)$	ν_9
530 (Q)	528	518 (Q)	ν(GeC)	ν_5
216.4	220	202.8	δ(Ge-C≡C)	ν_{10}

Vapor pressure data of $Ge(C{\equiv}CH)H_3$ in the range 212 to 276 K fit the equation log (p/Torr) = 7.744 − 1345.8/T; this gives a boiling point of 277 K and the heat of vaporization ΔH_v = 6.16 kcal/mol [6].

Thermodynamic functions of $Ge(C{\equiv}CH)H_3$ and $Ge(C{\equiv}CH)D_3$ were evaluated for the ideal gaseous state at atmospheric pressure for the range 100 to 1000 K, assuming an anharmonic and rigid rotor model of the molecule [18]. Earlier calculations assumed a rigid harmonic oscillator model [7].

$Ge(C{\equiv}CCl)H_3$ (Table **50**, No. **6**). The fundamental modes observed in the IR spectrum were assigned assuming C_{3v} symmetry [16]. An analysis of the rotational-fine structure of the IR-active perpendicular bands associated with motions of the GeH_3 group was reported. Rotation and Coriolis constants were given [17].

The UV spectrum exhibits a strong band starting near 43000 cm^{-1} with intensity increasing to beyond 50000 cm^{-1} attributed to the alkyne $\pi \rightarrow \pi^+$ transition [16].

The He(I) photoelectron spectrum has the following vertical ionization potentials (in eV): 10.19 (3e, CC π-Cl np), 12.45 ($5a_1$, GeH, GeC), 12.75 (2e, GeH), 13.83 (1e, Cl np + CC π), 17.1 ($4a_1$, CCl), 18.8 ($3a_1$, GeC, GeH) [16].

The mass spectrum showed a group of peaks (because of Ge isotopes and ready Ge-H bond cleavage) in a region expected for the molecular ion. Major breakdown paths involve

References on p. 312

loss of halogen followed by Ge-C bond cleavage; an alternative path involving a metastable ion leads to extrusion of C_2 leaving $[Ge(Cl)H_n]^+$ [16].

Ge(C≡CCF$_3$)H$_3$ (Table **50**, No. **7**). A boiling point of 294 ± 6 K and the vaporization parameters $\Delta H_v = 32.7 \pm 0.6$ kJ/mol and $\Delta S_v = 111.2$ J · mol^{-1} · K^{-1} were derived from the equation log (p/cm Hg) = 7.705 − 1710/T [21].

All frequencies of the IR spectra of $Ge(C{\equiv}CCF_3)H_3$ and $Ge(C{\equiv}CCF_3)D_3$ were listed, including bands assigned to overtones and combinations (IR spectrum of No. 7 in an Ar matrix also listed) [21]. The fundamental vibrations are compiled in Table 53. Complete IR and Raman spectra were also reported for $Ge(C{\equiv}CCF_3)D_3$; selected IR data: ν_{as}(GeD) 1547(vs), ν_s(GeD) 1524(vs), δ_{as}(GeD) 623.7(m), δ_s(GeD) 597.5(vs), and $\rho(GeD_3)$ 489(s) cm^{-1} [21]. The rotational fine-structure of the IR-active perpendicular bands associated with motions of the GeH_3 group was analyzed and the rotation and Coriolis constants were reported [17].

Table 53
Vibrational Spectra of $Ge(C{\equiv}CCF_3)H_3$; Fundamental Modes (C_{3v} Symmetry) [21].
Wavenumbers in cm^{-1}.

IR gas	Raman liquid	assignment	
2219(m)	2225(m,p)	ν(C≡C)	ν_1
2133.6(s)		ν_{as}(GeH)	ν_8
2128.1(s)	2135(vs,p)	ν_s(GeH)	ν_2
1262.5(vs)	1255(w,p)	ν_s(CF)	ν_3
1173(vs)	1160(w,dp)	ν_{as}(CF)	ν_9
886.0(vs)	885(s)	$\delta_{as}(GeH_3)$	ν_{10}
862(s)	865(m)	$\nu(C\text{-}CF_3)$	ν_5
821.8(vs)	825(s,p)	$\delta_s(GeH_3)$	ν_4
685(m)	660(w,p)	$\delta_s(CF_3)$	ν_6
626.8(vs)		$\rho(GeH_3)$	ν_{11}
	615(w,dp)	$\delta_{as}(CF_3)$	ν_{12}
474(w)	460(m,dp)	$\rho(CF_3)$	ν_{13}
264(m)	270(s,p)	ν(GeC)	ν_7
235(w)	245(vs,p?)	skeletal bend	ν_{14}
76(vw)	90(m,dp)	skeletal bend	ν_{15}

References:

[1] Brinckman, F. E.; Stone, F. G. A. (J. Inorg. Nucl. Chem. **11** [1959] 24/32).
[2] Satgé, J. (Ann. Chim. [Paris] [13] **6** [1961] 519/73).
[3] Ponomarenko, V. A.; Zueva, G. Ya.; Andreev, N. S. (Izv. Akad. Nauk SSSR Ser. Khim. **1961** 1758/62; Bull. Acad. Sci. USSR Div. Chem. Sci. [Engl. Transl.] **1961** 1639/43).
[4] Thomas, E. C.; Laurie, V. W. (J. Chem. Phys. **44** [1966] 2602/4).
[5] Mazerolles, P.; Manuel, G. (Bull. Soc. Chim. Fr. **1967** 2511/5).
[6] Lovejoy, R. W.; Baker, D. R. (J. Chem. Phys. **46** [1967] 658/65).
[7] Rao, D. V. R. A.; Rai, D. K. (Curr. Sci. **37** [1968] 41/2).
[8] Ramaswamy, K.; Balasubramanian, V. (Indian J. Phys. **43** [1969] 735/41).
[9] Parker, J.; Ladd, J. A. (Trans. Faraday Soc. **66** [1970] 1907/16).
[10] Durig, J. R.; Turner, J. B. (Spectrochim. Acta A **27** [1971] 1623/32).

[11] Yarandina, V. N.; Sverdlov, L. M. (Zh. Fiz. Khim. **47** [1973] 267; Russ. J. Phys. Chem. USSR [Engl. Transl.] **47** [1973] 153).
[12] Ramalingam, S. K.; Soundararajan, S. (J. Organomet. Chem. **72** [1974] 59/63).
[13] Durig, J. R.; Kizer, K. L.; Li, Y. S. (J. Am. Chem. Soc. **96** [1974] 7400/4).
[14] Jenkins, R. L.; Kedrowski, R. A.; Elliott, L. E.; Tappen, D. C.; Schlyer, D. J.; Ring, M. A. (J. Organomet. Chem. **86** [1975] 347/57).
[15] Rivière, P.; Satgé, J.; Castel, A. (C. R. Seances Acad. Sci. C **284** [1977] 395/8).
[16] Cradock, S.; Ebsworth, E. A. V.; Green, A. R. (J. Chem. Soc. Dalton Trans. **1978** 759/63).
[17] Cradock, S.; Green, A. R.; Duncan, J. L. (J. Mol. Spectrosc. **77** [1979] 385/401).
[18] Sabapathy, K.; Ramasamy, R.; Venkateswarlu, K. (Indian J. Pure Appl. Phys. **18** [1980] 625/6).
[19] McKean, D. C.; Torto, I.; Mackenzie, M. W.; Morrisson, A. R. (Spectrochim. Acta A **39** [1983] 387/98).
[20] McKean, D. C.; Torto, I.; Mackenzie, M. W. (Spectrochim. Acta A **39** [1983] 399/408).

[21] Anderson, D. W. W.; Cradock, S.; Ebsworth, E. A. V.; Green, A. R.; Rankin, D. W. H.; Robiette, A. G. (J. Organomet. Chem. **271** [1984] 235/47).
[22] Payami, F.; Kuttiappan, P.; Mohan, S. (J. Mol. Struct. **147** [1986] 315/9).
[23] Burshtein, K. Ya.; Shorygin, P. P. (Izv. Akad. Nauk SSSR Ser. Khim. **1987** 2764/9; Bull. Acad. Sci. Div. Chem. Sci. [Engl. Transl.] **1987** 2563/7).
[24] Burshtein, K. Ya.; Shorygin, P. P. (Izv. Akad. Nauk SSSR Ser. Khim. **1988** 1330/5; Bull. Acad. Sci. USSR Div. Chem. Sci. [Engl. Transl.] **1988** 1169/73).
[25] Dewar, M. J. S.; Jie, C. (Organometallics **8** [1989] 1544/7).
[26] Burshtein, K. Ya.; Isaev, A. N.; Shorygin, P. P. (J. Organomet. Chem. **361** [1989] 21/5).
[27] Dakkouri, M. (J. Am. Chem. Soc. **113** [1991] 7109/14).
[28] Gunasekaran, S.; Marshell, J. (Indian J. Phys. **65** B [1991] 367/70).
[29] Luo, Y.-R.; Holmes, J. L. (J. Phys. Chem. **96** [1992] 9568/71).

1.3.3.7 $GeRH_3$ Compounds with R = Cycloalkenyl

The compounds in this section are listed in Table 54. Further information on each compound is given at the end of the table.

Table 54
$GeRH_3$ Compounds with R = Cycloalkenyl.
Explanations, abbreviations, and units are given on p. X.

No.	R group	formation (yield) properties and remarks
1	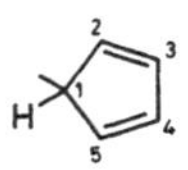	GeH_3Br + excess KC_5H_5 (or NaC_5H_5 [7]) at −196 to +20°C followed by trap-to-trap fractionation; purified by repeated vacuum condensation (76%) [5], (73%) [4]; see also [1] heating $Ge(C_5H_5\text{-}1)H_3$ at 64 to 133°C gave $Ge(C_5H_5\text{-}2)H_3$ and $Ge(C_5H_5\text{-}3)H_3$; heating at 64°C for 94 h yielded a 78:15:7 mixture of the three isomers, identified by ^1H NMR [4] m.p. 200 K [7]

Table 54 (continued)

No.	R group	formation (yield) properties and remarks
1 (continued)		vapor pressure (in Torr): 10.8 at 0.1 °C and 31.8 at 20.4 °C [4] ^{1}H NMR (neat liquid): 3.02 (br, H-1), 3.33 (d, HGe, ^{3}J(H, H) = 1.7), 5.50 (m, H-2,3,4,5) at −60 °C; 3.33 (HGe), 5.56 (br, C_5H_5) at 35 °C; 3.33 (HGe), 5.41 to 5.50 (sext, C_5H_5, J(H, H) = 0.45) at 100 °C [4]; see also [1]; 3.75 (HGe) and 6.05 (C_5H_5) in $CDCl_3$ at 35 °C [4]; for the ^{1}H NMR spectra of isomers, see p. 316 IR (gas): ν_{as}(GeH) 2120(s,sh), ν_s(GeH) 2099, 2088(vs), ν(C=C) 1635(w), ν(CC) 1001, 996(s), $\delta_{as}(GeH_3)$, 894(R), 887(Q), 881(P)(s); $\delta_s(GeH_3)$ 830(R), 823(Q), 820, 815(P), 811(vvs), $\rho(GeH_3)$ 520(m); other bands given and assigned [4]; see also [5] Raman (liquid): ν_s(GeH) 2053(vs,p), ν(C=C) 1600(m,p), 1565(w,p), ν(CC) 948(vs,p), $\delta_{as}(GeH_3)$ 877(s,dp), $\delta_s(GeH_3)$ 824(s,p), 820(m,sh,dp), ν(GeC) 369(vs,p), δ(GeC) 194(w,dp), 115(m,dp); other bands given and assigned [4] MS: $[GeC_5H_n]^+$ (33.6), $[GeC_4H_n]^+$ (1.8), $[GeC_3H_n]^+$ (11.2), $[GeC_2H_n]^+$ (5.3), $[GeCH_n]^+$ (1.6), $[GeH_n]^+$ (5.0), $[C_5H_n]^+$ (33.0), $[C_4H_n]^+$ (0.6), $[C_3H_n]^+$ (7.3), $[C_2H_n]^+$ (0.7) [4]
2	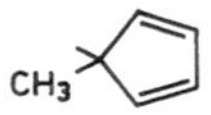 two isomers	prepared like No. 1 except using $KC_5H_4CH_3$ (ca. 45%) [4] m.p. ca. −100 °C [4] vapor pressure (in Torr): 3.3 at 0.2 °C and 10.3 at 19.1 °C [4] ^{1}H NMR (neat liquid): 1.61 (CH_3), 2.71, 2.95 (HCGe), 3.27, 3.30 (d's, HGe, J = 1.1, 1.0), 5.63 to 6.06 (HC=) at −60 °C; 1.63 (s, CH_3), 3.32 (s, HGe), 5.70 (v br, HC=) at 35 °C; 1.63 (CH_3), 3.32 (quint, HGe, J = 0.55), 5.08 to 5.20 (br, HC=) at 100 °C [2, 4] IR (gas): selected strong bands at 2102, 2078, 880, 820, 670; complete spectrum given [4] Raman (liquid): ν(GeH) 2075(vs,p), $\delta(GeH_3)$ 831(w), 810(w,p), $\nu(C(CH_3)$-C)(?) 405(s,p), ν(GeC) 377(vs,p); $\delta(C(CH_3)$-C)(?) 262(m,p); δ(Ge-C) 160(m,dp), 110(m,p) [4] UV (gas): λ_{max} = 260 nm [4] MS: $[GeC_6H_n]^+$ (17.0), $[GeC_5H_n]^+$ (3.2), $[GeC_4H_n]^+$ (2.5), $[GeC_3H_n]^+$ (4.1), $[GeC_2H_n]^+$ (6.8), $[C_6H_n]^+$ (42.4), $[GeH_n]^+$ (4.7), $[C_4H_n]^+$ (9.5), $[C_3H_n]^+$ (6.1), $[C_2H_x]^+$ (3.6) [4]
3		prepared like No. 1 except using KC_9H_7 at −30 to +20 °C (30 min) followed by fractionation ^{1}H NMR (neat liquid, 35 °C): 3.00 (HGe, ^{3}J(H, H) = 2.2), 3.22 (H-1), 5.90, 6.28 (m, H-2,3), 6.56 to 6.98 (m, C_6H_4) IR (liquid film): selected strong bands at 2065, 810, 752; complete spectrum given Raman (liquid): ν(GeH) 2078(s,p), $\delta(GeH_3)$ 815(w,p), ν(GeC) 391(vs,p) MS: major peaks correspond to $[GeC_9H_n]^+$ (23.1), $[C_9H_n]^+$ (53.8), $[C_5H_n]^+$ (5.6) [4]

Table 54 (continued)

No.	R group	formation (yield) properties and remarks
4	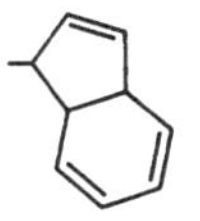	$GeH_3Br + KC_9H_9$ (1 : 1.2 mole ratio) in monoglyme at −63 °C (3 h) followed by fractionation; ca. 4 : 1 mixture of exo and endo isomers (indicated by ^1H NMR) colorless liquid ^1H NMR ($Si(CH_3)_4$): 1.4 to 3.4 (m, v br, CH-), 3.49 (d, HGe, J = 3.3) and 3.60 (d, HGe, J = 3.0) (ca. 1 : 4 ratio), 5.3 to 5.9 (m's, strongest component at 5.58, CH=) for the mixture of endo and exo isomers (spectrum at 28 °C illustrated) ^{13}C NMR ($CDCl_3$): 30.0, 38.1, 38.5, 42.1, 43.8, 44.1 (aliphatic C), 121.1, 121.4, 122.0, 122.9, 125.5, 126.0, 128.4, 128.8, 129.6, 129.9, 133.4, 134.8 (olefinic C) for mixture of exo and endo isomers (spectrum at 28 °C illustrated) IR (liquid film): selected strong bands at 2064, 874, 818, 782, 697, 560; complete spectrum given [6]
5	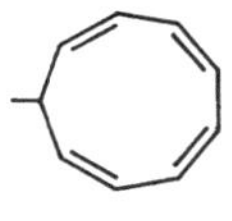	prepared in situ (for NMR measurements) from GeH_3Br and KC_9H_9 (1 : 1 mole ratio) in monoglyme at −196 to −50 °C (6 h) in a sealed tube; formed as an intermediate during the preparation of No. 4 (indicated by NMR) ^1H NMR ($Si(CH_3)_4$, −45 °C): 3.20 (aliphatic H), 3.73 (d, HGe, $^3J(H, H)$ = 2.5), 5.48 to 5.75 (olefinic H); (spectrum illustrated) ^{13}C NMR (DME, −45 °C): 28.0 (aliphatic C), 125.5, 127.1, 127.3, 128.9 (olefinic C); (spectrum illustrated) [6]

* Further information:

$Ge(C_5H_5)H_3$ (Table **54**, No. **1**). The ^1H NMR spectrum (illustrated) is consistent with fluxional behavior at 35 °C. A static configuration was adopted below −20 °C as indicated by the appearance of the resonance δ(H-1) and the splitting of the GeH resonance into a doublet [1, 4]. On heating towards a fast-exchange limit, coupling between the ring and germyl protons was observed to give a sextet at 100 °C (illustrated) consistent with an intramolecular, metallotropic shift as the mechanism of rearrangement [4].

The molecular structure of $Ge(C_5H_5)H_3$ (see **Fig. 6**) was determined for the solid (at 160 K) and the gas by X-ray and electron diffraction, respectively. Principal bond lengths for the gas: Ge-C(1) = 1.969(5), C(1)-C(2) = 1.478(13), C(2)-C(3) = 1.350(5), and C(3)-C(4) =

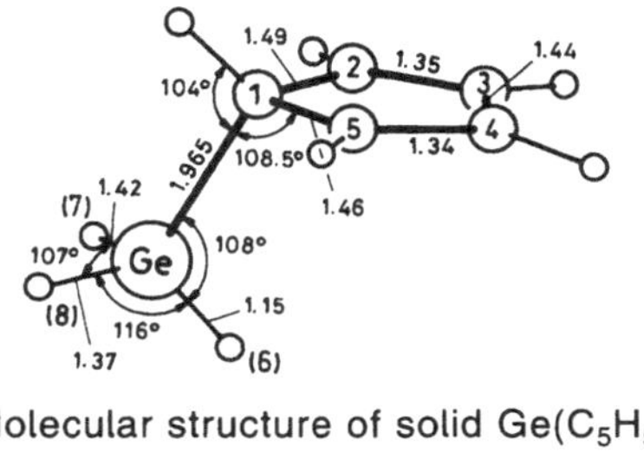

Fig. 6. Molecular structure of solid $Ge(C_5H_5)H_3$ [7].

References on p. 316

1.468(22) Å. Crystals of the compound are monoclinic with a = 4.90, b = 9.54, c = 13.11 Å, and β = 98.4°; space group $P2_1/c-C^5_{2h}$; Z = 4 gives d_c = 1.54 g/cm^3. The ring is essentially planar in both phases with ring angles close to 109° at all carbon atoms except C(1), where the angle is 100(3)° for the gas and 104.0(7)° for the solid [7].

The compound is thermally quite stable; 31% of it withstood 94 h of heating at 64 °C in a sealed tube. The ^{1}H NMR spectrum revealed that partial isomerization of the initially pure $Ge(C_5H_5\text{-}1)H_3$ to the 2-germyl isomer (15%) and 3-germyl isomer (7%) had taken place [4].

No. 1 reacted with an excess of dry HCl at 20 °C (1 h) to give GeH_3Cl and traces of C_5H_6. HCOOH and CH_3SH had no effect. Deprotonation to obtain $[Ge(C_5H_4)H_3]^-$ with K in ether or with LiC_4H_9 in C_5H_{12} failed [4]. The compound was converted at room temperature to Ge_2H_6, Ge_3H_8, and C_5H_6 in the presence of NH_2R' (R' = CH_3, C_2H_5, or $CH_2C_6H_5$) or NHR'_2 (R' = C_2H_5 or n-C_4H_9) [3]. The reaction with $Sn(CH_3)_3H$ at 60 °C (12 h) yielded GeH_4 and a small amount of $Sn(CH_3)_3C_5H_5$, probably contaminated with $GeH_3Sn(CH_3)_3$ (based on ^{1}H NMR) [4].

$Ge(C_5H_4CH_3)H_3$ (Table **54**, No. **2**). Temperature-dependent ^{1}H NMR spectra showed that the compound was fluxional at 35 °C, undergoing rapid metallotropic rearrangement which involved the isomers I, II and III. On cooling, two different GeH and GeCH resonances appeared at ca. −50 °C, attributed to the isomers II and III in almost equal concentrations. The spectrum at 100 °C, which showed a splitting of the GeH resonance into a quintet (J = 0.55 Hz), was interpreted like that of No. 1 [2, 4].

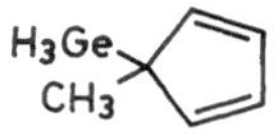

I

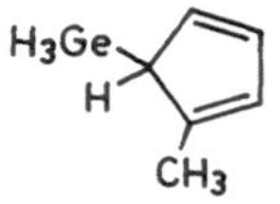

II

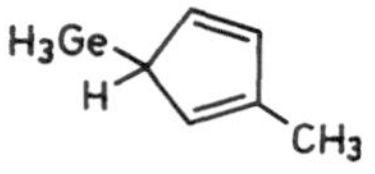

III

$Ge(C_9H_7)H_3$ (Table **54**, No. **3**). The ^{1}H NMR spectrum observed at 35 °C is consistent with the static configuration displayed in Table 54. Between 175 and 195 °C partial coalescence of the C_5-ring pattern could be seen, corresponding to the onset of fluxional exchange on the ^{1}H NMR time scale [4].

$Ge(C_9H_9)H_3$ (Table **54**, No. **4**) is a thermally stable liquid. It appeared to be unaffected by briefly exposing it to ambient atmosphere [6].

Mass spectrum: The polyisotopic parent-ion family accounted for 8.6% of the ion current, and $[C_9H_n]^+$ fragments were very prominent (64.4%). Other important ions were $[GeH_n]^+$ and $[C_7H_n]^+$ [6].

$Ge(C_9H_9)H_3$ (Table **54**, No. **5**). 9-germyl-all-cis-cyclonona-1,3,5,7-tetraene is stereochemically rigid on the NMR time scale at −45 °C; isomerization to No. 4 at higher temperatures prevented observation of fluxional behavior. Nonreversible isomerization occurred between −45 °C and room temperature [6].

References:

[1] Stobart, S. R. (J. Organomet. Chem. **33** [1971] C11/C12).
[2] Stobart, S. R. (J. Organomet. **43** [1972] C26/C28).
[3] Angus, P. C.; Stobart, S. R. (J. Chem. Soc. Chem. Commun. **1973** 127).
[4] Angus, P. C.; Stobart, S. R. (J. Chem. Soc. Dalton Trans. **1973** 2374/80).

[5] Angus, P. C.; Stobart, S. R. (Inorg. Synth. **17** [1977] 176/8).
[6] Bonny, A.; Stobart, S. R. (J. Chem. Soc. Dalton Trans. **1979** 786/91).
[7] Barrow, M. J.; Ebsworth, E. A. V.; Harding, M. M.; Rankin, D. W. H. (J. Chem. Soc. Dalton Trans. **1980** 603/6).

1.3.3.8 $GeRH_3$ Compounds with R = Aryl and Heterocycle

The compounds described in this chapter are listed in Table 55. The radical I was mentioned in connection with semiempirical MO calculations of $[E(\overline{CSCH_2CH_2CH_2S})H_3^{\cdot}]^+$ radical cations (E = Si, Ge, Sn) to compare two-center energies of their metal-carbon bonds; they become weaker in the series silyl > germyl > stannyl [50].

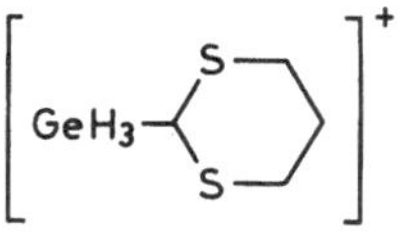

I

Table 55
$GeRH_3$ Compounds with R = Aryl and Heterocycle.
An asterisk indicates further information at the end of the table.
Explanations, abbreviations, and units are given on p. X.

No.	R group	formation (yield) properties and remarks
*1	C_6H_5	$Ge(C_6H_5)Cl_3$ + excess $LiAlH_4$ in refluxing ether (2 h) followed by evaporation, addition of petroleum ether, and vacuum distillation of the concentrated filtrate (75%) [4]; purified by trap-to-trap distillation [36]; see also [2, 47] b.p. 41 to 43°C/12 [4], 40 to 43°C/20 [31], 40°C/22 [2, 5], 50°C/35 [23] d^{20} = 1.2371; n_D^{20} = 1.5353 [2, 5, 23] (= n_D^{25} [31]) ^1H NMR: 4.20 (s, HGe) in C_5H_{12} [5]; 4.22 in CCl_4 [40]; 4.25 [2, 5, 24], 7.18 (C_6H_5) [24] for the neat compound; 4.29, 7.33 in $CDCl_3$ [47]; 5.52 (?) in [31] ^{13}C NMR (THF): 129.04 (C-3), 129.70 (C-4), 131.60 (C-1), 136.06 (C-2) [36] IR (liquid): ν(GeH) 2072(vs,br), ν(GeC) 1097(vs), $\delta_{as}(GeH_3)$ 874(vs,br), $\delta_s(GeH_3)$ 823(vs,br), $\rho(GeH_3$, out-of-plane) 585(vs,sh), $\rho(GeH_3$, in-plane) 575(vs) [6]; see also [1]; other bands given and assigned for the liquid [6]; ν(GeH) 2086 for the gas [37]; 2074 in C_7H_{16}; 2072 in CCl_4 [8] Raman (liquid): ν(GeH) 2072(100,p), ν(GeC) 1096(5,p), $\delta_{as}(GeH_3)$ 878(14,dp), $\delta_s(GeH_3)$ 836(18,p), $\rho(GeH_3$, in-plane) 580(2,dp); other bands given and assigned [6] with a slight excess of Na or $NaNH_2$ (see also [9]) in liquid NH_3 $\rightarrow$ $Ge(C_6H_5)H_2Na$ [36] with HX gas (X = Br or I) $\rightarrow$ GeH_3X [11]

Table 55 (continued)

No.	R group	formation (yield) properties and remarks
*1 (continued)		with GeF_2 → unstable insertion product [19] with $HgCl_2$ (1:0.8 mole ratio) at 25°C (4 h) → $Ge(C_6H_5)H_2Cl$ [5] with CH_3OCH_2Cl (ca. 1:1 mole ratio) at 60°C (30 min) in the presence of $AlCl_3$ → $Ge(C_6H_5)H_2Cl$; same reaction but ca. 1:2 mole ratio → $Ge(C_6H_5)(H)Cl_2$ [5]; see also [2] with N-bromosuccinimide (1:1 mole ratio) in petroleum ether at 60°C (3 h) → $Ge(C_6H_5)H_2Br$ (80%) [5]; see also [2]; similarly with N-iodosuccinimide at 45°C (2 h) → $Ge(C_6H_5)H_2I$ (20%), $Ge(C_6H_5)I_3$ (5%), and GeI_4 (30%) [5] with CH_2N_2 (50% excess) in ether and UV-irradiated (6 h) → $Ge(CH_3)(C_6H_5)H_2$ and $Ge(CH_3)_2(C_6H_5)H$ [2] with $C_4H_9C{\equiv}CH$ (1:3 mole ratio) at 80°C (4 h) in the presence of AIBN (sealed tube) → $Ge(CH{=}CHC_4H_9)_3C_6H_5$ [33] with $Co_2(CO)_8$ in C_6H_{14} at 20°C → $Ge(C_6H_5)Co_3(CO)_{11}$ [10]
*2	C_6F_5	$Ge(C_6F_5)Br_3$ + $LiAlH_4$ in ether-C_6H_{14} at ca. −25°C followed by hydrolysis with aqueous HCl at 20°C and fractionation of the organic layer (58%) b.p. 130 to 133°C/740 d^{20} = 1.702; n_D^{20} = 1.4420 IR: ν(GeH) 2120; other bands at 880, 840 [20]
*3	$C_6H_4CH_3$-4	$Ge(C_6H_4CH_3\text{-}4)Cl_3$ + excess $LiAlH_4$ in refluxing ether (4 h) followed by hydrolysis on ice and fractionation of the ether layer (ca. 40%) b.p. 166°C; n_D^{20} = 1.5322 [18]
*4	$C_6H_2(CH_3)_3$-2,4,6	$Ge(C_6H_2(CH_3)_3\text{-}2,4,6)Cl_3$ (from $GeCl_4$ + RMgBr in ether-THF) + $LiAlH_4$ in refluxing ether (6 h) followed by hydrolysis with aqueous HCl, extraction with ether, and distillation (55%), along with GeR_2H_2 and GeR_3H (small amount) [41]; see also [40] $Ge(C_6H_2(CH_3)_3\text{-}2,4,6)F_3$ (from $GeR(OCH_3)_3$ + HF) + excess $LiAlH_4$ in ether at 0°C followed by hydrolysis and GC of the ether phase (82%) [48] b.p. 50°C/0.05 [41, 48]; n_D^{20} = 1.5495 [41] ^1H NMR: 2.23 (s, CH_3-4), 2.34 (s, CH_3-2,6), 4.13 (s, HGe), 6.75 (s, C_6H_2) in CCl_4 [41, 48]; 2.20, 2.29, 4.06, 6.72 in THF-d_8 [51] ^{13}C NMR (THF-d_8): 21.12 (CH_3-4), 24.17 (CH_3-2,6), 127.31 (C-1), 128.74 (C-3), 139.27 (C-4), 144.08 (C-2) [48, 51] IR: ν(GeH) 2077 [41] stable under N_2 [41] with CCl_4 at 90°C in the presence of AIBN → $Ge(C_6H_2(CH_3)_3\text{-}2,4,6)Cl_3$ [40] with CH_3OCH_2Cl (1:2 mole ratio) at 40°C in the presence of $AlCl_3$ → $Ge(CH_2OCH_3)Cl_3$ [40]

Table 55 (continued)

No.	R group	formation (yield) properties and remarks
		with CH_3SSCH_3 in C_6H_6 at 10 °C (2 h) in the presence of AIBN (sealed tube) → $Ge(C_6H_2(CH_3)_3$-2,4,6)$(SCH_3)_3$ [48]; see also [41]
5	$C_6H_2(C_4H_9$-t$)_3$-2,4,6	$Ge(C_6H_2(C_4H_9$-t$)_3$-2,4,6)$(OCH_3)_3$ + $LiAlH_4$ in refluxing ether for 2 h (90%); recrystallized from C_5H_{12} at −30 °C long white needles; m.p. 124 °C ^{1}H NMR (C_6D_6): 1.38 (s, t-C_4H_9-4), 1.60 (s, t-C_4H_9-2,6), 4.82 (s, HGe), 7.60 (s, C_6H_2) IR (KBr): ν(GeH) 2100, 2060 with excess CCl_4 in the presence of AIBN at 80 °C in a sealed tube (3 h) → $Ge(C_6H_2(C_4H_9$-t$)_3$-2,4,6)Cl_3 (30%) [57]
*6	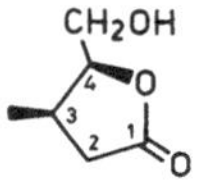	threo-Ge(CH($\overline{CHOC(CH_3)_2OC}H_2$)$CH_2COOC_2H_5$)$H_3$ in dioxane-H_2O (4 : 1) + Amberlyst 15 at 80 °C (2 h) followed by filtration, evaporation, and flash-chromatography of the residue over SiO_2 in C_6H_{14}-$CH_3COOC_2H_5$ (1 : 1) (48%) [46]; see also [44] m.p. 75.5 to 76.5 °C (from C_6H_{12}-$CH_3COOC_2H_5$) $[\alpha]_D$ = +32.1° (c = 1.12) in CH_3OH at 20 °C ^{1}H NMR (C_6D_6-D_2O): 1.447 (m, H-3), 2.052 (dd, H-2′, ^2J = 17.2, J(H-2′,3) = 9.7), 2.264 (dd, H-2, ^2J = 17.2, J(H-2,3) = 11.5), 3.173 (dd, H-5′, ^2J = 12.5, J(H-4,5′) = 3.3), 3.390 (dd, H-5, ^2J = 12.5, J(H-4,5) = 3.3), 3.486 (d, HGe, J = 2.9), 3.940 (dt, H-4, J(H-3,4) = 8.4, J(H-4,5) = 3.3) ^{13}C NMR (C_6D_6): 19.9 (C-3), 33.9 (C-2), 63.7 (C-5), 83.0 (C-4), 177.2 (C-1) IR (KBr): ν(GeH) 2090(s), ν(C=O) 1760(s,br) MS (CI): $[M - 1]^+$ [46]
*7	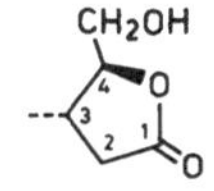	like No. 5 using the corresponding erythro compound (54%) m.p. 60.0 to 62.0 °C (from C_6H_{14}-$CH_3COOC_2H_5$) $[\alpha]_D$ = +59.4° (c = 1.10) in CH_3OH at 20 °C ^{1}H NMR (C_6D_6-D_2O): 1.904 (m, H-3), 2.088 (dd, H-2′, ^2J = 17.4, J(H-2′,3) = 12.5), 2.472 (dd, H-2, ^2J = 17.4, J(H-2,3) = 8.8), 3.499 (d, HGe, J = 2.6), 3.544 (dd, H-5′, ^2J = 12.5, J(H-4,5′) = 4.8), 3.802 (dd, H-5, ^2J = 12.8, J(H-4,5) = 2.9), 4.224 (ddd, H-4, J(H-3,4) = 10.3) ^{13}C NMR (C_6D_6): 18.6 (C-3), 34.6 (C-2), 62.8 (C-5), 86.2 (C-4), 177.3 (C-1) IR (KBr): ν(GeH) 2100(s), ν(C=O) 1770(br,s) MS (CI): $[M - 1]^+$ [46]
8	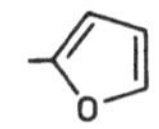	Ge($\overline{C=CHCH=CHO}$)Br_3 + solid $LiAlH_4$ (1 : 2 ratio [45] or 1 : 6 ratio [39]) in C_6H_6 at 8 °C (1 h) or at 25 °C (6 h) in the presence of the phase-transfer catalyst $[N(C_2H_5)_3CH_2C_6H_5]Cl$ (81 and 70%, respectively) [39, 45]

References on p. 324

Table 55 (continued)

No.	R group	formation (yield) properties and remarks
*9	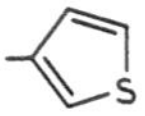	$\overline{C(Br)=CHSCH=C}H + LiC_4H_9$ in ether at −70 °C followed by condensation of GeH_3Cl, addition of H_2O under vacuum at 20 °C (after 3 h), extraction with ether, and GC (90%) colorless liquid; d^{20} = 1.48 ^1H NMR ($CDCl_3$): 4.23 (s, HGe), 7.16 (dd, H-4), 7.42 (dd, H-5, J(H-4,5) = 4.62), 7.46 (d, H-2, J(H-2,4) = 0.95, J(H-2,5) = 2.69) ^{13}C NMR ($CDCl_3$): 126.03 (C-5, ^1J(C, H) = 185.7), 127.89 (C-3), 132.54 (C-2, ^1J(C, H) = 180.6), 132.59 (C-4, ^1J(C, H) = 165.0) IR (film): ν(GeH) 2075(vs); other strong bands at 848, 822, 770, 641; general assignments for 3-substituted thiophenes given UV (CH_3CN): λ_{max} = 238 (log ε = 3.51) MS: $[C_4H_6]^+$ (100.0), $[^{74}Ge(C_4H_3S)H_3]^+$ (54.1), $[C_4H_3SH_2]^+$ (52.3), $[C_3H_3]^+$ (45.7), $[^{74}Ge]^+$ (25.8), $[CHS]^+$ (15.8) [52]

*Further information:

$Ge(C_6H_5)H_3$ (Table **55**, No. **1**) was obtained (14% yield) in addition to $Ge(CH_2CH=CH_2)(C_6H_5)H_2$ by reducing $Ge(CH_2CH=CH_2)(C_6H_5)Cl_2$ (from $Ge(C_6H_5)Cl$ and $CH_2=CHCH_2Cl$) with $LiAlH_4$ at −20 °C [13] or by reducing the insertion product $Ge(OCHR'X)C_6H_5$ (X = $N(CH_3)_2$ or $P(C_2H_5)_2$) resulting from $Ge(C_6H_5)X$ and R'CHO [25].

$Ge(C_6H_5)H_3$ was one of the products of the decomposition of $Ge_2(C_6H_5)_2H_4$ at room temperature in daylight [15]. Cleaving $Ge_2(C_6H_5)_2H_4$ with Li-Hg in THF (20 h) yielded 44% No. 1. $Ge(C_6H_5)H_3$ was also obtained when $Ge_2(C_6H_5)_2H_4$ in THF was treated with catalytic amounts of either Li in HMPT for 24 h (43% yield) or $Ge(C_6H_5)H_2Li$ in THF or LiC_4H_9 in C_6H_{14}. In each case $(Ge(C_6H_5)H)_n$ was also formed [23].

The compound was formed along with $Ge(C_6H_5)(H)Cl_2$ and $Ge(C_6H_5)H_2Cl$ by reducing $Ge(C_6H_5)Cl_3$ with $Sn(C_4H_9)_3H$ at 20°C (ca. 5 h) in the presence or absence of $CH(=N_2)COOC_2H_5$ (12 and 2% yield, respectively) [35]. UV irradiation of $Ge(C_6H_5)H_2OCH_3$ in C_6H_6 at 30 °C (12 h) yielded 42% of No. 1 (monitored by ^1H NMR). Photolysis of $Ge(C_6H_5)H_2SCH_3$ or $Ge(C_6H_5)(H)(SCH_3)_2$ at 30 °C (5 h) gave $Ge(C_6H_5)H_3$ along with $Ge(C_6H_5)(SCH_3)_3$ and $Ge(C_6H_5)(H)(SCH_3)_2$ or $Ge(C_6H_5)H_2SCH_3$, respectively [21].

Treating $[Ge_2(C_6H_5)_2Hg_3]_n$ with CH_3OH (10% in C_6D_6) regenerated $Ge(C_6H_5)H_3$ (indicated by ^1H NMR) from which the starting material was initially prepared; $Ge(C_6H_5)H_2OCH_3$ was the other product [34].

The deuterated compounds $Ge(C_6H_5)D_3$ (from $Ge(C_6H_5)Cl_3$ and $LiAlD_4$), $Ge(C_6D_5)H_3$ (from $Ge(C_6D_5)Cl_3$ and $LiAlH_4$) and $Ge(C_6D_5)D_3$ (from $Ge(C_6D_5)Cl_3$ and $LiAlD_4$) were isolated by hydrolysis with D_2O and distillation in a short-path apparatus, distillation at 33 to 35 °C/13 Torr, or at 136 to 137 °C, respectively (56% yield of $Ge(C_6D_5)D_3$) [6].

$Ge(C_6H_5)H_2T$ was prepared by adding tritiated water to $Ge(C_6H_5)H_3$ in CH_3OH containing CH_3ONa and DMSO, refluxing (20 h), and work-up by adding $C_6H_5CH_3$ and neutralizing with HCl [31].

The heat of formation of $Ge(C_6H_5)H_3$, $\Delta H_f^\circ = -11.8 \pm 1.0$ kcal/mol, was listed together with data of various C_6H_5X compounds [53].

The microwave spectra of $^{74}Ge(C_6H_5)H_3$, $^{72}Ge(C_6H_5)H_3$, $^{70}Ge(C_6H_5)H_3$, and $^{74}Ge(C_6H_5)D_3$ were measured in the range 12.4 to 26.5 GHz at ca. −30 °C. Ab initio calculations gave a plausible structure when the ab initio geometry was corrected to fit the experimental rotational constants: r(Ge-C) = 1.9429, r(Ge-H) = 1.5157 Å, and <(C-Ge-H) = 109.90° [54].

The dipole moment of $Ge(C_6H_5)H_3$ was measured in C_6H_{12} at 20 °C: $\mu = 0.68 \pm 0.03$ D [9].

A complete analysis of the ^{1}H NMR spectrum of $Ge(C_6H_5)H_3$ recorded in CCl_4 at various concentrations and infinite dilution was reported [9]. A discussion of possible factors influencing the chemical shifts of the ring protons revealed that the electric dipole effect mainly accounts for the shifts and that any pertubation of the π electrons by π(p-d) bonding was almost nonexistent [9]. ^{1}H NMR studies on organohydrides of Si, Ge, and Sn including $Ge(C_6H_5)H_3$ indicated weak association in the pure state and the formation of labile complexes of various compositions in aprotic donor solvents, corresponding to 5- and 6-coordinate central atoms [24]. Correlations between $\delta(C_6H_5)$ and δ(HE) values (E = C, Si, Ge) and the diamagnetic susceptibilities of various solvents were pointed out. Some double-bond character of the E-C bond (E = Si, Ge) was suggested, causing a fractional negative charge on the germyl and silyl hydrogen atoms [28].

The ^{13}C NMR spectra of $Ge(C_6H_5)_nH_{4-n}$ and $Ge(C_6H_5)_nH_{3-n}Na$ compounds (n = 1 to 3) were compared with those of analogous Group V derivatives [36].

A complete analysis of the IR and Raman spectra of liquid $Ge(C_6H_5)H_3$, $Ge(C_6H_5)D_3$, $Ge(C_6D_5)H_3$, and $Ge(C_6D_5)D_3$ (all illustrated) is given in [6]. The fundamental vibrations of the phenyl ring were assigned based on C_{2v} symmetry. A consideration of the rotational fine structure of the vapor phase revealed a free rotation of the phenyl group about the Ge-C axis at room temperature [6]. A theoretical study considering the effects of internal rotation on the IR spectra of $E(C_6H_5)H_3$ and $E(C_6H_5)D_3$ (E = Si, Ge; ν(GeH), $\delta(GeH_3)$, $\rho(GeH_3)$, and $\rho(GeD_3)$ regions at high resolution were illustrated) showed that internal rotation in these compounds was hindered by only a very small barrier (25 ± 10 cal/mol for $Ge(C_6H_5)H_3$) [7].

The IR spectra of phenyl derivatives of Si, Ge, and Sn were studied with respect to the frequency ranges of the E-C stretching vibrations. Thus, vibrations at ca. 360 to 290 cm^{-1} were assigned to ν(GeC), whereas on the basis of the integral intensities of the bands at ca. 1100 cm^{-1} the absorptions in the 1000 to 1120 cm^{-1} region are mostly related to the δ(CCH) and δ(CCC) vibrations of the C_6H_5 ring (cf. assignment in [6]; see Table 55). A dependence of the number of the C_6H_5 groups at the E atom on the intensities of these bands was found [29]. The steric effects of phenyl, xylyl, and mesityl substituents on the frequency and integrated intensity of the ν(GeH) band were studied for a series of Si and Ge hydrides including $Ge(C_6H_5)CH_3$ [32].

Studies of isolated ν(GeH) frequencies revealed that for alkyl germanes ν(GeH) correlated well with the sum of Taft σ^* parameters of the substituents at Ge; but no single correlation was possible for compounds with C_6H_5 ligands [37]; see also [8, 38]. The ν(GeH) and $\nu(GeC_6H_5)$ bands of the neat compound (2064 and 1093 cm^{-1}, respectively) were shifted in CH_3COCH_3 (2080 and 1060 cm^{-1}) because of association between the hydride and the solvent as a result of Ge←O coordination [26].

The mass spectrum showed peaks with m/e = 156 to 147 corresponding to all ions from $[^{76}Ge(C_6H_5)H_3]^+$ to $[^{70}GeC_6H_5]^+$. Fragments with lower mass numbers result from breakup of the Ge-C bond and the C_6H_5 ring [9]. The complete spectrum is given in [47].

References on p. 324

^1H NMR spectroscopy was used to determine the acidities of some aryl-substituted germanes in liquid NH_3. Increasing aryl substitution lowers the acidity. Factors such as solvation, ion pairing, and structural effects were discussed to explain the order [16]. The fact that phenylgermanes are weaker acids than germane was attributed to the absence of any significant π(p-p) interaction between the C_6H_5 ring and the Ge atom [9, 14].

On heating, $Ge(C_6H_5)H_3$ partially decomposed to give $Ge(C_6H_5)_2H_2$, GeH_4, Ge, and H_2, amounting to 6% after 12 h at 150 °C (sealed tube) and to 20% at 200 °C. 5% of the compound decomposed under UV irradiation at 30 °C during 12 h [5].

Nanosecond laser flash photolysis and monitoring of the electronic spectrum was used to study the formation and the decay of phenyl-substituted germyl radicals (UV absorption of the $Ge(C_6H_5)H_2^{\bullet}$ radical at λ_{max} = 310 nm). The rate constant for hydrogen abstraction by t-$C_4H_9O^{\bullet}$ radicals was determined in C_6H_6 at ca. 295 K to be $k = (4.4 \pm 0.3) \times 10^8\ M^{-1} \cdot s^{-1}$ for $Ge(C_6H_5)H_3$. The first-order decay was attributed to the addition of the $Ge(C_6H_5)H_2^{\bullet}$ radical to the C_6H_5 ring of $Ge(C_6H_5)H_3$, proceeding with the rate $k = (1.2 \pm 0.1) \times 10^6\ M^{-1} \cdot s^{-1}$ [47]. Photolysis by the ArF excimer laser (193 nm) of a supersonic free jet of $Ge(C_6H_5)H_3$ and $Ge(C_6H_5)D_3$ in argon produced GeH_2 and GeD_2 radicals whose laser-induced fluorescence excitation and dispersed fluorescence spectra could be observed and interpreted [56].

The Ge-H bond dissociation energies of $Ge(C_6H_5)_nH_{4-n}$ compounds (n = 1 to 3) were determined in C_6H_6 or isooctane at 23 °C by the technique of laser-induced photoacoustic calorimetry. They remain unaffected by increasing the phenyl substitution and are in the range 79.2 to 80.2 kcal/mol [47].

First-order rate constants were determined for the loss of tritium from $Ge(C_6H_5)(T)H_2$ and other tritiated Ge hydrides in CH_3OH-CH_3ONa at 20 to 40 °C. The relative rate referred to that of $Ge(C_6H_5)_3H$ is $k_{rel} = 1.5 \times 10^{-2}$ at 30 °C. Rates of the base-catalyzed hydrogen exchange decrease from $Ge(C_6H_5)_3T$ to $Ge(C_6H_5)H_2T$, indicating an order of kinetic acidities opposite to the order reported [3] for equilibrium acidities in liquid NH_3 [31].

The influence of various donor solvents (N bases such as 2-picoline or pyridine) on the rate of the reactions of No. 1 with CH_3COOH, $CH_2ClCOOH$, C_6H_5COOH, or 3-$NO_2C_6H_4$-COOH at 110 °C in the presence of Cu (replacement of hydrogen by the corresponding anion) was studied. The overall reaction rate depends not only on the electron-donor properties of the solvent, but also on steric factors [27].

$Ge(C_6H_5)H_3$ reacted with CH_3SSCH_3 (1:1 mole ratio) at 100°C (2 h) to give $Ge(C_6H_5)H_2SCH_3$, $Ge(C_6H_5)(H)(SCH_3)_2$, and $Ge(C_6H_5)(SCH_3)_3$. A 87% yield of $Ge(C_6H_5)$-$(SCH_3)_3$ was obtained with a 1:3 mole ratio at 100°C after 4 h [21].

Addition of LiC_4H_9-t in C_5H_{12} to $Ge(C_6H_5)H_3$ (1:1 mole ratio) in THF at −78°C (40 min), followed by treatment with excess CH_3I (20°C for 1 h) and hydrolysis, gave $Ge(CH_3)(C_6H_5)H_2$, $Ge(C_4H_9\text{-t})(C_6H_5)H_2$, $Ge(C_4H_9\text{-t})_2(C_6H_5)H$, and $Ge(C_6H_5)H_3$ in a 69:9:8:14 ratio [51]. The reaction with LiC_4H_9 in THF to yield $Ge(C_6H_5)H_2Li$ (45%), $Ge(C_4H_9)(C_6H_5)H_2$ (15%), and $Ge(C_4H_9)_2(C_6H_5)H$ (4%) was briefly mentioned in [23].

$Ge(C_6H_5)H_3$ reacted with $Hg(C_4H_9)_2$ (1:1 mole ratio) in C_6H_6 at 0°C to yield polymeric $(Ge_2(C_6H_5)_2Hg_3)_n$ [34]. Treatment with $Hg(C_2H_5)_2$ (1:3 mole ratio) gave $(Ge(C_6H_5)Hg_{1.5})_n$ which reacted with $CH_2{=}C(CH_3)C(CH_3){=}CH_2$ to give the digermane II [30].

Insertion of $Ge(C_6H_5)Cl$ into a Ge-H bond of $Ge(C_6H_5)H_3$ (1:1 mole ratio) at 60 °C (12 h in a sealed tube) was indicated by the formation of $Ge(C_6H_5)(H)(Cl)Ge(C_6H_5)H_2$ which thermally decomposed to yield $Ge(C_6H_5)H_2Cl$; further reaction with $Ge(C_6H_5)Cl$ led to $Ge(C_6H_5)(H)(Cl)Ge(C_6H_5)(H)Cl$ and $Ge(C_6H_5)(Cl)_2Ge(C_6H_5)H_2$ [17]. The formation of

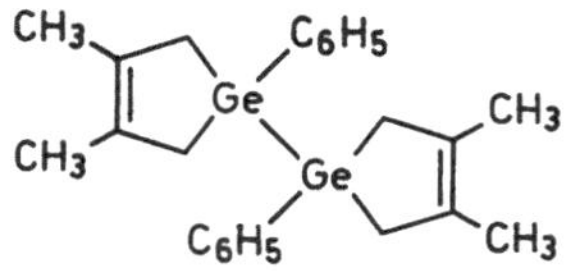

II

$Ge_2(C_6H_5)_2H_4$ from $Ge(C_6H_5)H_3$ and $Ge(C_6H_5)(OCH_3)H_2$ at 120 °C (2 h) in the presence of CH_3ONa-CH_3OH was reported to proceed via insertion of $Ge(C_6H_5)H$, formed from the unstable $Ge(C_6H_5)(OCH_3)H_2$ [12].

Long-time reactions of $(Co_2(CO)_7)_2(\mu_4$-Ge) (Formula III) with excess $Ge(C_6H_5)H_3$ in C_6H_{14} at 20 to 30 °C gave $Co_4(CO)_{13}(\mu_4$-Ge)(μ-Ge(C_6H_5)H) (Formula IV, R = C_6H_5) and Co_4-$(CO)_{12}(\mu_4$-Ge)(μ-Ge(C_6H_5)H$)_2$ (Formula V, R = R′ = C_6H_5) along with H_2 and CO. Increasing amounts of $Ge(C_2H_5)H_3$ and slightly elevated temperatures favor the formation of compound V. Under similar conditions, treatment of compound IV (R = C_2H_5) with $Ge(C_6H_5)H_3$ yielded compound V (R = C_2H_5 and R′ = C_6H_5) [55].

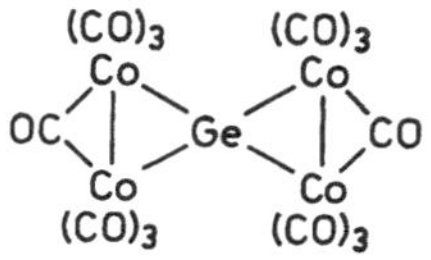

III

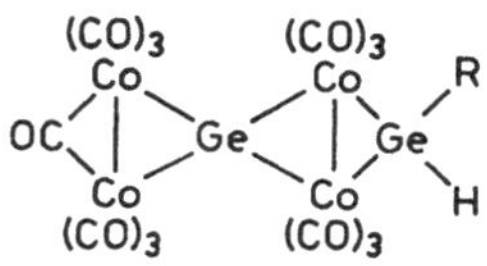

IV

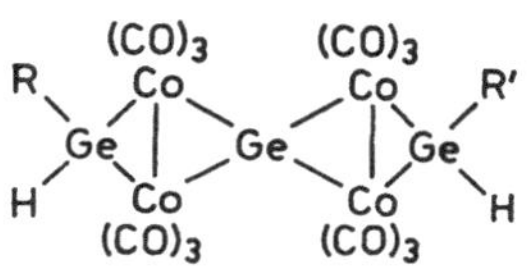

V

$Ge(C_6H_5)H_3$ polymerized to a three-dimensional gel in the presence of catalytic amounts of dimethyltitanocene; stepwise oligomerization occurred with vanadocene as the catalyst (at 50 °C) [42]; see also [43].

$Ge(C_6F_5)H_3$ (Table **55**, No. **2**) reacted with excess sulfur at 150 to 170 °C (1 h) with evolution of H_2S to give the adamantane-like compound VI [22].

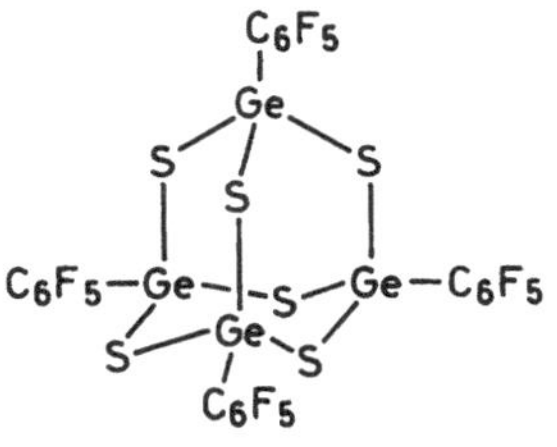

VI

The reaction with $Rh(P(C_6H_5)_3)_2R'$ (R′ = 3,6-di-t-butyl-o-benzoquinone or 3,6-di-t-butyl-4,5-dimethoxy-o-benzoquinone) in $C_6H_5CH_3$ at −60 °C yielded an adduct. Analysis of the EPR spectrum suggested a primary trans-addition of $Ge(C_6H_5)H_3$; this adduct rearranged irreversibly to the cis-isomer when the temperature was raised [49].

$Ge(C_6H_4CH_3$-4$)H_3$ (Table **55**, No. **3**). A comparison of the dipole moments of silyl and germyl substituted toluenes revealed that the GeH_3 group had a higher +I effect than the SiH_3 group, indicating the electronegativity of Ge to be lower than that of Si. The dipole moment of $Ge(C_6H_4CH_3$-4$)H_3$ in C_6H_6 at 25 °C is μ = 0.77 D. The σ_p constant was estimated from the dipole moment to be σ_p = 0.01 [18].

References on p. 324

$Ge(C_6H_2(CH_3)_3$-2,4,6)H_3 (Table **55**, No. **4**) was formed along with other Ge-containing compounds, when $Ge(C_6H_2(CH_3)_3$-2,4,6)H_2Li was either kept in THF at 20°C (72 h) or treated with $HgCl_2$, $Ge(C_6H_2(CH_3)_3$-2,4,6)$_3Cl$, $(C_6H_2(CH_3)_3$-2,4,6)COCl, or $(C_6H_5CO)_2O$ in THF at temperatures between −78 and +20°C (40°C for $HgCl_2$) [51].

The influence of mesityl substituents on the IR frequency and integrated intensity of ν(EH) bands (E = Si, Ge) was shown for $Ge(C_6H_2(CH_3)_3$-2,4,6)H_3 and related compounds to be determined not only by the inductive effect and π(p-d) interaction, but also by the steric effects of these groups [32].

Adding LiC_4H_9-t in C_5H_{12} to compound No. 4 in THF (1:1 mole ratio) at −60°C (40 min) and warming to +20°C gave $Ge(C_6H_2(CH_3)_3$-2,4,6)H_2Li (indicated by ^1H NMR). A similar reaction with LiC_4H_9-t (1:2 mole ratio) followed by treatment with excess CH_3I yielded $Ge(CH_3)(C_6H_2(CH_3)_3$-2,4,6)H_2 (36%) and $Ge(CH_3)_2(C_6H_2(CH_3)_3$-2,4,6)H (35%) [51].

Treating No. 4 with N-chlorosuccinimide (ca. 1:1 mole ratio) in THF at 70°C (3 h) in a sealed tube gave the monosubstitution product $GeRH_2Cl$; using a ca. 1:1.6 mole ratio (60°C, 6 h) yielded $GeRH_2Cl$, $GeR(H)Cl_2$, and $GeRCl_3$ (R = $C_6H_2(CH_3)_3$-2,4,6). Reactions with N-bromosuccinimide (30°C, 2 h) proceeded similarly [40].

Threo- and **erythro-Ge($\overline{CHCH_2C(O)OC}HCH_2OH)H_3$** (Table **55**, Nos. **6** and **7**). The absolute configurations were determined to be (3R,4R) and (3S,4R), respectively. Structural assignments were established by decoupling ^1H NMR experiments, and the optical purity was determined by HPLC analysis of the (+)- and (−)-MTPA esters of the compound (MTPA = α-methoxy-α-(trifluoromethyl)phenylacetic acid) [46].

The threo compound at 12.5 μg/mL showed significant inhibition of Mycobacterium tuberculosis ohtsuka [44].

Ge($\overline{C{=}CHSCH{=}C}H)H_3$ (Table **55**, No. **9**) was polymerized by electrooxidation at a constant potential of 1.90 V (reference electrode: silver wire; onset potential: E_a = 1.76 V) on indium-doped tin oxide glass electrodes in 0.2 M $[N(C_4H_9)_4]PF_6$-$C_6H_5NO_2$ solution to yield a free-standing polymer film containing a greatly reduced amount of Ge compared with that present before oxidation. The polymer was electrochromic and electroactive with a conductivity of Λ = 0.05 Ω/cm [52].

References:

[1] Cross, R. J.; Glockling, F. (J. Organomet. Chem. **3** [1965] 146/55).
[2] Satgé, J.; Rivière, P. (Bull. Soc. Chim. Fr. **1966** 1773/4).
[3] Birchall, T.; Jolly, W. L. (Inorg. Chem. **5** [1966] 2177/80).
[4] Kühlein, K.; Neumann, W. P. (Justus Liebigs Ann. Chem. **702** [1967] 17/23).
[5] Rivière, P.; Satgé, J. (Bull. Soc. Chim. Fr. **1967** 4039/46).
[6] Durig, J. R.; Sink, C. W.; Turner, J. B. (J. Chem. Phys. **49** [1968] 3422/41).
[7] Fleming, J. W.; Banwell, C. N. (J. Mol. Spectrosc. **31** [1969] 318/40).
[8] Mathis, R.; Barthelat, M.; Mathis, F. (Spectrochim. Acta A **26** [1970] 1993/2000).
[9] Birchall, T.; Drummond, I. (J. Chem. Soc. A **1970** 1401/5).
[10] Ball, R.; Bennett, M. J.; Brooks, E. H.; Graham, W. A. G.; Hoyano, J.; Illingworth, S. M. (J. Chem. Soc. Chem. Commun. **1970** 592/3).
[11] Mironov, V. F.; Kalinina, L. N.; Berliner, E. M.; Gar, T. K. (Zh. Obshch. Khim. **40** [1970] 2597/601; J. Gen. Chem. USSR [Engl. Transl.] **40** [1970] 2590/4).
[12] Massol, M.; Satgé, J.; Rivière, P.; Barrau, J. (J. Organomet. Chem. **22** [1970] 599/610).
[13] Massol, M.; Barrau, J.; Rivière, P.; Satgé, J. (J. Organomet. Chem. **30** [1971] 27/41).

[14] Jolly, W. L. (Inorg. Chem. **10** [1971] 2364/5).
[15] Rivière, P.; Satgé, J. (Synth. Inorg. Met.-Org. Chem. **2** [1972] 57/63).
[16] Birchall, T.; Drummond, I. (Inorg. Chem. **11** [1972] 250/2).
[17] Rivière, P.; Satgé, J. (Helv. Chim. Acta **55** [1972] 1164/73).
[18] Vo-Kim-Yen; Papouskova, Z.; Schraml, J.; Chvalovsky, V. (Collect. Czech. Chem. Commun. **38** [1973] 3167/75).
[19] Satgé, J.; Rivière, P.; Boy, A. (C. R. Seances Acad. Sci. C **278** [1974] 1309/12).
[20] Bochkarev, M. N.; Maiorova, L. P.; Korneva, S. P.; Bochkarev, L. N.; Vyazankin, N. S. (J. Organomet. Chem. **73** [1974] 229/36).
[21] Rivière, P.; Dousse, G.; Satgé, J. (Synth. React. Inorg. Met.-Org. Chem. **4** [1974] 281/93).
[22] Bochkarev, M. N.; Maiorova, L. P.; Vyazankin, N. S.; Razuvaev, G. A. (J. Organomet. Chem. **82** [1974] 65/71).
[23] Rivière, P.; Satgé, J.; Soula, D. (J. Organomet. Chem. **72** [1974] 329/38).
[24] Ivanov, V. A.; Reikhsfel'd, V. O.; Saratov, I. E. (Zh. Obshch. Khim. **45** [1975] 2036/40; J. Gen. Chem. USSR [Engl. Transl.] **45** [1975] 1999/2002).
[25] Riviére, P.; Rivière-Baudet, M.; Satgé, J. (J. Organomet. Chem. **97** [1975] C37/C40).
[26] Reikhsfel'd, V. O.; Ivanov, V. A.; Saratov, I. E. (Zh. Obshch. Khim. **45** [1975] 2243/5; J. Gen. Chem. USSR [Engl. Transl.] **45** [1975] 2202/4).
[27] Saratov, I. E.; Ivanov, V. A.; Reikhsfel'd, V. O. (Zh. Obshch. Khim. **46** [1976] 1052/7; J. Gen. Chem. USSR [Engl. Transl.] **46** [1976] 1048/52).
[28] Saratov, I. E.; Reikhsfel'd, V. O.; Ivanov, V. A. (Zh. Obshch. Khim. **47** [1977] 1776/81; J. Gen. Chem. USSR [Engl. Transl.] **47** [1977] 1625/9).
[29] Minaeva, N. A.; Nadtochii, Yu. G.; Chumaevskii, N. A. (Zh. Prikl. Spektrosk. **27** [1977] 276/83; J. Appl. Spectrosc. [Engl. Transl.] **27** [1977] 1020/6).
[30] Rivière, P. (unpublished results from G. Dousse; J. Satgé; Helv. Chim. Acta **60** [1977] 1381/7).
[31] Eaborn, C.; Singh, B. (J. Organomet. Chem. **177** [1979] 333/48).
[32] Skobeleva, S. E.; Egorochkin, A. N.; Korshev, S. Ya.; Ratushnaya, S. Kh.; Rivière, P.; Satgé, J.; Richelme, S.; Cazes, A. (J. Organomet. Chem. **182** [1979] 1/7).
[33] Sennikov, P. G.; Skobeleva, S. E.; Kuznetsov, V. A.; Egorochkin, A. N.; Rivière, P.; Satgé, J.; Richelme, S. (J. Organomet. Chem. **201** [1980] 213/9).
[34] Rivière, P.; Castel, A.; Satgé, J. (J. Organomet. Chem. **212** [1981] 351/67).
[35] Castel, A.; Rivière, P.; Satgé; J. (J. Organomet. Chem. **232** [1982] 137/46).
[36] Batchelor, R. J.; Birchall, T. (J. Am. Chem. Soc. **105** [1983] 3848/52).
[37] McKean, D. C.; Torto, I.; Mackenzie, M. W. (Spectrochim. Acta A **39** [1983] 399/408).
[38] McKean, D. C.; Torto, I.; Mackenzie, M. W.; Morrisson, A. R. (Spectrochim. Acta A **39** [1983] 387/98).
[39] Gevorgyan, V. N.; Ignatovich, L. M.; Lukevics, E. (J. Organomet. Chem. **284** [1985] C31/C32).
[40] Rivière, P.; Rivière-Baudet, M.; Castel, A.; Satgé, J.; Lavabre, A. (Synth. React. Inorg. Met.-Org. Chem. **17** [1987] 539/57).
[41] Rivière, P.; Rivière-Baudet, M.; Satgé, J. (Organomet. Synth. **4** [1988] 545/8).
[42] Aitken, C.; Harrod, J. F.; Malek, A.; Samuel, E. (J. Organomet. Chem. **349** [1988] 285/91).
[43] Harrod, J. F. (ACS Symp. Ser. **360** [1988] 89/100).
[44] Takahashi, Y.; Kakimoto, N.; Asai Germanium Research Institute (Jpn. Kokai Tokkyo Koho 02-131490 [90-131490] [1988/90]; C.A. **113** [1990] No. 212348).
[45] Lukevics, E.; Gevorgyan, V. (Chem. Technol. Silicon Tin. Proc. Asian Network Anal. Inorg. Chem. 1st Int. Chem. Conf. Silicon Tin, Kuala Lumpur 1989 [1992], pp. 165/77).

[46] Takahashi, Y.; Kakimoto, N. (J. Organomet. Chem. **399** [1990] 47/51).
[47] Clark, K. B.; Griller, D. (Organometallics **10** [1991] 746/50).
[48] Rivière, P.; Rivière-Baudet, M.; Castel, A.; Desor, D.; Abdennadher, C. (Phosphorus Sulfur Silicon Relat. Elem. **61** [1991] 189/99).
[49] Pankratov, L. V.; Nevodchikov, V. I.; Chersakov, V. K.; Bochkarev, M. N. (Metalloorg. Khim. **4** [1991] 516/20; Organomet. Chem. USSR [Engl. Transl.] **4** [1991] 247/9).
[50] Narasaka, K.; Okauchi, T.; Arai, N. (Chem. Lett. **1992** 1229/32).

[51] Castel, A.; Rivière, P.; Satgé, J.; Desor, D. (J. Organomet. Chem. **433** [1992] 49/61).
[52] Ritter, S. K.; Noftle, R. E. (Chem. Mater. **4** [1992] 872/9).
[53] Luo, Y.-R.; Holmes, J. L. (J. Phys. Chem. **96** [1992] 9568/71).
[54] Caminati, W.; Damiani, D.; Dakkouri, M.; Zeeb, S. (J. Mol. Struct. **296** [1993] 79/84).
[55] Lee, S. K.; Mackay, K. M.; Nicholson, B. K. (J. Chem. Soc. Dalton Trans. **1993** 715/22).
[56] Saito, K.; Obi, K. (Chem. Phys. Letters **215** [1993] 193/8).
[57] Richelme, S.; Andrianarison, M.; Couret, C.; Escudié, J.; Satgé, J. (Main Group Met. Chem. **10** [1987] 69/76).

Empirical Formula Index

In the following index the compounds are listed by their empirical formulas in the order of increasing carbon content (first column). Salts are identified by the empirical formula of the germanium-containing cation or anion.

The second column contains the substance formulas, wherein cyclic ligands are partly denoted also by their empirical formula or a linear sequence of ring atoms. In many cases the formulas given in the handbook had to be modified. This became necessary in order to make the index data compatible with the Gmelin Formula Index.

In the third column, page references are printed in ordinary types, table numbers in bold face, and compound numbers within the tables in italics.

For example:

$CH_3GeO_2^-$	$K[(H)_3Ge\text{-}CO_2]$	298/9
$C_5H_{12}Ge$	$(H)_2Ge[\text{-}CH_2CH_2CH_2CH_2CH_2\text{-}]$	262/3, **40**, *3*

Ligand Formula Index

Ligands containing carbon atoms can be used to locate a compound. These ligands are listed in the Ligand Formula Index in the order of increasing number of carbon atoms. They are generally not further characterized by linearized formulas or names unless this is necessary to distinguish between isomers. The number of identical ligands in a compound and the nature of bonding are not taken into consideration. Thus, several compounds may be listed at the same position. Compounds having two or more different carbon-containing ligands occur at more than one position.

The variable organic ligands are placed in the first three columns, while inorganic ligands appear in the fourth column and in the third one, if necessary.

In the fifth column, page references are printed in ordinary type, table numbers in bold face, and compound numbers within the tables in italics.

The following examples illustrate the arrangement:

For $HGe(C_4H_9\text{-}n)_2\text{-}CH_2CH_2\text{-}OH$ (p. 167, Table 24, No. 11):

C_2H_5O	C_4H_9	–	H	167, **24**, *11*
C_4H_9	C_2H_5O	–	H	167, **24**, *11*

CH_2Br	–	–	H	288, **45**, *4*
CH_2Br	CH_3	–	H	147, **22**, *4*
CH_2Cl	–	–	D	287/8, **45**, *3*
CH_2Cl	–	H	D	287/8, **45**, *3*
CH_2Cl	–	–	H	287/8, **45**, *3*
CH_2Cl	CH_3	–	H	147, **22**, *2*
				248, **37**, *2*
CH_2Cl	CH_3	C_6H_5	H	192, **28**, *3*
CH_2Cl	C_2H_5	–	H	158, **23**, *2*
				250, **37**, *13*
CH_2Cl	C_4H_9	–	H	252, **37**, *23*
CH_2F	–	–	D	287, **45**, *1*
CH_2F	–	H	D	287, **45**, *1*
CH_2F	–	–	H	287, **45**, *1*
CH_2I	–	–	H	288, **45**, *5*
CH_3				
$^{13}CH_3$	–	–	H	268
CD_3	–	–	D	4
				270
CD_3	–	–	H	4
				216
				268/71
CD_3	CF_3	–	H	153
				184
CD_3	CF_3	CH_3	H	153
CD_3	CH_3	–	H	5
CHD_2	–	–	H	218
				268
				270/1
CHD_2	CH_3	–	H	5
CH_2D	–	–	H	268
CH_3	–	–	D	1/2
				4/5
				7
				216
				218
				268/73
CH_3	–	H	D	218
				268
				270/1
CH_3	–	–	H	1/14
				216/22
				267/77

CH_3	CD_3	CF_3	H	153
CH_3	CF_3	–	H	147, **22**, *1*
				180, **26**, *1*
				248, **37**, *1*
CH_3	$CHCl_2$	–	H	147, **22**, *3*
CH_3	$CHCl_2$	C_6H_5	H	192, **28**, *4*
CH_3	CH_2Br	–	H	147, **22**, *4*
CH_3	CH_2Cl	–	H	147, **22**, *2*
				248, **37**, *2*
CH_3	CH_2Cl	C_6H_5	H	192, **28**, *3*
CH_3	CH_3O	–	H	147/8, **22**, *5*
CH_3	CH_3O	C_6H_5	H	192, **28**, *5*
CH_3	$C_2H_2F_3O_3S$	–	H	148, **22**, *6*
CH_3	C_2H_3	–	H	151, **22**, *30*
CH_3	C_2H_5	–	D	154
CH_3	C_2H_5	H	D	249, **37**, *3*
CH_3	C_2H_5	–	H	149, **22**, *13*
				249, **37**, *3*
CH_3	C_2H_5	C_3H_6Cl	H	192, **28**, *1*
CH_3	C_2H_5	C_6H_5	H	192, **28**, *2*
CH_3	C_2H_5O	C_6H_5	H	193, **28**, *8*
CH_3	$C_3H_5O_2$	–	H	149, **22**, *18*
CH_3	$C_3H_5O_2$	C_6H_5	H	192/3, **28**, *6*
CH_3	C_3H_7	–	H	149, **22**, *16*
				167, **24**, *7*
				249, **37**, *4*
CH_3	C_3H_9Si	–	H	148, **22**, *10*
CH_3	C_4H_3O	–	H	152, **22**, *41*
CH_3	C_4H_3S	–	H	153, **22**, *42*
CH_3	C_4H_5O	–	H	152, **22**, *40*
				183, **26**, *22*
				249, **37**, *8*
CH_3	C_4H_7	–	H	151, **22**, *31*
CH_3	C_4H_7O	–	H	152, **22**, *39*
CH_3	$C_4H_7O_2$	–	H	149/50, **22**, *19*
CH_3	C_4H_8	–	H	202, **30**, *5*
CH_3	C_4H_9	–	H	150, **22**, *23*
				167, **24**, *8*
CH_3	$C_4H_{10}N$	–	H	148, **22**, *7*
CH_3	$C_4H_{11}Si$	–	H	148, **22**, *9*
				180, **26**, *3*
CH_3	C_5H_5	–	H	249, **37**, *5*
CH_3	C_5H_7O	–	H	151, **22**, *33*

CH_3	C_5H_9	–	H	152, **22**, *34*
CH_3	C_5H_9O	–	H	151, **22**, *32*
CH_3	C_5H_{10}	–	H	202, **30**, *9*
				204, **30**, *16*
CH_3	C_5H_{11}	–	H	149, **22**, *17*
				182, **26**, *12*
CH_3	C_6H_5	–	H	152, **22**, *37*
				172/3, **25**, *1*
				249, **37**, *6*
CH_3	C_6H_5	C_7H_7	H	193, **28**, *7*
CH_3	C_6H_5	$C_{10}H_7$	H	193, **28**, *10*
CH_3	C_6H_5	$C_{10}H_{14}N$	H	193, **28**, *9*
CH_3	C_6H_{11}	–	H	152, **22**, *35*
CH_3	C_6H_{12}	–	H	205, **30**, *20*
				206, **30**, *23*
CH_3	$C_6H_{12}N$	–	H	148, **22**, *8*
CH_3	C_7H_4NO	–	H	184, **26**, *23*
				250, **37**, *9*
CH_3	C_7H_4NS	–	H	184, **26**, *24*
				250, **37**, *10*
CH_3	C_7H_7	–	H	149, **22**, *12*
				181, **26**, *6*
CH_3	C_7H_{13}	–	H	152, **22**, *36*
CH_3	C_7H_{14}	–	H	207, **30**, *27*
CH_3	C_8H_7NS	–	H	179/80
CH_3	C_8H_9	–	H	149, **22**, *15*
CH_3	$C_8H_{10}P$	–	H	149, **22**, *14*
CH_3	C_8H_{12}	–	H	203, **30**, *12*
CH_3	$C_9H_8FeO_3$	–	H	200/1
CH_3	C_9H_{11}	–	H	150, **22**, *21*
				152, **22**, *38*
				182, **26**, *14*
				249, **37**, *7*
CH_3	$C_9H_{12}P$	–	H	150, **22**, *20*
CH_3	$C_{10}H_{13}$	–	H	150, **22**, *24*
				150, **22**, *25*
CH_3	$C_{10}H_{15}$	–	H	151, **22**, *29*
CH_3	$C_{10}H_{18}$	–	H	211, **30**, *41*
CH_3	$C_{10}H_{20}$	–	H	210, **30**, *39*
CH_3	$C_{10}H_{20}O$	–	H	210/1, **30**, *40*
CH_3	$C_{10}H_{27}Si_3$	–	H	148/9, **22**, *11*
CH_3	$C_{11}H_{15}$	–	H	150, **22**, *26*
				151, **22**, *28*

CH_3	$C_{12}H_{17}$	–	H	151, **22**, *27*
CH_3	$C_{12}H_{18}O$	–	H	208, **30**, *30*
CH_3	$C_{15}H_{15}$	–	H	150, **22**, *22*
CH_3	$C_{15}H_{23}$	–	H	183, **26**, *20*
CH_3Cl_2Si	–	–	H	290, **45**, *14*
CH_3O	–	–	H	288, **45**, *6*
CH_3O	CH_3	–	H	147/8, **22**, *5*
CH_3O	CH_3	C_6H_5	H	192, **28**, *5*
CH_3O	C_6H_5	–	H	173, **25**, *2*
CH_4ClSi	–	–	H	290, **45**, *13*
CH_4N	–	–	H	289, **45**, *11*
CH_5Si	–	–	H	233/4, **34**, *2*
				289/90, **45**, *12*
CO_2	–	–	H	298/9
C$_2$Cl	–	–	H	308, **50**, *6*
C_2H	–	–	D	308, **50**, *5*
C_2H	–	–	H	308, **50**, *5*
C_2H_2	–	–	H	261/2
$C_2H_2F_3O_3S$	CH_3	–	H	148, **22**, *6*
$C_2H_2F_3O_3S$	C_6H_5	–	H	173, **25**, *4*
C_2H_3	–	–	D	307, **50**, *1*
C_2H_3	–	–	H	234, **34**, *6*
				307, **50**, *1*
C_2H_3	CH_3	–	H	151, **22**, *30*
C_2H_3	C_2H_5	–	H	161, **23**, *30*
$C_2H_3O_2$	–	–	H	289, **45**, *9*
C_2H_4	–	–	H	261
C_2H_4F	–	–	H	290, **45**, *17*
C_2H_5				
CD_2-CHD_2	–	–	D	279
CD_2-CH_3	–	–	H	279
CHD-CD_3	–	–	D	279
CH_2-CD_3	–	–	H	278/9
C_2H_5	–	–	D	18
				22
				56
				223/5
				278/9
				281
C_2H_5	–	H	D	224/5
				279
C_2H_5	–	–	H	14/72

				223/8
				278/83
C_2H_5	–	–	T	19
				55
C_2H_5	CF_3	–	H	158, **23**, *1*
				250, **37**, *12*
C_2H_5	$CHCl_2$	–	H	158, **23**, *3*
				251, **37**, *14*
C_2H_5	CH_2Cl	–	H	158, **23**, *2*
				250, **37**, *13*
C_2H_5	CH_3	–	D	154
C_2H_5	CH_3	H	D	249, **37**, *3*
C_2H_5	CH_3	–	H	149, **22**, *13*
				249, **37**, *3*
C_2H_5	CH_3	C_3H_6Cl	H	192, **28**, *1*
C_2H_5	CH_3	C_6H_5	H	192, **28**, *2*
C_2H_5	C_2H_3	–	H	161, **23**, *30*
C_2H_5	C_2H_5O	–	H	159, **23**, *5*
				251, **37**, *16*
C_2H_5	C_2H_5S	–	H	159, **23**, *6*
C_2H_5	C_3H_5O	–	H	161, **23**, *31*
C_2H_5	$C_3H_5O_2$	–	H	160, **23**, *14*
C_2H_5	C_3H_6Cl	–	H	159, **23**, *9*
				182, **26**, *11*
				251, **37**, *17*
C_2H_5	C_3H_7	–	H	159, **23**, *8*
				166, **24**, *1*
C_2H_5	C_3H_7O	–	H	159, **23**, *10*
				159, **23**, *11*
				251, **37**, *18*
C_2H_5	C_4H_7	–	H	161, **23**, *32*
				161, **23**, *33*
				251/2, **37**, *20*
C_2H_5	C_4H_7O	–	H	160, **23**, *13*
				161, **23**, *25*
C_2H_5	$C_4H_7O_2$	–	H	160, **23**, *15*
C_2H_5	C_4H_9	–	H	160, **23**, *19*
				160, **23**, *21*
				167, **24**, *10*
C_2H_5	$C_4H_9Cl_2Si$	–	H	160, **23**, *17*
C_2H_5	C_4H_9O	–	H	159, **23**, *12*
				160, **23**, *22*
C_2H_5	$C_4H_{10}N$	–	H	161, **23**, *27*

C_2H_5	$C_4H_{11}Si$	–	H	160, **23**, *16*
C_2H_5	C_5H_4O	–	H	209/10, **30**, *35*
C_2H_5	C_5H_8	–	H	203, **30**, *10*
C_2H_5	C_5H_9	–	H	162, **23**, *34*
				162, **23**, *35*
C_2H_5	C_5H_9O	–	H	161, **23**, *26*
				161, **23**, *29*
				162, **23**, *37*
C_2H_5	C_5H_{11}	–	H	161, **23**, *28*
				251, **37**, *19*
C_2H_5	$C_5H_{11}O$	–	H	161, **23**, *23*
C_2H_5	C_6F_5	–	H	182, **26**, *13*
C_2H_5	C_6H_5	–	H	162/3, **23**, *40*
				174, **25**, *9*
				252, **37**, *21*
C_2H_5	C_6H_5	–	T	178
C_2H_5	C_6H_5	$C_{10}H_7$	H	194, **28**, *11*
C_2H_5	C_6H_{12}	–	H	162, **23**, *36*
C_2H_5	C_6H_{13}	–	H	160, **23**, *20*
C_2H_5	$C_6H_{13}O$	–	H	161, **23**, *24*
C_2H_5	C_7H_7	–	H	158/9, **23**, *4*
				181, **26**, *7*
				251, **37**, *15*
C_2H_5	$C_8H_{10}P$	–	H	159, **23**, *7*
C_2H_5	C_8H_{13}	–	H	162, **23**, *38*
C_2H_5	C_9H_{14}	–	H	210, **30**, *36*
C_2H_5	$C_{10}H_{16}O$	–	H	208, **30**, *28*
C_2H_5	$C_{10}H_{17}$	–	H	162, **23**, *39*
C_2H_5	$C_{11}H_{10}$	–	H	210, **30**, *37*
C_2H_5	$C_{12}H_{12}O$	–	H	209, **30**, *33*
C_2H_5	$C_{12}H_{18}O$	–	H	208/9, **30**, *31*
C_2H_5	$C_{16}H_{19}Si$	–	H	160, **23**, *18*
C_2H_5O				
$CH(CH_3)$-OH	–	–	H	291, **45**, *19*
$CH(CH_3)$-OH	C_6H_5	–	H	256, **39**, *6*
CH_2-OCH_3	–	–	H	288/9, **45**, *7*
CH_2-OCH_3	C_6H_5	–	H	180, **26**, *2*
				255, **39**, *1*
CH_2CH_2-OH	–	–	H	290/1, **45**, *18*
CH_2CH_2-OH	CH_3	C_6H_5	H	193, **28**, *8*
CH_2CH_2-OH	C_2H_5	–	H	159, **23**, *5*
				251, **37**, *16*
CH_2CH_2-OH	C_3H_7	–	H	166, **24**, *2*

CH_2CH_2-OH	C_4H_9	–	H	167, **24**, *11*
CH_2CH_2-OH	C_6H_5	–	H	174, **25**, *10*
				256, **39**, *5*
C_2H_5S	C_2H_5	–	H	159, **23**, *6*
C_2H_6N	–	–	H	291, **45**, *21*
C_2H_7Si	–	–	H	291, **45**, *22*
C$_3F_3$	–	–	D	312
C_3F_3	–	–	H	308, **50**, *7*
$C_3H_4F_3$	–	–	H	292, **45**, *26*
C_3H_4N	C_3H_7	–	H	167, **24**, *6*
C_3H_4N	C_4H_9	–	H	168, **24**, *19*
C_3H_4N	C_6H_5	–	H	174, **25**, *12*
C_3H_5				
CH_2-CH=CH_2	–	–	D	307, **50**, *2*
CH_2-CH=CH_2	–	–	H	307, **50**, *2*
CH_2-CH=CH_2	C_4H_9	–	H	170, **24**, *29*
CH_2-CH=CH_2	C_6H_5	–	H	258, **39**, *16*
C_3H_5-c	–	–	D	302/3, **48**, *1*
C_3H_5-c	–	H	D	304
C_3H_5-c	–	–	H	302/3, **48**, *1*
C_3H_5O				
CH=CH-CH_2OH	C_2H_5	–	H	161, **23**, *31*
CH_2-C(=O)-CH_3	C_3H_7	–	H	167, **24**, *4*
CH_2CH_2-CH=O	C_6H_5	–	H	174, **25**, *11*
$C_3H_5O_2$				
C(=O)O-C_2H_5	–	–	H	289, **45**, *10*
CH_2-C(=O)-OCH_3	–	–	H	96
CH_2-C(=O)-OCH_3	C_3H_7	–	H	166/7, **24**, *3*
CH_2-C(=O)-OCH_3	C_4H_9	–	H	167/8, **24**, *12*
CH_2-O-C(=O)-CH_3	CH_3	C_6H_5	H	192/3, **28**, *6*
CH_2-O-C(=O)-CH_3	C_6H_5	–	H	173, **25**, *3*
CH_2CH_2-CO_2H	–	–	H	293, **45**, *33*
CH_2CH_2-CO_2H	CH_3	–	H	149, **22**, *18*
CH_2CH_2-CO_2H	C_2H_5	–	H	160, **23**, *14*
CH_2CH_2-CO_2H	C_3H_7	–	H	167, **24**, *5*
C_3H_6	C_4H_9	–	H	201, **30**, *1*
C_3H_6Cl	–	–	H	234, **34**, *4*
				292, **45**, *27*
C_3H_6Cl	CH_3	C_2H_5	H	192, **28**, *1*
C_3H_6Cl	C_2H_5	–	H	159, **23**, *9*
				182, **26**, *11*
				251, **37**, *17*

C_3H_6F	–	–	H	292, **45**, *25*
C_3H_6O	C_6H_5	–	H	201, **30**, *3*
C_3H_7				
C_3H_7-i	–	–	D	78
C_3H_7-i	–	H	D	229, **33**, *2*
				285
C_3H_7-i	–	–	H	77/9
				229, **33**, *2*
				283, **44**, *2*
C_3H_7-i	CH_3	–	H	149, **22**, *16*
				167, **24**, *7*
				249, **37**, *4*
C_3H_7-i	C_6H_5	–	H	174, **25**, *14*
C_3H_7-i	C_6H_5	$C_{10}H_7$	H	194, **28**, *12*
C_3H_7-n	–	–	D	283, **44**, *1*
C_3H_7-n	–	–	H	72/7
				229, **33**, *1*
				283, **44**, *1*
C_3H_7-n	C_2H_5	–	H	159, **23**, *8*
				166, **24**, *1*
C_3H_7-n	C_2H_5O	–	H	166, **24**, *2*
C_3H_7-n	C_3H_4N	–	H	167, **24**, *6*
C_3H_7-n	C_3H_5O	–	H	167, **24**, *4*
C_3H_7-n	$C_3H_5O_2$	–	H	166/7, **24**, *3*
				167, **24**, *5*
C_3H_7-n	C_4H_8	–	H	202, **30**, *6*
C_3H_7-n	C_4H_9	–	H	168, **24**, *14*
				252, **37**, *22*
C_3H_7O				
$C(CH_3)_2$-OH	–	–	H	292, **45**, *24*
CH_2-$CH(CH_3)$-OH	C_2H_5	–	H	159, **23**, *11*
				251, **37**, *18*
CH_2-$CH(CH_3)$-OH	C_6H_5	–	H	257, **39**, *9*
CH_2CH_2-OCH_3	–	–	H	291, **45**, *20*
$CH_2CH_2CH_2$-OH	–	–	H	292, **45**, *28*
$CH_2CH_2CH_2$-OH	C_2H_5	–	H	159, **23**, *10*
$CH_2CH_2CH_2$-OH	C_4H_9	–	H	168, **24**, *15*
$CH_2CH_2CH_2$-OH	C_6H_5	–	H	256/7, **39**, *8*
C_3H_8N	–	–	H	294, **45**, *39*
C_3H_8N	C_4H_9	–	H	168, **24**, *18*
C_3H_9Si				
CH_2-$SiH(CH_3)_2$	CH_3	–	H	148, **22**, *10*
$CH_2CH_2CH_2$-SiH_3	–	–	H	294, **45**, *41*

C_4H_3O	–	–	H	142, **21**, *20*
				245, **36**, *7*
				319, **55**, *8*
C_4H_3O	CH_3	–	H	152, **22**, *41*
C_4H_3S				
2-SC_4H_3	CH_3	–	H	153, **22**, *42*
3-SC_4H_3	–	–	H	320, **55**, *9*
C_4H_5	C_6H_5	–	H	175, **25**, *18*
C_4H_5O	CH_3	–	H	152, **22**, *40*
				183, **26**, *22*
				249, **37**, *8*
C_4H_6N	C_6H_5	–	H	174, **25**, *13*
C_4H_7				
CH_2-$C(CH_3)$=CH_2	C_4H_9	–	H	170, **24**, *30*
CH_2-CH=CH-CH_3	–	–	H	307/8, **50**, *3*
CH_2-CH=CH-CH_3	C_2H_5	–	H	161, **23**, *33*
				251/2, **37**, *20*
CH_2CH_2-CH=CH_2	CH_3	–	H	151, **22**, *31*
CH_2CH_2-CH=CH_2	C_2H_5	–	H	161, **23**, *32*
C_4H_7-c	–	–	H	303, **48**, *2*
C_4H_7O				
$CH(CH_3)$-CH_2-CH=O	C_5H_{11}	–	H	170, **24**, *34*
CH(OH)-CH=CH-CH_3	C_6H_5	–	H	258, **39**, *17*
CH_2-$CH(CH_3)$-CH=O	C_2H_5	–	H	160, **23**, *13*
CH_2CH_2-C(=O)-CH_3	C_2H_5	–	H	161, **23**, *25*
CH_2CH_2-C(=O)-CH_3	C_4H_9	–	H	169, **24**, *27*
CH_2CH_2-C(=O)-CH_3	C_6H_5	–	H	175, **25**, *17*
2-OC_4H_7	CH_3	–	H	152, **22**, *39*
$C_4H_7O_2$				
$CH(CH_3)$-CH_2-CO_2H	–	–	H	293, **45**, *34*
CH_2-C(=O)O-C_2H_5	C_4H_9	–	H	168, **24**, *13*
CH_2-$CH(CH_3)$-CO_2H	–	–	H	294, **45**, *35*
CH_2CH_2-C(=O)-OCH_3	CH_3	–	H	149/50, **22**, *19*
CH_2CH_2-C(=O)-OCH_3	C_2H_5	–	H	160, **23**, *15*
CH_2CH_2-C(=O)-OCH_3	C_4H_9	–	H	168, **24**, *17*
$C_4H_7S_2$	–	–	H	317
C_4H_8				
-CH_2-$CH(CH_3)$-CH_2-	C_4H_9	–	H	201, **30**, *2*
-$CH_2CD_2CD_2CH_2$-	C_4H_9	–	H	202, **30**, *7*
-$CH_2CD_2CD_2CH_2$-	C_6H_5	–	H	202, **30**, *8*
-$CH_2CH_2CH_2CH_2$-	–	–	D	265/6
-$CH_2CH_2CH_2CH_2$-	–	–	H	262, **40**, *1*
-$CH_2CH_2CH_2CH_2$-	CH_3	–	H	202, **30**, *5*

$-CH_2CH_2CH_2CH_2-$	C_3H_7	–	H	202, **30**, *6*
$-CH_2CH_2CH_2CH_2-$	C_4H_9	–	H	202, **30**, *7*
$-CH_2CH_2CH_2CH_2-$	C_6H_5	–	H	202, **30**, *8*
C_4H_8O	C_6H_5	–	H	201/2, **30**, *4*
C_4H_9				
C_4H_9-n	–	–	D	79
				86
C_4H_9-n	–	–	H	79/92
				229/30, **33**, *3*
				283/4, **44**, *3*
C_4H_9-n	$CHCl_2$	–	H	252, **37**, *24*
C_4H_9-n	CH_2Cl	–	H	252, **37**, *23*
C_4H_9-n	CH_3	–	H	150, **22**, *23*
				167, **24**, *8*
C_4H_9-n	C_2H_5	–	H	160, **23**, *19*
				167, **24**, *10*
C_4H_9-n	C_2H_5O	–	H	167, **24**, *11*
C_4H_9-n	C_3H_4N	–	H	168, **24**, *19*
C_4H_9-n	C_3H_5	–	H	170, **24**, *29*
C_4H_9-n	$C_3H_5O_2$	–	H	167/8, **24**, *12*
C_4H_9-n	C_3H_6	–	H	201, **30**, *1*
C_4H_9-n	C_3H_7	–	H	168, **24**, *14*
				252, **37**, *22*
C_4H_9-n	C_3H_7O	–	H	168, **24**, *15*
C_4H_9-n	C_3H_8N	–	H	168, **24**, *18*
C_4H_9-n	C_4H_7	–	H	170, **24**, *30*
C_4H_9-n	C_4H_7O	–	H	169, **24**, *27*
C_4H_9-n	$C_4H_7O_2$	–	H	168, **24**, *13*
				168, **24**, *17*
C_4H_9-n	C_4H_8	–	H	201, **30**, *2*
				202, **30**, *7*
C_4H_9-n	$C_4H_9Cl_2Si$	–	H	169, **24**, *21*
C_4H_9-n	C_4H_9O	–	H	168, **24**, *16*
C_4H_9-n	$C_4H_{11}Si$	–	H	168, **24**, *20*
C_4H_9-n	C_6H_5	–	D	178
C_4H_9-n	C_6H_5	–	H	170, **24**, *32*
				174, **25**, *15*
				252, **37**, *25*
C_4H_9-n	C_6H_9	–	H	170, **24**, *31*
C_4H_9-n	C_7H_7	–	H	167, **24**, *9*
				181, **26**, *8*
C_4H_9-n	C_8H_7	–	H	169, **24**, *28*
C_4H_9-n	$C_9H_{21}Si$	–	H	169, **24**, *24*

C_4H_9-n	$C_{11}H_{24}ClSi$	–	H	169, **24**, *26*
C_4H_9-n	$C_{11}H_{25}Si$	–	H	169, **24**, *25*
C_4H_9-n	$C_{12}H_{27}Si$	–	H	169, **24**, *22*
C_4H_9-n	$C_{16}H_{19}Si$	–	H	169, **24**, *23*
C_4H_9-t	–	–	H	230, **33**, *4*
				284/5, **44**, *9*
C_4H_9-t	C_2H_5	–	H	160, **23**, *21*
C_4H_9-t	C_5H_{10}	–	H	204, **30**, *17*
C_4H_9-t	C_6H_5	–	H	170, **24**, *33*
				252, **37**, *26*
C_4H_9-t	C_6H_{12}	–	H	205, **30**, *21*
				206/7, **30**, *24*
C_4H_9-t	C_9H_{18}	–	H	207, **30**, *26*
C_4H_9-t	$C_{18}H_{29}$	–	H	253, **37**, *27*
$C_4H_9Cl_2Si$	C_2H_5	–	H	160, **23**, *17*
$C_4H_9Cl_2Si$	C_4H_9	–	H	169, **24**, *21*
C_4H_9O				
$CH(CH_3)$-CH_2CH_2-OH	–	–	H	292/3, **45**, *30*
$CH(CH_3)$-CH_2CH_2-OH	C_6H_5	–	H	257/8, **39**, *12*
CH(OH)-C_3H_7-i	–	–	H	292, **45**, *29*
CH(OH)-C_3H_7-i	C_6H_5	–	H	257, **39**, *10*
CH(OH)-C_3H_7-n	–	–	H	294/5, **45**, *42*
CH_2-$CH(CH_3)$-CH_2-OH	C_2H_5	–	H	159, **23**, *12*
CH_2-$CH(CH_3)$-CH_2-OH	C_4H_9	–	H	168, **24**, *16*
CH_2CH_2-$CH(CH_3)$-OH	C_2H_5	–	H	160, **23**, *22*
CH_2CH_2-$CH(CH_3)$-OH	C_6H_5	–	H	175, **25**, *16*
$CH_2CH_2CH_2$-OCH_3	–	–	H	293, **45**, *31*
$CH_2CH_2CH_2CH_2$-OH	C_6H_5	–	H	175, **25**, *19*
$C_4H_{10}N$				
CH_2-$N(CH_3)$-C_2H_5	CH_3	–	H	148, **22**, *7*
CH_2CH_2-CH_2CH_2-NH_2	–	–	H	234, **34**, *5*
CH_2CH_2-CH_2CH_2-NH_2	C_2H_5	–	H	161, **23**, *27*
$C_4H_{11}Si$				
CH_2-$Si(CH_3)_3$	–	–	H	96
CH_2-$Si(CH_3)_3$	CH_3	–	H	148, **22**, *9*
				180, **26**, *3*
$CH_2CH_2CH_2$-SiH_2-CH_3	C_2H_5	–	H	160, **23**, *16*
$CH_2CH_2CH_2$-SiH_2-CH_3	C_4H_9	–	H	168, **24**, *20*
C$_5H_4O$	C_2H_5	–	H	209/10, **30**, *35*
C_5H_5	–	–	H	234, **34**, *7*
				313/4, **54**, *1*
C_5H_5	CH_3	–	H	249, **37**, *5*

C_5H_7O	CH_3	–	H	151, **22**, *33*
$C_5H_7O_3$	–	–	H	319, **55**, *6*
				319, **55**, *7*
C_5H_8	–	–	H	264, **40**, *10*
C_5H_8	C_2H_5	–	H	203, **30**, *10*
C_5H_9				
CH_2-$C(CH_3)$=CH-CH_3	C_2H_5	–	H	162, **23**, *35*
CH_2-$CH(CH_3)$-CH=CH_2	C_2H_5	–	H	162, **23**, *34*
$CH_2CH_2CH_2$-CH=CH_2	CH_3	–	H	152, **22**, *34*
C_5H_9-c	–	–	D	303, **48**, *3*
C_5H_9-c	–	H	D	305
C_5H_9-c	–	–	H	93
				303, **48**, *3*
C_5H_9O				
CH=CH-$C(CH_3)_2$-OH	C_2H_5	–	H	162, **23**, *37*
CH_2-$CH(CH_3)$-C(=O)-CH_3	C_2H_5	–	H	161, **23**, *26*
CH_2-CH(OH)-$C(CH_3)$=CH_2	CH_3	–	H	151, **22**, *32*
C_5H_8-2-OH	C_2H_5	–	H	161, **23**, *29*
$C_5H_9O_2$	–	–	H	294, **45**, *36*
C_5H_{10}				
-$CH(CH_3)$-$CH_2CH_2CH_2$-	–	–	H	262, **40**, *2*
-$CH(CH_3)$-$CH_2CH_2CH_2$-	CH_3	–	H	202, **30**, *9*
-$CH_2CH_2CH_2CH_2CH_2$-	–	–	H	262/3, **40**, *3*
-$CH_2CH_2CH_2CH_2CH_2$-	CH_3	–	H	204, **30**, *16*
-$CH_2CH_2CH_2CH_2CH_2$-	C_4H_9	–	H	204, **30**, *17*
-$CH_2CH_2CH_2CH_2CH_2$-	C_5H_{11}	–	H	204, **30**, *18*
-$CH_2CH_2CH_2CH_2CH_2$-	C_6H_5	–	H	204/5, **30**, *19*
C_5H_{11}				
CH_2-C_4H_9-t	–	–	H	96
CH_2-C_4H_9-t	CH_3	–	H	149, **22**, *17*
				182, **26**, *12*
C_5H_{11}-i	–	–	H	93
				284, **44**, *5*
C_5H_{11}-i	C_2H_5	–	H	161, **23**, *28*
				251, **37**, *19*
C_5H_{11}-n	–	–	H	92/3
				230/1, **33**, *5*
				284, **44**, *4*
C_5H_{11}-n	C_4H_7O	–	H	170, **24**, *34*
C_5H_{11}-n	C_5H_{10}	–	H	204, **30**, *18*
$C_5H_{11}O$				
CH_2-$CH(CH_3)$-$CH(CH_3)$-OH	C_2H_5	–	H	161, **23**, *23*
CH_2CH_2-CH_2CH_2-OCH_3	–	–	H	295, **45**, *43*

$C_5H_{11}O_2$	–	–	H	293, **45**, *32*
C$_6Cl_5$	–	–	H	142, **21**, *18*
C_6F_5	–	–	H	132/9
				245, **36**, *6*
				318, **55**, *2*
C_6F_5	C_2H_5	–	H	182, **26**, *13*
C_6H_4Cl				
C_6H_4-2-Cl	–	–	H	139, **21**, *3*
C_6H_4-2-Cl	–	–	T	142/3
C_6H_4-3-Cl	–	–	H	139, **21**, *4*
C_6H_4-3-Cl	–	–	T	142/3
C_6H_4-3-Cl	C_6H_5	–	H	176, **25**, *26*
C_6H_4-4-Cl	–	–	H	139, **21**, *5*
C_6H_4-4-Cl	–	–	T	142/3
C_6H_4F				
C_6H_4-3-F	–	–	H	139, **21**, *1*
C_6H_4-4-F	–	–	H	139, **21**, *2*
C_6H_4-4-F	C_6H_5	–	H	176, **25**, *25*
C_6H_4-4-F	C_6H_5	–	T	179
$C_6H_4NO_2$	C_6H_5	–	H	176, **25**, *28*
$C_6H_4NO_2$	C_6H_5	–	T	179
C_6H_5				
C_6D_5	–	–	D	320/1
C_6D_5	–	–	H	320/1
C_6H_5	–	–	D	100
				102
				106
				110
				112
				123
				238/9
				320/2
C_6H_5	–	H	D	238/9
C_6H_5	–	–	H	100/32
				237/43
				317/8, **55**, *1*
C_6H_5	–	T	H	238
				240
				320
				322
C_6H_5	–	–	T	106
				120

C_6H_5	$CHCl_2$	CH_3	H	192, **28**, *4*
C_6H_5	CH_2Cl	CH_3	H	192, **28**, *3*
C_6H_5	CH_3	–	H	152, **22**, *37*
				172/3, **25**, *1*
				249, **37**, *6*
C_6H_5	CH_3	CH_3O	H	192, **28**, *5*
C_6H_5	CH_3	C_2H_5	H	192, **28**, *2*
C_6H_5	CH_3	C_2H_5O	H	193, **28**, *8*
C_6H_5	CH_3	$C_3H_5O_2$	H	192/3, **28**, *6*
C_6H_5	CH_3	C_7H_7	H	193, **28**, *7*
C_6H_5	CH_3	$C_{10}H_7$	H	193, **28**, *10*
C_6H_5	CH_3	$C_{10}H_{14}N$	H	193, **28**, *9*
C_6H_5	CH_3O	–	H	173, **25**, *2*
C_6H_5	$C_2H_2F_3O_3S$	–	H	173, **25**, *4*
C_6H_5	C_2H_5	–	H	162/3, **23**, *40*
				174, **25**, *9*
				252, **37**, *21*
C_6H_5	C_2H_5	–	T	178
C_6H_5	C_2H_5	$C_{10}H_7$	H	194, **28**, *11*
C_6H_5	C_2H_5O	–	H	174, **25**, *10*
				180, **26**, *2*
				255, **39**, *1*
				256, **39**, *5*
				256, **39**, *6*
C_6H_5	C_3H_4N	–	H	174, **25**, *12*
C_6H_5	C_3H_5	–	H	258, **39**, *16*
C_6H_5	C_3H_5O	–	H	174, **25**, *11*
C_6H_5	$C_3H_5O_2$	–	H	173, **25**, *3*
C_6H_5	C_3H_6O	–	H	201, **30**, *3*
C_6H_5	C_3H_7	–	H	174, **25**, *14*
C_6H_5	C_3H_7	$C_{10}H_7$	H	194, **28**, *12*
C_6H_5	C_3H_7O	–	H	256/7, **39**, *8*
				257, **39**, *9*
C_6H_5	C_4H_5	–	H	175, **25**, *18*
C_6H_5	C_4H_6N	–	H	174, **25**, *13*
C_6H_5	C_4H_7O	–	H	175, **25**, *17*
				258, **39**, *17*
C_6H_5	C_4H_8	–	H	202, **30**, *8*
C_6H_5	C_4H_8O	–	H	201/2, **30**, *4*
C_6H_5	C_4H_9	–	D	178
C_6H_5	C_4H_9	–	H	170, **24**, *32*
				170, **24**, *33*
				174, **25**, *15*

				252, **37**, *25*
				252, **37**, *26*
C_6H_5	C_4H_9O	–	H	175, **25**, *16*
				175, **25**, *19*
				257, **39**, *10*
				257/8, **39**, *12*
C_6H_5	C_5H_{10}	–	H	204/5, **30**, *19*
C_6H_5	C_6H_4Cl	–	H	176, **25**, *26*
C_6H_5	C_6H_4F	–	H	176, **25**, *25*
C_6H_5	C_6H_4F	–	T	179
C_6H_5	$C_6H_4NO_2$	–	H	176, **25**, *28*
C_6H_5	$C_6H_4NO_2$	–	T	179
C_6H_5	C_6H_{10}	–	H	203, **30**, *11*
C_6H_5	C_6H_{11}	–	H	171, **24**, *35*
				258, **39**, *15*
C_6H_5	$C_6H_{11}O$	–	H	175, **25**, *21*
C_6H_5	C_6H_{12}	–	H	206, **30**, *22*
				207, **30**, *25*
C_6H_5	C_6H_{13}	–	H	175, **25**, *22*
C_6H_5	$C_6H_{13}O$	–	H	258, **39**, *13*
C_6H_5	$C_6H_{14}N$	–	H	257, **39**, *11*
C_6H_5	C_7H_4N	–	H	176, **25**, *27*
C_6H_5	C_7H_4N	–	T	179
C_6H_5	C_7H_5O	–	H	173, **25**, *7*
C_6H_5	C_7H_7O	–	H	173, **25**, *5*
				256, **39**, *2*
C_6H_5	C_7H_{12}	–	H	210, **30**, *38*
C_6H_5	$C_7H_{13}O_2$	–	H	175, **25**, *20*
C_6H_5	C_8H_{12}	–	D	213
C_6H_5	C_8H_{12}	–	H	203, **30**, *13*
C_6H_5	C_8H_{17}	–	H	175, **25**, *23*
C_6H_5	$C_{10}H_{11}O$	–	H	173, **25**, *8*
C_6H_5	$C_{10}H_{16}O$	–	H	208, **30**, *29*
C_6H_5	$C_{11}H_{21}O$	–	H	258, **39**, *18*
C_6H_5	$C_{11}H_{22}O$	–	H	211, **30**, *42*
C_6H_5	$C_{11}H_{23}O$	–	H	258, **39**, *14*
C_6H_5	$C_{12}H_{12}O$	–	H	209, **30**, *34*
C_6H_5	$C_{12}H_{18}O$	–	H	209, **30**, *32*
C_6H_5	$C_{13}H_{11}O$	–	H	173, **25**, *6*
C_6H_5	$C_{13}H_{12}N$	–	H	256, **39**, *3*
C_6H_5	$C_{13}H_{12}NO$	–	H	256, **39**, *4*
C_6H_5	$C_{14}H_{13}O$	–	H	256, **39**, *7*
C_6H_5	$C_{23}H_{20}NOPRe$	–	H	176, **25**, *24*

C_6H_5	$C_{27}H_{59}Si_6$	–	H	176, **25**, *29*
				259, **39**, *19*
C_6H_5	$C_{28}H_{20}$	–	H	204, **30**, *14*
C_6H_7	–	–	H	314, **54**, *2*
C_6H_9	C_4H_9	–	H	170, **24**, *31*
C_6H_{10}	–	–	H	264, **40**, *11*
C_6H_{10}	C_6H_5	–	H	203, **30**, *11*
C_6H_{11}				
CH_2CH_2-CH_2CH_2-CH=CH_2	CH_3	–	H	152, **22**, *35*
C_6H_{11}-c	–	–	D	94
C_6H_{11}-c	–	–	H	93/4
				303, **48**, *4*
C_6H_{11}-c	C_6H_5	–	H	171, **24**, *35*
				258, **39**, *15*
$C_6H_{11}O$				
CH(CH_2CH_2OH)-CH=CH-CH_3	–	–	H	308, **50**, *4*
$CH_2CH_2CH_2CH_2$-C(=O)-CH_3	C_6H_5	–	H	175, **25**, *21*
C_6H_{12}				
CH_2-CH(CH_3)=C(CH_3)$_2$	C_2H_5	–	H	162, **23**, *36*
-CH(CH_3)-$CH_2CH_2CH_2CH_2$-	–	–	H	263, **40**, *4*
-CH_2-CH(CH_3)-$CH_2CH_2CH_2$-	–	–	H	263, **40**, *5*
-CH_2-CH(CH_3)-$CH_2CH_2CH_2$-	CH_3	–	H	205, **30**, *20*
-CH_2-CH(CH_3)-$CH_2CH_2CH_2$-	C_4H_9	–	H	205, **30**, *21*
-CH_2-CH(CH_3)-$CH_2CH_2CH_2$-	C_6H_5	–	H	206, **30**, *22*
-CH_2CH_2-CH(CH_3)-CH_2CH_2-	–	–	H	263, **40**, *6*
-CH_2CH_2-CH(CH_3)-CH_2CH_2-	CH_3	–	H	206, **30**, *23*
-CH_2CH_2-CH(CH_3)-CH_2CH_2-	C_4H_9	–	H	206/7, **30**, *24*
-CH_2CH_2-CH(CH_3)-CH_2CH_2-	C_6H_5	–	H	207, **30**, *25*
$C_6H_{12}N$	CH_3	–	H	148, **22**, *8*
C_6H_{13}				
CH_2-CH(CH_3)-C_3H_7-i	C_2H_5	–	H	160, **23**, *20*
CH_2CH_2-C_4H_9-t	C_9H_{11}	–	H	183, **26**, *18*
C_6H_{13}-n	–	–	H	93
				231, **33**, *6*
				284, **44**, *6*
C_6H_{13}-n	C_6H_5	–	H	175, **25**, *22*
$C_6H_{13}O$				
CH_2-CH(OH)-C_4H_9-t	C_2H_5	–	H	161, **23**, *24*
CH_2CH_2-CH_2CH_2-CH(CH_3)-OH	C_6H_5	–	H	258, **39**, *13*
CH_2CH_2-O-C_4H_9-n	C_7H_{15}	–	H	181, **26**, *10*
$C_6H_{14}N$	–	–	H	294, **45**, *40*
$C_6H_{14}N$	C_6H_5	–	H	257, **39**, *11*

$C_7H_4F_3$	–	–	H	141, **21**, *13*
C_7H_4N	C_6H_5	–	H	176, **25**, *27*
C_7H_4N	C_6H_5	–	T	179
C_7H_4NO	CH_3	–	H	184, **26**, *23*
				250, **37**, *9*
C_7H_4NS	CH_3	–	H	184, **26**, *24*
				250, **37**, *10*
C_7H_5O	C_6H_5	–	H	173, **25**, *7*
C_7H_5O	C_9H_{11}	–	H	182, **26**, *17*
C_7H_7				
CH_2-C_6H_5	–	–	D	95
				234, **34**, *3*
CH_2-C_6H_5	–	–	H	94/5
				234, **34**, *3*
				290, **45**, *16*
CH_2-C_6H_5	CH_3	–	H	149, **22**, *12*
				181, **26**, *6*
CH_2-C_6H_5	CH_3	C_6H_5	H	193, **28**, *7*
CH_2-C_6H_5	C_2H_5	–	H	158/9, **23**, *4*
				181, **26**, *7*
				251, **37**, *15*
CH_2-C_6H_5	C_4H_9	–	H	167, **24**, *9*
				181, **26**, *8*
CH_2-C_6H_5	C_7H_{15}	–	H	181, **26**, *9*
C_6H_4-2-CH_3	–	–	H	140, **21**, *10*
C_6H_4-2-CH_3	–	–	T	142/3
C_6H_4-3-CH_3	–	–	H	140/1, **21**, *11*
C_6H_4-3-CH_3	–	–	T	142/3
C_6H_4-4-CH_3	–	–	D	143/4
C_6H_4-4-CH_3	–	–	H	141, **21**, *12*
				243, **36**, *1*
				318, **55**, *3*
C_6H_4-4-CH_3	–	–	T	142/3
C_7H_7O				
$CH(OH)$-C_6H_5	C_6H_5	–	H	173, **25**, *5*
				256, **39**, *2*
$CH(OH)$-C_6H_5	C_9H_{11}	–	H	182, **26**, *15*
				259, **39**, *20*
C_6H_4-2-OCH_3	–	–	H	140, **21**, *6*
C_6H_4-2-OCH_3	–	–	T	142/3
C_6H_4-3-OCH_3	–	–	H	140, **21**, *7*
C_6H_4-4-OCH_3	–	–	H	140, **21**, *8*
C_6H_4-4-OCH_3	–	–	T	142/3

C_7H_{12}	C_6H_5	–	H	210, **30**, *38*
C_7H_{13}	CH_3	–	H	152, **22**, *36*
$C_7H_{13}O_2$	C_6H_5	–	H	175, **25**, *20*
C_7H_{14}				
$-CH_2-C(CH_3)_2-CH_2CH_2CH_2-$	–	–	H	263/4, **40**, *9*
$-CH_2-C(CH_3)_2-CH_2CH_2CH_2-$	CH_3	–	H	207, **30**, *27*
$-CH_2-CH(CH_3)-CH_2-CH(CH_3)-CH_2-$	–	–	H	263, **40**, *8*
C_7H_{15}	–	–	H	94
				231, **33**, *7*
				284, **44**, *7*
C_7H_{15}	$C_6H_{13}O$	–	H	181, **26**, *10*
C_7H_{15}	C_7H_7	–	H	181, **26**, *9*
$C_7H_{15}O$	–	–	H	295, **45**, *44*
$C_7H_{19}Si_2$	$C_{13}H_9$	–	H	180/1, **26**, *4*
$C_7H_{19}Si_2$	$C_{14}H_{11}$	–	H	181, **26**, *5*
C$_8H_7$	C_4H_9	–	H	169, **24**, *28*
C_8H_7NS	CH_3	–	H	179/80
C_8H_9				
$CH_2CH_2-C_6H_5$	CH_3	–	H	149, **22**, *15*
$C_6H_3-3,4-(CH_3)_2$	–	–	H	141, **21**, *15*
$C_8H_{10}P$	CH_3	–	H	149, **22**, *14*
$C_8H_{10}P$	C_2H_5	–	H	159, **23**, *7*
C_8H_{12}	–	–	H	264, **40**, *12*
C_8H_{12}	CH_3	–	H	203, **30**, *12*
C_8H_{12}	C_6H_5	–	D	213
C_8H_{12}	C_6H_5	–	H	203, **30**, *13*
C_8H_{13}	C_2H_5	–	H	162, **23**, *38*
C_8H_{17}	–	–	H	94
				231, **33**, *8*
				284, **44**, *8*
C_8H_{17}	C_6H_5	–	H	175, **25**, *23*
C$_9H_7$	–	–	H	314, **54**, *3*
$C_9H_8FeO_3$	CH_3	–	H	200/1
C_9H_9				
2,4,6,8-cyclononatetraen-1-yl	–	–	H	315, **54**, *5*
3a,7a-dihydro-1H-inden-1-yl	–	–	H	315, **54**, *4*
C_9H_{11}				
$CH_2CH_2CH_2-C_6H_5$	CH_3	–	H	150, **22**, *21*
$C_6H_2-2,4,6-(CH_3)_3$	–	–	D	246
$C_6H_2-2,4,6-(CH_3)_3$	–	–	H	141/2, **21**, *16*
				244, **36**, *4*

				318/9, **55**, *4*
C_6H_2-2,4,6-$(CH_3)_3$	CH_3	–	H	152, **22**, *38*
				182, **26**, *14*
				249, **37**, *7*
C_6H_2-2,4,6-$(CH_3)_3$	C_6H_{13}	–	H	183, **26**, *18*
C_6H_2-2,4,6-$(CH_3)_3$	C_7H_5O	–	H	182, **26**, *17*
C_6H_2-2,4,6-$(CH_3)_3$	C_7H_7O	–	H	182, **26**, *15*
				259, **39**, *20*
C_6H_2-2,4,6-$(CH_3)_3$	$C_{10}H_{13}O$	–	H	259, **39**, *21*
C_6H_2-2,4,6-$(CH_3)_3$	$C_{10}H_{15}OSi$	–	H	182, **26**, *16*
C_6H_2-2,4,6-$(CH_3)_3$	$C_{13}H_9$	–	H	183, **26**, *19*
				183, **26**, *21*
				260, **39**, *23*
C_6H_2-2,4,6-$(CH_3)_3$	$C_{27}H_{59}Si_6$	–	H	259/60, **39**, *22*
$C_9H_{12}N$	–	–	H	140, **21**, *9*
$C_9H_{12}P$	CH_3	–	H	150, **22**, *20*
C_9H_{14}	C_2H_5	–	H	210, **30**, *36*
C_9H_{18}	–	–	H	263, **40**, *7*
C_9H_{18}	C_4H_9	–	H	207, **30**, *26*
$C_9H_{21}Si$	C_4H_9	–	H	169, **24**, *24*
C$_{10}H_7$	–	–	H	142, **21**, *19*
$C_{10}H_7$	CH_3	C_6H_5	H	193, **28**, *10*
$C_{10}H_7$	C_2H_5	C_6H_5	H	194, **28**, *11*
$C_{10}H_7$	C_3H_7	C_6H_5	H	194, **28**, *12*
$C_{10}H_{11}O$	C_6H_5	–	H	173, **25**, *8*
$C_{10}H_{13}$				
CH_2-$C(CH_3)_2$-C_6H_5	–	–	H	96
$CH_2CH_2CH(CH_3)$-C_6H_5	CH_3	–	H	150, **22**, *25*
$CH_2CH_2CH_2CH_2$-C_6H_5	CH_3	–	H	150, **22**, *24*
C_6H-2,3,5,6-$(CH_3)_4$	–	–	H	142, **21**, *17*
C_6H_3-2,6-$(C_2H_5)_2$	–	–	H	243, **36**, *2*
$C_{10}H_{13}O$	C_9H_{11}	–	H	259, **39**, *21*
$C_{10}H_{14}N$	CH_3	C_6H_5	H	193, **28**, *9*
$C_{10}H_{15}$	–	–	H	303, **48**, *5*
$C_{10}H_{15}$	CH_3	–	H	151, **22**, *29*
$C_{10}H_{15}OSi$	C_9H_{11}	–	H	182, **26**, *16*
$C_{10}H_{16}O$	C_2H_5	–	H	208, **30**, *28*
$C_{10}H_{16}O$	C_6H_5	–	H	208, **30**, *29*
$C_{10}H_{17}$	C_2H_5	–	H	162, **23**, *39*
$C_{10}H_{17}O_4$	–	–	H	294, **45**, *37*
				294, **45**, *38*
$C_{10}H_{18}$	CH_3	–	H	211, **30**, *41*

$C_{10}H_{20}$	CH_3	–	H	210, **30**, *39*
$C_{10}H_{20}O$	CH_3	–	H	210/1, **30**, *40*
$C_{10}H_{27}Si_3$	–	–	H	290, **45**, *15*
$C_{10}H_{27}Si_3$	CH_3	–	H	148/9, **22**, *11*
C$_{11}H_{10}$	C_2H_5	–	H	210, **30**, *37*
$C_{11}H_{15}$				
CH_2CH_2-$C(CH_3)_2$-C_6H_5	CH_3	–	H	150, **22**, *26*
$CH_2CH_2CH_2$-CH_2CH_2-C_6H_5	CH_3	–	H	151, **22**, *28*
$C_{11}H_{21}O$	C_6H_5	–	H	258, **39**, *18*
$C_{11}H_{22}O$	C_6H_5	–	H	211, **30**, *42*
$C_{11}H_{23}O$	C_6H_5	–	H	258, **39**, *14*
$C_{11}H_{24}ClSi$	C_4H_9	–	H	169, **24**, *26*
$C_{11}H_{25}Si$	C_4H_9	–	H	169, **24**, *25*
C$_{12}H_9$	–	–	H	141, **21**, *14*
$C_{12}H_9$	–	–	T	142/3
$C_{12}H_{12}O$	C_2H_5	–	H	209, **30**, *33*
$C_{12}H_{12}O$	C_6H_5	–	H	209, **30**, *34*
$C_{12}H_{17}$				
CH_2CH_2-$C(CH_3)_2$-C_6H_4-4-CH_3	CH_3	–	H	151, **22**, *27*
C_6H_3-2,6-$(C_3H_7\text{-}i)_2$	–	–	H	244, **36**, *3*
$C_{12}H_{18}O$	CH_3	–	H	208, **30**, *30*
$C_{12}H_{18}O$	C_2H_5	–	H	208/9, **30**, *31*
$C_{12}H_{18}O$	C_6H_5	–	H	209, **30**, *32*
$C_{12}H_{27}Si$	C_4H_9	–	H	169, **24**, *22*
C$_{13}H_9$	$C_7H_{19}Si_2$	–	H	180/1, **26**, *4*
$C_{13}H_9$	C_9H_{11}	–	H	183, **26**, *19*
				183, **26**, *21*
				260, **39**, *23*
$C_{13}H_{11}O$	C_6H_5	–	H	173, **25**, *6*
$C_{13}H_{12}N$	C_6H_5	–	H	256, **39**, *3*
$C_{13}H_{12}NO$	C_6H_5	–	H	256, **39**, *4*
C$_{14}H_{11}$	$C_7H_{19}Si_2$	–	H	181, **26**, *5*
$C_{14}H_{13}O$	–	–	H	291/2, **45**, *23*
$C_{14}H_{13}O$	C_6H_5	–	H	256, **39**, *7*
C$_{15}H_{15}$	CH_3	–	H	150, **22**, *22*
$C_{15}H_{23}$	–	–	H	245, **36**, *5*
$C_{15}H_{23}$	CH_3	–	H	183, **26**, *20*
$C_{15}H_{23}$	$C_{27}H_{59}Si_6$	–	H	260, **39**, *24*

C$_{16}H_{19}Si$	C_2H_5	–	H	160, **23**, *18*
$C_{16}H_{19}Si$	C_4H_9	–	H	169, **24**, *23*
C$_{18}H_{28}$	$C_{18}H_{29}$	–	H	204, **30**, *15*
$C_{18}H_{29}$	–	–	H	319, **55**, *5*
$C_{18}H_{29}$	C_4H_9	–	H	253, **37**, *27*
$C_{18}H_{29}$	$C_{18}H_{28}$	–	H	204, **30**, *15*
C$_{23}H_{20}NOPRe$	C_6H_5	–	H	176, **25**, *24*
C$_{27}H_{59}Si_6$	C_6H_5	–	H	176, **25**, *29* 259, **39**, *19*
$C_{27}H_{59}Si_6$	C_9H_{11}	–	H	259/60, **39**, *22*
$C_{27}H_{59}Si_6$	$C_{15}H_{23}$	–	H	260, **39**, *24*
C$_{28}H_{20}$	–	–	H	264, **40**, *13*
$C_{28}H_{20}$	C_6H_5	–	H	204, **30**, *14*

Physical Constants and Conversion Factors

Avogadro constant N_A (or L) = 6.02214×10^{23} mol^{-1}
Faraday constant F = 9.64853×10^{4} C/mol
molar gas constant R = 8.31451 $J \cdot mol^{-1} \cdot K^{-1}$
molar volume (ideal gas) V_m = 2.24141×10^{1} L/mol
(273.15 K, 101325 Pa)

Planck constant h = 6.62608×10^{-34} $J \cdot s$
elementary charge e = 1.60218×10^{-19} C
electron mass m_e = 9.10939×10^{-31} kg
proton mass m_p = 1.67262×10^{-27} kg

1 kg = 2.205 pounds
1 m = 3.937×10^{1} inches = 3.281 feet
1 m^3 = 2.642×10^{2} gallons (U.S.)
1 m^3 = 2.200×10^{2} gallons (Imperial)

Force	N	dyn	kp
1 N	1	10^{5}	1.019716×10^{-1}
1 dyn	10^{-5}	1	1.019716×10^{-6}
1 kp	9.80665	9.80665×10^{5}	1

Pressure	Pa	bar	kp/m²	at	atm	Torr	lb/in²
1 Pa = 1 N/m²	1	10^{-5}	1.019716×10^{-1}	1.019716×10^{-5}	9.86923×10^{-6}	7.50062×10^{-3}	1.450378×10^{-4}
1 bar = 10^{6} dyn/cm²	10^{5}	1	1.019716×10^{4}	1.019716	9.86923×10^{-1}	7.50062×10^{2}	1.450378×10^{1}
1 kp/m² = 1 mm H_2O	9.80665	9.80665×10^{-5}	1	10^{-4}	9.67841×10^{-5}	7.35559×10^{-2}	1.422335×10^{-3}
1 at (technical)	9.80665×10^{4}	9.80665×10^{-1}	10^{4}	1	9.67841×10^{-1}	7.35559×10^{2}	1.422335×10^{1}
1 atm = 760 Torr	1.01325×10^{5}	1.01325	1.033227×10^{4}	1.033227	1	7.60×10^{2}	1.469595×10^{1}
1 Torr = 1 mmHg	1.333224×10^{2}	1.333224×10^{-3}	1.359510×10^{1}	1.359510×10^{-3}	1.315789×10^{-3}	1	1.933678×10^{-2}
1 lb/in² = 1 psi	6.89476×10^{3}	6.89476×10^{-2}	7.03069×10^{2}	7.03069×10^{-2}	6.80460×10^{-2}	5.17149×10^{1}	1

Work, Energy, Heat	J	kW·h	kcal	Btu	eV
1 J = 1 W·s = 1 N·m = 10^7 erg	1	2.778×10^{-7}	2.39006×10^{-4}	9.4781×10^{-4}	6.242×10^{18}
1 kW·h	3.6×10^{6}	1	8.604×10^{2}	3.41214×10^{3}	2.247×10^{25}
1 kcal	4.1840×10^{3}	1.1622×10^{-3}	1	3.96566	2.6117×10^{22}
1 Btu (British thermal unit)	1.05506×10^{3}	2.93071×10^{-4}	2.5164×10^{-1}	1	6.5858×10^{21}
1 eV	1.602×10^{-19}	4.450×10^{-26}	3.8289×10^{-23}	1.51840×10^{-22}	1

1 cm^{-1} = 1.239842×10^{-4} eV
1 hartree = 27.2114 eV
1 Hz = 4.135669×10^{-15} eV
1 eV ≙ 23.0578 kcal/mol

Power	kW	hp	$kp \cdot m \cdot s^{-1}$	kcal/s
1 kW = 10^3 J/s	1	1.35962	1.01972×10^{2}	2.39006×10^{-1}
1 hp (horsepower, metric)	7.3550×10^{-1}	1	7.5×10^{1}	1.7579×10^{-1}
1 $kp \cdot m \cdot s^{-1}$	9.80665×10^{-3}	1.333×10^{-2}	1	2.34384×10^{-3}
1 kcal/s	4.1840	5.6886	4.26650×10^{2}	1

References:

Mills, I. (Ed.), International Union of Pure and Applied Chemistry, Quantities, Units and Symbols in Physical Chemistry, Blackwell Scientific Publications, Oxford 1988.
The International System of Units (SI), National Bureau of Standards Spec. Publ. 330 [1972].
Landolt-Börnstein, 6th Ed., Vol. II, Pt. 1, 1971, pp. 1/14.
ISO Standards Handbook 2, Units of Measurement, 2nd Ed., Geneva 1982.
Cohen, E. R., Taylor, B. N., Codata Bulletin No. 63, Pergamon, Oxford 1986.

Key to the Gmelin System of Elements and Compounds

HCl

$ZnCl_2$

System Number	Symbol	Element
1		Noble Gases
2	H	Hydrogen
3	O	Oxygen
4	N	Nitrogen
5	F	Fluorine
6	**Cl**	**Chlorine**
7	Br	Bromine
8	I	Iodine
8a	At	Astatine
9	S	Sulfur
10	Se	Selenium
11	Te	Tellurium
12	Po	Polonium
13	B	Boron
14	C	Carbon
15	Si	Silicon
16	P	Phosphorus
17	As	Arsenic
18	Sb	Antimony
19	Bi	Bismuth
20	Li	Lithium
21	Na	Sodium
22	K	Potassium
23	NH_4	Ammonium
24	Rb	Rubidium
25	Cs	Caesium
25a	Fr	Francium
26	Be	Beryllium
27	Mg	Magnesium
28	Ca	Calcium
29	Sr	Strontium
30	Ba	Barium
31	Ra	Radium
32	**Zn**	**Zinc**
33	Cd	Cadmium
34	Hg	Mercury
35	Al	Aluminium
36	Ga	Gallium

$CrCl_2$

$ZnCrO_4$

System Number	Symbol	Element
37	In	Indium
38	Tl	Thallium
39	Sc, Y La–Lu	Rare Earth Elements
40	Ac	Actinium
41	Ti	Titanium
42	Zr	Zirconium
43	Hf	Hafnium
44	Th	Thorium
45	Ge	Germanium
46	Sn	Tin
47	Pb	Lead
48	V	Vanadium
49	Nb	Niobium
50	Ta	Tantalum
51	Pa	Protactinium
52	**Cr**	**Chromium**
53	Mo	Molybdenum
54	W	Tungsten
55	U	Uranium
56	Mn	Manganese
57	Ni	Nickel
58	Co	Cobalt
59	Fe	Iron
60	Cu	Copper
61	Ag	Silver
62	Au	Gold
63	Ru	Ruthenium
64	Rh	Rhodium
65	Pd	Palladium
66	Os	Osmium
67	Ir	Iridium
68	Pt	Platinum
69	Tc	Technetium[1]
70	Re	Rhenium
71	Np, Pu . . .	Transuranium Elements

Material presented under each Gmelin System Number includes all information concerning the element(s) listed for that number plus the compounds with elements of lower System Number.

For example, zinc (System Number 32) as well as all zinc compounds with elements numbered from 1 to 31 are classified under number 32.

[1] A Gmelin volume titled "Masurium" was published with this System Number in 1941.

A Periodic Table of the Elements with the Gmelin System Numbers is given on the Inside Front Cover